Linear Programming

Linear Programming

Katta G. Murty
The University of Michigan

John Wiley & Sons
New York Chichester Brisbane Toronto Singapore

Library of Congress Cataloging in Publication Data:

Murty, Katta G., 1936–
Linear programming.

Rev. ed. of: Linear and combinatorial programming. c1976.
Includes index.
1. Linear programming. I. Title.
T57.74.M87 1983 519.7'2 83-7012
ISBN 0-471-09725-X

Printed in the United States of America

10 9 8 7 6 5 4 3 2 1

Foreword

Few realize that linear programming is a revolutionary development that permits us, for the first time in our long evolutionary history, to make decisions about the complex world in which we live that can approximate, in some sense, the optimal or best decision. Before the postwar period (1947), decisions were being made without gathering the relevant facts, organizing them in an applicable way, or considering the infinite number of alternative ways to achieve desired goals that were technologically possible and did not exceed the available resources. Of course, before the advent of computers, such a model-building approach would have been unthinkable. Even with the most modern methods, it is still unthinkable to sift through all the ways of assigning, say, 70 individuals to 70 jobs and selecting the best assignment for each person. However, by combining the computer with methods such as those discussed in this book, problems many times more complex are routinely solved every day all over the world.

Perhaps this is why there was so much excitement at a certain gathering in 1949 at the University of Chicago in which young mathematicians, economists, and statisticians presented their research on a new tool for attacking complex planning problems of the modern world. Linear programming was that new tool. Since then, four of those present have received the Nobel Prize for research influenced by that conference. With linear programming, one can *abstract* the underlying essential similarities in the management of such seemingly disparate systems as industrial production, the flows of resources in the economy, crop rotation, assignment of individuals to jobs, or the design of complex engineering systems (e.g., transport, water, energy, refinery scheduling).

In my opinion, Katta Murty has succeeded in his aim of providing students with an up-to-date, in-depth coverage of all the important practical, computational, and mathematical aspects of linear programming. He has done so in a clear, intuitively appealing, understandable way while maintaining an appropriate standard of rigor. Only minimal mathematical background is required.

I am honored to have been asked by Katta Murty to write this foreword. It was my privilege to have had him as a graduate student many years ago at Berkeley and to be able to follow his important contributions to the field since that time, both as a theoretician and as an expositor.

George B. Dantzig
Professor of Operations Research
Stanford University

Preface

INTRODUCTION

I am grateful for the enthusiastic reception given to my book *Linear and Combinatorial Programming* published by Wiley in 1976. I received constructive suggestions for improving it from readers all over the world, which proved to be extremely useful as I started preparing the material for this book. As the Telugu poet J. S. Sarma said:

> **రచనకు బాధ్యత కావలె,**
> **పుస్తకము వ్రాసెడి వేళన్, రుచి**
> **నెంచు సద్విమర్శకుడ చిరముగా**
> **మంచి మిత్రుడౌ రచయితకున్**

(A good critic is an author's best friend). Most of the people pointed out that, though the book was quite comprehensive and the coverage extensive, there were still several important topics not discussed. They suggested that in the second edition an effort should be made to make the coverage complete. Besides, several new developments have occurred since 1976. This rapid growth was made possible by the large number of very talented researchers who were attracted to work in these areas. There are reasons for this. One important reason is the existence of the very efficient simplex method developed by George B. Dantzig in 1947 for solving linear programs, as well as the wide availability of software packages for it. Another is the practical applicability of these areas. Today, linear programming and the related developments in network flow algorithms and integer and combinatorial programming make it possible for us to solve many of the complex optimization problems in our modern society, industry, and business with the use of computers. Another reason is the professional recognition given to these areas by the Nobel Prize Committee when it awarded the 1975 Nobel Prize in Economics to T. C. Koopmans and L. Kantarovitch for developing the application of linear programming to the economic problem of allocating resources.

THE THREE BOOKS

I tried to expand the coverage of the book to include all these new developments, as well as all other important topics not discussed in the first edition. It soon became clear that these additions would create a very large book. So it became essential to divide it into three books—one containing the material on linear programming, the second containing the material on network and combinatorial programming, and the third on linear complementarity. These divisions also make practical sense, since most universities are offering separate courses in each of these areas. This is the book on linear programming. I have included in this book all the material that is normally offered in courses with the words "linear programming" in their title.

WHAT IS NEW?

The chapters on the ellipsoid algorithm, the vector minima and iterative methods for solving linear inequalities, and linear programming problems are new. Even though many of the other chapter titles remain the same as those in the 1976 book, all of them have been very thoroughly revised to make the coverage complete, and to include many new illustrations and exercises.

THE OBJECTIVES

1 To provide an in-depth and clear coverage of all the important practical, technical, computational, and mathematical aspects of linear programming and the transportation problem.

2 To discuss and illustrate very clearly the methods used to model problems as linear programs and to help develop skill in modeling.
3 To discuss the theory and the geometry of linear programming systematically in an elementary but rigorous manner so that a reader without much mathematical background can easily understand it.
4 To discuss clearly the algorithms for solving linear programs and transportation problems, to present their efficient implementations for the computer, and to discuss their computational complexity.
5 To help develop skill in using algorithms intelligently to solve practical problems.

BACKGROUND NEEDED

The background required to study this book is some familiarity with matrix algebra (which is gained in an undergraduate course), especially the concept of linear independence, and the matrix reduction methods for solving systems of linear equations. All this essential background material is reviewed in Sections 3.1 to 3.3 of the book for the sake of completeness.

SUMMARY OF CHAPTER CONTENTS

The book begins with a section entitled Notation, in which all the symbols and several terms are defined. It is strongly recommended that the reader peruse this section first and refer to it whenever there is a question about the meaning of some symbol or term.

Chapter 1 presents various methods used for formulating problems as linear programs. Many illustrative examples are used. Section 1.1 discusses very elementary methods. Section 1.2 deals with the more complicated methods based on piecewise linear approximations. This section also presents applications of linear programming in parameter estimation during modeling. Section 1.4 briefly presents the various approaches for handling multiobjective linear programming problems. Section 1.5 discusses the practical aspects of modeling, including, estimation of the data in the model.

Chapter 2 discusses the simplex method for solving linear programs using canonical tableaux. This is the originally developed version of the simplex method. It is computationally inefficient, but ideally suited for solving small linear programs by hand computation and for introducing the basic concepts of the simplex method to the beginning student. We also clearly show that the simplex method is a version of a reduced-gradient method when the problem is viewed as one involving the nonbasic variables only, with the basic variables eliminated using the equality constraints in the problem. It is possible to branch off from this and discuss the reduced-gradient and the generalized reduced-gradient methods of nonlinear programming as being generalizations of the simplex algorithm.

Chapter 3 presents the necessary mathematical, geometrical, and linear algebra concepts for studying linear programs, the structure of their sets of feasible solutions, and how the simplex method operates on them. This chapter is heavily rooted in linear algebra, and it discusses the mathematical theory of linear programming. Related algorithms in linear algebra (e.g., the algorithm for testing linear independence, the algorithm for computing a maximal linearly independent subset of a given set of vectors, etc.) are all discussed here. Results on convex polyhedra that are used in studying linear programming are also presented. Sections 3.1 and 3.2 form a review of the relevant concepts from matrix algebra and *n*-dimensional geometry. Section 3.3 provides a review of the matrix-reduction methods for solving systems of linear equations. Sections 3.4 to 3.10 contain the basic mathematical results in linear programming. Sections 3.11 and 3.12 discuss the simplex method using matrix operations; the results in these sections are used later in Chapters 5 and 7 to develop computationally efficient implementations of the simplex method. Section 3.18 discusses the very important concept of *equivalence* among linear programming problems. Section 3.19 presents algorithms for ranking the extreme points of a convex polyhedron in increasing order of a linear objective function. Section 3.20 discusses methods for solving fractional linear programming problems.

Chapter 4 deals with the duality theory of linear programming. The economic arguments behind duality are clearly discussed. The dual variables are clearly shown to be the marginal values associated with the items in the original linear programming model when the dual problem has a unique optimum solution. In this chapter we also discuss the theorems of alternatives for linear equality or inequality systems, the stability of the linear programming model, necessary and sufficient conditions for a linear program to have a unique optimum solution, and zero-sum two-person matrix games.

Chapter 5 discusses the revised simplex method with both the explicit form and the product form of the inverses. Here, we also present the infeasibility analysis of linear programming, and the column generation procedure in modeling that the revised simplex format makes possible. Chapter 6 considers the dual simplex method using both the canonical tableaux and the inverse tableaux. In the chapter discussion, we provide a geometrical illustration of the paths taken by the primal simplex and the dual simplex algorithms on the same problem, so that the reader can compare and contrast these two algorithms. Chapter 7 discusses numerically stable implementations of the simplex algorithm based on matrix factorizations. These implementations improve the numerical accuracy and help to take advantage of sparsity.

Chapters 8 and 9 discuss the various types of sensitivity-postoptimality analysis and ranging in linear programming. In Chapter 8 we also discuss the computational complexity of the parametric linear programming problem and the properties of the optimum objective value function in a linear program when it is treated as a function of the right-hand-side constants, or the original cost coefficients. When the dual problem has alternate optimum solutions, the optimum objective value treated as a function of the right-hand-side constants vector b is not differentiable at that b. In this case, an optimum dual solution cannot be interpreted as the vector of marginal values associated with the items. As pointed out in reference [4.2], most textbooks on linear programming do not discuss this fact in their chapters on marginal analysis. At such points $b = (b_i)$, there are two marginal values for each i: one is the rate of change in the optimum objective value per unit increase in b_i, and the other is the rate of change in the same per unit decrease in b_i, known as the positive and negative marginal values. In Section 8.15 we discuss how to compute both the positive and the negative marginal values efficiently, using the set of optimum solutions of the dual problem.

Chapter 10 considers degeneracy and the problem of cycling in degenerate problems, which can prevent the (primal or dual) simplex algorithm from terminating in a finite number of steps. The geometry of degeneracy is clearly discussed, and various methods for preventing cycling under degeneracy are presented.

Chapter 11 deals with efficient versions of the primal and dual simplex methods for handling bounded variable linear programs. The generalization of this to the "generalized upper bounding" (or GUB) constraints is presented very clearly. This naturally leads to "structured linear programming," an area that deals with efficient implementations of the simplex method for solving large-scale linear-programming models, which exploit the special structure of the model. Chapter 12 presents the decomposition principle of linear programming. Chapter 13 deals with the special methods available for handling transportation problems. All the properties of this problem based on its special structure are clearly derived. Efficient implementations of the primal and dual simplex methods, using the tree corresponding to the basis, and the triple label representation for storing and manipulating the trees are presented very clearly. Using this implementation, very large transportation problems are being solved with little computer time in practical applications.

Chapter 14 discusses the worst case and the average computational complexity of the simplex method. Chapter 15 presents the polynomially bounded ellipsoid algorithm for solving system of linear inequalities, or linear programs.

Most books on linear programming usually ignore iterative methods. One reason for this might be the tremendous practical success of the simplex method, leading to the view that it is futile to present other approaches. However, for tackling large-scale unstructured linear programming models, iterative methods may provide a potentially useful alternative. Therefore, we present a brief survey of various iterative methods in Chapter 16. Chapter 17 deals with methods for computing the vector minima in a multiobjective linear programming problem.

HOW TO USE THE EXERCISES

The exercises in the book are of several types.

1. *Formulation problems.* An effort is made to include in Chapter 1 formulation problems from many of the areas of application that use linear programming. The instructor can formulate some of these problems in the class to illustrate the wide applicability of linear programming and assign some others as homework exercises.
2. *Numerical problems.* These require the application of the algorithms discussed in the book. Solving these problems will give insight into the ways in which algorithms work.
3. *Proofs and other problems.* These require a deeper understanding of the material discussed in the text.
4. *True or false questions.* Several of these questions in Chapters 3 and 4 contain a statement and ask the reader to mark whether it is true or false and to justify the answer by either a simple proof or a counterexample. These questions were very popular in the 1976 book, and the beginning students seemed to learn a lot by doing them. The instructor could give a quiz to the class based on these questions several times during the term, and they can be used to promote class discussion as well.

Exercises that are listed in the middle of a chapter can normally be solved using the methods discussed in the section in which they appear. A new sequence of exercise numbers begins with each chapter (e.g., Exercise 6.2 refers to exercise number 2 in Chapter 6).

WHAT DISTINGUISHES THIS BOOK FROM THE OTHERS IN THE AREA

This book's main distinguishing feature is its completeness and comprehensiveness. Most important topics in linear programming are discussed clearly and in depth. The practical, modeling, mathematical, geometrical, algorithmic, and computational aspects of linear programming are all covered very carefully in complete detail. Many of the books in the area usually specialize in only a few of these aspects. For example, this is one of the few books that discusses the implementation of the primal simplex algorithm for solving transporation problems using tree labels. Algorithms for solving different types of transporation models (those involving equality constraints only, or those involving inequality constraints, or the interval type of constraints, bounds on the variables, etc.) are also all discussed separately and clearly. It is possible for the instructor to assign the material for reading before the class and then spend the class time going over the main points, leaving the details to be read from the book. This feature, and the large number of exercises of various types make this book very convenient for either self-study or for an instructor to teach from with very little additional effort on his or her part.

HOW TO USE THE BOOK IN A COURSE

The material in Chapters 1 and 2; Sections 3.1 to 3.14 of Chapter 3; Sections 4.1 to 4.6 in Chapter 4; Chapters 5, 6, 7, 8, 9, 10, and 13 form the basic core of a one-term first-year graduate-level course in linear programming, and a course can be covered in this order. To teach the course in a mathematical style, I recommend emphasis on the mathematical theory discussed in Chapters 3 and 4. The same course can be taught to emphasize applications of the material, especially the formulation, algorithmic, and computational aspects of linear programming. The remaining material in the book can be the basis for a second course on linear programming. Or anyone who has taken the first course can study this material through self-study fairly easily. By maintaining a fast pace, it is also possible for a well-motivated class to cover the entire book in a single term.

An undergraduate course on "introduction to optimization" can be taught using selected material from this book and the companion book *Network and Combinatorial Programming* (reference [1.94], abbreviated here as NCP) in the following order: Sections 1.1 of this book covering elementary methods of formulation of linear programming problems; Sections 3.1 to 3.3 of this book to review the background material of matrix algebra; Sections 4.1 to 4.4 and Sections 4.5.5 and 4.5.6 of this book, covering duality, marginal analysis (in particular, how to write down the dual of a linear program in standard form), and the relationships of the primal and dual solutions associated with an optimum basis for the problem; Section 3.1 of NCP, discussing the Hungarian method for solving the assignment problem (this provides a special algorithm for a highly structured linear programming problem, which is very easy to understand); Section 13.2 of this book, discussing the simplified version of the primal algorithm for solving the balanced transportation problems; Chapter 2 of this book (when the algorithms are discussed in this order, the course moves from simple, elegant algorithms for specially structured problems to somewhat complicated algorithms for general problems that lack structure); Sections 9.1 to 9.8 and 13.5, discussing some sensitivity analysis and including general linear programming and transportation problems; Chapter 12 of NCP, covering some simple formulations of integer programming problems; Sections 15.1 to 15.7 of NCP, covering the branch-and-bound approach; and some supplemental material on applications of dynamic programming and nonlinear programming.

Of course, a well-motivated person can also use this book for self-study to learn linear programming. Operations researchers and people who use optimization in their professional work can use this book as a reference for details of the methods or for developing improved algorithms.

REFERENCES

References are listed at the end of each chapter, and a list of books on linear programming is provided at the end of Chapter 1. Because of the lack of space, the list of references had to be kept very brief. However, most references used in the preparation of this material have been cited.

Acknowledgments

In preparing this book I have received advice and encouragement from several people. For this I am heavily indebted to Mustafa Akgül, John Bartholdi, Pat Carstensen, R. Chandrasekharan, Sung Jin Chung, Yahya Fathi, Akli Gana, David M. Gay, Jack L. Goldberg, Ikuyo Kaneko, Olvi L. Mangasarian, Diana P. O'leary, M. H. Partovi, Clovis Perin, Mike Plantholt, Loren Platzman, Steve Pollock, Romesh Saigal, Art J. Schwartz, Bob Smith, W. Allen Spivey, Klaus Truemper, Layne Watson, and Chris Witzgall.

The final version of the manuscript was completed during 1981–1982. I was on sabbatical leave from the University of Michigan then and spent part of that year at the School of Management and Administration, University of Texas at Dallas; the remaining part of the year I was at the Indian Statistical Institute in Delhi, India, on a project sponsored by the grant, INT-8109825 from the U.S.-India Cooperative Science Program, NSF Special Foreign Currency Program. I thank the University of Michigan for this sabbatical, the University of Texas at Dallas (in particular, R. Chandrasekharan of the School of Management and Administration) for inviting me to spend a term with them, and the NSF (in particular, Osman Shinaishin, Program Director of US-India Cooperative Science Program) for making it possible for me to visit the Indian Statistical Institute.

From 1978 to 1982, the research carried out by my graduate students and myself in optimization was partially supported by the Directorate of Mathematical and Information Sciences of the U.S. Air Force Office of Scientific Research under Grant AFOSR 78-3646. Several results from this research appear throughout the book. I thank the AFOSR (in particular, our program manager, Joseph Bram) for this support and opportunity.

The manuscript was reviewed by Donovan Young of Georgia Institute of Technology. I thank him for his excellent comments. They were very useful in the final revision.

I am deeply indebted to George B. Dantzig for making it possible for me to study operations research, and for graciously agreeing to write the foreword for this book.

Most of the typing of the manuscript was done by Geraldine Cox and Jane Outslay at the Department of Industrial and Operations Engineering, The University of Michigan. I thank them for their patience and untiring effort.

I would also like to thank Bill Stenquist, Engineering Editor at Wiley, for his encouragement.

Finally, I thank my wife Vijaya, my daughters Vani and Madhusri, and Mother Adilakshmi for their patience and support. This book is dedicated to them.

Katta G. Murty

Contents

Notations

M A large positive number.

$\mathcal{N}$ Set of points in a graph or a network.

$\mathcal{A}$ Set of lines (arcs or edges) in a network.

G A graph or a network.

a An assignment.

e A vector in which all coordinates are equal to 1. e_r is a vector like this in **R**.

e The base of natural logarithms. It is approximately equal to 2.7.

(*i*, *j*) This denotes an ordered pair (the comma in the middle indicates that it is an ordered pair). It is used to denote an arc from node *i* to node *j*. in a network, or the cell in the *i*th row and the *j*th column of a tableau or an array.

(*i*; *j*) This is an unordered pair (the semicolon in the middle indicates that it is an unordered pair) denoting an edge connecting the nodes *i* and *j* in a graph or a network.

w, *z* Unless otherwise defined, these are usually the Phase I and Phase II objective functions, respectively, in the simplex method.

x, *y* Vectors or variables.

E, *A*, *B*, *X*, *P*, *F*, *D*, *U*, *L* Unless otherwise specified, these symbols usually denote matrices or vectors.

B Unless otherwise specified, this symbol usually denotes a basis for a linear program in which the constraints are a system of linear equations in nonnegative variables.

B^{-1} Inverse of the matrix *B*.

$\boldsymbol{\beta}$ Given a basis *B*, this symbol is used to denote the inverse $B^{-1} = (\beta_{ij})$.

I Unit matrix or identity matrix. It is a square matrix in which all diagonal entries are 1 and all off-diagonal entries are 0. See Section 3.2.

M, N, E, P, Q, F, Γ, D, I, U, H, J These symbols usually denote sets that are defined in that section or chapter.

B Augmented basis in a linear program when an additional constraint is introduced. See Sections 9.3, 9.4.

$\mathbf{K}^{\Delta}$ The convex hull of the extreme points of a convex polyhedron **K**. See Section 3.7.

$\mathbf{K}^{\mathrm{L}}$ The cone of homogeneous solutions corresponding to a convex polyhedron **K**. See Section 3.7.

c This symbol usually denotes the vector of original cost coefficients.

x_B The vector of basic variables associated with a basis *B* for a linear program.

d_B, c_B The row vectors of the Phase I and Phase II cost coefficients, respectively, associated with a basis *B* for a linear program.

b Unless otherwise specified, this symbol usually denotes the right-hand side constant vector in a linear program.

$\bar{a}_{ij}$, $\bar{b}_i$ Updated entries in a canonical tableau for a linear program.

$\bar{d}_j$, $\bar{c}_j$ Unless otherwise specified, these are usually the Phase I and Phase II relative cost coefficients, respectively, in the simplex method.

π Unless otherwise specified, this denotes the vector of dual variables.

u_i, v_j In a transportation model, these denote the dual variables associated with the ith source and the jth demand center, respectively. In other models these denote slack variables or other variables, as defined there.

P(j), S(j), EB(j), YB(j) Indices used in the transportation algorithm. See Section 13.4.

m, n In Chapter 13, these denote the number of sources and demand centers, respectively, in a transportation model.

$\mathscr{B}$ A basic set of cells in a transportation array for a balanced transportation problem.

$\mathbf{G}_{\mathscr{B}}$ The rooted tree corresponding to a basic set of cells $\mathscr{B}$ in a transportation problem.

$\mathbf{G}_{\boldsymbol{\Delta}}$ The graph corresponding to a set of cells, Δ, in a transportation array.

$\mathscr{A}_{\boldsymbol{\Delta}}$ The set of edges in the graph $G_{\boldsymbol{\Delta}}$.

$\mathscr{A}_{\mathscr{B}}$ The set of edges in the graph $G_{\mathscr{B}}$.

$\lceil\alpha\rceil$ Defined only for real numbers α. It represents the smallest integer that is greater than or equal to α, and is often called the *ceiling of* α. For example $\lceil -4.3 \rceil = -4$, $\lceil 4.3 \rceil = 5$.

$\lfloor\alpha\rfloor$ Defined only for real numbers α. It represents the largest integer less than or equal to α, and is often called the *floor of* α. For example $\lfloor -4.3 \rfloor = -5$, $\lfloor 4.3 \rfloor = 4$.

$\sum$ Summation sign.

∞ Infinity.

$\in$ Set inclusion symbol. If **F** is a set, "$\mathrm{F}_1 \in \mathbf{F}$" means that "$\mathrm{F}_1$ is an element of **F**." Also "$\mathrm{F}_2 \notin \mathbf{F}$" means that "$\mathrm{F}_2$ is not an element of **F**."

$\subset$ Subset symbol. If **E**, $\boldsymbol{\Gamma}$ are two sets, "$\mathbf{E} \subset \boldsymbol{\Gamma}$" means that "**E**" is a subset of $\boldsymbol{\Gamma}$, or that every element in **E** is also an element of $\boldsymbol{\Gamma}$."

$\cup$ Set union symbol. If **D**, **H** are two sets, $\mathbf{D} \cup \mathbf{H}$ is the set of all elements that are either in **D** or in **H** or in both **D** and **H**.

$\cap$ Set intersection symbol. If **D** and **H** are two sets, $\mathbf{D} \cap \mathbf{H}$ is the set of all elements that are in both **D** and **H**.

$\varnothing$ The empty set. The set containing no elements.

$\backslash$ Set difference symbol. If **D** and **H** are two sets, $\mathbf{D}\backslash\mathbf{H}$ is the set of all elements of **D** that are not in **H**.

$\{\ \}$ Set brackets. The notation $\{x\colon \text{some property}\}$ represents the set of all elements, x, satisfying the property mentioned after the ":".

Minimum $\{\ \}$ The minimum number among the set of numbers appearing inside the set brackets. Maximum $\{\ \}$ has a similar meaning. If the set is empty we will adopt the convention that the minimum in it is $+\infty$ and the maximum in it is $-\infty$.

$\mathbf{R}^n$ Real Euclidean n-dimensional vector space. It is the set of all *ordered* vectors $(x_1, \ldots, x_n)$ where each x_j is a real number, with the usual operations of addition and scalar multiplication defined on it.

$|\alpha|$ Absolute value of the real number α.

$|\mathbf{F}|$ If **F** is a set, this symbol denotes its cardinality, that is, the number of distinct elements in the set **F**.

■ This symbol indicates the end of a proof.

$\|x\|$ Euclidean norm of a vector $x \in \boldsymbol{R}^n$. If $x = (x_1, \ldots, x_n)$, $\|x\| = +\sqrt{x_1^2 + \cdots + x_n^2}$

$\geqq, \geq, >$ The symbols representing order relationships among vectors in $\boldsymbol{R}^n$. See Section 3.1.

$\succ$ Lexicographically greater than. See Section 10.2.3.

$A_{i.}$ The ith row vector of the matrix A.

$A_{.j}$ The jth column vector of the matrix A.

$\log_2 x$ Defined only for positive real numbers x. It is the logarithm of the positive real number x, with 2 as the base (or radix).

$\mathbf{x}_1^+, \mathbf{x}_1^-$ See Section 1.2.3.

iff If and only if.

LP Linear program.

BFS Basic feasible solution.

θ-loop A minimal linearly dependent set of cells in a transportation array. See section 13.1.6.

(*i.j*) This refers to the *j*th equation in the *i*th chapter. Equations are numbered serially in each chapter.

Section *i.j*; *i.j.k* The sections are numbered serially in each chapter. "*i.j*" refers to section *j* in Chapter *i*. "*i.j.k*" refers to subsection *k* in section *i.j*.

Figure *i.j* The *j*th figure in Chapter *i*. The figures are numbered serially in this manner in each chapter.

Reference [*i, j*] The *j*th reference in the list of references given at the end of the chapter *i*. References given at the end of each chapter are numbered serially.

Exercise *i.j* The *j*th exercise in Chapter *i*. Exercises are numbered serially in each chapter.

$n!$ n factorial. Defined only for nonnegative integers, $0! = 1$. And $n!$ is the product of all the positive integers from 1 to n, whenever n is a positive integer.

$\binom{n}{r}$ Defined only for positive integers $n \geqq r$. It is the number of distinct subsets of r objects from a set of n distinct objects. It is equal to $(n!)/((r!)(n-r)!)$.

Bounded set A subset $\mathbf{S} \subset \mathbf{R}^n$ is *bounded* if there exists a finite real number α such that $\|x\| \leqq \alpha$, for all $x \in \mathbf{S}$.

Superscript^{T} Denotes transposition. See Section 3.2. A^{T} is the transpose of the matrix A. If x is a column vector, x^{T} is the same vector written as a row vector and vice versa. Column vectors are printed as transposes of row vectors to conserve space in the text.

Superscripts We use superscripts to enumerate vectors or matrices or elements in any set. When considering a set of vectors, in $\mathbf{R}^n$, x^r may be used to denote the rth vector in the set, and it will be the vector $(x_1^r, \ldots, x_n^r)$. In a similar manner, while considering a sequence of matrices, the symbol P^r may be used to denote the rth matrix in the sequence. Superscripts should not be confused with exponents and these are distinguished by different type styles.

Exponents In the symbol $\varepsilon^{\mathbf{r}}$, $\mathbf{r}$ is the exponent. $\varepsilon^{\mathbf{r}} = \varepsilon \times \varepsilon \times \cdots \times \varepsilon$, where there are $\mathbf{r}$ ε's in this product. Notice the difference in type style between superscripts and exponents.

Proper subset If $\mathbf{E}$ is a subset of a set $\boldsymbol{\Gamma}$, $\mathbf{E}$ is said to be a proper subset of $\boldsymbol{\Gamma}$ if $\mathbf{E} \neq \boldsymbol{\Gamma}$, that is if $\boldsymbol{\Gamma} \backslash \mathbf{E} \neq \emptyset$.

Linear function A linear function of the variables (or parameters) $\lambda_1, \ldots, \lambda_r$ is a function of the form $c_1\lambda_1 + \cdots + c_r\lambda_r$, where $c_1, \ldots, c_r$ are given real numbers.

Affine function An affine function of the variables (or parameters) $\lambda_1, \ldots, \lambda_r$, is a function of the form $c_0 + c_1\lambda_1 + \cdots + c_r\lambda_r$, where $c_0, c_1, \ldots, c_r$ are given real numbers. If $c_0 = 0$, the affine function is a linear function. In general, an affine function is a function that is equal to a constant plus a linear function.

Consistent system of equations A system of simultaneous linear equations, say $Ax = b$, for which at least one solution, x, exists. The system is *inconsistent*, if it has no solutions. See Section 3.3.5.

Redundant equations In a system of simultaneous linear equations, a redundant equation is one that can be obtained as a linear combination of the others. See Sections 2.6.2 and 3.3.5.

Basis for R^n A set of n vectors in $\mathbf{R}^n$ that is a linearly independent set.

Basis, basic vector, for an LP These terms are *only* defined for LPs in which the constraints are a system of equations in nonnegative variables. Before talking about a basis or a basic vector for an LP, first transform it so that the constraints in it are a system of equations in nonnegative variables (see Section 2.2). Eliminate all the redundant equations from the system. Let the resulting system of constraints be $Ax = b$, $x \geqq 0$. In this system the variable x_j is associated with the column $A_{.j}$. Let A be of order $m \times n$. Any nonsingular square submatrix of A of order m is called a *basis* for this LP. A *basic vector* for this LP is the vector of variables associated with the columns in a basis. See Sections 2.3.2 and 3.5.1.

Pos$\{A_1, \ldots, A_k\}$ If $A_1, \ldots, A_k$ are vector in $\mathbf{R}^n$, then $\text{Pos}\{A_1, \ldots, A_k\} = \{y: y = \alpha_1 A_1 + \cdots + \alpha_k A_k, \alpha_1 \geqq 0, \alpha_2 \geqq 0, \ldots, \alpha_k \geqq 0\}$.

Cardinality Defined only for sets. The cardinality of a set is the number of elements in it.

Tight or slack constraints Let $\bar{x}$ be a point in $\mathbf{R}^n$ satisfying an inequality constraint $f(x) \geqq 0$. If $f(\bar{x}) > 0$, the constraint "$f(x) \geqq 0$" is said to be *slack* at $\bar{x}$. If $f(\bar{x}) = 0$, the constraint "$f(x) \geqq 0$" is said to be *tight* or *active* at $\bar{x}$.

Maximization In Chapter 2 and the following chapters, whenever a function $f(x)$ has to be maximized subject to some conditions, we look at the equivalent problem of minimizing $-f(x)$ subject to the same conditions. Both problems have the same set of optimum solutions, and the maximum value of $f(x) = -$minimum value of $(-f(x))$. Hence the algorithms discussed there are stated in terms of solving minimization problems.

Activity Each decision variable in a linear-programming model can be interpreted as the level at which an associated activity is carried out by the decision maker. Thus every possible activity that the decision maker can perform corresponds to a decision variable in the linear-programming model and vice versa.

Item Any material or resource on which there is either a requirement or a limit on its availability or use, thus leading to a constraint in the model. Each item leads to a constraint in the model and each constraint in the model is the material balance equation or inequality associated with an item. See Section 1.1.4.

Linear program in standard form This is a linear program in which all the variables are restricted to be nonnegative, all other constraints in the variables are linear equality constraints, and the objective function is a linear function that is required to be minimized.

Feasible solution A numerical vector that satisfies all the constraints and restrictions in the problem.

Optimum solution, or Optimum feasible solution A feasible solution that optimizes (i.e., either maximizes or minimizes as required) the objective value among all feasible solutions.

Algorithm The word comes from the last name of the Persian scholar Abu Ja'far Mohammed ibn Mûsâ alkhowârizmî whose textbook on arithmetic (about A.D. 825) had a significant influence on the development of these methods. An *algorithm* is a set of rules for getting a required output from a specific input in which each step is so precisely defined that it can be translated into computer language and executed by machine.

Necessary conditions, sufficient conditions, necessary and sufficient conditions When studying a property of a system, a condition is said to be a *necessary condition* for that property if that condition is satisfied whenever the property holds. A condition is said to be a *sufficient condition* for the property if the property holds whenever the condition is satisfied. A *necessary and sufficient condition* for the property is a condition that is both a necessary condition and a sufficient condition for that property.

Karush–Kuhn–Tucker necessary conditions for optimality Let $f(x)$, $g_i(x)$, $h_t(x)$ be real-valued differentiable functions defined on $\mathbf{R}^n$, for all i, t. Consider the following mathematical program: minimize $f(x)$, subject to $g_i(x) \geqq 0$ for $i = 1$ to m, $h_t(x) = 0$ for $t = 1$ to p. The Karush–Kuhn–Tucker Lagrangian for this problem is $L(x, \pi, \mu) = f(x) - \sum_{i=1}^{m} \pi_i g_i(x) - \sum_{t=1}^{p} \mu_t h_t(x)$, where π_i, μ_t are the Lagrange multipliers associated with the constraints. The Karush–Kuhn–Tucker (KKT) necessary optimality conditions are: $[\partial L(x, \pi, \mu)/\partial x] = \nabla f(x) - \sum_{i=1}^{m} \pi_i \nabla g_i(x) - \sum_{t=1}^{p} \mu_t \nabla h_t(x) = 0$, $g_i(x) \geqq 0$ for $i = 1$ to m, $h_t(x) = 0$ for $t = 1$ to p, $\pi_i g_i(x) = 0$ for $i = 1$ to m, $\pi_i \geqq 0$ for all i; where $\nabla f(x)$, etc., are the vectors of partial derivatives. If $\bar{x}$ is a local minimum for this problem, under fairly general conditions, it can be shown that there exist multipliers $\bar{\pi}$, $\bar{\mu}$ such that $\bar{x}$, $\bar{\pi}$, $\bar{\mu}$ satisfy the KKT conditions. In the literature these conditions are usually called *first-order necessary optimality conditions,* or *Kuhn–Tucker conditions.* But it has been found recently that Karush was the first to discuss them. Hence nowadays the name *Karush–Kuhn–Tucker necessary optimality conditions* is coming into vogue.

Size The size of an optimization problem is a parameter that measures how large the problem is. Usually it is the number of digits in the data in the optimization problem, when it is encoded in binary form.

$O(n^r)$ A finitely terminating algorithm for solving an optimization problem is said to be of order n^r or $O(n^r)$, if the computational effort required by the algorithm in the worst case, to solve a version of the problem of size n, grows as αn^r, where α, r are numbers that are independent of the size n and the data in the problem.

Polynomially bounded algorithm An algorithm is said to be polynomially bounded if it can be proved that the computational effort required by it is bounded above by a fixed polynomial in the size of the problem.

The class P of problems This is the class of all problems for solving which there exists a polynomially bounded algorithm.

chapter 7

Formulation of Linear Programs

chapter 1

1.1 FORMULATION OF LINEAR PROGRAMS

1.1.1 Introduction

Optimization, or *mathematical programming* is the branch of mathematics dealing with methods for optimizing (i.e., either maximizing or minimizing) an objective function of *n* decision variables subject to specified constraints on these variables. We begin this book with an incident related to optimization. Once I was planning for a trip to India. At that time my brother, who lives in India, had a four-year-old daughter and a boy who was just born. The question of what gift to take for my brother's family arose, and my wife suggested that I take a baby scale, which they could use to keep track of the baby's weight. We got a really fancy one, and my brother became so excited on seeing it that he wanted to check the baby's weight immediately. We set it up and he put the baby on the pan. His four-year-old daughter was watching us with great curiosity, but she could not figure out what was going on, so she asked her father what he was doing. Just to tease her, he said, "You see, honey, your uncle from America likes your brother so much that he agreed to buy him from us at a cost of $1 per ounce of his weight. So I am weighing the baby to see how much money we will get by selling him to your uncle!" On hearing this, the little girl looked quite shocked, and with a sad face, she asked, "You are not going to sell him now, are you, daddy?" From the worry in her face it was clear that she loved her brother very much; this pleased her father, and in order to console her, he said, "Don't worry, honey, I will never give your brother away to your uncle. I said it only to tease you." Rolling her eyes in surprise, the little girl then said, "That is not what I meant, daddy. You see the boy is very small now, but he is gaining weight rapidly every week. If we wait for some more weeks, he will be much heavier, and will fetch much more money . . . ." The attitude of the little girl in this story is one of optimization, and it is almost universal. Interest in optimization methods and their applications is very widespread now.

In the days before digital computers, only very simple optimization models involving few variables and few constraints could be solved by the tools then available. So, in those days, a great deal of emphasis was placed on keeping the model *very small and compact,* as otherwise it could not be solved. In the last 50 years, a large number of algorithms to solve a great variety of optimization problems have been developed, and large digital computers for implementing and executing these algorithms have become widely available. Consequently, many types of optimization models can now be solved efficiently; therefore, practitioners are building many large-scale optimization models and solving them almost routinely. The emphasis in constructing an optimization model these days is to keep it *computable* or *tractable.* One such model is the *linear programming model,* in which a linear objective function is to be optimized subject to linear equality and inequality constraints and sign restrictions (or lower and/or upper bounds) on the decision variables. We now have very efficient algorithms for solving linear programming problems, and very high-quality software for these algorithms is available at minimal cost. Many large-scale linear programming problems (involving a few thousand constraints in several thousands of variables) are being solved daily very satisfactorily using this software.

In this book we discuss linear programming in all its depth. In this chapter we discuss techniques for constructing linear programming models for practical problems. To formulate a real life problem as a linear program is an art in itself. Even though there are excellent methods for solving a problem once it is formulated as a linear program, there is little theory to help in formulating the problems in this way. In a real life problem, several approximations may have to be made before it can be modeled as a linear program. The basic principles will be illustrated by considering a simple example. We will use the abbreviation LP for linear program.

1.1.2 Product Mix Problem

A company makes two kinds of fertilizers, called Hi-phosphate and Lo-phosphate. Three basic raw materials are used in manufacturing these fertilizers as in the table given at the bottom.

The raw material supply comes from the company's own quarry. The net profit is the selling price minus the labor and other production costs. How much of each fertilizer should the company manufacture to maximize its total net profit? This problem will be formulated using two different approaches, the *direct approach* and the *input-output approach.*

1.1.3 The Direct Approach

Step 1 Prepare a list of all the decision variables in the problem. This list must be complete in the sense that if an optimum solution providing the values of each of the variables is obtained, the decision maker should be able to translate it into an optimum policy that can be implemented. In the problem of Section 1.1.2 the variables are:

$x_1 =$ the tons of Hi-phosphate to be manufactured

$x_2 =$ the tons of Lo-phosphate to be manufactured

Associated with each variable in the problem is an *activity* that the decision maker can perform. In this problem there are two such activities:

Activity 1: to manufacture 1 ton of Hi-phosphate

Activity 2: to manufacture 1 ton of Lo-phosphate

The variables in the problem just define the *levels* at which these activities are carried out. When the constraints binding the variables are written, it will be clear that there are additional variables (called *slack variables* in what follows) imposed on the problem by the inequality constraints. Slack variables will be discussed in greater detail later. Formulation of the problem as an LP requires the following assumptions:

Proportionality Assumption

It requires 2 tons of raw material 1 to manufacture 1 ton of Hi-phosphate. The *proportionality assumption* implies that $2x_1$ tons of raw material 1 are required to manufacture x_1 tons of Hi-phosphate for any $x_1 \geqq 0$. In general, the proportionality assumption guarantees that if a_{ij} units of the ith item are consumed (or produced) in carrying out activity j at unit level, then $a_{ij}x_j$ units of this item are consumed (or produced) in carrying out activity j at level x_j; and if $\$c_j$ of profit are earned in carrying out activity j at unit level, $\$c_jx_j$ is the contribution of activity j to the net profit; for any $x_j \geqq 0$.

Additivity Assumption

It requires 2 tons of raw material 1 to manufacture 1 ton of Hi-phosphate and 1 ton of the same raw material to manufacture 1 ton of Lo-phosphate. The *additivity assumption* implies that $2x_1 + x_2$ tons of raw material 1 is required to manufacture x_1 tons of Hi-phosphate and x_2 tons of Lo-phosphate for any $x_1 \geqq 0$, $x_2 \geqq 0$. The additivity assumption generally implies that the total consumption (or production) of an item is equal to the sum of the various quantities of the item consumed (or produced) in carrying out each individual activity at its specified level. The additivity assumption implies that the objective function is *separable in the variables*; that is, if the variables in the model are $x_1, \ldots, x_n$,

	Tons of Raw Material Required to Manufacture 1 Ton		
Raw Material	*Hi-Phosphate*	*Lo-Phosphate*	*Maximum Amount of Raw Materials Available per Month (tons)*
1	2	1	1500
2	1	1	1200
3	1	0	500
Net profit per ton manufactured	\$15	\$10	

and the objective function is $z(x_1, \ldots, x_n) = z(x)$, then $z(x)$ can be written as a sum of n functions, each of which involves only one variable in the model [i.e., as $z_1(x_1) + \cdots + z_n(x_n)$, where $z_j(x_j)$ is the contribution of the variable x_j to the objective function].

The proportionality and the additivity assumptions together are known as the *linearity assumptions.*

The Implications of the Proportionality and Additivity Assumptions In most real life problems, the proportionality and additivity assumptions may not hold exactly. For example, the selling price per unit normally decreases in a nonlinear fashion as the number of units purchased increases. Thus, in formulating real life problems as LPs, several approximations may have to be made. It has been found that most real life problems can be approached very closely by suitable linear approximations. This, and the relative ease with which LPs can be solved, have made it possible for linear programming to find a vast number of applications. The proportionality and the additivity assumptions automatically imply that all the constraints in the problem are either linear equations or inequalities. They also imply that the objective function is linear.

Continuity of Variation Assumption

It is assumed that each variable in the model can take all the real values in its range of variation.

In real life problems, some variables may be restricted to take only integer values (e.g., if the variable represents the number of empty buses transported from one location to another). Such restrictions make the problem an *integer programming problem.* Integer programs are much harder to solve than continuous variable LPs. They are not discussed in this book. See [1.71, 1.92, 1.94, 1.104, 1.109] for integer programming.

Step 2 Write down all the constraints and the objective function in the problem.

Nonnegativity Restrictions

The variables x_1, x_2 have to be nonnegative to make any practical sense. In linear programming models in general the nonnegativity restriction on the variables is a natural restriction that occurs because certain activities (manufacturing a product, etc.) can only be carried out at nonnegative levels.

Items and the Associated Constraints

There may be other constraints on the variables, imposed by lower or upper bounds on certain products that are either inputs to the production process or outputs from it. Any such product that leads to a constraint in the model is called an *item.* Each item leads to a constraint on the decision variables, and conversely every constraint in the model is associated with an item. Make a list of all the *items* that lead to constraints in the problem.

In the fertilizer problem of Section 1.1.2 each raw material leads to a constraint; for example, the amount of raw material 1 used is $2x_1 + x_2$ tons, and this cannot exceed 1500 tons. This imposes the constraint $2x_1 + x_2 \leqq 1500$. Since this inequality just compares the amount of raw material 1 used to the amount available, it is called a *material balance inequality.* The material balance equations or inequalities corresponding to the various items in the problem are the constraints of the problem.

Each material balance inequality in the model contains in itself the definition of another nonnegative variable known as a *slack variable.* For example, the constraint imposed by the restriction on the amount of raw material 1 available, namely $2x_1 + x_2 \leqq 1500$, can be written as $1500 - 2x_1 - x_2 \geqq 0$. If we define $x_3 = 1500 - 2x_1 - x_2$, then x_3 is a slack variable representing the amount of raw material 1 remaining unutilized, and the constraint can be written in the equivalent form $2x_1 + x_2 + x_3 = 1500$, where $x_3 \geqq 0$.

Objective Function

Now write down the objective function. By the proportionality and additivity assumptions, this is guaranteed to be a linear function of the variables. This completes the formulation of the problem as an LP. It is as follows:

		Item
Maximize	$z(x) = 15x_1 + 10x_2$	
Subject to	$2x_1 + x_2 \leqq 1500$	Raw 1
	$x_1 + x_2 \leqq 1200$	Raw 2
	$x_1 \leqq 500$	Raw 3
	$x_1 \geqq 0 \quad x_2 \geqq 0$	

(1.1)

Example 1.1 A Blending Problem

A refinery takes four raw gasolines, blends them, and produces three types of fuel. Here are the data.

Raw Gas Type	*Octane Rating*	*Available Barrels per Day*	*Price per Barrel*
1	68	4000	$31.02
2	86	5050	33.15
3	91	7100	36.35
4	99	4300	38.75

Fuel Blend Type	*Minimum Octane Rating*	*Selling Price ($/barrel)*	*Demand Pattern*
1	95	45.15	At most 10,000 barrels/day
2	90	42.95	Any amount can be sold
3	85	40.99	At least 15,000 barrels/day

The company sells raw gasolines not used in making fuels at $38.95/barrel if its octane rating is over 90 and at $36.85/barrel if its octane rating is less than 90. How can the company maximize its total daily profit? In formulating this problem, we find that the variables are

x_{ij} = barrels of raw gasoline of type i used in making fuel type j per day for $i = 1$ to 4 and $j = 1, 2, 3$

y_i = barrels of raw gasoline type i sold as it is in the market

The octane rating of fuel type 1 should be $\geqq 95$. Assuming that the proportionality and additivity assumptions hold for the octane rating of the blend, the octane rating of this fuel, if a positive quantity of this fuel is made, is $(68x_{11} + 86x_{21} + 91x_{31} + 99x_{41})/(x_{11} + x_{21} + x_{31} + x_{41})$. Thus, if a positive quantity of this fuel is made, the constraint on its octane rating requires $(68x_{11} + 86x_{21} + 91x_{31} + 99x_{41})/(x_{11} + x_{21} + x_{31} + x_{41}) \geqq 95$, or, equivalently, $68x_{11} + 86x_{21} + 91x_{31} + 99x_{41} - 95(x_{11} + x_{21} + x_{31} + x_{41}) \geqq 0$. This inequality will also hold when the amount of this fuel manufactured is zero. Hence, the octane rating constraint on this fuel may be represented by this linear inequality. If we proceed in a similar manner, we obtain the formulation given at the bottom.

By defining the variables as the amounts of various ingredients in the blend, blending problems can be formulated in this manner. However, if the variables are defined to be the proportions of the various ingredients in the blend, remember to include in the model the constraint that the sum of the proportions of the various ingredients in a blend should be equal to one.

$$\begin{aligned}
\text{Maximize}\quad & 45.15(x_{11} + x_{21} + x_{31} + x_{41}) + 42.95(x_{12} + x_{22} + x_{32} + x_{42}) \\
& + 40.99(x_{13} + x_{23} + x_{33} + x_{43}) + y_1(36.85 - 31.02) + y_2(36.85 - 33.15) \\
& + y_3(38.95 - 36.35) + y_4(38.95 - 38.75) - 31.02(x_{11} + x_{12} + x_{13}) \\
& - 33.15(x_{21} + x_{22} + x_{23}) - 36.35(x_{31} + x_{32} + x_{33}) - 38.75(x_{41} + x_{42} + x_{43}) \\
\text{Subject to}\quad & 68x_{11} + 86x_{21} + 91x_{31} + 99x_{41} - 95(x_{11} + x_{21} + x_{31} + x_{41}) \geqq 0 \\
& 68x_{12} + 86x_{22} + 91x_{32} + 99x_{42} - 90(x_{12} + x_{22} + x_{32} + x_{42}) \geqq 0 \\
& 68x_{13} + 86x_{23} + 91x_{33} + 99x_{43} - 85(x_{13} + x_{23} + x_{33} + x_{43}) \geqq 0 \\
& x_{11} + x_{12} + x_{13} + y_1 = 4000 \\
& x_{21} + x_{22} + x_{23} + y_2 = 5050 \\
& x_{31} + x_{32} + x_{33} + y_3 = 7100 \\
& x_{41} + x_{42} + x_{43} + y_4 = 4300 \\
& x_{11} + x_{21} + x_{31} + x_{41} \leqq 10{,}000 \\
& x_{13} + x_{23} + x_{33} + x_{43} \geqq 15{,}000 \\
& x_{ij}, y_i \geqq 0 \qquad \text{for all } i \text{ and } j
\end{aligned}$$

1.1.4 The Input-Output Approach

In this procedure the lists of all the possible *activities* and the *items* defining the constraints are first prepared. The list of *activities* should include all the possible actions that the decision maker can perform in the problem. Units should be fixed for measuring the level at which each activity is carried out. The levels at which the activities are performed define the variables in the problem.

An *item* in the problem is any material or resource on which there is either a requirement or a limit on its availability or use. Each item leads to a constraint in the problem. The list of items should include an item called *dollar* that defines the objective function. If money is required to carry out an activity at unit level, dollar is an *input* for carrying out this activity. If money is produced as a result of carrying out an activity, dollar is an *output from* it. Even if the optimized item is different from money, we will use the term *dollar* to describe it.

To write the constraints in tabular form, define x_j as the level at which the jth activity is carried out, and assign the jth column in the tableau to correspond to this activity. Likewise associate a row of the tableau with each item. If a_{ij} units of item i are required as an input for carrying out the jth activity at unit level, enter $+a_{ij}$ in the (i, j) position (i.e., in the ith row and jth column) of the tableau. On the other hand, if a_{ij} units of item i are produced as an output by carrying out the jth activity at unit level, enter $-a_{ij}$ in the (i, j) position of the tableau. Thus, inputs to activities are entered with a "+" sign, and outputs from activities are entered with a "−" sign ($\xrightarrow{+}\square\xrightarrow{-}$).

In the objective row corresponding to the item "dollar" all costs are inputs and, hence, are entered with a "+" sign, and all profits are outputs and are considered as *negative costs* and are entered with a "−" sign. The constraints are completed by entering the limits on the availability or requirements of the items, on a right-hand side column of the tableau. The objective row gives the coefficients of the objective function that is to be minimized. When formulated in this manner, the fertilizer product mix problem leads to the tableau given at the bottom. The blank entries in the tableau are zeros.

This is equivalent to the model formulated earlier. x_3, x_4, x_5 are the slack variables corresponding to the inequality constraints obtained in the earlier model.

Example 1.2 A Multiperiod Problem

A nationalized company can either manufacture capital goods or consumer goods. The company's production is being planned over a five-period horizon. At the beginning of period 1 the company has 100 units of capital goods and 50 units of consumer goods in stock. Each unit of capital goods can be used to produce *either* three new units of capital goods at the expense of three units of consumer goods to be provided from stock *or* 10 units of consumer goods, over an uninterrupted two-period interval, at the end of which the unit of capital goods again becomes available for use. Once capital goods are produced they can be used for production in subsequent periods. If necessary, capital goods can also be left idle. The company will sell everything in its stock at the beginning of period 6 and close up. The company can also sell either capital goods or consumer goods in any period according to the following price schedule. Which production plan will best maximize the total returns?

	Activities and Levels						
	To Make 1 Ton		*To Leave 1 Ton Unutilized of Raw Material*				
	Hi-phosphate	*Lo-phosphate*	*1*	*2*	*3*		*Right-Hand Side Constants*
Items	x_1	x_2	x_3	x_4	x_5		
Raw material 1	2	1	1			=	1500
Raw material 2	1	1		1		=	1200
Raw material 3	1				1	=	500
Dollars cost	−15	−10				=	$z(x)$ minimize

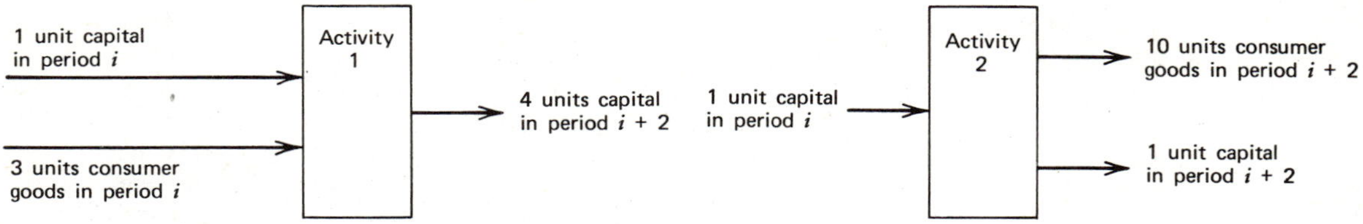

Figure 1.1

	Price in Monetary Units per Unit Goods During the Period	
Period	*Capital Goods*	*Consumer Goods*
1	4	0.1
2	6	0.15
3	8	0.15
4	6	0.15
5	3	0.1
6	3	0.1

We will formulate this problem by using the input-output approach. During each period the company can possibly engage in the following activities.

1 Assign one unit of capital goods to manufacture three units of capital goods over the next two periods.
2 Assign one unit of capital goods to manufacture 10 units of consumer goods over the next two periods.
3 Leave one unit of capital goods idle until next period.
4 Sell one unit of capital goods.
5 Sell one unit of consumer goods from stock.
6 Carry over one unit of consumer goods in stock until next period.

Of course the company cannot engage in activities 1 or 2 during period 5 because the company has to close down at the beginning of period 6. Let x_{ij} be the level at which activity j is carried out during the ith period.

The *items* are the capital goods and the consumer goods in stock in the various periods. Here is the list of items: Capital goods available at the beginning of period i, Consumer goods available at the beginning of period i, for various values of i, and dollar.

Clearly the item *capital goods at the beginning of period* 3 is an input into activities 1, 2, 3, 4 during period 3 and it is an output from activity 1 during period 1. Also one unit of capital goods assigned to produce new capital goods at the beginning of period 1 finishes this job by the end of period 2 and becomes available for use again at the beginning of period 3. Thus the output from the activity of assigning one unit of capital goods at the beginning of period 1 to produce capital goods is four units of capital goods (itself and the three new units of capital goods it has produced) at the beginning of period 3. Figure 1.1 represents some of the input-output features in this problem. From these considerations we arrive at the formulation of the problem in Tableau 1.1. Each of the constraints in the problem is a material balance equation; this formulation could also have been obtained by direct arguments.

Example 1.3 A Diet Problem

In the diet model a list of possible foods that can be included in the diet is provided, together with the nutrient content and the cost per unit weight of each food. A *diet* is a vector that specifies how much of each food is to be consumed per day. It is a *feasible diet* if the amount of each nutrient contained in it is greater than or equal to the specified minimum daily requirement (MDR) for that nutrient. The problem is to find a minimum cost feasible diet. In the following diet problem the nutrients are starch, protein, and vitamins, and the foods are two types of grain with data given below.

	Nutrient Units/Kilogram of Grain Type		
Nutrient	*1*	*2*	*MDR of Nutrient in Units*
Starch	5	7	8
Protein	4	2	15
Vitamins	2	1	3
Cost ($/kg) of food	0.60	0.35	

Tableau 1.1

	Activities and their levels																														
	Period 1						*Period 2*						*Period 3*						*Period 4*						*Period 5*				*Beginning of period 6*		
Item	*1* x_{11}	*2* x_{12}	*3* x_{13}	*4* x_{14}	*5* x_{15}	*6* x_{16}	*1* x_{21}	*2* x_{22}	*3* x_{23}	*4* x_{24}	*5* x_{25}	*6* x_{26}	*1* x_{31}	*2* x_{32}	*3* x_{33}	*4* x_{34}	*5* x_{35}	*6* x_{36}	*1* x_{41}	*2* x_{42}	*3* x_{43}	*4* x_{44}	*5* x_{45}	*6* x_{46}	*3* x_{53}	*4* x_{54}	*5* x_{55}	*6* x_{56}	*4* x_{64}	*5* x_{65}	
Capital in 1	1	1	1	1																											= 100
Consumer goods in 1	3				1	1																									= 50
Capital in 2			−1				1	1	1	1																					= 0
Consumer goods in 2						−1	3				1	1																			= 0
Capital in 3	−4	−1							−1				1	1	1	1															= 0
Consumer goods in 3		−10										−1	3				1	1													= 0
Capital in 4							−4	−1							−1				1	1	1	1									= 0
Consumer goods in 4								−10										−1	3				1	1							= 0
Capital in 5													−4	−1							−1				1	1					= 0
Consumer goods in 5														−10										−1			1	1			= 0
Capital at beginning of 6																			−4	−1					−1				1		= 0
Consumer goods at beginning of 6																				−10								−1		1	= 0
\$ = objective to be minimized				−4	−0.1					−6	−0.15					−8	−0.15					−6	−0.15			−3	−0.1		−3	−0.1	

The activities and their levels in this model are for $j = 1, 2$: Activity j: to include 1 kg of grain type j in the diet. Associated level $= x_j$. So x_j is the amount in kilograms of grain j included in the daily diet, $j = 1, 2$, and the vector $x = (x_1, x_2)^T$ is the diet. The items in this model are the various nutrients, each of which leads to a constraint in the model. For example, the amount of starch contained in diet x is $5x_1 + 7x_2$, which must be $\geqq 8$ for feasibility. This leads to the following formulation.

Minimize	$z(x) = 0.60x_1 + 0.35x_2$	Item
Subject to	$5x_1 + 7x_2 \geqq 8$	Starch
	$4x_1 + 2x_2 \geqq 15$	Protein
	$2x_1 + x_2 \geqq 3$	Vitamins

and

$$x_1, x_2 \geqq 0$$

Comment 1.1: A diet problem involving 77 different foods, to meet the minimum requirements of nine different nutrients at minimum cost was formulated by G. J. Stigler in "The Cost of Subsistence," *Journal of Farm Economics*, vol. 27, 1945. The formulation led to an LP in 77 nonnegative variables, with nine inequality constraints. Stigler did not know of any method for solving the LP at that time, but he obtained an approximate solution using a trial and error procedure that led to a diet that met the minimum requirements of the nine nutrients considered in the model at an annual cost of \$39.93 at 1939 prices. After Professor George B. Dantzig developed the simplex algorithm for solving LPs in 1947, Stigler's diet problem was probably one of the first large problems to be solved by the simplex method, and it gave the true optimum diet for this model, with an annual cost of \$39.67 at 1939 prices. So the trial and error solution of Steigler was very close to the optimum. Stigler's original diet model contained no constraints to guarantee that the diet is palatable, and does not allow much room for day-to-day variations in diet that contribute to eating pleasure, and hence is very hard to implement. The basic model can be modified by including additional constraints to make sure that the solution obtained leads to a tasteful diet with ample scope for variety. This sort of modification of the model after looking at the optimum solution to determine its reasonableness and implementability, solving the modified model, and even repeating this whole process several times, is fairly typical in practical applications of linear programming.

Human beings have the philosophy of live to eat as well as eat to live. And, if we can afford it, we do not really bother about the cost of the food. It is also impossible to make a human being eat the same diet every day. For all these reasons, it is not practical to determine human diet using an optimization model.

However, it is much easier to make cattle and fowl consume the diet that is determined as being optimal for them. For this reason, the diet model is used extensively in farms for determining optimal cattle feed, chicken feed, turkey feed, etc. Since the availability and prices of the foods keep varying all the time, large-scale farms run their diet model several times each year to determine the optimal feed mix with current data.

Example 1.4 The Transportation Problem

In the transportation model we have a set of nodes or places called *sources*, which have a commodity available for shipment, and another set of places called *demand centers*, which require this commodity. The amount of commodity available at each source (called the *availability*) and the amount required at each demand center (called the *requirement*) are specified, as well as the cost per unit of transporting the commodity from each source to each demand center. The problem is to determine the quantity to be transported from each source to each demand center, so as to meet all the requirements at minimum total shipping cost. Consider the problem where the commodity is iron ore, the sources are mines 1, 2, where the ore is produced, and the demand centers are three steel plants that require the ore, with data as given below.

	Unit Cost of Shipping Ore from Mine to Steel Plant (cents per ton)			*Amount of Ore Available at Mine (tons)*
	1	*2*	*3*	
Mine 1	9	16	28	103
Mine 2	14	29	19	197
Tons of ore required at steel plant	71	133	96	

The activities in the transportation model are: to ship one unit of the commodity from source i to demand center j, over i, j. It is convenient to represent the variable corresponding to the level at which this activity is carried out by the double subscripted symbol x_{ij}. Thus, x_{ij} represents the amount of iron ore in tons shipped from mine i to steel plant j. The items in this model are the ore at various locations. Consider ore at mine 1. There are only 103 tons of it available, and the amount of ore shipped out of mine 1, which is $x_{11} + x_{12} + x_{13}$, cannot exceed the amount available, leading to the constraint $x_{11} + x_{12} + x_{13} \leqq 103$. Likewise, considering ore at steel plant 1, there are at least 71 tons of it required, and the amount of ore shipped to steel plant 1 has to be greater than or equal to this amount, leading to the constraint $x_{11} + x_{21} \geqq 71$. This leads to the formulation given at the bottom.

The Special Structure of the Transportation Problem

As an LP, the transportation problem has a very special structure. It can be very compactly represented in the form of an array, in which row i corresponds to source i, and column j corresponds to demand center j. The cell in row i and column j of the array, denoted by (i, j), corresponds to the shipping route from source i to demand center j. Inside this cell record the decision variable x_{ij}, which represents the amount of commodity shipped along this route, and enter the cost per unit shipped on this route in the lower right-hand corner. The objective function in this model is the sum of the variables in the array multiplied by the cost coefficient in the corresponding cell. Record the availabilities at the sources in a column on the right-hand side of the array; and similarly the requirements at the demand centers, in a row at the bottom of the array. Then each constraint in the model is a constraint on the sum of all the variables either in a row or a column of the array, and it can be read off easily from the array as shown below.

Array Representation of the Transportation Problem

	Steel Plant			
	1	2	3	
Mine 1	x_{11} 9	x_{12} 16	x_{13} 28	$\leqq 103$
Mine 2	x_{21} 14	x_{22} 29	x_{23} 19	$\leqq 197$
	$\geqq 71$	$\geqq 133$	$\geqq 96$	

$x_{ij} \geqq 0$ for all i, j. Minimize cost.

Any LP, whether it comes from a transportation or a different context, that can be represented in this special form of an array, is called a *transportation problem*. In general, the constraints may be equations or inequalities.

In a general LP, even when all the data are integer valued, there is no guarantee that there will be an optimum integer solution. However, the special structure of the transportation problem makes the following theorem possible.

Integer Property in the Transportation Model

In a transportation model, if all the availabilities and requirements are positive integers, and if the problem has a feasible solution, then it has an optimum solution in which all the decision variables x_{ij} assume only integer values.

Caution: The above statement does not claim that an optimum solution for the transportation problem with positive integral availabilities and requirements must satisfy the integer property. There may

$$
\begin{array}{llll}
\text{Minimize} & z(x) = 9x_{11} + 16x_{12} + 28x_{13} + 14x_{21} + 29x_{22} + 19x_{23} & & \text{Item} \\
\text{Subject to} & x_{11} + x_{12} + x_{13} & \leqq 103 & \text{Ore at Mine 1} \\
& x_{21} + x_{22} + x_{23} & \leqq 197 & \text{Ore at Mine 2} \\
& x_{11} + x_{21} & \geqq 71 & \text{Ore at Plant 1} \\
& x_{12} \qquad x_{22} & \geqq 133 & \text{Ore at Plant 2} \\
& x_{13} \qquad x_{23} & \geqq 96 & \text{Ore at Plant 3}
\end{array}
$$

and

$$x_{ij} \geqq 0 \quad \text{for all } i = 1, 2, \quad j = 1, 2, 3$$

be many alternate optimum solutions for the problem, and the statement only says that at least one of these optimum solutions satisfies the integer property.

This integer property is proved in Chapter 13, and using it, a special algorithm is developed for transportation problems.

Example 1.5 The Assignment Problem

This problem, known as the *marriage problem*, was proposed as an application of linear programming to sociology in the early 1950s. We consider a club of sociologists consisting of five men and five women who know each other very well. Based on their personal opinions, we are told that the amount of happiness that the ith man and the jth woman derive by spending a period of time x_{ij} together is $c_{ij}x_{ij}$, where the c_{ij} are tabulated below.

$c = (c_{ij}) =$

$i \backslash j$	1	2	3	4	5
1	78	−16	19	25	83
2	99	98	87	16	92
3	86	19	39	88	17
4	−20	99	88	79	65
5	67	98	90	48	60

These are the rates at which happiness is acquired by the various possible couples, per unit time spent together. If c_{ij} is positive, the couple is happy when together. If c_{ij} is negative, the couple is unhappy when together, and so acquire unhappiness only.

Determine how much time each man in the club should spend with each woman, so as to maximize the overall happiness derived by all the members of the club. To keep the model simple, we assume that the remaining lifetimes of all club members are equal, and measure time in units of this lifetime. There are 25 activities in this model, and these are for $i, j = 1$ to 5: Activity: man i and woman j to spend one unit of time together. Associated level $= x_{ij}$. Thus x_{ij} is the fraction of their lifetime that man i and woman j spend together, for $i, j = 1$ to 5. The items in this model are the various lifetimes of each member of the club. For example, man 1's lifetime leads to the constraint that the sum of the fractions of his lifetime he spends with each woman should be equal to 1, that is, $x_{11}, + x_{12} + x_{13} + x_{14} + x_{15} = 1$. This marriage problem is the following transportation problem.

Man \ Woman	1	2	3	4	5	
1	x_{11} 78	x_{12} −16	x_{13} 19	x_{14} 25	x_{15} 83	= 1
2	x_{21} 99	x_{22} 98	x_{23} 87	x_{24} 16	x_{25} 92	= 1
3	x_{31} 86	x_{32} 19	x_{33} 39	x_{34} 88	x_{35} 17	= 1
4	x_{41} −20	x_{42} 99	x_{43} 88	x_{44} 79	x_{45} 65	= 1
5	x_{51} 67	x_{52} 98	x_{53} 90	x_{54} 48	x_{55} 60	= 1
	= 1	= 1	= 1	= 1	= 1	

$x_{ij} \geqq 0$ for all i, j; maximize objective

It is a special transportation problem in which the number of sources equals the number of demand centers, all the availabilities and requirements are 1, and all the constraints are equality constraints. Since all variables are $\geqq 0$, and the sum of all the variables in each row and column is required to be 1, all variables in this problem have to lie between 0 and 1. From the integer property mentioned earlier, this problem has an optimum solution in which all x_{ij} take integer values, and in such a solution all x_{ij} are equal to 0 or 1. One such solution, for example, is

$$x = (x_{ij}) = \begin{pmatrix} 0 & 0 & 0 & 0 & 1 \\ 1 & 0 & 0 & 0 & 0 \\ 0 & 0 & 0 & 1 & 0 \\ 0 & 1 & 0 & 0 & 0 \\ 0 & 0 & 1 & 0 & 0 \end{pmatrix} \tag{1.2}$$

In this solution man 1 (corresponding to row 1) spends all his lifetime with woman 5 (corresponding to column 5). So in this solution we can think of man 1 *being assigned to* woman 5, etc. Hence such an integer solution is *known* as an *assignment*, and the problem of finding an optimum solution, which is an assignment, is called the *assignment problem*. In the optimum assignment, each man lives happily ever after with the woman he is assigned to and vice versa, and there is never any divorce.

When I took the linear programming course from Professor George B. Dantzig at the University of California, Berkeley, he told us the following story about this result. He was delivering a lecture on linear programming one day in the city of Hollywood, the home of movie stars and glamorous fashion models, the city which probably has the highest divorce rate in the world. That was in the early days of the simplex algorithm, and the lecture was given wide publicity as a talk on a very significant new development. The city's newspaper, the *Hollywood Daily*, sent one of their seasoned reporters to cover the talk. The reporter sat through the entire lecture, but could not understand any of the theory presented. When Dantzig looked at him at the end, he was sitting with a glum face, which seemed to convey his dilemma of how to write this up as an exciting news story. To cheer him up, Dantzig told him, "Let me tell you an application of linear programming that has sex appeal." "Now you are talking," the reporter said with a smile. Dantzig continued, "Our work with linear programming models leads us to a mathematical proof that it is possible for all of us to live happily married without any divorce by choosing our mates optimally." "You say without any divorce?" the reporter asked, and shaking his head in the negative, continued, "Man, you have been working with the wrong kind of models."

The conclusion that there exists an optimum marriage policy that maximizes the overall club's happiness without any divorce is very interesting. Extending this logic to the whole society itself, one can argue that there exists a solution pairing each man with a woman in society that maximizes the society's happiness without any divorce. Natural systems have a tendency to move toward optimum solutions, and if such a divorceless optimum solution exists, one would expect it to manifest itself in nature. Why, then, is there so much divorce going on in society, and why

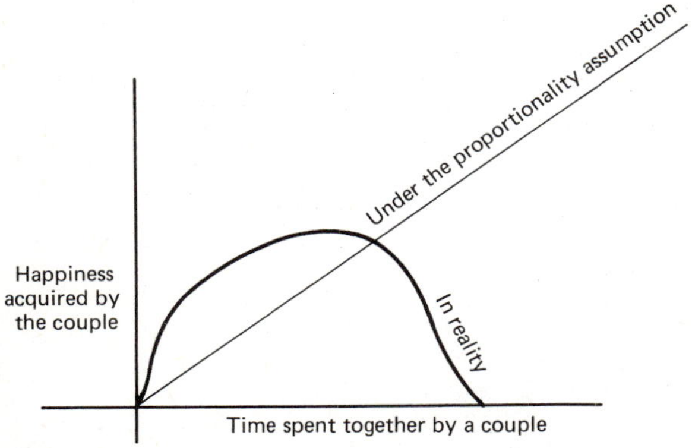

Figure 1.2 The inappropriateness of the proportionality assumption in the marriage problem.

is the frequency of divorce increasing rather than declining? This seems to imply that the conclusion obtained from the model—that there exists an optimal marriage policy that maximizes the society's happiness without any divorce—is false. If it is false, some of the assumptions on which the model is constructed must be invalid. The only assumptions used in constructing the model are the proportionality and the additivity assumptions. Let us examine each of them carefully.

The proportionality assumption states that the happiness acquired by a couple is proportional to the time they spend together, that is, it behaves as in the thin line in Figure 1.2. In practice, though, a couple may begin their life together in utter bliss, but develop a mutual dislike as they get to know each other over time. After all, the proverb says: "Familiarity breeds contempt." The actual happiness acquired by the couple as a function of the time spent together often behaves as the thick nonlinear curve in Figure 1.2. Thus the proportionality assumption is not reasonable in the marriage problem.

The additivity assumption states that the society's happiness is the sum of the happiness acquired by the various members in it. In particular, this states that a person's unhappiness cancels with another person's happiness. In reality these things are quite invalid. History has many instances of major social upheavals just because there was one single unhappy person. The additivity assumption is quite inappropriate for determining the society's happiness as a function of the happiness of its members.

Finally, the choice of the objective of maximizing society's happiness itself is quite inappropriate. In determining their marriage partners, most people are guided by the happiness they expect to acquire, and probably do not care what impact it will have on society. It is extremely hard to force people to do something just because it is good for society. Political idealogies based on societal objective functions that ignore individual objectives are not really acceptable to many people. This is illustrated by the following definition that appeared in a big city newspaper one day: "The government released a statement that "the current level of unemployment in the country is acceptable". What it really means is that the government economist who prepared this statement is confident that his job is secure and that he is not worried about joining the list of unemployed."

In summary, for studying the marriage problem, the linearity assumptions and the choice of the objective of maximizing the club's happiness seem very inappropriate, and hence models for this problem based on these assumptions can lead to very unnatural or even wrong conclusions. This problem is discussed here mainly to provide an example where the linearity assumptions are totally inappropriate.

Comment 1.2: The first formulations of the transportation problem date back to the late 1930s and the early 1940s. See L. V. Kantorovitch, *Mathematical Methods in the Organization and Planning of Production*, Publication House of the Leningrad State University, 1939, translated in *Management Science,* vol. 6, pp. 366–422, 1960; F. L. Hitchcock, "The Distribution of a Product from Several Sources to Numerous Localities," *Journal Mathematics and Physics*, vol. 20, pp. 224–230, 1941; and T. C. Koopmans, "Optimum Utilization of the Transportation System," *Econometrica*, vol. 17, nos. 3 and 4 (Supplement), 1949. For a detailed coverage on the early history of linear programming developments, see the book by G. B. Dantzig, [1.61] and his paper [1.8].

What Information Can be Derived from a Linear Programming Model?

In Section 1.2, Chapter 2, and in later chapters we present algorithms for solving LPs. Sometimes there may be alternate optimum solutions for an LP. In Section 3.10 we present methods for computing all of them. Using these methods, a suitable optimum solution (one that satisfies some conditions that may not have been included in the model, but which may be important) can be selected for implementation.

An *infeasible* LP is one that has no feasible solution. When this happens, there must be a subset of conditions in the model that are mutually contradictory (maybe the requirements of some items are so high that they cannot be met with the available resources), and these conditions have to be modified in order to make the model feasible. The methods discussed in Sections 5.4 and 6.2 help in identifying such a subset of conditions. After making the necessary modifications, the new model can be solved.

It will be very useful for decision making if we can find out the rate of change in the optimum objective value per unit change in the requirements or the availabilities of each item (e.g., in the fertilizer problem, what is the effect on the maximum net profit, of an increase of 1 ton in the availability of raw material 1?) These rates are called the *marginal values* associated with the items, or the *dual variables*, or the *shadow prices of the items.* These are the variables in another linear programming problem that is in *duality relationship* with the original problem. In this context the original problem is called the *primal problem* and the other problem is called the *dual problem.* The derivation of the dual problem is discussed in Chapter 4. The use of the dual solution in a *marginal-value analysis* (or *marginal analysis*) is discussed in Sections 4.6.3 and 8.15. When the original problem is solved by the revised simplex method (Chapter 5), the optimum dual solution is also obtained automatically as a by-product, without any additional effort. By performing a marginal analysis using the optimum dual solution, the decision maker can determine what the most critical resources are and how the requirements or resource availabilities can be modified to arrive at much better objective values than those possible under the existing requirements or resource availabilities. By providing this kind of information, the linear programming model becomes a very valuable planning tool.

Exercises

Formulate the following problems as LPs.

1.1 A nonferrous metals corporation manufactures four different alloys from two basic metals. The requirements are given below. Determine the optimal product mix to maximize gross revenue.

Metal	*Proportion of the Metal in Alloy* 1	2	3	4	*Total supply of Metal per Day*
1	0.5	0.6	0.3	0.1	6 tons
2	0.5	0.4	0.7	0.9	4 tons
Selling price of alloy per ton	$10	15	18	40	

Data	Paint Type				Thinner Type	
	1	2	3	4	1	2
Cost ($/gal)	9	7	5.57	4	3	1.85
Viscosity (CP)	900	780	620	375	2	25
Vapor pressure (PSI)	0.2	0.4	0.6	0.8	12.0	8.0
Brilliance content (gs/gal)	30	20	50	10	0	0
Durability content (gs/gal)	2000	1500	1000	500	0	0

Property	*Requirement*	*Reason*
Viscosity	$\geqq 400$	So that a single coat would do.
Brilliance content	Between 15 and 30	So that surface is properly reflective
Vapor pressure	Between 2 and 4 PSI	For proper drying in time
Durability content	$\geqq 575$	For weather resistance.

1.2 *BLENDING PROBLEM.* Determine an optimum blend for the paint to paint a huge skyscraper. The paint to be used can be obtained by blending four raw paints and two thinners. Data on them and requirements on the blend are given above.

1.2 FORMULATIONS USING PIECEWISE LINEAR FUNCTIONS

1.2.1 Convex and Concave Functions

Let $f(x)$ be a real-valued function defined over points $x = (x_1, \ldots, x_n)^{\mathrm{T}} \in \mathbf{\Gamma} \subset \mathbf{R}^n$, where $\mathbf{\Gamma}$ is either $\mathbf{R}^n$ or a convex subset (see Chapter 3 for a definition of this term) of $\mathbf{R}^n$. Then $f(x)$ is said to be a *convex function* iff for any $x^1 = (x_1^1, \ldots, x_n^1)^{\mathrm{T}}$, $x^2 = (x_1^2, \ldots, x_n^2)^{\mathrm{T}} \in \mathbf{\Gamma}$ and $0 \leqq \alpha \leqq 1$, we have

$$f(\alpha x^1 + (1 - \alpha)x^2) \leqq \alpha f(x^1) + (1 - \alpha)f(x^2) \quad (1.3)$$

The inequality (1.3), which defines the convexity of a function, is called *Jensen's inequality* after the Danish mathematician who first discussed it. The important property of convex functions is that when you join two points on the surface of the function by a *chord*, the function itself lies underneath the chord on the interval joining these points (see Figure 1.3). Similarly, if $g(x)$ is a real-valued function defined on the convex subset $\mathbf{\Gamma}$ of $\mathbf{R}^n$, it is said to be a *concave function* iff for any x^1, $x^2 \in \mathbf{\Gamma}$, and $0 \leqq \alpha \leqq 1$, we have

$$g(\alpha x^1 + (1 - \alpha)x^2) \geqq \alpha g(x^1) + (1 - \alpha)g(x^2) \quad (1.4)$$

Clearly a function is concave iff its negative is convex. Also, a concave function lies above the chord on any interval (see Figure 1.4). Convex and concave functions figure prominently in optimization. Some of the optimization algorithms can be mathematically proved to work only under suitable convexity assumptions on the objective and constraint

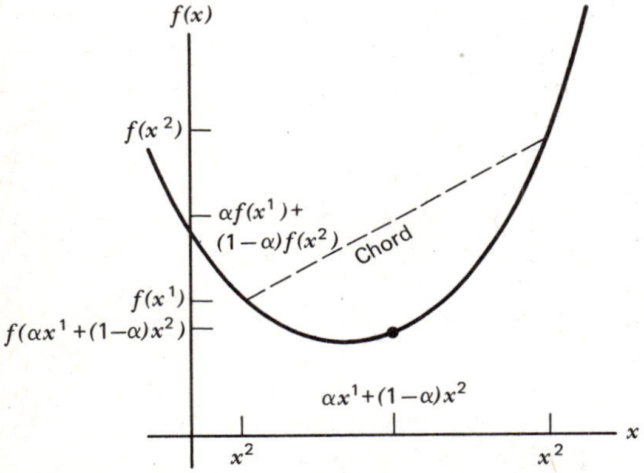

Figure 1.3 A convex function defined on the real line.

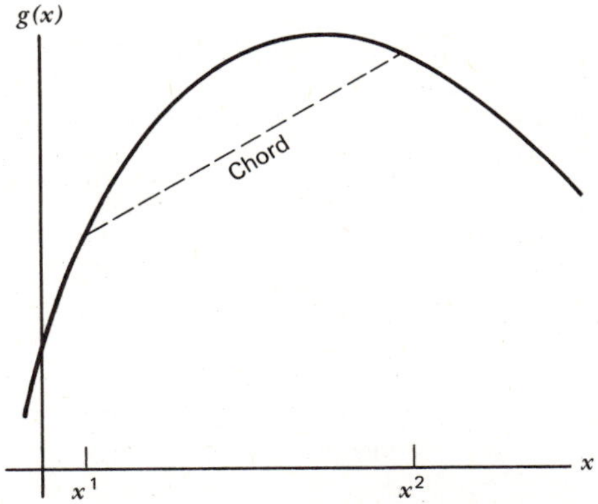

Figure 1.4 A concave function defined on the real line.

functions. The reader can verify the following properties of convex and concave functions, from the definitions given above.

1 A real-valued function defined on the real line (i.e., a function of one real variable) is convex iff its slope, or the first derivative, is monotone nondecreasing with the variable. It is a concave function iff its slope is monotone decreasing with the variable.

2 Let $f_1(x), f_2(x), \ldots, f_r(x)$ be all real-valued functions defined on the convex subset $\Gamma \subset \mathbf{R}^n$. The function $\theta(x) = \alpha_1 f_1(x) + \cdots + \alpha_r f_r(x)$ is a linear combination of these functions. If $f_1(x), \ldots, f_r(x)$ are all convex and $\alpha_1, \ldots, \alpha_r \geqq 0$, $\theta(x)$ is convex. If $f_1(x), \ldots, f_r(x)$ are all concave and $\alpha_1, \ldots, \alpha_r \geqq 0$, $\theta(x)$ is concave. $\theta(x)$ is convex if, for every t satisfying $\alpha_t > 0$, $f_t(x)$ is convex, and for every t satisfying $\alpha_t < 0$, $f_t(x)$ is concave.

In optimization literature, mathematical optimality conditions have been proved, and nice algorithms constructed, for optimization problems in which a convex function has to be minimized, or a concave function has to be maximized over a convex set. Hence such problems are considered to be nice problems and mathematical programmers refer to them as *convex programming problems.* If an optimization problem requires the minimization of a nonconvex function (or equivalently, the maximization of a nonconcave function), it is a hard problem and, in general, the only known algorithms that may be able to handle such problems might be enumerative algorithms, which could be quite inefficient, particularly when the problem size is large. In mathematical programming literature, these hard problems are referred to as *nonconvex programming problems.*

The reader can verify that the objective function in an LP is both convex and concave. So LPs belong to the class of nice convex programming problems. We have nice algorithms for solving LPs, and these are discussed in later chapters.

A real-valued function $\theta(x)$ defined on a convex subset $\Gamma \subset \mathbf{R}^n$ is said to be an *affine function* iff it is both convex and concave; that is, if it satisfies $\theta(\alpha x^1 + (1-\alpha)x^2) = \alpha\theta(x^1) + (1-\alpha)\theta(x^2)$ for every pair of points x^1, x^2 in Γ and $0 \leqq \alpha \leqq 1$. Given the affine function $\theta(x)$, it can be shown that there exist real numbers $c_0, c_1, \ldots, c_n$ such that $\theta(x) = c_0 + c_1x_1 + \cdots + c_nx_n$.

Piecewise Linear Functions Defined on the Real Line

Let $\theta(\lambda)$ be a real-valued function defined on the interval $\underline{\lambda} \leqq \lambda \leqq \bar{\lambda}$ of the real line. It is said to be a *piecewise linear function* if this interval can be partitioned into subintervals such that in each subinterval $\theta(\lambda)$ is affine: that is, if there exist values $\lambda_1, \lambda_2, \ldots, \lambda_r$ such that

$$\theta(\lambda) = \delta_t + k_t\lambda \qquad \text{for } \lambda_{t-1} \leqq \lambda < \lambda_t, \quad t = 1 \text{ to } r+1 \tag{1.5}$$

where $\lambda_0 = \underline{\lambda}$, $\lambda_{r+1} = \bar{\lambda}$ and $\delta_1, \ldots, \delta_{r+1}$; $k_1, \ldots, k_{r+1}$ are known constants. This $\theta(\lambda)$ is a *continuous piecewise linear function* if $\delta_t + k_t\lambda_t = \delta_{t+1} + k_{t+1}\lambda_t$ for each $t = 1$ to r. The constant k_t is the *slope* of this piecewise linear function in the interval λ_{t-1} to λ_t, for $t = 1$ to $r+1$. For a continuous piecewise linear function, the points where its slope changes are known as *breakpoints.* It can be verified that $\theta(\lambda)$ defined in (1.5) is convex iff it is continuous and its slope is nondecreasing with λ (i.e., $k_1 \leqq k_2 \leqq k_3 \leqq \cdots \leqq k_{r+1}$). $\theta(\lambda)$ defined in (1.5) is concave iff it is continuous and its slope is nonincreasing with λ (i.e., $k_1 \geqq k_2 \geqq \cdots + k_{r+1}$) (see Figures 1.6 and 1.7).

1.2.2 Modeling with Separable Piecewise Linear Objective Functions

The real-valued function $z(x)$ defined over $x = (x_1, \ldots, x_n)^T \in \mathbf{R}^n$ is said to be a separable piecewise linear function if we can write $z(x)$ as $z_1(x_1) + \cdots + z_n(x_n)$, the sum of n functions where each involves one variable only, and each $z_j(x_j)$ is piecewise linear as defined above, for $j = 1$ to n. In such a function $z_j(x_j)$ is the contribution of the variable x_j to the objective function $z(x)$. In the optimization models discussed thus far, we assumed that the contribution of a variable, say x_1, to the objective function varies linearly with x_1, as the term c_1x_1 (Figure 1.5). However, in most real life problems this assumption is too rigid. It is probably more realistic to assume that the contribution of the variable x_1 to the cost function varies in a piecewise linear manner. If this is a piecewise linear convex function in x_1, it can be depicted as in Figure 1.6 (see page 17).

Suppose the constraints are all linear, and the cost function, which is to be minimized, is separable in the variables and piecewise linear. Then, in constructing the objective function in the model, the additivity assumption holds, but the proportionality assumption is violated. However, if the separable

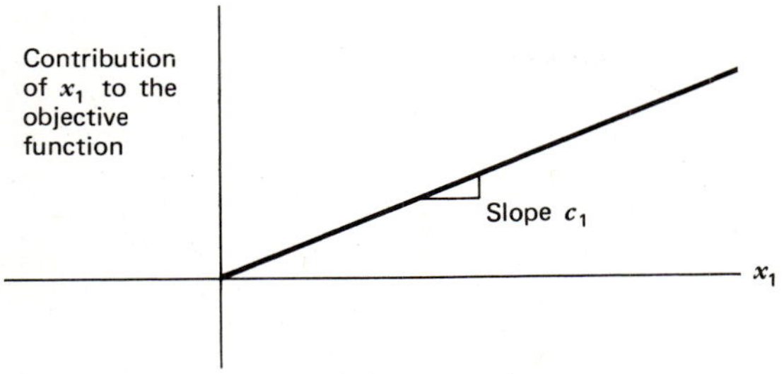

Figure 1.5 A linear function.

and piecewise linear objective function to be minimized is convex, the problem can still be modeled as LP. In this section we discuss how this can be done.

Let $z_1(x_1)$ denote the contribution of x_1 to the objective function. Let $0 < x_1^1 < x_1^2 < \cdots < x_1^r < \infty$ be the points at which $z_1(x_1)$ changes slope, and let the slope in the interval $x_1^{t-1} \leqq x_1 \leqq x_1^t$ be c_1^t for $t = 1$ to $r + 1$, where $x_1^0 = 0$, $x_1^{r+1} = \infty$. The intervals within which $z_1(x_1)$ is linear are 0 to x_1^1, x_1^1 to $x_1^2, \ldots,$ x_1^r to ∞. Let y_t be the portion of x_1 lying in the tth interval, x_1^{t-1} to x_1^t, (i.e., y_t is the length of the overlap of the interval 0 to x_1 with the interval x_1^{t-1} to x_1^t), $t = 1$ to $r + 1$. When defined in this manner, the new variables $y_1, \ldots, y_{r+1}$ partition x_1 as $y_1 + \cdots + y_{r+1}$. They are subject to the constraints:

$$\begin{aligned} &0 \leqq y_1 \leqq x_1^1 \\ &0 \leqq y_2 \leqq x_1^2 - x_1^1 \\ &\quad\vdots \\ &0 \leqq y_r \leqq x^r - x_1^{r-1} \\ &0 \leqq y_{r+1} \end{aligned} \qquad (1.6)$$

and

for every t, if $y_t > 0$, then each of y_j is equal to its upper bound $x_1^j - x_1^{j-1}$, for all $j < t$. (1.7)

When the variables $y_1, \ldots, y_{r+1}$ are defined in this manner, $z_1(x_1)$ is clearly equal to $c_1^1 y_1 + \cdots + c_1^{r+1} y_{r+1}$.

Example 1.6

Consider the case where $z_1(x_1)$ has the following slopes:

Interval of x_1	*Slope*
0–10	3
10–25	5
25–∞	7

In this case, x_1 will be partitioned as $y_1 + y_2 + y_3$, where $0 \leqq y_1 \leqq 10$, $0 \leqq y_2 \leqq 15$, $0 \leqq y_3$, and

$$\begin{aligned} y_2 > 0 \quad &\text{implies} \quad y_1 = 10 \\ y_3 > 0 \quad &\text{implies} \quad y_1 = 10 \text{ and } y_2 = 15 \end{aligned}$$

Then $z_1(x_1)$ can be expressed as $3y_1 + 5y_2 + 7y_3$.

The variable x_1 can now be eliminated from the model by substituting $x_1 = y_1 + \cdots + y_{r+1}$ wherever it appears in the model. In the objective function $z_1(x_1)$ is replaced by $c_1^1 y_1 + \cdots + c_1^{r+1} y_{r+1}$. Notice that the new objective function is linear in the new variables. The constraints (1.6) on the new variables are included with the other constraints in the model. The following facts guarantee that in any optimum solution of the transformed model, the constraints (1.7) are automatically satisfied: (1) We are trying to minimize the overall objective function; and (2) the slopes discussed satisfy the conditions $c_1^1 < c_1^2 < \cdots < c_1^{r+1}$.

If the slopes do not satisfy condition 2, and the objective function has to be minimized, then the constraints (1.7) on the new variables have to be specifically included in the model; since these constraints are not linear constraints, the transformed model is not a linear programming model.

Example 1.7

Consider the function $z_1(x_1)$ defined in Example 1.6. Let x_1 be any nonnegative number. Keep x_1 fixed and consider the following optimization problem

$$\begin{aligned} &\text{Minimize} \quad 3y_1 + 5y_2 + 7y_3 \\ &\text{Subject to} \quad y_1 + y_2 + y_3 = x_1 \qquad (1.8) \\ &0 \leqq y_1 \leqq 10 \quad 0 \leqq y_2 \leqq 15 \quad 0 \leqq y_3 \end{aligned}$$

In (1.8), the slopes of the variables y_1, y_2, y_3 in the objective function are strictly increasing in that order. This implies that the optimum solution for (1.8) is

$$\begin{aligned} &y_1 = x_1 \quad y_2 = y_3 = 0; \quad &&\text{if } 0 \leqq x_1 \leqq 10 \\ &y_1 = 10 \quad y_2 = x_1 - 10 \quad y_3 = 0; \quad &&\text{if } 10 \leqq x_1 \leqq 25 \\ &y_1 = 10 \quad y_2 = 15 \quad y_3 = x_1 - 25; \quad &&\text{if } 25 \leqq x_1 \end{aligned}$$

This implies that the optimum objective value in (1.8) is $z_1(x_1)$ specified in Example 1.6, for any $x_1 \geqq 0$, and that condition (1.7) holds automatically in the optimum solution for (1.8). Hence, in any objective function to be minimized, in which $z_1(x_1)$

appears with a positive coefficient, we can replace x_1 by $y_1 + y_2 + y_3$ and $z_1(x_1)$ by $3y_1 + 5y_2 + 7y_3$, where the variables y_1, y_2, y_3 are constrained by the bounds given in (1.8).

If the contribution of some other variable to the objective function is also piecewise linear, partition that variable too in a similar manner. If the original separable, piecewise linear objective function to be minimized is convex, constraints on the new variables of the type (1.7) can be ignored in the transformed model, and the transformed model is a linear programming model in terms of the new variables.

The same technique can be used for transforming any problem in which a separable, piecewise linear objective function is to be maximized, subject to linear constraints, into an LP, provided the objective function is concave (i.e., if the slope of each component function in the objective is monotone decreasing with the variable).

Example 1.8

A company manufactures three different products, 1, 2, 3, using limestone as the basic raw material. The company has its own limestone quarries, which can produce up to 250 units of limestone per day at a cost of \$2/unit. If the company needs additional limestone, it can buy it from a supplier at a cost of \$5/unit.

The regional electric utility has recently adopted a modern stepwise rate system to discourage wastage. It charges the company \$30 per unit for the first 1000 units of electricity used daily, \$45 per unit for 500 units per day beyond the initial 1000 units, and a hefty \$75 per unit for any amount beyond the initial 1500 units of electricity used per day.

The region's water distribution authority charges at the rate of \$6 per unit of water used per day up to 800 units, and \$7 per unit for any amount used beyond 800 units per day. The company buys fuel from a supplier at the rate of \$4/unit, but the energy conservation laws restrict the company from using more than 3000 units of fuel per day.

The company's labor force provide 640 man-hours of labor per day during regular working hours, and the regular wages for these laborers are paid directly by the company's parent organization and do not cost the company itself anything directly. However, if the company needs more than 640 man-hours of labor per day, they can get up to a maximum of 160 more man-hours per day by asking the laborers to work overtime, for which the company has to pay itself at the rate of \$12 per man-hour. The remaining data in the problem are tabulated at the bottom.

We will now formulate the problem of determining how much of each product to produce daily, so as to maximize the company's daily net profit. First, let us examine the structure of the objective function. Let $y_L, y_e, y_w, y_f, y_\ell$ denote the number of units of limestone, electricity, water, fuel, labor, respectively, used by the company per day. We denote by the symbols $z_L(y_L)$, $z_e(y_e)$, etc., the costs to the company (\$/day) of these respective inputs. See Figures 1.6 and 1.7.

The fuel cost and the sales revenue from the sale of products 2 and 3 are linear. The costs of limestone, electricity, water, and labor are all piecewise linear and convex, but all these things enter the objective function of net profit to the company with a -1 coefficient. The sales revenue from the sale of product 1 is piecewise linear and concave, and this enters the net profit function with a coefficient of $+1$. Thus the overall objective function, net profit, is

Product	*Units of Inputs Needed to Produce One Unit of Product*					*Selling Price*
	Limestone	*Electricity*	*Water*	*Fuel*	*Labor*	
1	1/2	3	1	1	2	\$300/unit for the first 50 units; \$250/unit beyond 50 units per day
2	1	2	1/4	1	1	\$350/unit to an upper limit of 100 units/day
3	3/2	5	2	3	1	\$450/unit

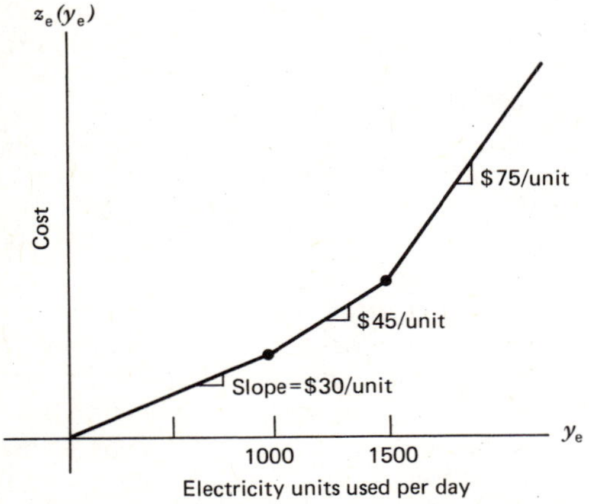

Figure 1.6 Cost to the company of electricity used daily. It is a convex piecewise linear function. Notice that the slope of $z_e(y_e)$ is monotone increasing with y_e.

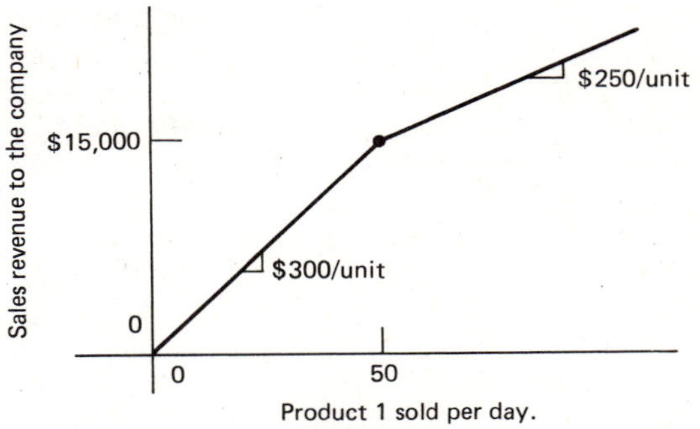

Figure 1.7 Sales revenue per day from production and sale of product 1. It is a piecewise linear concave function. Notice that its slope is monotone decreasing.

piecewise linear concave, and since this is to be maximized, this problem can be modeled as an LP.

The primary decision variables in this model are clearly x_j = number of units of product j manufactured and sold per day, $j = 1, 2, 3$. The limestone used per day is $y_L = (x_1/2) + x_2 + (3x_3/2)$, in terms of the primary decision variables. Since the cost of limestone has a single breakpoint at 250 units, we define the variables y_{L1}, y_{L2} by

$$\begin{aligned} \frac{x_1}{2} + x_2 + \frac{3x_3}{2} &= y_L = y_{L1} + y_{L2} \\ 0 \leqq y_{L1} \leqq 250 &\qquad 0 \leqq y_{L2} \\ z_L(y_L) &= 2y_{L1} + 5y_{L2} \end{aligned} \tag{1.9}$$

Here y_{L1}, y_{L2} refer to the amounts of limestone used per day in the two intervals for limestone within which its cost is linear. In (1.9), the variable y_L can easily be eliminated. In the following model, the variables y_{e1}, y_{e2}, y_{e3}; y_{w1}, y_{w2}; $y_{\ell 1}, y_{\ell 2}$; x_{11}, x_{12} have similar interpretations.

Maximize

$$(300x_{11} + 250x_{12}) + 350x_2 + 450x_3 - (2y_{L1} + 5y_{L2}) - (30y_{e1} + 45y_{e2} + 75y_{e3}) - (6y_{w1} + 7y_{w2}) - 4y_f - 12y_{\ell 2}$$

Subject to

$$\frac{x_1}{2} + x_2 + \frac{3x_3}{2} - y_{L1} - y_{L2} = 0$$

$$3x_1 + 2x_2 + 5x_3 - y_{e1} - y_{e2} - y_{e3} = 0$$

$$x_1 + \frac{x_2}{4} + 2x_3 - y_{w1} - y_{w2} = 0$$

$$x_1 + x_2 + 3x_3 - y_f = 0$$

$$2x_1 + x_2 + x_3 - y_{\ell 1} - y_{\ell 2} = 0$$

$$x_1 - x_{11} - x_{12} = 0$$

$$0 \leqq y_{L1} \leqq 250,\ 0 \leqq y_{L2}$$

$$0 \leqq y_{e1} \leqq 1000,\ 0 \leqq y_{e2} \leqq 500,\ 0 \leqq y_{e3}$$

$$0 \leqq y_{w1} \leqq 800,\ 0 \leqq y_{w2}$$

$$0 \leqq y_f \leqq 3000$$

$$0 \leqq y_{\ell 1} \leqq 640,\ 0 \leqq y_{\ell 2} \leqq 160$$

$$0 \leqq x_{11} \leqq 50,\ 0 \leqq x_{12}$$

$$0\ \ x_2 \leqq 100,\ 0 \leqq x_3$$

Exercises

1.3 A firm manufactures four products called P_1, P_2, P_3, P_4. Product P_1 can be sold at a profit of $10 per ton up to a quantity of 10 tons. Quantities of P_1 over 10 tons but not more than 25 tons can be sold at a profit of $7 per ton. Quantities beyond 25 tons earn a profit of only $5 per ton.

Product P_2 yields a profit of $8 per ton up to 7 tons. Quantities of P_2 above 7 tons yield a profit of only $4 per ton. Everyone who buys P_2 also buys P_4 to go along with it. This is described later. Both products P_1, P_2 can be sold in unlimited amounts.

P_3 is a by-product obtained while producing P_1. Up to 10 tons of P_3 can be sold at $2 profit per ton. However, beyond 10 tons there is no market for P_3, and since it cannot be stored, it has to be disposed of at a cost to the firm of $3 per ton.

P_4 is a by-product obtained while producing P_2. Also, P_4 can be produced independently. Every customer who buy θ tons of P_2 has to buy $\theta/2$ tons of P_4 to go along with it for every $\theta \geqq 0$. Also, P_4 has an independent market in unlimited quantities. One ton of P_4 yields a profit of $3 per ton if it is sold along with P_2. One ton of P_4 sold independently yields a profit of $2.50 per ton.

Production of 1 ton of P_1 requires 1 h of machine 1 time plus 2 h of machine 2 time. One ton of P_2 requires 2 h of machine 1 time plus 3 h of machine 2 time. Each ton of P_1 produced automatically delivers 3/2 tons of P_3 as a by-product without any additional work. Each ton of P_2 produced yields 1/4 tons of P_4 as a by-product without any additional work. To produce 1 ton of P_4 independently requires 3 h of machine 3 time.

The company has 96 h of machine 1 time, 120 h of machine 2 time, and 240 h of machine 3 time available. The company wishes to maximize its total net profit. Formulate this problem as an LP and justify your formulation.

1.2.3 Minimizing the Maximum of Several Linear Functions

Consider an optimization problem that is like an LP with the exception that there are k different linear or affine cost functions instead of one. Suppose the objective is to minimize the maximum of these k affine cost functions. Let $x = (x_1, \ldots, x_n)^T$ represent the column vector of variables in the problem and let $c_0^1 + c^1x, \ldots, c_0^k + c^kx$ be the k affine cost functions, where $c^r = (c_1^r, c_2^r, \ldots, c_n^r)$ is the row vector of cost coefficients in the rth function. We have to find an x that minimizes

$$f(x) = \text{maximum}\{c_0^1 + c^1x, c_0^2 + c^2x, \ldots, c_0^k + c^kx\}$$

subject to the constraints in the problem. We do not have $f(x)$ given by an explicit formula, but for any given vector x, $f(x)$ can be computed easily from the above equation. That is why the function $f(x)$ defined in this manner is known as the *pointwise maximum (or supremum) of the affine functions* $c_0^r + c^rx$, $r = 1$ to k. This function is not in general separable, but it can be shown to be piecewise linear and convex on $\mathbf{R}^n$ (see Section 8.14 where this is proved). Since $f(x)$ is piecewise linear and convex, the problem of minimizing $f(x)$ subject to linear constraints can be transformed into an LP.

Problems like this often appear in practice. For example, suppose we want to finish a project in minimal time. There may be k different agencies working independently on segments of this project. Let x be the vector of variables in the problem. $c_0^r + c^rx$ may represent the time taken by the rth agency by adopting the solution vector x. Then the overall time taken for the project is $f(x)$, defined above.

While dealing with an uncertain situation, a method in which decisions are taken so as to minimize the maximum cost that may be incurred (or to maximize the minimum profit that may be made) is known as a *worst case analysis* method. Worst case analysis often leads to models of this type.

To transform this into an LP, define a new variable x_{n+1} and introduce these constraints.

$$\begin{gathered} x_{n+1} - c_0^1 - c^1x \geqq 0 \\ \vdots \\ x_{n+1} - c_0^k - c^kx \geqq 0 \end{gathered}$$

Let $X = (x_1, \ldots, x_n, x_{n+1})^T$. Augment the constraints in the original problem with these constraints. Then minimize $z(X) = x_{n+1}$, subject to all the constraints. The new objective function is linear in the variables and, hence, the new problem is an LP. If $\hat{X} = (\hat{x}_1, \ldots, \hat{x}_n, \hat{x}_{n+1})^T$ is an optimum solution of this new LP, $\hat{x} = (\hat{x}_1, \ldots, \hat{x}_n)^T$ is an optimum solution, of the original problem and $\hat{x}_{n+1} = f(\hat{x})$ is the minimum value of $f(x)$ in the original problem.

Exercises

1.4 Justify the preceding statement.

1.5 Consider the problem in which ore has to be transported from m mines to n plants. The amount of ore available at the ith mine is a_i tons and the amount of ore required at the jth plant is b_j tons. $\sum_i a_i = \sum_j b_j$. Exactly one truck can be used to make these shipments, and the truck is restricted to making no more than one trip from a mine to a plant. What is the minimum capacity truck that can do this job? What is the optimal shipping schedule? Formulate this as an LP.

Example 1.9

Consider the problem:

$$\begin{array}{ll} \text{Minimize} & \\ & f(x) = \text{maximum}\ (2x_1 - 3x_2, -7x_1 + 8x_2) \\ \text{Subject to} & x_1 - 3.5x_2 \geqq 8 \\ & 9x_1 + 15x_2 \geqq 100 \\ & x_1 \text{ and } x_2 \geqq 0 \end{array}$$

When transformed as above, this problem becomes the LP

$$\begin{array}{ll} \text{Minimize} & x_3 \\ \text{Subject to} & x_3 - 2x_1 + 3x_2 \geqq 0 \\ & x_3 + 7x_1 - 8x_2 \geqq 0 \\ & x_1 - 3.5x_2 \geqq 8 \\ & 9x_1 + 15x_2 \geqq 100 \\ & x_1 \text{ and } x_2 \geqq 0 \end{array}$$

If $c_0^1 + c^1x, \dots, c_0^r + c^rx$ are given affine functions, the function defined by $g(x) = \text{minimum}\ \{c_0^1 + c^1x, \dots, c_0^r + c^rx\}$, is known as the *pointwise minimum (or infimum) of the functions* $c_0^1 + c^1x, \dots, c_0^r + c^rx$. A problem of maximizing the pointwise minimum of several given linear functions, subject to linear constraints, can also be transformed into an LP using exactly similar arguments. The function $g(x)$ defined above is not in general separable, but it is piecewise linear and concave. Suppose it is required to maximize this function subject to some linear constraints. Define a new variable x_{n+1} and introduce the additional constraints, $x_{n+1} - c_0^1 - c^1x \leqq 0, \dots, x_{n+1} - c_0^r - c^rx \leqq 0$. The problem of maximizing x_{n+1} subject to these additional constraints, and the linear constraints in the original problem, is an LP that is an equivalent problem.

Illustration of the Worst Case Analysis

Consider the fertilizer product mix problem discussed in Section 1.1.2. Suppose we are determining the product mix for next year. Let the net profits per ton of Hi-phosphate and Lo-phosphate fertilizers be $\$c_1$ and $\$c_2$, respectively. In Section 1.1.2, we had $(c_1, c_2) = (15, 10)$. Suppose we do not know at this stage what the exact net profits from these fertilizers will be next year. However, assume that it is known from market analysis that the vector (c_1, c_2) will be either (15, 10) or (10, 15) or (12, 12). Assume that the remaining data in the problem will be the same next year as in Section 1.1.2. If x_1, x_2 are the amounts (in tons) of Hi-phosphate and Lo-phosphate fertilizer, respectively, manufactured next year, the total net profit will be $15x_1 + 10x_2$, or $10x_1 + 15x_2$, or $12x_1 + 12x_2$, depending on the fertilizer net profit vector at that time. So, the minimum net profit will be $h(x) = \text{minimum}\ \{15x_1 + 10x_2,\ 10x_1 + 15x_2,\ 12x_1 + 12x_2\}$. Under this uncertainty, the worst case analysis dictates that we choose (x_1, x_2) to maximize $h(x)$ subject to the constraints in (1.1). Transforming this into an LP leads to:

$$\begin{array}{ll} \text{Maximize} & x_3 \\ \text{Subject to} & x_3 - 15x_1 - 10x_2 \leqq 0 \\ & x_3 - 10x_1 - 15x_2 \leqq 0 \\ & x_3 - 12x_1 - 12x_2 \leqq 0 \\ & 2x_1 + x_2 \leqq 1500 \\ & x_1 + x_2 \leqq 1200 \\ & x_1 \leqq 500 \\ & x_1 \geqq 0, \quad x_2 \geqq 0 \end{array} \tag{1.10}$$

If $(\bar{x}_1, \bar{x}_2, \bar{x}_3)$ is an optimum solution of this LP, then $\bar{x} = (\bar{x}_1, \bar{x}_2)$ is an optimum solution of the worst case problem, and $\bar{x}_3 = h(\bar{x})$ is the optimum objective value in it.

1.2.4 Minimizing a Nonnegative Weighted Sum of Absolute Values

In some cases we may have to solve an optimization problem of the form: Find $(x_1, \dots, x_n)$ to

$$\begin{array}{lll} \text{Minimize} & V(x) = \sum_{j=1}^{n} c_j|x_j| & \\ \text{Subject to} & \sum_{j=1}^{n} a_{ij}x_j = b_i & \text{for } i = 1 \text{ to } m \end{array} \tag{1.11}$$

where $|x_j|$ is the absolute value of x_j and all the c_j are given nonnegative numbers. The absolute value function $|x_j|$ is a piecewise linear convex function with its only breakpoint at $x_j = 0$ (see

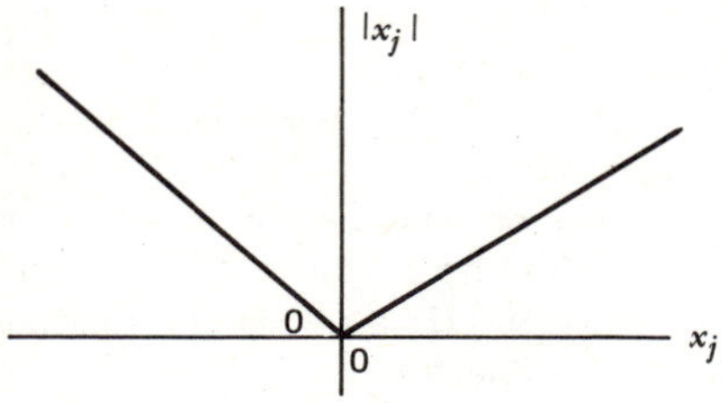

Figure 1.8 The absolute value function is piecewise linear and convex.

Figure 1.8). So when all $c_j \geqq 0$, the objective function in (1.11) is separable, piecewise linear, and convex, and hence problems such as this can be transformed into LPs. The variable x_j is unrestricted in sign in the problem. Every real number can be expressed as the difference of two nonnegative numbers. Hence, we can express the variable x_j as the difference of two nonnegative variables, as in

$$x_j = x_j^+ - x_j^- \qquad x_j^+ \geqq 0 \quad x_j^- \geqq 0 \tag{1.12}$$

By defining

$$\begin{aligned} x_j^+ &= 0 && \text{if } x_j \leqq 0 \\ &= x_j && \text{if } x_j > 0 \\ x_j^- &= 0 && \text{if } x_j \geqq 0 \\ &= -x_j && \text{if } x_j < 0 \end{aligned} \tag{1.13}$$

we can even guarantee that

$$x_j^+ x_j^- = 0 \tag{1.14}$$

that is, at least one of x_j^+ and x_j^- is always zero. When defined in this manner, x_j^+ is known as the *positive part* and x_j^- as the *negative part* of the unrestricted variable x_j. If x_j^+ and x_j^- satisfy both (1.12) and (1.13), clearly $|x_j| = x_j^+ + x_j^-$. By using this equation, (1.11) is transformed into the problem. Find $x^+ = (x_1^+, \ldots, x_n^+)$ and $x^- = (x_1^-, \ldots, x_n^-)$ to

$$\text{Minimize} \quad U(x^+, x^-) = \sum_{j=1}^{n} c_j(x_j^+ + x_j^-)$$

$$\begin{aligned} \text{Subject to} \quad & \sum_{j=1}^{n} a_{ij}(x_j^+ - x_j^-) = b_i \qquad \text{for } i = 1 \text{ to } m \\ & x_j^+ \geqq 0, x_j^- \geqq 0 \qquad \text{for all } j \end{aligned} \tag{1.15}$$

$$x_j^+ x_j^- = 0 \qquad \text{for all } j \tag{1.16}$$

Let $\tilde{x}^+$, $\tilde{x}^-$ be any feasible solution to (1.15) that may or may not satisfy (1.16). Let $\bar{v}_j = \min(\tilde{x}_j^+, \tilde{x}_j^-)$, $\hat{x}_j^+ = \tilde{x}_j^+ - \bar{v}_j$, $\hat{x}_j^- = \tilde{x}_j^- - \bar{v}_j$ for all j. Clearly $(\hat{x}^+, \hat{x}^-)$ is also a feasible solution to (1.15), it satisfies (1.16), and since all $c_j \geqq 0$,

$$U(\hat{x}^+, \hat{x}^-) \leqq U(\tilde{x}^+, \tilde{x}^-) \tag{1.17}$$

Thus the property that $c_j \geqq 0$ for all j implies that if the original problem (1.11) has an optimum solution, then there is an optimum solution to (1.15) that automatically satisfies (1.16). When the LP (1.15) is solved by the simplex method, an optimum solution for it that satisfies (1.16) automatically will be obtained, provided (1.15) has a feasible solution. From the optimum solution of (1.15) an optimum solution of (1.11) is obtained by using (1.12). Hence (1.11) is equivalent to the LP (1.15).

Example 1.10

This example illustrates (1.17). Consider $z(x_1) = 6|x_1|$. Transforming this as above, we get $U(x_1^+, x_1^-) = 6(x_1^+ + x_1^-)$, where

$$x_1^+ - x_1^- = x_1 \qquad x_1^+ \geqq 0, \quad x_1^- \geqq 0 \tag{1.18}$$

Give a specific value to x_1, say $x_1 = -20$. Then in order to satisfy the constraints in (1.18), we need $(x_1^+ = \alpha,\ x_1^- = 20 + \alpha)$, for any $\alpha \geqq 0$. In this general solution to (1.18), $U(x_1^+, x_1^-)$ has the value of $6(20 + 2\alpha)$, and the minimum value for $U(x_1^+, x_1^-)$ is attained by setting $\alpha = 0$, leading to the unique solution $(x_1^+ = 0,\ x_1^- = 20)$, which satisfies (1.16). The same phenomenon holds for any other real value of x_1.

Exercises

1.6 If some $c_j < 0$, show that (1.17) may not hold, and hence the transformation of (1.11) into the LP (1.15) will not work.

Note 1.1: It is possible to handle problems slightly more general than (1.11) by the methods discussed here. Let $c_0^t + \sum_{j=1}^{n} c_j^t x_j$, $t = 1$ to r be r given affine functions. Consider the problem of minimizing the general objective function

$$\theta(x) = \sum_{j=1}^{n} d_j x_j + \sum_{j=1}^{n} f_j |x_j| + \sum_{t=1}^{r} g_t \left| c_0^t + \sum_{j=1}^{n} c_j^t x_j \right|$$

subject to linear constraints on the decision variables x_j, $j = 1$ to n. If f_j, g_t are nonnegative for all j, t, this problem can be transformed into an equivalent LP by the methods discussed above. For this, define each of the affine functions $c_0^t + \sum_{j=1}^n c_j^t x_j$ as a new variable, say y_t, by including additional linear constraints

$$c_0^t + \sum_{j=1}^n c_j^t x_j - y_t = 0 \qquad t = 1 \text{ to } r \qquad (1.19)$$

in the system of constraints for the problem. In terms of x_j and y_t, the objective function

$$\theta(x) = \psi(x, y) = \sum_{j=1}^n d_j x_j + \sum_{j=1}^n f_j|x_j| + \sum_{t=1}^r g_t|y_t|$$

If x_j is restricted to be nonnegative in the system of constraints, then $|x_j|$ can be replaced by x_j; otherwise, express x_j as the difference of two nonnegative variables, $x_j^+ - x_j^-$, and y_t similarly as $y_t^+ - y_t^-$. Assuming that $f_j \geqq 0$ for all $j = 1$ to n, $g_t > 0$ for all $t = 1$ to r, replace $|x_j|$ by $x_j^+ + x_j^-$, replace x_j in the linear part of the objective function and in the constraints by $x_j^+ - x_j^-$, replace $|y_t|$ by $y_t^+ + y_t^-$ in the objective function, and replace y_t in the constraints (1.19) by $y_t^+ - y_t^-$ (see Example 1.11, below).

1.2.5 Applications of Linear Programming in Curve Fitting

Consider the following problem. In a laboratory a chemical reaction was conducted at various temperatures and the yield was measured. The data are given below.

Temperature, t	-5	-3	-1	0	1
Yield, $y(t)$	80	92	96	98	100

It is expected that the yield $y(t)$ can be approximated by a cubic expression in terms of the temperature t, namely, $a_0 + a_1 t + a_2 t^2 + a_3 t^3$. Let $a = (a_0, a_1, a_2, a_3)$, and $f(a, t) = a_0 + a_1 t + a_2 t^2 + a_3 t^3$. The coefficients a_0, a_1, a_2, a_3 are the parameters in the model to be estimated using the data. The *curve-fitting problem,* or the *parameter estimation problem,* is the problem of determining the optimum values for these parameters that make the value of the fit $f(a, t)$ as close as possible to the observed yield $y(t)$ over the value of t used in the experiment. In almost all practical applications, the functional form of the curve or surface to be fitted is determined from practical considerations or theoretical reasoning, and the values of the parameters in the

Example 1.11

Consider the following problem.

$$\begin{aligned} \text{Minimize} \quad & \theta(x) = -3x_1 - 4x_2 + 5x_3 + 3|x_2| + 2|x_3| + 9|-13 - 12x_1 + 5x_2 - 7x_3| + 13|15 - 3x_1 - 4x_2 + 6x_3| \\ \text{Subject to} \quad & 3x_1 - 2x_2 + 13x_3 = -9 \\ & -8x_1 + 3x_2 + 3x_3 = 10 \end{aligned} \qquad (1.20)$$

First we define two new variables $y_1 = -13 - 12x_1 + 5x_2 - 7x_3$, $y_2 = 15 - 3x_1 - 4x_2 + 6x_3$. Since all the variables are unrestricted, we then substitute $x_j = x_j^+ - x_j^-$ for $j = 1$ to 3 and $y_t = y_t^+ - y_t^-$ for $t = 1, 2$, where $x_j^+, x_j^-, y_t^+, y_t^-$ are all restricted to be nonnegative. The transformed problem is:

$$\begin{aligned} \text{Minimize} \quad & -3(x_1^+ - x_1^-) - 4(x_2^+ - x_2^-) + 5(x_3^+ - x_3^-) + 3(x_2^+ + x_2^-) \\ & + 2(x_3^+ + x_3^-) + 9(y_1^+ + y_1^-) + 13(y_2^+ + y_2^-) \\ \text{Subject to} \quad & 3(x_1^+ - x_1^-) - 2(x_2^+ - x_2^-) + 13(x_3^+ - x_3^-) = -9 \\ & -8(x_1^+ - x_1^-) + 3(x_2^+ - x_2^-) + 3(x_3^+ - x_3^-) = 10 \\ & -12(x_1^+ - x_1^-) + 5(x_2^+ - x_2^-) - 7(x_3^+ - x_3^-) - (y_1^+ - y_1^-) = 13 \\ & -3(x_1^+ - x_1^-) - 4(x_2^+ - x_2^-) + 6(x_3^+ - x_3^-) - (y_2^+ - y_2^-) = -15 \\ & x_j^+, x_j^- \geqq 0 \qquad \text{for } j = 1 \text{ to } 3 \\ & y_t^+, y_t^- \geqq 0 \qquad \text{for } t = 1, 2 \end{aligned} \qquad (1.21)$$

functional form are then estimated using the data to give a close fit.

Given specific values for the parameters a, the deviation at temperature t of the fit $f(a, t)$ and the observed yield $y(t)$ is $f(a, t) - y(t)$. This may be positive or negative or zero. Let $\mathbf{S} = \{-5, -3, -1, 0, 1\}$. The overall deviation of $f(a, t)$ from $y(t)$ over $t \in \mathbf{S}$ is a function of the parameter vector a, and there are several ways of measuring it. Notice that the actual sum of deviation, that is $\sum_{t \in \mathbf{S}} (f(a, t) - y(t))$, is not a good measure of the overall deviation, since in this sum positive deviations at some values of $t \in \mathbf{S}$ will get cancelled by negative deviations at other values of $t \in \mathbf{S}$. Three different measures are commonly used for the overall deviation of the fit. They are

(i) $L_1(a) = \sum_{t \in \mathbf{S}} |f(a, t) - y(t)|$

(ii) $L_\infty(a) = \text{maximum } \{|f(a, t) - y(t)| : t \in \mathbf{S}\}$

(iii) $L_2(a) = \sum_{t \in \mathbf{S}} (f(a, t) - y(t))^2$

The first measure of deviation, $L_1(a)$, is known as the L_1-*norm*. The second measure of deviation, $L_\infty(a)$, is known as the L_∞-*norm* or the *uniform norm*, or the *Chebyshev norm* after the Russian mathematician P.L. Chebyshev who first considered it in the context of curve fitting. The third measure of deviation, $L_2(a)$, is known as the L_2-*norm*, or the *Euclidean norm*, or the *sum of squares of deviations.* Regardless of which measure of deviation is used, the best approximation is obtained by choosing those values for the parameters a that minimize that measure of deviation. Generally, there will be different best approximate solutions for different measures of deviation, and even for a particular measure of deviation there may be a large set of alternate best approximate solutions. In a practical curve-fitting problem, the choice of a measure of deviation is often made by the ease with which we can minimize that measure of deviation.

Best L_1-Approximation

Let us consider the problem of finding the optimum values for the parameters $a = (a_0, a_1, a_2, a_3)$ in our model that minimize $L_1(a)$. This problem is

$$\text{Minimize} \quad L_1(a) = \sum_{t \in \mathbf{S}} |a_0 + a_1 t + a_2 t^2 + a_3 t^3 - y(t)| \tag{1.22}$$

over all possible real values for a_0, a_1, a_2, a_3. The actual deviation at $t = -5$, is $a_0 - 5a_1 + 25a_2 - 125a_3 - 80$. This may be positive, negative, or zero. It can be expressed as a difference of two nonnegative numbers, as in: $a_0 - 5a_1 + 25a_2 - 125a_3 - 80 = u_1 - v_1$, $u_1 \geqq 0$, $v_1 \geqq 0$. In addition, if we restrict u_1 and v_1 to satisfy $u_1 v_1 = 0$, then we have $|a_0 - 5a_1 + 25a_2 - 125a_3 - 80| = u_1 + v_1$. By this and the discussion in Section 1.2.4, (1.22) is equivalent to the LP:

$$\text{Minimize} \quad \sum_{i=1}^{5} u_i + v_i$$

Subject to

$$\begin{aligned}
a_0 - 5a_1 + 25a_2 - 125a_3 - u_1 + v_1 &= 80 \\
a_0 - 3a_1 + 9a_2 - 27a_3 - u_2 + v_2 &= 92 \\
a_0 - a_1 + a_2 - a_3 - u_3 + v_3 &= 96 \\
a_0 \qquad\qquad\qquad - u_4 + v_4 &= 98 \\
a_0 + a_1 + a_2 + a_3 - u_5 + v_5 &= 100 \\
u_i, v_i \geqq 0 \quad \text{for all } i,\ a_0, a_1, a_2, a_3 &\text{ unrestricted}
\end{aligned} \tag{1.23}$$

If $(\bar{a}_0, \bar{a}_1, \bar{a}_2, \bar{a}_3; \bar{u}_i, \bar{v}_i : i = 1 \text{ to } 5)$ is an optimum solution of the LP (1.23), then $\bar{a} = (\bar{a}_0, \bar{a}_1, \bar{a}_2, \bar{a}_3)$ is an optimum parameter vector, and $f(\bar{a}, t) = \bar{a}_0 + \bar{a}_1 t + \bar{a}_2 t^2 + \bar{a}_3 t^3$ is the best fit under this measure. The minimum objective value in the LP (1.23) gives the minimum value of the measure of deviation $L_1(a)$, and depending on its magnitude, one can determine whether the cubic fit obtained is a reasonably good fit or not. If it is not considered a good fit, a different functional form or model can be tried.

Chebyshev Approximation

Now let us consider the problem of determining the values of the parameters $a = (a_0, a_1, a_2, a_3)$ that minimize the measure of deviation $L_\infty(a)$. This is the problem

$$\begin{aligned}
&\text{minimize } L_\infty(a) \\
&= \text{maximum } \{|a_0 + a_1 t + a_2 t^2 + a_3 t^3 - y(t)| : t \in \mathbf{S}\}
\end{aligned} \tag{1.24}$$

over all possible real values for a_0, a_1, a_2, a_3. By the same arguments as those used earlier, and the discussion in Sections 1.2.3 and 1.2.4, it is clear that this problem is equivalent to the LP

Minimize z

Subject to $z - u_i - v_i \geqq 0$ for all $i = 1$ to 5

$$\begin{aligned}
a_0 - 5a_1 + 25a_2 - 125a_3 - u_1 + v_1 &= 80 \\
a_0 - 3a_1 + 9a_2 - 27a_3 - u_2 + v_2 &= 92 \\
a_0 - a_1 + a_2 - a_3 - u_3 + v_3 &= 96 \qquad (1.25) \\
a_0 \qquad\qquad - u_4 + v_4 &= 98 \\
a_0 + a_1 + a_2 + a_3 - u_5 + v_5 &= 100 \\
u_i, v_i \geq 0 \quad \text{for all } i, a_0, a_1, a_2, a_3 &\text{ unrestricted}
\end{aligned}$$

As before, if, $(\tilde{a}_0, \tilde{a}_1, \tilde{a}_2, \tilde{a}_3; \tilde{z}; \tilde{u}_i, \tilde{v}_i: i = 1 \text{ to } 5)$ is an optimum solution of the LP (1.25), $f(\tilde{a}, t) = \tilde{a}_0 + \tilde{a}_1 t + \tilde{a}_2 t^2 + a_3 t^3$ is known as a *minimax solution* for the curve-fitting problem (because it minimizes the maximum deviation of the cubic fit from the observed yields at values of t used in the experiment), or a *Chebyshev approximation*.

Least Squares Approximation

The problem of determining the values of the parameters $a = (a_0, a_1, a_2, a_3)$ that minimize the measure of deviation $L_2(a)$ is

minimize $L_2(a)$

$$= \sum_{t \in S} (a_0 + a_1 t + a_2 t^2 + a_3 t^3 - y(t))^2 \qquad (1.26)$$

over all possible real values for a_0, a_1, a_2, a_3. For obvious reasons, problem (1.26) is known as a *least squares problem* and this method is known as the *method of least squares.* Problem (1.26) can be solved by setting each of the partial derivatives of $L_2(a)$ with respect to each a_i. to zero, since there are no constraints on the parameters a_i.

Curve-Fitting Problems with Constraints on the Values of the Parameters

In most practical curve-fitting problems, there may be constraints on parameter values. In that case, the parameter values should be determined to minimize the chosen measure of deviation subject to the constraints.

Comparison of the Various Measures of Deviation

The L_1- and the L_∞-norms are not differentiable at points in the parameter space where some deviation term is zero, and this term changes sign depending on which direction we move in the parameter space from that point. The L_2-norm does not have these problems and it is in general differentiable everywhere. Before the development of the simplex algorithm for solving linear programs in 1947, most optimization methods were calculus based and relied exclusively on the use of partial derivatives. Hence in statistics and other applied sciences, traditionally the method of least squares had a historical advantage, and it has become well established as the standard tool for solving curve-fitting problems. However, when there are known constraints on parameter values, the least squares problem becomes a constrained nonlinear programming problem. Traditionally this is handled by ignoring the constraints. If the resulting unconstrained least squares solution violates the constraints on the parameter values, that solution is somehow modified to another solution which satisfies the constraints. This procedure is in general unsatisfactory, and since most practical curve-fitting problems tend to have constraints on parameter values, it becomes very hard to use the method of least squares to solve them satisfactorily.

If the parameters to be determined in the model occur linearly in it (as in the example discussed above), and if the constraints on parameter values are either linear equality or inequality constraints, the problem of determining the best L_1- or L_∞-approximation can be transformed into an LP and solved efficiently by the simplex method (see Chapter 2 and subsequent chapters). Nowadays, in most computing centers all over the world, excellent and reliable computer codes to solve linear programs are available, and they work very well even on fairly large-size problems; hence the use of an L_1- or L_∞-approximation (as opposed to the least squares approximation) in linear curve-fitting problems is becoming widespread. If the parameters to be determined appear nonlinearly in the model (for example, if we wish to approximate the yield $y(t)$ by a function of the form $g(t) = \alpha \cos(\beta(\log t)^\gamma)$, where α, β, γ are the parameters), or if there are nonlinear constraints on parameter values, the use of a least squares approximation with some nonlinear programming algorithm is preferred (see Reference [1.30]).

Uses in Solving Practical Optimization Problems

To determine an optimum solution in a practical problem, we first construct a mathematical model of the problem, and then find the optimum solution by solving the model. If the mathematical model involves only linear or affine functions of the decision variables, the model will be an LP and can be solved by the methods to solve LPs discussed in the sequel. However, the model itself might involve several parameters that may have to be estimated using actual data. In constructing linear models, this parameter estimation itself can be done by using L_1- or L_∞-approximations and a linear programming formulation for finding these approximations as discussed above. For example, in optimizing the operations of a chemical plant we may have a constraint that the yield y in the plant should be above a certain specified minimum level. Suppose the yield depends on the temperature x_1, flow rate x_2, and pressure x_3. Suppose we want to approximate the yield by the affine function $a_0 + a_1x_1 + a_2x_2 + a_3x_3$, where a_0, a_1, a_2, a_3 are the parameters in this model. The best values for these parameters can be estimated from data on the yield using L_1- or L_∞-approximations and a linear programming formulation of it as discussed above.

1.2.6 Modeling with Excesses and Shortages

In many practical problems (e.g., inventory modeling), we are required to compare a linear function of the problem variables, say, $\sum a_jx_j$ (which in inventory models may be *the amount of production*) with a known constant, say, b (which will correspond to the *demand* in inventory models). If $\sum a_jx_j > b$, there is an *excess situation* and a cost is incurred, which is equal to $c_1(\sum a_jx_j - b)$. In inventory models this will be the *holding cost*. On the other hand, if $\sum a_jx_j < b$, there is a *shortage situation* and a penalty or cost is incurred, which is equal to $c_2(b - \sum a_jx_j)$. In inventory models this will be the *penalty for shortage*. There is no cost due to excess in a shortage situation; and likewise, there is no penalty for shortage in an excess situation. Assume that the problem is to minimize the total cost. We can express the real number $\sum a_jx_i - b$ as a difference of two nonnegative numbers as in Section 1.2.4 and clearly the excess is the positive part of $(\sum a_jx_j - b)$ and the shortage is its negative part. Hence, if y_1 and y_2 are such that

$$\sum a_jx_j - y_1 + y_2 = b$$
$$y_1 \geqq 0 \qquad y_2 \geqq 0 \tag{1.27}$$

$$y_1y_2 = 0 \tag{1.28}$$

then y_1 is the excess and y_2 is the shortage. Hence the cost incurred is $c_1y_1 + c_2y_2$. Since the aim is to minimize the total cost, using arguments similar to those in Section 1.2.4, it can be shown that the constraint (1.28) will be satisfied automatically by an optimum solution to the problem obtained by the simplex method, if $c_1 + c_2 \geqq 0$. Assuming that $c_1 + c_2 \geqq 0$, we can therefore represent the total contribution of the excess and shortage costs as $c_1y_1 + c_2y_2$ where y_1 and y_2 are subject to the constraints (1.27). This models the problem correctly as long as $c_1 + c_2 \geqq 0$. If $c_1 + c_2 < 0$, it is not possible to model this problem as an LP.

Exercises

1.7 Justify the above model.

Example 1.12

A firm has m cement manufacturing plants and n cement selling markets. During a season the ith plant can manufacture at most a_i tons of cement during *regular production time* at a cost of r_i dollars per ton for $i = 1$ to m. In addition to *regular production*, the ith plant can manufacture an amount of at most b_i tons of cement by running *overtime* at a cost of s_i dollars per ton (where $s_i > r_i$) for $i = 1$ to m. Note that a plant cannot operate on *overtime* unless the *regular time* is fully used up. The transportation cost from the ith plant to the jth market is c_{ij} dollars per ton of cement. All the cement manufactured during a season can be transported to the market in the same season. At the jth market there is a *demand* of d_j tons of cement for $j = 1$ to n. There is no restriction that the *demand* should be met; however, each ton of *demand met* in market j fetches a price of α_j dollars and each ton of demand left unsatisfied at market j results in a penalty of δ_j dollars to the firm. If the total amount of cement shipped to market j exceeds the demand at that market, the excess can be sold to a warehouse at a price of γ_j dollars per ton (note $\gamma_j < \alpha_j$), for $j = 1$ to n, in unlimited quantities. The problem is to determine the optimal production-distribution-marketing pattern that minimizes the firm's net cost during a season.

We will now formulate this problem as an LP. The variables in the problem are:

x_{ij} = tons of cement shipped from the ith plant to the jth market

x_i = total tons of cement manufactured at the ith plant

y_{1i} = total tons of cement manufactured at the ith plant in regular time

Machine Type	Capacity: Production per Day	Production Costs per Day If Used to Manufacture Purity			
		1	2	3	4
1	50 tons	\$900	1000	1250	1500
2	40 tons	600	750	1000	1050
3	25 tons	600	700	800	900
Monthly demand for chemical at purity level (tons)		150	300	90	175
Penalty per ton short		40	60	75	90

y_{2i} = total tons of cement manufactured at the ith plant in overtime

u_j = excess amount in tons (amount supplied above the demand) at the jth market

v_j = shortage of cement in tons (amount of unfulfilled demand) at the jth market

The production cost for producing x_i tons of cement at the ith plant is a piecewise linear convex function. As in Section 1.2.2 we formulate this by expressing x_i as $y_{1i} + y_{2i}$, with the restrictions, $0 \leqq y_{1i} \leqq a_i$, $0 \leqq y_{2i} \leqq b_i$, and the production cost will be $r_i y_{1i} + s_i y_{2i}$. Similarly the shortage and excess amounts at the jth market are obtained by expressing the difference between the total supply and demand for cement at this market as the difference of excess and shortage, both of which are nonnegative numbers, that is, $\sum_{i=1}^{m} x_{ij} - d_j = u_j - v_j$, and then the net cost at this market can be expressed as $\delta_j v_j - \alpha_j(d_j - v_j) - \gamma_j u_j$. Hence eliminating the constant terms from the objective function, the overall problem is

Minimize

$$\sum_{i=1}^{m} (r_i y_{1i} + s_i y_{2i}) + \sum_{j=1}^{n} ((\delta_j + \alpha_j)v_j - \gamma_j u_j) + \sum_{i=1}^{m} \sum_{j=1}^{n} c_{ij} x_{ij}$$

Subject to

$$\sum_{j=1}^{n} x_{ij} - y_{1i} - y_{2i} = 0 \qquad \text{for } i = 1 \text{ to } m$$

$$\sum_{i=1}^{m} x_{ij} - u_j + v_j = d_j \qquad \text{for } j = 1 \text{ to } n$$

$$y_{1i} \leqq a_i \qquad \text{for } i = 1 \text{ to } m$$

$$y_{2i} \leqq b_i \qquad \text{for } i = 1 \text{ to } m$$

$$x_{ij}, y_{1i}, y_{2i}, u_j, v_j \geqq 0 \qquad \text{for all } i, j$$

Exercises

1.8 There are three types of machines that can be used to make a chemical. The chemical can be manufactured at four different purity levels. Data are given in the table at the top. The chemical is used by the company internally; if the company does not manufacture enough of it, it can be bought at a price, which is called the penalty for shortage. Assume that there are 30 production days per month at the company. Formulate the problem of minimizing the overall cost as an LP.

1.3 SOLVING LINEAR PROGRAMS IN TWO VARIABLES

LPs involving only two variables can be solved by drawing a diagram on the Cartesian plane. Given an LP, a *feasible solution* is a vector that specifies a value for each variable in the problem, which when substituted satisfies all the constraints and sign restrictions. An *optimum solution* is a feasible solution that either maximizes the objective function or minimizes it (depending on what is required to be done) on the set of feasible solutions. To solve an LP involving only two variables, the set of feasible solutions is first identified on the two-dimensional Cartesian plane. The optimum solution is then identified by tracing the line corresponding to the set of feasible solutions that give a specific value to the objective function and then moving this line parallel to itself.

As an example consider the fertilizer product mix problem (1.1). The constraint $2x_1 + x_2 \leqq 1500$ requires that any feasible solution (x_1, x_2) to the problem should be on one side of the line $2x_1 + x_2 = 1500$, the side that contains the origin (because the origin makes $2x_1 + x_2 = 0 < 1500$). This side is indicated by an arrow on the line in Figure 1.9. Likewise, all the constraints can be represented on the diagram. The set of feasible solutions is the set of points on the plane satisfying all the constraints and sign restrictions; this is the shaded region in Figure 1.9.

For any real value of z_0, $z(x) = z_0$ represents all the points on a straight line in the two-dimensional plane. Changing the value of z_0 translates the entire line to another parallel line. Pick an arbitrary value of z_0 and draw the line $z(x) = z_0$ on the diagram. If this line intersects the set of feasible solutions, the points of intersection are the feasible solutions that give the value z_0 to the objective function. If the line does not intersect the set of feasible solutions, check whether increasing z_0, or decreasing it, translates it to a parallel line that is closer to the set of feasible solutions. Change the value of z_0 appropriately until a value for z_0, say z_0', is found such that the line $z(x) = z_0'$ has a nonempty intersection with the set of feasible solutions. Then go to the next step. In the fertilizer problem, if $z_0 =$

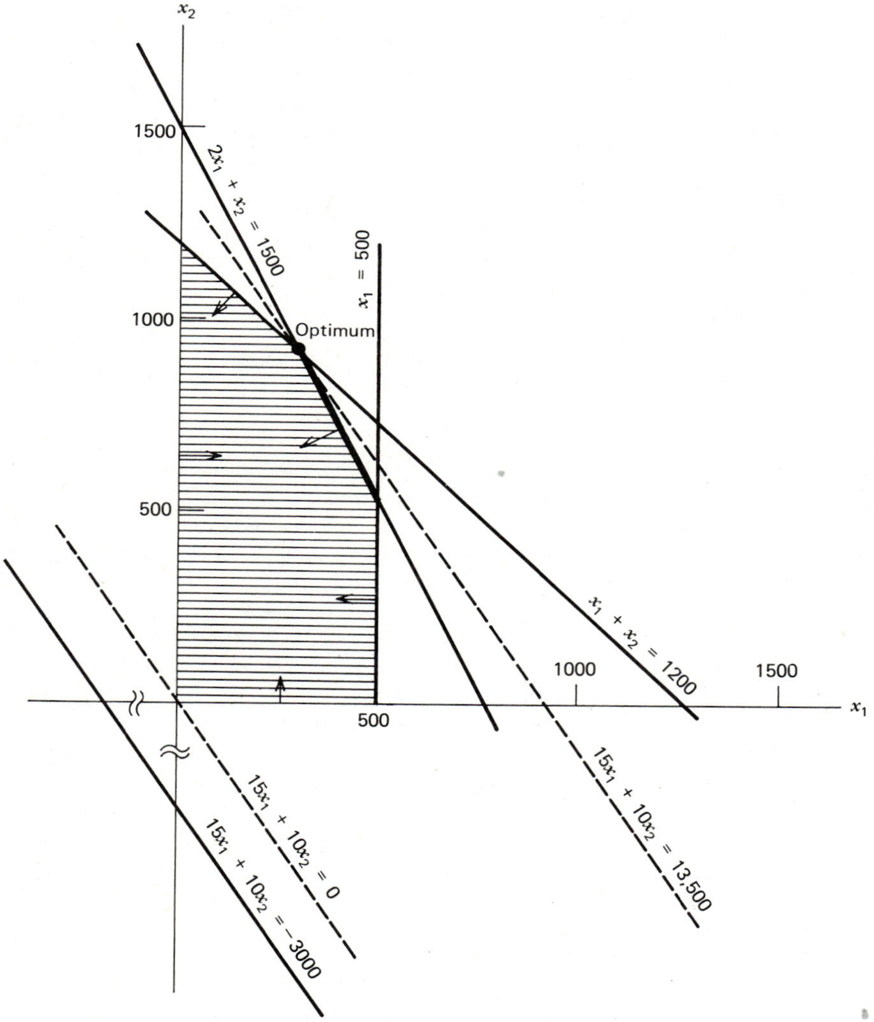

Figure 1.9 Fertilizer product mix problem.

-3000, the line $z(x) = -3000$ does not intersect the set of feasible solutions at all. Also, the line $z(x) = z_0$ moves closer to the set of feasible solutions as z_0 increases from -3000, and when $z_0 = 0$, the line $z(x) = 15x_1 + 10x_2 = 0$, has a nonempty intersection with the set of feasible solutions.

If $z(x)$ has to be maximized, try to move the line $z(x) = z_0$ in a parallel fashion by increasing the value of z_0 steadily from z_0' as far as possible while still intersecting the set of feasible solutions. If $\hat{z}_0$ is the maximum value of z_0 obtained in this process, it is the maximum value of $z(x)$ in the problem and the set of optimum feasible solutions is the set of feasible solutions that lie on the line $z(x) = \hat{z}_0$.

On the other hand, if the line $z(x) = z_0$ has a nonempty intersection with the set of feasible solutions for every $z_0 \geqq z_0'$, then $z(x)$ is *unbounded above* on that set. In this case $z(x)$ can be made to diverge to $+\infty$ and the problem has no finite optimum solution.

If the aim is to minimize $z(x)$, then steadily decrease the value of z_0 from z_0' and apply the same kind of arguments.

For the fertilizer problem, as z_0 is increased from 0, the line $15x_1 + 10x_2 = z_0$ moves up until $\hat{z}_0 = 13{,}500$. For any value of $z_0 > 13{,}500$ the line $z(x) = z_0$ does not intersect the set of feasible solutions. Thus,the optimum objective value is \$13,500 and the optimum solution of the problem is $(x_1, x_2) = (300, 900)$.

Exercises

1.9 Solve the following LPs by drawing diagrams.

(a) Maximize $30x_1 + 20x_2$

Subject to
$$\begin{aligned} x_1 + x_2 &\geqq 1 \\ x_1 - x_2 &\geqq -1 \\ 3x_1 + 2x_2 &\leqq 6 \\ x_1 - 2x_2 &\leqq 1 \\ x_1 \geqq 0 \quad x_2 &\geqq 0 \end{aligned}$$

(b) Minimize $2x_1 - x_2$

Subject to
$$\begin{aligned} x_1 + x_2 &\geqq 10 \\ -10x_1 + x_2 &\leqq 10 \\ -4x_1 + x_2 &\leqq 20 \\ x_1 + 4x_2 &\geqq 20 \\ x_1 \geqq 0 \quad x_2 &\geqq 0 \end{aligned}$$

(c) Minimize $x_1 + 3x_2$

Subject to
$$\begin{aligned} 2x_1 + x_2 &\geqq 10 \\ -x_1 + x_2 &\leqq 20 \\ x_1 - 2x_2 &\leqq 10 \\ x_1 + x_2 &\leqq 30 \\ x_1 \geqq 0 \quad x_2 &\geqq 0 \end{aligned}$$

(d) Minimize $10x_1 + 3x_2$

Subject to
$$\begin{aligned} x_1 + x_2 &\geqq 20 \\ x_1 &\leqq 6 \\ x_1 &\geqq 2 \\ x_2 &\leqq 12 \\ x_2 &\geqq 1 \end{aligned}$$

(e) Maximize $2x_1 - 2x_2$

Subject to
$$\begin{aligned} -2x_1 + x_2 &\leqq 2 \\ x_1 - x_2 &\leqq 1 \\ x_1 \geqq 0 \quad x_2 &\geqq 0 \end{aligned}$$

(f) In the following problem show that the variables x_3, x_4, x_5 play the roles of slack variables. Show that the constraints in the problem can be transformed into inequality constraints in the variables x_1, x_2, by eliminating these slack variables. Use this and solve this problem by drawing a diagram.

Minimize $x_1 - 3x_2$

Subject to
$$\begin{aligned} 2x_1 + x_2 + x_3 \qquad\qquad &= 20 \\ x_1 \qquad\quad - x_4 \qquad &= 7 \\ x_2 \qquad\qquad - x_5 &= 17 \\ x_j \geqq 0 \quad \text{for all } j = 1 \text{ to } 5 & \end{aligned}$$

Note 1.2: LPs in higher dimensional spaces ($\geqq 3$) cannot be solved by drawing diagrams, but the simplex algorithm of Chapter 2 provides a systematic procedure for solving them. However, we will gain a lot of intuition on how the simplex algorithm works by referring to and visualizing the concepts of geometry.

1.4 MODELING MULTIPLE OBJECTIVE PROBLEMS

In all the problems discussed so far, there was a single, well-defined objective function to be optimized. In some problems the decision maker may want to optimize two or more objective functions simultaneously; these are called *multiple objective problems*.

Example 1.13

Consider the fertilizer manufacturer's problem formulated in Section 1.1.3. The fertilizer manufacturer clearly wants to maximize its profit $z^1(x) = 15x_1 + 10x_2$, which is in dollars. Now, suppose that in the world of fertilizers the prestige of a company goes up with the amount of Hi-phosphate fertilizer it manufactures. Our manufacturer may consider it very important to maintain the image of its company as a prestigious company, and in this case it should also want to maximize the amount of Hi-phosphate fertilizer produced by its company, which is $z^2(x) = x_1$, in tons. In this case, we have two different objective functions $z^1(x) = 15x_1 + 10x_2$ and $z^2(x) = x_1$, both of which the manufacturer would like to see maximized simultaneously, subject to the constraints derived in Section 1.1.3. This is a multiobjective problem.

In a multiobjective problem, if there exists a feasible solution x^1, which simultaneously optimizes every one of the objective functions, x^1 is clearly the optimum solution for the problem. Unfortunately, this phenomenon is quite rare. There is usually no single feasible solution that simultaneously provides the best value for each objective function. As an illustration, in Example 1.13, $z^1(x)$ is maximized by the unique feasible solution $\bar{x} = (300, 900)^T$, yielding a value to $z^1(x) = \$13,500$. The problem of maximizing $z^2(x)$ has alternate optima, with the points $x(\alpha) = (500, \alpha)$ $0 \leqq \alpha \leqq 500$, all being optimal to it, giving a value to $z^2(x)$ of 500 tons. Clearly, there is no single feasible solution in this problem that simultaneously optimizes (here maximizes) both $z^1(x)$ and $z^2(x)$.

Very often, in multiobjective problems, the various objective functions conflict with each other. An optimum solution for one objective function might turn out to be a very undesirable solution for another objective function. This conflict among the various objective functions in a multiobjective problem is illustrated by the cartoon in Figure 1.10.

There is no theoretically satisfactory concept of optimality yet in multiobjective problems in which the objective functions conflict with each other, but many practically reasonable approaches have been developed for handling them. We discuss these approaches here briefly. We begin with a few definitions first.

Let **K** denote the set of feasible solutions (i.e., set of solutions satisfying all the constraints and restrictions on the decision variables), and $f^1(x)$, $f^2(x), \ldots, f^t(x)$, the various objective functions in a multiobjective problem. We assume $t \geqq 2$, as otherwise it is a single objective problem. Then the vector $f(x) = (f^1(x), f^2(x), \ldots, f^t(x))$ is known as the *vector-valued objective* (or *criterion*) *function*. In this problem. A feasible solution $\bar{x} \in \mathbf{K}$ is said to be a

vector minimum, if there exists no other feasible solution $x \in \mathbf{K}$ that satisfies $f^r(x) \leqq f^r(\bar{x})$, for $r = 1$ to t, with the inequality holding strictly for at least one r;

vector maximum, if there exists no other feasible solution $x \in \mathbf{K}$ that satisfies $f^r(x) \geqq f^r(\bar{x})$, for $r = 1$ to t, with the inequality holding strictly for at least one r.

So, given a vector minimum $\bar{x}$, any move from $\bar{x}$ that strictly decreases an objective function from its value at $\bar{x}$, automatically leads to a strict increase in another objective function from its value at $\bar{x}$. Thus a vector-minimum solution may be a desirable solution to seek when it is required to minimize each objective function in the problem. A similar argument shows that a vector-maximum solution may be a desirable solution to seek when it is required to maximize each objective function in the problem.

Example 1.14

The set of feasible solutions of the fertilizer problem formulated in Section 1.1.3 is drawn in Figure 1.9. For the two-objective version of this problem discussed in Example 1.13 above, it can be verified that every point on the thick line segment joining $x^1 = (300, 900)$ and $x^2 = (500, 500)$ is a vector-maximum feasible solution.

A vector maximum over **K** for the vector function $f(x)$, is a vector minimum over **K** for the vector function $-f(x)$ and vice versa. We now discuss some of the approaches that can be used to handle multiobjective problems.

1.4.1 The Vector-Minimum Approach

Let $f(x) = (f^1(x), \ldots, f^t(x))$ be the vector of objective functions. If it is required to make one of these

Figure 1.10 A multiple objective problem with conflicting objective functions. (Drawing by Ziegler, *copyright* © *1980* The New Yorker Magazine, Inc.).

functions as large as possible, replace it by its negative in the vector. After this modification, every function in the vector $f(x)$ is required to be made as low as possible. Then we could take a vector minimum of $f(x)$ over the set of feasible solutions as a solution to this multiobjective problem. In this case, the vector minimum is also known as a *pareto optimal solution* or *efficient solution* or a *nondominated solution*, and this method of choosing a vector minimum as the solution is known as the *pareto optimality criterion*.

One drawback of this approach is that even though the concept of vector minimum is mathematically very appealing, usually there are many vector minima, and there exists no satisfactory theory for comparing and choosing the best among them.

If all the constraints are linear, and each objective function in $f(x)$ is affine, the problem of finding a vector minimum for $f(x)$ over the feasible region is known as a *linear vector-minimization problem*. In linear problems a vector minimum can be found very efficiently, see Chapter 17 or references [1.37–1.48], and efficient methods even exist for computing all the vector minima. However, so far, these methods have remained as mathematical tools that have not found many uses in practical applications.

1.4.2 Approach Based on Ordering the Objective Functions in Priority Order

Let $f(x) = (f^1(x), \ldots, f^t(x))$ be the vector of objective functions, each of which has to be made as low as possible. This approach can be used if there exists a clear priority order among the objective functions, so that one of the objective functions is recognized as the most important among all, another is recognized as being the most important among all but the first, etc. Let $f^1(x), \ldots, f^t(x)$ be the objective functions in this priority order. When such universally recognized priority order exists among the objective functions, proceed in stages as follows. First, in stage 1 minimize $f^1(x)$ (the most important objective function) over the set of feasible solutions. If the optimum solution for this problem is unique, accept it as the optimum solution for the multiobjective problem and terminate. Otherwise let $\mathbf{K}_1$ denote the set of alternate optima, each of which minimizes $f^1(x)$ over the original set of feasible solutions. Go to stage 2. In general, if the method does not terminate in stage r, we let $\mathbf{K}_r$ denote the set of alternate optimum solutions of the stage r problem. Then in stage $r + 1$, minimize $f^{r+1}(x)$ over $\mathbf{K}_r$. If the optimum solution of this problem is unique, terminate by accepting it as the optimum solution of the multiobjective problem. Otherwise let $\mathbf{K}_{r+1}$ denote the set of alternate optimum solutions of the stage $r + 1$ problem. Go to stage $r + 2$ and continue in the same way. If all the constraints are linear, and each objective function is linear, this approach can be implemented very efficiently using the simplex algorithm for solving linear programs (see Section 3.10 and Exercise 3.72). If a natural priority order exists among the objective functions, this is the best approach to use. This is known as a *sequential procedure*, and it can be shown (see Chapter 17, in particular Exercise 17.6) that the solution obtained under this procedure is a vector minimum for this multiobjective problem.

Requirements for Using the Other Approaches

All the other approaches for handling multiobjective problems discussed below require that each objective function be expressed in some common comparable units. For example, in Example 1.13, $z^1(x) = 15x_1 + 10x_2$ measures the profit in dollars, and $z^2(x) = x_1$ measures the Hi-phosphate fertilizer production in tons, and these two objective functions are not comparable. However, we can change the second objective function into the profit derived in dollars from Hi-phosphate fertilizer production, $f^2(x) = 15x_1$. Maximizing $z^2(x)$ is the same as maximizing $f^2(x)$ and vice versa. When the vector of objective functions for this problem is taken to be $z^1(x)$ and $f^2(x)$, both objective functions are in comparable units of dollars. Using similar ideas, it is often possible to express all the objective functions in a multiobjective problem in common units. We assume that this is done for using any of the approaches discussed below.

1.4.3 The Worst Case Approach

Let $f(x) = (f^1(x), \ldots, f^t(x))$ be the vector of objective functions, each of which is required to be minimized, all the objective functions being in common units. Define for each x, $\theta(x) = \text{maximum}\ \{f^1(x), \ldots, f^t(x)\}$, the pointwise supremum function of all the objective functions. This approach takes the feasible solution that minimizes $\theta(x)$ as an optimum solution of the multiobjective problem. If all the constraints and the objective functions are linear, the problem of minimizing $\theta(x)$ can be transformed into an LP as discussed in Section. 1.2.3. This approach tries to minimize the maximum of the different objective values, and hence can be interpreted as a worst case approach.

1.4.4 Approach Based on Constructing a Weighted Average of the Different Objective Functions

Let $f(x) = (f^1(x), \ldots, f^t(x))$ be the vector of objective functions each of which is required to be minimized, all of them being in common units. For $r = 1$ to t, determine a *weight*, α_r, for the rth objective function, $0 < \alpha_r \leqq 1$, depending on the importance of this objective function in the vector, where this weight is given higher values for more important objective functions. Then $\theta(x) = \sum_{r=1}^{t} \alpha_r f^r(x)$ is a positive weighted combination of all the objective functions in the multiobjective problem. By choosing the weights properly, $\theta(x)$ can be viewed as the result of balancing the various objective functions in the problem into a single function. Find the feasible solution that minimizes $\theta(x)$, and take it as the optimum solution of the multiobjective problem. If the solution obtained does not look very reasonable, a new $\theta(x)$ can be generated by altering the weights and the process repeated until a reasonable solution is obtained. It is proved in Chapter 17, that the solution obtained using this approach is a vector minimum, and conversely for every vector minimum $\bar{x}$ there exists a positive vector of weights $\alpha = (\alpha_1, \ldots, \alpha_t)$, such that $\bar{x}$ minimizes $\sum_{r=1}^{t} \alpha_r f^r(x)$ over the set of feasible solutions.

1.4.5 The Goal Programming Approach

Let $f^r(x) = c^r x$, $r = 1$ to t, be all the objective functions in the problem, where $c^r = (c_1^r, c_2^r, \ldots, c_n^r)$ is the row vector of cost coefficients in the rth objective function. Suppose the constraints on the decision variables are $Ax = b$, $x \geqq 0$. In this approach, instead of trying to optimize each objective function, the decision maker is asked to specify a *goal* or a *target value* for each objective function that is most desirable to attain. Let g_r denote the specified goal for the rth objective function, $r = 1$ to t. Consider the following LP:

$$\begin{aligned} \text{Minimize} \quad & z(x, u, v) = \sum_{r=1}^{t} (\alpha_r u_r + \beta_r v_r) \\ \text{Subject to} \quad & c^r x - u_r + v_r = g_r \qquad r = 1 \text{ to } t \qquad (1.29) \\ & Ax = b \\ & x \geqq 0 \quad u_r, v_r \geqq 0 \qquad \text{for all } r = 1 \text{ to } t \end{aligned}$$

where the coefficients α_r, β_r are determined by the following: If the decision maker likes to make $f^r(x)$ as large as possible, he or she may not mind a value for $f^r(x)$ larger than g_r, but if $f^r(x)$ is going to have a value less than g_r, the decision maker may like to see the difference $g_r - f^r(x)$ be as small as possible. In this case, make $\alpha_r = 0$ and give β_r a positive value. On the other hand, if he or she wants $f^r(x)$ to have as low a value as possible, make $\beta_r = 0$, and give α_r a positive value. As opposed to these two possibilities, the decision maker may just be interested in having the value of $f^r(x)$ be as close to the goal g_r as possible; the deviation between these, whether positive or negative, being considered undesirable. In this case, make the values of both α_r and β_r positive. When one or both of these coefficients α_r and or β_r is to be made positive, the actual values chosen for them can be made to depend on the relative importance of the objective function in comparison with the others. If $(\bar{x}, \bar{u}, \bar{v})$ is an optimum solution of (1.29), $\bar{x}$ can be taken as the optimum solution of this multiobjective problem, obtained by this goal programming approach. The LP (1.29) can be solved for different sets of values of the weights α_r, β_r, until at some stage, the optimum solution obtained for (1.29) is a reasonable solution for the multiobjective problem. Also, the goals could be altered and (1.29) solved again. Exploring with the optimum solutions for (1.29) obtained in this manner, one can expect to get a practically satisfactory solution to the multiobjective problem.

Exercises

1.10 In any optimum solution $(\bar{x}, \bar{u}, \bar{v})$ for the LP (1.29), prove that at least one of $\bar{u}_r$ or $\bar{v}_r$ will be 0 for each $r = 1$ to t. (Use the same arguments as in Section 1.2.4).

1.11 We have the following system of constraints:

$$\begin{aligned} x_1 \qquad\qquad & + x_4 - 2x_5 + 7x_6 + x_7 = 18 \\ x_2 \qquad & - x_4 + 3x_5 - 8x_6 + 2x_7 = 13 \\ & x_3 + 2x_4 - 2x_5 + 2x_6 - x_7 = 19 \\ & x_j \geqq 0 \qquad \text{for all } j \end{aligned}$$

We also have three objective functions: $f^1(x) = 13x_1 - 17x_2 + 8x_3 - 4x_4 + 3x_5 - 7x_6$, $f^2(x) = 12x_2 - 3x_3 + 4x_4 + 5x_5 + 7x_6 + 18x_7$, $f^3(x) = \sum_{j=1}^{7} x_j$. Objective function $f^1(x)$ should be made as large as possible, but preferably above its target value of 80; objective function $f^2(x)$ should be made as low as possible, but preferably below its target value of 100; and objective function $f^3(x)$ should be as close to its target value of 25 as possible. Formulate the problem of finding a feasible solution to this system that satisfies these requirements, as an LP.

1.4.6 Approach Based on Goal Setting Using Marginal Values

If any objective function in the list is required to be maximized, replace it by its negative. After this change, we want every objective function to be minimized. In this approach, one of the objective functions, which the decision maker would particularly like minimized, is singled out as the *cost function*. In most practical applications, this problem of identifying a special objective function is usually easy. It is done by taking the cost of implementing the solution (or the negative of the net profit from implementing the solution) as the special function. Let $f^0(x) = c^0 x$ be this special function, and let $f^r(x) = c^r x$, $r = 1$ to t be the other objective functions in the model. Suppose the constraints on the decision variables are $Ax = b$, $x \geqq 0$, where A is a given matrix of order $m \times n$. Take an initial feasible solution, say $\bar{x}$, preferably a vector minimum for $\{f^1(x), \ldots, f^r(x)\}$

that can be found by using the methods discussed in Chapter 17. Now solve the LP.

$$\begin{aligned} \text{Minimize} \quad & f^0(x) = c^0 x \\ \text{Subject to} \quad & Ax = b \\ & c^r x = g_r \qquad r = 1 \text{ to } t \\ & x \geqq 0 \end{aligned} \tag{1.30}$$

where $g_r = c^r \bar{x}$, for $r = 1$ to t. An optimum solution of (1.30) minimizes the cost of achieving the initial goal of $g_r = c^r \bar{x}$,for the rth objective function, for $r = 1$ to t. As discussed in Section 1.1.4 (and in Chapter 4 and Section 8.15 in greater detail), the dual optimum solution associated with (1.30) provides the marginal values, the rates of change in the optimum objective value when the right-hand side constants in (1.30) are perturbed. Given the present goals g_r and the marginal values μ_r, the decision maker can usually select a new set of goals and easily estimate how much it would cost to make this change. Replace g_r in (1.30) with these newly selected goal values, solve the modified problem, and repeat the whole process until a satisfactory set of goals is reached. To obtain the new optimum solution for (1.30) efficiently when the values of g_r are changed, the dual simplex algorithm discussed in Chapter 6 can be used. The practical usefulness of this approach stems from the fact that it provides the marginal values (rates of change in cost per unit change in goals) associated with the goals of the various objective functions, and thereby provides estimates for the change in the cost for effecting required changes in the goal values.

1.5 CONSTRUCTING LINEAR PROGRAMMING MODELS IN PRACTICAL APPLICATIONS

We have already mentioned that the linearity assumptions may not hold exactly in practical applications, but they may lead to a reasonably good approximation of the actual problem.

A major task in constructing an LP model for a practical problem is that of obtaining all the data in the model. It may be hard to obtain the exact values of the coefficients in all the constraints and the objective function, and the right-hand-side constants. Some of them might even be random variables whose values change over time stochastically. The area of *stochastic linear programming* deals with methods for solving LPs in which some or all the coefficients are subject to random fluctuations. Unfortunately stochastic programming methods are complicated, are hard to use for large-scale models, and thus have not come into popular usage. Because of this, practitioners tend to construct a deterministic model that is a reasonable approximation for the problem by choosing the most likely or the closest estimated values for the coefficients in the model. If a particular coefficient is actually a random variable, they can replace it by its average value or some other measure of central tendency. As will be shown later, LP models in general tend to be stable, so small errors in the data lead to small errors in the optimum solution. Also practitioners never really accept the optimum solution from a single run of the model as the solution to be implemented. They usually solve the model several times, with different likely sets of values for the data, and watch how the optimum solution changes. This usually reveals which coefficients are critical, and they can then try to obtain more precise estimates for those coefficients and run the model again. Algorithms for solving LP models are quite efficient in practice, and very high-quality software for these algorithms is available everywhere. So it is quite feasible to run several versions of the model, even with a modest computing budget. The optimum solutions obtained under the various runs provide a guideline for choosing a final solution that is the best one to implement, using the practitioner's practical knowledge about the problem. In the rest of the book we develop the theory and computational algorithms for solving LP models, given the values of all the data elements in them, and also methods for analyzing what changes may occur in the output when small perturbations take place in the data. But one has to keep the above discussion in mind while trying to solve practical problems using these methods.

1.6 WHAT IS MEANT BY SOLVING A LINEAR PROGRAM?

Consider the LP in the following form: find $x = (x_1, \ldots, x_n)^{\mathrm{T}}$ to

$$\begin{aligned} \text{Minimize} \quad & z(x) = cx = \sum_{j=1}^{n} c_j x_j \\ \text{Subject to} \quad & Ax = b \\ & x \geq 0 \end{aligned}$$

where A is a given matrix of order $m \times n$. In Chapter 2 we show that every LP can be transformed into this form. The variables $x_1, \ldots, x_n$ are the *decision variables* in this problem and x is the column vector of decision variables. The entries in A, c, b are the *data* in this problem. A *feasible solution* for this problem is a numerical vector x that satisfies all the constraints and sign restrictions. An *optimum feasible solution* (or *an optimum solution*) is a feasible solution that minimizes the objective value $z(x)$ among all feasible solutions. Other types of solutions for this problem called *basic solutions* and *basic feasible solutions* (BFSs) are defined later in Chapters 2 and 3. There are three distinct possibilities for this problem. They are:

1. There may not be any feasible solution for the problem. In this case, this problem is said to be *infeasible*.
2. Feasible solutions for the problem may exist (in this case the problem is said to be *feasible*), but no finite optimum feasible solution may exist because the objective function $z(x)$ is unbounded below on the set of feasible solutions. In this case, the minimum value of $z(x)$ in the problem is $-\infty$, and the LP is said to be *unbounded below*.
3. Feasible solutions for the problem may exist and the optimum objective value may be finite. In this case, the problem has an optimum feasible solution.

An *instance* of this LP is obtained by specifying particular numerical values to m, n, and all the data elements in the problem. Any algorithm for solving this LP should be able to determine which of the three possibilities (1), (2), (3) occurs, when applied on any instance of the problem. Also, if possiblity (3) occurs, it should terminate with an optimum feasible solution for that instance. In addition to these, under possibility (2), the *simplex method* for solving LPs discussed in Chapter 2 and later chapters obtains a *half-line*, which begins at a point and moves in a fixed direction along a straight line, satisfying the property that every point on it is feasible and that the objective value $z(x)$ decreases indefinitely as you continue to move along this half-line.

Using digital arithmetic we can carry out exact arithmetical operations only on rational numbers. So, for computational purposes, we assume that all the data are rational. All the theorems proved in the sequel remain valid when the data are arbitrary real numbers, but statements about computational complexity, etc., always refer to rational problems. After reading Chapters 2, 3, and 4, the reader can verify that when the data are all rational, all basic solutions (both primal and dual) are rational vectors, and on such problems, the simplex method executed with exact arithmetic will obtain an optimum solution (if one exists) that is a BFS, a rational vector.

In solving large LPs, practical limitations (such as the limit on the amount of memory space we can use) may prevent us from carrying out the arithmetical operations in the algorithm exactly, even with rational data. Roundoff errors are inevitably introduced when carrying out arithmetical operations on finite precision machines. In this case, if the LP has an optimum solution, by using a robust implementation of the simplex method (or any other method for solving LPs), we can find an approximation to an optimum solution for it to a reasonable degree of accuracy. This is what we mean by solving an LP.

Exercises

Formulate Exercises 1.12–1.34 as LPs.

1.12 A forestry company has four sites on which they grow trees. They are considering four species of trees, the pines, spruces, walnuts, and other hardwoods. Data on the problem are given below. How much area should the company devote to the growing of various species in the various sites?

Site Number	Area Available at Site (ka)	Expected Annual Yield from Species (m^3/ka)				Expected Annual Revenue from Species (money units per ka)			
		Pine	Spruce	Walnut	Hardwood	Pine	Spruce	Walnut	Hardwood
1	1500	17	14	10	9	16	12	20	18
2	1700	15	16	12	11	14	13	24	20
3	900	13	12	14	8	17	10	28	20
4	600	10	11	8	6	12	11	18	17
Minimal required expected annual yield (km^3)		22.5	9	4.8	3.5				

1.13 A farmer has to purchase the following quantities of fertilizer from four different shops, subject to the following capacities and prices. How can he fulfill his requirements at minimal cost?

Fertilizer Type	*Minimum Required (tons)*
1	185
2	50
3	50
4	200
5	185

Shop Number	*Maximum (all types combined) They Can Supply*
1	350 tons
2	225
3	195
4	275

	Price in Money Units per Ton of Fertilizer Type				
At Shop	*1*	*2*	*3*	*4*	*5*
1	45.0	13.9	29.9	31.9	9.9
2	42.5	17.8	31.0	35.0	12.3
3	47.5	19.9	24.0	32.5	12.4
4	41.3	12.5	31.2	29.8	11.0

1.14 A product can be made in three sizes, large, medium, and small, which yield a net unit profit of $12, 10, and 9, respectively. The company has three centers where this product can be manufactured and these centers have a capacity of turning out 550, 750, and 275 units of the product per day, respectively, regardless of the size or combination of sizes involved. Manufacturing this product requires cooling water and each unit of large, medium, and small sizes produced require 21, 17, and 9 gallons of water, respectively. The centers 1, 2, and 3 have 10,000, 7000, and 4200 gallons of cooling water available per day, respectively. Market studies indicate that there is a market for 700, 900, and 450 units of the large, medium, and small sizes, respectively, per day. By company policy, the fraction (scheduled production)/(center's capacity) must be the same at all the centers. How many units of each of the sizes should be produced at the various centers in order to maximize the profit?

1.15 A housewife wants to fix a naturally vitamin-rich cake made out of blended fruit for supper. Data on the available types of fruit are given in the table at the bottom. To keep the cake palatable she has to make sure that the use of the various fruits is within the bounds specified. Formulate the problem of finding the minimum cost blend that satisfies all of these constraints. By simple linear transformations on the variables, show that this LP can be transformed into an equivalent LP in which the lower bounds on all the variables are zero.

1.16 *THREE DIMENSIONAL TRANSPORTATION PROBLEM.* A firm transports wood logged from its forest stands through one of its own depots to a customer's depot. They have m forest stands, p depots, and n customers, c_{ijk} is the cost of transporting logged wood from the ith forest stand to the kth customer through the jth depot in dollars/cubic meter, for $i = 1$ to m,

Fruit Type	*Number of Units of Nutrient/kg of Fruit* — *Vitamin A*	*Vitamin C*	*Cost ($/kg) of Fruit*	*Restrictions on Fruit Use (kg)* — *Minimum*	*Maximum*
1	2	0	0.25	2	10
2	0	3	0.31	3	6
3	2	4	0.48	0	7
4	1	2	0.21	5	20
5	3	2	0.19	0	5
Minimum nutrient requirement (units)	41	80			

Process	Requirements per Unit Time			Output per Unit Time	
	Raw Material 1 (units)	Raw Material 2 (units)	Fuel (units)	Chemical 1 (units)	Chemical 2 (units)
1	9	5	50	9	6
2	6	8	75	7	10
3	4	11	100	10	6
Amount available	200	400	1850		

$j = 1$ to p, $k = 1$ to n; a_i is the annual production of wood in the ith forest stand in cubic meters; b_j is the amount of wood that the jth depot can handle in cubic meters per year; and d_k is the demand for wood by the kth customer in cubic meters per year. It is required to minimize the overall annual transportation bill subject to all the restrictions.

1.17 A company makes a blend consisting of two chemicals, 1 and 2, in the ratio 5:2 by weight. These chemicals can be manufactured by three different processes using two different raw materials and a fuel. Production data are given in the table at the top. For how much time should each process be run in order to maximize the total amount of *blend* manufactured?

1.18 In constructing a hydrological model using the data in the table given at the bottom, it is required to obtain the expected runoff, denoted by R_i during the ith period, as a linear function of the observed precipitation. From hydrological considerations the expected runoff depends on the precipitation during that period and the previous two periods. So the model for expected runoff is $R_i = b_0p_i + b_1p_{i-1} + b_2p_{i-2}$, where p_i equals precipitation during the ith period; and b_0, b_1, b_2 are the coefficients that are required to be estimated. These coefficients have to satisfy the following constraints from hydrological considerations: $b_0 + b_1 + b_2 = 1$, where $b_0 \geqq b_1 \geqq b_2 \geqq 0$. Obtain the best estimates for b_0, b_1, b_2, if the objectives are

(i) to minimize the sum of absolute deviations $\sum_i |R_i - b_0p_i - b_1p_{i-1} - b_2p_{i-2}|$; and

(ii) to minimize the maximum absolute deviation $\max_i |R_i - b_0p_i - b_1p_{i-1} - b_2p_{i-2}|$.

1.19 Each circle in Figure 1.11 is a city. Material can be shipped from city i to city j only if there is directed arc from city i to city j as:

$$(i) \text{---} \ell_{ij}, k_{ij}, c_{ij} \rightarrow (j).$$

The three numbers on the arc are the least amount you can ship along this arc, the maximum tonnage you can ship along this arc, and the cost in dollars/ton shipped along this arc, respectively. There are 27 tons of material at city 1, and all of it should be shipped to city 6 at minimal total cost. All material should originate from city 1 and end up in city 6. At any of the intermediate cities the amount of material reaching should be equal to the amount of material leaving.

Period	1	2	3	4	5	6	7	8	9	10	11	12
Precipitation (inch hours)	3.8	4.4	5.7	5.2	7.7	6.0	5.4	5.7	5.5	2.5	0.8	0.4
Runoff (acre feet)	0.05	0.35	1.0	2.1	3.7	4.2	4.3	4.4	4.3	4.2	3.6	2.7

—(*R. Deininger*)

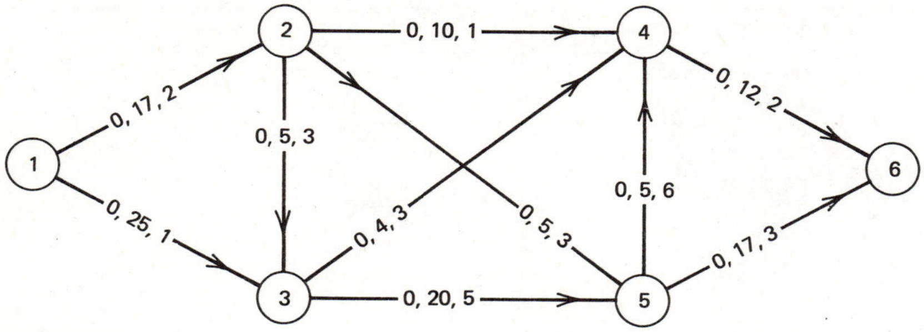

Figure 1.11

1.20 A European company has three coke oven plants, code named 1,2,3. The coal comes from four different sources; USA, Ruhr, Lorraine, and Saar. The plants produce coke, which may be classified into two categories, metallurgical coke and coke screenings; they also produce coke oven gas. The coke oven plants are operated by heating them with either blast furnace gas or coke oven gas. The process flow chart is given in Figure 1.12. The production of coke oven gas and coke depends on the coal used. The proportion of coke screenings in the coke produced depends on the coal used and the plant where it is used. Metallurgical coke is what is left in the coke after coke screenings are separated. Processing 1 ton of coal requires the heat equivalent of 0.611 *k*th of coke oven gas.[1] One unit of blast furnace gas is equivalent to 0.927 *k*th of coke oven gas. The annual processing capacities of the three plants are 9×10^5 tons, 7×10^5 tons, and 3×10^5 tons of coal, respectively.

Saar coal cannot be used in plants 1 and 2. USA coal and Lorraine coal cannot be used in plant 3. The percentage of Lorraine coal in the coal used at plant 1 cannot exceed 30. The percentage of Lorraine coal in the coal used at plant 2 cannot exceed 35. The percentage of Saar coal in the coal used at plant 3 cannot exceed 40. Coke oven gas can be bought or sold in any amounts at $11/*k*th. Coke screenings can be sold in any amounts at $98/ton. Blast furnace gas can be purchased in any amounts at $8/unit. The prices of coal are given below. All the coke screenings produced are sold. All the coke oven gas produced is either used up at the plants or sold. It is required to produce a total of 10^6 tons of metallurgical coke at minimal cost.

Coal Source	*Coke Oven Gas Produced kth/Ton of Coal*	*Coke Tons/ Ton of Coal*
USA	1.08	0.88
Ruhr	1.10	0.80
Saar	1.25	0.74
Lorraine	1.45	0.72

Coal Source	
USA	$80.39/ton
Ruhr	$80.90/ton at plant 1 and 2 $83.93/ton at plant 3
Saar	$80.14/ton
Lorraine	$68.70/ton

Proportion of Coke Screenings in the Coke Produced

At Plant	*Using Coal from* USA	Ruhr	Saar	Lorraine
1	0.10	0.09	0.10	0.15
2	0.08	0.08	0.08	0.11
3	0.07	0.07	0.07	0.10

—(*J. F. Collard*)

1.21 A farmer is planning his operations over a three-year period. At the beginning of the period he has two bushels of grain. At the beginning of years 1, 2, 3 he has to decide how much he will plant. A bushel of grain planted at the beginning of a year yields λ bushels by the

[1] The amount of coke oven gas produced is measured by its heat content. *k*th is a kilotherm, where therm is the amount of heat needed to raise the temperature of 1 ton of water by 1 degree centigrade.

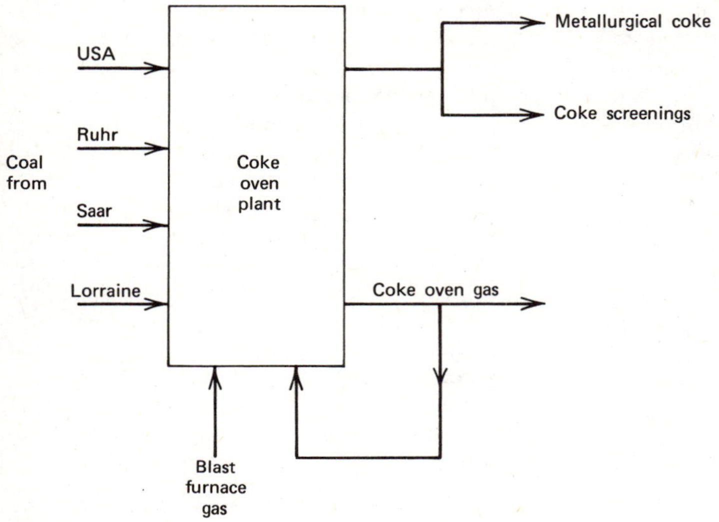

Figure 1.12

end of that year. The profit per bushel of grain sold at the beginning of year i is expected to be p_i dollars for $i = 1, 2, 3, 4$. The farmer will sell all the grain he has available at the end of the third year and close down his farm. Determine an optimal selling-planting program that will maximize his total profit. Use the input-output approach.

1.22 A company manufactures a product, the demand for which varies from month to month. The raw material and labor availability exhibit seasonal variations. During the months 10, 11, 12, 1, 2, 3 the company can hire at most enough labor to produce 1200 and 600 tons per month during regular time and overtime, respectively. In months 4, 5, 6, 7, 8, 9 these labor capacities are 800 and 500 tons, respectively. The product manufactured during a month can be sold anytime during the next month or later. Storage costs are \$1.00/ton from one month to the next for the product. Raw material cannot be

	Cost of Labor (\$/Ton of Production) During		*Limit on Raw Material Availability (enough to make tons of product)*	*Demand (tons)*	*Selling price (\$/ton)*
Month	*Regular Time*	*Overtime*			
1	\$4	\$6	600	400	18
2	during these		450	700	18
3	months		425	600	18
4	\$6	\$9	1200	900	25
5	during these		1300	900	25
6	months		1600	900	25
7			1600	800	25
8			1500	600	25
9			1300	800	25
10	\$4	\$6	500	1200	30
11			500	1100	30
12			500	1400	30

stored. It has to be used up in the month in which it is obtained. Operations begin in month 1 with a stock of 50 tons of the product. At the end of month 12 the company should have a stock of at least 50 tons of the product. Determine an optimum production schedule.

1.23 An agency controls the operation of a system consisting of two water reservoirs with one hydroelectric power generation plant attached to each. The planning horizon for the system is a year, divided into six periods. Reservoir 1 has a capacity for holding 3500 kilo acre-ft (ka-ft) of water and reservoir 2 has a capacity of 5500 ka-ft. At any given instant of time, if the reservoir is at its full capacity, the additional inflowing water is spilled over a spillway. Spilled water does not produce any electricity.

During each period some specified minimal amount of water must be released from the reservoirs to meet the downstream requirements for recreation, irrigation, and navigation purposes. However, there is no upper limit on the amount of water that can be released from the reservoirs. Any unreleased water is stored (up to the capacity of the reservoir) and can be used for release in subsequent periods. All water released from the reservoirs (even though it is released for recreation and other purposes) produces electricity.

It can be assumed that during each period, the water inflows and releases occur at a constant rate. Also, on an average 1 a-ft of water released from reservoir 1 produces 310 kWh of electricity and 1 a-ft released from reservoir 2 produces 420 kWh. At the beginning of the year reservoir 1 contains 1800 ka-ft of water and reservoir 2 contains 2500 ka-ft of water. The same amounts of water must be left in the respective reservoirs at the end of the year.

The electricity produced can either be sold to a local firm (called a class I customer) or to class II customers. Class I customers buy electricity on an annual basis; they require that specified percentages of it should be supplied in the various periods. They pay $10 per 1000 kWh. Class II customers buy electricity on a period by period basis. They will purchase any amount of electricity in any period at $5 per 1000 kWh. The other data for the problem are given in the table at the bottom. Operate the system to maximize the total annual revenue from the sale of electricity.

1.24 A company has a permit to operate for five seasons. It can manufacture only during the first four seasons and in the fifth period it is only allowed to sell any leftover products. It can manufacture two types of products. One unit of product 1 requires five man-hours in preparatory shop and three man-hours in finishing shop. Each unit of product 2 requires six man-hours in preparatory shop and one man-hour in finishing shop. During each season the company has at most 12,000 man-hours in preparatory and 15,000 man-hours in the finishing shop (only during the first four seasons). The product manufactured during some season can be sold anytime from the next season onward. However, selling requires some marketing effort and it is expected that 0.1 and 0.2 man-hours of marketing effort are required to sell 10 units of

Period	*Inflows in Kilo Acre-Feet into Reservoir* 1	2	*Minimum Release from Reservoir* 1	2	*Percentage of Annual Energy Sold to Class I Customer to Be Delivered in Period*
1	547	2616	200	304	10
2	1471	2335	200	578	12
3	982	1231	200	995	15
4	146	731	200	1495	32
5	32	411	200	558	21
6	159	497	200	392	10

—(*S. Parikh*)

Season	Normal Rate per Man-Hour	Maximum Man-Hours at Normal Rate	Overtime Rate per Man-Hour	Expected Selling Price per Unit of Product 1	Expected Selling Price per Unit of Product 2
2	$2	400	$20	$20	$45
3	$4	300	$20	25	40
4	$1	600	$20	30	40
5	$10	1000	$20	15	30

products 1 and 2, respectively. Man-hours for marketing effort can be hired at the rates given in the table at the top. There is no limit on the number of man-hours that can be hired for marketing effort at the overtime rate. If a unit of product is available for sale during a season, but is not sold in that season, the manufacturer has to pay carryover charges of $2 per unit to put it up for sale again in the next season. The selling prices in the various seasons are given in the table above. How should the company operate in order to maximize its total profit?

1.25 The table at the bottom gives the composition of various foods used in making cereals. The other material in each food is fiber, water, etc. The company blends these food materials and makes two kinds of cereals. In the process of blending, 3% of protein, 5% of starch, and 10% of minerals and vitamins are completely lost from the mix. For each 100 kg of foods added in the blend, the blending process adds 5 kg of other material (mainly water and fat).

Cereal type 1 sells for $1.50/kg. It should contain at least 22% of protein, 2% of minerals and vitamins, and at most 30% of starch by weight. Cereal type 2 sells for $1.00/kg. It should contain at least 30% starch by weight. What is the optimal product mix for the company?

1.26 A contractor is working on a project, work on which is expected to last for a period of T weeks. It is estimated that during the jth week, the contractor will need u_j man-hours of labor, $j = 1$ to T, for this project. The contractor can fulfill these requirements either by hiring laborers over the entire T week horizon (called *steady labor*) or by hiring laborers on a weekly basis each week (called *casual labor*) or by employing a combination of both. One man-hour of steady labor costs c_1 dollars; the cost is the same each week. However, the cost of casual labor may vary from week to week, and it is expected to be c_{2j} dollars/man-hour, during week j, $j = 1$ to T. How can he fulfill his labor requirements at minimal cost?

—(*S. Kedia and G. Ponce Compos*)

1.27 There are m refineries with the ith refinery having the capacity to supply a_i gal of fuel. There are n cities with a demand for this fuel, and the demand in the jth city is b_j gal. f_{ij} is the fraction of a gal of fuel consumed in transportation when 1 gal. of fuel leaves refinery i to city j by a delivery truck. Find a feasible shipping schedule that minimizes the total amount of fuel consumed by the delivery trucks.

1.28 A farmer has three farms. He can grow three different crops on them. Data for the coming

Food	Percentage Content (by Weight): Protein	Starch	Minerals, Vitamins, etc.	Other Material	Price ($/kg)	Availability per Day (kg)
1	45	12	4	39	0.68	1500
2	7	38	1	54	0.27	500
3	12	25	2	61	0.31	1000
4	27	40	3	30	0.45	2000

Farm	Usable Acreage	Water Available in a-ft at \$25/a-ft	Cost per a-ft of Water Beyond Amounts in Previous Column
1	400	1200	\$50
2	600	2200	\$60
3	450	1100	\$70

Crop	Maximum Acreage Farmer Can Plant	Water Consumption (a-ft/a)	Yield units/a	Selling Price per Unit
A	500	6	50	\$25/unit up to 10,000 units \$20/unit beyond 10,000 units
B	700	5	100	\$15/unit up to 40,000 units \$8/unit beyond 40,000 units
C	350	4	75	\$45/unit up to 20,000 units \$42/unit beyond 20,000 units

season are given in the table at the top. To maintain a uniform work load among the three farms, the farmer adopts the policy that the percentage of usable acreage planted must be the same at each of the three farms. However, any combination of the crops may be grown at any of the farms. The only expenses the farmer counts are for the water used at the various farms. How much acreage at each farm should be devoted to the various crops to maximize the farmer's *net profit*?

1.29 A farmer can lease land up to maximum of 1000 a. He has to pay \$5 per acre per year if he leases up to 600 a. Beyond 600 a, he can lease at \$8 per acre per year. He grows corn on the land. He can grow corn at the *normal* level or at an *intense* level (more fertilizer, frequent irrigation, etc.) Normal level yields 70 bushels per acre. Intense level yields 100 bushels per acre. The requirements are given in the table at the bottom. Harvesting requires 0.5 man-hours of labor per bushel harvested. The farmer can sell corn at the rate of \$2.50 per bushel in the wholesale market. He can also raise poultry. Poultry is measured in poultry units. To raise one poultry unit requires 25 bushels of corn, 20 man-hours of labor, and 25 ft^2 of shed floor space. He can either use the corn that he has grown himself or buy corn from the retail market. He gets corn at the rate of \$3.50 per bushel from the retail market. He can sell at the price of \$175 per poultry unit in the wholesale market up to 200 units. Any amount of poultry over 200 units sells for \$160 per unit. He has only one shed for raising poultry with 15,000 ft^2 of floor space. He and his family can contribute 4000 man-hours of labor per year at no cost. If he needs more labor, he can hire it at \$3 per man-hour up to 3000 man-hours. For any amount of labor hired over 3000 man-hours, he has to pay \$6 per man-hour. Maximize his net profit.

1.30 A company makes products 1 and 2. Machines 1, 2, and 3 are required in the manufacture of

Requirements per Acre per Year	Normal Level	Intense Level
Labor (man-hours)	6	9
Materials (seed, fertilizer, water etc.)	\$20	\$35

these products. The relevant data are tabulated below. The manufactured products may be within specification or not. Product i, which is within specifications, is denoted by W_i, and the same product that is outside the specifications is denoted by R_i, $i = 1, 2$. Both W_i and R_i can be sold in the market. The market price of R_i is c_{2i}/unit for $i = 1, 2$. W_i can be sold either as W_i or it can also be sold as R_i. If W_i is sold as W_i, it sells for c_{1i}/unit, and if it is sold as R_i, it is sold at the same price as R_i. However, R_i cannot be sold as W_i.

When product i is manufactured, an average proportion p_i comes out as W_i, $i = 1, 2$. L_i, ℓ_i are the minimum demands for W_i and R_i, respectively. K_i and k_i are the maximum amount of W_i and R_i, respectively, that can be sold. Determine a production and marketing schedule for the company that maximizes its total sales revenue subject to all the constraints.

	Machine-Hours Required to Manufacture One Unit of Product on Machine		
Product	*1*	*2*	*3*
1	3	9	2
2	7	4	5
Machine-hours available on the machine	350	970	1420

1.31 There are 14 food ingredients available. They can be classified into seven groups. It is required to mix them into a minimum cost diet satisfying the following constraints: (i) Protein content should be at least 20%; (ii) fat content should be at least 3%; (iii) fiber content should be at most 12%; (iv) Ca content should be between 1% and 2%; (v) Ph content should be between 0.6% and 2%; (vi) Ca content should be greater than or equal to the Ph content. Data and other constraints are given below.

1.32 A company manufactures products 1, 2, and 3 with labor and material as inputs. Data are given in the table at the top of page 42. The material required in this process is obtained from outside. There are two suppliers for this material. The amount of material available, called x_4 in the table, depends on the supplier chosen. The small supplier can provide at most 25 tons of material per month. The big supplier will only supply if at least 35 tons of the material is taken per month. By company regulations, the company has to choose one of these two suppliers and get the material only from it. The company would like to adopt the policy that maximizes its monthly profit. Can this be found by solving one LP? If not, discuss how you can solve this problem.

		Percent of Nutrient Content							
Group	*Food Ingredient*	*Protein*	*Fat*	*Fiber*	*Ca*	*Ph*	*Cost per Ton*	*Range of Ingredient (%)*	*Range of Group (%)*
1	1 Dried beet pulp	9	0.5	20	0.7	0.05	$64	4 to 20	5 to 20
	2 Dried citrus pulp	6	3	16	2	0.1	35	1 to 20	
2	3 Ground yellow corn	8.5	4	2.5	0.02	0.25	55	1 to 25	20 to 35
	4 Ground oats	12	4.5	12	0.1	0.4	54	1 to 25	
3	5 Corn molasses	3.5			0.6	0.1	19	5 to 14	5 to 14
4	6 Wheat middlings	16	4	8	0.1	0.9	64	5 to 30	10 to 30
	7 Wheat bran	16	4	10.5	0.1	1.2	62	5 to 30	
5	8 Distiller grains	26	8.5	9	0.15	0.6	77	5 to 15	2 to 25
	9 Corn gluten feed	24	2	8	0.3	0.65	66	1 to 25	
6	10 Cottonseed meal	41	1.5	13	0.1	1.2	74	1 to 35	3 to 35
	11 Linseed oil meal	34	1	8	0.35	0.8	85	1 to 35	
	12 Soybean oil meal	45	0.5	6.5	0.2	0.6	108	1 to 35	
7	13 Ground limestone				36	0.5	10	0 to 2	
	14 Ground phosphate				32	14	66	1 to 2	

(*G. Brigham*)

Item	*Items Requirement to Produce 1 Ton of Product* 1	2	3	*Amount of Item Available per Month*
Labor (man-hours)	7	2	4	46
Material (tons)	4	5	5	x_4
Profit ($/ton)	4	1	5	

1.33 The following table gives some of the data collected during a ballistic evaluation of a new armor plate material, t is the thickness of the target armor plate in inches and v is the striking velocity of the projectile (in feet per second) required to penetrate the plate. v is the control parameter. Estimate t (the thickness of the plate that a projectile with striking velocity v would penetrate) as a function of v. Use the data to fit the following curve: $t = a_3 v + a_1(a_3 v)^{\mathbf{a_2}}$, where a_1, $\mathbf{a_2}$, a_3 are the coefficients to be estimated. a_1, $\mathbf{a_2}$, a_3 should be all nonnegative and $\mathbf{a_2}$ should lie between 0 and 1. Discuss a practical method for obtaining the values of a_1, $\mathbf{a_2}$, a_3 that give the best fit.

v	2300	2800	2850	2900	3200
t	4.087	4.596	4.646	4.697	5.002

1.34 *SEPARABLE CONVEX PROGRAMMING PROBLEM.* Consider the following problem; where $f_i(x_i) = c_i x_i^{-\mathbf{a}_i}$ and c_i, $\mathbf{a_i}$, ℓ_i, k_i are given positive numbers, and $0 \leqq \ell_i < k_i$ for all i. $f_i(x_i)$ is a nonlinear, convex function for all i. Transform this into a problem in which a convex nonlinear objective function has to be minimized subject to linear constraints on the variables. Discuss how an approximate solution of this problem can be obtained by using a piecewise linear approximation for the objective function. How can the precision of the solution obtained be improved?

$$\begin{aligned} \text{Minimize} \quad & \textstyle\sum f_i(x_i) \\ \text{Subject to} \quad & \textstyle\sum x_i^2 = 1 \\ k_i \geqq x_i \geqq \ell_i \quad & \text{for all } i \end{aligned}$$

1.35 *THE ALLOY-MAKING PROBLEM.* Suppose we want to make an alloy consisting of metals 1 to m in proportions $p_1, \ldots, p_m$, respectively, by weight, where $p_1, \ldots, p_m$ are all given positive numbers satisfying $p_1 + \cdots + p_m = 1$. We have t different raw materials that can be used in the manufacture of this alloy. The jth raw material contains a proportion a_{ij} of metal i by weight, where a_{ij} are all given nonnegative numbers satisfying $\sum_{i=1}^{m} a_{ij} = 1$, for each $j = 1$ to t. For $j = 1$ to t, c_j is the cost of raw material j in dollars per ton. In addition, each metal can be obtained in very pure form, and the cost for the ith metal is d_i in dollars per ton. Model the problem of determining the minimum cost mix of raw materials and pure metals, to manufacture the alloy, as an LP.

Can the variables representing the amounts of pure metals included in the mix be considered as slack variables in this model? Why?

1.36 Let $w = (w_1, \ldots, w_n) > 0$ be a given vector of positive weights. For each vector $y = (y_1, \ldots, y_n) \in \mathbf{R}^n$, define the function L($y$) by L($y$) $=$ maximum $\{w_i|y_i|\colon i = 1 \text{ to } n\}$. Given a vector $a = (a_1, \ldots, a_n) \in \mathbf{R}^n$, it is required to find a vector $x = (x_1, \ldots, x_n)$ to minimize L($a - x$), subject to $x_i - x_{i+1} \leqq 0$, $i = 1$ to $n - 1$. Formulate this problem as an LP. A problem of this type arises when trying to estimate the failure rate of an item as a function of its age. An experiment can be run with several new items put in use at time 0. Time can be divided into unit intervals, and we observe the number still working at the beginning of each interval. If f_i is the observed number of items still working at time point i, $f_i - f_{i+1}$ is the number of items that failed in the ith interval, and $((f_i - f_{i+1})/f_i) = a_i$ is the observed failure rate at the age of i time units. Suppose from practical and engineering considerations we know that the failure rate increases with age (this is known as the IFR, or the increasing failure rate property). The actual failure rates observed $a_1, \ldots, a_n$ may not be increasing due to random fluctuations in the observations. An optimum solution x of the above problem provides a vector satisfying the IFR property that is closest to the observed vector of failure rates a, and hence x can be taken as the estimate of the vector of failure rates of the item at various ages. How does the formulation change if each x_i is required to be nonnegative, in addition, and also less than or equal to one?

—(*V. A. Ubhaya* [1.32, 1.33])

1.37 *COAL DISTRIBUTION PROBLEM.* Formulate this problem as an LP. A region contains a large number of small individual coal mines. The quality of coal varies from mine to mine considerably because of the heterogeneous nature of coal production in the area. The major coal market in the region is for thermal generation of electricity that requires large guaranteed deliveries of coal, consisting of relatively uniform grades. The objective is to indicate mine operating levels to the coal producers, such that their net delivered price of the make-up coal on a BTU basis is maximized, subject to the condition that the mixed coal product must satisfy the utility requirements. Let x_{ij} denote the tons of coal shipped from mine i (total of m mines) to the jth collection point (n central collection points). We are also given the values of the following data elements; q_i = daily production capacity (tons) of mine i, where $i = 1$, to m; b = projected daily demand (tons) by the utility; a_i, s_i, w_i, v_i, f_i, c_i = percentage (by weight) of ash, sulfur, moisture, volatile matter, ash fusion, and fixed carbon, respectively, in the coal from the jth mine; a, s, w = permissible maximum ash, sulpher, and moisture percentage (by weight), respectively, in the mix; v, f = permissible minimum volatile matter and ash fusion percentage (by weight), respectively, in the mix; c_l, c_u = permissible lower and upper bounds for percentage (by weight) of fixed carbon in the mix; λ_j = delivered price of coal at collection point j per ton; γ_{ij} = shipping charges per ton of coal from mine i to collection point j.

—(*G. Manula and Y. Kim* [1.24])

1.38 *PARTICULATE BLENDING OF CHEMICALS.* There are t batches of chemicals available. The following data are available: w_k = weight (tons) of material in the kth batch; $a_{1k}, a_{2k}, \ldots, a_{mk}$ ($\sum_{i=1}^{m} a_{ik} = 1$) is the fraction of material (by weight) in the kth batch in the various particle sizes; $b_1, b_2, \ldots, b_m (\sum b_i = 1)$ is the desired fractions of blend (by weight) in the various particle sizes; w = weight (tons) of the blend to be produced. It is desired to blend material from the various batches, so as to produce a blend with fractional size-weight distribution as close to the desired as possible. Formulate this problem as an LP. Discuss how the formulation changes if it is required to minimize the maximum deviation from the desired fraction in the blend over all the sizes. In particular, formulate the problem with the data at the bottom of the page.

1.39 *THE BETTING PROBLEM.* A linear programmer is planning to spend a day at the race course. He will have \$$b$ available for betting. On the day of his proposed visit, there is only one race planned in which k horses numbered 1, 2, . . . , k are competing. If he bets a dollar on the ith horse, his gross payoff from this bet will be 0 if the ith horse does not come first in the race, or \$$\alpha_i$ if the ith horse comes first in the race, for $i = 1$ to k. The numbers $\alpha_1, \ldots, \alpha_k$ are known positive integers. He is, of course, allowed to bet any nonnegative amount on any number of horses. His problem is to determine how much to bet on each of the horses in the

	Fraction by Weight of Various Sizes				
Batch	*1*	*2*			*Tons Available*
1	0.10	0.40	0.30	0.20	100
2	0.20	0.10	0.50	0.20	250
3	0.30	0.58	0.02	0.10	50
4	0.40	0.15	0.30	0.15	500
5	0.60	0.17	0.18	0.05	200
Desired fraction in blend	0.45	0.30	0.15	0.10	800 tons required

—(*R. R. Klimpel* [1.20])

race (i.e., how to divide his available $\$b$ for bets on the various horses) so as to maximize his minimum net gain, irrespective of whichever horse comes first in the race. Formulate this as an LP. Write down the special case of this problem when $b = 100$, $k = 5$, and $(\alpha_1, \alpha_2, \alpha_3, \alpha_4, \alpha_5) = (2, 5, 1.5, 7, 3)$.

—(*S. Vajda* [1.110])

1.40 *A PRODUCTION PLANNING PROBLEM.* A company has to make deliveries of a product to its dealers at the end of each month. At the beginning of the year the company knows that they have to deliver d_i tons of this material at the end of the ith month in that year, $i = 1$ to 12. Material produced during a month can be delivered either that same month itself or stored in inventory and delivered in some other month. It costs the company $\$c_1$/ton to carry the material in inventory from one month to the next. The year begins with 0 tons in inventory and is required to end with 0 tons in inventory at the end of month 12. The company has to determine how much material should be produced each month of the year. If the company decides to produce x_i tons in month i and x_{i+1} tons in month $i+1$, it incurs a cost of $c_2|x_i - x_{i+1}|$ to adjust the production levels at the beginning of month $i+1$. It is required to determine how much to produce in each of the 12 months of the year so as to minimize the total of inventory and production-level adjustment costs over the year. Formulate this problem as an LP. Write down the problem for the specific case when $d = (30, 40, 20, 70, 80, 90, 100, 30, 150, 200, 150, 150)$, and $c_1 = 10$, $c_2 = 4$.

—(*A. J. Hoffman and W. W. Jacobs* [1.16])

1.41 Consider the following system of linear inequalities. It is required to find a feasible solution x to this system that makes each of the four constraints to be satisfied as an equation as nearly as possible. Formulate this problem as an LP.

$$\begin{aligned}
x_1 - 2x_2 + x_3 - x_4 - x_5 + x_6 &\leqq 13\\
-2x_1 + x_2 - 12x_3 + 2x_4 + 3x_5 + 4x_6 &\leqq 1\\
3x_1 + 13x_2 + 18x_3 + 17x_4 + 25x_5 + 12x_6 &\leqq 3\\
5x_1 + 3x_2 - 8x_3 + 13x_4 + 8x_5 - 7x_6 &\leqq 5\\
x_j \geqq 0 \quad \text{for all } j &= 1 \text{ to } 6
\end{aligned}$$

1.42 Consider the following system of linear equations. It is required to find a vector x that makes the left-hand side in this system as close to the right-hand side componentwise as possible. Formulate this problem as an LP.

$$\begin{aligned}
x_1 + 2x_2 - 3x_3 + x_4 - 2x_5 &= 13\\
2x_1 - 3x_2 + 4x_3 + 5x_4 - 4x_5 &= -18\\
- x_2 + 15x_3 - 10x_4 - 8x_5 &= 22\\
17x_1 + 13x_3 + 22x_4 - 17x_5 &= -10\\
13x_1 - 12x_2 - 10x_3 - 2x_4 &= -8\\
2x_1 + x_2 + 3x_3 - 12x_5 &= 13\\
-x_1 - 13x_2 + 11x_3 + 8x_4 + 9x_5 &= 11
\end{aligned}$$

1.43 AGGREGATE PLANNING PROBLEM IN A STEEL ROLLING MILL. This application arises in the production planning process of a large steel rolling mill, where approximately 4000 to 6000 cast rolls of 300 different types, each requiring an average of 10 operations, are machined in a large job shop each year. The order book is first balanced in an aggregate manner using the linear programming model discussed here, such that monthly or other periodic demands are approximately in line with the manufacturing capabilities. Detailed production scheduling and planning is then carried out using heuristic methods. We are given the following data and definitions of the decision variables for the aggregate planning problem. The constraints are the machine-time availability constraints and supply-demand constraints (the number of rolls of a given group machined and shipped in a period cannot exceed the demand for the group in that period). The objective is to maximize the total tonnage of all rolls machined and shipped during the planning period. Ignoring the integer requirement on the variables x_{ij}, formulate this problem as an LP. Data given are: N = the number of months (or periods) into the future for which planning is being done; M = the number of groups of rolls under consideration during the planning period; K = the number of different types of machines; D_{ij} = the number of rolls of group i demanded in period j; P_{ik} = the processing time (hours) required by each roll in group i on machines of type k; A_{ij} = the processing time available in hours in period j on machines of type k; W_{ij} = the average

Bidder	Shipping Point	Maximum Quantity for Day Desired from Shipping Point	Bonus Offered per Barrel
Company C	A	10,000 barrels	$0.10
	B	10,000 barrels	$0.09
Company D	A	10,000 barrels	$0.20
	B	10,000 barrels	$0.15

—(*B. L. Jackson and J. M. Brown* [1.17])

shipping weight of a roll in group i demanded in period j. The variables are: x_{ij} = the number of rolls in group i that should be machined and shipped in period j.

—(*S. K. Jain, K. L. Scott, E. G. Vasold* [1.18])

1.44 *CRUDE OIL SALES.* This problem arose in the sale of crude oil produced at the Naval Petroleum Reserves at the Elk Hills (A) and Buena Vista (B) fields in California. A maximum of 10,000 barrels per day is up for sale from each of these fields. Two companies, C and D, bid for the purchase of this crude oil according to the offers tabulated at the top. The bonus here is the amount the company agrees to pay over an announced minimum base price per barrel. The U.S. Government is constrained by a law not to sell more than 15,000 barrels per day from both these fields together to any single purchaser. It is required to determine how much crude oil to sell to each of companies C and D from each of the fields A and B, so as to maximize the total bonus obtained, subject to these constraints. Formulate this problem as an LP.

1.45 *AIRPLANE FUELING PROBLEM.* An airplane has a route to fly that takes it from city r to $r + 1$, $r = 1$ to $n - 1$, and then at the end from city n to city 1. For $r = 1$ to n, w_r is the expected weight of the plane in units (including the body plus crew, passengers, and baggage, but excluding fuel, estimated by using the known passenger inputs and departures at the various cities) when in flight from city r to $r + 1$. We also have the following data for each $r = 1$ to n: c_r = the cost of aviation fuel per unit weight at city r, in dollars; k_r = the maximum amount of aviation fuel that can be purchased in city r in weight units. Let F units be the maximum weight of fuel that can be loaded into the plane. By FAA regulations the plane must carry a specified amount of γ units of reserve fuel, when it takes off from any city on the route, beyond what it needs to complete the immediate flight segment and landing at the next city on the route. Denote by x_r the amount of fuel in weight units purchased and loaded into the plane at city r, $r = 1$ to n. Let W_r denoted the total combined weight of the plane, including fuel, at takeoff from city r on the route $r = 1$ to n.

1 Assume that the amount of fuel (in weight units) needed to take off from city r and fly to and land in the next city, city $r + 1$. in the route (define city $n + 1$ = city 1) is $\alpha_r W_r$, $r = 1$ to n (where α_r are given positive numbers determined by using the flying distance and time between cities r and $r + 1$. etc). Formulate the problem of determining $x_1, \ldots, x_n$ optimally so as to minimize the total fuel cost for completing this trip, as an LP. Also, write down the special case of this model when the data are $n = 5$, $\gamma = 25$, F = 225.

r	c_r	k_r	w_r	α_r
1	1.5	200	350	0.17
2	2.5	75	300	0.1
3	0.7	∞	310	0.07
4	0.5	∞	400	0.1
5	3.5	35	425	0.15

2 For $r = 1$ to n, let t_r be the expected duration of the flight from city r to city $r + 1$. In (1) we gave simple linear formulas for determining the fuel consumption in the various flight segments in the route. If you think that this only gives a crude approximation, consider the following. Suppose that the fuel consumption is to be determined by a more accurate procedure described below. When the plane takes off and keeps flying, fuel will be burnt and the weight of the plane continuously decreases by the amount of spent fuel. At any point of time t, when the plane is in the air, let $V(t)$ denote the combined weight of the plane, including the fuel at that time point. Suppose the rate at which fuel is burnt at that time point is $\lambda V(t)$ dt, where λ is a given positive constant. Use this to determine the amount of fuel spent by the plane to fly the various flight segments in the route. With this, construct a model for determining the optimum values of $x_1, \ldots, x_n$, so as to minimize the total fuel cost for completing this trip.

3 An airline operates a large number of routes with a fleet of airplanes. Discuss how the model discussed in (1) can be extended to determine which plane should buy how much fuel in which city and when, so as to minimize the total fuel bill for the airline.

—(*M. Queyranne* [1.29])

1.46 *A RESOURCE ALLOCATION PROBLEM.* The federal government has agreed to provide funds for a project to transform some untrained and unemployable people into occupationally trained employable individuals through a process of training. If a participant completes the training period for a particular program, there is a positive benefit as a result of the cost incurred in the training. If the participant does not complete the training, there is no benefit, but a fractional cost of training is incurred anyway, depending on how much of the period he or she remained in training. The problem is to allocate the funds and participants among the various programs to maximize the total expected benefit, subject to the budget and other constraints. Available data is tabulated at the bottom of this page (some of the values are estimated from input information for a community of about 700,000 people consisting primarily of disadvantaged persons, of whom a maximum of 500 are expected to apply for a training program of 12 weeks duration). The

	Training Program			
	$i = 1$	2	3	4
Description	*Work Experience*	*On-the-Job Training*	*Classroom Training*	*Public Service Job*
c_i = cost of program in dollars per participant accepted to take it	960	2400	2880	3600
b_i = the increased annual income in dollars that a participant can expect to receive, if he or she completes the program successfully	8320	10,400	10,400	7800
ϕ_i = the probability that an applicant will qualify to be accepted as a participant in the program	0.8	0.5	0.7	0.6
π_i = the probability that a participant will complete the program successfully if he or she begins it	0.4	0.72	0.33	0.38
θ_i = the probability that a client who completes the program will succeed in obtaining a job (giving him or her the expected increased annual income)	0.34	0.7	0.4	0.4
t_i = fractional cost incurred per participant beginning in the program but not completing it successfully	0.25	0.13	0.27	0.26
N_i = maximum number of applicants that can be screened by the program	150	315	20	50
x_i = decision variable, the number of applicants sent to the ith program	x_1	x_2	x_3	x_4
y_i = decision variables, dollars of budget allocated to the ith program	y_1	y_2	y_3	y_4

—(*J. K. Bandyopadhyay, S.K. Sungupta, and R. K. Bogh* [1.3])

total budget available is \$500,000. The expected benefit is the expected increased annual income earned by successful participants. Ignoring the integer requirements on the x_i, formulate the problem of determining the x, y to maximize the expected total benefit as an LP.

1.47 Transform the following mathematical program into an LP. Justify clearly that the solution obtained from the transformed LP solves this problem. In this problem, if there is an additional constraint $x_1 - 2x_2 + 3x_3 - 30| \geqq 5$, could you still transform the whole problem into a single LP? Why? Discuss briefly an approach for solving the problem with this additional constraint.

$$
\begin{aligned}
\text{Minimize} \quad & z(x) = 6x_1 + 5x_2 - 3x_3 \\
\text{Subject to} \quad & x_1 + 2x_2 - 7x_3 \leqq 1 \\
& |3x_1 - 5x_2 - 20| \leqq 4 \\
& x_j \geqq 0 \quad \text{for all } j
\end{aligned}
$$

REFERENCES

On Applied Linear Programming

1.1 S. Advani, "A Linear Programming Approach to Air Cleaner Design," *Operations Research* (March/April 1974), 295–297.

1.2 L. R. Arnold and D. Botkin, "Portfolios to Satisfy Damage Judgements: A Linear Programming Approach," *Interfaces 8*, 2 (February 1978), 38–42.

1.3 J. K. Bandopadhyay, S. K. Sengupta, and R. K. Bogh, "A Resource Allocation Model for an Employability Planning System," *Interfaces 10*, 5 (October 1980), 90–94.

1.4 F. M. Bass and R. T. Lonsdale, "An Exploration of Linear Programming in Media Selection," *Journal of Marketing Research 3* (May 1966), 179–188.

1.5 R. G. Bland, "The Allocation of Resources by Linear Programming," *Scientific American 244*, 6 (June 1981), 126–144.

1.6 J. Byrd, Jr., and L. T. Moore, "The Application of a Product Mix Linear Programming Model in Corporate Policy Making," *Management Science 24*, 13 (September 1978), 1342–1350.

1.7 A. Charnes, C. Colantoni, W. W. Cooper, and K. O. Kortanek, "Economic Social and Enterprise Accounting and Mathematical Models," *The Accounting Review 47*, (January 1972), 85–108.

1.8 G. B. Dantzig, "Reminiscences About the Origins of Linear Programming," *Operations Research Letters 1*, 2 (April 1982), 43–48.

1.9 R. Dutton, G. Hinman, and C. B. Millham, "The Optimal Location of Nuclear Power Facilities in the Pacific Northwest," *Operations Research* (May/June 1974), 478–487.

1.10 J. F. Engel and R. Warshaw, "Allocating Advertising Dollars by Linear Programming," *Journal of Advertising Research 4*, 3 (September 1964), 42–48.

1.11 A. Geoffrion, "The purpose of Mathematical Programming is Insight, not Numbers," *Interfaces 7*, 1, 1976, pp. 81–92.

1.12 C. R. Glassey and V. K. Gupta, "A Linear Programming Analysis of Paper Recycling," *Management Science* (December 1974), 392–408.

1.13 J. J. Glen, "A Mathematical Programming Approach to Beef Feedlot Optimization," *Management Science 26*, 5 (May 1980), 524–535.

1.14 P. Gray and C. Cullinan-James, "Applied Optimization—A Survey," *Interfaces 6*, 3 (May 1976), 24–29.

1.15 R. V. Hartley, "Linear Programming: Some Implications for Management Accounting," *Management Accounting 51* (November 1969), 48–51.

1.16 A. J. Hoffman and W. W. Jacobs, "Smooth Patterns of Production," *Management Science 1*, 1954, pp. 86–91.

1.17 B. L. Jackson and J. M. Brown, "Using LP for Crude Oil Sales at Elk Hills: A Case Study," *Interfaces 10*, 3 (June 1980), 65–70.

1.18 S. K. Jain, K. L. Scott, and E. G. Vasold, "Orderbook Balancing Using a Combination of Linear Programming and Heuristic Techniques," *Interfaces 9*, 1 (November 1978), 55–67.

1.19 W. J. Kennedy and J. E. Gentle, *Statistical Computing,* Dekker, New York, 1980.

1.20 R. R. Klimpel, "Operations Research in the Chemical Industry, Parts I and II," *Chemical Engineering 80*, 9 and 10, 1973, pp. 103–108 and 87–94.

1.21 D. S. Kochhar and G. Allora-Abbondi, "Channel Allocation on a Cable Distribution System," *Computers and Industrial Engineering 4*, 1980, pp. 173–184.

1.22 D. B. Kotak, "Application of Linear Programming to Plywood Manufacturer," *Interfaces 7*, 1 (November 1976), 56–68.

1.23 R. F. Love, "An Application of a Facilities Location Model in a Prestressed Concrete Industry," *Interfaces 6*, 4 (August 1976), 45–49.

1.24 G. Manula and Y. Kim, "A Linear Programming Simulator for Coal Distribution Problems," *Proceedings of the Symposium in Operations Research in the Mineral Industries*, University Park, P., 1966.

1.25 W. Marcuse, L. Bodin, E. Cherniavsky, and S. Yasuko, "A Dynamic Time Dependent Model for the Analysis of Alternative Energy Policies," In *Operational Research*, North-Holland, Amsterdam, The Netherlands 1975.

1.26 J. J. Moder and S. E. Elmaghraby (Eds.), *Handbook of Operations Research Models and Applications*, Van Nostrand Reinhold, New York, 1978.

1.27 E. F. Peter Newson, (Ed.), *Management Science and the Manager, A Case Book*, Prentice-Hall, Englewood Cliffs, N.J., 1980.

1.28 R. H. Philipson and A. Ravindran, "Applications of Mathematical Programming to Metal Cutting," *Mathematical Programming Study*, 11 (1979), 116–134.

1.29 M. Queyranne, "The Tankering Problem", CBA Working Paper series 1982, University of Houston, Houston, Texas.

1.30 J. R. Rice, *The Approximation of Functions: Linear Theory*, Vol. 1, Addison-Wesley, Reading, Mass., 1964.

1.31 E. L. Summers, "The Audit Staff Assignment Problem: A Linear Programming Analysis," *The Accounting Review 47* (July 1972), 443–453.

1.32 V. A. Ubhaya, "Isotone Optimization, I," *Journal of Approximation Theory 12*, 2 (October 1974), 146–159.

1.33 V. A. Ubhaya, "Isotone Optimization, II," *Journal of Approximation Theory 12*, 4 (December 1974), 315–331.

1.34 H. M. Weingartner *Mathematical Programming and the Analysis of Capital Budgeting Problems,* Prentice-Hall Englewood Cliffs, N.J., 1963.

1.35 T. K. Zierer, A. W. Mitchell, and T. R. White, "Practical Applications of Linear Programming to Shell's Distribution Problems," *Interfaces 6*, 4 (August 1976), 13–26.

1.36 *Communications in Statistics, Part B*, (*Special Issue on Computations for Least Absolute Values Estimation*) *B6*, 4, 1977.

On Multiobjective Programming

1.37 J. L. Arthur and A. Ravindran, "PAGP, A Partitioning Algorithm for (Linear) Goal Programming Problems," *ACM Transactions on Mathematical Software 6*, 3 (September 1980), 378–386.

1.38 A. Charnes and W. W. Cooper, "Goal Programming and Multiple Objective Optimizations, Part I," *European Journal of Operations Research 1*, 1, 1977, pp. 39–54.

1.39 J. L. Cochrane and M. Zeleny (Eds.), *Multiple Criteria Decision Making*, University of South Carolina Press, Columbia, S. C., 1973.

1.40 J. L. Cohon, *Multiobjective Programming and Planning*, Academic, New York, 1978.

1.41 G. Fendel and T. Gal (Eds.), *Multiple Criteria Decision Making, Theory and Application*, Springer, Berlin, 1980.

1.42 C. L. Hwang and A. S. Masud, *Multiple Objective Decision Making Methods and Applications: A State of the Art Survey*, Springer, New York, 1979

1.43 J. S. H. Kornbluth, "A Survey of Goal Programming," *Omega 1*, 2, 1973, pp. 193–205.

1.44 S. M. Lee, *Goal Programming for Decision Analysis*, Auerbach, Philadelphia, P., 1972.

1.45 J. Sponk, *Interactive Multiple Goal Programming: Applications to Financial Management*, Martinus Nijhoff, Boston, Mass., 1981.

1.46 M. Starr and M. Zeleny, (Eds.), *Multiple Criteria Decision Making*, North-Holland/TIMS Studies in Management Science, vol. 6, Amsterdam, The Netherlands, 1977.

1.47 S. Zionts and J. Wallenius, "An Interactive Programming Method for Solving the Multiple Criteria Problems," *Management Science* 22, 6, 1976, pp. 652–663.

1.48 S. Zionts (Ed.), *Proceedings of the Conference on Multiple Criteria Problem Solving* (Buffalo, N.Y. 1977), Springer Verlag, New York, 1978.

Selected Books on Linear Programming and Related Areas

1.49 J. R. Aronofsky, J. M. Dutton, and M. T. Tayyabkhan, *Managerial Planning with Linear Programming in Process Industry Operations*, Wiley-Interscience, New York, 1978.

1.50 T. S. Arthanari and Y. Dodge *Mathematical Programming in Statistics*, Wiley, New York, 1981.

1.51 M. L. Balinski (Ed.), *Mathematical Programming Study 1, Pivoting and Extensions: In Honor of A. W. Tucker*, North-Holland, Amsterdam, The Netherlands, 1974.

1.52 M. L. Balinski and E. Hellerman (Eds.), *Mathematical Programming Study 4, Computational Practice in Mathematical Programming*, North-Holland, Amsterdam, The Netherlands, 1976.

1.53 M. Bazaraa and J. J. Jarvis, *Linear Programming and Network Flows*, Wiley, New York, 1977.

1.54 E. M. L. Beale, *Mathematical Programming*, Isaac Pitman, London, 1968.

1.55 S. P. Bradley, A. C. Hax, and T. L. Magnanti, *Applied Mathematical Programming*, Addison-Wesley, Reading, Mass., 1977.

1.56 A. Charnes and W. W. Cooper, *Management Models and Industrial Applications of Linear Programming*, Vols. I and II, Wiley, New York, 1961.

1.57 A. Charnes, W. W. Cooper, and A. Henderson, *An Introduction to Linear Programming*, Wiley, New York, 1953.

1.58 N. Christofides, *Graph Theory, An Algorithmic Approach*, Academic, New York, 1975.

1.59 V. Chvátal, *Linear Programming*, W. H. Freeman & Co., San Francisco, California, 1983.

1.60 L. Cooper and D. Steinberg, *Introduction to Methods of Optimization*, Saunders, Philadelphia, 1970.

1.61 G. B. Dantzig, *Linear Programming and Extensions*, Princeton University Press, Princeton, N.J., 1963.

1.62 G. B. Dantzig, M. A. H. Dempster, and M. Kallio (Eds.), *Large-Scale Linear Programming* (Proceedings of an IIASA Workshop, CP-81-S1-2 Volumes, June 2–6, 1980), The International Institute for Applied Systems Analysis, Laxenburg, Austria.

1.63 R. Dorfman, P. A. Samuelson, and R. M. Solow, *Linear Programming and Economic Analysis*, McGraw-Hill, New York, 1958.

1.64 N. J. Driebeck, *Applied Linear Programming*, Addison-Wesley, Reading, Mass., 1969.

1.65 S. E. Elmghraby, *Some Network Models in Management Science*, Springer, New York, 1970.

1.66 G. D. Eppen and F. J. Gould, *Quantitative Concepts for Management: Decision Making Without Algorithms*, Prentice-Hall, Englewood Cliffs, N.J., 1979.

1.67 S. Even, *Graph Algorithms*, Computer Science Press, Potomac, Md., 1979.

1.68 L. R. Ford, Jr., and D. R. Fulkerson, *Flows in Networks*, Princeton University Press, Princeton N.J., 1962.

1.69 D. Gale, *Theory of Linear Economic Models*, McGraw-Hill, New York, 1960.

1.70 M. R. Garey and D. S. Johnson, *Computers and Intractability: A Guide to the Theory of NP-Completeness*, Freeman, San Francisco, 1979.

1.71 R. S. Garfinkel and G. C. Nemhauser, *Integer Programming*, Wiley-Interscience, New York, 1972.

1.72 W. W. Garvin, *Introduction to Linear Programming*, McGraw-Hill, New York, 1960.

1.73 S. I. Gass, *Linear Programming: Methods and Applications*, 4th ed., McGraw-Hill, New York, 1975.

1.74 D. P. Gaver and G. L. Thompson, *Programming and Probability Models in Operations*

Research, Brookes/Cole Publishing Co., Monterey, Calif., 1973.

1.75 R. L. Graves and P. Wolfe (Eds.), *Recent Advances in Mathematical Programming*, McGraw-Hill, New York, 1963.

1.76 M. R. Greenberg, *Applied Linear Programming for the Socioeconomic and Environmental Sciences*, Academic, New York, 1978.

1.77 G. Hadley, *Linear Programming*, Addison-Wesley, Reading, Mass., 1962.

1.78 F. S. Hillier and G. J. Lieberman, *Introduction to Operations Research*, 3rd ed., Holden-Day, San Francisco, 1980.

1.79 T. C. Hu, *Integer Programming and Network Flows*, Addison-Wesley, Reading, Mass., 1970.

1.80 P. A. Jensen and J. W. Barnes, *Network Flow Programming*, Wiley, New York, 1980.

1.81 J. F. Kennington and R. V. Helgason, *Algorithms for Network Programming*, Wiley-Interscience, New York, 1980.

1.82 T. C. Koopmans (Ed.), *Activity Analysis of Production and Allocation*, Wiley, New York, 1959.

1.83 H. W. Kuhn and A. W. Tucker (Eds.), *Linear Inequalities and Related Systems*, Princeton University Press, Princeton, N.J., 1956.

1.84 L. S. Lasdon, *Optimization Theory for Large Systems*, Macmillan, New York, 1970.

1.85 E. L. Lawler, *Combinatorial Optimization—Networks and Matroids*, Holt, Rinehart and Winston, New York, 1976.

1.86 R. I. Levin and R. P. Lamone, *Linear Programming for Management Decisions*, Irwin, Homewood, Ill., 1969.

1.87 R. W. Llewellyn, *Linear Programming*, Holt, Rinehart and Winston, New York, 1964.

1.88 D. G. Luenberger, *Introduction to Linear and Non-Linear Programming*, Addison-Wesley, Reading, Mass., 1973.

1.89 R. E. Machol, *Elementary Systems Mathematics: Linear Programming for Business and the Social Sciences*, McGraw-Hill, New York, 1976.

1.90 E. Minieka, *Optimization Algorithms for Networks and Graphs*, Dekker, New York, 1978.

1.91 B. A. Murtagh, *Advanced Linear Programming*, McGraw-Hill, New York, 1981.

1.92 K. G. Murty, *Linear and Combinatorial Programming*, Wiley, New York, 1976.

1.93 K. G. Murty, *Linear Complementarity*, to be published.

1.94 K. G. Murty, *Network and Combinatorial Programming*, to be published.

1.95 B. Noble, *Applied Linear Algebra*, Prentice-Hall, Englewood Cliffs, N.J., 1969.

1.96 W. Orchard-Hays, *Advanced Linear Programming Computing Techniques*, McGraw-Hill, New York, 1968.

1.97 D. P. Phillips, A. Ravindran, and J. Solberg, *Operations Research: Principles and Practice*, Wiley, New York, 1976.

1.98 P. H. Randolph and H. D. Meeks, *Applied Linear Optimization*, Grid, Columbus, Ohio, 1978.

1.99 S. S. Rao, *Optimization Theory and Applications*, Wiley Eastern Ltd, New Delhi, 1978.

1.100 R. I. Rothberg, *Linear Programming*, North Holland, New York, 1979.

1.101 T. L. Saaty, *Mathematical Methods of Operations Research*, McGraw-Hill, New York, 1959.

1.102 M. Sakarovitch, *Notes on Linear Programming*, Van Nostrand Reinhold, New York, 1971.

1.103 H. M. Salkin and J. Saha, *Studies in Linear Programming*, North-Holland, Amsterdam, The Netherlands, 1975.

1.104 H. M. Salkin, *Integer Programming*, Addison-Wesley, Reading, Mass., 1975.

1.105 D. M. Simmons, *Linear Programming for Operations Research*, Holden-Day, San Francisco, 1972.

1.106 M. Simonnard, *Linear Programming* (Translated from French by W. S. Jewell), Prentice-Hall, Englewood Cliffs, N.J., 1966.

1.107 W. A. Spivey and R. M. Thrall, *Linear Optimization*, Holt, Rinehart and Winston, New York, 1970.

1.108 J. E. Strum, *Introduction to Linear Programming*, Holden-Day, San Francisco, 1972.

1.109 H. A. Taha, *Integer Programming: Theory, Applications, and Computations*, Academic Press, New York, 1975.

1.110 S. Vajda, *Mathematical Programming*, Addison-Wesley, Reading, Mass., 1961.

1.111 C. Van de Panne, *Linear Programming and Related Techniques*, North-Holland, Amsterdam, The Netherlands, 1971.

1.112 C. Van de Panne, *Methods for Linear and Quadratic Programming*, North-Holland, Amsterdam, The Netherlands, 1975.

1.113 H. M. Wagner, *Principles of Operations Research with Applications to Managerial Decisions*, Prentice-Hall, Englewood Cliffs, N.J., 2nd ed., 1975.

1.114 H. P. Williams, *Model Building in Mathematical Programming*, Wiley-Interscience, New York, 1978.

1.115 N. Wu and R. Coppins, *Linear Programming and Extensions*, McGraw-Hill, New York, 1981.

1.116 M. Zeleny, *Linear Multiobjective Programming*, Springer, New York, 1974.

1.117 S. Zionts, *Linear and Integer Programming*, Prentice-Hall, Englewood Cliffs, N.J., 1974.

1.118 S. I. Zukhovitskiy and L. I. Avdeyeva, *Linear and Convex Programming*, Saunders, Philadelphia, 1966.

chapter 2 The Simplex Method

2.1 INTRODUCTION

In an LP we have to optimize the objective function subject to linear equality and/or inequality constraints, and sign restrictions on the variables. If there are no sign restrictions on the variables and no inequality constraints, the LP will be of the form:

$$\begin{aligned} \text{Minimize} \quad & z(x) = cx \\ \text{Subject to} \quad & Ax = b \end{aligned}$$

where A, b, c, are given matrices or vectors. It can be shown (see Exercise 3.17) that in this case, if there is a feasible solution, $z(x)$ is either a constant on the set of feasible solutions, or it can be made to diverge to $-\infty$ on this set. Hence, such LPs are trivial, and not of practical interest. Hereafter, we will assume that the LPs being discussed consist of at least one inequality constraint or a variable restricted in sign.

As a naive approach for solving such problems it may be tempting to use the familiar technique in elementary calculus of trying to solve the system of equations obtained by putting the partial derivatives of the objective function equal to zero. This technique does not work here because of the constraints. Even the standard Lagrange multiplier technique does not apply directly to this problem because we have to deal with inequality constraints, and the Lagrange multiplier technique only handles equality constraints. We take recourse to the Karush–John–Kuhn–Tucker theories to handle inequality constraints. Unfortunately these theories provide only certain necessary conditions that every optimum solution must satisfy, and they do not provide any systematic method for solving these conditions. In any case, it is clear that any method that can find an optimum solution of an LP must contain in itself a method for finding a feasible solution satisfying a system of linear equality and inequality constraints. It can be shown that this problem of finding an initial feasible solution can itself be posed as another LP. Solving such a linear programming formulation seems to be the only practicable approach for finding a feasible solution, satisfying all the constraints in the problem. Thus, none of the classical methods in calculus or linear algebra offer any approaches for solving LPs or for solving systems of linear inequalities. Special techniques had to be developed to tackle these problems. This chapter discusses one of these special techniques, known as the *simplex method* due to George B. Dantzig (1947).

Recently, a special direct method for solving systems of linear inequalities (which does not need the transformation of this problem into an LP) known as the *ellipsoid method* has been developed. The ellipsoid method belongs to a class of algorithms known as *variable metric methods* in nonlinear programming literature. By applying the ellipsoid method to solve the system of necessary and sufficient optimality conditions corresponding to an LP, we can solve the LP. The ellipsoid method is polynomially bounded (in the sense that the worst case computational effort needed to solve an LP or a system of linear inequalities with integer data, is bounded above by a polynomial in the size of the problem). The simplex method is a *pivotal method* (based on carrying out a sequence of pivot operations on the matrix of coefficients of the system of constraints) and it is not polynomially bounded in the worst case. However, for solving practical LP models, the simplex method seems to be much more efficient than the ellipsoid method. In this chapter we discuss the simplest form of the simplex method to establish notation and develop ideas. Other forms of the simplex method are discussed in later chapters (see, for example, Chapters 5, 6, 7, 10, and 11). The ellipsoid method is discussed in detail in Chapter 15.

It would be convenient if the final answer to a problem can be obtained either as a simple formula, in terms of the data, or in some tabular form from which the answers can be looked up easily. It will be shown in Chapter 3 that this is generally impossible for LPs. LPs can only be solved by means of an *algorithm*, which is a step-by-step procedure that leads to an answer to the problem in a finite number of steps.

2.2 TRANSFORMING THE PROBLEM INTO THE STANDARD FORM

Before applying the simplex method on an LP, all the constraints in it on which pivot operations are carried out must be transformed into equality constraints. The reasons for this are described below. Also one may want to either maximize or minimize the objective function. It becomes very cumbersome if we have to discuss versions of the simplex method for each of these different types of LP models separately. Instead, we discuss the simplex method for solving LPs in a particular form known as the *standard form*, and show here how every type of LP model can be transformed into this standard form.

In an LP model, sign restrictions on a variable would require that the variable be either nonnegative or nonpositive, as specified. We refer to such conditions as *restrictions* in the LP model. In the simplex method, pivot operations are not carried out on these restrictions. They are held separately and the method tries to satisfy these restrictions implicitly. All the other conditions in an LP model (whether it involves one or many variables) will be referred to in this chapter as *constraints.* Constraints are all the conditions on the decision variables in the model which are included in the pivot operations when the simplex method is used to solve the problem. For the version of the simplex method that we discuss here, all conditions other than sign restrictions on individual variables, are called constraints.

In Chapter 11, we discuss a special efficient implementation of the simplex method known as the *bounded-variable simplex method*, in which both upper bound and lower bound conditions on individual variables are treated as restrictions, and constraints refer to only those conditions that involve two or more decision variables. In that method, pivot operations are only carried out on the system of constraints in the model, each of which involves two or more decision variables.

To transform an LP into the *standard form*, carry out the following steps in the order listed.

Transform Lower Bound Constraints into Nonnegativity Restrictions

If there is any variable in the problem that has a required lower bound other than zero, for example, $x_1 \geqq \ell_1$, where ℓ_1 is some specified number, eliminate this variable from the model by substituting $x_1 = y_1 + \ell_1$ wherever it occurs in the model. Clearly the constraint $x_1 \geqq \ell_1$ is equivalent to requiring $x_1 = y_1 + \ell_1$, $y_1 \geqq 0$. In this manner all lower bounded variables in the problem can be replaced by variables with lower bounds of zero. The lower bound constraint on x_1 gets replaced by the nonnegativity restriction on the variable y_1.

Transforming all the Constraints into Equality Constraints

Convert all the remaining inequality constraints (including upper bound constraints on variables, if any) into equations by introducing appropriate slack variables. A constraint of the form,

$$\sum_{j=1}^{n} a_{1j}x_j \leqq b_1 \text{ becomes}$$

$$\sum_{j=1}^{n} a_{1j}x_j + x_{n+1} = b_1, \qquad x_{n+1} \geqq 0$$

$$\text{and } \sum_{j=1}^{n} a_{2j}x_j \geqq b_2 \text{ becomes}$$

$$\sum_{j=1}^{n} a_{2j}x_j - x_{n+2} = b_2, \qquad x_{n+2} \geqq 0$$

Here, x_{n+1}, x_{n+2} are the slack variables. The coefficient of the slack variable in the transformed equality constraint depends on whether the original inequality constraint was a $\leqq$ or a $\geqq$ inequality. Slack variables are always nonnegative, and represent unutilized capacities, etc. For example, in the above constraints, the slack variables are $x_{n+1} = b_1 - \sum_{j=1}^{n} a_{1j}x_j$ and $x_{n+2} = \sum_{j=1}^{n} a_{2j}x_j - b_2$. In some textbooks, the name *slack variable* is only used for variables such as x_{n+1} above; and the variables such as x_{n+2} are called *surplus variables.* Here we use the term *slack variable* for both these kinds of variables. The slack variables are as much a part of the original problem as the variables used in modeling the constraints. Slack variables remain in the problem throughout, and their values in an optimal feasible solution of the problem give significant information.

Note 2.1: Each inequality constraint in the original problem leads to a different slack variable.

Nonnegative Variables

In the standard form we require all the variables to be nonnegative. If the original problem contains some unrestricted variables, they can be handled by two methods. In one method each unrestricted variable is transformed into the difference of two nonnegative variables, as in Section 1.2.4. For example, let the variable x_1 in the problem be unrestricted in sign. Any real variable can be expressed as a difference of two nonnegative variables. So we can substitute $x_1 = x_1^+ - x_1^-$, where $x_1^+ \geqq 0$ and $x_1^- \geqq 0$. If $x_1^+ - x_1^-$ is substituted for x_1 wherever it occurs in the original LP and the new variables x_1^+ and x_1^- are restricted to be nonnegative, the resulting LP is *equivalent* to the original one. Each unrestricted variable can be handled in a similar manner. This method handles unrestricted variables conveniently, but it is not used much in practice because it increases the number of variables in the problem. Also if this method is used, the resulting LP will invariably be *unstable* (see Section 4.8).

The other method for handling unrestricted variables takes advantage of their unrestricted nature and eliminates them from the problem. Suppose x_1 is an unrestricted variable in an LP. Since x_1 is a variable appearing in the problem, the coefficient of x_1 in at least one of the constraints must be nonzero. Suppose it is the ith constraint, which is $a_{i1}x_1 + a_{i2}x_2 + \cdots + a_{in}x_n = b_i$, where $a_{i1} \neq 0$. This constraint is equivalent to $x_1 = (b_i - a_{i2}x_2 - \cdots - a_{in}x_n)/a_{i1}$. Since there are no further restrictions on the variable x_1, once the optimal values of the other variables $x_2, \ldots, x_n$ are obtained, the associated optimum value of x_1 can be obtained by substituting the values of x_2 to x_n in the above expression. Eliminate x_1 from the problem by storing this expression for x_1 somewhere, by substituting this expression for x_1 in all the other constraints and the objective function, and by eliminating the ith constraint from further consideration.

All the unrestricted variables in the problem can be eliminated from the problem in this manner one by one. Each unrestricted variable eliminated in this way reduces the number of variables in the LP by one and reduces the number of constraints by one; this makes the order of the problem smaller and hence easier to solve.

Exercises

2.1 Explain why variables that must be nonnegative cannot be eliminated from the problem in the above manner.

If there is a variable in the model that must be nonpositive, say $x_1 \leq 0$, let $x_1 = -y_1$ and eliminate x_1 from the model by using this equation; y_1 is a nonnegative variable. After these transformations, all the variables in the transformed model are nonnegative variables. All restrictions in the model are nonnegativity restrictions on the variables.

Minimization Form

Eliminate constant terms in the objective function. This does not change the set of optimum solutions for the problem. Then express the objective function in the *minimization form.* In the original statement of the problem, if the objective is maximize $z'(x) = \sum_{j=1}^n c'_j x_j$, change the problem with the new objective as minimize $z(x) = -z'(x) = \sum_{j=1}^n - c'_j x_j$. This change does not affect the set of optimal feasible solutions of the problem, and the equation [maximum value of $z'(x)$] = $-$[minimum value of $z(x)$], can be used to get the maximum value of the original objective function.

Nonnegative Right-Hand-Side Constants Now the LP has been transformed into the following form:

Minimize	$z(x) = \sum_{j=1}^n c_j x_j$		
Subject to	$\sum_{j=1}^n a_{ij} x_j = b_i$	$i = 1$ to m	These are the constraints in the problem.
and	$x_j \geqq 0$	$j = 1$ to n	These are the sign restrictions on the variables.

Multiply the ith equation on both sides by -1, if b_i is negative. As a result of this, all the right-hand constants in the equality constraints become nonnegative.

Standard Form

The problem is now in standard form. It may be written in *detached coefficient tabular form*.

Tableau 2.1 Original Tableau

x_1	$x_2 \ldots x_j \ldots x_n$	$-z$	b
a_{11}	$a_{12} \ldots a_{1j} \ldots a_{1n}$	0	b_1
$\vdots$	$\vdots \quad \vdots \quad \vdots$	$\vdots$	$\vdots$
a_{i1}	$a_{i2} \ldots a_{ij} \ldots a_{in}$	0	b_i
$\vdots$	$\vdots \quad \vdots \quad \vdots$	$\vdots$	$\vdots$
a_{m1}	$a_{m2} \ldots a_{mj} \ldots a_{mn}$	0	b_m
c_1	$c_2 \ldots c_j \ldots c_n$	1	0

$x_j \geqq 0$ for all j; minimize z

Each row in the tableau (except the last row) represents a constraint in the problem. The ith row represents the constraint $a_{i1}x_1 + a_{i2}x_2 + \cdots + a_{in}x_n = b_i$. The last row in the tableau represents the equation $c_1x_1 + \cdots + c_nx_n - z = 0$. This row is called the *cost row*, or the *objective row*. The c_j are known as the *original cost coefficients*; the numbers $b_1, \ldots, b_m$ are known as the *original right-hand side constants*; and the numbers a_{ij} are known as the *original input-output coefficients*. The column vector associated with x_j in Tableau 2.1 $(a_{ij}, \ldots, a_{mj}, c_j)^{\mathrm{T}}$ is known as the *original column vector of* x_j in this problem. The variables in the original tableau $x_1, \ldots, x_n$ are all the variables in the model, including slack variables (excepting possibly some unrestricted variables that may have been eliminated from the problem). A *feasible solution* is any vector $x = (x_1, \ldots, x_n)$ that satisfies all the constraints and the sign restrictions.

Why Inequality Constraints Should be Transformed into Equality Constraints Before Applying the Simplex Method

In any system of equations we can **(1)** multiply all the coefficients on both sides of an equation by any nonzero number and **(2)** multiply an equation by a real number and add it to another equation. These transformations do not change the set of feasible solutions of the system. For example, consider the system:

$$3x_1 - 7x_2 = -1$$
$$x_1 + x_2 = 13$$

Add two times the second equation to the first. Then multiply the second equation by four. This leads to the following system, which can be verified to have the same solution.

$$5x_1 - 5x_2 = 25$$
$$4x_1 + 4x_2 = 52$$

When transformations of types 1 and 2 are carried out on a system of inequalities, the set of feasible solutions is normally altered. For example, consider the system of inequalities $x_1 \geqq 0$, $x_2 \geqq 0$ in $\mathbf{R}^2$. (As discussed above, these are treated as sign restrictions in the simplex method, and not included in the pivot operations. However, we choose this simple system purely to illustrate what can happen when operations of types 1 and 2 are performed on a system of linear inequalities.) The set of feasible solutions of this system is the nonnegative quadrant in $\mathbf{R}^2$, the shaded region in Figure 2.1. Now transform the system by adding two times the second inequality to the first. The new system and its set of feasible solutions are given in Figure 2.2. Clearly

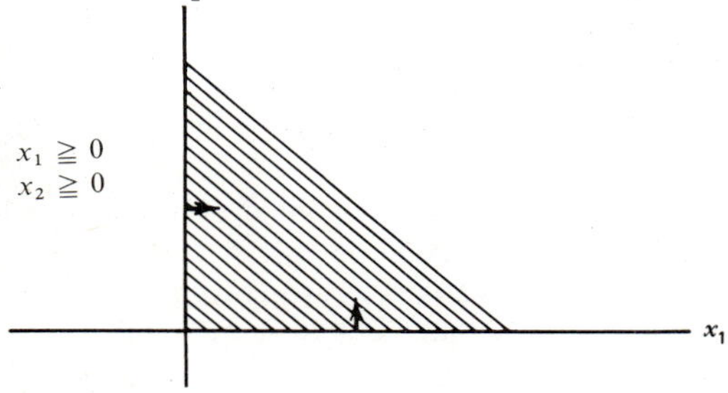

Figure 2.1 System of inequalities.

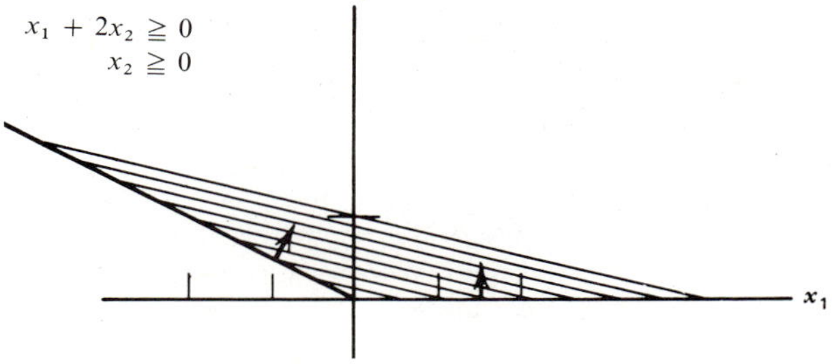

Figure 2.2 Transformed system.

the transformed system does not have the same set of feasible solutions as the original system.

However, when a system of inequality constraints is transformed into a system of equations by introducing the appropriate slack variables, then transformations of types 1 and 2 can be carried out on the resulting system of equations, and these will not change the set of feasible solutions of it.

The simplex method uses transformations of types 1 and 2 on the system of constraints to find an optimum solution. Hence it is essential that the system of constraints of an LP on which pivot operations are performed be transformed into a system of equations by introducing appropriate slack variables before applying the simplex method on it.

Example 2.1 Transforming a Problem into Standard Form

Consider the following LP

Maximize

$$-x_1 + 2x_2 - x_3 + y_4 + 3y_5 - y_6 + y_7$$

Subject to

$$\begin{aligned}
3x_2 + 2x_3 + y_4 - y_5 + y_6 + y_7 &\geqq -4 \\
2x_1 - x_3 - y_4 - 2y_6 + y_7 &\leqq -58 \\
-x_2 + 2y_4 + 2y_5 + y_6 &= 12 \\
-x_1 - x_2 + y_4 - y_5 + 2y_6 + y_7 &= 7 \\
y_4 &\geqq -5 \\
x_1 &\leqq 9
\end{aligned}$$

$$x_1, x_2, x_3 \geqq 0 \qquad y_5 \leqq 0 \qquad y_4, y_6, y_7 \text{ unrestricted}$$

First the lower bounded variable y_4 is eliminated by substituting $y_4 = -5 + x_4$, where x_4 is a nonnegative variable, in all the constraints and the objective function. Replace the constraint $y_4 \geqq -5$ by the nonnegativity restriction on x_4. The transformed first constraint is $3x_2 + 2x_3 + x_4 - y_5 + y_6 + y_7 \geqq -4 + 5 = 1$. Introducing the slack variable s_1, this constraint is equivalent to $3x_2 + 2x_3 + x_4 - y_5 + y_6 + y_7 - s_1 = 1$; $s_1 \geqq 0$. From this constraint, we derive $y_6 = 1 - 3x_2 - 2x_3 - x_4 + y_5 - y_7 + s_1$. Substitute this expression for y_6 in the objective function and all the constraints, thereby eliminating y_6 from the system. These transformations change the fourth constraint into $-x_1 - 7x_2 - 4x_3 - x_4 + y_5 - y_7 + 2s_1 = 10$. From this we have $y_7 = -10 - x_1 - 7x_2 - 4x_3 - x_4 + y_5 + 2s_1$. Substitute this expression for y_7 in all the remaining constraints and the objective function, thus eliminating y_7 from the system. The original constraints 1 and 4, from which the expressions for the unrestricted variables y_6, y_7 are obtained, are eliminated from the system.

Slack variables are now introduced to transform the remaining inequality constraints into equations. Transform the nonpositive variable y_5 by substituting $y_5 = -x_5$, where x_5 is a nonnegative variable. Eliminate constant terms from the transformed objective function, and multiply it by -1 to transform it into the minimization form. In the resulting system, the first constraint is $-x_1 - 15x_2 - 9x_3 - 2x_4 - x_5 + 4s_1 + s_2 = -31$. Multiply both sides of this by -1, to make the right-hand side constant positive. The resulting problem in detached coefficient tableau form is as follows:

x_1	x_2	x_3	x_4	x_5	s_1	s_2	s_3	$-z$	b
1	15	9	2	1	−4	−1	0	0	31
1	3	2	2	−2	−1	0	0	0	11
1	0	0	0	0	0	0	1	0	0
3	9	7	0	4	3	0	0	1	0

$x_j \geqq 0$ for all j; $s_1, s_2, s_3 \geqq 0$; z to be minimized.

2.3 CANONICAL TABLEAUX

2.3.1 Nonsingular Linear Transformations

Each row of the tableau represents a constraint in the problem. From linear algebra it is well known that nonsingular linear transformations, that is transformations such as the following, transform the system into an *equivalent system*.

1 Multiplying all the entries in a row of the tableau by a nonzero constant, which is equivalent to multiplying both sides of the corresponding constraint by that nonzero constant.

2 Adding a constant multiple of a row of the tableau to another row, which is equivalent to adding a constant multiple of a constraint to another one.

Since they operate on the rows of the tableau, transformations of this type are known as *elementary row operations*. Given the equivalent system obtained after making several elementary row operations, a reverse series of elementary row operations can be performed on the equivalent system to yield the original system again. Neither the set of feasible solutions nor the set of optimum solutions of the

Row 1	a_{11} ... a_{1j} ... a_{1n}	b_1
Row 2	a_{21} ... a_{2j} ... a_{2n}	b_2
α(Row 1) + Row 2	$\alpha a_{11}+a_{21}$ ... $\alpha a_{1j}+a_{2j}$ ... $\alpha a_{1n}+a_{2n}$	αb_1+b_2
β(Row 2)	βa_{21} ... βa_{2j} ... βa_{2n}	βb_2

problem is changed by making such transformations. Hence we take the liberty to perform such transformations. The main approach of the simplex algorithm is to move from one basic feasible solution to another by making such transformations until an optimum solution is obtained.

Note 2.2: Row operations are always performed element by element using corresponding elements in the respective rows. The tableau at the top illustrates how a row operation is performed using two rows, rows 1 and 2. The third row in this tableau gives α(row 1) + row 2, where α is any real number. The fourth row in this tableau is β(row 2), where β is any nonzero real number. A numerical example is given below.

Row 1	1	0	3	−5	3
Row 2	0	1	2	4	1
−5(Row 1) + Row 2	−5	1	−13	29	−14

Pivot Operation

A *pivot operation* or *a pivot step* on a system of equality constraints, or on the tableau representing the system in detached coefficient form, is uniquely specified by specifying the *pivot element* for the operation. *The pivot element has to be nonzero*. (It cannot be zero, because the pivot operation requires divisions with the pivot element as the denominator.) Consider the pivot operation on the system in Tableau 2.2, with the element a_{rs} as the pivot element. The row containing the pivot element a_{rs} (i.e., the rth constraint in Tableau 2.2 here) is called the *pivot row*, and the column containing the pivot element (i.e., the sth column in Tableau 2.2 here) is called the *pivot column* in this pivot operation. The pivot operation consists of a series of elementary row operations on Tableau 2.2, to transform the pivot column into one containing an entry of +1 in the pivot row, and an entry of 0 in all the other rows. It requires the following:

Tableau 2.2 Pivot Operations on a System of Equality Constraints, With the Pivot Element Circled

x_1	x_2 ... x_s ... x_n	
a_{11}	a_{12} ... a_{1s} ... a_{1n}	$= b_1$
⋮	⋮ ⋮ ⋮	⋮
a_{r1}	a_{r2} ... $\textcircled{a_{rs}}$ ... a_{rn}	$= b_r$ Pivot row
⋮	⋮ ⋮ ⋮	⋮
a_{m1}	a_{m2} ... a_{ms} ... a_{mn}	$= b_m$

Pivot column

i For each $i \neq r$, subtract a suitable multiple of the pivot row from the ith row so that the entry in the pivot column in the ith row becomes zero.

ii Divide the pivot row by the pivot element.

Since this is a system of equations, these operations lead to an equivalent system of equations. The equivalent system obtained after this pivot operation is given in Tableau 2.3, where for $j=1$ to n: $\bar{a}_{rj}=a_{rj}/a_{rs}$, $\bar{a}_{ij}=a_{ij}-a_{is}a_{rj}/a_{rs}$, for $i \neq r$; $\bar{b}_r=b_r/a_{rs}$, $\bar{b}_i=b_i-a_{is}b_r/a_{rs}$, for $i \neq r$.

See Example 2.11 below for an illustration of the pivot operation. The simplex method is called a *pivot method*, since it solves an LP by carrying out a sequence of pivot operations on the system of constraints in the LP.

Tableau 2.3 Equivalent System Obtained After the Pivot Operation

x_1	x_2 ... x_s ... x_n	
$\bar{a}_{11}$	$\bar{a}_{12}$ ... 0 ... $\bar{a}_{1n}$	$\bar{b}_1$
⋮	⋮ ⋮ ⋮	⋮
$\bar{a}_{r1}$	$\bar{a}_{r2}$ ... 1 ... $\bar{a}_{rn}$	$\bar{b}_r$
⋮	⋮ ⋮ ⋮	⋮
$\bar{a}_{m1}$	$\bar{a}_{m2}$ ... 0 ... $\bar{a}_{mn}$	$\bar{b}_m$

2.3.2 Canonical Form

The original tableau, Tableau 2.1, is in canonical form if there exists a unit matrix (see Section 3.2) of order m among the first m rows of the original tableau (after the problem is written in standard form). When this occurs it is easy to read out a feasible solution of the problem from the tableau. This is a special type of feasible solution known as a *basic feasible solution* (BFS). The significance of BFSs will be explained in Chapter 3, where it is shown that they correspond to *extreme* or *corner points* of the set of feasible solutions. A significant result is that if an LP in standard form has an optimum feasible solution, it has an optimum feasible solution that is a BFS. The simplex method draws on this result heavily, and it searches only among BFSs for an optimum feasible solution.

In defining a BFS for the problem, the variables are partitioned into two sets called *nonbasic* (or *independent*) variables and *basic* (or *dependent*) variables. The nonbasic variables are all equal to zero in the BFS; the basic variables are such that the set of their column vectors in the tableau forms a column basis for the matrix of input-output coefficients.

If the problem is in canonical form, rearrange the numbers of the variables, if necessary, so that the column vectors corresponding to the variables $x_1, \ldots, x_m$ in the tableau form a unit matrix of order m among the first m rows. (This step of rearranging the numbers of the variables is really unnecessary. It is done purely for the convenience of referring to the basic variables as $x_1, \ldots, x_m$.) Select the variables associated with the column vectors of the unit matrix as the *basic variables*. All the other variables in the problem become *nonbasic variables*. The column vectors in the tableau associated with the basic variables are called the present *basic column vectors*. All the other column vectors in the tableau are called the *nonbasic column vectors*.

The basic variable that corresponds to the first column vector of the unit matrix is called the *first basic variable* or the *basic variable in the first row*. (For convenience, we assume that this variable is called x_1.) In general, the basic variable that corresponds to the rth column vector of the unit matrix is called the rth *basic variable* or the *basic variable in the* rth *row*. The *basic vector* is the vector of basic variables in their proper order. Record the basic variables in their order on the left-hand side of the tableau. When this is done, this tableau is of the form seen in Tableau 2.4.

Tableau 2.4

Basic Variables	$x_{m+1} \ldots x_n$	x_1	$x_2 \ldots x_m$	$-z$	b
x_1	$a_{1,m+1} \ldots a_{1n}$	1	$0 \ldots 0 \ldots$	0	b_1
x_2	$a_{2,m+1} \ldots a_{2n}$	0	$1 \ldots 0$	0	b_2
$\vdots$	$\vdots \qquad \vdots$	$\vdots$	$\vdots \qquad \vdots$	$\vdots$	$\vdots$
x_m	$a_{m,m+1} \ldots a_{mn}$	0	$0 \ldots 1$	0	b_m
	$c_{m+1} \ldots c_n$	c_1	$c_2 \ldots c_m$	1	0

In putting the problem in standard form, the right-hand side constants have all been made nonnegative. The solution given by setting the ith basic variable equal to b_i for $i = 1$ to m, and all nonbasic variables equal to zero, is a feasible solution of the problem. It satisfies all the constraints and is nonnegative as required.

2.3.3 Pricing Out Routine

Subtract suitable multiples of the first m rows from the cost row, so that the entries in it under all the basic column vectors become zero. This is called the *pricing out operation* or the *cost updating operation*. This gives Tableau 2.5, known as the *canonical tableau with respect to the present basic vector*. In the canonical tableau with respect to a given basic vector, the basic column vectors in their proper order should form the unit matrix, and they should all be priced out.

Here is the initial BFS:

$$\begin{aligned} \text{All nonbasic variables} &= 0 \\ i\text{th basic variable} &= b_i \qquad i = 1 \text{ to } m \qquad (2.1) \\ \text{Objective value} &= z_0 \end{aligned}$$

The simplex algorithm always requires a BFS like this to start with. In this case, the basic vector $(x_1, \ldots, x_m)$ is known as a *feasible basic vector*, or a *primal feasible basic vector* for this LP. The $\bar{c}_j$ are called the *relative* (or *updated*) *cost coefficients with respect to the basic vector* $(x_1, \ldots, x_m)$, "relative" because their values depend on the choice of the basic vector. They are obtained by "pricing out" all the basic column vectors. It is explained later that a unit change in a nonbasic variable x_j from its

Tableau 2.5 Initial Canonical Tableau

Basic Variables	x_{m+1} ... x_j ... x_n	x_1 ... x_i ... x_m	$-z$	b
x_1	$a_{1,m+1}$ ... a_{1j} ... a_{1n}	1 ... 0 ... 0	0	b_1
⋮	⋮ ⋮ ⋮	⋮ ⋮ ⋮	⋮	⋮
x_i	$a_{i,m+1}$ ... a_{ij} ... a_{in}	0 ... 1 ... 0	0	b_i
⋮	⋮ ⋮ ⋮	⋮ ⋮ ⋮	⋮	⋮
x_m	$a_{m,m+1}$ ... a_{mj} ... a_{mn}	0 ... 0 ... 1	0	b_m
$-z$	$\bar{c}_{m+1}$ ... $\bar{c}_j$ ... $\bar{c}_n$	0 ... 0 ... 0	1	$-z_0$

present value of zero, while retaining feasibility, results in a change of $\bar{c}_j$ units in the objective value.

Words such as *basis* and *basic feasible solution* have been used without being defined adequately. Rigorous definitions of these terms are given in Chapter 3.

In some problems the original input-output coefficient matrix may not contain a full unit matrix of order m as a submatrix, see, for instance, Example 2.3. In such problems there is no way of directly obtaining a BFS from the original tableau. From the original tableau we may not even be able to decide whether a feasible solution exists. For such problems there is a method of introducing *artificial variables* and transforming the problem of finding a feasible solution into another LP. This augmented LP is formulated in such a way that it has an initial BFS (which is artificial) that can be obtained directly. Starting with this BFS, the augmented problem can be solved by applying the simplex algorithm. The augmented LP is known as the *Phase I problem.* An optimum solution of the Phase I problem either provides a BFS of the original problem or a proof that the original problem has no feasible solution at all. If a BFS of the original problem is obtained at the termination of the Phase I problem, the original problem is then solved by applying the simplex algorithm starting with it. This latter part of the computational effort is known as the *Phase II problem.*

To solve an LP, the simplex algorithm always requires a starting BFS. The *simplex method* for solving a general LP applies the simplex algorithm in two phases as discussed above. If the original tableau of the problem in standard form contains a full unit matrix as a submatrix, a BFS can be obtained directly without working on Phase I. Therefore, in this case we proceed directly to Phase II of the simplex method.

Example 2.2 Initial Basic Vector Selection for an LP in Canonical Form

Consider the following LP in detached coefficient tableau form.

x_1	x_2	x_3	x_4	x_5	x_6	x_7	$-z$	b
1	0	1	−2	0	1	1	0	3
1	0	0	1	1	−2	0	0	4
−1	1	0	1	0	3	0	0	5
0	4	3	20	6	5	−2	1	0

$x_j \geqq 0$ for all j; minimize z.

Here both the variables x_3, x_7 correspond to the first unit column vector, and between them suppose we select x_7 as the first basic variable. The only possible choices for basic variables in rows 2 and 3 are x_5, x_2, respectively. Hence our initial basic vector is (x_7, x_5, x_2). Pricing out requires the following operation on the original tableau: (row 4) − (−2) (row 1) − 6 (row 2) − 4 (row 3). This leads to the canonical tableau with respect to this basic vector. The corresponding BFS is $(x_1, x_2, x_3, x_4, x_5, x_6, x_7) = (0, 5, 0, 0, 4, 0, 3)$ with an objective value of $z = 38$.

Canonical Tableau

Basic Variables	x_1	x_2	x_3	x_4	x_5	x_6	x_7	$-z$	b
x_7	1	0	1	−2	0	1	1	0	3
x_5	1	0	0	1	1	−2	0	0	4
x_2	−1	1	0	1	0	3	0	0	5
$-z$	0	0	5	6	0	7	0	1	−38

2.4 PHASE I WITH A FULL ARTIFICIAL BASIS

Before discussing the simplex algorithm we discuss how to set up the Phase I problem for obtaining an initial BFS of an LP if the original tableau corresponding to it (in standard form with nonnegative right-hand side constants) is not in canonical form.

Let the original tableau for the problem be Tableau 2.1, and suppose the first m rows of this tableau (the constraint rows) do not contain the unit matrix of order m as a submatrix. Augment the tableau with the *artificial variables* $x_{n+1}, \ldots, x_{n+m}$, whose column vectors form the unit matrix of order m. Let the cost coefficient of each of these artificial variables in the original objective row be equal to zero. Now the augmented problem has an initial BFS that is obtained directly: $x_i = 0$, $i = 1, \ldots, n$; $x_{n+i} = b_i, i = 1, \ldots, m$. Therefore, the augmented problem is in canonical form with the artificial vector $(x_{n+1}, \ldots, x_{n+m})$ as the basic vector. Any feasible solution of the augmented problem, $(\hat{x}_1, \ldots, \hat{x}_{n+1}, \ldots, \hat{x}_{n+m})$, in which all the artificial variables $\hat{x}_{n+1}, \ldots, \hat{x}_{n+m}$ are zero yields the feasible solution $(\hat{x}_1, \ldots, \hat{x}_n)$ of the original problem. This is the main idea exploited in Phase I of the simplex method. Starting with the initial BFS, try to get a feasible solution of the augmented problem in which all the artificial variables are zero. This may be achieved by restricting all the artificial variables to be nonnegative and minimizing their sum, that is, Minimize $w = x_{n+1} + \cdots + x_{n+m}$, where w is known as the *Phase I objective function* or the *Phase I cost function* or the *infeasibility form*. The augmented tableau is Tableau 2.6.

The last row in Tableau 2.6 is the *Phase I objective row*. As long as the simplex method is in Phase I, the objective is to minimize w. In this context the original objective row is known as the *Phase II objective row*. This problem of minimizing w, starting with a full artificial basis, is called the *Phase I problem*.

Price out the basic column vectors of the augmented tableau by subtracting suitable multiples of the first m rows from the Phase I objective row and the Phase II objective row so that all the entries under the basic columns in these rows become zero. This gives the initial canonical tableau for the Phase I problem. The Phase I problem is an LP in canonical form, with an initial BFS. Therefore, the simplex algorithm can be applied to solve it.

Tableau 2.7 Initial Canonical Tableau for the Phase I Problem

Basic Variables	$x_1 \ldots x_n$	$x_{n+1} \ldots x_{n+m}$	$-z$	$-w$	b
x_{n+1}	$a_{11} \ldots a_{1n}$	$1 \ldots 0$	0	0	b_1
$\vdots$	$\vdots \quad \vdots$	$\vdots \quad \vdots$	$\vdots$	$\vdots$	$\vdots$
x_{n+m}	$a_{m1} \ldots a_{mn}$	$0 \ldots 1$	0	0	b_m
$-z$	$\bar{c}_1 \ldots \bar{c}_n$	$0 \ldots 0$	1	0	$-z_0$
$-w$	$\bar{d}_1 \ldots \bar{d}_n$	$0 \ldots 0$	0	1	$-w_0$

Interpretation

The original constraints are

$$\sum_{j=1}^{n} a_{ij}x_j = b_i \qquad i = 1 \text{ to } m$$

$$x_j \geqq 0 \qquad \text{for all } j$$

The constraints of the corresponding Phase I problem are

$$\sum_{j=1}^{n} a_{ij}x_j + x_{n+i} = b_i \qquad i = 1 \text{ to } m \tag{2.2}$$

$$x_j \geqq 0 \qquad \text{for all } j \text{ and } x_{n+i} \geqq 0 \qquad \text{for } i = 1 \text{ to } m$$

The following conclusions can be drawn:

1 If $(\hat{x}_1, \ldots, \hat{x}_n)$ is a feasible solution of the original problem, $(\hat{x}_1, \ldots, \hat{x}_n, \hat{x}_{n+1}, \ldots, \hat{x}_{n+m})$, where

Tableau 2.6 Augmented Tableau with a Full Artificial Basis

Basic Variables	$x_1 \ldots x_n$	$x_{n+1} \ldots x_{n+m}$	$-z$	$-w$	b
x_{n+1}	$a_{11} \ldots a_{1n}$	$1 \ldots 0$	0	0	b_1
$\vdots$	$\vdots \quad \vdots$	$\vdots \quad \vdots$	$\vdots$	$\vdots$	$\vdots$
x_{n+m}	$a_{m1} \ldots a_{mn}$	$0 \ldots 1$	0	0	b_m
$-z$	$c_1 \ldots c_n$	$0 \ldots 0$	1	0	0
$-w$	$0 \ldots 0$	$1 \ldots 1$	0	1	0

$\hat{x}_{n+1}, \ldots, \hat{x}_{n+m}$ are all equal to zero, is a feasible solution of the Phase I problem [i.e., it satisfies (2.2)].

2 Conversely, if $(\bar{x}_1, \ldots, \bar{x}_{n+1}, \ldots, \bar{x}_{n+m})$ is a feasible solution of the Phase I problem in which $\bar{x}_{n+1}, \ldots, \bar{x}_{n+m}$ all happen to be zero, $(\bar{x}_1, \ldots, \bar{x}_n)$ is a feasible solution to the original problem.

3 In (2.2) all the artificial variables $x_{n+1}, \ldots, x_{n+m}$ are restricted to be nonnegative. The sum of nonnegative numbers is always nonnegative. Therefore the Phase I objective value is greater than or equal to zero on the set of feasible solutions of the Phase I problem.

4 If the original problem has a feasible solution, by 1, a feasible solution to the Phase I problem can be constructed from it by setting all the artificial variables equal to zero. This feasible solution makes the value of w equal to zero. Since w is nonnegative on the set of feasible solutions of the Phase I problem, any feasible solution that makes w equal to zero must be an optimum solution for the Phase I problem. Therefore, if the original problem has a feasible solution, the minimum value of w in the Phase I problem is zero.

5 Conversely, if the minimum value of w in the Phase I problem is zero, there must exist a feasible solution $(\bar{x}_1, \ldots, \bar{x}_n; \bar{x}_{n+1}, \ldots, \bar{x}_{n+m})$ that makes w equal to zero. Therefore $\bar{x}_{n+1} + \cdots + \bar{x}_{n+m} = 0$. Since each of $\bar{x}_{n+1}, \ldots, \bar{x}_{n+m}$ is nonnegative, their sum can be zero only if each of them is zero. Hence $\bar{x}_{n+1}, \ldots, \bar{x}_{n+m}$ are all equal to zero. So $(\bar{x}_1, \ldots, \bar{x}_m)$ is a feasible solution of the original problem. Therefore, if the minimum value of w in the Phase I problem is zero, the original problem has a feasible solution.

6 From these arguments we conclude that the original problem has a feasible solution iff the minimum value of w in the Phase I problem is zero.

7 When the Phase I problem is solved, two things can happen.
(a) The minimum value of w is greater than zero. By (6) this implies that the original problem cannot have a feasible solution. Therefore the original constraints and the sign restrictions together must be inconsistent. An example of such a system is $-x_1 - x_2 = 1$, where $x_1 \geqq 0$ and $x_2 \geqq 0$.
(b) The minimum value of w is zero. From an optimum solution the Phase I problem, obtain a feasible solution to the original problem as in 2.

Note 2.3: If the original tableau contains some column vectors of the unit matrix of order m, the variables corresponding to these column vectors can be picked as basic variables in the initial basic vector. It is necessary to augment the tableau only with artificial variables whose column vectors are the column vectors of the unit matrix that do not appear in the original tableau. If this is done, the initial basic vector to the augmented problem consists of some original problem variables and some artificial variables. *The Phase I objective function is always the sum of the artificial variables that were introduced into the problem.* Hence the original Phase I cost coefficient of any original problem variable is zero, and of any artificial variable is one. The original Phase II cost coefficient of any artificial variable is zero. The initial canonical tableau for the Phase I problem is obtained by pricing out all the basic variables in both the objective rows.

Note 2.4: In Section 3.14 we discuss a method for formulating the Phase I problem using one artificial variable only. In computational tests, the Phase I formulation discussed in Section 3.14 performed much more efficiently than the Phase I formulation using a full artificial basis discussed here (see Reference [3.33]).

Example 2.3 Phase I Formulation

Consider the following LP.

Original Tableau

x_1	x_2	x_3	x_4	x_5	$-z$	b
-2	1	1	0	1	0	3
-2	1	0	1	0	0	2
1	1	0	1	0	0	7
3	-2	-2	4	2	1	0

$x_j \geqq 0$ for all j; minimize z.

There are no variables corresponding to the second and third unit vectors in the original tableau, and hence it is not in canonical form. Introducing the artificial variables x_6, x_7, x_8, the original tableau for the Phase I problem with a full artificial basis is first given. This is followed by the canonical tableau with respect to the initial artificial basic vector (x_6, x_7, x_8).

Original Tableau for the Phase I Problem

x_1	x_2	x_3	x_4	x_5	x_6	x_7	x_8	$-z$	$-w$	b
−2	1	1	0	1	1	0	0	0	0	3
−2	1	0	1	0	0	1	0	0	0	2
1	1	0	1	0	0	0	1	0	0	7
3	−2	−2	4	2	0	0	0	1	0	0
0	0	0	0	0	1	1	1	0	1	0

x_6, x_7, x_8 are the artificials.

First Canonical Tableau for the Phase I Problem

Basic Variable	x_1	x_2	x_3	x_4	x_5	x_6	x_7	x_8	$-z$	$-w$	b
x_6	−2	1	1	0	1	1	0	0	0	0	3
x_7	−2	1	0	1	0	0	1	0	0	0	2
x_8	1	1	0	1	0	0	0	1	0	0	7
$-z$	3	−2	−2	4	2	0	0	0	1	0	0
$-w$	3	−3	−1	−2	−1	0	0	0	0	1	−12

2.5 THE SIMPLEX ALGORITHM

2.5.1 The Initial Stage

This is an algorithm that can be used to solve any LP in standard form for which a starting feasible basic vector is known. Suppose the problem is of the form:

$$\text{Minimize} \quad z(x) = \sum_{j=1}^{n} c_j x_j$$

$$\text{Subject to} \quad \sum_{j=1}^{n} a_{ij} x_j = b_i \quad i = 1, \ldots, m \quad (2.3)$$

$$x_j \geqq 0 \quad j = 1, \ldots, n$$

Suppose the initial basic vector is $(x_1, \ldots, x_m)$. Let the canonical tableau with respect to this basic vector be:

Tableau 2.8 Canonical Tableau

Basic Variables	$x_{m+1} \ldots x_n$	$x_1 \ldots x_m$	$-z$	Basic Values
x_1	$\bar{a}_{1,m+1} \ldots \bar{a}_{1,n}$	$1 \ldots 0$	0	$\bar{b}_1$
$\vdots$	$\vdots \quad \vdots$	$\vdots \quad \vdots$	$\vdots$	$\vdots$
x_m	$\bar{a}_{m,m+1} \ldots \bar{a}_{m,n}$	$0 \ldots 1$	0	$\bar{b}_m$
$-z$	$\bar{c}_{m+1} \ldots \bar{c}_n$	$0 \ldots 0$	1	$-\bar{z}$

In every canonical tableau, the column vectors of the basic variables in their proper order form a unit matrix. Also, the cost row has zero entries under all the basic column vectors. The entries in the canonical tableau are called *updated entries* or *entries in the canonical tableau with respect to the present basic vector*, in order to distinguish them from the entries of the original tableau. The column vector associated with the variable x_j in Tableau 2.8 is known as the *updated column vector of x_j with respect to the present basic vector*. The entries $\bar{a}_{ij}$ are known as the *updated input-output coefficients* with respect to the present basic vector. The BFS corresponding to this basic vector is given below. The updated right-hand side constants in the canonical tableau will always represent the values of the current basic variables in the corresponding BFS.

$$\begin{array}{lll} \text{Nonbasic variables} & x_i = 0 & i = m+1, \ldots, n \\ \text{Basic variables} & x_i = \bar{b}_i & i = 1, \ldots, m \qquad (2.4) \\ \text{Objective value} & z = \bar{z} & \end{array}$$

2.5.2 The Fundamental Optimality Criterion

An *optimum feasible solution* for this problem is a feasible solution that gives the objective function its minimum value among all feasible solutions. Hence, a feasible solution $\tilde{x} = (\tilde{x}_1, \ldots, \tilde{x}_n)$ is an optimum feasible solution if $z(\tilde{x}) \leqq z(x)$ for all other feasible solutions x. This is the fundamental criterion for checking whether a particular feasible solution is optimal or not.

2.5.3 Termination Criteria

The termination criteria provide a computationally efficient means to check whether the fundamental optimality criterion is satisfied in any step of the algorithm.

Optimality

The present BFS is optimal if all the relative cost coefficients with respect to this basic vector are nonnegative, that is, $\bar{c}_j \geqq 0$, for all j. Of course, $\bar{c}_j$ is zero if x_j is a basic variable. The criterion requires that the relative cost coefficients of all nonbasic variables be nonnegative.

Note 2.5: The optimality criterion requires that the relative cost coefficients $\bar{c}_j$ with respect to the present basic vector (and not the original cost coefficients c_j) should be nonnegative.

THEOREM 2.1 If the optimality criterion stated above is satisfied, the feasible solution in that step is an optimum feasible solution for that LP.

Proof: Let the LP be (2.3) and let Tableau 2.8 be the canonical tableau for the LP in that step. The present BFS of the problem makes the objective function equal to $\bar{z}$. From the last row of the present canonical tableau we see that

$$\sum_{j=m+1}^{n} \bar{c}_j x_j - z(x) = -\bar{z}$$

that is,

$$z(x) = \bar{z} + \sum_{j=m+1}^{n} \bar{c}_j x_j$$

x_j must be nonnegative for all j in every feasible solution. Hence, if $\bar{c}_j$ is nonnegative for all j, $\sum_{j=m+1}^{n} \bar{c}_j x_j$ must be nonnegative. Hence, $z(x)$ must be greater than or equal to $\bar{z}$ at every feasible solution x. Thus, the present BFS must be optimal to the problem. ∎

Example 2.4 Illustration of the Optimality Criterion of the Simplex Algorithm

Consider the LP in Example 2.2. The optimality criterion is satisfied in the canonical tableau because all the relative cost coefficients are nonnegative. From the last row of the canonical tableau, we have $5x_3 + 6x_4 + 7x_6 - z = -38$, that is, $z(x) = 38 + (5x_3 + 6x_4 + 7x_6)$. Since all the variables are nonnegative in every feasible solution, $5x_3 + 6x_4 + 7x_6 \geqq 0$, at every feasible solution. Hence from the above expression we conclude that $z(x) \geqq 38$ at all feasible solutions x to this problem. Hence the present BFS $x = (x_1, \ldots, x_7) = (0, 5, 0, 0, 4, 0, 3)$ is an optimum solution of the problem, with an optimum objective value of $z(x) = 38$.

Mathematically, the optimality criterion stated above is a *sufficient optimality condition*, because, when it is satisfied, we are guaranteed that the present BFS is an optimum feasible solution for the problem.

Unboundedness

The objective function is unbounded below (i.e., there exists a class of feasible solutions along which it diverges to $-\infty$), if there exists an s such that, in a canonical tableau with respect to some feasible basic vector,

$$\bar{c}_s < 0 \qquad \text{and } \bar{a}_{is} \leqq 0 \qquad \text{for all } i = 1, \ldots, m$$

where $\bar{c}_s$ is the relative cost coefficient of x_s with respect to this basic vector, and $\bar{a}_{is}$, $i = 1$ to m are the entries in the updated column vector of x_s.

Discussion Consider the canonical tableau with respect to some feasible basic vector. For convenience in referring to it, assume that the basic vector is $(x_1, \ldots, x_m)$. Suppose the column vector of x_s satisfies the unboundedness criterion in the present canonical tableau. Rearranging the column vectors, if necessary, let the canonical tableau be Tableau 2.9.

Tableau 2.9 Canonical Tableau

Basic Variables	*Other Nonbasic Variables*	x_s	$x_1 \ldots x_m$	$-z$	b
x_1	...	$\bar{a}_{1s}$	$1 \ldots 0$	0	$\bar{b}_1$
$\vdots$	$\vdots$	$\vdots$	$\vdots \quad \vdots$	$\vdots$	$\vdots$
x_m	...	$\bar{a}_{ms}$	$0 \ldots 1$	0	$\bar{b}_m$
$-z$	...	$\bar{c}_s$	$0 \ldots 0$	1	$-\bar{z}$

The present BFS denoted by $\bar{x}$ is given in (2.4). Since this is a feasible solution, all the $\bar{b}_i$ are nonnegative. In $\bar{x}$, all the current nonbasic variables, including x_s, are zero. Try to increase the value of x_s from zero to some nonnegative value, say λ, keeping all the remaining nonbasic variables fixed at zero. From the results in Section 2.3.1, any vector that satisfies the system of constraints represented by the present canonical tableau also satisfies the original system of constraints and vice versa. Hence, if the equality constraints in the problem are to be satisfied, the values of the basic variables have to be changed such that the following equations hold.

$$\begin{aligned}
&\bar{a}_{1s}x_s + x_1 && = \bar{b}_1 \\
&\bar{a}_{2s}x_s \quad + x_2 && = \bar{b}_2 \\
&\vdots \qquad\qquad \ddots && \vdots \\
&\bar{a}_{ms}x_s \qquad\qquad + x_m && = \bar{b}_m \\
&\bar{c}_s x_s && -z = -\bar{z}
\end{aligned} \tag{2.5}$$

This implies that the new solution is

$$\begin{aligned}
i\text{th basic variable, } x_i &= \bar{b}_i - \bar{a}_{is}\lambda \\
i &= 1 \text{ to } m \\
x_s &= \lambda \\
\text{All other nonbasic variables} &= 0 \\
z &= \bar{z} + \bar{c}_s\lambda
\end{aligned} \tag{2.6}$$

Since the column vector of x_s satisfied the unboundedness criterion, $\bar{a}_{is} \leqq 0$ for all i, and $\bar{c}_s < 0$. In the solution given in (2.6), the values of all the variables x_j remain nonnegative for every $\lambda \geqq 0$. Hence, it is a feasible solution for every $\lambda \geqq 0$. However, as λ is made larger and larger, the objective value corresponding to this solution, which is equal to $\bar{z} + \bar{c}_s\lambda$, decreases indefinitely, since $\bar{c}_s < 0$. By choosing a λ arbitrarily large, we get from (2.6) a feasible solution to the problem at which the value of $z(x)$ is less than any arbitrary real number. Hence in this case the objective function $z(x)$ is unbounded below on the set of feasible solutions.

The set of feasible solutions obtained by giving λ values from 0 to ∞ in (2.6) is known as an *extreme half-line* of the set of feasible solutions for this LP. The objective values diverges to $-\infty$ as we travel along this extreme half-line by varying λ from 0 to ∞ in (2.6). See Section 3.7 for details.

Note 2.6: An LP in which the objective function $z(x)$ has to be minimized is said to be *unbounded below* if $z(x)$ diverges to $-\infty$ on the set of feasible solutions of the problem. In a similar manner an LP in which an objective function $z'(x)$ has to be maximized is said to be *unbounded above* if $z'(x)$ diverges to $+\infty$ on the set of feasible solutions.

Example 3.5 Illustration of the Unboundedness Criterion in the Simplex Algorithm

Here is the canonical tableau for an LP:

Canonical Tableau

Basic Variables	x_1	x_2	x_3	x_4	x_5	x_6	$-z$	b
x_3	0	0	1	1	−1	−5	0	17
x_1	1	0	0	−1	−1	−3	0	19
x_2	0	1	0	−1	−1	0	0	11
$-z$	0	0	0	9	3	−4	1	53

The column vector of x_6 satisfies the unboundedness criterion. Making the value of the nonbasic variable x_6 equal to λ, and keeping x_4, x_5 equal to zero, leads to the feasible solution

$$\begin{aligned}
&(x_3, x_1, x_2, x_6, x_4, x_5) \\
&\quad = (17, 19, 11, 0, 0, 0) + \lambda(5, 3, 0, 1, 0, 0) \\
&z = -53 - 4\lambda
\end{aligned}$$

Notice that this solution remains feasible for all $\lambda \geqq 0$. As λ tends to ∞, its objective value tends to $-\infty$.

Terminal Canonical Tableau

For an LP in standard form, a feasible basic vector is said to be a *terminal primal feasible basic vector* if either the optimality criterion or the unboundedness criterion is satisfied in the canonical tableau corresponding to that basic vector. The canonical tableau corresponding to a terminal primal feasible basic vector for an LP in standard form is known as a *terminal canonical tableau* for the LP in the primal simplex algorithm.

2.5.4 Improving a Nonoptimal Basic Feasible Solution

Interpretation of the Relative Cost Coefficients of the Nonbasic Variables

Let Tableau 2.4 be the original tableau for an LP in standard form. So we have $b_i \geqq 0$ for all $i = 1$ to m.

The original objective function is $z(x)=c_1x_1+\cdots+c_nx_n$, as a function of the original variables in the problem. So the original cost coefficient vector $c=(c_1, \ldots, c_n)$ is $[\partial z(x)]/\partial x$, the vector of partial derivatives of the objective function with respect to the original variables, or the *gradient vector* of $z(x)$. Select $x_B=(x_1, \ldots, x_m)$ as the initial basic vector in Tableau 2.4. Then $x_D=(x_{m+1}, \ldots, x_n)$ is the corresponding nonbasic vector. The BFS corresponding to this choice is $\bar{x}$ given in (2.1). For $i=1$ to m, we obtain from the ith constraint in Tableau 2.4 $x_i=b_i-a_{i,m+1}x_{m+1}-\cdots-a_{in}x_n$. These equations give the expressions for the basic variables in x_B as functions of the nonbasic variables in x_D in any feasible solution of the problem. That is why the nonbasic variables are also known as *independent variables* and the basic variables are known as *dependent variables*. Using these equations the basic variables can be totally eliminated from the problem, and the problem expressed purely in terms of the nonbasic (or independent) variables only. It can be verified that this leads to $z(x)=z_0+\bar{c}_{m+1}x_{m+1}+\cdots+\bar{c}_nx_n$, where $z_0=z(\bar{x})$, and changes the problem into:

$$\begin{aligned}
&\text{Minimize} && z_0+\bar{c}_{m+1}x_{m+1}+\cdots+\bar{c}_nx_n \\
&\text{Subject to} && b_i-a_{i,m+1}x_{m+1}-\cdots-a_{in}x_n \geqq 0 \\
& && i=1 \text{ to } m \qquad\qquad (2.7) \\
& && x_{m+1}\geqq 0, \ldots, x_n \geqq 0
\end{aligned}$$

where the $\bar{c}_j$ are the relative cost coefficients from the canonical Tableau 2.5. Representation (2.7) is the *representation* of the original problem in terms of the nonbasic variables in x_D. In the present BFS $\bar{x}$, all the nonbasic variables $x_{m+1}, \ldots, x_n$ are zero. Hence the current BFS $\bar{x}$ corresponds to the feasible point $x_D=(x_{m+1}, \ldots, x_n)=0$ in the representation (2.7).

The vector of partial derivatives of the objective function in the representation (2.7), is $(\bar{c}_{m+1}, \ldots, \bar{c}_n)$. Hence the vector of relative cost coefficients of the nonbasic variables in a canonical tableau is equal to the partial derivative vector of the objective function in a representation of the original problem in terms of these nonbasic variables only, and is therefore called *the reduced gradient vector* of the objective function with respect to these nonbasic variables. For $j=m+1$ to n, clearly the relative cost coefficient $\bar{c}_j$ in Tableau 2.5 is *the rate of change in the objective-value, per unit change in the value of the nonbasic variable x_j from its present value of zero in the present BFS*. This implies that in order to decrease the objective function from its present value, we should increase the values of the nonbasic variables, whose relative cost coefficients are strictly negative, from their present value of zero, while leaving the values of all other nonbasic variables equal to zero. From the representation (2.7), it is also clear that if $\bar{c}_{m+1}, \ldots, \bar{c}_n$ are all nonnegative, then the current feasible solution $(x_{m+1}, \ldots, x_n)=0$ for it (which corresponds to the current BFS $\bar{x}$ for the problem in terms of the original variables) is an optimum solution.

The choice of the nonbasic vector is not necessarily unique. For example, in Tableau 2.4, we can take any vector of m variables corresponding to column vectors in Tableau 2.4 that form a linearly independent set (see Chapter 3 for definitions of these terms and further details) and choose that as a basic vector. Actually, in the simplex algorithm, the basic vector is changed by one variable in each step. Whenever we obtain a canonical tableau with respect to a new basic vector, we obtain the representation of the original LP in terms of the corresponding nonbasic vector. However, all these representations are equivalent to the original problem.

Change in the Basic Vector

If neither the optimality nor the unboundedness criteria are satisfied in the canonical tableau with respect to the present basic vector, the simplex algorithm moves to a better feasible solution [a feasible solution $\hat{x}$ is said to be better than the present feasible solution $\bar{x}$ if $z(\hat{x}) \leqq z(\bar{x})$] from the present one. The simplex algorithm does this by changing the value of one judiciously selected nonbasic variable from its present value of 0 to some nonnegative value. The important thing to remember is that *in each step, the simplex algorithm tries to change the value of only one nonbasic variable.* The nonbasic variable for this change is selected so that the objective value decreases as a result of this change. It was shown above that this will happen only if the nonbasic variable selected for the change has a negative relative cost coefficient in the present canonical tableau. If x_s is the selected variable, it must satisfy

$$\bar{c}_s < 0 \qquad (2.8)$$

After some simple calculations, the nonnegative value that this nonbasic variable can take, denoted by θ, is determined. When x_s is made equal to θ, the values of the present basic variables have to be recomputed so that the resulting solution satisfies all the constraints. The modified value of at least one of the basic variables turns out to be zero. One such basic variable is selected and dropped from the basic vector, and x_s is made a basic variable in its place. This gives a new basic vector. By making the necessary transformations of the type discussed in Section 2.3.1 the column vector of x_s is transformed into the appropriate column vector of the unit matrix, and this leads to the canonical tableau with respect to the new basic vector. This whole operation is called *bringing the nonbasic variable* x_s *into the basic vector*; x_s is known as the *entering variable* in this operation.

Any nonbasic variable with a strictly negative relative cost coefficient is a candidate for being the entering variable, and such variables are called *nonbasic variables eligible to enter the basic vector*, or *eligible variables* in short. The entering variable can be chosen to be any one of the nonbasic variables that is eligible to enter. The column vector of the entering variable in the present canonical tableau is called the *pivot column*. When there are several eligible variables, the entering variable should be selected from among them, in such a way that the solution of the problem can be completed with the least computational effort. However, at present, no such selection method is known. One good rule that is commonly used and that seems to work well in practice, is to select the nonbasic variable x_s as the entering variable where

$$\bar{c}_s = \text{minimum } \{\bar{c}_j, j = 1 \text{ to } n\} \tag{2.9}$$

Break ties in (2.9) arbitrarily. This rule has the rationale of choosing the variable with the largest per unit capability of reducing the value of z. (In this sense, it may be viewed as a *steepest descent rule*.) A rule such as this to decide which nonbasic variable should be selected as the entering variable is known as a *pivot column choice rule* or *entering variable choice rule*. See Section 2.5.7 for a discussion of several entering variable choice rules.

Assume that the present basic vector is $(x_1, \ldots x_m)$ and let the canonical tableau with respect to this basic vector be Tableau 2.9. All the nonbasic variables excepting the entering variable x_s remain equal to zero in the next solution. After we substitute the value zero for those variables, the remaining system of constraints is (2.5). Hence if x_s is given the value λ, the new solution is the one in (2.6).

From these equations it is clear that if x_s is increased from its present value of 0 to the nonnegative value λ, the objective value decreases from its present value of $\bar{z}_0$, because $\bar{c}_s < 0$. *The change in the objective value is* $\bar{c}_s$ *units per unit change in the value of* x_s.

This is the reason for requiring that the relative cost coefficient of the entering variable should be negative. If the relative cost coefficient of the entering variable is zero, the objective value remains unchanged; if the relative cost coefficient of the entering variable is positive, the objective value increases.

Since $\bar{c}_s < 0$, the maximum decrease in objective value in this step can be achieved by giving λ the maximum possible value. However, the value of λ should be such that the new values of the present basic variables are all nonnegative. The new value of the ith basic variable is $\bar{b}_i - \bar{a}_{is}\lambda$, and it should be nonnegative, for all $i = 1$ to m. But we know that $\bar{b}_i \geqq 0$, because it is the value of the ith basic variable in the present BFS. Also, $\lambda \geqq 0$. Hence, if $\bar{a}_{is} \leqq 0$, the above restriction is satisfied automatically for any $\lambda \geqq 0$. On the other other hand, if $\bar{a}_{is} > 0$, then $\bar{b}_i - \bar{a}_{is}\lambda \geqq 0$ implies $\lambda \leqq \bar{b}_i/\bar{a}_{is}$. Thus the maximum value that we can give to λ is

$$\begin{aligned} \theta &= \text{minimum} \left\{ \frac{\bar{b}_i}{\bar{a}_{is}} : i \text{ such that } \bar{a}_{is} > 0 \right\} \\ &\qquad \text{if at least one } \bar{a}_{is} > 0 \\ &= \infty \qquad \text{if } \bar{a}_{is} \leqq 0 \qquad \text{for all } i = 1 \text{ to } m \end{aligned} \tag{2.10}$$

The second alternative, $\theta = \infty$, happens only when $\bar{a}_{is} \leqq 0$ for all i (see the unboundedness criterion, Section 2.5.3). If the unboundedness criterion is not satisfied, then the maximum value that x_s can have is finite, and it is the θ obtained from (2.10), known as the *minimum ratio* (or specifically the *primal simplex minimum ratio*) in this step. The operation of computing θ is known as the *minimum ratio test* (or specifically the *primal simplex minimum ratio test*). In this case, let $i = r$ be an index that ties for the minimum in (2.10), that is, $\theta = \bar{b}_r/\bar{a}_{rs}$. When x_s is given the value θ, x_r, the present rth basic variable becomes equal to zero. Replace the present rth basic variable by x_s. The rth row in the present canonical tableau is called the *pivot row*. The element in the pivot row and the pivot column, that is, $\bar{a}_{rs}$, is known as the *pivot element*. The present rth basic variable is the *leaving variable,* or the *dropping variable.* So if r ties for the minimum in (2.10), the rth basic variable is called a variable *eligible to drop* (or *eligible to leave*) from the basic vector, or a *blocking variable.* A rule that helps to determine the dropping variable among those in the present basic vector that tie for the minimum in (2.10) is known as a *dropping variable choice rule,* or a *pivot row choice rule* in the primal simplex algorithm. See Section 2.5.8 for a discussion on dropping variable choice rules.

All the basic vectors for this LP contain the same number of variables. Giving λ the maximum value that it can take (subject to nonnegativity restrictions on the variables) is also necessary to drive one of the present basic variables to zero and, thus, retain the same number of variables in the new basic vector.

The purpose of the minimum ratio test is to guarantee that the new basic solution obtained when x_s is brought into the basic vector satisfies the nonnegativity restrictions on the variables.

2.5.5 Pivot Operation

Bringing the nonbasic variable x_s into the basic vector as the rth basic variable produces a new canonical tableau with respect to the new basic vector by transforming the present column vector of x_s into the rth column vector of the unit matrix, using a *pivot operation* with the column vector of x_s in the present canonical tableau as the *pivot column* and the present rth row as the *pivot row* and the entry contained in both of them, that is, $\bar{a}_{rs}$ as the *pivot element.* This leads to a new canonical tableau in which x_s is the rth basic variable.

2.5.6 Change in Cost Due to Pivoting

The objective value corresponding to the new BFS is:

$$\begin{matrix}\text{The new} \\ \text{objective value}\end{matrix} = \begin{matrix}\text{the previous} \\ \text{objective value}\end{matrix} + \theta\bar{c}_s \quad (2.11)$$

This follows from the objective row equation from the new tableau. Since $\theta \geqq 0$ and $\bar{c}_s < 0$, the new basic solution will be better than the present one.

2.5.7 Discussion of the Unboundedness Criterion and the Entering Variable Choice Rules

Suppose the optimality criterion is not satisfied in a canonical tableau for an LP. To check whether the unboundedness criterion is satisfied in the present tableau, one has to look at the updated column of each eligible variable and examine whether it contains a positive entry or not. This often requires examining most of the data in the present canonical tableau, which is computationally burdensome, particularly when solving a large problem on the computer. Hence, in practice, when the optimality criterion is not satisfied, one does not waste time checking whether the unboundedness criterion is satisfied. Instead, an entering variable is selected from the current set of eligible variables, using some entering variable choice rule. If x_s is the entering variable, the minimum ratio corresponding to this pivot column choice, which we conveniently denote by θ_s, is defined as in (2.10). If $\theta_s = +\infty$, the pivot column has no positive entries and hence the objective function is unbounded below in this LP. The extreme half-line along which the objective function diverges to $-\infty$ can be constructed as in Section 2.5.3 using the information from the pivot column and the updated right-hand side constants vector. Then we terminate. If θ_s is finite, we perform the pivot, obtain the next canonical tableau, and continue.

When this procedure is used, it is possible that the unboundedness criterion is satisfied in the present canonical tableau, and we may not notice it until a pivot for which the entering variable chosen happens to correspond to a column in which the unboundedness criterion is actually satisfied. The algorithm will continue pivoting until the pivot column has no positive entries.

Example 2.6

Consider the following LP.

Basic Variables	x_1	x_2	x_3	x_4	x_5	x_6	$-z$	$\bar{b}$	*Ratios*
x_1	1	0	0	−1	1	−1	0	2	
x_2	0	1	0	(1)	−1	−1	0	1	1/1 Minimum
x_3	0	0	1	1	−2	−1	0	5	5/1
$-z$	0	0	0	−1	0	−1	1	3	

$x_j \geqq 0$ for all j; z to be minimized.

The unboundedness criterion is satisfied in this tableau in the column of x_6. However, suppose x_4 has been chosen as the entering variable. The ratios are computed, the minimum ratio is 1, and the pivot row is row 2. Performing the pivot leads to the next canonical tableau.

Basic Variables	x_1	x_2	x_3	x_4	x_5	x_6	$-z$	$\bar{b}$
x_1	1	1	0	0	0	-2	0	3
x_4	0	1	0	1	-1	-1	0	1
x_3	0	-1	1	0	-1	0	0	4
$-z$	0	1	0	0	-1	-2	1	4

In this tableau, x_5, x_6 are the eligible variables. Clearly, the unboundedness of the objective function in this LP will be recognized in this step, regardless of which of the eligible variables x_5 or x_6 is chosen as the entering variable.

Now, we discuss various entering variable choice rules.

Steepest Descent Rule

In this rule the entering variable is chosen in such a way that the rate of decrease in the objective value, per unit change in the value of the entering variable from its present value of zero, is the highest among all eligible variables. By the results in Section 2.5.4, this is achieved by choosing x_s as the entering variable, where s is determined to satisfy (2.9). Hence this rule is also known as *the most negative relative cost coefficient choice rule*. It is also known as *Dantzig's pivot column choice rule*, after G. B. Dantzig, who suggested it. To implement this rule, all the entries in the updated cost row must be examined and the most negative among them located. It seems to perform very well on an average, in practice.

Example 2.7

Consider the following canonical tableau for an LP.

Basic Variables	x_1	x_2	x_3	x_4	x_5	x_6	$-z$	$\bar{b}$	Ratios If x_4 Enters	Ratios If x_6 Enters
x_1	1	0	0	10	-1	-1	0	2	2/10 min	
x_2	0	1	0	-2	1	0	0	2		
x_3	0	0	1	20	6	1	0	30	30/20	30/1 min
$-z$	0	0	1	-15	-5	-1	1	-100		

$x_j \geqq 0$ for all j; z to be minimized

The entering variable chosen according to this rule is x_4, with a relative cost coefficient of $\bar{c}_4 = -15$. When x_4 enters, the minimum ratio is $\theta_4 = 2/10$, and the actual change in the objective value will be $\bar{c}_4\theta_4 = (-15)(2/10) = -3$. On the other hand, if the entering variable is x_6 with $\bar{c}_6 = -1$, the objective value undergoes a change of -30.

Leftmost of the Eligible Variables Rule

In this rule the entering variable is chosen as x_s, where $s = \text{minimum } \{j : j \text{ such that } x_j \text{ is an eligible variable in this stage}\}$.

Greatest Decrease in the Objective Value Rule

In this rule, the actual change in the objective value that will occur if x_j is chosen as the entering variable is computed for each eligible variable x_j in this stage. This is, of course, $\bar{c}_j\theta_j$, where $\bar{c}_j$ is the present relative cost coefficient of x_j, and θ_j is the minimum ratio if x_j is chosen as the entering variable. Then the entering variable is chosen as that eligible variable that corresponds to the greatest decrease in objective value. Hence the entering variable is x_s, where s is such that $\bar{c}_s\theta_s = \text{minimum } \{\bar{c}_j\theta_j\text{: } j, \text{ such that } x_j \text{ is an eligible variable in this stage}\}$. For example, in the canonical tableau for the LP discussed in Example 2.7, x_6 will be chosen as the entering variable. Mathematically, this rule is very appealing. But to make the entering variable choice, the minimum ratio corresponding to each eligible variable must be computed. This is very burdensome, since it requires examining almost all the data in the canonical tableau. One feels that if this rule is used, the number of pivot steps required for solving the problem is minimized. Unfortunately, there is no theoretical guarantee that this is true (see Exercise 2.38). Because of the computational burden involved in making the entering variable choice, this rule is rarely used by practitioners.

Bland's Rule

This entering variable choice rule is a generalization of the "leftmost of the eligible variables rule" discussed above, suggested by R. G. Bland. This rule requires that the variables be arranged in some specific order before the algorithm is initiated. This order can be arbitrary, but once it is selected, it is fixed during the entire algorithm. In each pivot step, this rule chooses the entering variable to be that eligible variable that is the first among all eligible variables in this step, in the specific order selected for the variables. So, if the specific order chosen for the variables is the natural order $x_1, x_2, \ldots, x_n$ in increasing order of the subscripts, this rule reduces to the "leftmost of the eligible variables rule" discussed above. As an example, consider the LP presented in Example 2.7. The eligible variables in that pivot step are x_4, x_5, x_6. If the specific order chosen for the variables is $x_1, x_6, x_2, x_5, x_3, x_4$, this rule would select x_6 as the entering variable in this pivot step in that LP. In empirical tests, Bland's rule turned out to be very inefficient in comparison with other rules. See reference [10.1].

LRC Rule

The LRC in this name is an abbreviation for *Least Recently Considered eligible variable.* This rule also requires that the variables be arranged in some specific order before the algorithm is initiated. This order can be arbitrary, but once it is selected, it is fixed during the entire algorithm. Suppose this order is $(x_{t_1}, \ldots, x_{t_n})$. In the very first pivot step of the algorithm, the entering variable can be chosen arbitrarily. In all subsequent pivot steps use the following: To select the entering variable in a pivot step, identify the entering variable chosen in the previous step. Suppose it is x_{t_r}, then the entering variable is the first eligible variable when you examine the variables in the order $x_{t_{r+1}}, x_{t_{r+2}}, \ldots, x_{t_n}, x_{t_1}, \ldots, x_{t_r}$. So, in this rule the variables are examined in a cyclical order, beginning with the entering variable in the previous step, in the specific order chosen for the variables, until the first eligible variable turns up, which is then chosen as the entering variable.

In implementing this rule the initial specific order for the variables is usually taken to be the natural order $x_1, x_2, \ldots, x_n$. This rule has some interesting properties that make it very attractive (see Chapter 14 and W. H. Cunnigham's paper [13.5]). This seems to be the rule frequently used in many commercial implementations of the simplex algorithm.

Programmers who write computer packages usually test different entering variable choice rules on several randomly generated problems, and choose the one that seems to give the best empirical performance.

2.5.8 Discussion of the Dropping Variable Choice Rules

If there is no tie in the minimum ratio test, the dropping variable (and hence the pivot row) is identified uniquely and unambiguously by the minimum ratio test. Thus, we need a dropping variable choice rule only in those pivot steps in which there are ties in the minimum ratio test. The following are some of the rules that can be used for this choice.

Topmost of the Eligible Dropping Variables Rule

In this rule, the dropping variable is chosen among those eligible, so that the pivot row is the topmost in the tableau, among the possible.

Lexico–Minimum Ratio Test

This rule selects the dropping variable in each pivot step uniquely and unambiguously. It has to be used in every pivot step of the algorithm from the beginning. If the usual minimum ratio rule discussed in this chapter identifies the dropping variable uniquely, that will be the dropping variable under this rule too. Whenever there are ties in the usual minimum ratio test, this rule selects the dropping variable among those tied by carrying out additional minimum ratio steps using the columns from the inverse of the present basis in place of the updated right-hand-side column. See Chapter 10 for a complete discussion of this rule. If this rule is used, the simplex method is guaranteed to terminate in a finite number of steps (it *resolves the problem of cycling under degeneracy*).

Bland's Dropping Variable Choice Rule

As in Bland's entering variable rule, here also you are required to select a specific ordering of the variables before initiating the algorithm. Once this order is selected, it is fixed during the entire algorithm. In each pivot step, this rule selects the dropping variable to be that blocking variable that is the first among all the blocking variables in the specific order chosen for the variables. It can be proved that the simplex algorithm executed using Bland's entering and dropping variable choice rules terminates after a finite number of pivot steps (see Chapter 10).

However, in empirical tests, these rules performed very poorly. See references [10.1].

See references [2.2–2.6] for studies comparing the various entering and dropping variable choice rules.

2.5.9 Nondegenerate or Degenerate Pivot Steps, Numerical Examples

Example 2.8 Nondegenerate Pivot Step

Consider canonical tableau, Tableau 2.10 for an LP. Suppose x_5 has been selected as the entering variable. Give x_5 a value of λ, keep the other nonbasic variables x_1, x_3, x_6 equal to zero, and reevaluate the values of the basic variables. This leads to the following feasible solution, which we will denote by $x(\lambda)$: $(x_4, x_7, x_8, x_2, x_5, x_1, x_3, x_6) - (0, 0, 5, 8, 0, 0, 0, 0) + \lambda(0, 1, -1, -2, 1, 0, 0, 0)$, $z = 20 - 2\lambda$. As λ increases from zero, the new values of the present basic variables remain nonnegative until λ becomes equal to $8/2 = 4$. When $\lambda = 4$, the value of x_2 becomes equal to zero and if $\lambda > 4$, x_2 will become negative in this solution. Hence the maximum value that λ can have is 4, which is the minimum ratio. Thus, when x_5 enters, x_2 leaves the basic vector. The new canonical tableau is obtained by performing a pivot with the updated column vector of x_5 as the pivot column and the fourth row as the pivot row, it is given in Tableau 2.11.

The new BFS can be verified to be the same one obtained by substituting λ = minimum ratio = 4 in the solution $x(\lambda)$. Also verify that the new objective value is $12 = 20 + 4(-2)$, as discussed in equation (2.11). The application of the algorithm should be continued in the same manner until one of the termination criteria are satisfied.

Note 2.7: In normal computation, the analysis of giving the entering variable a value of λ, writing down the new solution $x(\lambda)$, etc., is not carried out. The moment an entering variable is selected, its updated column vector in the present canonical tableau is identified as the pivot column. If the pivot column has no positive entries, termination occurs with the unboundedness criterion being satisfied. If there are some positive entries in the pivot column, the ratios are computed on the right-hand side of the tableau (as in Tableau 2.10), the minimum ratio is identified, the row in which the minimum ratio occurs is chosen as the pivot row, and the pivot is performed, which leads to the new canonical tableau.

A pivot step in the simplex algorithm in which the minimum ratio is strictly positive is known as a *nondegenerate pivot step.* In every nondegenerate pivot

Tableau 2.10

Basic Variables	x_1	x_2	x_3	x_4	x_5	x_6	x_7	x_8	$-z$	b	Ratios for x_5 Entering
x_4	2	0	-2	1	0	1	0	0	0	0	
x_7	1	0	-1	0	-1	1	1	0	0	0	
x_8	5	0	1	0	1	0	0	1	0	5	5/1
x_2	8	1	2	0	(2)	4	0	0	0	8	8/2
$-z$	-1	0	3	0	-2	-6	0	0	1	-20	

$x_j \geqq 0$ for all j; z to be minimized.

Tableau 2.11

Basic Variables	x_1	x_2	x_3	x_4	x_5	x_6	x_7	x_8	$-z$	b
x_4	2	0	-2	1	0	1	0	0	0	0
x_7	5	$\frac{1}{2}$	0	0	0	3	1	0	0	4
x_8	1	$-\frac{1}{2}$	0	0	0	-2	0	1	0	1
x_5	4	$\frac{1}{2}$	1	0	1	2	0	0	0	4
$-z$	7	1	5	0	0	-2	0	0	1	-12

step, the simplex algorithm obtains a new feasible solution, with a strict decrease in its objective value.

Example 2.9 Degenerate Pivot Step

Consider Tableau 2.10 for the LP discussed in Example 2.8. Suppose x_1 is selected as the entering variable. The minimum ratio for this step is minimum $\{0/2, 0/1, 5/5, 8/8\} = 0$. Giving the entering variable x_1 the value λ, and leaving x_3, x_5, x_6 equal to zero, leads to the solution $(x_4, x_7, x_8, x_2, x_1, x_3, x_5, x_6) = (0, 0, 5, 8, 0, 0, 0, 0) + \lambda(-1, -1, -5, -8, 1, 0, 0, 0)$, $z = 20 - \lambda$. Hence in this step, x_1 enters the present basic vector replacing either of the zero-valued basic variables x_4 or x_7 from it, and x_1 itself will be a zero-valued basic variable in the new basic vector. Notice that when λ is made equal to the minimum ratio 0, the new solution is exactly the same as the present solution. Hence in this pivot step, no change occurs either in the present solution or the present objective value, but the algorithm obtains a new basic vector corresponding to the same solution.

A pivot step of the simplex algorithm in which the minimum ratio is zero is known as a *degenerate pivot step*. In every degenerate pivot step, no change occurs in the objective value or the feasible solution, but a new basic vector is obtained.

2.5.10 Reduced Gradient Method

In Section 2.5.4 it was pointed out that the vector of relative cost coefficients of the nonbasic variables in a canonical tableau is the *reduced gradient* of the objective function in terms of the nonbasic (independent) variables only. Since the simplex algorithm uses it, the simplex algorithm is sometimes referred to as a reduced gradient method. Actually, the term *reduced gradient method* usually refers to a nonlinear programming method that minimizes a nonlinear objective function subject to linear equality constraints and sign restrictions on the variables, which is based on an extension of the ideas used in the simplex algorithm. A further extension of these ideas has led to the *generalized reduced gradient method* (GRG), which can be used to minimize a nonlinear objective function subject to general (possibly nonlinear) equality constraints and bounds on the variables. See books on nonlinear programming (for example, [2.7, 2.8, 2.10, 2.12–2.15, 2.18, 2.20]) for detailed discussions of these nonlinear programming methods.

2.5.11 Salient Features of the Primal Simplex Algorithm

For solving an LP, the (primal) simplex algorithm always needs an initial feasible solution of a special type known as a basic feasible solution (BFS). It goes through a sequence of pivot steps, each of which may be either nondegenerate or degenerate. In every nondegenerate pivot step, it moves from one BFS to another distinct BFS, leading to a strict improvement in objective value (i.e., a strict decrease in objective value in minimization problems). In every degenerate pivot step, there is no change in the objective value, or even the BFS, but the basic vector changes. In each pivot step (whether degenerate or not), a basic variable in the present basic vector is replaced by a nonbasic variable. In each pivot step, the (primal simplex) minimum ratio test used to determine the dropping variable guarantees that the solution obtained at the end will be a feasible solution (i.e., satisfies all the constraints and sign restrictions on the variables) for the problem. The criteria used to choose the entering nonbasic variable guarantee that the objective value either undergoes a strict improvement or stays the same in each pivot step.

The main property of the simplex algorithm of moving from one feasible basic vector to another through a sequence of feasible basic vectors, each differing from the preceding in exactly one variable, is similar to the requirements in the following word game: Change the word "winter" into "summer" via a series of English words, each only one letter different from the preceding word (one solution is winter, sinter, sinner, sunner, sumner, summer).

2.6 OUTLINE OF THE SIMPLEX METHOD FOR A GENERAL LINEAR PROGRAM

An LP in standard form can be solved by the simplex algorithm directly if the original tableau for it is in canonical form (see Section 2.3.2). Otherwise we must first apply the simplex algorithm to solve a Phase I problem. If the minimum Phase I objective value is zero, the original LP is solved in Phase II, beginning with the feasible basic vector for it obtained at the end of Phase I. The combined Phase I, II approach to solve an LP (requiring the use of the simplex algorithm twice) is known as the *simplex method*. The simplex algorithm (method) discussed in this chapter is specifically known as the *primal simplex algorithm* (*method*), to distinguish it from the dual simplex algorithm (method) discussed in Chapter 6.

2.6.1 Case I

If the original tableau is in canonical form, an initial feasible basic vector associated with the unit matrix is chosen and the simplex algorithm applied directly from there. Termination may occur in two different ways.

1. If the final tableau satisfies the optimality criterion, the final BFS is an optimal feasible solution.
2. If the final tableau satisfies the unboundedness criterion, the objective function $z(x)$ is unbounded below, and there is no finite optimal solution (see Section 2.5.3).

2.6.2 Case II

If the original tableau is not in canonical form, the simplex method involves two phases. If the original problem in standard form is the one in Tableau 2.1, the original Phase I tableau with a full aritificial basis is Tableau 2.6, where $x_{n+1}, \ldots, x_{n+m}$ are the artificial variables and w, the Phase I objective function, is their sum. The choice of $x_{n+1}, \ldots, x_{n+m}$ as the initial basic vector for the Phase I problem, and the pricing out of the basic column vectors in both the Phase I and II cost rows leads to the initial canonical tableau.

Suppose in some general stage of the Phase I problem we find the canonical tableau to be as arranged in Tableau 2.12.

The $\bar{d}_j$ are the *Phase I relative cost coefficients* of the original problem variables. Phase I terminates in this stage if $\bar{d}_j \geqq 0$ for $j = 1$ to n. (See Section 2.6.3. for an explanation as to why the Phase I relative cost coefficients of the artificial variables are not considered in this criterion.)

Infeasibility Criterion

At Phase I termination, if the minimum value of $w = \bar{w}$ is positive, the original problem has no feasible solution. See Section 2.4.

Switching Over to Phase II

If $\bar{w} = 0$ at Phase I termination, a BFS of the original problem is obtained from the final tableau of the Phase I problem by suppressing the artificial variables. The algorithm should be continued in such a way that all subsequent solutions obtained are feasible to the original problem. This requires that the value of w must be kept equal to zero in all the subsequent steps. Thus any nonbasic artificial variable at this stage will never again be considered as a candidate to enter the basic vector. All such variables and their column vectors are deleted from the tableau. In normal computation, artificial variables are deleted from the tableau the moment they leave the basic vector during Phase I (see Section 2.6.3). From the $m + 2$th row of the terminal Phase I tableau we get

$$w = \bar{w} + \sum_{j=1}^{n} \bar{d}_j x_j + \sum_{j=1}^{m} (-\sigma_j) x_{n+j} \tag{2.12}$$

where $\bar{w}$ is 0 according to the assumption made here. Since we want to keep w equal to 0 in all subsequent iterations, any artificial variable x_{n+i} that is still in the tableau must remain equal to zero in all subsequent steps. Basic artificial varibles can be kept in the basic vector until they are replaced by some original problem variables during Phase II iterations, but they must be equal to zero. From (2.12) it is also clear that if any original problem variable x_j is such that $\bar{d}_j > 0$, that variable must be equal to zero in all subsequent steps, if w is to be kept equal to zero. Hence such variables will never be considered eligible to enter the basic vector during Phase II. Any variable x_j such that

Tableau 2.12

Basic Variables	$x_1 \ldots x_n$	$x_{n+1} \ldots x_{n+m}$	$-z$	$-w$	b
	$\bar{a}_{11} \ldots \bar{a}_{1n}$	$\beta_{11} \ldots \beta_{1m}$	0	0	$\bar{b}_1$
	$\vdots \quad \vdots$	$\vdots \quad \vdots$	$\vdots$	$\vdots$	$\vdots$
	$\bar{a}_{m1} \ldots \bar{a}_{mn}$	$\beta_{m1} \ldots \beta_{mm}$	0	0	$\bar{b}_m$
	$\bar{c}_1 \ldots \bar{c}_n$	$-\pi_1 \ldots -\pi_m$	1	0	$-\bar{z}$
	$\bar{d}_1 \ldots \bar{d}_n$	$-\sigma_1 \ldots -\sigma_m$	0	1	$-\bar{w}$

Basic Variables	x_1	x_2	x_3	x_4	x_5	x_7	$-z$	$-w$	b
x_3	3	-1	1	1	0	0	0	0	6
x_7	0	0	0	0	0	1	0	0	0
x_5	2	1	0	2	1	0	0	0	5
$-z$	-2	-4	0	-3	0	0	1	0	-10
$-w$	0	0	0	0	0	0	0	1	0

$\bar{d}_j > 0$ in the final Phase I tableau is permanently set equal to zero and its column vector is deleted from the tableau. Thus, in all subsequent iterations, all the variables that are candidates for entering into the basic vector will be original problem variables x_j for which $\bar{d}_j = 0$. From (2.12) this automatically guarantees that the value of w remains equal to zero and hence that any artificial variable that is still in the basic vector will have a value of zero. The Phase I objective is not needed any more and it is deleted from the tableau. Thus, switching over to Phase II requires the following steps:

1. All artificial variables and their column vectors are deleted from the tableau when they leave the basic vector during Phase I (see Section 2.6.3).
2. All original problem variables x_j for which the final Phase I relative cost coefficient $\bar{d}_j$ is positive are set equal to zero and their column vectors are deleted from the tableau.
3. The Phase I objective row is deleted from the tableau.
4. Any artificial variable that is still in the basic vector can be left in it until it is replaced by some original problem variable during Phase II. However the values of all the artificial variables will be zero in all solutions obtained during Phase II.
5. During Phase II, the Phase II objective row is used for checking optimality, unboundedness, etc.

Redundant Equations

The original system of constraints is $\sum_{j=1}^{n} a_{ij}x_j = b_j$, where $i = 1$ to m. If one of these equations can be obtained as a linear combination of the others, that equation is redundant and it can be deleted from the system. For example, consider this system:

$$\begin{aligned} x_1 + x_2 + x_3 + x_4 + x_5 &= 5 \\ x_1 + x_2 + 2x_3 + 2x_4 + 2x_5 &= 8 \\ x_1 + x_2 &= 2 \\ x_3 + x_4 + x_5 &= 3 \end{aligned}$$

In this system the first equation is the sum of the third and fourth equations, hence it can be eliminated. Again, the second equation is the sum of the third equation and two times the fourth equation, and hence it can be eliminated too. Thus the original system is equivalent to the system of the last two equations only. See Section 3.3.5 for a complete discussion of redundant equations in systems of linear equations.

If the system of equations is consistent, the number of nonredundant equations in the system is the rank of the coefficient matrix $A = (a_{ij})$. If *A has rank m*, there are no redundant equations in the system. The existence of redundant equations can be discovered during Phase I of the simplex method. Assume that all the nonbasic artificial column vectors have been deleted from the tableau. In the remaining final Phase I canonical tableau, if there is a row in which all the entries are zero except a single "1" entry in an artificial column vector, that row corresponds to a redundant equation in the original system. All such rows and the corresponding artificial column vectors can be deleted from the tableau before entering Phase II.

Example 2.10

The tableau given at the top is the canonical tableau obtained at Phase I termination of an LP with three constraints. x_1 to x_5 are original problem variables. Variable x_7 is the only artificial variable left in the tableau at Phase I termination. Notice that the second row in this tableau represents a redundant constraint. Hence delete it from the tableau. Thus initiate phase II with the following canonical tableau.

Basic Variables	x_1	x_2	x_3	x_4	x_5	$-z$	b
x_3	3	-1	1	1	0	0	6
x_5	2	1	0	2	1	0	5
$-z$	-2	-4	0	-3	0	1	-10

Eliminating the Artificial Variables at the End of Phase I

After eliminating the rows corresponding to redundant equations from the terminal Phase I canonical tableau, there may still be some artificial variables that are in the basic vector. Suppose the ith basic variable is an artificial variable. The updated right-hand side constant in the ith row $\bar{b}_i$ must be zero, since all the artificial variables are zero in the present BFS. Since this is not a redundant row, there must be some j such that $\bar{a}_{ij} \neq 0$. Pick any j such that x_j is an original problem variable, and for which $\bar{a}_{ij} \neq 0$. The artificial variable can be replaced from the basic vector by x_j. To do this, a pivot operation is performed with the jth column vector as the pivot column, and the ith row as the pivot row. After the pivot operation the artificial column vector can be deleted from the tableau. Each artificial variable in the basic vector can be deleted in this manner if desired. However, it is not necessary to remove artificial variables from the basic vector before going over to Phase II. They can be left in the basic vector until they are replaced from it during Phase II.

Improving the Phase I Objective Function

If the Phase I termination criterion is not satisfied in the present canonical tableau, the Phase I part of the simplex method is continued. An original problem variable, whose Phase I relative cost coefficient $\bar{d}_j$ is negative, is selected as the entering variable. If several $\bar{d}_j$ are negative, a convenient pivot column choice rule is to select the variable x_s as the entering variable where s is such that

$$\bar{d}_s = \text{minimum } \{\bar{d}_j : j = 1 \text{ to } n\}$$

After the pivot, repeat the same process. During Phase I, entering variables are selected using the Phase I relative cost coefficients in the updated Phase I objective row (this is the last row in the Phase I canonical tableau, if the tableau is arranged as discussed here). The Phase II objective row is not used at all during Phase I, and it can actually be deleted from the tableau and reintroduced into the tableau at the end of Phase I, just before going over to Phase II (in this case, you have to price out the basic column vectors in the Phase II objective row before going over to Phase II).

Note 2.8: Any of the entering variable choice rules similar to those discussed in section 2.5.7, but using the Phase I relative cost coefficients, can be used for entering variable choice during Phase I.

2.6.3 Why Is It All Right to Delete Nonbasic Artificial Variables from the Phase I Canonical Tableau?

Let the LP for which we are trying to find a BFS be the one in Tableau 2.1 in standard form. So $b_i \geqq 0$ for all $i = 1$ to m. The purpose of Phase I is to find a BFS for this LP if one exists, or to conclusively establish that there exists no feasible solution. There are many ways in which a Phase I problem can be constructed to do this. One is to use a Phase I with a full artificial basis by augmenting Tableau 2.1 with artificial variables $x_{n+1}, \ldots, x_{n+m}$ whose columns form the unit matrix of order m, and then minimizing the sum of these m artificial variables. This leads to the Phase I problem given in Tableau 2.6. If Tableau 2.1 contains some columns of the unit matrix of order m, another way of formulating a Phase I problem is to introduce only artificial variables whose column vectors are the remaining columns of the unit matrix of order m, and then minimize the sum of the artificial variables introduced. Whichever Phase I formulation is used, by using the arguments of Section 2.4, we conclude that the minimum Phase I objective value is zero iff the original problem in Tableau 2.1 has a feasible solution.

Now suppose a Phase I pivot step is performed on Tableau 2.6 in which an artificial variable, say x_{n+1}, is replaced by the original problem variable, say x_1. After this pivot step, and deleting x_{n+1}, suppose we are lead to Tableau 2.13.

Let (P) refer to the system consisting of only the columns of $x_1, \ldots, x_n$ and that of b and the first m constraints in Tableau 2.13. This (P) is clearly obtained by performing a pivot step with a_{11} as the pivot element in Tableau 2.1, and so is equivalent to it. Since the column of x_1 in (P) is the first column of the unit matrix, a Phase I problem corre-

Tableau 2.13

Basic Variables	x_1	$x_2 \;\ldots\; x_n$	$x_{n+2} \ldots x_{n+m}$	$-w$	b
x_1	1	$a'_{12} \ldots a'_{1n}$	$0 \;\ldots\; 0$	0	b'_1
x_{n+2}	0	$a'_{22} \ldots a'_{2n}$	$1 \;\ldots\; 0$	0	b'_2
$\vdots$	$\vdots$	$\vdots \quad \vdots$	$\vdots \quad \vdots$	$\vdots$	$\vdots$
x_{n+m}		$a'_{m2} \ldots a'_{mn}$	$0 \;\ldots\; 1$	0	b'_m
$-w$	0	$d'_2 \;\ldots\; d'_n$	$0 \;\ldots\; 0$	1	$-w'$

sponding to (P) can be constructed by introducing $m-1$ artificial variables, call them $x_{n+2}, \ldots, x_{n+m}$, associated with the remaining $m-1$ columns of the unit matrix of order m, and it can be verified that Tableau 2.13 is exactly the canonical tableau of this Phase I problem with respect to the basic vector $(x_1, x_{n+2}, \ldots, x_{n+m})$. So the minimum value of w in Tableau 2.13 is zero iff the original problem in Tableau 2.1 has a feasible solution. We can save the computational effort required to maintain the column of x_{n+1} in canonical tableaux by switching over to the new Phase I problem in Tableau 2.13 at this stage.

Repeating the same argument in subsequent steps, we conclude that during Phase I, the column vector of an artificial variable can be deleted from the canonical tableau whenever that variable drops out of the basic vector.

The operations involved in solving the Phase I problem can be viewed as those for creating a unit matrix of order m on the left-hand side of Tableau 2.1 through elementary row operations, while maintaining the right-hand-side constants vector b nonnegative throughout. If we do not insist on maintaining the b vector nonnegative, it is possible to create a unit matrix of order m on the left-hand side of Tableau 2.1 (if there are no redundant constraints) by performing at most m pivot steps on it. However, a unit matrix of order m on the left-hand side of Tableau 2.1 is not useful in constructing a *feasible solution* (i.e., one in which all the variables are nonnegative) unless the b vector is nonnegative in that same tableau.

***Note 2.9:* On Feasibility** Normally infeasibility results when the resources available for carrying out the activities are not adequate to meet the requirements of the items. An infeasible model can be made feasible by eliminating some constraints (see Chapters 5 and 6 for a discussion on how to identify which constraints should be relaxed to make the problem feasible), or by keeping all the constraints but modifying the right-hand-side constants vector. There are usually many ways in which the right-hand-side constants vector in an infeasible model can be modified to make it feasible. One such modification is discussed in Exercise 2.40. Also see Section 5.4.

2.6.4 Alternate Optimum Solutions

Suppose an optimum solution is obtained when an LP is solved. If some of the nonbasic variables have zero relative cost coefficients in the final optimum canonical tableau, alternate optimum solutions for the problem can be obtained by bringing into the basic vector a nonbasic variable with a zero relative cost coefficient. See Section 3.10.

2.6.5 Numerical Examples

Example 2.11 Phase II Applied Directly

Minimize

$$z = -14\,G_1 - 18\,x_3 - 16\,x_4 - 80\,x_5$$

Subject to

$$\begin{aligned} 4.5\,G_1 + 8.5\,x_3 + 6\,x_4 + 20\,x_5 &\leqq 6000 \\ G_1 + x_3 + 4\,x_4 + 40\,x_5 &\leqq 4000 \\ x_3, \quad x_4, \quad x_5 &\geqq 0 \end{aligned}$$

G_1 unrestricted in sign

Replace the unrestricted variable G_1 by $x_1 - x_2$, where x_1, x_2 are two nonnegative variables. Convert the inequalities into equality constraints by adding the slack variables x_6, x_7.

The original tableau is in canonical form with respect to the basic vector (x_6, x_7). Start Phase II directly. Looking at $\bar{c}_j$, the most negative $\bar{c}_j$ is $\bar{c}_5 = -80$. Therefore bring x_5 into the basic vector. By the minimum ratio test x_7 is the leaving variable. The pivot element is 40, which has been circled. The pivot operation requires the following elementary

Original Tableau

Basic Variables	x_1	x_2	x_3	x_4	x_5	x_6	x_7	$-z$	*Basic Values*	*Ratio*
x_6	4.5	−4.5	8.5	6	20	1	0	0	6000	300
x_7	1	−1	1	4	(40)	0	1	0	4000	100 min
$-z$	−14	14	−18	−16	−80	0	0	1	0	
					↑		↓			

Basic Variables	x_1	x_2	x_3	x_4	x_5	x_6	x_7	$-z$	Basic Values	Ratio
x_6	4	-4	⑧	4	0	1	$-\frac{1}{2}$	0	4000	500 min
x_5	$\frac{1}{40}$	$-\frac{1}{40}$	$\frac{1}{40}$	$\frac{1}{10}$	1	0	$\frac{1}{40}$	1	100	4000
$-z$	-12	12	-16	-8	0	0	2	0	8000	
x_3	$\frac{1}{2}$ (circled)	$-\frac{1}{2}$	1	$\frac{1}{2}$	0	$\frac{1}{8}$	$-\frac{1}{16}$	0	500	1000 min
x_5	$\frac{1}{80}$	$-\frac{1}{80}$	0	$\frac{7}{80}$	1	$-\frac{1}{320}$	$\frac{17}{640}$	1	$\frac{350}{4}$	7000
$-z$	-4	4	0	0	0	2	1	0	16,000	

Optimum Tableau

Basic Variables	x_1	x_2	x_3	x_4	x_5	x_6	x_7	$-z$	Basic Values
x_1	1	-1	2	1	0	$\frac{1}{4}$	$-\frac{1}{8}$	0	1000
x_5	0	0	$-\frac{1}{40}$	$\frac{3}{40}$	1	$-\frac{1}{160}$	$\frac{9}{320}$	0	75
$-z$	0	0	8	4	0	3	$\frac{1}{2}$	1	20,000

row operations in the order listed: Row 1 − (20/40)(Row 2), Row 3 + (80/40)(Row 2), (1/40)(Row 2). This gives the new canonical tableau given above. We continue in the same manner. In each step, the pivot element is circled.

From the final optimum tableau, the optimum solution is $(x_1, x_2, x_3, x_4, x_5, x_6, x_7) = (1000, 0, 0, 75, 0, 0)$ with the minimum objective value $= -20{,}000$. In terms of the original problem variables, the optimum solutions is $G_1 = 1000$, $x_5 = 75$, $x_3 = x_4 = 0$.

Example 2.12 Illustration of Phase I and Phase II

$$\begin{aligned} \text{Maximize} \quad & x_1 + 2x_2 + 3x_3 - x_4 \\ \text{Subject to} \quad & x_1 + 2x_2 + 3x_3 = 15 \\ & -2x_1 - x_2 + 5x_3 = -20 \\ & x_1 + 2x_2 - x_3 + x_4 = 10 \\ & x_1, \quad x_2, \quad x_3, \quad x_4 \geqq 0 \end{aligned}$$

After putting the problem in standard form, we get the original tableau:

x_1	x_2	x_3	x_4	$-z$	b
1	2	3	0	0	15
2	1	5	0	0	20
1	2	1	1	0	10
-1	-2	-3	1	1	0

This tableau is not in canonical form. Artificial variables x_5, x_6, x_7 are introduced and the original Phase I tableau is given below. After pricing out the basic column vectors in it, we get the Phase I initial tableau.

Original Phase 1 Tableau with Full Artificial Basis

x_1	x_2	x_3	x_4	x_5	x_6	x_7	$-z$	$-w$	b
1	2	3	0	1	0	0	0	0	15
2	1	5	0	0	1	0	0	0	20
1	2	1	1	0	0	1	0	0	10
-1	-2	-3	1	0	0	0	1	0	0
0	0	0	0	1	1	1	0	1	0

(x_5, x_6, x_7) constitute an artificial basic vector.

Phase I Initial Tableau

Basic Variables	x_1	x_2	x_3	x_4	x_5	x_6	x_7	$-z$	$-w$	*Basic Values*	*Ratio*
x_5	1	2	3	0	1	0	0	0	0	15	5
x_6	2	1	⑤	0	0	1	0	0	0	20	4 min
x_7	1	2	1	1	0	0	1	0	0	10	10
$-z$	-1	-2	-3	1	0	0	0	1	0	0	
$-w$	-4	-5	-9	-1	0	0	0	0	1	-45	

The most negative $\bar{d}_j$ is $\bar{d}_3$. Thus, x_3 is brought into the basic vector. By the minimum ratio test x_6 is the leaving variable. The new canonical tableau is given below. We continue in the same way. Pivot elements

Basic Variables	x_1	x_2	x_3	x_4	x_5	x_7	$-z$	$-w$	*Basic Values*	*Ratio*
x_5	$-\frac{1}{5}$	$\left(\frac{7}{5}\right)$	0	0	1	0	0	0	3	$\frac{15}{7}$ min
x_3	$\frac{2}{5}$	$\frac{1}{5}$	1	0	0	0	0	0	4	20
x_7	$\frac{3}{5}$	$\frac{9}{5}$	0	1	0	1	0	0	6	$\frac{30}{9}$
$-z$	$\frac{1}{5}$	$-\frac{7}{5}$	0	1	0	0	1	0	12	
$-w$	$-\frac{2}{5}$	$-\frac{16}{5}$	0	-1	0	0	0	1	-9	
x_2	$-\frac{1}{7}$	1	0	0		0	0	0	$\frac{15}{7}$	
x_3	$\frac{3}{7}$	0	1	0		0	0	0	$\frac{25}{7}$	
x_7	$\frac{6}{7}$	0	0	①		1	0	0	$\frac{15}{7}$	$\frac{15}{7}$
$-z$	0	0	0	1		0	1	0	15	
$-w$	$-\frac{30}{35}$	0	0	-1		0	0	1	$-\frac{15}{7}$	
x_2	$-\frac{1}{7}$	1	0	0			0	0	$\frac{15}{7}$	
x_3	$\frac{3}{7}$	0	1	0			0	0	$\frac{25}{7}$	$\frac{25}{3}$
x_4	$\left(\frac{6}{7}\right)$	0	0	1			0	0	$\frac{15}{7}$	$\frac{15}{6}$ min
$-z$	$-\frac{6}{7}$	0	0	0			1	0	$\frac{90}{7}$	
$-w$	0	0	0	0			0	1	0	

are circled and nonbasic artificial variables are deleted.

Therefore at this point Phase I terminates and the last artificial variable has left the basic vector. Proceed with Phase II. The entering variable is x_1.

Basic Variables	x_1	x_2	x_3	x_4	$-z$	Basic Values
x_2	0	1	0	$\frac{1}{6}$	0	$\frac{5}{2}$
x_3	0	0	1	$-\frac{1}{2}$	0	$\frac{35}{14}$
x_1	1	0	0	$\frac{7}{6}$	0	$\frac{15}{6}$
$-z$	0	0	0	1	1	15

This is an optimum tableau because all c_j are now nonnegative. The optimum feasible solution is $(x_1, x_2, x_3, x_4) = (15/6, 5/2, 35/14, 0)$. The optimum objective value is -15. The optimum value of the original objective function is $-(\min z) = -(-15) = +15$.

Example 2.13 An Infeasible LP

$$\begin{aligned} \text{Minimize} \quad & x_1 + x_2 \\ \text{Subject to} \quad & x_1 + x_2 \leqq -1 \\ & x_1 - x_2 \geqq 0 \\ & x_1 \geqq 0 \qquad x_2 \geqq 0 \end{aligned}$$

Introducing slack variables x_3, x_4 and artificial variables x_5, x_6, the original Phase I tableau is:

x_1	x_2	x_3	x_4	x_5	x_6	$-z$	$-w$	b
-1	-1	-1	0	1	0	0	0	1
1	-1	0	-1	0	1	0	0	0
1	1	0	0	0	0	1	0	0
0	0	0	0	1	1	0	1	1

Since $\bar{d} \geqq 0$, for all $j = 1$ to 4, Phase I terminates with this tableau and, since the value of $w = -(-1) = 1$, the original problem is infeasible. The constraints and sign restrictions of the original problem when represented on the x_1, x_2 Cartesian plane are as in Figure 2.3. The line $x_1 + x_2 = -1$ is indicated on the plane. Any point satisfying the constraint $x_1 + x_2 \leqq -1$ lies on the side of the arrow drawn on the line

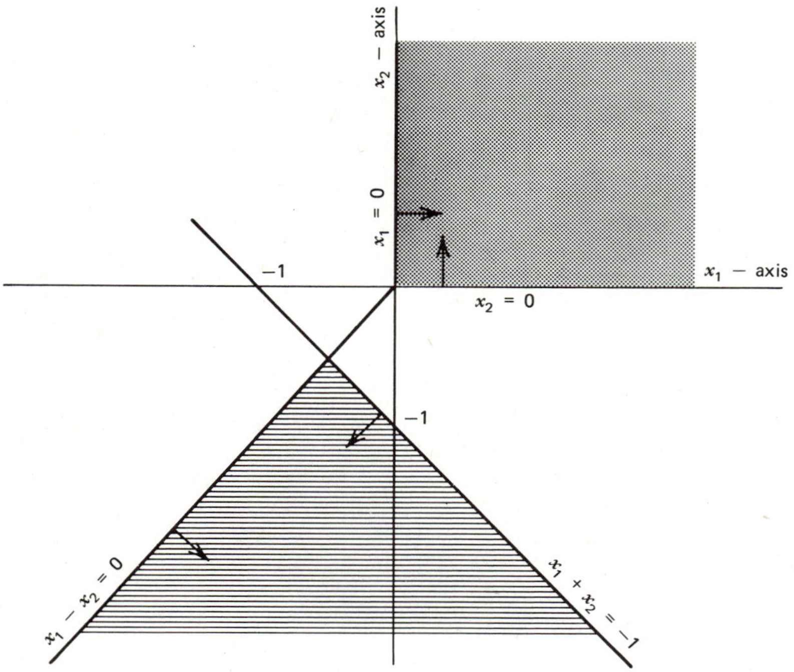

Figure 2.3

Initial Phase I Canonical Tableau

Basic Variables	x_1	x_2	x_3	x_4	x_5	x_6	$-z$	$-w$	*Basic Values*
x_5	−1	−1	−1	0	1	0	0	0	1
x_6	1	−1	0	−1	0	1	0	0	0
$-z$	1	1	0	0	0	0	1	0	0
$-w$	0	2	1	1	0	0	0	1	−1

$x_1 + x_2 = -1$. In a similar manner all the constraints and sign restrictions are indicated by the corresponding lines and arrows pointing from them. Clearly, if there is a feasible solution to the problem, it must lie in the intersection of the dotted and shaded regions of the plane. Obviously no such point exists, since the regions have an empty intersection.

Example 2.14 Illustration of Unboundedness

$$\begin{aligned} \text{Minimize} \quad & -x_1 - x_2 \\ \text{Subject to} \quad & x_1 + x_2 \geqq 1 \\ & x_1 - x_2 \geqq 0 \\ x_1 \geqq 0 \quad & x_2 \geqq 0 \end{aligned}$$

After introducing the slack variables x_3, x_4 and artificial variables, x_5, x_6, we get the original Phase I tableau:

x_1	x_2	x_3	x_4	x_5	x_6	$-z$	$-w$	b
1	1	−1	0	1	0	0	0	1
1	−1	0	−1	0	1	0	0	0
−1	−1	0	0	0	0	1	0	0
0	0	0	0	1	1	0	1	0

Pricing out the artificial column vectors, we get the initial Phase I canonical tableau given at the bottom.

Discussion of Degeneracy

A BFS of an LP in standard form is said to be *degenerate* if the value of at least one of the basic variables is zero in it. In a degenerate BFS, if we perform the standard pivot operation to improve the objective value, the minimum ratio could turn out to be zero, as it happened here. When this happens, the pivot operation just changes the basic vector, but the BFS corresponding to it and the objective value remain unchanged. The same thing might happen in the next pivot step and, after a series of such pivot steps, it might return to the basic vector that started these degenerate pivot steps. In this case, we say that the *simplex algorithm has cycled due to degeneracy* and unless something is done, the algorithm can cycle among this set of degenerate basic vectors indefinitely without ever moving out of the cycle and reaching a terminal basic vector. In a nondegenerate BFS the values of all the basic variables are positive. Hence, the minimum ratio in a pivot step from a nondegenerate BFS is always positive, and such a step in the course of the simplex algorithm is guaranteed to result in a strict decrease in the objective value. Since the objective value never increases in the course of the algorithm, once a strict decrease in it occurs, we will never go back to the previous basic vector in subsequent steps. Thus cycling can never begin with a nondegenerate BFS. There are special techniques for resolving the cycling problem under degeneracy. Among these, the lexico–minimum ratio rule will be discussed later. (See Chapter 10.) However, in most practical applications of linear programming, it has been found that these special techniques for

Basic Variables	x_1	x_2	x_3	x_4	x_5	x_6	$-z$	$-w$	*Basic Values*	*Ratio*
x_5	1	1	−1	0	1	0	0	0	1	1
x_6	(1)	−1	0	−1	0	1	0	0	0	0 min
$-z$	−1	−1	0	0	0	0	1	0	0	
$-w$	−2	0	1	1	0	0	0	1	−1	

Basic Variables	x_1	x_2	x_3	x_4	x_5	$-z$	$-w$	Basic Values	Ratios
x_5	0	②	-1	1	1	0	0	1	$\frac{1}{2}$ min
x_1	1	-1	0	-1	0	0	0	0	
$-z$	0	-2	0	-1	0	1	0	0	
$-w$	0	-2	1	-1	0	0	1	-1	
x_2	0	1	$-\frac{1}{2}$	$\frac{1}{2}$		0	0	$\frac{1}{2}$	
x_1	1	0	$-\frac{1}{2}$	$-\frac{1}{2}$		0	0	$\frac{1}{2}$	
$-z$	0	0	-1	0		1	0	1	
$-w$	0	0	0	0		0	1	0	

resolving degeneracy are unnecessary. When a degenerate pivot has to be made, we continue as usual as if nothing has happened; after a small number of degenerate pivots, the algorithm practically always resolves the degeneracy and continues on its way to a terminal basis. Continuing our numerical example, we obtain the tableau given at the top.

Terminate Phase I and begin Phase II. The column vector of x_3 satisfies the unboundedness criterion in this canonical tableau. Actually the solution

$$(x_2, x_1, x_3, x_4) = (\tfrac{1}{2}, \tfrac{1}{2}, 0, 0) + \lambda(\tfrac{1}{2}, \tfrac{1}{2}, 1, 0)$$
$$z = -1 - \lambda \qquad (2.13)$$

is a feasible solution of the original problem for all $\lambda \geqq 0$. And as λ tends to $+\infty$, the objective value tends to $-\infty$. The set of feasible solutions is the shaded region in Figure 2.4. For each $\lambda \geqq 0$, the point represented by (2.13) is on the thick line and, as λ tends to $+\infty$, it moves on this line in the direction of the arrow. This thick line is called an *extreme half-line* (to be defined in Chapter 3) of the set of feasible solutions, and the objective function $z(x)$ diverges to $-\infty$ along this line.

Comment 2.1: The version of the simplex algorithm discussed here is the version developed originally. It is not computationally efficient. All the computations to be performed in the simplex algorithm can be carried out much more efficiently by using the inverse of the basis in each step. These improved methods require matrix theoretic manipulations. In Chapter 3 the necessary matrix algebra is reviewed; the improved methods of carrying out the simplex algorithm are discussed in Chapters 3, 5, and 7. The study of the original simplex algorithm using canonical tableaux is very useful to introduce the basic concepts of linear programming and the simplex algorithm; for this reason, it has been discussed here in great detail.

Note 2.10: A particular relative cost coefficient may be positive at the beginning of a pivot step, and yet its value may become negative at the end of that pivot step. (See Exercises 2.32 and 2.33.) In general, it is not possible to execute the simplex algorithm to satisfy the property that once a relative cost coeffi-

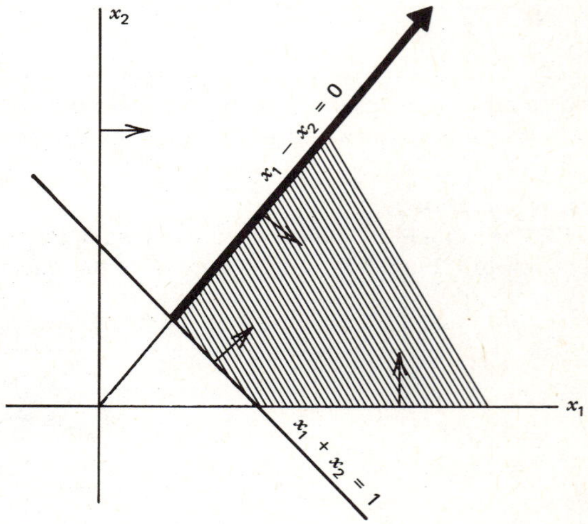

Figure 2.4

cient attains a nonnegative value in some iteration, it stays nonnegative in all subsequent iterations. (See Exercise 2.33.) Also a particular variable may enter the basic vector in some iteration of the simplex algorithm, and yet it may drop out of the basic vector in a subsequent iteration, and the same feature may repeat later on. In general, it is not possible to execute the simplex algorithm to satisfy the property that once a variable is entered into the basic vector, it stays in the basic vector in all subsequent iterations. (See Exercise 2.34.) In the same manner, a variable may drop out of the basic vector in some iteration of the simplex algorithm, and the same variable may be chosen as the entering variable into the basic vector in a subsequent iteration (the only exceptions to this are the artificial variables used in the simplex method for initiating Phase I, see Exercise 2.35).

2.7 THE BIG-*M* METHOD

To solve a general LP, the simplex algorithm is applied twice, once in Phase I, and again in Phase II. There have been several attempts to combine these two phases into a single problem. *The Big-**M** method* is one of them. *In this section, the symbol **M** will be used to denote an arbitrarily large positive number.* Consider the LP in standard form (2.3). Suppose a full artificial basis is introduced. Let the artificial variables be called $t_1, \ldots, t_m$. Consider the augmented problem:

$$\text{Minimize} \qquad Z(x, t) = \sum c_j x_j + \boldsymbol{M}(t_1 + \cdots + t_m)$$

$$\text{Subject to} \quad \sum_{j=1}^{n} a_{ij}x_j + t_i = b_i \qquad i = 1 \text{ to } m \tag{2.14}$$

$$x_j, t_i \geqq 0 \qquad \text{for all } i, j$$

It is not necessary to give a specific value to $\boldsymbol{M}$, but it is treated as a parameter that is very much larger than any number with which it is compared. The problem is now an LP with $(t_1, \ldots, t_m)$ as an initial feasible basic vector. Hence the simplex algorithm can be applied directly to solve it.

During the application of the simplex algorithm, the relative cost coefficient of x_j is of the form $\bar{c}_j + \boldsymbol{M}\bar{c}_j^*$, where $\bar{c}_j$ is the constant term [coming from the original objective function, $z(x)$] and the $\bar{c}_j^*$ is the coefficient of the parameter $\boldsymbol{M}$. If $\bar{c}_j^* < 0$, this relative cost coefficient is negative, whatever $\bar{c}_j$ may be (since $\boldsymbol{M}$ is very large). If $\bar{c}_j^* > 0$, this relative cost coefficient is positive. If $\bar{c}_j^* = 0$, the sign of this relative cost coefficient is the sign of $\bar{c}_j$. To make the computations easier, the $\bar{c}_j$ and the $\bar{c}_j^*$ are kept in two different rows of the tableau. The actual relative cost coefficient of x_j is $\bar{c}_j + \boldsymbol{M}\bar{c}_j^*$. If x_j is the entering variable, it should be priced out in both the $\bar{c}$ row and the $\bar{c}_j^*$ row. When applied to this problem, the simplex algorithm might terminate in several ways.

1 If the simplex algorithm leads to an optimum solution $(\tilde{x}, \tilde{t})$ in which $\tilde{t} = 0$, then $\tilde{x}$ is an optimum feasible solution of the original problem.

2 If the simplex algorithm leads to an optimum solution $(\tilde{x}, \tilde{t})$ in which $\tilde{t} \neq 0$, the optimality criterion being satisfied independently of how large a value the parameter $\boldsymbol{M}$ has, the original problem has no feasible solution.

3 If the unboundedness criterion is satisfied independently of how large a value the parameter $\boldsymbol{M}$ has, then $Z(x, t)$ is unbounded below in (2.14) for all sufficiently large values of $\boldsymbol{M}$. In this case, $z(x)$ is unbounded below in (2.3) if it is feasible. A proof of this is discussed in Section 4.5.7. The Big-**M** method terminates here.

   In this case, the feasibility of (2.3) can only be determined by continuing the algorithm after changing the original objective function, $\sum c_j x_j$ to 0. With this change, the objective function of (2.14) becomes $Z^1(x, t) = \boldsymbol{M}(t_1 + \cdots + t_m)$. So $Z^1(x, t) \geqq 0$ at all feasible solutions (x, t) of (2.14); hence, it is bounded below. Making this change in the objective function is equivalent to making all the entries in the $\bar{c}$ row equal to zero while leaving the $\bar{c}^*$ row as it is in the present tableau. From now on all the relative cost coefficients are treated to be equal to $\boldsymbol{M}\bar{c}_j^*$ for all j and, hence, have the same sign as $\bar{c}_j^*$. The application of the simplex algorithm is continued, starting again from the present basis. Let the optimum solution obtained at termination be $(\hat{x}, \hat{t})$. If $\hat{t} = 0$, (2.3) is feasible and $z(x)$ is unbounded below on it. If $\hat{t} \neq 0$, (2.3) has no feasible solution.

Therefore if (2.3) has an optimum feasible solution, the Big-**M** method will find it after the application of the simplex algorithm once. If (2.3) is infeasible or if $z(x)$ is unbounded below on it, we may have to apply the simplex algorithm twice before completely determining the status of the problem.

It is impossible to conclude theoretically whether the Big-**M** method solves LPs with less computational effort than the standard Phase I, II approach. Practical experience seems to indicate that both approaches take approximately the same amount of work.

Example 2.15

Consider the LP:

x_1	x_2	x_3	x_4	x_5	$-z$	b
1	−1	0	1	−2	0	1
0	1	0	−2	2	0	4
0	−1	1	2	1	0	6
3	−12	5	23	−9	1	0 (z minimize)

$x_j \geqq 0$ for all j.

Since the column vectors of x_1, x_3 are unit vectors, they can be included in the initial basic vector. Only one artificial variable need be introduced. As explained earlier, the objective function for the Big-***M*** problem is entered in two rows in the following tableau. The artificial variable is t.

Original Tableau for the Big-***M*** Problem

	x_1	x_2	x_3	x_4	x_5	t	$-Z$	b
	1	−1	0	1	−2	0	0	1
	0	1	0	−2	2	1	0	4
	0	−1	1	2	1	0	0	6
$\bar{c}$ row	3	−12	5	23	−9	0	1	0
$\bar{c}^*$ row	0	0	0	0	0	1		0

Picking (x_1, t_1, x_3) as the initial basic vector and pricing out in both the cost rows leads to the initial canonical tableau.

Initial Canonical Tableau

Basic Variables	x_1	x_2	x_3	x_4	x_5	t	$-Z$	b
x_1	1	−1	0	1	−2	0	0	1
t	0	(1)	0	−2	2	1	0	4
x_3	0	−1	1	2	1	0	0	6
$\bar{c}$ row	0	−4	0	10	−8	0	1	−33
$\bar{c}^*$ row	0	−1	0	2	−2	0		− 4

Remember that the actual relative cost coefficient is the entry in $\bar{c}$ row plus ***M*** times the entry in $\bar{c}^*$ row. Thus the relative cost coefficient of x_2 is $-4-\boldsymbol{M}$, and it is negative when ***M*** is large. Hence x_2 is a candidate to enter the basic vector. The value of $-Z$ in the present solution is $-33-4\boldsymbol{M}$. Bringing x_2 into the basic vector leads to

Basic Variables	x_1	x_2	x_3	x_4	x_5	t	$-Z$	b
x_1	1	0	0	−1	0	1	0	5
x_2	0	1	0	−2	2	1	0	4
x_3	0	0	1	0	3	1	0	10
$\bar{c}$ row	0	0	0	2	0	4	1	−17
$\bar{c}^*$ row	0	0	0	0	0	1		0

This is a terminal basis. Clearly the solution $(x_1, x_2, x_3, x_4, x_5) = (5, 4, 10, 0, 0)$ is optimal to the original problem with an objective value of $z(x) = 17$.

Comment 2.2: G. B. Dantzig's first published paper "Programming in a Linear Structure" appeared in *Econometrica*, vol. 17, 1949, pp. 73–74. On seeing the title of this paper, T. C. Koopmans is reported to have suggested the name *linear programming* for the subject. According to G. B. Dantzig, the name *simplex method* has been suggested by T. S. Motzkin. See Section 3.13 for a mathematical explanation of this name. The 1975 Nobel Prize in Economics was jointly awarded to L. V. Kantorovich of Russia and T. C. Koopmans of the United States, and the citation for the prize mentions their contributions on the application of linear programming to the economic problem of allocating resources. See "The Nobel Prizes" in the "Science and the Citizen" section of *Scientific American*, vol. 233, no. 6, December 1975, pp. 48–49, and the article by H. E. Scarf, "The 1975 Nobel Prize in Economics: Resource Allocation," *Science*, vol. 190, no. 4215, November 14, 1975, pp. 649, 710, and 712. In 1976, G. B. Dantzig was awarded the National Medal of Science of the United States, by President G. R. Ford at a White House ceremony, ". . . for inventing linear programming and discovering methods that led to wide-scale scientific and technical applications to important problems in logistics, scheduling, and network optimization, and to the use of computers in making efficient use of mathematical theory." See "George Dantzig Awarded National Medal of Science," *OR/MS Today*, vol. 4, no. 1, January 1977, last page. Even though other methods such as the ellipsoid method have been developed for solving linear programs, the simplex method (using efficient implementations of it discussed in Chapters 5 and 7) continues to be the most efficient method for solving practical linear programming models. Not only did George B. Dantzig develop the simplex method, his leadership role over the last 35 years is primarily responsible for making linear programming, and the simplex method for solving it, the vital planning tool it is today.

Exercises

2.2 Consider the following LP:

x_1	x_2	x_3	x_4	x_5	x_6	x_7	$-z$	b
0	1	0	α	1	0	3	0	β
0	0	1	-2	2	Δ	-1	0	2
1	0	0	0	-1	2	1	0	3
0	0	0	δ	3	γ	ξ	1	0

$x_j \geq 0$ for all j; z to be minimized.

The entries α, β, γ, δ, Δ, ξ in the tableau are parameters. *Each of the following questions is independent, and they all refer to this original problem.* Clearly state the ranges of values of the various parameters that will make the conclusions in the following questions true. B_1 is the basis for this problem corresponding to the basic vector $x_{B_1} = (x_2, x_3, x_1)$.

1 The present tableau is such that Phase II of the simplex method can be applied using this as an initial tableau.

2 Row 1 in the present tableau indicates that the problem is infeasible.

3 B_1 is a feasible but nonoptimal basis for this problem.

4 B_1 is a feasible basis for the problem, and the present tableau indicates that z is unbounded below.

5 B_1 is a feasible basis, x_6 is a candidate to enter the basic vector, and when x_6 is the entering variable, x_3 leaves the basic vector.

6 B_1 is a feasible basis, x_7 is a candidate to enter the basic vector, but when x_7 is the entering variable, the solution and the objective value remain unchanged.

2.3 Consider the LPs

$$\begin{aligned} \text{Minimize} \quad & z(x) = cx \\ \text{Subject to} \quad & Ax = b \\ & x \geqq 0 \end{aligned}$$

and

$$\begin{aligned} \text{Minimize} \quad & z(x) = (\mu c)x \\ \text{Subject to} \quad & Ax = (\lambda b) \\ & x \geqq 0 \end{aligned}$$

where λ and μ are strictly positive real numbers, and A, b, c are the same matrices or vectors in both the problems. Obtain a relationship between the optimum solutions of the two problems. Explain briefly why this relationship fails when either λ or μ is negative.

2.4 Solve the following LP: If $z(x)$ is unbounded below in this problem, find an extreme half-line along which it diverges to $-\infty$. Find out a feasible solution on this half-line that corresponds to an objective value of -415.

$$\begin{aligned} \text{Minimize} \quad z(x) = \; & x_1 - 2x_2 - 4x_3 + 4x_4 \\ \text{Subject to} \quad & -x_2 + 2x_3 + x_4 \leqq 4 \\ & -2x_1 + x_2 + x_3 - 4x_4 \leqq 5 \\ & x_1 - x_2 + 2x_4 \leqq 3 \\ & x_j \leqq 0 \quad \text{for all } j \end{aligned}$$

2.5 Let Tableau 2.8 be the canonical tableau obtained in the course of solving the LP (2.3). In this problem it is known that every feasible solution satisfies $\sum x_j \leqq \alpha$, where α is a known positive number. Suppose the present canonical tableau does not satisfy the optimality criterion. Let $\bar{c}_s = \text{minimum } \{\bar{c}_1, \ldots, \bar{c}_n\}$. Prove that $\bar{z} + \alpha\bar{c}_s$ is a lower bound for the minimum objective value in this Exercise.

Solve LPs 2.6 to 2.22.

2.6
$$\begin{aligned} \text{Minimize} \quad & 12x_1 - 10x_2 - 30x_3 \\ \text{Subject to} \quad & -3x_1 + 2x_2 + 8x_3 \leqq 17 \\ & -x_1 + x_2 + 3x_3 \leqq 9 \\ & -2x_1 + x_2 + 8x_3 \leqq 16 \\ & x_j \geqq 0 \quad \text{for all } j \end{aligned}$$

2.7
$$\begin{aligned} \text{Minimize} \quad & -x_1 - 4x_2 - 5x_3 \\ \text{Subject to} \quad & x_1 + 2x_2 + 3x_3 \leqq 2 \\ & 3x_1 + x_2 + 2x_3 \leqq 2 \\ & 2x_1 + 3x_2 + x_3 \leqq 4 \\ & x_j \geqq 0 \quad \text{for all } j \end{aligned}$$

Use only one pivot step.

2.8

$$\begin{aligned}
\text{Minimize} \quad & -3x_1 - 11x_2 - 9x_3 + x_4 + 29x_5 \\
\text{Subject to} \quad & x_2 + x_3 + x_4 - 2x_5 \leqq 4 \\
& x_1 - x_2 + x_3 + 2x_4 + x_5 \geqq 0 \\
& x_1 + x_2 + x_3 - 3x_5 \leqq 1 \\
& x_j \geqq 0 \quad \text{for all } j \neq 1 \\
& x_1 \text{ unrestricted}
\end{aligned}$$

2.9

$$\begin{aligned}
\text{Minimize} \quad & -17x_1 + 13x_2 + 2x_3 - 8x_4 \\
\text{Subject to} \quad & 2x_1 - x_2 + x_4 \leqq 4 \\
& -x_1 + x_3 - x_4 \geqq -3 \\
& x_1 + 3x_2 - 2x_3 \leqq 4 \\
& x_j \geqq 0 \quad \text{for all } j
\end{aligned}$$

2.10

$$\begin{aligned}
\text{Minimize} \quad & x_6 \\
\text{Subject to} \quad & 2x_2 + x_3 + x_4 - x_5 = 0 \\
& 3x_1 - 2x_2 - x_3 + 2x_4 \leqq 15 \\
& x_1 + x_3 \leqq 5 \\
& -9x_1 + 8x_2 + x_3 - 2x_4 - x_6 = 0 \\
& x_j \geqq 0 \quad \text{for all } j \neq 6 \\
& x_6 \text{ unrestricted}
\end{aligned}$$

2.11

$$\begin{aligned}
\text{Minimize} \quad & x_1 \\
\text{Subject to} \quad & x_1 + x_2 + x_3 + x_4 + x_5 \geqq 4 \\
& x_1 - 2x_2 - x_3 + 2x_4 \geqq -2 \\
& 5x_1 - 4x_2 - x_3 + 8x_4 + 2x_5 \leqq 1 \\
& x_j \geqq 0 \quad \text{for all } j \neq 1 \\
& x_1 \text{ unrestricted}
\end{aligned}$$

2.12

$$\begin{aligned}
\text{Maximize} \quad & x_5 \\
\text{Subject to} \quad & -x_1 + 5x_2 + x_4 \geqq x_5 \\
& 2x_1 - 2x_3 + x_4 \geqq x_5 \\
& -2x_2 + 2x_3 + x_4 \geqq x_5 \\
& -x_1 - 2x_2 - x_3 - x_4 = -4 \\
& x_j \geqq 0 \quad \text{for all } j
\end{aligned}$$

2.13

$$\begin{aligned}
\text{Minimize} \quad & x_1 + 3x_2 - x_3 + x_4 - x_5 + x_6 \\
\text{Subject to} \quad & x_1 - x_3 + x_4 + 2x_5 - x_6 + x_7 = 6 \\
& x_1 + x_2 + x_5 + x_6 + 4x_7 = 14 \\
& 2x_1 - 2x_2 - 3x_3 + 3x_4 + 3x_5 - 4x_6 + 2x_7 = 22 \\
& x_j \geqq 0 \quad \text{for all } j
\end{aligned}$$

2.14

$$\begin{aligned}
\text{Minimize} \quad & -5x_1 - 4x_3 - 13x_4 + 3x_5 + 11x_6 + x_7 \\
\text{Subject to} \quad & -x_1 + 2x_2 + 5x_3 - x_4 + 5x_5 + 4x_6 + x_7 = 2 \\
& x_1 + x_2 + 4x_3 + 4x_4 + x_5 - x_6 - x_7 = 7 \\
& x_j \geqq 0 \quad \text{for all } j
\end{aligned}$$

2.15

$$\begin{aligned}
\text{Minimize} \quad & -x_2 - 4x_3 + 2x_5 - 2x_7 + 10 \\
\text{Subject to} \quad & 2x_1 - x_2 - x_3 + x_4 - 2x_5 + x_6 - x_7 = 4 \\
& 2x_1 + 2x_2 - 3x_3 + x_4 + 2x_5 + 2x_6 + x_7 = 6 \\
& 6x_1 + 2x_2 - 3x_3 + 4x_4 + x_5 + 2x_6 = 12 \\
& 2x_1 + x_2 + x_3 + 2x_4 + x_5 - x_6 = 2 \\
& x_j \geqq 0 \quad \text{for all } j
\end{aligned}$$

2.16

$$\begin{aligned}
\text{Minimize} \quad & -x_1 + 2x_2 - x_3 - x_4 + 2x_5 - x_6 + 2x_7 + 3x_8 - 5x_9 \\
\text{Subject to} \quad & x_1 + 5x_2 - 3x_3 + 4x_4 - 3x_5 + 2x_8 - 8x_9 = -3 \\
& -5x_2 - x_3 - 5x_4 + 4x_5 - x_6 + 2x_7 + x_9 = -4 \\
& x_1 + 2x_2 + 5x_4 + x_5 + x_6 + x_7 - 2x_8 + 3x_9 = 9 \\
& x_j \geqq 0 \quad \text{for all } j
\end{aligned}$$

2.17

$$\begin{aligned}
\text{Minimize} \quad & -12x_1 - x_2 - 2x_3 - 5x_4 - 4x_5 + 6x_6 \\
\text{Subject to} \quad & 8x_1 + x_2 + 3x_3 + 2x_4 + 3x_5 - 3x_6 = 17 \\
& -5x_1 - x_2 - x_3 - x_4 - 2x_5 + 2x_6 = -12 \\
& -5x_1 - x_3 - 2x_4 - x_5 + 4x_6 \leqq -8 \\
& x_j \geqq 0 \quad \text{for all } j
\end{aligned}$$

2.18

$$\begin{aligned}
\text{Minimize} \quad & 34x_1 + 5x_2 + 19x_3 + 9x_4 \\
\text{Subject to} \quad & -2x_1 - x_2 - x_3 - x_4 \leqq -9 \\
& 4x_1 - 2x_2 + 5x_3 + x_4 \leqq 8 \\
& -4x_1 + x_2 - 3x_3 - x_4 \leqq -5 \\
& x_j \geqq 0 \quad \text{for all } j
\end{aligned}$$

2.19

$$\begin{aligned}
\text{Minimize} \quad & x_1 + x_2 + x_3 + x_4 + x_5 + x_6 + x_7 \\
\text{Subject to} \quad & x_1 + 2x_3 + 3x_4 - 2x_5 - x_6 + x_7 = 1 \\
& 3x_1 + 2x_2 + 2x_3 + 5x_4 - 2x_5 + x_6 + 3x_7 = 5 \\
& -2x_1 + x_2 + x_4 - x_5 + x_6 - 2x_7 = 4 \\
& x_j \geqq 0 \quad \text{for all } j
\end{aligned}$$

2.20

$$\begin{aligned}
\text{Minimize} \quad & 2x_1 \\
\text{Subject to} \quad & x_1 - x_4 = 3 \\
& x_1 - x_2 - 2x_5 = 1 \\
& 2x_1 + x_3 + x_5 = 7 \\
& x_j \geqq 0 \quad \text{for all } j
\end{aligned}$$

2.21

$$\begin{aligned}
\text{Maximize} \quad & x_8 \\
\text{Subject to} \quad & -x_1 - x_3 + x_4 + x_5 = -2 \\
& x_1 - 2x_3 - 3x_4 + x_6 - x_8 = 4 \\
& x_1 - 2x_4 + x_7 = 3 \\
& 2x_1 + x_2 - x_8 = 6 \\
& x_j \geqq 0 \quad \text{for all } j \neq 8 \\
& x_8 \text{ unrestricted}
\end{aligned}$$

Use at most one artificial variable.

2.22 (a) Minimize

$$-3x_1 + 5x_2 + 3x_3 + 9x_4 + 4x_5$$

Subject to

$$\begin{aligned} -x_1 + x_2 + x_3 + 2x_4 + x_5 \phantom{{}+x_6} &= 0 \\ x_3 + 3x_4 + 3x_5 + x_6 &= 5 \\ -2x_1 + 2x_2 + x_3 + x_4 - x_5 - x_6 &= -5 \\ x_j \geqq 0 \quad \text{for all } j \end{aligned}$$

(b) Minimize

$$-x_1$$

Subject to

$$\begin{aligned} x_1 + x_2 - x_3 + x_4 - x_5 + 2x_6 &= 2 \\ 2x_1 - x_2 - x_3 - 2x_4 + x_5 - x_6 &= 3 \\ 3x_1 - 2x_3 - x_4 + x_6 &= 5 \\ x_j \geqq 0 \quad \text{for all } j \end{aligned}$$

2.23 Find the optimum solution or the extreme ray along which z diverges to $-\infty$ in each of the following LPs.

	x_1	x_2	x_3	x_4	x_5	x_6	$-z$	$= b$
	1	0	2	1	2	−1	0	3
(a)	2	1	2	0	1	−1	0	6
	−1	0	2	0	−3	0	1	10
	0	1	0	1	2	1	0	2
(b)	1	−3	−1	1	1	0	0	0
	0	−1	−1	2	2	0	1	−4
	1	0	−1	−2	−7	−9	0	5
(c)	0	1	−2	−1	−8	−10	0	4
	0	0	3	0	6	10	1	6

$x_j \geqq 0$ for all j in each problem.

2.24 Solve the following LP:

Minimize

$$3x_1 + 5x_2 + 7x_3 + 9x_4 + 11x_5$$

Subject to

$$9x_1 + 25x_2 + 28x_3 + 18x_4 + 66x_5 = 105$$

$$x_j \geqq 0 \quad \text{for all } j$$

2.25 Consider the one constraint LP:

$$\begin{aligned} \text{Minimize} \quad & \textstyle\sum c_j x_j \\ \text{Subject to} \quad & \textstyle\sum a_j x_j = b \\ x_j \geqq 0 \quad & \text{for all } j \end{aligned}$$

(a) Develop a simple test for checking the feasibility of this problem.

(b) Develop a simple test for checking unboundedness.

(c) Develop a simple method for obtaining an optimum solution directly.

2.26 Find a feasible solution of each of the following systems:

(a)
$$\begin{aligned} x_1 + x_2 + x_3 + x_4 + x_5 &= 2 \\ -x_1 + 2x_2 + x_3 - 3x_4 + x_5 &= 1 \\ x_1 - 3x_2 - 2x_3 + 2x_4 - 2x_5 &= -4 \\ x_j \geqq 0 \quad \text{for all } j \end{aligned}$$

(b)
$$\begin{aligned} -4x_1 + x_2 + 4x_3 &= 3 \\ 2x_1 + x_2 + 5x_3 &= 2 \\ x_j \geqq 0 \quad \text{for all } j \end{aligned}$$

(c)
$$\begin{aligned} -5x_1 + x_2 - 3x_3 + 3x_4 + 8x_5 &\leqq -3 \\ 3x_1 - x_2 + 2x_3 - x_4 - 5x_5 &\leqq 2 \\ -2x_1 + x_2 - x_3 + 3x_5 &\leqq 2 \\ x_j \geqq 0 \quad \text{for all } j \end{aligned}$$

(d)
$$\begin{aligned} x_1 - 2x_2 + x_3 + 2x_4 - x_5 &= 5 \\ -x_1 - 3x_2 + 5x_3 + 6x_4 - x_5 &= 10 \\ -3x_1 + x_2 + 3x_3 + 2x_4 + x_5 &= 3 \\ x_j \geqq 0 \quad \text{for all } j \end{aligned}$$

2.27 Use a linear programming formulation to show that the following constraints imply $x_1 + 2x_2 \leqq 8$.

$$\begin{aligned} 4x_1 + x_2 &\leqq 4 \\ 2x_1 - 3x_2 &\leqq 6 \\ x_1, \quad x_2 &\geqq 0 \end{aligned}$$

2.28 Show that none of the feasible solutions of the following system satisfies $-6x_1 + 8x_2 + 7x_3 - 9x_4 - 5x_5 \leqq -18$.

$$\begin{aligned} 2x_1 - x_2 - x_3 + 2x_4 + x_5 &\leqq 3 \\ -3x_1 + x_2 + 4x_3 - 5x_4 - 2x_5 &\leqq -4 \\ x_j \geqq 0 \quad \text{for all } j \end{aligned}$$

2.29 Consider the LP

$$\begin{aligned} \text{Minimize} \quad & z(x) = cx \\ \text{Subject to} \quad & Ax = b \\ & x \geqq 0 \end{aligned}$$

where A is a matrix of order $m \times n$ and rank m. It is required to transform this into an LP of the following form where F is a matrix of order $(m+1) \times n$. Show how this can be done.

$$\begin{aligned} &\text{Minimize} \quad z(x) = cx \\ &\text{Subject to} \quad Fx \geqq f \\ &\qquad x \geqq 0 \end{aligned}$$

2.30 Consider the following LP where A, b, c are given $m \times n$, $m \times 1$, $1 \times n$ matrices, respectively, and l_j, k_j are given constants for each $j = 1$ to n; l_j and k_j are finite, and $l_j \leqq k_j$ for all j. Some of the l_j and even k_j may be negative. Show how to transform this problem into an LP in n or less nonnegative variables, in which each variable is restricted to be less than or equal to 1.

$$\begin{aligned} &\text{Minimize} \quad z(x) = cx \\ &\text{Subject to} \quad Ax = b \\ &\text{And} \quad l_j \leqq x_j \leqq k_j \qquad j = 1 \text{ to } n \end{aligned}$$

2.31 Solve the following LPs by the Big-***M*** method.

(a) Minimize

$$x_1$$

Subject to

$$\begin{aligned} x_1 + x_2 + x_3 + 2x_4 - x_5 - x_6 &\geqq 2 \\ -x_1 - 2x_3 + x_4 - x_5 - 2x_6 &\geqq 3 \\ x_2 - 3x_3 + 4x_4 - 3x_5 - 5x_6 &\leqq 7 \\ x_j \geqq 0 \quad \text{for all } j \end{aligned}$$

(b) Minimize

$$-2x_2 + 2x_3 + x_4$$

Subject to

$$\begin{aligned} x_1 + x_3 + x_4 - x_5 + x_6 + 2x_7 &= 6 \\ x_2 + x_4 - x_5 + x_6 &= 5 \\ -x_2 + x_3 - x_4 + x_5 + x_7 &= -3 \\ x_j \geqq 0 \quad \text{for all } j \end{aligned}$$

(c) Minimize

$$x_1 + x_2 + x_3 - 3x_4 + 6x_5 + 4x_6$$

Subject to

$$\begin{aligned} x_1 + x_2 + 3x_4 - x_5 + 2x_6 &= 6 \\ x_2 + x_3 - x_4 + 4x_5 + x_6 &= 3 \\ x_1 + x_3 - 2x_4 + x_5 + 5x_6 &= 5 \\ x_j \geqq 0 \quad \text{for all } j \end{aligned}$$

(d) Minimize

$$3x_1 - 2x_3 + x_4$$

Subject to

$$\begin{aligned} x_1 + 2x_2 - 2x_3 - x_4 + x_5 &\geqq 5 \\ 5x_1 + 6x_2 - 3x_3 - x_4 - x_5 &\leqq 10 \\ 3x_1 + 2x_2 + x_3 + x_4 - 3x_5 &\geqq 3 \\ x_j \geqq 0 \quad \text{for all } j \end{aligned}$$

2.32 Construct a numerical example to illustrate the fact that a particular relative cost coefficient may be positive at the beginning of a pivot step of the simplex algorithm, and yet its value may become negative at the end of that pivot step.

2.33 Consider the application of the simplex algorithm to solve the following LP starting with (x_1, x_2, x_3) as the initial basic vector. Using this tableau, show that, in general, it is not possible to execute the simplex algorithm to satisfy the property that once a relative cost coefficient becomes nonnegative, it stays nonnegative in all subsequent iterations.

x_1	x_2	x_3	x_4	x_5	x_6	x_7	$-z$	b
1	0	0	1	−1	8	−1	0	3
0	1	0	1	−2	−6	1	0	5
0	0	1	1	3	7	−1	0	7
0	0	0	3	0	5	−1	1	0

$x_j \geqq 0$ for all j; z to be minimized.

2.34 Consider the following LP:

x_1	x_2	x_3	x_4	x_5	x_6	$-z$	b
1	0	0	10	−1	2	0	3
0	1	0	10	2	−2	0	5
0	0	1	−10	1	−1	0	7
−3	3	6	−110	0	−3	1	1

$x_j \geqq 0$ for all j; z to be minimized.

Apply the simplex algorithm for solving it, starting with (x_1, x_2, x_3) as the initial basic vector, and the steepest descent entering variable choice rule. Show that x_4 enters the basic vector, but drops off in subsequent iterations, and is not contained in the optimal basic vector for this LP.

2.35 In Exercise 2.34, when x_4 enters the basic vector (x_1, x_2, x_3), x_1 drops out of the basic vector. And yet verify that x_1 comes back into the basic vector in subsequent iterations, and is contained in the optimal basic vector for that LP.

2.36 While solving an LP by the simplex algorithm, suppose that Tableau 2.8 is obtained in some iteration, and assume that in this tableau $\bar{c}_j \geqq 0$ for all $j = m + 1$ to $n - 1$, and $\bar{c}_n < 0$. So x_n is the only possible choice for the entering variable into the basic vector in this iteration. Assume that the pivot step of bringing x_n into the basic vector in this iteration is a nondegenerate pivot step. Under these conditions, prove that x_n can never drop out of the basic vector in all subsequent iterations of the algorithm. Also in this case, prove that every optimum feasible basic vector for this LP must include x_n as a basic variable.

Are these results true even if the pivot step of bringing x_n into the basic vector in Tableau 2.8 is a degenerate pivot step? Why? Illustrate with the following problem:

x_1	x_2	x_3	x_4	$-z$	b
1	0	−1	2	0	0
0	1	0	0	0	1
0	0	1	−1	1	0

$x_j \geqq 0$ for all j; minimize z.

—(*R. Chandrasekharan, private communication*)

2.37 Consider the LP (2.3). If there exist numbers $\pi_1, \ldots, \pi_m$ such that $c_j + \sum_{i=1}^{m} \pi_i a_{ij} \geqq 0$ for all j, prove that the unboundedness criterion cannot be satisfied while solving this LP.

2.38 Consider the following LP:

$$\begin{aligned}
\text{Minimize} \quad & x_2 \\
\text{Subject to} \quad & x_1 + x_2 \leqq 17 \\
& 8x_1 - x_2 \leqq 64 \\
& x_1 + 3x_2 \geqq 8 \\
& 2x_1 + 3x_2 \geqq 13 \\
& 2x_1 + x_2 \geqq 7 \\
& x_1 \geqq 1 \\
& 2x_1 - 7x_2 \geqq -47 \\
& x_1, x_2 \text{ unrestricted in sign}
\end{aligned} \tag{2.15}$$

1 Plot the set of feasible solutions of this LP on the x_1, x_2-Cartesian plane and identify the optimum solution.

2 Introduce the slack variable y_i to make the ith constraint in the LP into an equation for $i = 1$ to 7. Eliminate the unrestricted variables x_1, x_2 from the model using the equations corresponding to the sixth and first constraints, respectively. Show that this transforms the LP into

$$\begin{aligned}
&\text{Minimize} \\
&-y_1 - y_6 \\
&\text{Subject to} \\
&y_1 + y_2 + 9y_6 = 72 \\
&3y_1 + y_3 + 2y_6 = 41 \\
&3y_1 + y_4 + y_6 = 37 \\
&y_1 + y_5 - y_6 = 11 \\
&7y_1 + 9y_6 - y_7 = 63 \\
&y_1 \text{ to } y_7 \geqq 0
\end{aligned} \tag{2.16}$$

3 Obtain the canonical tableau corresponding to the basic vector $(y_2, y_3, y_4, y_5, y_6)$ for (2.16). Show that this is a feasible basic vector. Find the BFS and the solution corresponding to it for the original problem (2.15). Mark this solution for (2.15) on its set of feasible solutions.

4 Solve (2.16) beginning with $(y_2, y_3, y_4, y_5, y_6)$, using the "greatest decrease in the objective value" entering variable choice rule. Mark the path traced on the set of feasible solutions for (2.15).

5 Choose y_7 as the entering variable into the initial basic vector $(y_2, y_3, y_4, y_5, y_6)$, and verify that the simplex algorithm finds the optimum solution of (2.16) with much less computational effort in this case. Interpret these results geometrically.

2.39 *RELATIONSHIP OF DEGENERACY TO TIES IN THE MINIMUM RATIO TEST* Let Tableau 2.8 be the canonical tableau for an LP. Suppose the nonbasic variable x_s is chosen as the entering variable in this tableau. The minimum ratio in this step θ is given by (2.10). Suppose θ is finite and there are several i, say $i_1, \ldots, i_t$, that tie for a minimum in (2.10). Among these suppose row i_1 has been selected as the pivot row for this step. Prove that in the BFS obtained after this pivot step, the basic variables in rows

$i_2, \ldots, i_t$ will all be equal to zero. Illustrate your argument with a numerical example. Using this fact prove that if ever there is a tie for the minimum ratio in a pivot step of the primal simplex algorithm, then the BFS obtained after that pivot step must be degenerate. If it is known that the LP in Tableau 2.8 has no degenerate BFSs, then prove that the minimum ratio test identifies the pivot row uniquely and unambiguously in every pivot step of the primal simplex algorithm when solving this LP.

2.40 *MODIFYING THE RIGHT-HAND-SIDE CONSTANTS IN AN INFEASIBLE LP TO MAKE IT FEASIBLE.* Consider the LP in Tableau 2.1. Suppose the corresponding Phase I problem is solved using Tableau 2.7 as the initial Phase I canonical tableau. Suppose this LP is infeasible. Let x_{n+i_u}, $u = 1$ to t, be all the artificial variables that remained as basic variables at Phase I termination. Let $\bar{x} = (\bar{x}_1, \ldots, \bar{x}_n)$ be the vector of values of the original problem variables in the final Phase I solution and let $\bar{b} = (\bar{b}_1, \ldots, \bar{b}_m)$ be the updated right-hand-side constants vector in the terminal Phase I canonical tableau. Define $b' = (b'_i)$ by

$$\begin{aligned} b'_i &= b_i && \text{if } i \notin \{i_1, \ldots, i_t\} \\ &= b_i - \bar{b}_{i_u} && i = i_u, \quad u = 1 \text{ to } t \end{aligned}$$

Prove that $\bar{x}$ can be treated as an approximate solution of the original LP in Tableau 2.1. What measure of deviation does $\bar{x}$ minimize?

Let the *modified LP* refer to the LP obtained by changing the right-hand-side constants vector in Tableau 2.1 from b to b'. Prove that the modified LP is feasible, and that, in fact, $\bar{x}$ is a feasible solution for it. In the final Phase I canonical tableau, change all the $\bar{b}_{i_1}, \ldots, \bar{b}_{i_t}$, $\bar{w}$ to zero. Show that this changes it into a canonical tableau for the modified LP. Apply this modification procedure to Exercise 2.13, and obtain an optimum solution of the modified LP.

2.41 Consider an LP in which the constraints in standard form are those in Tableau 2.1. Let Tableau 2.6 be the corresponding Phase I problem. In Phase I of the simplex method, the artificial variables are discarded from the tableau whenever they drop out of the basic vector. Let $\bar{w}^1$ denote the value of w at Phase I termination.

Now consider Tableau 2.6 as an LP by itself, in which the decision variables are the nonnegative variables $x_1, \ldots, x_n$, $x_{n+1}, \ldots, x_{n+m}$. Make no distinction between the problem variables $x_1, \ldots, x_n$ and the artificial variables $x_{n+1}, \ldots, x_{n+m}$. Solve it by applying the simplex algorithm, beginning with the feasible basic vector $(x_{n+1}, \ldots, x_{n+m})$ for it. Let $\bar{w}^2$ denote the value of w at termination.

Prove that $\bar{w}^1 > 0$ iff $\bar{w}^2 > 0$. Prove that $\bar{w}^1 = 0$ iff $\bar{w}^2 = 0$. Is the claim $\bar{w}^1 = \bar{w}^2$ true? Why? If not, what relationship exists between $\bar{w}^1$ and $\bar{w}^2$? Illustrate with numerical examples.

2.42 Discuss how to solve the following LP directly by the simplex algorithm without changing the objective into minimization form: maximize $z^1(x) = c^1 x$, subject to $Ax = b$, $x \geqq 0$.

REFERENCES

2.1 G. B. Dantzig, "Expected Number of Steps of the Simplex Method for a Linear Program with a Convexity Constraint," Technical Report SOL 80-3R, Systems Optimization Laboratory, Stanford University, Stanford, Calif., October 1980.

2.2 A. J. Hoffman, M. Mannos, D. Sokolowsky, and N. Wiegmann, "Computational Experience in Solving Linear Programs," *Journal of the Society for Industrial and Applied Mathematics*, *1*, 1, 1953, pp. 17–33.

2.3 H. W. Kuhn and R. E. Quandt, "An Experimental Study of the Simplex Method," *Proceedings of the Symposia in Applied Mathematics*, vol. XV, Americal Mathematical Society, 1963.

2.4 T. M. Liebling, "On the Number of Iterations of the Simplex Method," *Methods of Operations Research*, *XVII*, *V* Oberwolfach-Tagung uber Operations Research, 13–19, (August 1977), 248–264.

2.5 A. Orden, "Computational Investigation and Analysis of Probabilistic Parameters of Convergence of a Simplex Algorithm," In *Progress in Operations Research*, Vol. II, A. Prekopa (Ed.), North-Holland, Amsterdam, The Netherlands, 1976, pp. 705–715.

2.6 P. Wolfe and L. Cutler, "Experiments in Linear Programming," In *Recent Advances in Mathematical Programming*, R. L. Graves and P. Wolfe (Eds.), McGraw-Hill, New York, 1963, pp. 177–200.

Selected Books in Nonlinear Programming

2.7 M. Avriel, *Nonlinear Programming: Analysis and Methods*, Prentice-Hall, Englewood Cliffs, N. J., 1976.

2.8 M. S. Bazaraa and C. M. Shetty, *Nonlinear Programming: Theory and Algorithms*, Wiley, New York, 1979.

2.9 A. V. Fiacco and G. P. McCormick, *Nonlinear Programming: Sequential Unconstrained Minimization Technique*, Wiley, New York, 1968.

2.10 R. Fletcher, *Practical Methods of Optimization*, Wiley, New York, 1980.

2.11 C. B. Garcia and W. I. Zangwill, *Pathways to Solutions, Fixed Points and Equilibria*, Prentice-Hall, Englewood Cliffs, N. J., 1981.

2.12 P. E. Gill, W. Murray, and M. H. Wright, *Practical Optimization*, Academic, New York, 1981.

2.13 P. E. Gill and W. Murray, *Numerical Methods for Constrained Optimization*, Academic, New York, 1974.

2.14 D. M. Himmelblau, *Applied Nonlinear Programming*, McGraw-Hill, New York, 1972.

2.15 H. P. Künzi and W. Krelle, *Nonlinear Programming*, Blaisdell, Waltham, Mass., 1966.

2.16 O. L. Mangasarian, *Nonlinear Programming*, McGraw-Hill, New York, 1969.

2.17 J. M. Ortega and W. C. Rheinboldt, *Iterative Solutions of Nonlinear Equations in Several Variables*, Academic, New York, 1970.

2.18 J. F. Shapiro, *Mathematical Programming: Structures and Algorithms*, Wiley, New York, 1979.

2.19 J. Stoer and C. Witzgall, *Convexity and Optimization In Finite Dimensions I*, Springer, Berlin, 1970.

2.20 W. I. Zangwill, *Nonlinear Programming, A Unified Approach*, Prentice-Hall, Englewood Cliffs, N. J., 1969.

The Geometry of the Simplex Method

chapter 3

3.1 EUCLIDEAN VECTOR SPACES: DEFINITIONS AND GEOMETRICAL CONCEPTS

Familiarity with the notion of the n-dimensional Euclidean space $\mathbf{R}^n$ as a linear vector space is assumed. Every point $x \in \mathbf{R}^n$ is an *ordered vector* of the form $x = (x_1, \ldots, x_n)$, where each of the x_j is a real number. If the order changes, the vector changes. For example, the vectors (1, 2) and (2, 1) are different. Each vector can be viewed as the coordinate vector of a point in the Euclidean vector space of appropriate dimension. When all the x_j are equal to zero, the vector is known as the zero vector and it is also denoted by the symbol "0." It will always be clear from the context whether "0" refers to the real number 0 or the zero vector in some Euclidean space.

The vector x is said to be *nonnegative* if each of its components x_j is greater than or equal to zero. Symbolically this is written as $x \geqq 0$. (Notice the two lines under the inequality sign.) Obviously $0 \geqq 0$.

The vector $x \in \mathbf{R}^n$ is said to be *semipositive* if each of the components x_j is greater than or equal to zero and at least one of them is strictly greater than zero. We denote this by $x \geq 0$. (Notice the single line under the inequality sign.) That is, $x \geq 0$ iff $x \geqq 0$ and $x \neq 0$ or, iff $x \geqq 0$ and $\sum x_j$ is strictly positive. Obviously $0 \ngeq 0$.

The vector x is said to be *positive* if each of its components x_j is strictly greater than zero. This is denoted by $x > 0$. If $x = (x_j)$, $y = (y_j) \in \mathbf{R}^n$,

$x \geqq y$ means that $x - y \geqq 0$; that is, $x_j \geqq y_j$ for each $j = 1$ to n.

$x > y$ means that $x - y > 0$; that is, $x_j > y_j$ for each $j = 1$ to n.

$x \geq y$ means that $x - y \geq 0$; that is, $x_j \geqq y_j$ for each j and $x_j > y_j$ for at least one j.

Translate

If $\mathbf{S} \subset \mathbf{R}^n$, $\hat{x} \in \mathbf{R}^n$, the *translate* of $\mathbf{S}$ to $\hat{x}$ is the set $\{x : x = \hat{x} + y, y \in \mathbf{S}\}$.

Linear, Affine, Convex, and Nonnegative Combinations

Let $x = (x_1, \ldots, x_n)$ be a vector in $\mathbf{R}^n$ and α a real number. Then $\alpha x = (\alpha x_1, \ldots, \alpha x_n)$. Thus multiplication of a vector by a real number is the multiplication of each coordinate in the vector by that real number.

If $x^1 = (x_1^1, \ldots, x_n^1) \neq 0$, for $0 \leqq \alpha \leqq 1$, the vector αx^1 represents a point between 0 (the origin) and the point x^1, on the line segment joining them. For $\alpha > 1$, the point αx^1 is a point beyond x^1 on the straight line obtained by starting at the origin and joining it to x^1 and continuing in that direction. For $\alpha < 0$, the vector αx^1 is a point beyond the origin on the straight line obtained by joining x^1 to 0 and continuing in that direction. See Figure 3.1.

Let $x^1 = (x_1^1, \ldots, x_n^1)$ and $x^2 = (x_1^2, \ldots, x_n^2)$ be two vectors in $\mathbf{R}^n$. Their sum $x^1 + x^2$ is obtained by adding the corresponding coordinates in x^1 and x^2. Thus $x^1 + x^2 = (x_1^1 + x_1^2, \ldots, x_n^1 + x_n^2)$. If x^1, x^2 are distinct from 0 and x^1 cannot be obtained by multiplying x^2 by a real number (i.e., the point represented by x^1 in $\mathbf{R}^n$ is not on the straight line joining 0 and x^2), the three points x^1, x^2, 0 determine a unique two-dimensional subspace of $\mathbf{R}^n$. The point $x^1 + x^2$ is the fourth vertex of the parallelogram in this two-dimensional subspace, whose other vertices are 0, x^1, x^2. This is known as the *parallelogram law of addition of vectors*. See Figure 3.2.

Linear Combinations and Linear Hull

Let $\{x^1, \ldots, x^k\}$ be a finite set of points in $\mathbf{R}^n$, where $x^r = (x_1^r, \ldots, x_n^r)$, for $r = 1$ to k. A *linear combination* of these points is any point x of the form $x = \alpha_1 x^1 + \cdots + \alpha_k x^k$, where $\alpha_1, \ldots, \alpha_k$ are all real numbers. The set of all linear combinations of $x^1, \ldots, x^k$ is the

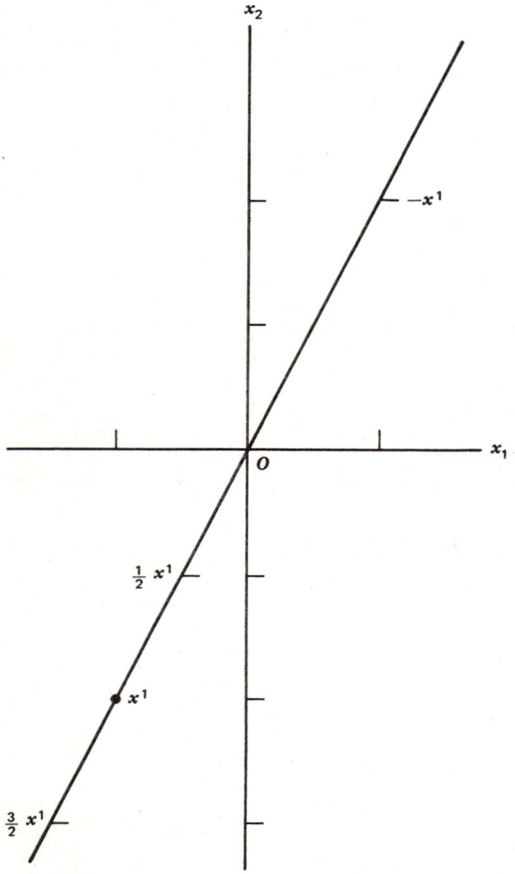

Figure 3.1 The linear hull of $\{x^1\}$ in $\mathbf{R}^2$.

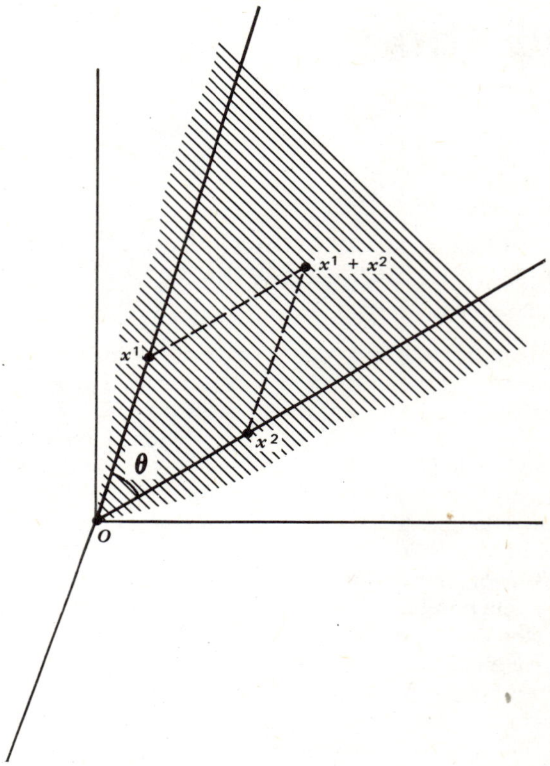

Figure 3.2 The linear hull of $\{x^1, x^2\}$ in $\mathbf{R}^3$ is the plane containing 0, x^1, x^2.

subspace of $\mathbf{R}^n$ of smallest dimension containing $x^1, \ldots, x^k$; it is called the *linear hull* of $\{x^1, \ldots, x^k\}$. See Figures 3.1 and 3.2.

A set of vectors $\{x^1, \ldots, x^k\} \subset \mathbf{R}^n$ is said to *span* all the vectors in a subset $\Gamma \subset \mathbf{R}^n$, if every vector in Γ can be expressed as a linear combination of vectors in the set $\{x^1, \ldots, x^k\}$, that is, iff $\Gamma \subset$ linear hull of $\{x^1, \ldots, x^k\}$. The linear hull of $\{x^1, \ldots, x^k\}$ is sometimes called the *span* or the *linear span* of $\{x^1, \ldots, x^k\}$. It is a *subspace* of $\mathbf{R}^n$. A subspace of $\mathbf{R}^n$ is a subset $\boldsymbol{\Gamma} \subset \mathbf{R}^n$ satisfying the property that if x, $y \in \boldsymbol{\Gamma}$, then every linear combination of x and y is also in $\boldsymbol{\Gamma}$.

Example 3.1

Let $x^1 = (1, 0, -1)$, $x^2 = (-2, 3, 17)$. Any point of the form $x = (\alpha_1 - 2\alpha_2, 3\alpha_2, -\alpha_1 + 17\alpha_2)$ is a linear combination of x^1 and x^2.

Example 3.2

Let $x^1 = (-1, -2)$. The linear hull of $\{x^1\}$ is the set of all points in $\mathbf{R}^2$ that are scalar multiples of x^1. It is the set of all points on the straight line joining x^1 and the origin in Figure 3.1.

Affine Combinations and Affine Hull

An *affine combination* of $x^1, \ldots, x^k$ is any point of the form

$$x = \alpha_1 x^1 + \cdots + \alpha_k x^k \tag{3.1}$$

where $\alpha_1, \ldots, \alpha_k$ are real numbers satisfying

$$\alpha_1 + \cdots + \alpha_k = 1 \tag{3.2}$$

The set of all affine combinations of $x^1, \ldots, x^k$ is known as the *affine hull* of $\{x^1, \ldots, x^k\}$. The affine hull of a set of points in $\mathbf{R}^n$ is a subset of its linear hull.

Clearly the point x in (3.1) can be expressed as $x = x^1 + (\alpha_1 - 1)x^1 + \alpha_2 x^2 + \cdots + \alpha_k x^k$, and by (3.2) this is $x^1 + [\alpha_2(x^2 - x^1) + \cdots + \alpha_k(x^k - x^1)]$. If $y = \alpha_2(x^2 - x^1) + \cdots + \alpha_k(x^k - x^1)$, y is a linear combination of $x^2 - x^1, \ldots, x^k - x^1$. As $\alpha_2, \ldots, \alpha_k$ assume all possible real values, y varies over the entire linear hull of $\{(x^2 - x^1), \ldots, (x^k - x^1)\}$. Therefore the affine hull of $\{x^1, \ldots, x^k\}$ is the translate of the linear hull of $\{(x^2 - x^1), \ldots, (x^k - x^1)\}$ to x^1.

An *affine space* is a subset $\Gamma \subset \mathbf{R}^n$ satisfying the property that if $x, y \in \Gamma$, then every affine combination of x and y is also in Γ. The affine hull of $\{x^1, \ldots, x^k\} \subset \mathbf{R}^n$ is an affine space.

Example 3.3

Let $x^1 = (1, 0, 0)$, $x^2 = (0, 1, 0)$, $x^3 = (0, 0, 1)$. An affine combination of x^1, x^2, x^3 is a point of the form $x = (\alpha_1, \alpha_2, \alpha_3)$, where $\alpha_1 + \alpha_2 + \alpha_3 = 1$. Thus the affine hull of x^1, x^2, x^3 is the hyperplane of dimension 2 containing x^1, x^2, and x^3. See Figure 3.3.

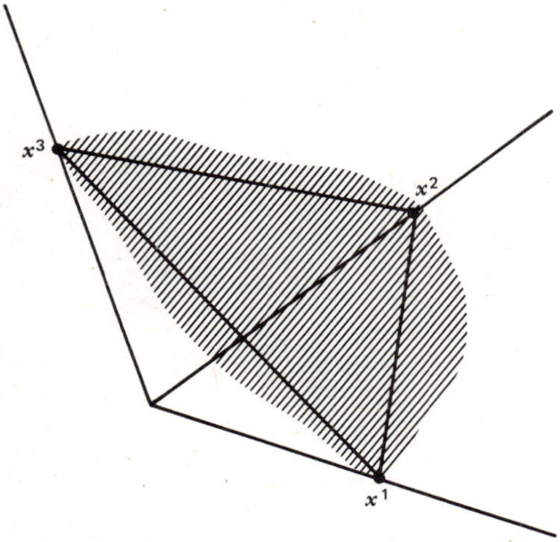

Figure 3.3 Affine hull of $\{x^1, x^2, x^3\}$ is the hyperplane containing x^1, x^2, and x^3.

Example 3.4

Let $x^1 = (3, 1)$, $x^2 = (4, 3)$. The affine hull of $\{x^1, x^2\}$ is the set of all points on the straight line joining x^1, x^2. Also it is the translate to x^1 of the linear hull of $\{(4, 3) - (3, 1) = (1, 2)\}$ in $\mathbf{R}^2$. See Figure 3.4.

In general, the affine hull of any pair of distinct points is the straight line joining these points.

Convex Combinations and Convex Hull

A *convex combination* of $x^1, \ldots, x^k$ is a point x of the form $x = \alpha_1 x^1 + \cdots + \alpha_k x^k$, where $\alpha_1, \ldots, \alpha_k$ are real numbers satisfying $\alpha_1 + \cdots + \alpha_k = 1$ and $\alpha_1 \geqq 0, \ldots, \alpha_k \geqq 0$. The set of all convex combinations of $x^1, \ldots, x^k$ is known as the *convex hull* of $\{x^1, \ldots, x^k\}$. It is a subset of the affine hull of the same set of points.

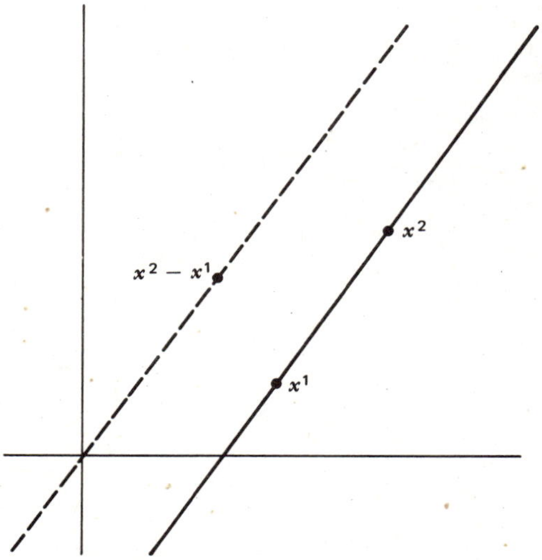

Figure 3.4 Affine hull of $\{x^1, x^2\}$ is the straight line through x^1 and x^2. It is a translate of the linear hull of $\{x^2 - x^1\}$ to x^1.

Example 3.5

If $x^1 = (1, 0)$, $x^2 = (0, 1)$, the convex hull of $\{x^1, x^2\}$ is the set of all points in $\mathbf{R}^2$ of the form (α_1, α_2) with $\alpha_1 \geqq 0$, $\alpha_2 \geqq 0$, $\alpha_1 + \alpha_2 = 1$. See Figures 3.5 and 3.6.

In general the convex hull of any two points in $\mathbf{R}^n$ is the set of all points on the *line segment* joining these two points.

Nonnegative (Linear) Combinations, Pos Cones

If $\{x^1, \dots, x^k\}$ is a subset of points from $\mathbf{R}^n$, a *nonnegative (linear) combination* of $\{x^1, \dots, x^k\}$ is any vector x of the form $x = \alpha_1 x^1 + \cdots \alpha_k x^k$, where $\alpha_1 \geqq 0, \dots, \alpha_k \geqq 0$. The set of all nonnegative combinations of $\{x^1, \dots, x^k\}$ is known as the *nonnegative hull* or the *Pos cone* of $\{x^1, \dots, x^k\}$, and is denoted by $\text{Pos}\{x^1, \dots, x^k\}$. For example, any vector of the form $(\alpha_1, \alpha_2)^T = \alpha_1(1, 0)^T + \alpha_2(0, 1)^T$, where $\alpha_1 \geqq 0$, $\alpha_2 \geqq 0$, is a nonnegative combination of $\{(1, 0)^T, (0, 1)^T\}$. The set of all these nonnegative combinations is the nonnegative orthant in $\mathbf{R}^2$, pictured in Figure 3.15(b). See Figure 3.17 where a Pos cone in $\mathbf{R}^3$, $\text{Pos}\{A_{.1}, A_{.2}, A_{.3}, A_{.4}\}$, is illustrated. Also see Example 3.8 and Figure 3.22 where the Pos cone of the four column vectors of a matrix A given there is illustrated.

Summary of Various Types of Combinations of Vectors

Let $\mathbf{\Gamma} = \{x^1, \dots, x^k\} \subset \mathbf{R}^n$.

The Usefulness of Various Types of Combinations of Vectors

A system of linear equations is said to be a *homogeneous system* if all the right hand side constants in it are zero. Thus a system such as "$Ax = 0$" is a homogeneous system of linear equations. If x^1, x^2 are feasible solutions of this system, we have $Ax^1 = 0$, $Ax^2 = 0$ and, hence, $A(\alpha_1 x^1 + \alpha_2 x^2) = 0$ for all α_1, α_2. Thus any linear combination of solutions of a homogeneous system of linear equations is also a solution of this system. Thus the operation of generating linear combinations of vectors is useful if we have to generate the set of all feasible solutions of a homogeneous system of linear equations. The set of feasible solutions of such a system is a subspace.

Consider a nonhomogeneous system of linear "$Ax = b$." Thus here $b \neq 0$. If x^1, x^2 are feasible solutions of this system, we have $Ax^1 = b$ and $Ax^2 = b$ and, hence, $A(\alpha_1 x^1 + \alpha_2 x^2) = (\alpha_1 + \alpha_2)b$. Thus the linear combination $\alpha_1 x^1 + \alpha_2 x^2$ is also a feasible solution of this system iff $\alpha_1 + \alpha_2 = 1$, that is, iff $\alpha_1 x^1 + \alpha_2 x^2$ is an affine combination of x^1, x^2. Similarly, it can be verified that any affine combination of solutions of a nonhomogeneous system of linear equations is also a solution of this system. Therefore the operation of generating affine combinations of vectors is useful if we have to generate the set of all feasible solutions of a nonhomogeneous system of linear equations. The set of feasible solutions of such a system is an affine space.

Consider a nonhomogeneous system of linear equations in nonnegative variables, "$Ax = b, x \geqq 0$."

Type of Combination	*Is*	*Name of the Set of All Such Combinations*
Linear Combination of $\mathbf{\Gamma}$	Any point of the form $\alpha_1 x^1 + \cdots + \alpha_k x^k$, where $\alpha_1, \dots, \alpha_k$ are real numbers	Linear hull or linear span of $\mathbf{\Gamma}$; it is a subspace of $\mathbf{R}^n$
Affine combination of $\mathbf{\Gamma}$	Any point of the form $\alpha_1 x^1 + \cdots + \alpha_k x^k$, where $\alpha_1, \dots, \alpha_k$ are real numbers satisfying $\alpha_1 + \cdots + \alpha_k = 1$	Affine hull of $\mathbf{\Gamma}$; it is a translate of a subspace of $\mathbf{R}^n$
Convex combination of $\mathbf{\Gamma}$	Any point of the form $\alpha_1 x^1 + \cdots + \alpha_k x^k$, where $\alpha_1, \dots, \alpha_k$ are real numbers satisfying $\alpha_1 + \cdots + \alpha_k = 1$ and $\alpha_1 \geqq 0, \dots \alpha_k \geqq 0$	Convex hull of $\mathbf{\Gamma}$
Nonnegative combination of $\mathbf{\Gamma}$	Any point of the form $\alpha_1 x^1 + \cdots + \alpha_k x^k$, where $\alpha_1 \geqq 0, \dots, \alpha_k \geqq 0$	Nonnegative hull or Pos cone of $\mathbf{\Gamma}$; or $\text{Pos}(\mathbf{\Gamma})$

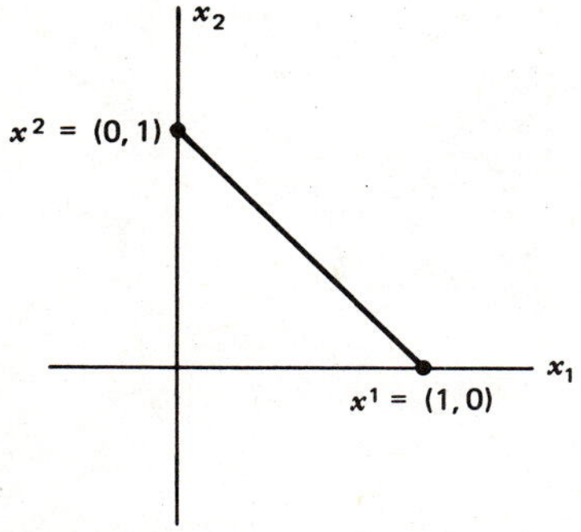

Figure 3.5 Convex hull of $\{x^1, x^2\}$.

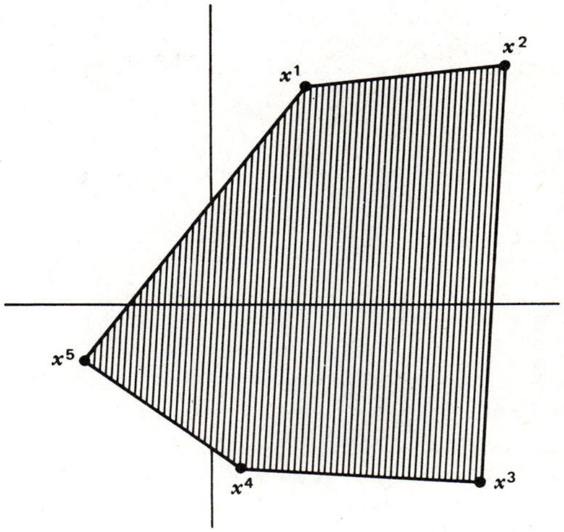

Figure 3.6 Convex hull of $\{x^1, \ldots, x^5\}$ in $\mathbf{R}^2$.

If x^1, x^2 are feasible solutions of this system, we have $Ax^1 = Ax^2 = b$, x^1, $x^2 \geqq 0$ and, hence, $A(\alpha_1 x^1 + \alpha_2 x^2) = b$, $\alpha_1 x^1 + \alpha_2 x^2 \geqq 0$ provided that $\alpha_1 + \alpha_2 = 1$, $\alpha_1, \alpha_2 \geqq 0$. Thus a convex combination of feasible solutions of this system is also feasible. In the same manner, it can be verified that any convex combination of feasible solutions of a system of linear equations and linear inequalities, is also a feasible solution to the system. Conversely, to guarantee that a linear combination of feasible solutions of a system of linear equations and linear inequalities is also feasible, in general, we need the condition that the linear combination is a convex combination. Thus the operation of generating convex combinations of vectors is useful if we have to generate the set of feasible solutions of a system of linear equations and inequalities.

Now consider a homogeneous system of linear inequalities, that is, a system of the form "$Ax \geqq 0$." It can be verified that any nonnegative combination of feasible solutions of this system is also feasible to this system. Thus the operation of generating nonnegative combinations of vectors is useful if we have to generate the set of feasible solutions of a homogeneous system of linear equations and inequalities.

Straight Lines, Line Segments, Rays, and Half-Lines in $\mathbf{R}^n$

A straight line in $\mathbf{R}^n$ is the locus of a point, all of whose coordinates are affine functions of a parameter θ

$$x = \begin{pmatrix} x_1 \\ \vdots \\ x_n \end{pmatrix} = \begin{pmatrix} a_1 \\ \vdots \\ a_n \end{pmatrix} + \theta \begin{pmatrix} b_1 \\ \vdots \\ b_n \end{pmatrix}$$

as the parameter assumes all real values, where at least one of the b_i is nonzero [i.e., $(b_1, \ldots, b_n) \neq 0$]. A straight line is also the affine hull of any two points on it. If x^1 and x^2 are two distinct points in $\mathbf{R}^n$, a general point on the straight line joining them is $x(\theta) = x^1 + \theta(x^2 - x^1)$, where θ is a real-valued parameter. As θ takes all real values, $x(\theta)$ generates all the points on the straight line joining x^1 and x^2.

In $\mathbf{R}^2$, a straight line is the set of all points satisfying a single linear equation. However, for $n > 2$, the set of all points satisfying a single linear equation in $\mathbf{R}^n$ will be a hyperplane of dimension $n - 1$, and not a straight line. So this parametric representation is the most convenient method for representing straight lines in $\mathbf{R}^n$ for $n \geqq 2$.

Example 3.6

Let $x^1 = (3, 1)$, $x^2 = (4, 4)$. Then the straight line joining x^1 and x^2 is the set of all points of the form

$$\left\{ x = \begin{pmatrix} x_1 \\ x_2 \end{pmatrix} = \begin{pmatrix} 3 \\ 1 \end{pmatrix} + \theta \begin{pmatrix} 4 - 3 \\ 4 - 1 \end{pmatrix} = \begin{pmatrix} 3 + \theta \\ 1 + 3\theta \end{pmatrix} : \theta \text{ real number} \right\}.$$

The straight line joining x^1 and x^2 is the affine hull of x^1, x^2. See Figure 3.4.

Exercises

3.1 As θ varies between 0 and 1, prove that $x(\theta) = x^1 + \theta(x^2 - x^1)$ generates all the points on the line segment joining x^1 and x^2.

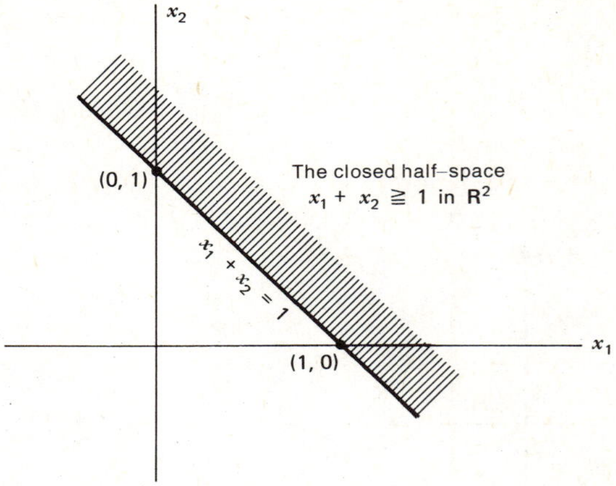

Figure 3.8

Rays and Half-Lines

Let $\tilde{x} \in \mathbf{R}^n$, $\tilde{x} \neq 0$. The *ray generated* by $\tilde{x}$ is the set $\{x\colon x = \lambda\tilde{x}, \lambda \geqq 0\}$. If $\hat{x} \in \mathbf{R}^n$, then the set $\{x\colon x = \hat{x} + \theta\tilde{x}, \theta \geqq 0\}$ is known as the *half-line* through $\hat{x}$ parallel to the ray generated by $\tilde{x}$. It starts at $\hat{x}$ (this point corresponds to $\theta = 0$), and as θ ranges from 0 to ∞, it traces the line through $\hat{x}$ parallel to the ray of $\tilde{x}$. See Figure 3.7. Every ray contains the origin, and a half-line is a translate of a ray.

Hyperplane

A hyperplane in $\mathbf{R}^n$ is the set of all points $x = (x_1, \ldots, x_n) \in \mathbf{R}^n$ satisfying a single linear equation, $a_1x_1 + \cdots + a_nx_n = b$, where $a_1, \ldots, a_n, b$ are given numbers and at least one of the a_i is nonzero, that is, $(a_1, \ldots, a_n) \neq 0$.

Half-Space

Consider an inequality constraint $a_{11}x_1 + \cdots + a_{1n}x_n \geqq b_1$, where $(a_{11}, \ldots, a_{1n}) \neq 0$. The set of all $x \in \mathbf{R}^n$ satisfying this inequality is the set of all points lying on one side of the hyperplane $a_{11}x_1 + \cdots + a_{1n}x_n = b_1$, and is known as a half-space, or a *closed half-space*, to be mathematically precise (see Figure 3.8). Any equality constraint of the form $a_{11}x_1 + \cdots + a_{1n}x_n = b_1$ is equivalent to the pair of inequality constraints $a_{11}x_1 + \cdots + a_{1n}x_n \geqq b_1$, and $a_{11}x_1 + \cdots + a_{1n}x_n \leqq b_1$, both of which must hold. In a general LP there may be some equality constraints, inequality constraints, and sign restrictions on the variables. Each sign restriction is actually an inequality constraint. Each equality constraint is equivalent to a pair of inequality constraints. Hence, all the constraints in any LP may be expressed as linear inequality constraints. The set of all points satisfying a linear inequality constraint is a closed half-space. Every feasible solution of the LP must satisfy all the constraints and, hence, it must be in each of the corresponding half-spaces. Hence in *linear programming problems, the set of feasible solutions is the intersection of a finite number of closed half-spaces.*

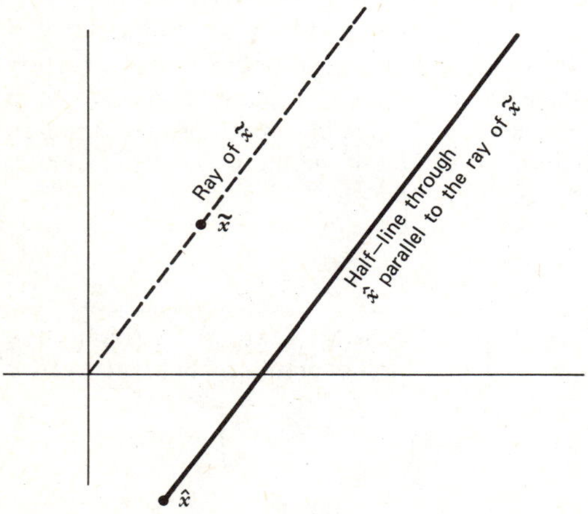

Figure 3.7

Convex Sets

A subset $\mathbf{K} \subset \mathbf{R}^n$ is said to be a *convex set* if every convex combination of any pair of points in $\mathbf{K}$ is also in $\mathbf{K}$. That is, if $\tilde{x} \in \mathbf{K}$, and $\hat{x} \in \mathbf{K}$, then $\alpha\tilde{x} + (1-\alpha)\hat{x} \in \mathbf{K}$ for all $0 \leqq \alpha \leqq 1$. Hence, if $\mathbf{K}$ is a convex set, the line segment joining any pair of points in $\mathbf{K}$ lies entirely in $\mathbf{K}$. Examples of convex sets in the plane are in Figure 3.9. Figure 3.10 contains examples of nonconvex sets.

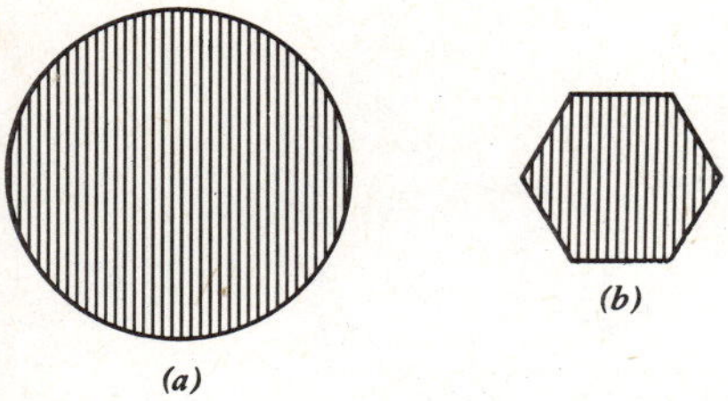

Figure 3.9 Convex sets. (*a*) All points inside or on the circle. (*b*) All points inside or on the polygon.

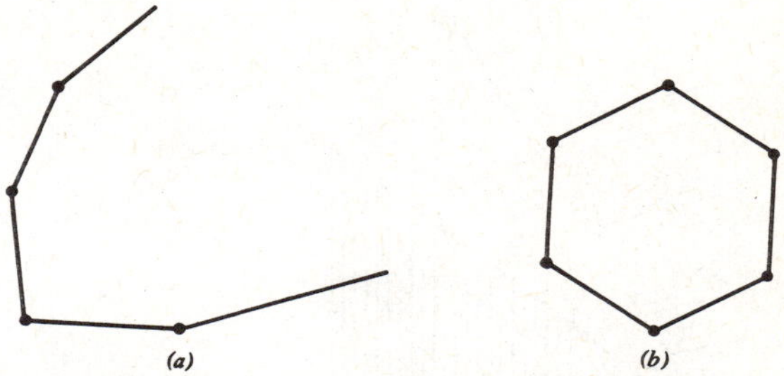

Figure 3.11 (*a*) A convex polyhedron that is not a polytope. (*b*) A convex polytope.

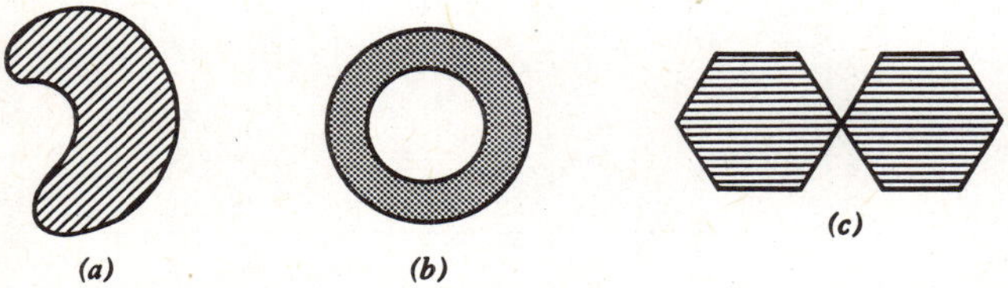

Figure 3.10 Non-convex sets. (*a*) All points inside or on the cashew nut. (*b*) All points on or between two circles. (*c*) All points on at least one of the two polygons.

Exercises

3.2 Prove that every half-space is a convex set.

3.3 Prove that if **K** is a convex set as defined above, and if $\{x^1, \ldots, x^r\}$ is a finite set of points in **K**, then the convex hull of $\{x^1, \ldots, x^r\}$ lies entirely in **K**.

3.4 Prove that the intersection of a family of convex sets is a convex set.

3.5 Prove that the set of feasible solutions of the following system of constraints is a convex set: $Ax + By = b$, $Ex + Fy \geqq d$, $x \geqq 0$, y unrestricted.

Convex Polyhedral Sets and Convex Polytopes

The intersection of a finite number of half-spaces is known as a *convex polyhedral set* or a *convex polyhedron*. From the discussion earlier, it is clear that the set of feasible solutions of an LP is a *convex polyhedron*. A convex polyhedron that is bounded is known as a *convex polytope* (Figure 3.11).

Cones, Convex Cones and Convex Polyhedral Cones

A subset $\mathbf{S} \subset \mathbf{R}^n$ is called a *cone* iff $x \in \mathbf{S}$ implies that $\alpha x \in \mathbf{S}$ for all $\alpha \geqq 0$; that is, the ray generated by any point in a cone lies entirely in the cone.

A *convex cone* is a cone that is also a convex set. Hence if **S** is a convex cone and $x \in \mathbf{S}$, $y \in \mathbf{S}$, all convex combinations of x, y must lie in **S**. Also all convex combinations of nonnegative multiples of x and y must also lie in **S**, since **S** is a cone. Therefore **S** is a convex cone iff $x \in \mathbf{S}$, $y \in \mathbf{S}$ implies $\alpha x + \beta y \in \mathbf{S}$ for all $\alpha \geqq 0$, $\beta \geqq 0$.

Exercises

3.6 Prove that **S** is a convex cone iff **S** contains every nonnegative linear combination of any finite number of points in **S**.

A convex cone **S** is a *convex polyhedral cone* if it is the intersection of a finite number of half-spaces. Thus every convex polyhedral cone is the set of feasible solutions of a system of constraints of the form $Ax \geqq 0$ (Figures 3.12 to 3.14).

Simplicial Cones

Let $\{A_{.1}, \ldots, A_{.r}\}$ be a linearly independent set of r vectors in $\mathbf{R}^n$ (see Section 3.3 for the definition of linear independence). Then

$$\text{Pos}\{A_{.1}, \ldots, A_{.r}\} = \{y : y = \alpha_1 A_{.1} + \cdots + \alpha_r A_{.r},\ \alpha_1 \geqq 0, \ldots, \alpha_r \geqq 0\}$$

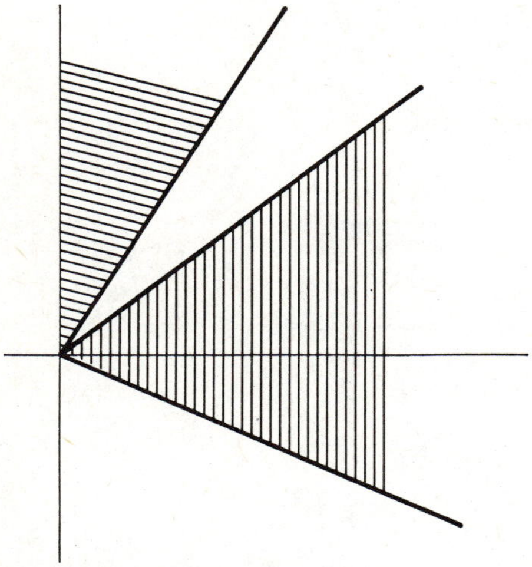

Figure 3.12 The cone in $\mathbf{R}^2$ part of which is shaded, is not a convex cone.

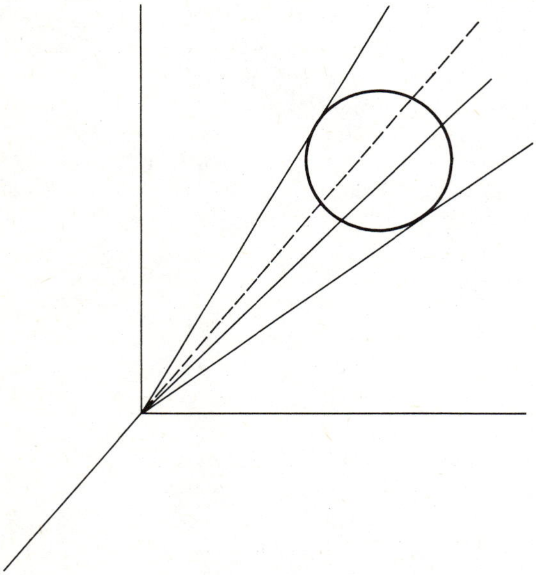

Figure 3.13 A convex cone in $\mathbf{R}^3$ that is not polyhedral.

is known as a *simplicial cone* of dimension r. If $\{A_{.1}, \ldots, A_{.r}\}$ is not a linearly independent set of vectors, the cone $\text{Pos}\{A_{.1}, \ldots, A_{.r}\}$ is not a simplicial cone. See Figures 3.15, 3.16, 3.17. The cone in Figure 3.17 is $\text{Pos}\{A_{.1}, A_{.2}, A_{.3}, A_{.4}\}$. From standard results in linear algebra it is well known that a

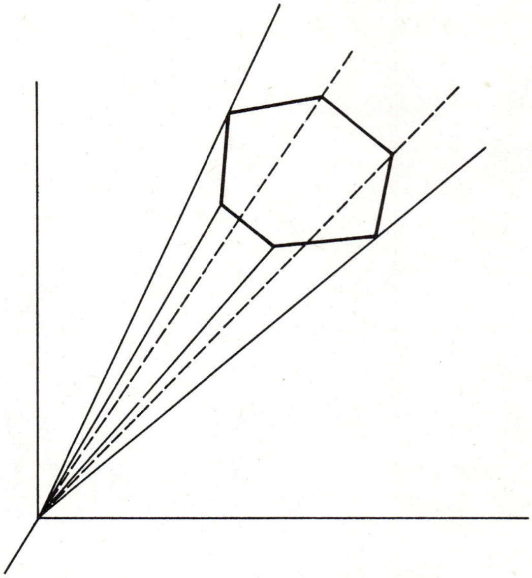

Figure 3.14 A convex polyhedral cone in $\mathbf{R}^3$.

set of vectors containing $n + 1$ or more vectors from $\mathbf{R}^n$ is linearly dependent for any $n \geqq 1$. Hence the cone $\text{Pos}\{A_{.1}, A_{.2}, A_{.3}, A_{.4}\}$ in Figure 3.17 is not a simplicial cone.

Simplex

Let $\{A_{.1}, A_{.2}, \ldots, A_{.r}\}$ be a set of r vectors from $\mathbf{R}^n$ satisfying the property that the set of vectors $\{A_{.2} - A_{.1}, \ldots, A_{.r} - A_{.1}\}$ is a linearly independent set. Then the convex hull $\{y: y = \alpha_1 A_{.1} + \cdots + \alpha_r A_{.r},$ where α's satisfy $\alpha_1 + \cdots + \alpha_r = 1,\ \alpha_1 \geqq 0, \ldots, \alpha_r \geqq 0\}$,

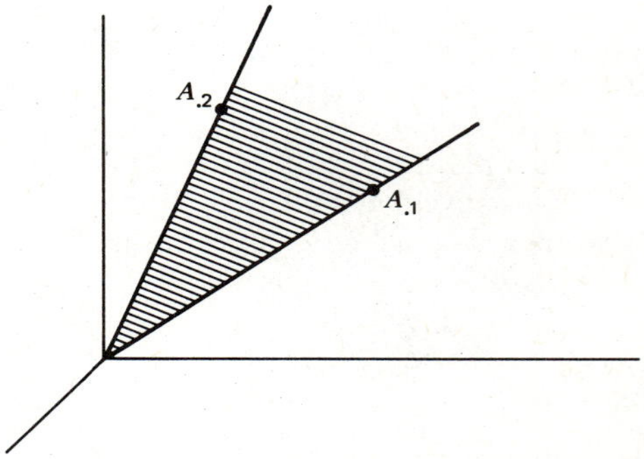

Figure 3.15(*a*) A two-dimensional simplicial cone $\text{Pos}\{A_{.1}, A_{.2}\}$ in $\mathbf{R}^3$.

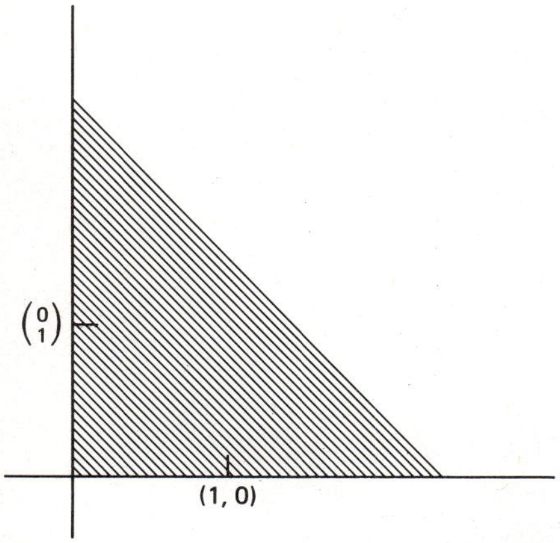

Figure 3.15(*b*) The nonnegative orthant in $\mathbf{R}^2$, Pos$\{\binom{1}{0}, \binom{0}{1}\}$ is a two-dimensional simplicial cone in $\mathbf{R}^2$.

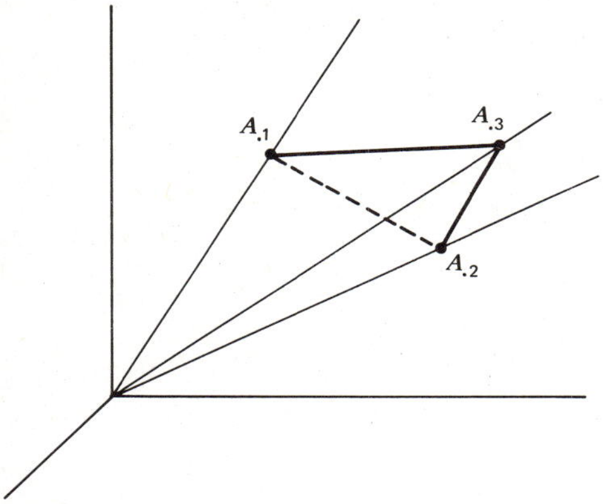

Figure 3.16 A simplicial cone of dimension 3.

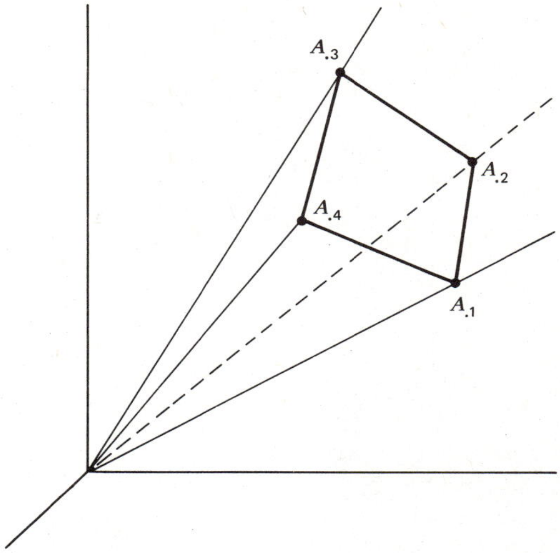

Figure 3.17 A convex polyhedral cone in $\mathbf{R}^3$, which is not a simplicial cone.

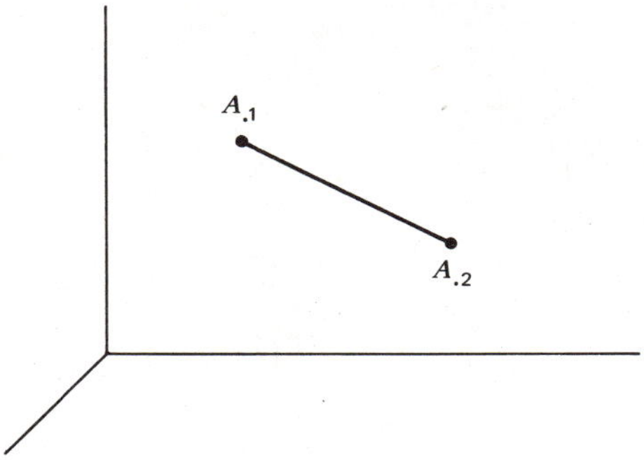

Figure 3.18 A simplex of dimension 1 is a line segment joining two distinct points.

is known as a *simplex* of dimension $r - 1$. It is well known from linear algebra that if four vectors $A_{.1}$, $A_{.2}$, $A_{.3}$, $A_{.4}$ lie in a two-dimensional plane, then the set $\{A_{.2} - A_{.1}, A_{.3} - A_{.1}, A_{.4} - A_{.1}\}$ cannot be linearly independent. This implies that the convex hull of $\{A_{.1}, A_{.2}, A_{.3}, A_{.4}\}$ in Figure 3.20 is not a simplex. See Figures 3.18, 3.19, 3.21 for simplexes.

Euclidean Norm

Let $x = (x_1, \ldots, x_n) \in \mathbf{R}^n$. The *Euclidean norm* of x is $\|x\| = +\sqrt{\sum_{j=1}^{n} x_j^2}$. If x and y are a pair of vectors in $\mathbf{R}^n$, the *Euclidean distance between x and y is* $\|x - y\|$.

Dot Product or Inner Product

Let $x = (x_1, \ldots, x_n)$ and $y = (y_1, \ldots, y_n)$ be two vectors in $\mathbf{R}^n$. Then the *dot product* (also called the *inner product* or the *scalar product*) of the vectors x and y is defined to be $\sum_{j=1}^{n} x_j y_j$ and is denoted by the product xy, where the vector x on the left of the product is written as a row vector, and the vector y on the right of the product is written as a column vector. In some books the dot product of x and y is also denoted by symbols like $x \cdot y$ or $\langle x, y \rangle$.

Let $x \neq 0$, $y \neq 0$ be two given vectors in $\mathbf{R}^n$. The angle θ between x and y is defined to be zero if y is a positive multiple of x (i.e., $y = \beta x$ for some $\beta > 0$;

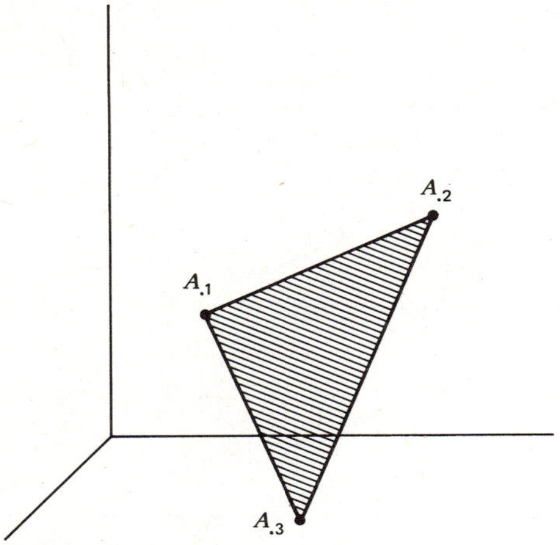

Figure 3.19 A simplex of dimension 2 is the convex hull of three distinct points that do not all lie on any straight line.

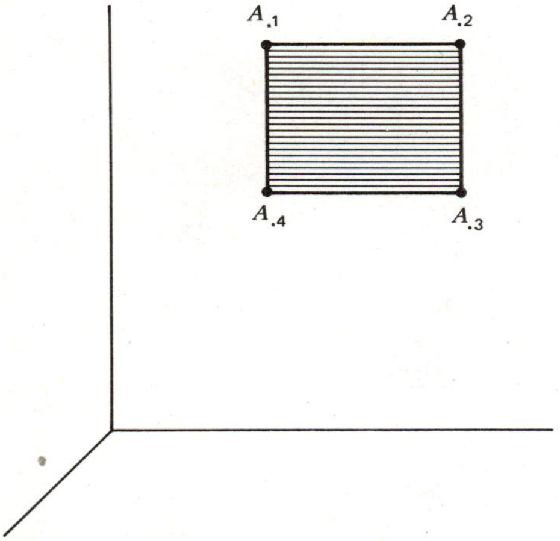

Figure 3.20 The convex hull of four points, all of which lie on a two-dimensional plane, is not a simplex.

in this case, the rays of x and y are the same, e.g., let $x = x^1$, $y = \frac{3}{2}x^1$, as in Figure 3.1). If y is a negative multiple of x (i.e., $y = \beta x$ for some $\beta < 0$), the rays of x and y are the two halves of the straight line that is the linear hull of $\{x\}$, (e.g., let $x = x^1$, $y = -x^1$, as in Figure 3.1), and in this case, the angle between x and y is defined to be 180° when measured in degrees. If y is not a multiple of x, since they are both nonzero, $\{x, y\}$ must be a linearly independent set. The linear hull of $\{x, y\}$ is a subspace of $\mathbf{R}^n$ of dimension 2. The angle between x and y is the angle θ between the rays of x and y in this subspace of dimension 2 (see

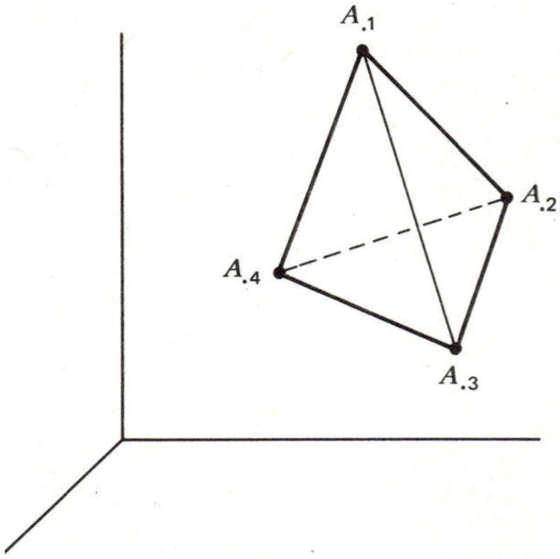

Figure 3.21 A simplex of dimension 3 is a tetrahedron.

Figure 3.2). If θ is this angle, then $\cos\theta =$ (dot product of x and $y)/(\|x\|\,\|y\|)$. So the rays of x and y are perpendicular to each other iff the dot product of x and y is zero.

3.2 MATRICES

A matrix is a rectangular array of numbers. Here is a matrix with m rows and n columns.

$$A = \begin{bmatrix} a_{11} & \cdots & a_{1j} & \cdots & a_{1n} \\ \vdots & & \vdots & & \vdots \\ a_{i1} & \cdots & a_{ij} & \cdots & a_{in} \\ \vdots & & \vdots & & \vdots \\ a_{m1} & \cdots & a_{mj} & \cdots & a_{mn} \end{bmatrix}$$

The entry in the ith row and the jth column of A is a_{ij}; it is known as the *(i, j)th entry in A*. The *order* of the matrix A is $m \times n$. When the order of A is understood, we will also denote A by (a_{ij}).

The coefficients of the variables in a system of linear equations is a matrix. For example, A is the matrix of coefficients of the variables in the system

$$\begin{array}{l} a_{11}x_1 + \cdots + a_{1n}x_n = b_1 \\ \vdots \qquad\qquad\quad \vdots \\ a_{m1}x_1 + \cdots + a_{mn}x_n = b_m \end{array}$$

If α is a real number, αA is the matrix obtained by multiplying each entry in A by α. Thus the (i, j)th entry in αA is αa_{ij}.

Two matrices $A = (a_{ij})$ and $B = (b_{ij})$ can only be added if they are of the same order. In this case $A + B = C$, where C is of the same order as A and B

and the (i, j)th entry in C is $c_{ij} = a_{ij} + b_{ij}$. Thus matrix addition consists of adding the entries in corresponding positions.

The transpose of the matrix A, denoted by A^{T}, is obtained by recording each row vector of A as a column. Thus

$$A^{\mathrm{T}} = \begin{bmatrix} a_{11} & \cdots & a_{i1} & \cdots & a_{m1} \\ \vdots & & \vdots & & \vdots \\ a_{1j} & \cdots & a_{ij} & \cdots & a_{mj} \\ \vdots & & \vdots & & \vdots \\ a_{1n} & \cdots & a_{in} & \cdots & a_{mn} \end{bmatrix}$$

Hence, if A is of order $m \times n$, A^{T} will be of order $n \times m$.

A row vector in $\mathbf{R}^n$ can be treated as a matrix of order $1 \times n$. Likewise a column vector in $\mathbf{R}^n$ can be treated as a matrix of order $n \times 1$.

A square matrix is a matrix that has the same number of rows and columns. The matrices

$$D = \begin{pmatrix} d_{11} & \cdots & d_{1m} \\ \vdots & & \vdots \\ d_{m1} & \cdots & d_{mm} \end{pmatrix}, \qquad I = \begin{pmatrix} 1 & 0 & 0 \\ 0 & 1 & 0 \\ 0 & 0 & 1 \end{pmatrix}$$

are square matrices. D is said to be of *order m*. The entries $d_{11}, d_{22}, \ldots, d_{mm}$ are the *diagonal entries* in D. They constitute its *principal diagonal*. All the other entries in D are its *off-diagonal* entries.

A *unit matrix* or *identity matrix* is a square matrix in which all diagonal entries are equal to one, and all off-diagonal entries are zero. Thus I given above is the unit matrix of order 3.

If $A = (a_{ij})$ is a matrix of order $m \times n$, we will denote by $A_{i.}$ the ith row vector of A, and by $A_{.j}$, the jth column vector of A. Thus

$$A_{i.} = (a_{i1}, \ldots, a_{in}). \qquad A_{.j} = (a_{1j}, \ldots, a_{mj})^{\mathrm{T}}$$

Let $\{i_1, \ldots, i_r\} \subset \{1, \ldots, m\}$ and $\{j_1, \ldots, j_s\} \subset \{1, \ldots, n\}$. Then the matrix

$$\begin{pmatrix} a_{i_1,j_1} & a_{i_1,j_2} & \cdots & a_{i_1,j_s} \\ a_{i_2,j_1} & a_{i_2,j_2} & \cdots & a_{i_2,j_s} \\ \vdots & \vdots & & \vdots \\ a_{i_r,j_1} & a_{i_r,j_2} & \cdots & a_{i_r,j_s} \end{pmatrix}$$

is known as a *submatrix* of the matrix A generated by the subset $\{i_1, \ldots, i_r\}$ of rows and the subset $\{j_1, \ldots, j_s\}$ of columns.

The rules for multiplication among matrices have evolved out of the study of linear transformations on systems of linear equations. Matrix multiplication is not commutative, that is, the order in which matrices are multiplied is very important. If A is a matrix of order $m \times n$ and B is a matrix of order $r \times s$, the product AB is defined only if $n = r$. In this case, the product AB is a matrix C of order $m \times s$, where the (i, j)th entry in C is $c_{ij} = \sum_{t=1}^{n} a_{it}b_{tj}$. Hence the (i, j)th entry in AB is the sum of the products of the entries in the ith row in A, by the corresponding entry in the jth column of B. Therefore, matrix multiplication is known as *row by column multiplication*, and the (i, j)th entry in the matrix product AB is the dot product of the ith row vector of A and the jth column vector of B.

Example 3.7

Let

$$A = \begin{pmatrix} 1 & -1 & -5 \\ 2 & -3 & -6 \end{pmatrix} \qquad B = \begin{pmatrix} -2 & -4 \\ -6 & -4 \\ 3 & -2 \end{pmatrix}$$

Then,

$$AB = \begin{pmatrix} -11 & 10 \\ -4 & 16 \end{pmatrix}$$

where, for example, the (1, 2)th element in AB, 10, is $A_{1.}B_{.2}$, etc.

Caution If $x = (x_i)$ and $y = (y_i)$ are column vectors in $\mathbf{R}^n$, the matrix product xy^{T} is not the dot product of x and y, but instead it is the matrix (x_iy_j) of order $n \times n$.

When the product AB exists, the product BA may not be defined, and even if it is defined, AB and BA may not be equal.

If A is a matrix of order $m \times n$, and π, x are vectors, the products πA and Ax exist only if π is a row vector $\mathbf{R}^m$ and x is a column vector in $\mathbf{R}^n$.

Pos (*A*)

Let A be a matrix of order $m \times n$. Let $A_{.j}$ denote the jth column vector of A, $j = 1$ to n. Then the Pos cone, $\mathrm{Pos}\{A_{.1}, \ldots, A_{.n}\} = \{Ay : y \geqq 0\}$ is also denoted by $\mathrm{Pos}(A)$.

Conditions for the Feasibility of a Linear Program in Standard Form

Let the constraints in an LP in standard form be

$$Ax = b \qquad x \geqq 0 \tag{3.3}$$

where A is a given matrix of order $m \times n$. If there is a feasible solution $\bar{x}$ for (3.3), then $\bar{x} \geqq 0$ and $A\bar{x} = b$, which implies that $b \in \text{Pos}(A)$. If $b \in \text{Pos}(A)$, then by the definition of Pos(A) there must exist an $\bar{x} \in \mathbf{R}^n$, $\bar{x} \geqq 0$ satisfying $A\bar{x} = b$, which implies that such $\bar{x}$ must be feasible to (3.3). Hence (3.3) has a feasible solution iff $b \in \text{Pos}(A)$. Even though this condition $b \in \text{Pos}(A)$ is easily stated, it is hard to verify directly when $m \geqq 3$, and we need an algorithm similar to Phase I of the simple method to check whether it holds.

Example 3.8

Consider the system

x_1	x_2	x_3	x_4	b
1	2	1	2	b_1
2	1	-1	-3	b_2

(3.4)

$x_j \geqq 0$ for all j.

Pos (A) in this case is the convex cone in $\mathbf{R}^2$, shown in Figure 3.22. The system (3.4) has a feasible solution iff the right-hand side constants vector $(b_1, b_2)^{\mathrm{T}}$ in (3.4) lies in Pos(A). If $b = (3, -1)^{\mathrm{T}}$, it can be verified that it is in the cone Pos(A) in Figure 3.22, and in this case (3.4) has a feasible solution. If $b = (-1, -1)^{\mathrm{T}}$, it can be verified that it does not lie in the cone Pos(A) in Figure 3.22, and in this case (3.4) has no feasible solution.

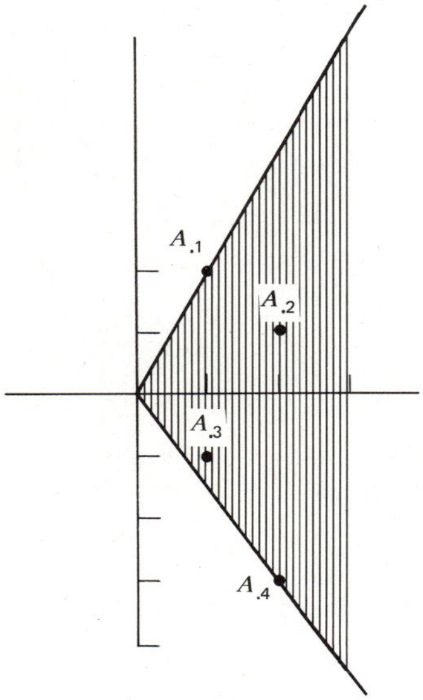

Figure 3.22 The cone Pos(A).

If the system (3.3) is infeasible, it can be made feasible by changing b into a point that lies in Pos(A). Hence, when (3.3) is infeasible, there are usually many ways in which the vector b in (3.3) can be modified to make it feasible.

3.3 LINEAR INDEPENDENCE OF A SET OF VECTORS AND SIMULTANEOUS LINEAR EQUATIONS

A set of vectors $\{A_{.1}, \ldots, A_{.k}\}$ in $\mathbf{R}^n$ is said to be *linearly dependent* iff it is possible to express the zero vector as a nonzero linear combination of vectors in the set; that is, iff there exist real numbers $\alpha_1, \ldots, \alpha_k$, not all zero such that $0 = \alpha_1 A_{.1} + \cdots + \alpha_k A_{.k}$; in this case, this equation is known as a linear *dependence relation* for this set of vectors. The set $\{A_{.1}, \ldots, A_{.k}\}$ is said to be *linearly independent* if it is not linearly dependent.

Note 3.1: From the definition it is clear that the empty set of vectors is linearly independent.

Thus a nonempty set of vectors $\{A_{.1}, \ldots, A_{.k}\}$ is linearly independent iff $\alpha_1 A_{.1} + \cdots + \alpha_k A_{.k} = 0$, which is a system of simultaneous linear equations in $\alpha_1, \ldots, \alpha_k$, has the unique solution, $\alpha_1 = \alpha_2 = \cdots = \alpha_k = 0$.

Exercises

3.7 Prove that any set of vectors containing the zero vector is linearly dependent.

3.8 Let D be a matrix that contains a row vector all the entries in which are zero, and let E be the matrix obtained by deleting that zero row vector from D. Prove that the set of column vectors of D is linearly independent iff the set of column vectors of E is linearly independent.

3.9 Let D be a matrix that contains a row vector or a column vector that contains a single nonzero entry, and let E be the matrix obtained by deleting the row and the column of that single nonzero entry from D. Prove that the set of column vectors of D is linearly independent iff the set of column vectors of E is linearly independent.

3.3.1 An Algorithm for Testing Linear Independence

Let $\{A_{.1}, \ldots, A_{.k}\}$ be a set of vectors in $\mathbf{R}^m$, where $A_{.j} = (a_{1j}, \ldots, a_{mj})^{\mathrm{T}}$ for $j = 1$ to k. Suppose it is required to test whether this set is linearly independent. Write each vector in the set as a row vector in a tableau.

Tableau

Vector	
$A_{.1}$	$a_{11} \ldots\ldots a_{m1}$
$\vdots$	$\vdots \qquad \vdots$
$A_{.k}$	$a_{1k} \ldots\ldots a_{mk}$

In the course of the algorithm the rows in the tableau will be altered. But at every step, each row in the tableau will be a nonzero linear combination of the vectors in the original set $\{A_{.1}, \ldots, A_{.k}\}$. The expression of each row in the current tableau as a nonzero linear combination of vectors in the set $\{A_{.1}, \ldots, A_{.k}\}$ will be indicated on the left-hand side of the tableau. A convenient method for keeping track of the expression is to store it in detached coefficient form. To do this, the original tableau is recorded as below.

Original Tableau

$A_{.1}$	$A_{.2} \ldots A_{.k}$		
1	$0 \ldots 0$	a_{11}	$a_{21} \ldots a_{m1}$
0	$1 \ldots 0$	a_{12}	$a_{22} \ldots a_{m2}$
$\vdots$	$\vdots \quad \vdots$	$\vdots$	$\vdots \quad \vdots$
0	$0 \ldots 1$	a_{1k}	$a_{2k} \ldots a_{mk}$

The matrix on the left-hand side here is the identity matrix of order k. The ith row in this left-hand portion of the tableau contains the coefficients of $A_{.1}, \ldots, A_{.k}$ in the expression for the vector in this row on the right-hand portion of the tableau. Whenever row operations are carried out, they are also performed on the left-hand portion of the tableau. In a general stage of the algorithm, if the entries in the ith row of the tableau are:

$\beta_{i1} \quad \beta_{i2} \ldots \beta_{ik}$	$\bar{a}_{1i} \quad \bar{a}_{2i} \ldots \bar{a}_{mi}$

this implies that the vector $(\bar{a}_{1i}, \ldots, \bar{a}_{mi})$ is equal to $\sum_{j=1}^{k} \beta_{ij} A_{.j}$. The algorithm starts with the original tableau.

Step 1 Pick row 1 for examining.

Step 2 If all the entries in the right-hand portion of the tableau in the row being examined are zero, the left-hand portion in this row provides an expression for the zero vector as a nonzero linear combination of the vectors $A_{.1}, \ldots, A_{.k}$; hence, the set is linearly dependent, terminate. If not, go to Step 3. The row that is being examined becomes the *pivot row* in Step 3.

Step 3 If the pivot row is the last row in the tableau, conclude that the set $\{A_{.1}, \ldots, A_{.k}\}$ is linearly independent and terminate. Otherwise, choose a nonzero element in the right-hand portion of the tableau in the pivot row as the pivot element, its column in the current tableau as the *pivot column*, and perform the pivot operation (see Note 3.2). From the nature of the pivot operation it is clear that each of the rows in the new tableau is a nonzero linear combination of the original vectors $A_{.1}, \ldots, A_{.k}$. In the new tableau pick the next row for examining and go back to Step 2.

Note 3.2: The phrase *pivot operation* in standard terminology requires computations (i) and (ii) defined in Section 2.3.1. In our application here, (ii) is actually unnecessary. Hence, whenever a pivot operation is called for in this algorithm, some computational effort can be saved by performing only (i) and omitting (ii).

After at most k pivot steps, the above algorithm will either conclude that the set $\{A_{.1}, \ldots, A_{.k}\}$ is linearly independent, or it finds an expression for the zero vector as a nonzero linear combination of the given vectors $\{A_{.1}, \ldots, A_{.k}\}$.

Example 3.9

Check whether the set of column vectors of the matrix A is a linearly independent set.

$$A = \begin{bmatrix} 0 & 1 & 1 & 2 \\ 1 & 0 & 1 & 2 \\ 2 & 1 & 1 & 4 \\ 0 & 1 & 1 & 2 \\ 1 & 0 & 1 & 2 \end{bmatrix}$$

Let $A_{.j}$ be the jth column vector of A. Here is the tableau.

Vector				Pivot Column					
$A_{.1}$	$A_{.2}$	$A_{.3}$	$A_{.4}$						
1	0	0	0	0	①	2	0	1	*Pivot Row*
0	1	0	0	1	0	1	1	0	
0	0	1	0	1	1	1	1	1	
0	0	0	1	2	2	4	2	2	

Pick the first row as the pivot row and the nonzero element "1" in the second column as the pivot element. The pivot operation leads to the next tableau and we continue. Pivot elements are circled.

Vector								
$A_{.1}$	$A_{.2}$	$A_{.3}$	$A_{.4}$					
1	0	0	0	0	1	2	0	1
0	1	0	0	①	0	1	1	0
−1	0	1	0	1	0	−1	1	0
−2	0	0	1	2	0	0	2	0

Vector								
$A_{.1}$	$A_{.2}$	$A_{.3}$	$A_{.4}$					
1	0	0	0	0	1	2	0	1
0	1	0	0	1	0	1	1	0
−1	−1	1	0	0	0	(−2)	0	0
−2	−2	0	1	0	0	−2	0	0

Vector								
$A_{.1}$	$A_{.2}$	$A_{.3}$	$A_{.4}$					
0	−1	1	0	0	1	0	0	1
$-\frac{1}{2}$	$\frac{1}{2}$	$\frac{1}{2}$	0	1	0	0	1	0
$\frac{1}{2}$	$\frac{1}{2}$	$-\frac{1}{2}$	0	0	0	1	0	0
−1	−1	−1	1	0	0	0	0	0

The last row in the right-hand side of the tableau has become the zero vector. So $-A_{.1} - A_{.2} - A_{.3} + A_{.4} = 0$, which exhibits the linear dependence of the set of column vectors $\{A_{.1}, A_{.2}, A_{.3}, A_{.4}\}$

Example 3.10

From the above example it is clear that the set of column vectors $\{A_{.1}, A_{.2}, A_{.3}\}$ is linearly independent.

Exercises

3.10 Is the set of column vectors of the following matrix D linearly independent?

$$D = \begin{bmatrix} 1 & 2 & 1 & 1 \\ 0 & 0 & 1 & 0 \\ 2 & 3 & 0 & 2 \\ 1 & 1 & 0 & 3 \\ 3 & 0 & 0 & 0 \end{bmatrix}$$

Before proceeding further, the reader should look at Section 3.21, in which the issue of precision in computation is discussed.

3.3.2 The Inverse of a Nonsingular Square Matrix

Inverse matrices are only defined for square matrices. If D is a square matrix of order m, its inverse is defined to be a square matrix E of order m satisfying $ED = DE = I$, where I is the unit matrix of order m. The inverse of D exists iff the set of row vectors of D is linearly independent, and such matrices are known as *nonsingular square matrices*. A square matrix in which the set of row vectors is linearly dependent is known as a *singular square matrix*. If D is a nonsingular square matrix, its inverse is normally denoted by D^{-1}.

A square matrix that can be transformed into the unit matrix by rearranging its row vectors is known as a *permutation matrix*. The matrix of order 5 in (1.2), Example 1.5 of Section 1.1.4 is a permutation matrix of order 5. If P is a permutation matrix of order m and A is any matrix of order $m \times n$, then PA is a matrix that is obtained by permuting the row vectors of A. That's why P is called a permutation matrix.

Exercises

3.11 If P is a permutation matrix and P^{T} is its transpose, prove that $PP^{\mathrm{T}} = P^{\mathrm{T}}P = I$, the identity matrix. Hence, if P is a permutation matrix, $P^{-1} = P^{\mathrm{T}}$.

Let B be a square matrix of order m, and b be a given column vector of order $m \times 1$. Consider the following problems: (1) To find B^{-1}. (2) To find the unique expression for b as a linear combination of the column vectors of B. We discuss a method for solving these problems by performing row operations. Let

I be the unit matrix of order m. Start with the following initial tableau.

Initial Tableau

B	I	b

Step 1 Pick the first row of the tableau to examine.

Step 2 The row being examined is known as the *pivot row*. If the pivot row consists of zero entries in all the first m columns of the current tableau, the set of row vectors of B is not linearly independent and B must be singular. Hence, B^{-1} does not exist. Terminate. Otherwise, pick a nonzero entry in the pivot row among the first m columns of the tableau and make it the pivot element. The column containing the pivot element is the *pivot column*. Perform the pivot operation.

Step 3 If the pivot row is the last row in the tableau, go to Step 4. Otherwise, pick the next row to examine and go back to Step 2.

Step 4 If all the rows in the tableau have been examined, the row operations performed so far would have transformed the matrix contained among the first m column vectors of the tableau (which was B originally) into a *permutation matrix*. Let this matrix be P. Rearrange the rows of the tableau so that P becomes the identity matrix. This operation is equivalent to multiplying every column in the tableau on the left by P^{T}.

The square matrix contained among the $m+1$ to $2m$ columns of the final tableau is B^{-1}. (Notice that the initial tableau contained the identity matrix in this same position.)

Let $B_{.j}$ denote the jth column vector of the original matrix, B. If the last column in the final tableau is $\bar{b} = (\bar{b}_1, \ldots, \bar{b}_m)^{\mathrm{T}}$, then $b = \sum \bar{b}_j B_{.j}$; this gives the unique expression of b as a linear combination of the column vectors of B. The column vector $\bar{b}$ is known as the *representation* of b as a linear combination of the column vectors of B. Actually, $\bar{b} = B^{-1}b$.

Example 3.11

Let

$$B = \begin{pmatrix} 0 & 2 & -1 & -2 \\ 1 & 0 & 1 & 1 \\ 2 & 1 & 0 & -1 \\ 3 & -1 & 2 & 1 \end{pmatrix}$$

Suppose it is required to compute B^{-1} and the expressions for the vectors $b^1 = (1, -2, 1, 0)$ and $b^2 = (-2, 2, 1, -1)$ as linear combinations of the column vectors of B. In the following, the pivot element in each step is circled.

								b^1	b^2
0	2	(−1)	−2	1	0	0	0	1	−2
1	0	1	1	0	1	0	0	−2	2
2	1	0	−1	0	0	1	0	1	1
3	−1	2	1	0	0	0	1	0	−1
0	−2	1	2	−1	0	0	0	−1	2
(1)	2	0	−1	1	1	0	0	−1	0
2	1	0	−1	0	0	1	0	1	1
3	3	0	−3	2	0	0	1	2	−5
0	−2	1	2	−1	0	0	0	−1	2
1	2	0	−1	1	1	0	0	−1	0
0	−3	0	(1)	−2	−2	1	0	3	1
0	−3	0	0	−1	−3	0	1	5	−5
0	4	1	0	3	4	−2	0	−7	0
1	−1	0	0	−1	−1	1	0	2	1
0	−3	0	1	−2	−2	1	0	3	1
0	(−3)	0	0	−1	−3	0	1	5	−5
0	0	1	0	$\frac{5}{3}$	0	−2	$\frac{4}{3}$	$-\frac{1}{3}$	$-\frac{20}{3}$
1	0	0	0	$-\frac{2}{3}$	0	1	$-\frac{1}{3}$	$\frac{1}{3}$	$\frac{8}{3}$
0	0	0	1	−1	1	1	−1	−2	6
0	1	0	0	$\frac{1}{3}$	1	0	$-\frac{1}{3}$	$-\frac{5}{3}$	$\frac{5}{3}$
1	0	0	0	$-\frac{2}{3}$	0	1	$-\frac{1}{3}$	$\frac{1}{3}$	$\frac{8}{3}$
0	1	0	0	$\frac{1}{3}$	1	0	$-\frac{1}{3}$	$-\frac{5}{3}$	$\frac{5}{3}$
0	0	1	0	$\frac{5}{3}$	0	−2	$\frac{4}{3}$	$-\frac{1}{3}$	$-\frac{20}{3}$
0	0	0	1	−1	1	1	−1	−2	6

Hence:

$$B^{-1} = \begin{pmatrix} -\frac{2}{3} & 0 & 1 & -\frac{1}{3} \\ \frac{1}{3} & 1 & 0 & -\frac{1}{3} \\ \frac{5}{3} & 0 & -2 & \frac{4}{3} \\ -1 & 1 & 1 & -1 \end{pmatrix}$$

It can be verified that $BB^{-1} = B^{-1}B = I$, where I is the unit matrix of order 4. Also we see that the representation of b^1 is $(1/3, -5/3, -1/3, -2)^{\mathrm{T}}$; that is, $b^1 = (1/3)B_{.1} + (-5/3)B_{.2} + (-1/3)B_{.3} + (-2)B_{.4}$ and $(1/3, -5/3, -1/3, -2)^{\mathrm{T}} = B^{-1}b^1$. Similarly, the representation of b^2 is $(8/3, 5/3, -20/3, 6)^{\mathrm{T}}$.

Example 3.12

Find the inverse of

$$D = \begin{pmatrix} 1 & 0 & 2 & 3 \\ 0 & 2 & 0 & 2 \\ -1 & 1 & 0 & 0 \\ 1 & -1 & 1 & 1 \end{pmatrix}$$

Applying the algorithm, we get the following. Pivot elements are circled.

(1)	0	2	3	1	0	0	0
0	2	0	2	0	1	0	0
−1	1	0	0	0	0	1	0
1	−1	1	1	0	0	0	1
1	0	2	3	1	0	0	0
0	(2)	0	2	0	1	0	0
0	1	2	3	1	0	1	0
0	−1	−1	−2	−1	0	0	1
1	0	2	3	1	0	0	0
0	1	0	1	0	$\frac{1}{2}$	0	0
0	0	(2)	2	1	$-\frac{1}{2}$	1	0
0	0	−1	−1	−1	$\frac{1}{2}$	0	1
1	0	0	1	0	$\frac{1}{2}$	−1	0
0	1	0	1	0	$\frac{1}{2}$	0	0
0	0	1	1	$\frac{1}{2}$	$-\frac{1}{4}$	$\frac{1}{2}$	0
0	0	0	0	$-\frac{1}{2}$	$\frac{1}{4}$	$\frac{1}{2}$	1

Since the last row has all zero entries in the left hand portion of the last tableau, this implies that D is a singular matrix. Hence, D^{-1} does not exist.

Exercises

3.12 Let

$$A = \begin{bmatrix} 0 & 2 & 3 & 1 \\ 1 & 0 & 2 & 2 \\ 0 & -1 & 0 & -3 \\ 1 & 0 & -2 & 0 \end{bmatrix} \quad B = \begin{bmatrix} 0 & 1 & -1 & -1 \\ 0 & 0 & 2 & 3 \\ 1 & 2 & 0 & 1 \\ -1 & 0 & 0 & 0 \end{bmatrix}$$

Find A^{-1} and the unique expressions for $(1, 2, 3, 4)^{\mathrm{T}}$ and $(-2, 1, 2, 1)^{\mathrm{T}}$ as linear combinations of the column vectors of A.

3.13 Apply the inverse finding algorithm on B given above.

3.3.3 The Rank of a Matrix

If $\mathbf{F} \subset \mathbf{R}^n$ is linearly independent, clearly every subset of $\mathbf{F}$ is also linearly independent. However, when a new vector is introduced into $\mathbf{F}$, the resulting set may not be linearly independent.

Let Γ be a subset of row vectors of A. Γ is said to be a *maximal linearly independent subset of row vectors of A* if it satisfies the following properties: (1) it is linearly independent and (2) either Γ contains all the row vectors of A, or, if it does not contain all row vectors of A, we cannot include any more of the remaining row vectors in Γ without losing the linear independence property. Clearly, if Γ is a maximal linearly independent subset of row vectors of A, every row vector of A can be expressed as a linear combination of row vectors in Γ. Here is a fundamental result from linear algebra: given a matrix A, every maximal linearly independent subset of row vectors of A contains the same number of row vectors in it. This number (the number of row vectors in any maximal linearly independent subset of row vectors of A) is known as the *rank* of the matrix A.

Algorithm for Finding the Rank of A Given Matrix

Given a matrix A of order $m \times n$, its rank can be found by performing pivots on it. The following procedure not only finds the rank of A, it also finds a maximal linearly independent subset of rows of A and an expression for each of the remaining rows of A as a linear combination of rows from this subset.

Step 1 Enter the matrix in the form of a tableau, and put a unit matrix of order m on its left-hand side. Label the ith column vector of this unit matrix by $A_{i.}$

$A_{1.}$	$A_{2.} \dots A_{i.} \dots A_{m.}$		
1	$0 \dots 0 \dots 0$	a_{11}	$a_{12} \dots a_{1n}$
0	$1 \dots 0 \dots 0$	a_{21}	$a_{22} \dots a_{2n}$
$\vdots$	$\vdots \quad \ddots \quad \vdots \quad\quad \vdots$	$\vdots$	$\vdots \quad\quad \vdots$
0	$0 \dots 1 \dots 0$	a_{i1}	$a_{i2} \dots a_{in}$
$\vdots$	$\vdots \quad\quad \vdots \quad \ddots \quad \vdots$	$\vdots$	$\vdots \quad\quad \vdots$
0	$0 \dots \quad \dots 1$	a_{m1}	$a_{m2} \dots a_{mn}$

As in Section 3.3.1, at any stage of the algorithm, the ith row on the left-hand side of the tableau contains the coefficients of $A_{1.}, \dots, A_{m.}$ in the expression for the vector in this row on the right-hand portion of this tableau. Select row 1 of the tableau for examining and go to Step 2.

Step 2 Let t be the number of the row being examined in the current tableau. If all the entries in the right-hand portion of the current tableau in row t are zero, go to Step 5 if $t = m$, or to Step 4 if $t < m$. If there are some nonzero entries in the right-hand portion of the current tableau in row t, go to Step 5 if $t = m$, or select this row as the pivot row and go to Step 3 if $t < m$.

Step 3 Select a nonzero entry in the right-hand portion of the tableau in the pivot row and make it the pivot element. The column in which the pivot element lies is the pivot column. Perform the pivot operation. The comments made in Note 3.2 apply here as well.

Step 4 Select the next row for examining and go to Step 2.

Step 5 Identify the rows in the tableau at this stage that have nonzero entries in the right-hand portion of the tableau. Suppose these are rows $i_1, \dots, i_r$. Then the rank of the matrix A is r, and $\{A_{i_1.}, A_{i_2.}, \dots, A_{i_r.}\}$ is a maximal linearly independent subset of row vectors of A.

If row t in the tableau at this stage has all zero entries in the right-hand portion, and if $\beta_{t1}, \dots, \beta_{tm}$ are the entries in the left-hand portion of the tableau at this stage in row t, then $\sum_{u=1}^{m} \beta_{tu} A_{u.} = 0$. β_{tt} will be equal to 1 and

$$\sum_{\substack{u=1 \\ u \neq t}}^{m} (-\beta_{tu}) A_{u.}$$

is an expression for $A_{t.}$ as a linear combination of row vectors from the maximal linearly independent subset obtained above.

Example 3.13

Find the rank of the matrix A given on the right-hand side of the first tableau below, a maximal linearly independent subset of row vectors of it, and an expression for each of its other rows as a linear combination of rows from its maximal set. The original tableau is the first tableau given below. In each tableau, the pivot element is circled.

$A_{1.}$	$A_{2.}$	$A_{3.}$	$A_{4.}$	$A_{5.}$						
1	0	0	0	0	①	2	−1	1	0	−2
0	1	0	0	0	2	4	−2	2	0	−4
0	0	1	0	0	0	0	3	1	2	2
0	0	0	1	0	2	4	1	3	2	−2
0	0	0	0	1	1	−1	−1	1	−2	1
1	0	0	0	0	1	2	−1	1	0	−2
−2	1	0	0	0	0	0	0	0	0	0
0	0	1	0	0	0	0	3	①	2	2
−2	0	0	1	0	0	0	3	1	2	2
−1	0	0	0	1	0	−3	−2	0	−2	3
1	0	−1	0	0	1	2	−4	0	−2	−4
−2	1	0	0	0	0	0	0	0	0	0
0	0	1	0	0	0	0	3	1	2	2
−2	0	−1	1	0	0	0	0	0	0	0
−1	0	0	0	1	0	−3	−2	0	−2	3

The nonzero rows in the right-hand portion of the terminal tableau are rows 1, 3, 5. Hence $\{A_{1.}, A_{3.}, A_{5.}\}$ is a maximal linearly independent subset of row vectors of A, and the rank of A is 3. From rows 2 and 4 in the final tableau, we have $-2A_{1.} + A_{2.} = 0$, and $-2A_{1.} - A_{3.} + A_{4.} = 0$. These equations give $A_{2.} = 2A_{1.}$ and $A_{4.} = 2A_{1.} + A_{3.}$, which are the required expressions.

In the above discussion we only used the row vectors of A and, hence, the rank defined there might appropriately be called the *row rank of the matrix A*. The *column rank of the matrix A* can be defined in a similar manner by replacing the word "row" in the above discussion by the word "column." This will be the same as the row rank of the matrix A^{T}. The *rank theorem* in linear algebra states that the row rank of any matrix is equal to its column rank. This number is therefore called the rank of the matrix. If A is of order $m \times n$, and rank r, clearly $r \leqq m$ and $r \leqq n$.

3.3.4 The Rank of a Given Set of Vectors from $\mathbf{R}^n$

Let $\mathbf{F} = \{A_1, \ldots, A_k\}$ be a nonempty set of vectors in $\mathbf{R}^n$. A subset $\mathbf{E} \subset \mathbf{F}$ is said to be a *maximal linearly independent subset of* $\mathbf{F}$ if (1) $\mathbf{E}$ is linearly independent, and (2) either $\mathbf{E} = \mathbf{F}$, or for every $A_j \in \mathbf{F} \backslash \mathbf{E}$, $\mathbf{E} \cup \{A_j\}$ is linearly dependent. A fundamental theorem in linear algebra states that all maximal linearly independent subsets of $\mathbf{F}$ have the same cardinality and this number is known as the *rank* of the set $\mathbf{F}$. See references [3.9, 3.11–3.14]. To find the rank of a given set of vectors $\mathbf{F}$ from $\mathbf{R}^n$, and to find a maximal linearly independent subset of $\mathbf{F}$, write down the vectors in $\mathbf{F}$ as the row vectors of a matrix and then apply the algorithm discussed above to compute the rank of this matrix.

Let $\Gamma \subset \mathbf{R}^n$ and let $\mathbf{F} \subset \Gamma$ satisfy the property that every vector in Γ can be expressed as a linear combination of vectors in $\mathbf{F}$. Then $\mathbf{F}$ is said to be a *spanning subset of* Γ. For example, in the set of vectors $\{(1, 0, 0), (0, 1, 0), (-2, -1, 0), (1, -2, 0)\}$, the subset $\{(1, 0, 0), (0, 1, 0)\}$ is a spanning subset.

If $\mathbf{F}$ is a spanning subset of $\Gamma \subset \mathbf{R}^n$, proper subsets of $\mathbf{F}$ may not have this property. However subsets obtained by including additional vectors of Γ into $\mathbf{F}$ all have the spanning property.

A *minimal spanning subset* of Γ is a subset $\mathbf{F} \subset \Gamma$ satisfying: (1) $\mathbf{F}$ is a spanning subset of Γ, and (2) no proper subset of $\mathbf{F}$ is a spanning subset of Γ. For example $\{(1, 0, 0), (0, 1, 0)\}$ is a minimal spanning subset of $\{(1, 0, 0), (0, 1, 0), (-2, -1, 0), (1, -2, 0)\}$.

A theorem in linear algebra states that all minimal spanning subsets of Γ have the same cardinality, and it is the same as the cardinality of any maximal linearly independent subset of Γ and, hence, the same as the rank of Γ. Actually, every minimal spanning subset of Γ is a maximal linearly independent subset of Γ and vice versa.

Let A be a given matrix of order $m \times n$. A subset of column vectors of A that is a maximal linearly independent subset and, thus, a minimal spanning subset of column vectors of A is called a *basis for A.* If the rank of A is m, then a basis for A is a set of m column vectors of A that is linearly independent, or the matrix that consists of these column vectors. In other words, when the rank of A is m, a basis for A is a square nonsingular submatrix of A of order m. Given a basis B for A, columns of A in the basis are called *basic columns* and the columns of A not in the basis are called *nonbasic columns.* Every nonbasic column can be expressed as a linear combination of the basic columns (by the spanning property of the basis) and this expression is unique (because the basis is a linearly independent set).

3.3.5 Systems of Simultaneous Linear Equations

Let A be an $m \times n$ matrix and b an $m \times 1$ column vector. The system

$$Ax = b \tag{3.5}$$

is a system of m linear equations in n unknowns. The ith equation in this system is $A_{i.}x = b_i$, for $i = 1$ to m. This system of equations can also be viewed as $\sum_{j=1}^{n} A_{.j}x_j = b$. Hence, if x is a solution of this system, it is the vector of coefficients in an expression of b as a linear combination of the column vectors of A. Thus (3.5) has a solution iff b can be expressed as a linear combination of the column vectors of A. This leads to the following standard result of linear algebra.

Result 3.1: (3.5) has a solution iff the rank of the augmented matrix $(A \vdots b)$ is equal to the rank of A. If rank $(A \vdots b) = 1 + \text{rank}\ (A)$, (3.5) has no solution.

In other words, (3.5) has a solution iff the column vector b lies in the linear hull of the set of column vectors of A. Thus, for studying the system of equations (3.5), the space of the column vectors of A plays a major role.

Note 3.3: The set of equations $\{p_1, \ldots, p_t\}$ from (3.5), where $t \geqq 2$ is said to form a *cycle of linearly dependent equations* if there exist real numbers $\alpha_1, \ldots, \alpha_t$, all of them different from zero, such that when the p_kth equation in (3.5) is multiplied on both sides by α_k, and summed over $k = 1$ to t, we get an equation in which the coefficients of all the variables and the right-hand side constant are all zero. An equation in system (3.5), which can be obtained as a linear combination of the other equations in the system, is known as a *redundant equation.* Thus each equation in a cycle of linearly dependent equations can be treated as a redundant equation, since it can be obtained as a linear combination of the other equations in the cycle, and conversely every redundant equation in (3.5) lies in some cycle of linearly dependent equations. If a cycle of linearly dependent equations is identified in (3.5), exactly one of the equations in the cycle can be selected arbitrarily and eliminated from the system. This does not change the set of solutions of the system. The same procedure can be repeated with the resulting system.

Exercises

3.14 If A is a square nonsingular matrix, prove that (3.5) has a solution and that this solution is unique.

3.15 Prove that (3.5) has a unique solution iff it has a solution and the set of column vectors of A is linearly independent; that is iff rank $(A \vdots b) =$ rank $(A) =$ number of columns in A.

The system of equations (3.5) is said to be *consistent* if it has a solution. Otherwise it is *inconsistent*. Let the rank of $A =$ rank of $(A \vdots b) = r$. If $r = m$, all the equations in the system (3.5) are said to be linearly independent. In this case, (3.5) is known as a *linearly independent system of equations*. This implies that there are no redundant equations in it. If $r < m$, system (3.5) has redundant equations, and exactly $m - r$ equations can be eliminated from it before it becomes a linearly independent system of equations. The algorithm that follows helps in identifying those equations that can be eliminated one after the other. After the elimination, (3.5) gets transformed into an equivalent system of r equations. Since r is the rank of the matrix A, $r \leqq m$ and $r \leqq n$. Hence, every consistent system of linear equations is equivalent to a system in which the number of equations is less than or equal to the number of unknowns.

The system of constraints (excluding the sign restrictions on variables) in an LP in standard form is a system of equations. Hence in discussing LPs in standard form, we can assume that the number of equality constraints is less than or equal to the number of variables.

The system of equations (3.5) can be solved by transforming it into *row echelon normal form*. The row echelon normal form represents an equivalent system of equations obtained by performing row operations on (3.5) and eliminating any redundant equations. The method for obtaining it is discussed below.

3.3.6 The Gauss–Jordan Elimination Method

Step 1 Express (3.5) in detached coefficient form as in the tableau below. Pick the first row of the tableau for examining and go to Step 2.

$x_1 \ldots x_j \ldots x_n$	b
$a_{11} \ldots a_{1j} \ldots a_{1n}$	$= b_1$
$\vdots \qquad \vdots \qquad \vdots$	$\vdots$
$a_{i1} \ldots a_{ij} \ldots a_{in}$	$= b_i$
$\vdots \qquad \vdots \qquad \vdots$	$\vdots$
$a_{m1} \ldots a_{mj} \ldots a_{mn}$	$= b_m$

Step 2 The row being examined in the present tableau is known as the *pivot row* for this stage. If all the entries in the pivot row on both sides of the tableau are zero, this row corresponds to a redundant equation in the system. Eliminate this row from the tableau and go to Step 3.

If all the entries in the pivot row in the present tableau on the left-hand side are zero, and the entry in the pivot row on the right-hand side is nonzero, this row represents the inconsistent equation, "0 = a nonzero number." Since this equation is obtained as a linear combination of equations from the original system, that system must be inconsistent. Hence, the system has no solution and we terminate.

If a nonzero entry exists in the pivot row on the left-hand side of the present tableau, pick any one of those entries as the pivot element. The column in which the pivot element lies is the *pivot column*. The variable corresponding to the pivot column is known as the *dependent variable* (or *basic variable* in linear programming terminology) *in the present pivot row*. Perform the pivot operation. (In linear algebra, this operation is known as *Gauss–Jordan elimination step* or a *Gauss–Jordan pivot step*).

Step 3 If there are some more rows in the tableau to be examined, pick the next row for examining and go to Step 2.

Suppose all the rows have been examined and the system is not inconsistent. For convenience in notation, assume that the dependent (basic) variables selected are $x_1, \ldots, x_r$ in that order. The columns of these variables in the final tableau constitute a permutation matrix. When these are arranged in proper order, it becomes the unit matrix, and the final tableau will be as below.

Dependent (basic) Variable in Row	$x_1 \dots x_r$	$x_{r+1} \dots x_j \dots x_n$	
x_1	$1 \dots 0$	$\bar{a}_{1,r+1} \dots \bar{a}_{1j} \dots \bar{a}_{1n}$	$\bar{b}_1$
$\vdots$	$\vdots \quad \vdots$	$\vdots \quad \vdots \quad \vdots$	$\vdots$
x_r	$0 \dots 1$	$\bar{a}_{r,r+1} \dots \bar{a}_{r,j} \dots \bar{a}_{rn}$	$\bar{b}_r$

The number of rows in this tableau, r, will be equal to m, if no redundant constraints were found in Step 2. Otherwise, $m - r$ is the number of redundant constraints eliminated from the system. The rank of the matrix A is r. All the variables other than the dependent variables are known as *independent variables* (*nonbasic variables* in linear programming terminology). A solution to the problem is obtained by giving arbitrary values to the independent variables, substituting these values in the final tableau, and then obtaining the values of the dependent variables. In particular, a solution is

$$\text{All independent variables} = 0$$
$$\text{Dependent variables in } i\text{th row} = \bar{b}_i \qquad i = 1 \text{ to } r$$

Note 3.4: This algorithm for solving a system of linear equations cannot guarantee that the solution obtained will be nonnegative. Hence this algorithm cannot be used directly to find a *feasible solution* (i.e., a nonnegative solution) of an LP in standard form.

Note 3.5: The algorithm (based on Gauss–Jordan elimination) discussed here for solving systems of linear equations may not perform very well from the numerical analysis point of view. Round-off errors can accumulate in this algorithm. The newer algorithms discussed in numerical analysis books reduce the round-off error accumulation. Most computer programs in use today are based on these newer algorithms. Some of these newer algorithms are discussed later in this text.

Note 3.6: Assume that (3.5) is a linearly independent system. Let the row-echelon normal form obtained for it be as above, with $r = m$. The column vector of the variable x_j in the row echelon normal form is $\bar{A}_{.j} = (\bar{a}_{1j}, \dots, \bar{a}_{mj})^T$, and in the original system it is $A_{.j} = (a_{1j}, \dots, a_{mj})^T$. Let B be the square matrix consisting of the column vectors of the basic variables in their proper order in the original system. Then from Section 3.3.2, it is clear that $\bar{A}_{.j} = B^{-1} A_{.j}$. Hence $\bar{A}_{.j}$ is the vector of coefficients in the expression of $A_{.j}$ as a linear combination of the column vectors of the basic variables in the original system; that is, it is the representation of $A_{.j}$ as a linear combination of the column vectors of B. The same result holds for the canonical tableaux obtained when an LP in standard form is solved by the simplex method. As an example consider the LP in Example 2.11 of Section 2.6.5. The column vector corresponding to x_4 in the optimum canonical tableau with respect to the basic vector (x_1, x_5) is $(1, 3/40)^T$. The column vectors corresponding to x_4, x_1, x_5 in the original tableau for this problem are $(6, 4)^T$, $(4.5, 1)^T$, $(20, 40)^T$, respectively. Clearly $(6, 4)^T = (4.5, 1)^T + (3/40)\ (20, 40)^T$, which verifies this result in this case.

Example 3.14 Illustration of the Gauss–Jordan Elimination Method

Consider the system of linear equations in the first tableau below. The various tableaux obtained when it is solved by the Gauss–Jordan elimination method are given below. The pivot element in each step is circled.

Dependent Variable in Row	x_1	x_2	x_3	x_4	x_5	b
	(1)	−2	2	−1	1	−8
	−1	0	4	−7	7	16
	0	−2	6	−8	8	6
x_1	1	−2	2	−1	1	−8
	0	(−2)	6	−8	8	8
	0	−2	6	−8	8	6
x_1	1	0	−4	7	−7	−16
x_2	0	1	−3	4	−4	−4
	0	0	0	0	0	−2

From the last row in the last tableau we conclude that this system is inconsistent. It has no solution.

Example 3.15 Illustration of the Gauss–Jordan Elimination Method

Consider the system of linear equations in the first tableau given below. Application of the Gauss–

$$
\begin{aligned}
d_{11}y_1 + d_{12}y_2 + d_{13}y_3 + \cdots + d_{1,m-1}y_{m-1} \quad &+ d_{1m}y_m &= d_1 \\
d_{22}y_2 + d_{23}y_3 + \cdots + d_{2,m-1}y_{m-1} \quad &+ d_{2m}y_m &= d_2 \\
d_{33}y_3 + \cdots + d_{3,m-1}y_{m-1} \quad &+ d_{3m}y_m &= d_3 \\
\vdots \qquad\qquad &\quad \vdots & \vdots \\
d_{m-1,m-1}y_{m-1} &+ d_{m-1,m}y_m &= d_{m-1} \\
&d_{mm}y_m &= d_m
\end{aligned} \tag{3.6}
$$

Jordan elimination method leads to the following tableaux. The pivot element in each step is circled.

Dependent Variable in Row	x_1	x_2	x_3	x_4	x_5	x_6	b
	0	(1)	−1	2	−2	1	−7
	−1	−1	−2	3	−1	−1	4
	−1	0	−3	5	−3	0	−3
	1	0	2	−1	2	−2	8
x_2	0	1	−1	2	−2	1	−7
	(−1)	0	−3	5	−3	0	−3
	−1	0	−3	5	−3	0	−3
	1	0	2	−1	2	−2	8
x_2	0	1	−1	2	−2	1	−7
x_1	1	0	3	−5	3	0	3
	0	0	0	0	0	0	0
	0	0	(−1)	4	−1	−2	5
x_2	0	1	0	−2	−1	3	−12
x_1	1	0	0	7	0	−6	18
x_3	0	0	1	−4	1	2	−5

In Step 3, the third equation became a redundant equation, and it was eliminated. The general solution for the system is

$$
\begin{aligned}
x_2 &= -12 - (-2x_4 - x_5 + 3x_6) \\
x_1 &= 18 - (7x_4 - 6x_6) \\
x_3 &= -5 - (-4x_4 + x_5 + 2x_6) \\
&x_4, x_5, x_6 \text{ arbitrary}
\end{aligned}
$$

In particular, giving the value 0 to all the independent variables, x_4, x_5, x_6 leads to the solution $x = (18, -12, -5, 0, 0, 0)^T$ of the system.

For solving (3.5) a modified method known as the *Gaussian elimination method* is much more numerically stable and also computationally more efficient; this method is discussed below. However, we need the results in Section 3.3.7 before we can discuss it.

3.3.7 Back-Substitution Method for Solving a Square Nonsingular, Upper-Triangular System of Linear Equations

A square nonsingular matrix $D = (d_{ij})$ is said to be an *upper-triangular matrix* if $d_{ij} = 0$ for all $i > j$. If D is upper triangular, the system of equations $Dy = d$ is of the form given at the top.

If D is upper triangular, its nonsingularity automatically implies that all its diagonal entries are nonzero. The value of y_m in the solution of the above system is obtained from the last equation to be d_m/d_{mm}. Substituting this value for y_m in the $(m-1)$th equation, the value of y_{m-1} can be obtained. In this manner, when the values of $y_m, \ldots, y_{t+1}$ are computed, they can be substituted in the tth equation and then the value of y_t in the solution can be obtained from it, for any $t = m-1, \ldots, 1$. This is called the *method of solving by back substitution.*

Example 3.16

Consider the following system:

y_1	y_2	y_3	y_4	
2	0	−2	1	−1
0	1	2	0	9
0	0	3	6	30
0	0	0	−2	−6

Applying back substitution, we get the values of the variables in the solution to be $(y_4, y_3, y_2, y_1) = (3, 4, 1, 2)$ in that order.

Let $D = (d_{ij})$ be a nonsingular upper triangular matrix of order m, and let $f = (f_1, \ldots, f_m)$ be a row vector in $\mathbf{R}^m$. Let $\pi = (\pi_1, \ldots, \pi_m)$ be a row vector of variables. Consider the system of linear equations

$$\pi D = f \tag{3.7}$$

In (3.7) the first equation determines π_1 to be f_1/d_{11}. Substituting the value for π_1 in the second equation we can compute the value of π_2 in the solution for (3.7). In a similar way we can continue and obtain the value of π_r from the rth equation by substituting the values of $\pi_1, \ldots, \pi_{r-1}$ obtained from earlier equations in (3.7).

In solving (3.6) by the back-substitution method, we obtained the values of the variables $y_m, y_{m-1}, \ldots, y_1$ in that order. In solving (3.7) by the similar method described above, we obtained the values of the variables $\pi_1, \pi_2, \ldots, \pi_m$ in that order. Hence in some books, the method of solving (3.7) discussed above is called the *forward substitution method*. Since both methods are similar, we will not make any distinction between them here, but refer to them for solving either (3.6) for y or (3.7) for π, when D is upper triangular, by the name *back-substitution method*.

A square nonsingular matrix $L = (\ell_{ij})$ is said to be *lower triangular* if L^T is upper triangular, that is, if $\ell_{ij} = 0$ for all $j > i$. If L is a nonsingular lower-triangular matrix, clearly the system of equations $Ly = a$, or the system $\pi L = b$, can both be solved by the back-substitution method of evaluating the values of the variables in the solution, one at a time, in some order.

3.3.8 The Gaussian Elimination Method for Solving a System of Linear Equations

In the Gauss–Jordan elimination method discussed earlier, the matrix of coefficients of the basic variables in their proper order is transformed into the unit matrix. In the *Gaussian elimination method*, the matrix of coefficients is merely transformed into an upper-triangular matrix. The change needed is that when the pivot is being performed with the tth row as the pivot row, suitable multiples of the pivot row are subtracted from the rows below it (i.e., the rows $t+1$, $t+2$, etc.) to transform all the entries in the pivot column in rows below the pivot row into zero. No changes are made in the pivot row, or the rows lying above it in the tableau. The rows of the tableau are examined serially from top to bottom, one after the other, as in the Gauss–Jordan elimination method. The inconsistency termination criterion and the redundant constraint criterion are also the same as under the Gauss–Jordan elimination method. Once the matrix of coefficients of the basic variables is transformed into an upper-triangular matrix, arbitrary values can be given to the nonbasic variables, their values substituted in the final tableau, the resulting constant terms transferred to the right-hand side, and the values of the basic variables obtained by back substitution from the resulting system.

Example 3.17

Consider the system in the first tableau below. Applying the Gaussian elimination method leads to the following tableaux. In each step, the pivot element is circled.

Dependent Variable in Row	x_1	x_2	x_3	x_4	x_5	x_6	b
	1	(−2)	1	1	1	−1	−4
	2	2	0	1	−1	−1	8
	2	−4	−2	0	−2	−2	−7
	0	0	−1	4	1	1	5
x_2	1	−2	1	1	1	−1	−4
	3	0	1	(2)	0	−2	4
	0	0	−4	−2	−4	0	1
	0	0	−1	4	1	1	5
x_2	1	−2	1	1	1	−1	−4
x_4	3	0	1	2	0	−2	4
	(3)	0	−3	0	−4	−2	5
	−6	0	−3	0	1	5	−3
x_2	1	−2	1	1	1	−1	−4
x_4	3	0	1	2	0	−2	4
x_1	3	0	−3	0	−4	−2	5
x_5	0	0	−9	0	(−7)	1	7

Hence the system of equations, after transforming the independent variables to the right-hand side and rearranging the dependent variables, is

x_2	x_4	x_1	x_5	
−2	1	1	1	$-4-(x_3-x_6)$
0	2	3	0	$4-(x_3-2x_6)$
0	0	3	−4	$5-(-3x_3-2x_6)$
0	0	0	−7	$7-(-9x_3+x_6)$

Giving both the independent variables x_3, x_6 the value zero leads to the solution $(x_1, x_2, x_3, x_4, x_5, x_6)^T$ $(1/3, 29/12, 0, 3/2, -1, 0)^T$.

A *Gauss–Jordan pivot step* transforms the pivot column into the column vector of the unit matrix, with a "1" entry in the pivot row and "0" entries in all the other rows. A *Gaussian pivot step* transforms the pivot column into upper-triangular form, with zero entries in all the rows below the pivot row. The usual simplex algorithm is based on the use of Gauss–Jordan pivots. Hence throughout the book, except Chapter 7, a pivot step always denotes a Gauss–Jordan pivot step. The simplex algorithm using *LU*-decomposition, discussed in Chapter 7, is based on the use of Gaussian pivots.

Improving the Numerical Stability of the Gaussian Elimination Method

Because of round-off errors whenever divisions or multiplications are performed, it is possible that the final solution $\bar{x}$ of (3.5) obtained by the Gaussian elimination method contains some errors and may fail to satisfy (3.5) exactly. Round-off errors cannot be completely avoided and, in practice, the method can be considered to be satisfactory if $\|A\bar{x} - b\|$ is sufficiently small.

When a sequence of arithmetical operations is performed, the round-off errors tend to accumulate. Special pivot element selection rules have been developed for use in the Gaussian elimination method that reduce this round-off error accumulation and improve its numerical stability. We discuss one of these rules here.

Examine the rows of the tableau serially from top to bottom, one after the other, as before. Consider the stage in which the tth row in the tableau is being examined. In this stage, the pivot column can be selected to be any column in which there is at least one nonzero entry in the rows from t to m. If there are no such columns, all the rows from t to m in the tableau correspond to redundant constraints (if the present entries in the right-hand-side constants column are all zero in these rows), or the system of equations is inconsistent (if there is at least one nonzero entry in the present right-hand-side constants column in rows t to m).

Suppose the sth column has been selected as the pivot column in this stage. Let the entries in this column be $\bar{a}_{1s}, \ldots, \bar{a}_{ms}$, in that order. So at least one of $\bar{a}_{ts}, \bar{a}_{t+1,s}, \ldots, \bar{a}_{ms}$ is nonzero. Now find r between t and m so that

$$|\bar{a}_{rs}| = \text{maximum } \{|\bar{a}_{is}| : i = t \text{ to } m\}$$

Break ties for r in the above equation arbitrarily. The special pivot element selection rule is to choose $\bar{a}_{rs}$ as the pivot element for the pivot to be performed in this step. Interchange the rth row and the tth row in the present tableau, and after the interchange perform a Gaussian pivot step with the sth column as the pivot column and the new tth row as the pivot row. After this pivot step continue in the same way.

In Gaussian elimination, this strategy of selecting the pivot element in the pivot column to be the updated element of maximum absolute value among those in the pivot column on or below the pivot row (and then interchanging rows if the selected element is strictly below the pivot row) is called the *partial pivoting strategy* in numerical analysis. An alternate strategy called the *complete pivoting strategy* first selects the pivot element as the maximum absolute value element among all the updated elements in nonbasic columns and in the pivot row or below, and then defines the pivot column to be the column containing the pivot element (it then performs row interchange if the pivot element is strictly below the pivot row and continues). For numerical stability the complete pivoting strategy would be better than the partial pivoting strategy, but it involves a lot more work for selecting the pivot element and is therefore not very popular. Computational experience indicates that partial pivoting strategy works almost as well as the complete pivoting strategy.

We will illustrate the application of the Gaussian elimination method with this special pivot element selection rule, on the following system.

x_1	x_2	x_3	x_4	x_5	x_6	b
2	−2	1	6	1	−1	−4
2	2	0	−5	−1	−1	8
2	−4	−2	10	−2	−2	−7
0	0	−1	4	1	1	5

We examine the first row and suppose the column of x_2 is chosen as the pivot column. The entries in this column are −2, 2, −4, 0, and the entry with maximum absolute value in this column occurs in row 3. So we interchange rows 1 and 3 leading to the following tableau, and then perform a Gaussian pivot step on it. Pivot elements are circled.

x_1	x_2	x_3	x_4	x_5	x_6	b
2	$\boxed{-4}$	-2	10	-2	-2	-7
2	2	0	-5	-1	-1	8
2	-2	1	6	1	-1	-4
0	0	-1	4	1	1	5

Basic Variable	x_1	x_2	x_3	x_4	x_5	x_6	b
x_2	2	-4	-2	10	-2	-2	-7
	3	0	-1	0	-2	-2	$\frac{9}{2}$
	1	0	2	1	2	0	$-\frac{1}{2}$
	0	0	-1	4	1	1	5

Now examine row 2 in this tableau. Choosing the column of x_4 as the pivot column, we notice that the entries in this column in the row being examined, and the rows below it, are 0, 1, 4, respectively. Among these, the entry in row 4 has maximum absolute value. So we interchange rows 2 and 4, which leads to the following tableau, and perform a Gaussian pivot step.

Basic Variables	x_1	x_2	x_3	x_4	x_5	x_6	b
x_2	2	-4	-2	10	-2	-2	-7
	0	0	-1	$\boxed{4}$	1	1	5
	1	0	2	1	2	0	$-\frac{1}{2}$
	3	0	-1	0	-2	-2	$\frac{9}{2}$
x_2	2	-4	-2	10	-2	-2	-7
x_4	0	0	-1	4	1	1	5
	1	0	$\boxed{\frac{9}{4}}$	0	$\frac{7}{4}$	$-\frac{1}{4}$	$-\frac{7}{4}$
	3	0	-1	0	-2	-2	$\frac{9}{2}$

Now examine row 3 in this tableau. Suppose the column of x_3 is chosen as the pivot column. The entries in this column in the row being examined and below it are 9/4 and 1. Among these, the one of maximum absolute value lies in the row being examined itself, so there is no need to interchange rows in this step. The pivot element is circled. Performing the Gaussian pivot leads to the following tableau.

Basic Variables	x_1	x_2	x_3	x_4	x_5	x_6	b
x_2	2	-4	-2	10	-2	-2	-7
x_4	0	0	-1	4	1	1	5
x_3	1	0	$\frac{9}{4}$	0	$\frac{7}{4}$	$-\frac{1}{4}$	$-\frac{7}{4}$
x_1	$\frac{31}{9}$	0	0	0	$-\frac{11}{9}$	$-\frac{19}{9}$	$\frac{67}{18}$

In the last row, suppose we select x_1 as the basic variable. Giving both the nonbasic variables x_5, x_6 the value zero leads to the solution

$$(x_1, x_2, x_3, x_4, x_5, x_6)^{\mathrm{T}} = (67/62,\ 847/124,\ -39/31,\ 97/62,\ 0,\ 0)^{\mathrm{T}}.$$

The operation that introduces error is the division by the pivot element. Suppose this element has the value a, which is stored with round-off error ε. The error in $1/a$ is $(1/a) - (1/(a-\varepsilon))$, which is approximately ε/a^2. So to minimize the error, $|a|$ should be as large as possible. This is the rationale behind this pivot element selection rule. See references [3.12, 3.13, 5.2], where a complete error analysis of Gaussian elimination method with this special pivot element selection rule is given. With this pivot element selection rule and the row interchanges that go along with it, the Gaussian elimination method becomes numerically stable and can be expected to produce accurate answers.

Here we provide an example to illustrate the effectiveness of this special pivot element selection rule in finite precision arithmetic. Suppose our computer maintains only three significant digits in the decimal representation of each number (i.e., it stores the first nonzero digit, the two digits following it, and the position of the decimal point in the number). Using this computer we want to solve the following system of two equations in two unknowns.

$$0.000100x_1 + 1.00x_2 = 1.00$$
$$1.00x_1 + 1.00x_2 = 2.00$$

In our computer, this system will be stored as follows in detached coefficient tableau form:

x_1	x_2	
$10^{-4}(1.00)$	1.00	1.00
1.00	1.00	2.00

The true rounded solution of this system is $\bar{x} = (\bar{x}_1, \bar{x}_2) = (1.00, 0.999)$. To solve this system, suppose we select the column of x_1 as the pivot column and row 1 as the pivot row. For this pivot step, we have to multiply the first equation by 10^4 and subtract from the second. The coefficient of x_2 and the right-hand side constant term in the updated second equation are -9999 and -9998, respectively. However, since our computer keeps only three significant digits, it will round off each of these numbers to $-10^4(1.00)$. So after the pivot step, the system becomes:

x_1	x_2	
$10^{-4}(1.00)$	1.00	1.00
0	$-10^4(1.00)$	$-10^4(1.00)$

This system leads to the solution $\hat{x} = (\hat{x}_1, \hat{x}_2) = (0.00, 1.00)$, which is quite far away from the true solution $\bar{x}$.

For pivoting in the column of x_1 in the original system, if we had selected the pivot element to be the element of maximum absolute value in the column, the pivot row would have been row 2. So we interchange rows 1 and 2, leading to:

x_1	x_2	
1.00	1.00	2.00
$10^{-4}(1.00)$	1.00	1.00

When the pivot step is performed with row 1 as the pivot row and the column of x_1 as the pivot column, the coefficient of x_2 and the right-hand side constant term in the second row become 0.9999 and 0.9998, respectively, both of which are rounded off to 1.00 by our computer. This leads to:

x_1	x_2	
1.00	1.00	2.00
0.00	1.00	1.00

which gives the solution $\tilde{x} = (\tilde{x}_1, \tilde{x}_2) = (1.00, 1.00)$. The solution $\tilde{x}$ is very close to the true solution $\bar{x}$, and is a reasonable answer for a machine of this word length.

We discuss the implementation of the simplex algorithm using Gaussian pivot steps in Chapter 7. In Section 7.2, whenever a Gaussian pivot step is performed, the pivot element in the pivot column is chosen according to this special pivot element selection rule, and row interchanges are performed whenever necessary.

3.3.9 Cramer's Rule

This rule gives the solution of a square nonsingular system of simultaneous linear equations in the form of a formula. Consider the system:

$x_1 \dots x_m$	
$a_{11} \dots a_{1,m}$	$= b_1$
$\vdots \quad \vdots$	
$a_{m1} \dots a_{mm}$	$= b_m$

Here $A = (a_{ij})$ is nonsingular. Let $\det(A_{.1}, \dots, A_{.j-1}, b, A_{.j+1}, \dots, A_{.m})$ be the determinant of a matrix whose column vectors are listed inside the brackets. Let $\bar{x}_j = \det(A_{.1}, \dots, A_{.j-1}, b, A_{.j+1}, \dots, A_{.m})/\det A$. Then $\bar{x} = (\bar{x}_1, \dots, \bar{x}_m)^{\mathrm{T}}$ is a solution of the above system. Since this involves the computation of $(m+1)$ determinants, solving the system of equations by this rule is computationally inefficient. However, Cramer's rule is theoretically quite important.

Exercises

3.16 Solve the following systems:

(a)
$$2x_1 - 3x_2 + 4x_3 = 8$$
$$6x_1 + 5x_2 - 7x_3 = 4$$

(b)
$$\begin{aligned} x_1 - x_2 + x_3 - x_4 + x_5 &= 3 \\ 2x_1 + 3x_3 + 2x_4 + 2x_5 &= -2 \\ 4x_1 - 2x_2 + 5x_3 + 4x_5 &= 2 \end{aligned}$$

(c)
$$\begin{aligned} x_1 + 2x_2 + 3x_3 &= 2 \\ -x_1 + x_2 - 2x_3 &= 1 \\ x_1 + 5x_2 + 4x_3 &= 5 \\ 3x_2 + x_3 &= 3 \\ -x_1 + 4x_2 - x_3 &= 4 \end{aligned}$$

3.17 Consider this LP in which there are no inequality constraints or sign restrictions: minimize $z(x) = cx$, subject to $Ax = b$. Prove that either $z(x)$ is a constant on the set of feasible solutions of this system

$$\left[\text{which happens when rank} \begin{pmatrix} A \\ \cdots \\ c \end{pmatrix} = \text{rank } (A)\right]$$

or that if there exist two feasible solutions that give different values to $z(x)$, then $z(x)$ is unbounded below on the set of feasible solutions of this system

$$\left[\text{which happens when rank} \begin{pmatrix} A \\ \cdots \\ c \end{pmatrix} = 1 + \text{rank}(A)\right]$$

3.4 BASIC FEASIBLE SOLUTIONS AND EXTREME POINTS OF CONVEX POLYHEDRA

3.4.1 Extreme Points

Let Γ be a convex subset of $\mathbf{R}^n$. A point $\bar{x} \in \Gamma$ is said to be an *extreme point* of Γ if it is impossible to express it as a convex combination of two other *distinct points* in Γ. That is, $\bar{x}$ is an extreme point of Γ iff $x^1 \in \Gamma$, $x^2 \in \Gamma$ and $0 < \lambda < 1$ such that

$$\bar{x} = \lambda x^1 + (1 - \lambda)x^2 \text{ implies that } x^1 = x^2 = \bar{x}$$

Another suggestive name for extreme points is *corner points* (see Figure 3.23). Extreme points play a very important role in solving problems related to convex polyhedra.

Given a point $x \in \Gamma$, it is very hard to computationally check whether it is an extreme point, using this geometric definition. If Γ is a convex polyhedron, it turns out that we can develop an algebraic characterization of extreme points in terms of the linear independence of a specified set of vectors. This algebraic characterization leads to an efficient computational method for checking whether a given point in a convex polyhedron is an extreme point or not, using the algorithm discussed in Section 3.3.1. In linear programming literature, extreme points defined through an algebraic characterization are usually called *basic feasible solutions*.

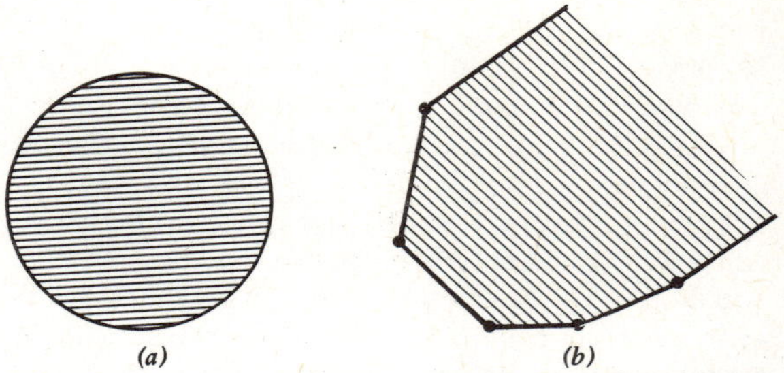

(a) (b)

Figure 3.23 (*a*) For this convex set, all boundary points are extreme points. (*b*) For this convex polyhedron, all bold corner points are extreme points.

3.4.2 Basic Feasible Solutions of a System of Linear Equations in Nonnegative Variables

Consider the system of linear equations in nonnegative variables

$$\begin{aligned} Ax &= b \\ x &\geqq 0 \end{aligned} \tag{3.8}$$

where A, b are matrices of orders $m \times n$, $m \times 1$, respectively, and $x \in \mathbf{R}^n$.

***Solutions and Feasible Solutions for* (3.8)**

Any vector x that satisfies $Ax = b$, (and may or may not satisfy $x \geqq 0$), is known as a *solution of* (*3.8*). A solution of (3.8) that satisfies $x \geqq 0$ is known as a *feasible solution of* (*3.8*).

Basic Feasible solutions

Let $\mathbf{K}$ be the set of feasible solutions of (3.8). If $\bar{x} \in \mathbf{K}$, the set of column vectors of A that $\bar{x}$ *uses* is $\{A_{.j}: j, \text{ such that } \bar{x}_j > 0\}$. The feasible solution $\bar{x} \in \mathbf{K}$ is said to be a *basic feasible solution* for (3.8) iff the set of column vectors of A that $\bar{x}$ *uses* is a linearly independent set; that is, the set $\{A_{.j}: j, \text{ such that } \bar{x}_j > 0\}$ is linearly independent. We use the abbretion BFS for basic feasible solution. The definition of a BFS given here applies only to a system of linear equality constraints in nonnegative variables. For a general system of constraints, including some inequality constraints or unrestricted variables, the definition of a BFS has to be modified so that every BFS is an extreme point of the set of feasible solutions, and vice versa.

A Mathematical Argument

Consider the system of constraints (3.8). Let $\bar{x}$ be a feasible solution for (3.8). Then $A\bar{x} = \sum_{j=1}^{n} \bar{x}_j A_{.j} = b$. Let $\{j_1, \ldots, j_r\} = \{j: j, \text{ such that } \bar{x}_j > 0\}$. Hence, $\bar{x}_j = 0$, for all $j \notin \{j_1, \ldots, j_r\}$. So we have $\bar{x}_{j_1} A_{.j_1} + \cdots + \bar{x}_{j_r} A_{.j_r} = b$. If $\{A_{.j_1}, \ldots, A_{.j_r}\}$ is linearly dependent, let the linear dependence relation among them be $\alpha_{j_1} A_{.j_1} + \cdots + \alpha_{j_r} A_{.j_r} = 0$, where $(\alpha_{j_1}, \ldots, \alpha_{j_r}) \neq 0$. Combining these two equations we get $(\bar{x}_{j_1} + \theta\alpha_{j_1})A_{.j_1} + \cdots + (\bar{x}_{j_r} + \theta\alpha_{j_r})A_{.j_r} = b$. Define the vector $x(\theta) = (x_1(\theta), \ldots, x_n(\theta))^{\mathrm{T}}$, where

$$\begin{aligned} x_j(\theta) &= \bar{x}_j + \theta\alpha_j && \text{for } j \in \{j_1, \ldots, j_r\} \\ &= 0 && \text{if } j \notin \{j_1, \ldots, j_r\} \end{aligned} \tag{3.9}$$

Then we see that $x(\theta)$ satisfies $Ax = b$. Define

$$\begin{aligned} \theta_1 &= \text{maximum } \{-\bar{x}_j/\alpha_j : j \in \{j_1, \ldots, j_r\}, \\ &\qquad \text{and such that } \alpha_j > 0\} \\ &= -\infty, \qquad \text{if } \alpha_j \leqq 0, \text{ for all } j \in \{j_1, \ldots, j_r\} \\ \theta_2 &= \text{minimum } \{-\bar{x}_j/\alpha_j : j \in \{j_1, \ldots, j_r\}, \\ &\qquad \text{and such that } \alpha_j < 0\} \\ &= +\infty, \qquad \text{if } \alpha_j \geqq 0, \text{ for all } j \in \{j_1, \ldots, j_r\} \end{aligned} \tag{3.10}$$

then $x(\theta) \geqq 0$ and hence is a feasible solution of (3.8) for all θ satisfying $\theta_1 \leqq \theta \leqq \theta_2$. Clearly θ_1 is either $-\infty$ or a strictly negative finite number; θ_2 is either $+\infty$ or a strictly positive finite number; and since $(\alpha_{j_1}, \ldots, \alpha_{j_r}) \neq 0$, at least one of the two numbers among θ_1, θ_2, must be finite.

Example 3.18

Consider the case where the solution $\bar{x}$ is $\bar{x}_j =$ 6, 17, 18, 5, 12, for $j = 2, 4, 7, 10, 13$, respectively, and 0 for $j \notin \{2, 4, 7, 10, 13\}$. So we have $6A_{.2} + 17A_{.4} + 18A_{.7} + 5A_{.10} + 12A_{.13} = b$. The rest of the data about the system of constraints are not relevant for this example, and hence we do not provide them here. Suppose the set of vectors $\{A_{.2}, A_{.4}, A_{.7}, A_{.10}, A_{.13}\}$ is linearly dependent in this example, and let a linear dependence equation among them be $-2A_{.2} + 3A_{.7} - A_{.10} + 6A_{.13} = 0$. Combining these we get $(6-2\theta)A_{.2} + 17A_{.4} + (18+3\theta)A_{.7} + (5-\theta)A_{.10} + (12+6\theta)A_{.13} = b$. Define the solution $x(\theta) = (x_j(\theta))$, where $x_j(\theta) = 6 - 2\theta, 17, 18 + 3\theta, 5 - \theta, 12 + 6\theta$, for $j = 2, 4, 7, 10, 13$, and 0, for $j \notin \{2, 4, 7, 10, 13\}$. Using (3.10), we find that in this case, $\theta_1 = \text{maximum } \{-18/3, -12/6\} = -2$, $\theta_2 = \text{minimum } \{6/2, 5/1\} = 3$. Hence the solution $x(\theta)$ here is a feasible solution for this example for all values of θ in $-2 \leqq \theta \leqq 3$.

THEOREM 3.1 Let **K** be the set of feasible solutions of (3.8). $x \in \mathbf{K}$ is an extreme point of **K** as defined in Section 3.4.1 iff it is a BFS of (3.8), as defined here.

Proof: Suppose $\bar{x}$ is a feasible solution of (3.8), which is not a BFS. As in the mathematical argument given above, we can construct a solution $x(\theta)$ for (3.8) as in (3.9), which remains feasible for all $\theta_1 \leqq \theta \leqq \theta_2$, where $\theta_1 < 0$ and $\theta_2 > 0$, using the linear dependence relation among the column vectors in the set $\{A_{.j} : j \text{ such that } \bar{x}_j > 0\}$. Choose ε satisfying $0 < \varepsilon < \text{minimum } \{|\theta_1|, |\theta_2|\}$. Then clearly $\bar{x} = \frac{1}{2}(x(-\varepsilon) + x(+\varepsilon))$, $x(-\varepsilon)$, and $x(+\varepsilon)$ are both feasible to (3.8), and $x(-\varepsilon) \neq x(+\varepsilon)$, since $(\alpha_j : j \in \{j_1, \ldots, j_r\}) \neq 0$ in (3.9). So $\bar{x}$ is not an extreme point of **K**.

Suppose $\bar{x}$ is a BFS of (3.8). As before, let $\mathbf{J} = \{j : \bar{x}_j > 0\}$. Suppose $\bar{x}$ is not an extreme point of **K**. Then there must exist feasible solutions $\tilde{x}$, $\hat{x}$ such that $\bar{x} = \lambda\tilde{x} + (1-\lambda)\hat{x}$, where $0 < \lambda < 1$ and $\tilde{x} \neq \hat{x}$. Since $\bar{x}_j = 0$ for all $j \notin \mathbf{J}$, this implies that $\tilde{x}_j = \hat{x}_j = 0$ for all $j \notin \mathbf{J}$. So $\sum_{j \in \mathbf{J}} A_{.j}\tilde{x}_j = \sum_{j \in \mathbf{J}} A_{.j}\hat{x}_j = b$. Since $\bar{x}$ is assumed to be a BFS, the set $\{A_{.j} : j \in \mathbf{J}\}$ is linearly independent. So the above equation implies that $\tilde{x}_j - \hat{x}_j = 0$ for all $j \in \mathbf{J}$. Therefore, $\tilde{x} = \hat{x}$, a contradiction. Hence $\bar{x}$ must be an extreme point of **K**. ∎

Example 3.19

Consider the system

x_1	x_2	x_3	x_4	x_5	b
1	0	−1	3	−1	1
0	1	1	4	2	4
0	0	0	−7	3	0

$x_j \geqq 0$ for all j.

For this system $x = (2, 3, 1, 0, 0)^T$ is not a BFS because it *uses* the column vectors of x_1, x_2, and x_3 and this set is not linearly independent, because (column vector x_1) − (column vector x_2) + (column vector x_3) = 0.

However, the feasible solution $x = (1, 4, 0, 0, 0)^T$ is clearly a BFS of this system.

3.4.3 Basic Feasible Solutions of a System of Equality and Inequality Constraints in Nonnegative Variables

The definition of a BFS for a system of linear constraints is always specified in such a way that a BFS is an extreme point and vice versa. While extreme points are defined in terms of geometric concepts, BFSs are defined in algebric terms using the concept of linear independence. For a BFS to correspond to an extreme point and vice versa, it is necessary to formulate the definition of a BFS taking into account the type of constraints in the system, the presence of inequality constraints, unrestricted variables, etc. Consider the system consisting of some equality constraints and some inequality constraints in nonnegative variables.

$$\begin{aligned} Ax &= b \\ Dx &\geqq d \\ x &\geqq 0 \end{aligned} \tag{3.11}$$

Transform this into an equivalent system of equality constraints in nonnegative variables, by introducing slack variables corresponding to the inequality constraints. It becomes

$$\begin{aligned} Ax + Os &= b \\ Dx - Is &= d \\ x \geqq 0 \qquad s &\geqq 0 \end{aligned} \tag{3.12}$$

Here O is the zero matrix of appropriate order, and $s = Dx - d$ is the vector of slack variables. A feasible solution x of (3.11) is said to be a *basic feasible solution* (BFS) of (3.11) iff the associated feasible solution $(x; s)$ (where $s = Dx - d$) of (3.12) is a BFS of (3.12) as defined in Section 3.4.2.

Let A be of order $m \times n$, and let D be of order $p \times n$. Then from the above definition, a feasible solution x of (3.11) is a BFS iff the set of column vectors

$$\left\{\begin{pmatrix} A_{.j} \\ \cdots \\ D_{.j} \end{pmatrix} : j, \text{ such that } x_j > 0\right\} \cup \left\{\begin{pmatrix} 0 \\ -I_{.t} \end{pmatrix} : t, \text{ such that } D_{t.}x - d_t > 0\right\}$$

is linearly independent, where the 0 in $(0, -I_{.t})$ is the column vector of zero entries in $\mathbf{R}^m$, and $-I_{.t}$ is the tth column vector of the negative of the unit matrix of order p. This definition guarantees that every BFS of (3.11) is an extreme point of the set of feasible solutions and vice versa.

Example 3.20

Consider the system of constraints

$$\begin{aligned} x_1 \qquad\qquad &\geqq 6 \\ x_2 + x_3 &\geqq 2 \\ x_1, x_2, \quad x_3 &\geqq 0 \end{aligned}$$

The point $\tilde{x} = (7, 2, 0)^{\mathrm{T}}$ is a feasible solution of the system, and the set of column vectors corresponding to positive x_j in this system is $\{(1, 0)^{\mathrm{T}}, (0, 1)^{\mathrm{T}}\}$, which is linearly independent. However, $\tilde{x}$ is not a BFS of this system according to the definition here. If we introduce the slack variables, the system becomes

$$\begin{aligned} x_1 \qquad\qquad - s_1 \qquad &= 6 \\ x_2 + x_3 \qquad - s_2 &= 2 \\ x_1, x_2, \quad x_3, s_1, \quad s_2 &\geqq 0 \end{aligned}$$

and $\tilde{x}$ corresponds to $(\tilde{x}, \tilde{s}) = (7, 2, 0, 1, 0)^{\mathrm{T}}$ and the set of column vectors corresponding to x_1, x_2, s_1 in this system is $\{(1, 0)^{\mathrm{T}}, (0, 1)^{\mathrm{T}}, (-1, 0)^{\mathrm{T}}\}$, which is linearly dependent. Actually, $\tilde{x} = \frac{1}{2}\ ((6, 2, 0)^{\mathrm{T}} + (8, 2, 0)^{\mathrm{T}})$, and both $(6, 2, 0)^{\mathrm{T}}$ and $(8, 2, 0)^{\mathrm{T}}$ are feasible solutions of the original system of constraints. Hence $\tilde{x}$ is not an extreme point.

Exercises

3.18 Prove that a BFS of (3.11) as defined here, is an extreme point of the set of feasible solutions of (3.11) and vice versa.

3.4.4 Basic Feasible Solutions of a System of Equality Constraints

Consider the system of equality constraints in variables, some of which are restricted in sign and some of which are not:

$$\begin{aligned} Dx + Ey &= b \\ x \geqq 0 \qquad y &\text{ unrestricted} \end{aligned} \tag{3.13}$$

where D is a matrix of order $m \times n_1$, and E is a matrix of order $m \times n_2$. Let the set of feasible solutions of this system be $\mathbf{F}$. Then a feasible solution $(\tilde{x}, \tilde{y})$ of (3.13) is called a *basic feasible solution* (BFS) iff the set of column vectors $\{D_{.j}: j \text{ such that } \tilde{x}_j > 0\} \cup \{E_{.j}: \text{for all } j\}$, which contains all the column vectors of E and all the column vectors of D, corresponding to positive $\tilde{x}_j$, is linearly independent.

If $n_1 = 0$, that is, if (3.13) contains only the unrestricted variables y, then a solution is a BFS only if the column vectors of E form a linearly independent set. In this case, if (3.13) has a solution, it must be unique. So a system of equality constraints in unrestricted variables has a basic solution iff it has a unique solution. If the solution is not unique, the set of solutions of such a system is a translate of a subspace and it cannot have any BFSs or extreme points.

Transforming an Unrestricted Variable into the Difference of Two Nonnegative Variables

As in Chapter 2, (3.13) can be transformed into a system of equality constraints in nonnegative variables by expressing each unrestricted variable as the difference of two nonnegative variables. Then the system becomes

$$\begin{aligned} Dx + Ey^{+} - Ey^{-} &= b \\ x \geqq 0 \qquad y^{+} \geqq 0 \qquad y^{-} &\geqq 0 \end{aligned} \tag{3.14}$$

The correspondence between feasible solutions of (3.13) and those of (3.14) is not one to one. For example, if $(\tilde{x}, \tilde{y}^+, \tilde{y}^-)$ is feasible to (3.14), then $(\tilde{x}, (\tilde{y}_i^+ + \delta), (\tilde{y}_i^- + \delta))$ obtained by adding a δ to each entry in $\tilde{y}^+$ and $\tilde{y}^-$ is also feasible to (3.14) for all $\delta \geqq 0$, and all these feasible solutions of (3.14) correspond to the feasible solution $(\tilde{x}, \tilde{y}^+ - \tilde{y}^-)$ of (3.13). *Every BFS of* (3.14) *need not correspond to a BFS of* (3.13).

Exercises

3.19 Prove that $(\tilde{x}, \tilde{y}) \in \mathbf{F}$ is an extreme point of $\mathbf{F}$ iff it is a BFS as defined above.

3.20 Consider the system $x_1 + x_2 = 1$. Prove that this system has no BFSs and the set of feasible solutions of this system has no extreme points. Now transform the system by expressing each variable x_1, x_2 as the difference of two nonnegative variables. Find all the BFSs of the transformed system.

Among the transformations on a linear programming model, discussed in Section 2.2, or those that transform one LP into an equivalent LP, discussed in Section 3.18, only the transformation of replacing an unrestricted variable in the system by the difference of two nonnegative variables creates new BFSs in the transformed system that do not correspond to BFSs of the original system.

3.4.5 Basic Feasible Solutions of a General Linear System of Constraints

Consider a system of linear constraints in which there may be some equality constraints and some inequality constraints, some unrestricted variables and some variables restricted to be nonnegative. Transform this system into a system of equality constraints by introducing slack variables corresponding to the inequality constraints. A feasible solution of the original system is called a *basic feasible solution* (BFS) of this system iff the corresponding solution (with the corresponding values of slack variables) is a BFS of the transformed system of equality constraints as defined in Sections 3.4.2 to 3.4.4. Also, see Exercise 3.74.

Exercises

3.21 Let Γ be the set of feasible solutions of the system given below. Prove that a point in Γ is an extreme point, iff it is a BFS as defined above.

$$\begin{aligned} Ax + Dy &= b \\ Ex + Fy &\geqq d \\ x \geqq 0, \quad & y \text{ unrestricted} \end{aligned}$$

3.22 Prove that 0 is the unique extreme point of the convex cone, which is the set of feasible solutions of $Ax \geqq 0$, where A is a matrix of order $m \times n$ and rank n.

Basic Feasible Solutions of a Bounded Variable Linear Program

As an illustrative example, here we characterize the BFSs of bounded-variable linear programs. The constraints are of the form:

$$\begin{aligned} Ax &= b \\ \ell_j \leqq x_j &\leqq U_j \qquad \text{for } j = 1 \text{ to } n \end{aligned} \tag{3.15}$$

where A is a matrix of order $m \times n$; ℓ_j could be $-\infty$ (in which case the lower bound constraint on x_j is redundant); U_j could be $+\infty$ (in which case the upper bound constraint on x_j is redundant). Also, $\ell_j \leqq U_j$ for each $j = 1$ to n. Bounded variable LPs are discussed in Chapter 11. Let $\mathbf{K}$ denote the set of feasible solutions of (3.15). Because the lower bound and upper bound restrictions are already included in the system of constraints in (3.15), each variable x_j is an unrestricted variable in that model. For each $j = 1$ to n, associate the slack variables s_j with the lower bound constraint on x_j if ℓ_j is finite; and the slack variable t_j with the upper bound constraint on x_j if U_j is finite. With these slacks, (3.15) gets transformed into the system of constraints in Tableau 3.1.

Tableau 3.1 Constraints in the Bounded Variable Linear Program

x_1	...	x_n	s_1	t_1	...	s_n	t_n	
a_{11}	...	a_{1n}	0	0	...	0	0	b_1
⋮		⋮	⋮	⋮		⋮	⋮	⋮
a_{m1}	...	a_{mn}	0	0	...	0	0	b_m
1	...	0	1	0	...	0	0	ℓ_1
1	...	0	0	-1	...	0	0	U_1
⋮		⋮	⋮	⋮		⋮	⋮	⋮
0	...	1	0	0	...	1	0	ℓ_n
0	...	1	0	0	...	0	-1	U_n

x_j unrestricted for all j; $s_j, t_j \geqq 0$ for each $j = 1$ to n.

Let $\bar{x}$ be a feasible solution of (3.15). Define $\bar{s}_j = \bar{x}_j - \ell_j, \bar{t}_j = U_j - \bar{x}_j, j = 1$ to n. We have the following:

If $\ell_j < \bar{x}_j < U_j$, then $\bar{s}_j > 0$, $\bar{t}_j > 0$

If $\bar{x}_j = \ell_j$, then $\bar{s}_j = 0$ and $\bar{t}_j > 0$ only if $U_j > \ell_j$

If $\bar{x}_j = U_j$, then $\bar{t}_j = 0$ and $\bar{s}_j > 0$ only if $U_j > \ell_j$

From the definition given above, $\bar{x}$ is a BFS of (3.15) iff the set {column of x_j in Tableau 3.1: $j = 1$ to n} $\cup$ {column of s_j in Tableau 3.1: j such that $\bar{s}_j > 0$} $\cup$ {column of t_j in Tableau 3.1: j such that $\bar{t}_j > 0$} is linearly independent. Let Q be the matrix whose columns are the columns in this set. If there exists a column vector of Q that contains only a single nonzero entry of $+1$ or -1 (these are columns corresponding to the slacks s_j or t_j in Tableau 3.1), or a row vector of Q that contains a single nonzero entry of $+1$ (this will be a row vector from the $(m + 1)$ to $(m + 2n)$th rows of Tableau 3.1, which means that the $+1$ entry in that row has to lie in a column of x_j that is such that $\bar{x}_j = \ell_j$ or U_j), then we will use the term *reduction* to refer to the operation of striking off the row and the column of that single $+1$ or -1 entry from the matrix Q. Also, if there exists a row in Q, all the entries in which are zero, that row is eliminated in the process of reduction. Obviously, the set of columns of Q form a linearly independent set, iff the set of columns of the matrix obtained after the reduction is linearly independent.

If j is such that $\ell_j < \bar{x}_j < U_j$, then the columns of both s_j and t_j in Tableau 3.1 appear in Q, and in the process of reduction, these columns and the rows of the $+1$ and -1 entries in them are eliminated from Q, although the column of x_j remains in Q. If j is such that $\bar{x}_j = \ell_j$ and $\bar{t}_j > 0$, then the column of t_j in Tableau 3.1 appears in Q, and in the process of reduction, this column of t_j and the row of the -1 entry in it are deleted from Q. In this case, since $\bar{s}_j = 0$, the column corresponding to s_j in Tableau 3.1 does not appear in Q; and the $+1$ entry in the column corresponding to x_j and the $(m + 2(j - 1) + 1)$th row of Tableau 3.1 is the only nonzero entry in its row in Q; and hence that row and the column corresponding to x_j get deleted from Q in the process of reduction. In the same way, if j is such that $\bar{x}_j = U_j$ and $\bar{s}_j > 0$, then the columns corresponding to x_j and s_j, and the rows corresponding to the $(m + 2(j - 1) + 1)$ and $(m + 2(j - 1) + 2)$ rows in Tableau 3.1, get deleted from Q in the process of reduction. Also, if j is such that $\ell_j = U_j$, then $\bar{x}_j = \ell_j = U_j$ and $\bar{s}_j = \bar{t}_j = 0$, in which case the columns corresponding to s_j and t_j in Tableau 3.1 do not appear in Q; and the column corresponding to x_j in Tableau 3.1 and the rows corresponding to the two bottommost nonzero entries of $+1$ in it get deleted from Q in the process of reduction.

Thus after the reduction is carried out on Q as often as possible, at the end we are left with a matrix whose columns are $A_{.j}$ for j, such that $\ell_j < \bar{x}_j < U_j$. Hence the columns of the original matrix Q form a linearly independent set iff the set $\{A_{.j}: j$, such that $\ell_j < \bar{x}_j < U_j\}$ is linearly independent. Hence, a feasible solution $\bar{x}$ of (3.15) is a BFS of (3.15), iff the set $\{A_{.j}: j$, such that $\ell_j < \bar{x}_j < U_j\}$ is linearly independent.

3.5 THE USEFULNESS OF BASIC FEASIBLE SOLUTIONS IN LINEAR PROGRAMMING

Every LP can be put in standard form by the transformations discussed in Chapter 2. The problem in standard form is an equivalent problem in which the constraints are all linear equations in nonnegative variables. Hence, for studying LPs, it is sufficient to restrict the discussion to systems of linear equations in nonnegative variables. In all subsequent sections of this chapter, we will only consider problems in which the constraints are linear equations in nonnegative variables.

3.5.1 Basis and its Primal Feasibility

Consider the system of constraints (3.8) where A is a matrix of order $m \times n$ and $x \in \mathbf{R}^n$. Suppose (3.8) does not contain any redundant constraints, that is, the rank of A is m. Any solution, x, of (3.8) is the vector of the coefficients of the column vectors of A in an expression for b as a linear combination of the column vectors of A. *Hence in studying the system (3.8), the vector space of the columns of matrix A plays an important role.*

Any nonsingular square submatrix of A of order m is known as a *basis* for this system. Suppose B is a basis. The column vectors of A and the variables in the problem can be partitioned into the *basic* and the *nonbasic parts* with respect to this basis B. Each column vector of A, which is in the basis B, is known as a *basic column vector*. All the remaining column vectors of A are called the *nonbasic column vectors.*

Let x_B be the vector of the variables associated with the basic column vectors. The variables in x_B are known as the *basic variables* with respect to the basis B, and x_B is the *basic vector*. Let x_D be the vector of the remaining variables, which are called the *nonbasic variables*. Let D be the matrix consisting of the column vectors associated with x_D. Rearranging the variables and column vectors, if necessary, (3.8) can be viewed as

$$Bx_B + Dx_D = b$$
$$x_B \geqq 0 \qquad x_D \geqq 0$$

The *basic solution of (3.8) corresponding to the basis B* is obtained by setting all the nonbasic variables equal to zero and then solving the remaining system of equations for the values of the basic variables. Since B is nonsingular, this leads to the solution

$$x_D = 0$$
$$x_B = B^{-1}b$$

B is said to be a *primal feasible basis* for (3.8) iff the basic solution corresponding to it satisfies the nonnegativity restrictions, that is, iff $B^{-1}b \geqq 0$.

3.5.2 (Primal) Nondegeneracy of a Basis or BFS for (3.8)

Suppose $\bar{x}$ is a BFS of (3.8). Let the rank of A be m. Since any set of $m+1$ or more column vectors in $\mathbf{R}^m$ is linearly dependent, the number of positive variables $\bar{x}_j$ cannot exceed m. $\bar{x}$ is said to be a *nondegenerate BFS* of (3.8) or *primal nondegenerate BFS* of (3.8), to be more specific, if exactly m of the $\bar{x}_j$ are positive. If the number of positive $\bar{x}_j$ is less than m, $\bar{x}$ is known as a *degenerate primal BFS* of (3.8).

There are two types of degeneracy associated with an LP. The concept discussed here is the degeneracy of the LP itself, specifically *primal degeneracy*. The other type of degeneracy associated with an LP is the degeneracy of the corresponding dual problem defined in Chapter 4, called *dual degeneracy*. In this chapter, the words *nondegenerate, degenerate* will mean *primal nondegenerate, primal degenerate,* respectively.

If $\bar{x}$ is a nondegenerate BFS of (3.8), the column vectors of A corresponding to positive $\bar{x}_j$ form a basis of (3.8) for which $\bar{x}$ is the associated basic solution.

If $\bar{x}$ is a degenerate BFS of (3.8), the set of column vectors of A corresponding to positive $\bar{x}_j$ is a linearly independent set, but does not consist of enough vectors to form a basis. It is possible to augment this set with some more column vectors of A and make it into a basis. However, there may be several ways of augmenting the set of column vectors of A *used* by $\bar{x}$ to make it into a basis. Every basis that includes all the column vectors of A corresponding to positive $\bar{x}_j$ is a basis that has $\bar{x}$ as its associated basic solution. The basic vectors corresponding to each of these bases differ in only the zero-valued basic variables.

In summary, every nondegenerate BFS, $\bar{x}$, is the basic solution associated with the basis consisting of the column vectors of A corresponding to positive $\bar{x}_j$. A degenerate BFS, $\bar{x}$, is the basic solution associated with any basis that contains all the column vectors of A corresponding to positive $\bar{x}_j$.

Hence a nondegenerate BFS has a unique basis associated with it. A degenerate BFS may have many bases associated with it.

Exercises

3.23 Find all the bases of the following system associated with the BFS, $x = (1, 4, 0, 0, 0)^T$.

x_1	x_2	x_3	x_4	x_5	b
1	0	−1	3	−1	1
0	1	1	4	2	4
0	0	0	−7	3	0
$x_j \geqq 0$ for all j.					

A primal feasible basis B of (3.8) is said to be a *primal nondegenerate basis* if $B^{-1}b > 0$. It is said to be a *primal degenerate basis* if at least one of the components in the vector $B^{-1}b$ is zero.

Primal Nondegeneracy of an LP in Standard Form

The LP is standard form

$$\begin{aligned} \text{Minimize} \quad & z(x) = cx \\ \text{Subject to} \quad & Ax = b \\ & x \geqq 0 \end{aligned} \tag{3.16}$$

where A is a matrix of order $m \times n$ and rank m is said to be a nondegenerate LP if every vector $\bar{x}$ that satisfies $A\bar{x} = b$ has at least m nonzero $\bar{x}_j$ in it. This happens iff b cannot be expressed as a linear combination of any set of $m-1$ or less column vectors of A; that is, iff $b \in \mathbf{R}^m$ does not lie in any subspace of $\mathbf{R}^m$ generated by a set of $m-1$ or less column vectors of A. In this case, the right-hand-side constants' vectors b in (3.16) is said to be *nondegenerate* in (3.16).

3.5.3 Finite Number of Basic Feasible Solutions

Since we assumed that the rank of A is m in (3.16), each basis for (3.16) consists of m column vectors of A. Hence the total number of distinct bases for (3.16) is less than or equal to $\binom{n}{m}$. We know that each BFS is associated with one or more bases for (3.16). Hence the total number of distinct BFSs of (3.16) cannot exceed $\binom{n}{m}$ and, hence, is finite.

3.5.4 Existence of a Basic Feasible Solution

THEOREM 3.2 If the system of linear equality constraints in nonnegative variables, (3.16) has a feasible solution, it has a BFS.

Proof:

Case 1 If $b = 0$, $x = 0$ is a BFS of (3.16) by definition.

Case 2 Let $b \neq 0$. In this case $x = 0$ is not feasible to (3.16). Therefore every feasible solution of (3.16) must be semipositive. Suppose $\tilde{x}$ is a feasible solution. Suppose $\{j: \tilde{x}_j > 0\} = \{j_1, \ldots, j_r\}$. Then $\{A_{.j_1}, \ldots, A_{.j_r}\}$ is the set of column vectors of A that the feasible solution $\tilde{x}$ *uses*. If $\{A_{.j_1}, \ldots, A_{.j_r}\}$ is linearly independent, $\tilde{x}$ is a BFS of (3.16) and we have proved the theorem. If not, there must exist real numbers $\alpha_1, \ldots, \alpha_r$, not all zero, such that $\alpha_1 A_{.j_1} + \cdots + \alpha_r A_{.j_r} = 0$. Using this linear dependence relation and $\tilde{x}$, define the vector $x(\theta)$ as in (3.9). Define θ_1, θ_2 as in (3.10). Then for all values of θ satisfying $\theta_1 \leqq \theta \leqq \theta_2$, $x(\theta)$ is a feasible solution of (3.16). Since $(\alpha_1, \ldots, \alpha_r) \neq 0$, at least one of the two numbers θ_1 or θ_2 must be finite. Let δ be either θ_1 or θ_2, which is finite. Clearly $x(\delta)$ is a feasible solution in which at most $(r - 1)$ of the variables are positive.

Hence, starting with a feasible solution $\tilde{x}$ in which r variables are strictly positive, we have constructed another feasible solution $x(\delta)$ in which at most $(r - 1)$ variables are strictly positive.

Either $x(\delta)$ is a BFS, in which case we are done, or we can apply the same procedure on it and construct another feasible solution of (3.16) in which the number of strictly positive variables is at least one less than the corresponding number for $x(\delta)$. When this procedure is repeated, we are guaranteed to find a BFS of (3.16) after at most $(r - 1)$ applications of the procedure, since in this case every feasible solution of (3.16) is semipositive. ■

Example 3.21 How to Find a Basic Feasible Solution Given a Feasible Solution

The above proof is constructive. It gives a procedure for constructing a BFS from a feasible solution. Consider the following system:

x_1	x_2	x_3	x_4	x_5	x_6	x_7	b
1	1	3	4	0	0	0	28
0	−1	−1	−2	1	0	0	−13
0	1	1	2	0	1	0	13
1	0	2	2	0	0	−1	15

$x_j \geqq 0$ for all j.

$\tilde{x} = (1, 2, 3, 4, 0, 0, 0)^T$ is a feasible solution of this system. Denote the column vector corresponding to x_j by $A_{.j}$. Then $\tilde{x}$ uses $\{A_{.1}, A_{.2}, A_{.3}, A_{.4}\}$. Using the algorithm of Section 3.3.1, we find that this set is linearly dependent and that $-2A_{.1} - A_{.2} + A_{.3} = 0$. Hence $\tilde{x}$ is not a BFS, of the system. Since $\tilde{x}$ is feasible $A_{.1} + 2A_{.2} + 3A_{.3} + 4A_{.4} = b = (28, -13, 13, 15)^T$. Using these two equations we get $(1 - 2\theta)A_{.1} + (2 - \theta)A_{.2} + (3 + \theta)A_{.3} + 4A_{.4} = b$. To keep the coefficients of $A_{.1}, \ldots, A_{.4}$ nonnegative, θ must satisfy $\theta_1 = -3 \leqq \theta \leqq \frac{1}{2} = \theta_2$. Putting $\theta = \theta_1 = -3$ leads to $7A_{.1} + 5A_{.2} + 4A_{.4} = b$. This implies that $\bar{x} = (7, 5, 0, 4, 0, 0, 0)^T$ is another feasible solution, using only the column vectors $\{A_{.1}, A_{.2}, A_{.4}\}$, which is a subset of the set of column vectors that $\tilde{x}$ used. Applying the test for linear independence to this set $\{A_{.1}, A_{.2}, A_{.4}\}$, we find that $-2A_{.1} - 2A_{.2} + A_{.4} = 0$. From these, we get $(7 - 2\theta)A_{.1} + (5 - 2\theta)A_{.2} + (4 + \theta)A_{.4} = b$. If the coefficients of $A_{.1}, A_{.2}, A_{.4}$ in this equation are to be nonnegative, θ must satisfy $-4 \leqq \theta \leqq 5/2$. Putting $\theta = 5/2$ in it leads to $2A_{.1} + (13/2)A_{.4} = b$, which implies that $\hat{x} = (2, 0, 0, 13/2, 0, 0, 0)^T$ is another feasible solution of the system. It is easily verified that $\hat{x}$ is a BFS.

Unfortunately the theorem proved here is not necessarily true for a general linear systems of constraints. Consider: $x_1 + x_2 \geqq 1$. The set of all feasible solutions of this system is a half-space. Obviously this set has no extreme points or BFSs. See Figure 3.8.

We will now discuss the importance of BFSs in LPs, which are expressed in standard form, and the way in which these solutions are used by the simplex algorithm.

$$\begin{array}{llcl}
\text{Minimize} & z(x) = x_1 + x_2 + x_3 + 2x_4 + 2x_5 + 2x_6 + 4x_7 \\
\text{Subject to} & x_1 + x_2 + 2x_4 + x_5 + x_7 = 16 \\
 & x_2 + x_3 + x_4 + 2x_5 + 5x_7 = 19 \\
 & x_1 + x_3 + x_4 + x_5 + 2x_6 + 2x_7 = 13 \\
 & x_j \geqq 0 \quad \text{for all } j
\end{array}$$

3.5.5 Existence of an Optimum Basic Feasible Solution

THEOREM 3.3 If an LP in standard form has an optimum feasible solution, then it has a basic feasible solution that is optimal.

Proof: Consider the LP in standard form (3.16). Let $\tilde{x}$ be an optimum feasible solution for it. Let $\{j_1, \ldots, j_r\} = \{j: \tilde{x}_j > 0\}$. If $\{A_{.j_1}, \ldots, A_{.j_r}\}$ is linearly independent, $\tilde{x}$ is a BFS, and we are done. Suppose this set is linearly dependent. So there exists $\alpha = (\alpha_1, \ldots, \alpha_r) \neq 0$ satisfying

$$\alpha_1 A_{.j_1} + \cdots + \alpha_r A_{.j_r} = 0 \tag{3.17}$$

We now show that the assumption $\tilde{x}$ is optimal to (3.16) implies that any $\alpha = (\alpha_1, \ldots, \alpha_r)$ satisfying (3.17) must also satisfy

$$\alpha_1 c_{j_1} + \cdots + \alpha_r c_{j_r} = 0 \tag{3.18}$$

Suppose not. As in the mathematical argument given in Section 3.4.2, construct the solution $\hat{x}(\theta) = (\hat{x}_j(\theta))$ for (3.16), where $\hat{x}_j(\theta) = \tilde{x}_j + \theta\alpha_j$ for $j \in \{j_1, \ldots, j_r\}$, 0 otherwise. Then $z(\hat{x}(\theta)) = z(\tilde{x}) + \theta(\alpha_1 c_{j_1} + \cdots + \alpha_r c_{j_r})$. Each of $\alpha_1, \ldots, \alpha_r$ are finite real numbers and $\tilde{x}_1, \ldots, \tilde{x}_r$ are all strictly positive. So it is clear that a finite positive number, $\varepsilon > 0$, can be found such that for all θ satisfying $-\varepsilon \leqq \theta \leqq \varepsilon$, $\hat{x}(\theta)$ it is a feasible solution. If $\alpha_1 c_{j_1} + \cdots + \alpha_r c_{j_r} > 0$, let $\gamma = -\varepsilon$ and if $\alpha_1 c_{j_1} + \cdots + \alpha_r c_{j_r} < 0$, let $\gamma = +\varepsilon$. Then $\hat{x}(\gamma)$ is a feasible solution of (3.16), $z(\hat{x}(\gamma)) < z(\tilde{x})$, which contradicts the assumption that $\tilde{x}$ is an optimum feasible solution of (3.16). Hence, (3.18) must hold. By using equation (3.17), we can obtain another feasible solution $\bar{x}$, such that the set of column vectors that $\bar{x}$ *uses* is a proper subset of the column vectors that $\tilde{x}$ uses, namely $\{A_{.j_1}, \ldots, A_{.j_r}\}$, as in Section 3.5.4. By (3.18), any such feasible solution $\bar{x}$ that we obtain must also satisfy $z(\bar{x}) = z(\tilde{x})$, and so $\bar{x}$ is also an optimum feasible solution. Hence, when this procedure is applied repeatedly, an optimal BFS of (3.16) will be obtained after at most r applications of the procedure. ■

Exercises

3.24 For the LP given at the top $x = (1, 2, 3, 4, 5, 0, 0)^T$ is known to be an optimum feasible solution. Obtain an optimum BFS from it.

Note 3.7: The theorem discussed here is true for LPs in which the constraints are all equality constraints in nonnegative variables. It may not be true for a problem in general form.

3.5.6 Finiteness of the Simplex Algorithm

Consider again the LP in standard form (3.16) where A is a matrix of order $m \times n$ and rank m. If the problem has an optimum feasible solution, then it has an optimum BFS by the results in Section 3.5.5. Hence it is enough to search among the finite number of BFSs of (3.16) for an optimum feasible solution. This is what the simplex algorithm does. All the solutions generated in the course of the simplex algorithm are basic solutions.

If (3.16) is a nondegenerate LP, every BFS of it has exactly m positive components and, hence, has a unique associated basis. In this case, when the simplex algorithm is applied to solve (3.16), the minimum ratio will turn out to be strictly positive in every pivot step. Therefore the objective value decreases strictly in each pivot step. Hence a basis that appears once in the course of the algorithm can never reappear. Also the total number of bases for (3.16) is finite. Hence the simplex algorithm must either terminate with an optimum basic vector or by satisfying the unboundedness criterion after a *finite number of pivot steps*.

Suppose (3.16) is a degenerate LP. While solving (3.16) by the simplex algorithm, a *degenerate pivot step* may occur. The next pivot step in the algorithm might also turn out to be a degenerate pivot step. Actually there might be a sequence of consecutive degenerate pivot steps. In all these steps the BFS of the problem and the objective value do not change at all. During these pivot steps, the algorithm is just moving among bases, all of which are associated with the same BFS. After several of these degenerate pivot steps, the algorithm might return to the basis that started this sequence of degenerate pivot steps, completing a *cycle*. See Chapter 10 for an example of this cycling. The algorithm can now go through the same cycle again and again indefinitely without ever reaching a basis satisfying a termination criterion. This phenomenon is known as *cycling under degeneracy*. Cycling does not seem to occur among the numerous degenerate LPs encountered in practical applications. However, the fact that it can occur has been established in specially constructed examples, one of which is discussed in Chapter 10.

Pivot row choice rules known as *techniques for resolving degeneracy* have been developed that guarantee that a basis once examined never reappears in the course of the simplex algorithm. The *lexicographic minimum ratio rule* discussed in Chapter 10 is one of them. When this is applied, the new basis is picked in such a way that there is a *strict lexico decrease* in the objective function. For the definitions of these terms and all other details, see Chapter 10. Since every pivot step results in a strict lexico decrease in the objective function, we conclude that a basis that appeared once in the course of the application of the simplex algorithm with the lexicographic pivot row choice rule, can never reappear. Hence after a finite number of pivot steps, the algorithm must terminate with a basis in which either the optimality criterion or the unboundedness criterion is satisfied. This leads to the following theorem:

THEOREM 3.4 Starting with a primal feasible basis, the simplex algorithm (with some technique for resolving degeneracy, if necessary) finds, after a finite number of pivot steps, one of the following:

1 An optimal feasible basis, the canonical tableau with respect to which satisfies the optimality criterion.

2 A primal feasible basis, the canonical tableau with respect to which satisfies the unboundedness criterion.

Exercises

3.25 Consider a general LP in which there is an unrestricted variable x_1. Suppose this problem is put in standard form, during which the unrestricted variable x_1 is expressed as the difference of two nonnegative variables as in $x_1 = x_1^+ - x_1^-$, $x_1^+ \geqq 0$, $x_1^- \geqq 0$. Let the transformed problem in standard form be called (P). Prove the following:

1 While solving (P) by the simplex method at least one of the variables in the pair (x_1^+, x_1^-) remains equal to zero always.

2 If (P) has an optimum feasible solution, it has an optimum feasible solution that satisfies $x_1^+ x_1^- = 0$, and the simplex method finds only an optimum feasible solution of (P) that satisfies this property.

3.26 *ON THE OPTIMALITY CRITERION IN THE SIMPLEX ALGORITHM.* Consider the LP (3.16) where A is a matrix of order $m \times n$ and rank m. Let x_B be a feasible basic vector for (3.16) associated with the BFS $\bar{x}$, and let $\bar{c}$ be the corresponding vector of relative cost coefficients.

1 If x_B is a nondegenerate feasible basic vector for (3.16), prove that $\bar{x}$ is an optimum feasible solution for (3.16) iff $\bar{c} \geqq 0$.

2 If x_B is a degenerate feasible basic vector for (3.16), $\bar{x}$ may be an optimum solution, even though the feasible basic vector x_B does not satisfy the optimality criterion. Illustrate with a numerical example.

3 Suppose $\tilde{x}$ is an optimum BFS of (3.16). Whether $\tilde{x}$ is degenerate or not, prove that there exists a feasible basic vector $x_{\tilde{B}}$ for (3.16) corresponding to $\tilde{x}$, such that in the canonical tableau of (3.16) with respect to $x_{\tilde{B}}$ the optimality criterion is satisfied.

3.6 THE EDGE PATH TRACED BY THE SIMPLEX ALGORITHM

In this section we study how the simplex algorithm walks on the set of feasible solutions of an LP before a termination criterion is satisfied.

3.6.1 A System of m Linearly Independent Equations in $m + 1$ Nonnegative Variables

Consider the system of linear equations in nonnegative variables

$$Ey = d$$
$$y \geqq 0 \tag{3.19}$$

where E is a matrix of order $m \times (m + 1)$, and rank m, and $y \in \mathbf{R}^{m+1}$. Since E is of rank m, there exists a square submatrix of E of order m that is nonsingular. Suppose it is $B = (E_{.1} \vdots \cdots \vdots E_{.m})$. Then the system of equations $Ey = d$ can be written as $B(y_1, \ldots, y_m)^{\mathrm{T}} + E_{.m+1}y_{m+1} = d$. That is, $(y_1, \ldots, y_m)^{\mathrm{T}} = B^{-1}(d - E_{.m+1}y_{m+1})$. Giving a value of θ to y_{m+1}, this is equivalent to

$$(y_1, \ldots, y_m)^{\mathrm{T}} = B^{-1}d - \theta(B^{-1}E_{.m+1})$$
$$y_{m+1} = \theta$$

As θ varies from $-\infty$ to $+\infty$ this generates a straight line in $\mathbf{R}^{m+1}$. The set of feasible solutions of (3.19) is the intersection of this straight line with the nonnegative orthant of $\mathbf{R}^{m+1}$. The following possibilities exist. (1) The straight line does not intersect the nonnegative orthant of $\mathbf{R}^{m+1}$. In this case the set of feasible solutions of (3.19) is empty. Figure 3.24 illustrates this situation in $\mathbf{R}^2$. (2) The intersection of this straight line and the nonnegative orthant of $\mathbf{R}^{m+1}$ is nonempty and bounded. In this case the set of feasible solutions of (3.19) is a line segment of the straight line. Hence it has exactly two extreme points, which are the end points of this line segment. See Figure 3.25. (3) The intersection of this straight line and the nonnegative orthant of $\mathbf{R}^{m+1}$ is nonempty and unbounded. In this case the set of feasible solutions of (3.19) is a half-line. Hence it has exactly one extreme point, which is the end point of this half-line. See Figure 3.26.

Hence, any linearly independent system of m equations in $m + 1$ nonnegative unknowns can have at most two BFSs. Notice that the linear independence of the system of constraints is essential for this result to hold.

Suppose we are solving the LP (3.16), where A is of order $m \times n$ and rank m, by the simplex algorithm. In each step, the simplex algorithm works with a basis, which is a nonsingular square submatrix of A consisting of m column vectors of A. The algorithm then picks exactly one of the nonbasic column vectors, called the *pivot column*, and tries to bring it into the basis. During this step of the algorithm, all the other $n - m - 1$ nonbasic variables are set equal to zero. Hence in this step, the simplex algorithm is dealing with the subset of the set of feasible solutions, in which only the present basic variables and the entering variable are free to vary, while all the other variables are zero. This subset corresponds to the *restricted system* of (3.16), which is obtained by

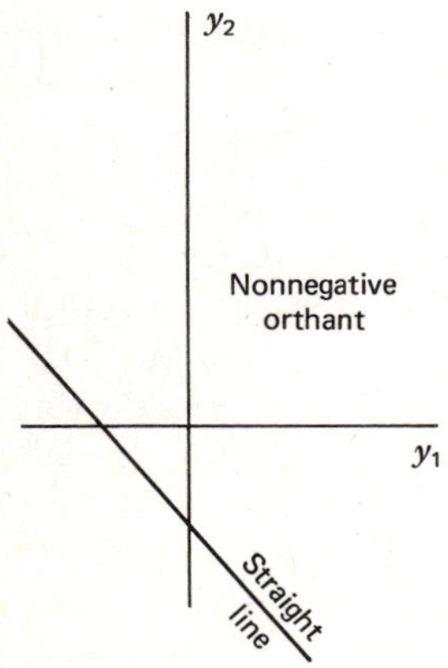

Figure 3.24

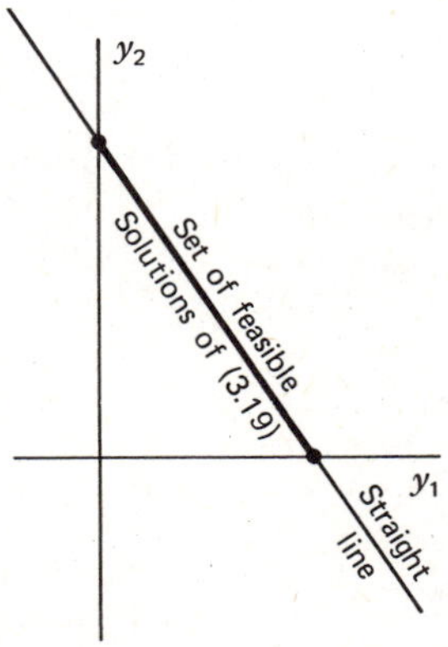

Figure 3.25

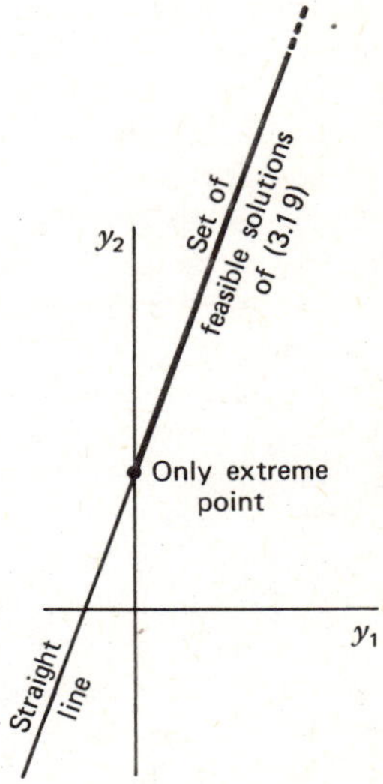

Figure 3.26

eliminating the $(n - m - 1)$ zero-valued nonbasic variables and their column vectors from (3.16). This restricted system is clearly of the form (3.19). Hence the results of this section are useful in determining what the simplex algorithm does in a pivot step. Before discussing this material, we review some concepts related to convex polyhedra.

3.6.2 Adjacency, Bounded Edges

Consider the convex polyhedron represented in Figure 3.27. Clearly every point on the line segment joining the extreme points x^1 and x^2 cannot be expressed as a convex combination of any pair of points in the convex polyhedron that are not on this line segment. This, however, is not true of points on the line segment joining the extreme points x^1 and x^3. Extreme points such as x^1 and x^2 are known as *adjacent extreme points* of the convex polyhedron. The extreme points x^1 and x^3 are not adjacent.

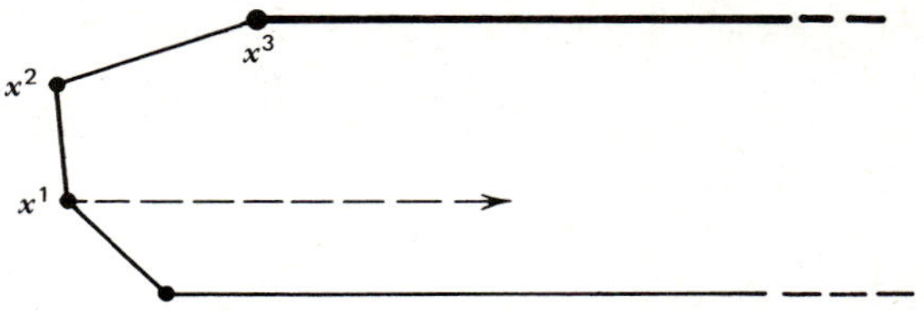

Figure 3.27

Definition: Geometric Characterization of Adjacency

Two extreme points $\tilde{x}$ and $\hat{x}$ of a convex polyhedron **K** are said to be *adjacent* iff every point $\bar{x}$ on the line segment joining them has the property that if $\bar{x} = \alpha x^1 + (1 - \alpha)x^2$, where $0 < \alpha < 1$ and $x^1 \in \mathbf{K}$, $x^2 \in \mathbf{K}$, then both x^1 and x^2 must themselves be on the line segment joining $\tilde{x}$ and $\hat{x}$.

The line segment joining a pair of adjacent extreme points of a convex polyhedron is called a *bounded edge* or an *edge* of the convex polyhedron.

Definition: Algebraic Characterization of Adjacency

When dealing with a convex polyhedron **K**, which is the set of feasible solutions of a system of linear equations in nonnegative variables, there is a simple characterization of adjacency of a pair of extreme points of **K** in terms of the rank of a set of vectors. Let **K** be the set of feasible solutions of (3.16). Two distinct extreme points of **K**, $\tilde{x}$ and $\hat{x}$, are adjacent extreme points of **K** iff the rank of the set $\{A_{.j}: j$, such that either $\tilde{x}_j$ or $\hat{x}_j$ or both are $> 0\}$ is one less than its cardinality; that is, this set is linearly dependent, and it contains a column vector such that when this column vector is deleted from this set, the remaining set is linearly independent (see reference [3.35]).

THEOREM 3.5 Let **K** be the set of feasible solutions of (3.16). Let $\tilde{x}$, $\hat{x}$ be two distinct BFSs of **K**, and $\mathbf{J} = \{j: j$, such that either $\tilde{x}_j$ or $\hat{x}_j$ or both are $>0\}$. Let $\mathbf{L} = \{x(\alpha): x(\alpha) = \alpha\tilde{x} + (1 - \alpha)\hat{x},\ 0 \leqq \alpha \leqq 1\}$. The following statements are equivalent:

1. If a point on **L** can be expressed as a convex combination of x^1, $x^2 \in \mathbf{K}$, both x^1, x^2 are themselves in **L**.
2. The rank of $\{A_{.j}: j \in \mathbf{J}\}$ is one less than its cardinality.

Proof: For notational convenience, assume that $\mathbf{J} = \{1, \ldots, r\}$. We will first prove that (1) implies (2). So $x_j(\alpha) \geqq 0$ for $j \in \mathbf{J}$, and $=0$ for $j \notin \mathbf{J}$. For each $0 \leqq \alpha \leqq 1$, $x(\alpha)$ is a solution to the system of equations: $\sum_{j \in \mathbf{J}} A_{.j}x_j = b$. Since this system has many solutions, the rank of $\{A_{.j}: j \in \mathbf{J}\}$ is $r - 1$ or less. If this rank is $r - 1$, (2) holds and we are done. So suppose the rank of $\{A_{.j}: j \in \mathbf{J}\}$ is $\leqq r - 2$. Let $\bar{x} = x(\frac{1}{2}) = \frac{1}{2}(\tilde{x} + \hat{x})$, $\mathbf{H} = \{x: x \in \mathbf{R}^n,\ x_j = 0$ for $j \notin \mathbf{J}$, and $\sum_{j \in \mathbf{J}} A_{.j}x_j = b\}$. By our rank assumption, **H** is a translate of a subspace of $\mathbf{R}^n$ of dimension $\geqq 2$. The BFSs $\tilde{x}$, $\hat{x}$ are both $\in \mathbf{H}$ and hence the straight line $\mathbf{L}_1 = \{x: x = \bar{x} + \lambda(\tilde{x} - \hat{x}),\ \lambda$ real$\} \subset \mathbf{H}$. Since the dimension of **H** is $\geqq 2$, there exists a $y \neq 0$ such that the straight line $\mathbf{L}_2 = \{x: x = \bar{x} + \lambda y,\ \lambda$ real$\} \subset \mathbf{H}$ and is distinct from $\mathbf{L}_1$. So $\mathbf{L}_1 \cap \mathbf{L}_2 = \{\bar{x}\}$ and y must satisfy $y_j = 0$ for $j \notin \mathbf{J}$, and $(y_j: j \in \mathbf{J}) \neq 0$. Since $\bar{x}_j > 0$ for all $j \in \mathbf{J}$, there exists an $\varepsilon > 0$ such that $\bar{x} + \lambda y \geqq 0$ for all $|\lambda| \leqq \varepsilon$. Since $\mathbf{L}_2 \subset \mathbf{H}$, this implies that $\bar{x} + \lambda y \in \mathbf{K}$ for all $-\varepsilon \leqq \lambda \leqq +\varepsilon$. Also $\bar{x} = \frac{1}{2}(\bar{x} - \varepsilon y) + \frac{1}{2}(\bar{x} + \varepsilon y)$, $\bar{x} - \varepsilon y$ and $\bar{x} + \varepsilon y$ are both $\in \mathbf{K}$, and since $\mathbf{L}_1 \cap \mathbf{L}_2 = \{\bar{x}\}$, neither $\bar{x} - \varepsilon y$ nor $\bar{x} + \varepsilon y$ is on the line segment **L**, a contradition to (1). So the rank of $\{A_{.j}: j \in \mathbf{J}\}$ must be $r - 1$, and (1) implies (2).

Now we prove that (2) implies (1). As before assume that $\mathbf{J} = \{1, \ldots, r\}$. Let the rank of $\{A_{.j}: j \in \mathbf{J}\}$ be $r - 1$. So that a column vector can be deleted from this set to make it linearly independent, suppose this is $A_{.r}$. Since either $\tilde{x}_r$ or $\hat{x}_r > 0$, and both $\tilde{x}$, $\hat{x}$ are BFSs of (3.16), we must have $A_{.r} \neq 0$. Consider the following parametric system of linear equations:

$$\begin{aligned} x_{r+1} = \cdots = x_n = 0 \\ x_1A_{.1} + \cdots + x_{r-1}A_{.r-1} = b - \lambda A_{.r} \end{aligned} \tag{3.20}$$

Since $\{A_{.1}, \ldots, A_{.r-1}\}$ is linearly independent, (3.20) has a unique solution for each λ, which we denote by $x^0(\lambda)$, and by the discussion in Section 3.6.1, $\mathbf{L}_3 = \{x^0(\lambda)\colon \lambda \text{ real}\}$ is a straight line in $\mathbf{R}^n$. From (3.20), $x^0(\lambda) \in \mathbf{K}$ if it is $\geqq 0$. Also $x^0(\tilde{x}_r) = \tilde{x}$ and $x^0(\hat{x}_r) = \hat{x}$. So $\tilde{x}$, $\hat{x} \in \mathbf{L}_3$, and hence $\mathbf{L} \subset \mathbf{L}_3$. Since $\tilde{x}$, $\hat{x}$ are both extreme points of $\mathbf{K}$, we must have $\mathbf{K} \cap \mathbf{L}_3 = \mathbf{L}$. From (3.20) this implies (1). Therefore (1) and (2) are equivalent. ■

The algebraic characterization of adjacency provides a convenient method for checking whether a given pair of BFSs of (3.16) are adjacent. It can be extended in an obvious manner to an algebraic characterization of adjacency of a pair of BFSs of a general system of linear constraints such as those considered in Sections 3.4.3, 3.4.4, and 3.4.5.

Exercises

3.27 Let **K** be the set of feasible solutions of a general system of linear constraints such as those considered in Sections 3.4.3, 3.4.4, or 3.4.5. Given a pair of distinct BFSs on **K**, develop a characterization for this pair of BFSs to be adjacent on **K**, in terms of the rank of some set of column vectors. Prove that your characterization for adjacency is equivalent to the geometric characterization for adjacency discussed above.

3.6.3 Adjacency in the Simplex Algorithm

Consider the LP (3.16) where A is a matrix of order $m \times n$ and rank m. Let **K** be the set of feasible solutions of (3.16). Suppose this problem is being solved by the simplex algorithm. Suppose B is the current basis and for convenience in referring to it, assume that $B = (A_{.1} \vdots \cdots \vdots A_{.m})$. Let $\tilde{x}$ be the BFS corresponding to the present basis B. Therefore $\tilde{x}_{m+1} = \tilde{x}_{m+2} = \cdots = \tilde{x}_n = 0$. Suppose in this step the entering variable is x_{m+1}. The pivot column in this step is $\bar{A}_{.m+1} = B^{-1}A_{.m+1} = (\bar{a}_{1,m+1}, \ldots, \bar{a}_{m,m+1})^{\mathrm{T}}$. We assume that $\bar{A}_{.m+1} \nleqq 0$. The case where $\bar{A}_{.m+1} \leqq 0$ corresponds to unboundedness; this is discussed in Section 3.7.

During this pivot step, the remaining nonbasic variables $x_{m+2}, \ldots, x_n$ are fixed at 0. Only the nonbasic variable x_{m+1} is changed from its present value of 0 to a nonnegative λ and the values of the basic variables are reevaluated to satisfy the equality constraints. By the discussion in Chapter 2, this leads to the solution $x(\lambda) = (\tilde{x}_1 - \bar{a}_{1,m+1}\lambda, \ldots, \tilde{x}_m - \bar{a}_{m,m+1}\lambda, \lambda, 0, \ldots, 0)^{\mathrm{T}}$. The maximum value that λ can take is the *minimum ratio* θ in this pivot step, and this is determined to keep $x(\lambda)$ nonnegative. After the pivot step the simplex algorithm moves to the BFS, $x(\theta)$. If the minimum ratio θ is equal to zero in this pivot step, this new BFS is $\tilde{x}$ itself. In this case the pivot step is a *degenerate pivot step*. Hence we have the following result.

1 In a degenerate pivot step, the simplex algorithm remains at the same BFS, but it obtains a new basis associated with it.

If the minimum ratio in this pivot step θ is strictly positive, the simplex algorithm moves to the new BFS, $x(\theta)$, which is distinct from $\tilde{x}$.

Let $\bar{x}$ be a point on the line segment joining $\tilde{x}$ and $x(\theta)$. Clearly the line segment joining $\tilde{x}$ and $x(\theta)$ is itself generated by varying λ in $x(\lambda)$ from 0 to θ. Hence $\bar{x}$ is $x(\lambda)$ for some value of λ between 0 and θ. Therefore, $\bar{x}_{m+2} = \cdots = \bar{x}_n = 0$. If $\bar{x}$ is expressed as a convex combination $\bar{x} = \alpha x^1 + (1-\alpha)x^2$, where $0 < \alpha < 1$ and x^1, x^2 are both distinct feasible solutions of (3.16), then $x^1_{m+2} = \cdots = x^1_n = x^2_{m+2} = \cdots = x^2_n = 0$. Hence from the arguments in Section 3.6.1, it is clear that both x^1, x^2 must be points of the form $x(\lambda)$ for some values of λ between 0 and θ. Therefore x^1, x^2 themselves lie on the line segment joining $\tilde{x}$ and $x(\theta)$. This implies that $\tilde{x}$ and $x(\theta)$ are themselves adjacent extreme points of **K**, which leads to the following result, that follows directly from Theorem 3.5.

2 During a nondegenerate pivot step, the simplex algorithm moves from one extreme point of the set of feasible solutions to an adjacent extreme point. The edge joining the extreme points is generated by giving the entering variable all possible values between 0 and the minimum ratio.

Example 3.22

Consider the following system of constraints:

Basic Variables	x_1	x_2	x_3	x_4	x_5	x_6	b
x_1	1	0	0	2	-1	-1	0
x_2	0	1	0	-3	1	-2	2
x_3	0	0	1	-1	2	0	1

$x_j \geqq 0$ for all j.

Let $A_{.j}$ denote the column vector corresponding to the variable x_j in this system for $j = 1$ to 6. Let $B = (A_{.1}, A_{.2}, A_{.3})$ be a basis. The BFS corresponding to this basis is $\tilde{x} = (0, 2, 1, 0, 0, 0)^T$. If the nonbasic variables x_4 is the entering variable, the minimum ratio turns out to be zero. This gives a new basis $(A_{.4}, A_{.2}, A_{.3})$, which is also associated with the same BFS $\tilde{x}$. This is a degenerate pivot step that results in a new basis, but no change in the actual BFS.

On the other hand, if the nonbasic variable x_5 is the entering variable, the minimum ratio is $\theta = \text{minimum}\left\{\frac{2}{1}, \frac{1}{2}\right\} = \frac{1}{2}$, and hence x_5 enters the basic vector (x_1, x_2, x_3), replacing x_3 from it. Keeping x_4, x_6 equal to zero, giving x_5 a value of λ, and evaluating the values of x_1, x_2, x_3 leads to the solution $x(\lambda) = (\lambda, 2 - \lambda, 1 - 2\lambda, 0, \lambda, 0)^T$. Therefore the new BFS is $\hat{x} = x(\frac{1}{2}) = (\frac{1}{2}, \frac{3}{2}, 0, 0, \frac{1}{2}, 0)^T$ and the new basis is $\hat{B} = (A_{.1}, A_{.2}, A_{.5})$, $\tilde{x}$ and $\hat{x}$ are adjacent BFSs. The line segment joining them is an edge of the set of feasible solutions of the system. Clearly, $x(\lambda) = ((\theta - \lambda)\tilde{x} + \lambda\hat{x})/\theta = ((\frac{1}{2} - \lambda)\tilde{x} + \lambda\hat{x})/(\frac{1}{2})$. As λ varies between 0 and θ, $x(\lambda)$ generates all the points on the edge joining $\tilde{x}$ and $\hat{x}$ (see Figure 3.28).

Hence, starting from a BFS, the simplex algorithm walks along the edges of the convex polyhedral set of feasible solutions (moving from an extreme point to an adjacent extreme point in every nondegenerate pivot step and sitting at the same extreme point in every degenerate pivot step) until termination occurs. Any path similar to that, which moves from an extreme point of a convex polyhedron **K** to another extreme point along the edges of **K**, is known as an *edge path*, or *adjacent extreme point path*, or just *adjacency path*. The simplex algorithm moves along an adjacency path on the set of feasible solutions of the problem, along which the objective value steadily improves.

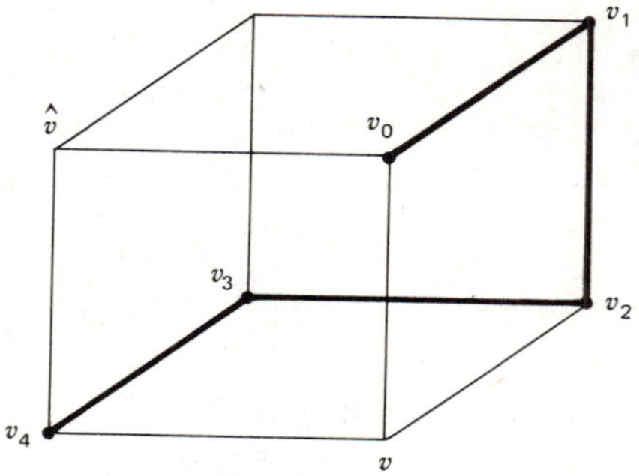

Figure 3.29 The path followed by the simplex algorithm.

Example 3.23

Consider an LP for which the set of feasible solutions is the cube in Figure 3.29. Let v_0 be the starting BFS. Adjacent extreme points of v_0 on the set of feasible solutions are v_1, $\bar{v}$, and $\hat{v}$. So from v_0 the simplex algorithm moves to either v_1, $\bar{v}$, or $\hat{v}$ in a nondegenerate pivot step. It cannot move from v_0 to an extreme point such as v_2, which is not adjacent to v_0, in one nondegenerate pivot step. The path taken by the simplex algorithm on the set of feasible solutions of this problem might be similar to the path indicated with thick edges, namely v_0 to v_1 (in one nondegenerate pivot step), v_1 to v_2, v_2 to v_3, and v_3 to v_4.

Definitions: Adjacent Basic Vectors, Adjacent Bases

Let x_{B_1} and x_{B_2} be two feasible basic vectors for (3.16). They are said to be *adjacent basic vectors* for (3.16) iff they differ in exactly one variable. Let B_1, B_2 be the bases of (3.16) corresponding to the basic vectors x_{B_1}, x_{B_2}. If x_{B_1} and x_{B_2} are adjacent basic vectors, then B_1, B_2 are known as *adjacent bases* for (3.16).

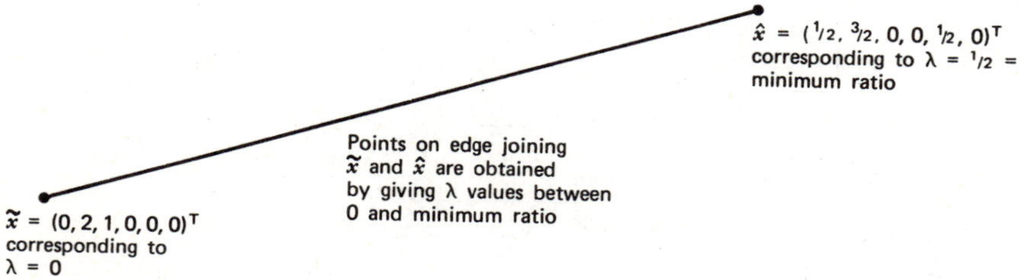

Figure 3.28

Exercises

3.28 *ADJACENCY OF A PAIR OF EDGES OF A CONVEX POLYHEDRON.* Let **K** be the set of feasible solutions of (3.16), where A is a given matrix of order $m \times n$, and b is a given column vector in $\mathbf{R}^m$. Two distinct edges of **K** are said to be adjacent on **K** if they meet at a common extreme point of **K**. Given two points x^1, x^2 in **K**, neither of which is an extreme point of **K**, and both of which are known to be on edges in **K**, discuss an efficient method to check whether the edges of **K** containing them are distinct edges that are adjacent on **K**.

3.7 HOMOGENEOUS SOLUTIONS, UNBOUNDEDNESS, RESOLUTION THEOREMS

Consider the system of equations in nonnegative variables in (3.8). A *homogeneous solution* corresponding to (3.8) is a vector y satisfying

$$Ay = 0$$
$$y \geqq 0$$

The set of all homogeneous solutions is a convex polyhedral cone. If $\tilde{x}$ is a feasible solution of (3.8), and $\hat{y}$ is a homogeneous solution corresponding to (3.8), then $\tilde{x} + \theta\hat{y}$ is also a feasible solution of (3.8) for any $\theta \geqq 0$. For example, consider the problem in Example 2.14 of section 2.6.5. For this problem in standard form, $\tilde{x} = (1/2, 1/2, 0, 0)^{\mathrm{T}}$ is a feasible solution and $\hat{y} = (1, 1, 2, 0)^{\mathrm{T}}$ is a homogeneous solution. For $\theta \geqq 0$, $\tilde{x} + \theta\hat{y}$ is a point on the thick half-line in the Figure 2.4 in Section 2.6.5.

THEOREM 3.6 RESOLUTION THEOREM 1 Every feasible solution of (3.8) can be expressed as the sum of (i) a convex combination of the BFSs of (3.8), and (ii) a homogeneous solution corresponding to (3.8).

Proof: Suppose $\tilde{x}$ is a feasible solution of (3.8). Let $\{j: \tilde{x}_j > 0\} = \{j_1, \ldots, j_k\}$. Hence the set of column vectors of A *used* by $\tilde{x}$ is $\{A_{.j_1}, \ldots, A_{.j_k}\}$. The proof of this theorem is based upon induction on the number k, the number of column vectors in this set.

Case 1 Suppose $k = 0$. This can only happen when $\tilde{x} = 0$, and since $\tilde{x}$ is assumed feasible, $b = 0$. In this case, $\tilde{x} = 0$ is itself a BFS of (3.8) and also a homogeneous solution corresponding to (3.8). Hence

$$\tilde{x} = \underbrace{0}_{\text{BFS}} + \underbrace{0}_{\substack{\text{Homogeneous}\\ \text{solution}}}$$

and hence the theorem holds in this case.

Case 2 $k \geqq 1$.

Induction Hypothesis Suppose the theorem holds for every feasible solution that *uses* a set of $k - 1$ or less column vectors of A.

We will now show that under the induction hypothesis, the theorem also holds for $\tilde{x}$, which uses a set of k column vectors of A. If $\{A_{.j_1}, \ldots, A_{.j_k}\}$ is linearly independent, then $\tilde{x}$ is a BFS of (3.8), and in this case

$$\tilde{x} = \underbrace{\tilde{x}}_{\text{BFS}} + \underbrace{0}_{\substack{\text{Homogeneous}\\ \text{solution}}}$$

Hence the theorem holds for $\tilde{x}$. On the other hand, if $\{A_{.j_1}, \ldots, A_{.j_k}\}$ is not linearly independent, there exist real numbers $\alpha_1, \ldots, \alpha_k$ not all zero, such that $\alpha_1 A_{.j_1} + \cdots + \alpha_k A_{.j_k} = 0$. Using this linear dependence relation and $\tilde{x}$, construct the feasible solution $\hat{x}(\theta)$ as in (3.9) of the mathematical argument given in Section 3.4.2. Define θ_1, θ_2 as in (3.10). At least one of θ_1 or θ_2 must be finite, call it θ_r. Then $\hat{x}(\theta_r)$ is a feasible solution of (3.8) that *uses* at most $k - 1$ column vectors of A. There are two possible subcases here:

1 Suppose all the real numbers $\alpha_1, \ldots, x_k$ are of the same sign. If all $\alpha_t \geqq 0$ for $t = 1$ to k, then $\theta_1 < 0$ is finite and $\theta_2 = +\infty$. On the other hand, if $\alpha_t \leqq 0$ for $t = 1$ to k, then $\theta_1 = -\infty$ and $\theta_2 > 0$ is finite. Hence if $\hat{y} = (\hat{y}_j)$, where $\hat{y}_{j_t} = |\theta_r \alpha_t|$ for $t = 1$ to k and $\hat{y}_j = 0$ for $j \notin \{j_1, \ldots, j_k\}$, from the above equations $\hat{x}(\theta_r) = \tilde{x} - \hat{y}$, so $\tilde{x} = \hat{x}(\theta_r) + \hat{y}$. Obviously $\hat{y}$ is a homogeneous solution corresponding to (3.8). Since $\hat{x}(\theta_r)$ is a feasible solution of (3.8) that *uses* a set of $k - 1$ or less column vectors of A, the theorem holds for it by the induction hypothesis and, hence, $\hat{x}(\theta_r) = x^1 + y^1$, where x^1 is a convex combination of BFSs of (3.8) and y^1 is homogeneous solution corresponding to (3.8). Therefore,

$$\tilde{x} = \underbrace{x^1}_{\substack{\text{Convex combination}\\ \text{of BFSs}}} + \underbrace{(y^1 + \hat{y})}_{\substack{\text{Homogeneous}\\ \text{solution}}}$$

$y^1 + \hat{y}$ is a homogeneous solution corresponding to (3.8) because both y^1 and $\hat{y}$ are homogeneous solutions and the sum of two homogeneous solutions is also a homogeneous solution. Hence the theorem holds for $\tilde{x}$.

2 Suppose the real numbers $\alpha_1, \ldots, \alpha_k$ are not all of the same sign. In this case both θ_1, θ_2 are finite. θ_1 is strictly negative and θ_2 is strictly positive. Therefore, $\tilde{x} = (\theta_2\hat{x}(\theta_1) + (-\theta_1)\hat{x}(\theta_2))/(\theta_2 - \theta_1) = \beta\hat{x}(\theta_1) + (1-\beta)\hat{x}(\theta_2)$, where $\beta = (\theta_2/(\theta_2 - \theta_1))$, $0 < \beta < 1$. Since both $\hat{x}(\theta_1)$ and $\hat{x}(\theta_2)$ are feasible solutions of (3.8) *using* a set of $k-1$ or less column vectors of A, the theorem holds for them by the induction hypothesis. Therefore $\hat{x}(\theta_1) = x^1 + y^1$, $\hat{x}(\theta_2) = x^2 + y^2$, where x^1, x^2 are both convex combinations of BFSs of (3.8) and y^1, y^2 are both homogeneous solutions corresponding to (3.8).

$$\tilde{x} = \underbrace{(\beta x^1 + (1-\beta)x^2)}_{\substack{\text{Convex combination}\\ \text{of BFSs}}} + \underbrace{(\beta y^1 + (1-\beta)y^2)}_{\substack{\text{Homogeneous}\\ \text{solution}}}$$

Hence the theorem holds for $\tilde{x}$. By Case 1 and by induction, the theorem therefore holds for k. Therefore the theorem is true in general. ■

Boundedness of Convex Polyhedra

If the set of feasible solutions of

$$Ax = b$$
$$x \geqq 0$$

is nonempty, by Resolution Theorem I, it is a convex polytope (i.e., is bounded) iff

$$Ay = 0$$
$$y \geqq 0$$

has a unique solution, namely, $y = 0$.

Extreme Homogeneous Solutions

A homogeneous solution corresponding to (3.8) is is called an *extreme homogeneous solution* iff it is a BFS of

$$Ay = 0$$
$$\sum_{j=1}^{n} y_j = 1 \qquad (3.21)$$
$$y_j \geqq 0 \qquad \text{for all } j$$

Thus there are only a finite number of distinct extreme homogeneous solutions. The constraint that the sum of all the variables should be equal to 1 eliminates 0 from consideration and normalizes the solution.

LEMMA 3.1 Either 0 is the unique homogenous solution, or every homogeneous solution corresponding to (3.8) can be expressed as a nonnegative linear combination of extreme homogeneous solutions.

Proof: Suppose 0 is not the only homogeneous solution corresponding to (3.8). Every nonzero homogeneous solution must be semipositive and, hence, it must satisfy

$$Ay = 0$$
$$\sum_{j=1}^{n} y_j = \alpha > 0$$
$$y_j \geqq 0 \qquad \text{for each } j$$

Hence it is equal to a feasible solution of (3.21) multiplied by α. Thus every homogeneous solution corresponding to (3.8), including zero, is a nonnegative multiple of some feasible solution of (3.21). Apply Resolution Theorem I to the system of constraints (3.21). The homogeneous system corresponding to (3.21) is

$$Ay = 0$$
$$\sum_{j=1}^{n} y_j = 0$$
$$y_j \geqq 0 \qquad \text{for each } j$$

$\sum y_j = 0$ and $y \geqq 0$ together imply that $y = 0$. Hence the only homogeneous solution corresponding to (3.21) is $y = 0$. Therefore by Resolution Theorem I, every feasible solution of (3.21) can be expressed as a convex combination of BFSs of (3.21), that is, extreme homogeneous solutions corresponding to (3.8). Hence every homogeneous solution corresponding to (3.8) can be expressed as a nonnegative linear combination of extreme homogeneous solutions corresponding to (3.8). ■

THEOREM 3.7 RESOLUTION THEOREM II FOR CONVEX POLYHEDRA Let **K** be the set of feasible solutions of (3.8).

1 If 0 is the unique homogeneous solution corresponding to (3.8), every feasible solution of (3.8) can be expressed as a convex combinations of BFSs of (3.8).

2 If 0 is not the only homogeneous solution corresponding to (3.8), every feasible solution of (3.8) is the sum of a convex combination of BFSs of (3.8), and a nonnegative combination of extreme homogeneous solutions corresponding to (3.8).

Proof: This theorem follows from Resolution Theorem I and Lemma 3.1. ■

Discussion Let $\mathbf{K}^{\Delta}$ be the convex hull of all the extreme points of $\mathbf{K}$, that is, of BFSs of (3.8). Let $\mathbf{K}^{\angle}$ be the cone of all homogeneous solutions corresponding to (3.8). If $\mathbf{\Gamma}_1$ and $\mathbf{\Gamma}_2$ are two subsets of $\mathbf{R}^n$, define $\mathbf{\Gamma}_1 + \mathbf{\Gamma}_2 = \{x + y : x \in \mathbf{\Gamma}_1, y \in \mathbf{\Gamma}_2\}$. Then Resolution Theorem II states that

$$\text{If } \mathbf{K}^{\angle} = \{0\} \qquad \mathbf{K} = \mathbf{K}^{\Delta}$$

$$\text{If } \mathbf{K}^{\angle} \neq \{0\} \qquad \mathbf{K} = \mathbf{K}^{\Delta} + \mathbf{K}^{\angle}$$

Let $x^1, \ldots, x^p$ be all the distinct BFSs of (3.8). If $\mathbf{K}^{\angle} = \{0\}$, then every feasible solution of (3.8) is of the form

$$x = \alpha_1 x^1 + \cdots + \alpha_p x^p \tag{3.22}$$

where $\alpha_1 + \cdots + \alpha_p = 1$ and $\alpha_t \geqq 0$ for all t.

If $\mathbf{K}^{\angle} \neq \{0\}$, let $y^1, \ldots, y^{\mathrm{L}}$ be all the extreme homogeneous solutions corresponding to (3.8). Then every feasible solution of (3.8) is of the form

$$x = \alpha_1 x^1 + \cdots + \alpha_p x^p + \beta_1 y^1 + \cdots + \beta_{\mathrm{L}} y^{\mathrm{L}} \tag{3.23}$$

where $\alpha_1 + \cdots + \alpha_p = 1$ and $\alpha_t \geqq 0$, $\beta_r \geqq 0$ for all t, r. This is a very useful theorem in the theory of linear programming. The *decomposition principle*, which helps in decomposing very large LPs into several smaller ones is based primarily on this theorem.

COROLLARY 3.1 If the LP (3.16) has a feasible solution, it has an optimum feasible solution iff $z(y) \geqq 0$ for every homogeneous solution y, corresponding to (3.16)

COROLLARY 3.2 Suppose there exists an extreme homogeneous solution y^r corresponding to (3.16), such that $cy^r < 0$. In this case, if (3.16) is feasible, $z(x)$ is unbounded below on the set of feasible solutions of (3.16).

Proofs of Corollaries 1 and 2: Let $x^1, \ldots, x^p$ be all the BFSs of (3.16). Suppose 0 is the unique homogeneous solution corresponding to (3.16). Then only Corollary 3.1 can hold. Let x be any feasible solution of (3.16). By Resolution Theorem II (Theorem 3.7), x can be expressed as in (3.22). Thus $cx = \alpha_1(cx^1) + \cdots + \alpha_p(cx^p)$, and therefore $\text{minimum}_{t=1 \text{ to } p} (cx^t) \leqq cx \leqq \text{maximum}_{t=1 \text{ to } p} (cx^t)$. Hence, the BFS that minimizes cx among the finite number of BFSs of (3.16) must be optimal to (3.16).

Now consider the case where 0 is not the only homogeneous solution corresponding to (3.16). Let $y^1, \ldots, y^{\mathrm{L}}$ be all the extreme homogeneous solutions corresponding to (3.16). If x is feasible solution of (3.16), by Resolution Theorem II, x can be expressed as in (3.23). Therefore,

$$cx = \sum_{t=1}^{p} \alpha_t(cx^t) + \sum_{r=1}^{\mathrm{L}} \beta_r(cy^r)$$

By Lemma 3.1, $z(y) \geqq 0$ for all homogeneous solutions y iff $cy^r \geqq 0$ for all extreme homogeneous solutions y^r. In this case, from (3.23), we conclude that $z(x)$ is minimized by the BFS, which gives the least value to $z(x)$ among $x^1, \ldots, x^p$.

On the other hand, if there exists a y satisfying $Ay = 0$, $y \geqq 0$, $cy < 0$, then by Lemma 3.1, there must exist an extreme homogeneous solution y^r such that $cy^r < 0$. By making β_r sufficiently large in (3.23) we can obtain a feasible solution here that makes $z(x)$ less than any given real number. ■

COROLLARY 3.3 If $z(x)$ is unbounded below on the set of feasible solutions of the LP (3.16), it remains unbounded below even if the b vector is changed as long as the problem remains feasible.

Proof: Since $z(x)$ is unbounded below on (3.16) by Corollaries 3.1 and 3.2, there exists a y^1 satisfying

$$Ay^1 = 0$$
$$cy^1 < 0$$
$$y^1 \geqq 0$$

Suppose b is changed to b^1; let x^1 be a feasible solution of the modified LP. Then $x^1 + \beta y^1$ is also a feasible solution of this LP for every $\beta \geqq 0$, and since $c(x^1 + \beta y^1) = cx^1 + \beta(cy^1)$ tends to $-\infty$ as β tends to $+\infty$ (because $cy^1 < 0$), $z(x)$ is unbounded below in this modified problem too. ■

COROLLARY 3.4 *Unboundedness of the set of feasible solutions*: Let **K** denote the set of feasible solutions of (3.16) and let $\mathbf{K}^{\angle}$ denote the set of homogeneous solutions corresponding to this system. **K** is unbounded iff $\mathbf{K} \neq \phi$ and $\mathbf{K}^{\angle}$ contains a nonzero point (that is, equivalently, the system $Ay = 0$, $\sum_{j=1}^{n} y_j = 1$, $y \geqq 0$ has a feasible solution y). Conversely, if $\mathbf{K}^{\angle} = \{0\}$, **K** is bounded.

Proof: Follows from Theorem 3.7. ■

COROLLARY 3.5 When (3.16) has a feasible solution, a necessary and sufficient condition for it to have a finite optimum feasible solution is that the optimum objective value in the LP: minimize cx, subject to $Ax = 0$, $x \geqq 0$; is zero.

Proof: Follows from Corollary 3.1. ■

Unbounded Edges or Extreme Half-Lines of Convex Polyhedra

Let **K** be the set of feasible solutions of (3.16). Suppose $\tilde{x}$ is a BFS of (3.16) and $\hat{y}$ an extreme homogeneous solution corresponding to (3.16). Then every point on the half-line

$$\{x: x = \tilde{x} + \theta\hat{y}, \theta \geqq 0\} \tag{3.24}$$

is a feasible solution for (3.16). This half-line is called an *unbounded edge* or an *extreme half-line* of **K** if every point $\bar{x}$ on it has the property that $\bar{x} = \alpha x^1 + (1-\alpha)x^2$, for some $0 < \alpha < 1$. $x^1 \in \mathbf{K}$ and $x^2 \in \mathbf{K}$ imply that both x^1 and x^2 are also on this half-line. Equivalently the half-line in (3.24) is an extreme half-line of **K** iff the set of column vectors $\{A_{.j}$: j, such that either $\tilde{x}_j$ or $\hat{y}_j$ or both are positive$\}$ is linearly dependent, and it contains a column vector such that when this column vector is deleted, the remaining set is linearly independent.

Exercises

3.29 Prove that the two definitions of an extreme half-line given here are equivalent.

3.30 Let $\bar{x} \in \mathbf{K}$, the set of feasible solutions of (3.16). If $\bar{x}$ is not an extreme point of **K**, prove that it lies on an extreme half-line of **K** iff properties (1) and (2) hold.

1 Let $\mathbf{S} = \{j: j$, such that $\bar{x}_j > 0\}$. There exists numbers α_j *such that* $\sum_{j \in \mathbf{S}} \alpha_j A_{.j} = 0$, $\alpha_j \geqq 0$ for all $j \in \mathbf{S}$ and $\alpha_j > 0$ for at least one $j \in \mathbf{S}$.

2 Let $\mathbf{U} = \{j: \alpha_j > 0\} \subset \mathbf{S}$, then the set of column vectors obtained by deleting any one of the column vectors $A_{.j}$ for $j \in \mathbf{U}$ from $\{A_{.j}: j \in \mathbf{S}\}$ is linearly independent.

Example 3.24

Consider Example 2.14 in Section 2.6.5. For this problem in standard form, let $\tilde{x} = (1/2, 1/2, 0, 0)^{\mathrm{T}}$ and $\hat{y} = (1/4, 1/4, 1/2, 0)^{\mathrm{T}}$. Then $\tilde{x}$ is a BFS of this system and $\hat{y}$ an extreme homogeneous solution corresponding to the system. The half-line $\{x: x = \tilde{x} + \theta\hat{y}, \theta \geqq 0\}$ is an extreme half-line of the set of feasible solutions. It is the thick half-line in Figure 2.4.

Example 3.25

In Figure 3.27 the thick half-line through x^3 is an extreme half-line. The parallel dashed half-line through x^1 is not an extreme half-line.

Example 3.26 An illustration of Resolution Theorems

We have discussed resolution theorems for convex polyhedra that are the sets of feasible solutions of systems of linear equations in nonnegative variables. Similar theorems can be proved for general convex polyhedra. While considering the set of feasible solutions of a general system of linear equations and inequalities, the associated homogeneous system is obtained by changing the right-hand-side constants in all the equations and inequalities to zero. For example, consider the general system:

System		*Corresponding Homogeneous System*	
$Ax + Dy = b$		$Ax + Dy = 0$	
$Ex + Fy \geqq d$		$Ex + Fy \geqq 0$	
$x \geqq 0$	y unrestricted	$x \geqq 0$	y unrestricted

We will illustrate resolution theorems by considering a system of linear inequalities in two variables, because in this case all the relevant sets can be drawn conveniently on the two-dimensional Cartesian plane. Let **K** be the set of feasible solutions of the system given below. The last constraint in this homogeneous system is automatically implied by the others and hence can be deleted. Let $\mathbf{K}^{\angle}$ be the set of feasible solutions of the homogeneous system.

System	Corresponding Homogeneous System
$x_1 - x_2 \geqq -5$	$x_1 - x_2 \geqq 0$
$x_1 - 3x_2 \leqq -9$	$x_1 - 3x_2 \leqq 0$
$x_1 \geqq 1$	$x_1 \geqq 0$
$x_2 \geqq 4$	$x_2 \geqq 0$
$x_1 + x_2 \geqq 6$	$x_1 + x_2 \geqq 0$

$\mathbf{K}^{\angle}$ is the dotted cone in Figure 3.30. The extreme homogeneous solutions associated with this system are $x^5 = (1/2, 1/2)$ and $x^6 = (3/4, 1/4)$. The rays of x^5, x^6 are the extreme rays of $\mathbf{K}^{\angle}$. The extreme points of **K** are x^1, x^2, x^3, x^4 marked in Figure 3.30. $\mathbf{K}^{\Delta}$ is their convex hull, and it is the part marked with dashed lines in **K**. Every vector in **K** can be expressed as the sum of a vector in $\mathbf{K}^{\Delta}$ and a vector in $\mathbf{K}^{\angle}$. This is Resolution Theorem I. Equivalently, Resolution Theorem II states that every vector in **K** can be expressed as $\alpha_1 x^1 + \alpha_2 x^2 + \alpha_3 x^3 + \alpha_4 x^4 + \beta_1 x^5 + \beta_2 x^6$, where $\alpha_1 + \alpha_2 + \alpha_3 + \alpha_4 = 1$, and $\alpha_1, \alpha_2, \alpha_3, \alpha_4, \beta_1, \beta_2 \geqq 0$.

K has two extreme half-lines. The extreme half line of **K** through x^1 is parallel to the ray of x^5. It is $\{x: x = x^1 + \lambda x^5, \lambda \geqq 0\}$. The extreme half line of **K** through x^4 is parallel to the ray of x^6. It is $\{x: x = x^4 + \lambda x^6, \lambda \geqq 0\}$.

While considering a convex polyhedron that is the set of feasible solutions of a general system of linear equations and inequalities, the definition of an extreme homogeneous solution and the statements of the resolution theorems have to be modified appropriately (see references [3.26, 3.36, 3.39, 3.40, 2.19]). For our study of linear programming, it is adequate to consider systems of linear equations in nonnegative variables, and hence we have only discussed this case.

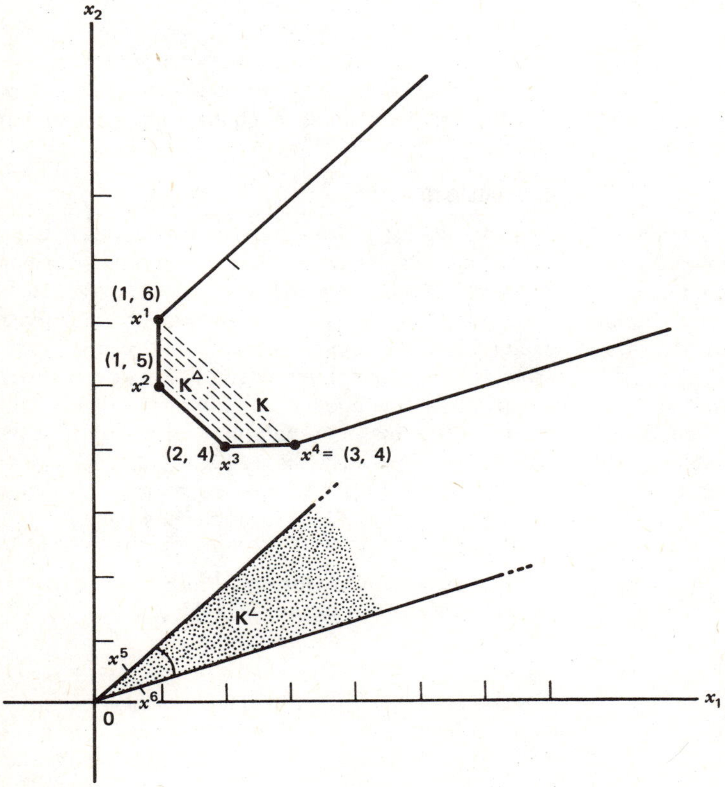

Figure 3.30

How to Check Whether a Given Feasible Solution Is on an Edge

Let **K** be the set of feasible solutions of (3.16). If $\bar{x} \in \mathbf{K}$, we provide here an algorithm to check whether it is on an edge of **K**, and if so, determine that edge. Let $\{j: \bar{x}_j > 0\} = \{j_1, \ldots, j_r\}$. If $\{A_{.j_1}, \ldots, A_{.j_r}\}$ is linearly independent, $\bar{x}$ is an extreme point of **K**. If $\{A_{.j_1}, \ldots, A_{.j_r}\}$ is not linearly independent, find its rank using the algorithm discussed in Section 3.3.3. If this rank is $\leqq r-2$, $\bar{x}$ does not lie on any edge of **K**. If this rank is $r-1$, let the linear dependence relation identified be $\sum_{t=1}^{r} \alpha_t A_{.j_t} = 0$. Using this linear dependence relation and the feasible solution $\bar{x}$, define the feasible solutions $x(\theta)$ as in (3.9) and compute θ_1, θ_2 as in (3.10). If both θ_1 and θ_2 are finite (which happens iff at least one $\alpha_t < 0$, and at least one $\alpha_t > 0$); $x(\theta_1)$, $x(\theta_2)$ are adjacent extreme points of **K** and $\bar{x}$ lies on the edge of **K** joining them. If only one of the numbers θ_1, θ_2 is finite (which happens iff all the α_t are of the same sign), $\bar{x}$ lies on an unbounded edge of **K**. This unbounded edge is $\{x(\theta): \theta \geqq \theta_1\}$, if θ_1 is finite, or $\{x(\theta): \theta \leqq \theta_2\}$, if θ_2 is finite.

Unboundedness in the Simplex Algorithm

Consider the LP (3.16). Suppose $z(x)$ is unbounded below on the set of feasible solutions of this problem. We will now show that at termination, the simplex algorithm provides an extreme half-line in this problem along which $z(x)$ diverges to $-\infty$. In solving this problem, the simplex algorithm will terminate with a canonical tableau in which the unboundedness criterion is satisfied. Suppose the terminal basis is the basis B. For convenience in referring to it, assume that the associated basic vector is $x_B = (x_1, \ldots, x_m)^{\mathrm{T}}$. Suppose the final canonical tableau is Tableau 2.8 of Section 2.5.1. In this canonical tableau, suppose the column vector of the nonbasic variable x_{m+1} satisfies the unboundedness criterion. Therefore

$$\bar{c}_{m+1} < 0 \qquad \bar{a}_{i,m+1} \leqq 0 \qquad \text{for all } i$$

$\bar{A}$ is obtained from A by doing a sequence of elementary row operations on A. Hence, if a vector $\bar{y}$ satisfies $\bar{A}\bar{y} = 0$, then $A\bar{y} = 0$ must hold. Define $\bar{y}$ by

$$\begin{aligned} \bar{y}_i &= -\bar{a}_{i,m+1} && \text{for } i = 1 \text{ to } m \\ &= 1 && \text{for } i = m+1 \\ &= 0 && \text{for } i \geqq m+2 \end{aligned}$$

By the above, $\bar{y} \geq 0$. From the final canonical tableau, clearly $\bar{A}\bar{y} = 0$. So $A\bar{y} = 0$, also. Normalizing $\bar{y}$ so that the sum of all the components in it becomes equal to 1 leads to $\hat{y} = \bar{y}(1/(1 - \sum_{i=1}^{m} \bar{a}_{i'm+1}))$. $\hat{y} \geqq 0$ and $\sum_{j=1}^{n} \hat{y}_j = 1$, and $A\hat{y} = 0$ by the above discussion. Clearly, $\hat{y}$ is an extreme homogeneous solution corresponding to (3.16). From the canonical tableau, we also verify that $z(\hat{y}) = c\hat{y} = \bar{c}_{m+1}/(1 - \sum_{i=1}^{m} \bar{a}_{i,m+1})$. Let the BFS of (3.16) associated with the terminal basis be $\bar{x}$. The half-line constructed in the simplex algorithm, along with $z(x)$ diverges to $-\infty$, is (see Section 2.5.3) $\{(\bar{b}_1 - \lambda\bar{a}_{1,m+1}, \ldots, \bar{b}_m - \lambda\bar{a}_{m,m+1}, \lambda, 0, \ldots, 0)^{\mathrm{T}}: \lambda \geqq 0\}$. This half-line is exactly the half-line $\{x: x = \bar{x} + \lambda\hat{y}, \lambda \geqq 0\}$. Every point on this half-line makes the variables $x_{m+2} = \cdots = x_n = 0$, conversely every feasible point x satisfying these conditions lies on this half-line. If x^0 is any point on this line and $x^0 = \alpha x^1 + (1-\alpha)x^2$ for some $0 < \alpha < 1$ and feasible solutions x^1 and x^2, then $x_{m+2}^1 = \cdots = x_n^1 = x_{m+2}^2 = x_n^2 = 0$. Hence x^1 and x^2 are on the half-line, too. Hence this half-line is an extreme half-line of the set of feasible solutions of (3.16).

Thus when the objective function in the LP being solved is unbounded, the simplex algorithm terminates by providing an extreme half-line of the set of feasible solutions along which the objective function diverges.

Example 3.27

Consider the LP of Example 2.14, Section 2.6.5. Consider the basic vector (x_2, x_1). This is a feasible basic vector. The associated BFS is $\tilde{x} = (1/2, 1/2, 0, 0)^{\mathrm{T}}$. The canonical tableau with respect to this basic vector is the final tableau obtained in Example 2.14. From the updated column vector of x_3, we construct the extreme homogeneous solutions $\hat{y} = (1/4, 1/4, 1/2, 0)^{\mathrm{T}}$. The extreme half-line along which $z(x)$ diverges to $-\infty$ is $\{x: x = (1/2, 1/2, 0, 0)^{\mathrm{T}} + \lambda(1/4, 1/4, 1/2, 0)^{\mathrm{T}}, \lambda \geqq 0\}$, which is the extreme half-line obtained in Section 2.6.5, Example 2.14.

3.7.1 Algorithm to Obtain All the Adjacent Extreme Points of a Given Extreme Point

Consider the LP (3.16) where A is a matrix of order $m \times n$ and rank m. Let **K** be the set of feasible solutions of this LP. Let $\bar{x}$ be an extreme point of **K**. In some applications, it may be necessary to do the following.

1 Generate all the adjacent extreme points of $\bar{x}$ on **K**.

2 Generate all the adjacent extreme points of $\bar{x}$ on **K** that make $z(x) \geqq z(\bar{x})$.

3 Generate all the bounded and unbounded edges of **K** containing $\bar{x}$.

Here we provide algorithms for doing these things. The computation involved is easy if $\bar{x}$ is a nondegenerate BFS and quite complicated if it is degenerate. We discuss these in two separate cases below.

Case 1 Suppose $\bar{x}$ is a nondegenerate BFS. Then exactly m of the $\bar{x}_j$ are positive. Let $\mathbf{J} = \{j_1, \ldots, j_m\} = \{j: j \text{ such that } \bar{x}_j > 0\}$ in this case. Then $x_B = (x_{j_1}, \ldots, x_{j_m})$ is the unique basic vector for (3.16) associated with the BFS $\bar{x}$. Obtain the canonical tableau for (3.16) with respect to the basic vector x_B.

Denote the entries in the canonical tableau by $\bar{a}_{ij}$, $\bar{b}_i$, $\bar{c}_j$, $-\bar{z}$, etc., as usual. Clearly $\bar{b}_i = \bar{x}_{j_i}$ for $i = 1$ to m, and $\bar{z} = z(\bar{x})$. By the nondegeneracy of $\bar{x}$, $\bar{b}_i > 0$ for all i. For each $j \notin \mathbf{J}$, compute θ_j = minimum $\{\bar{b}_i/\bar{a}_{ij}: i, \text{ such that } \bar{a}_{ij} > 0\}$, or $+\infty$ if $\bar{a}_{ij} \leqq 0$ for all i. θ_j is the minimum ratio when x_j enters the basic vector x_B. For each $s \notin \mathbf{J}$, define $x^s(\lambda) = (x_1^s(\lambda), \ldots, x_n^s(\lambda))^{\mathrm{T}}$, where

$$x_{j_i}^s(\lambda) = \bar{b}_i - \bar{a}_{is}\lambda \qquad \text{for } i = 1 \text{ to } m$$

$$x_s^s(\lambda) = \lambda$$

$$x_j^s(\lambda) = 0 \qquad \text{for all } j \notin \mathbf{J} \cup \{s\}$$

Then the set $\{x^s(\theta_s): s \notin \mathbf{J} \text{ such that } \theta_s \text{ is finite}\}$ is the set of adjacent extreme points of $\bar{x}$ in **K**. The set of adjacent extreme points of $\bar{x}$ on **K** at which $z(x) \geqq z(\bar{x})$ is $\{x^s(\theta_s): s \notin \mathbf{J} \text{ such that } \bar{c}_s \geqq 0 \text{ and } \theta_s \text{ is finite}\}$. The set of unbounded edges of **K** through $\bar{x}$ is $\{\{x^s(\lambda): \lambda \geqq 0\}: s \notin \mathbf{J} \text{ such that } \theta_s = +\infty\}$. The set of bounded edges of **K** through $\bar{x}$ is $\{\{x^s(\lambda): 0 \leqq \lambda \leqq \theta_s\}: s \notin \mathbf{J} \text{ such that } \theta_s \text{ is finite}\}$.

Example 3.28

Consider the following LP:

Basic Variables	x_1	x_2	x_3	x_4	x_5	x_6	x_7	x_8	x_9	$-z$	b
x_1	1	0	0	1	−2	2	−1	0	−1	0	2
x_2	0	1	0	−1	1	2	−1	−1	−1	0	4
x_3	0	0	1	2	1	2	0	−1	−1	0	6
$-z$	0	0	0	2	−6	0	2	−2	0	1	−10

$x_j \geqq 0$ for all j.

Let $\bar{x} = (2, 4, 6, 0, 0, 0, 0, 0, 0)^{\mathrm{T}}$. $x_B = (x_1, x_2, x_3)$ is the unique basic vector associated with the nondegenerate BFS, $\bar{x}$. The tableau at the bottom is the canonical tableau for this LP with respect to the basic vector x_B. $\mathbf{J} = \{1, 2, 3\}$. We compute $(\theta_4, \theta_5, \theta_6 \; \theta_7, \theta_8, \theta_9) = (2, 4, 1, +\infty, +\infty, +\infty)$. Let

$$x^4(\lambda) = (2 - \lambda, 4 + \lambda, 6 - 2\lambda. \lambda, 0, 0, 0, 0, 0)^{\mathrm{T}}$$

$$x^5(\lambda) = (2 + 2\lambda, 4 - \lambda, 6 - \lambda, 0, \lambda, 0, 0, 0, 0)^{\mathrm{T}}$$

$$x^6(\lambda) = (2 - 2\lambda, 4 - 2\lambda, 6 - 2\lambda, 0, 0, \lambda, 0, 0, 0)^{\mathrm{T}}$$

$$x^7(\lambda) = (2 + \lambda, 4 + \lambda, 6, 0, 0, 0, \lambda, 0, 0)^{\mathrm{T}}$$

$$x^8(\lambda) = (2, 4 + \lambda, 6 + \lambda, 0, 0, 0, 0, \lambda, 0)^{\mathrm{T}}$$

$$x^9(\lambda) = (2 + \lambda, 4 + \lambda, 6 + \lambda, 0, 0, 0, 0, 0, \lambda)^{\mathrm{T}}$$

Then $\{x^4(2), x^5(4), x^6(1)\}$ are the adjacent extreme points of $\bar{x}$ on **K**, the set of feasible solutions of this problem. $\{x^4(2), x^6(1)\}$ are the adjacent extreme points of $\bar{x}$ on **K** which make $z(x) \geqq z(\bar{x}) = 10$. The set of unbounded edges of **K** containing $\bar{x}$ is $\{\{x^s(\lambda): \lambda \geqq 0\}: s = 7, 8, 9\}$. The set of bounded edges of **K** containing $\bar{x}$ is $\{\{x^4(\lambda): 0 \geqq \lambda \geqq 2\}, \{x^5(\lambda): 0 \leqq \lambda \leqq 4\}, \{x^6(\lambda): 0 \leqq \lambda \leqq 1\}\}$.

Case 2 Suppose $\bar{x}$ is a degenerate BFS. In this case, let $\mathbf{J} = \{j_1, \ldots, j_r\} = \{j: \bar{x}_j > 0\}$. Then $r < m$, and $(x_{j_1}, \ldots, x_{j_r})$ is not a basic vector for (3.16), but it can be made into a basic vector by including in it $m - r$ variables (whose values are zero in $\bar{x}$) such that the column vectors of those variables, together with the column vectors $A_{.j_1}, \ldots, A_{.j_r}$ form a linearly independent set. A subset of $m - r$ variables $\{x_{p_1}, \ldots, x_{p_{m-r}}\} \subset \{x_j: j \notin \mathbf{J}\}$ is said to be a *zero set* for $\bar{x}$, if $\{A_{.j_1}, \ldots, A_{.j_r}, A_{.p_1}, \ldots, A_{.p_{m-r}}\}$ is a linearly independent set. Each basic vector for (3.16) associated with the BFS $\bar{x}$, has a zero set for $\bar{x}$ as its subset.

To get the set of all adjacent extreme points of $\bar{x}$ in **K**, we have to obtain the canonical tableau with respect to each of the basic vectors for (3.16) associated with $\bar{x}$, and obtain all the adjacent extreme points of $\bar{x}$ that can be obtained from that canonical tableau by bringing exactly one nonbasic variable into that basic vector. The computation can be organized efficiently by adopting the following procedure. Maintain a stack of zero sets called stack 1. Each zero set in stack 1 is either scanned or unscanned. When all the zero sets in stack 1 are scanned, the algorithm terminates. The algorithm also maintains the following stacks.

Stack 2: Adjacent extreme points of $\bar{x}$ in **K**.

Stack 3: Adjacent extreme points of $\bar{x}$ in **K** satisfying $z(x) \geqq z(\bar{x})$

Stack 4: Bounded edges of **K** through $\bar{x}$.

Stack 5: Unbounded edges of **K** through $\bar{x}$.

In the beginning, all of stacks 2 to 5 are empty. As new things are obtained during the algorithm, they are put in their respective stacks. The procedure begins by obtaining a zero set for $\bar{x}$, which is generated by completing $\{A_{.j_1}, \dots, A_{.j_r}\}$ into a maximal linearly independent set of column vectors of A. For this, write the matrix of order $n \times m$ with its rows in proper order as $A^{\mathrm{T}}_{.j_1}, \dots, A^{\mathrm{T}}_{.j_r}$, and then $A^{\mathrm{T}}_{.j}$ for $j \notin \mathbf{J}$ in some order, and apply the algorithm discussed in Section 3.3.3.

The Procedure

Step 1 Let $\{x_{p_1}, \dots, x_{p_{m-r}}\}$ be the initial zero set. In the beginning this is the only zero set in stack 1, and it is unscanned. Select this zero set for scanning, and go to Step 2.

Step 2 Let $\{x_{p_1}, \dots, x_{p_{m-r}}\}$ be the current zero set for $\bar{x}$, selected for scanning. Obtain the canonical tableau for (3.16) with respect to the basic vector $x_B = (x_{j_1}, \dots, x_{j_r}, x_{p_1}, \dots, x_{p_{m-r}})$. Let $\mathbf{J}_B = \{j_1, \dots, j_r, p_1, \dots, p_{m-r}\}$. For each $j \notin \mathbf{J}_B$, compute the minimum ratio θ_j in this canonical tableau, treating column j as the entering column. If $s \notin \mathbf{J}_B$ and $\theta_s > 0$, define $x^s(\lambda)$ as in Case 1. For each $s \notin \mathbf{J}_B$ such that $\theta_s > 0$ and is finite, $x^s(\theta_s)$ is an adjacent extreme point of $\bar{x}$ in **K** with an objective value of $z(\bar{x}) + \bar{c}_s\theta_s$, and if this extreme point is not already in stack 2, add it to stack 2. Also if $\bar{c}_s \geqq 0$, and $x^s(\theta_s)$ is not yet in stack 3, add it to stack 3. For each $s \notin \mathbf{J}_B$ such that $\theta_s > 0$ and is finite, $\{x^s(\lambda): 0 \leqq \lambda \leqq \theta_s\}$ is a bounded edge of **K** through $\bar{x}$, and if this edge is not already in stack 4, add it to stack 4. For each $s \notin \mathbf{J}_B$ such that $\theta_s = +\infty$, $\{x^s(\lambda): \lambda \geqq 0\}$ is an unbounded edge of **K** through $\bar{x}$, and if this unbounded edge is not already in stack 5, add it in stack 5. For each $j \notin \mathbf{J}_B$, and for u such that the updated column vector of x_j in the current canonical tableau has a nonzero entry in the row in which the zero basic variable x_{p_u} is the basic variable, $\{x_{p_1}, \dots, x_{p_{u-1}}, x_j, x_{p_{u+1}}, \dots, x_{p_{m-r}}\}$ is a zero set for $\bar{x}$, and if this zero set is not in stack 1 already, include it in stack 1 as an unscanned zero set.

After all these operations, declare the current zero set $\{x_{p_1}, \dots x_{p_{m-r}}\}$ as a scanned zero set, and go to Step 3.

Step 3 Go to stack 1 and look for an unscanned zero set from it at this stage. If it contains some unscanned zero sets, select one of them for scanning and with it as the current zero set go to Step 2. If there are no unscanned zero sets in stack 1 at this stage, go to Step 4.

Step 4 Terminate. Stacks 2, 3, 4, 5 at this stage are complete.

Example 3.29

Consider the following LP:

Basic Variables	x_1	x_2	x_3	x_4	x_5	x_6	x_7	x_8	$-z$	
x_1	1	0	0	1	0	−1	1	2	0	4
x_2	0	1	0	−1	1	−1	2	1	0	8
x_3	0	0	1	1	−1	0	0	0	0	0
$-z$	0	0	0	2	−1	2	−1	1	1	−10

$x_j \geqq 0$ for all j.

Let $\bar{x} = (4, 8, 0, 0, 0, 0, 0, 0)^T$. $\bar{x}$ is a degenerate BFS of this LP. $\mathbf{J} = \{j: j, \text{ such that } \bar{x}_j > 0\} = \{1, 2\}$. Clearly $\{x_3\}$ is a zero set for $\bar{x}$, and (x_1, x_2, x_3) is a basic vector associated with $\bar{x}$, and we begin the procedure with this as the initial basic vector. The procedure can be verified to yield the following.

$$x^1(\lambda) = (4, 8 - \lambda, \lambda, 0, \lambda, 0, 0, 0)^T$$
$$x^2(\lambda) = (4 + \lambda, 8 + \lambda, 0, 0, 0, \lambda, 0, 0)^T$$
$$x^3(\lambda) = (4 - \lambda, 8 - 2\lambda, 0, 0, 0, 0, \lambda, 0)^T$$
$$x^4(\lambda) = (4 - 2\lambda, 8 - \lambda, 0, 0, 0, 0, 0, \lambda)^T$$
$$x^5(\lambda) = (4 - \lambda, 8, 0, \lambda, \lambda, 0, 0, 0)^T$$

The set of adjacent extreme points of $\bar{x}$ in **K** that satisfy $z(x) \geqq z(x)$ is $\{x^4(2), x^5(4)\}$. There is only one unbounded edge of **K** through $\bar{x}$, and it is $\{x^2(\lambda): \lambda \geqq 0\}$. The set of bounded edges of **K** containing $\bar{x}$ is $\{\{x^1(\lambda): 0 \leqq \lambda \leqq 8\}, \{x^3(\lambda): 0 \leqq \lambda \leqq 4\}, \{x^4(\lambda): 0 \leqq \lambda \leqq 2\}, \{x^5(\lambda): 0 \leqq \lambda \leqq 4\}\}$.

The algorithm described here for obtaining all the adjacent extreme points of a given degenerate extreme point of the set of feasible solutions of (3.16) is discussed briefly in [3.34]; see [3.32] for some computational results. There are other methods for obtaining all the adjacent extreme points of a given degenerate extreme point of the set of feasible solutions of (3.16), but computationally all these methods seem to be quite complicated too (see references [3.19, 3.20, 3.37]).

3.8 THE MAIN GEOMETRICAL ARGUMENT IN THE SIMPLEX ALGORITHM

From the results discussed in the previous sections, we derive the following result, which is the main mathematical result in the simplex algorithm: Let **K** be a convex polyhedron in $\mathbf{R}^n$, and let $z(x)$ be a linear objective function defined on it.

1 If x^* is any extreme point of **K**, either x^* minimizes $z(x)$ on **K** or there exists an edge (bounded or unbounded) of **K** through x^*, such that $z(x)$ decreases strictly as we move along this edge from x^*.

2 If $z(x)$ is bounded below on **K**, and $\tilde{x}$ is any extreme point of **K**, either $\tilde{x}$ minimizes $z(x)$ on **K**, or there exists an adjacent extreme point of $\tilde{x}$ on **K**, say $\hat{x}$, such that $z(\hat{x}) < z(\tilde{x})$.

Adjacent Vertex Methods

For minimizing a linear objective function $z(x)$ on a convex polyhedron **K**, the simplex algorithm starts at an extreme point of **K**. In each step it moves from the present extreme point of **K** along an edge of **K** incident at this extreme point, such that $z(x)$ decreases as the algorithm travels along this edge. If the edge is an unbounded edge, $z(x)$ must be unbounded below on **K**. If the edge is bounded, the algorithm moves to the adjacent extreme point at the other end of this edge and continues in the same manner until either (1) an extreme point is reached with the property that $z(x)$ does not decrease by moving from this extreme point along every edge of **K** containing this extreme point, or (2) an unbounded edge is found along which $z(x)$ diverges to $-\infty$.

Algorithms of this type are known as *adjacent vertex methods*. The class of optimization problems that can be solved by adjacent vertex methods includes LPs, fractional linear programming problems (those in which a function of the form $z(x) = (dx + \alpha)/(fx + \beta)$ has to be minimized subject to equality constraints in nonnegative variables) and some other class of problems in which a nonlinear objective function $\theta(x)$ satisfying certain monotonicity properties has to be minimized subject to linear equations in nonnegative variables. See Section 3.20 and references [3.2, 3.7, 3.8].

3.9 THE DIMENSION OF A CONVEX POLYHEDRON

A subset $\mathbf{S} \subset \mathbf{R}^n$ is said to be a *subspace* of $\mathbf{R}^n$ if $x^1 \in \mathbf{S}$ and $x^2 \in \mathbf{S}$ implies $\alpha x^1 + \beta x^2 \in \mathbf{S}$ for all real values of α, β; that is, the linear hull of any pair of points in a subspace is in the subspace. All subspaces contain the origin 0.

A subset $\{x^1, \ldots, x^k\}$ of points in a subspace **S** is said to be a *maximal linearly independent subset of points* in **S** if $\{x^1, \ldots, x^k\}$ is linearly independent and $\{x^1, \ldots, x^k, x\}$ is linearly dependent for every $x \in \mathbf{S}$. It can be shown that all maximal linearly independent subsets of a given subspace of $\mathbf{R}^n$, contain the same number of points in them. This number is called the *dimension* of that subspace.

Also, if $\{x^1, \ldots, x^k\}$ is a maximal linearly independent subset of points in a subspace **S**, then **S** is the linear hull of $\{x^1, \ldots, x^k\}$.

A subset $\Gamma \subset \mathbf{R}^n$ is said to be an *affine space* if $x^1 \in \Gamma$ and $x^2 \in \Gamma$ implies $\alpha x^1 + (1 - \alpha)x^2 \in \Gamma$ for all real values of α. Let Γ be an affine space and let $x^0 \in \Gamma$. Then $\{x : x = y - x^0, y \in \Gamma\}$ is a subspace of $\mathbf{R}^n$. The *dimension* of Γ is defined as the dimension of this subspace. Hence the dimension of Γ is the maximum value of r such that there exist points $x^1, \ldots, x^r$ in Γ satisfying $\{x^1 - x^0, \ldots, x^r - x^0\}$ is a linearly independent set.

Let A be a matrix of order $m \times n$ and rank r. Then the set of solutions of $Ax = 0$ is a subspace of $\mathbf{R}^n$ of dimension $n - r$ (see references [3.9, 3.11, 3.12, 3.14]. The set of solutions of $Ax = b$ is either the empty set or an affine space of dimension $n - r$.

Let **K** be a convex polyhedral subset of $\mathbf{R}^n$. If $\mathbf{K} \neq \emptyset$, its dimension is defined to be the dimension of the smallest dimension affine space containing it. Equivalently, let x^0 be any point in **K**. The dimension of **K** is the maximum number r such that there exist points $x^1, \ldots, x^r$ in **K** satisfying $\{x^1 - x^0, \ldots, x^r - x^0\}$ is a linearly independent set.

The dimension of a convex polyhedron that is the set of feasible solutions of a system of linear equations in nonnegative variables is easily specified, under nondengeneracy. Let **K** be the set of feasible solutions of (3.16) where A is a matrix of order $m \times n$ and rank m. Let b be nondegenerate (i.e., every solution for the system has at least m nonzero coordinates, or equivalently, b does not lie in the linear hull of any set containing $m - 1$ or less columns of A). Then dimension of **K** is $n - m$. If b is degenerate, the dimension of **K** is $n - m$ or less. See the example below, and Exercises 3.67, 3.68.

Degenerate Example 3.30

Consider the system

$$x_1 - x_3 = 0$$
$$x_2 + x_3 = 0$$
$$x_1, x_2, x_3 \geqq 0$$

Here $n = 3$, $m = 2$. Hence $n - m = 1$. However, the system has the unique feasible solution $x = 0$, and since the set of feasible solutions contains a single point, its dimension is zero.

Let **K** be a convex polyhedral subset of $\mathbf{R}^n$. Let x^0 be an extreme point of **K**. Let $\mathbf{E}_1, \ldots, \mathbf{E}_r$ be all the edges of **K** containing x^0. Let x^t be a point on the edge $\mathbf{E}_t$ distinct from x^0, for $t = 1$ to r. Then it can be shown that the affine hull of $\{x^0, x^1, \ldots, x^r\}$ is the smallest dimension affine space containing **K**. Also the dimension of **K** is the dimension of this affine space, and it is equal to the rank of the set $\{x^1 - x^0, \ldots, x^r - x^0\}$.

We have the following result concerning the dimension of convex polyhedra that are the sets of feasible solutions of systems of linear equations in nonnegative variables.

THEOREM 3.8 Let **K** be the set of feasible solutions of $Ax = b$, where $x \geqq 0$, and where A is a matrix of order $m \times n$ and rank m. If this system has a nondegenerate BFS, then the dimension of **K** is $n - m$.

Proof: Let $\bar{x}$ be a nondegenerate BFS of this system corresponding to the basic vector $(x_1, \ldots, x_m)$. Let $\bar{A}_{.j} = (\bar{a}_{1j}, \ldots, \bar{a}_{mj})^{\mathrm{T}}$ be the updated column of x_j, and $\bar{b} = (\bar{b}_1, \ldots, \bar{b}_m)^{\mathrm{T}}$ be the updated right-hand-side constants vector in the canonical tableau for the system with respect to the basic vector $(x_1, \ldots, x_m)$. By nondegeneracy, $\bar{b} > 0$. So the solution obtained by making the nonbasic variable x_j equal to a parameter λ, leaving all the other nonbasic variables equal to zero, and then evaluating the values of $(x_1, \ldots, x_m)$ to satisfy the equality constraints, remains nonnegative as long as λ is positive and sufficiently small. Let $\varepsilon_j > 0$ be such a value of λ, and let $x^j = (x_1^j, \ldots, x_n^j)^{\mathrm{T}}$ denote this solution when $\lambda = \varepsilon_j$. We have

$$\begin{aligned} x_i^j &= \bar{b}_i - \varepsilon_j \bar{a}_{ij} && \text{for } i = 1 \text{ to } m \\ &= 0 && \text{for } i \neq j, \text{ and } i \text{ between } m + 1 \text{ to } n \\ &= \varepsilon_j && \text{for } i = j \end{aligned}$$

By doing this for $j = m + 1$ to n, we get the distinct feasible solutions $x^{m+1}, \ldots, x^n$ to this problem. The matrix consisting of column vectors $x^j - \bar{x}$ is

$$\begin{pmatrix} -\varepsilon_{m+1}\bar{a}_{1,m+1} & -\varepsilon_{m+2}\bar{a}_{1,m+2} & \cdots & -\varepsilon_n\bar{a}_{1n} \\ \vdots & \vdots & & \vdots \\ -\varepsilon_{m+1}\bar{a}_{m,m+1} & -\varepsilon_{m+2}\bar{a}_{m,m+2} & \cdots & -\varepsilon_n\bar{a}_{mn} \\ \varepsilon_{m+1} & 0 & \cdots & 0 \\ 0 & \varepsilon_{m+2} & \cdots & 0 \\ \vdots & \vdots & & \vdots \\ 0 & 0 & \cdots & \varepsilon_n \end{pmatrix}$$

Basic Variables	x_1	x_2	x_3	x_4	x_5	x_6	x_7	$-z$	b
x_1	1	0	0	0	1	1	0	0	3
x_2	0	1	0	1	1	-2	3	0	0
x_3	0	0	1	-2	1	1	7	0	2
$-z$	0	0	0	0	6	0	8	1	-10

$x_j \geqq 0$ for all j; z to be minimized.

and since $\varepsilon_{m+1}, \varepsilon_{m+2}, \ldots, \varepsilon_n$ are all positive, the set of column vectors of this matrix is a linearly independent set. Hence, $\{x^{m+1} - \bar{x}, \ldots, x^n - \bar{x}\}$ is a linearly independent set. Hence the dimension of **K** is $n - m$. ■

3.10 ALTERNATE OPTIMUM FEASIBLE SOLUTIONS AND FACES OF CONVEX POLYHEDRA

Consider the LP (3.16) where A is of order $m \times n$ and rank m. Suppose B_1 is an optimum feasible basis of this problem. Let the canonical tableau with respect to this basis be Tableau 2.8 of Section 2.5.1, with on optimum objective value of $\bar{z}$, and relative cost coefficients $\bar{c}_j$. Therefore, by the fundamental optimality criterion any feasible solution x that makes $z(x) = \bar{z}$ is an optimum feasible solution of this problem and vice versa. Hence the set of all optimum feasible solutions of (3.16) is the set of feasible solutions of

$$Ax = b$$
$$z(x) = cx = \bar{z}$$
$$x \geqq 0$$

This leads to the following results.

1 The set of optimum feasible solutions of an LP is itself a convex polyhedral set. Consequently,

2 every convex combination of optimum feasible solutions is also optimal.

Since B_1 is an optimum basis, all the relative cost coefficients $\bar{c}_j$ must be nonnegative. The cost row in the canonical tableau leads to

$$z(x) = \bar{z} + \sum_{\substack{j \text{ such that} \\ x_j \text{ is nonbasic}}} \bar{c}_j x_j$$

From this equation it is clear that every feasible solution of (3.16) in which the variables x_j corresponding to strictly positive $\bar{c}_j$ are zero, is an optimum feasible solution. Hence, if $\Gamma = \{j : j, \text{such that } \bar{c}_j > 0\}$, the set of optimum feasible solutions of (3.16) is the set of feasible solutions of (3.16) in which $x_j = 0$ for all $j \in \Gamma$.

Thus, if x_s is a nonbasic variable in the final optimum canonical tableau, and $\bar{c}_s = 0$, and if the operation of bringing x_s into the basic vector involves a nondegenerate pivot step, this step yields an alternate optimum BFS. Alternate optimal bases are obtained by bringing into an optimal basic vector any nonbasic variable with a zero relative cost coefficient. These considerations lead us to the following result, which is a *sufficient* (but not necessary) *condition* for (3.16) to have a unique optimum feasible solution.

Result 3.2: Consider the LP (3.16). If there exists an optimum basic vector for it, in which the relative cost coefficients of all the nonbasic variables are strictly positive, the corresponding BFS is the unique optimum feasible solution of this LP.

Example 3.31

Consider the canonical tableau given at the top.

An optimum basic vector of this problem is (x_1, x_2, x_3). The optimum solution corresponding to it is $x^{\mathrm{T}} = (3, 0, 2, 0, 0, 0, 0)^{\mathrm{T}}$ with objective value $= 10$. When x_6 is brought into this basic vector, it drives x_3 out of the basic vector and leads to a new optimum BFS, $x^1 = (1, 4, 0, 0, 0, 2, 0)$.

Exercises

3.31 Obtain all the optimum BFSs to the LP in Example 3.31.

3.32 In the LP in Example 3.31 prove that every feasible solution in which $x_5 = x_7 = 0$ is an optimum feasible solution and vice versa.

3.33 Write down a general expression that represents an optimum feasible solution to the LP in Example 3.31.

Example 3.32

Consider the canonical tableau given at the bottom. (x_1, x_2, x_3) is an optimum basic vector for this problem. The relative cost coefficient of the nonbasic variable x_4 is zero. So the condition mentioned in Result 3.2 is not satisfied, and yet, it can be verified that the present BFS $x = (0, 2, 2, 0, 0, 0)^{\mathrm{T}}$ is the unique optimum feasible solution of this LP, since this is the only feasible solution of the problem in which both the variables x_5, x_6 are zero.

Basic Variables	x_1	x_2	x_3	x_4	x_5	x_6	$-z$	b
x_1	1	0	0	1	-1	-1	0	0
x_2	0	1	0	1	0	0	0	2
x_3	0	0	1	0	2	4	0	2
$-z$	0	0	0	0	5	10	1	0

$x_j \geqq 0$ for all j; z to be minimized.

See Exercise 3.34 for necessary and sufficient conditions for an LP in standard form to have a unique optimum solution, and Section 4.7 for a complete discussion of necessary and sufficient conditions for an LP in general form to have a unique optimum feasible solution.

Exercise

3.34 *NECESSARY AND SUFFICIENT CONDITIONS FOR UNIQUE OPTIMUM SOLUTION.* Consider the LP (3.16) is standard form where A is a matrix of order $m \times n$ and rank m. Let $\bar{x}$ be a given optimum feasible solution for it. If $\bar{x}$ is not a BFS, prove that (3.16) has alternate optimum solutions besides $\bar{x}$. Suppose $\bar{x}$ is a BFS, and let x_B be a basic vector for (3.16) corresponding to $\bar{x}$, which satisfies the optimality criterion. Let $\bar{c}$ denote the vector of relative cost coefficients with respect to x_B. Let $\mathbf{J} = \{j: j, \text{ such that } \bar{c}_j > 0\}$, $\mathbf{L} = \{j: j, \text{ such that } \bar{c}_j = 0 \text{ and } x_j \notin x_B\}$. If $\bar{x}$ is nondegenerate and $\mathbf{L} \neq \emptyset$, prove that (3.16) has alternate optimum solutions besides $\bar{x}$. If $\mathbf{L} = \emptyset$, prove that $\bar{x}$ is the unique optimum solution of (3.16), whether $\bar{x}$ is degenerate or not. If $\bar{x}$ is degenerate and $\mathbf{L} \neq \emptyset$, consider the LP

$$\begin{aligned} \text{Maximize} \quad & f(x) = \sum_{j \in \mathbf{L}} x_j \\ \text{Subject to} \quad & Ax = b \\ & x_j = 0 \quad \text{for } j \in \mathbf{J} \\ & x_j \geqq 0 \text{ for } j \notin \mathbf{J} \end{aligned}$$

In this case, $\bar{x}$ is the unique optimum solution of (3.16) iff the maximum value of $f(x)$ in this LP is zero.

Faces of a Convex Polyhedron

Let $\mathbf{K} \subset \mathbf{R}^n$ be a convex polyhedron. Suppose $d \neq 0$ is a row vector of order $1 \times n$. Let $\mathbf{H} = \{x: dx = d_0\}$ for some real number d_0. If $dx \geqq d_0$ for all $x \in \mathbf{K}$, the hyperplane $\mathbf{H}$ is said to have the convex polyhedron $\mathbf{K}$ on one of its sides. If $\mathbf{H}$ has $\mathbf{K}$ on one of its sides and, in addition, if there is a point $\tilde{x}$ such that $\tilde{x} \in \mathbf{K} \cap \mathbf{H}$, that is, if $\tilde{x} \in \mathbf{K}$ and $d\tilde{x} = d_0$, then $\mathbf{H}$ is said to be a *supporting hyperplane* for the convex polyhedron $\mathbf{K}$ at the point $\tilde{x} \in \mathbf{K}$. See Figure 3.31.

A *face* of a convex polyhedron $\mathbf{K}$ is either the convex polyhedron itself, the empty set, or the intersection of the convex polyhedron with a supporting hyperplane. A *facet* of a convex polyhedron is a face of it, whose dimension is one less than the dimension of the convex polyhedron (Figure 3.32).

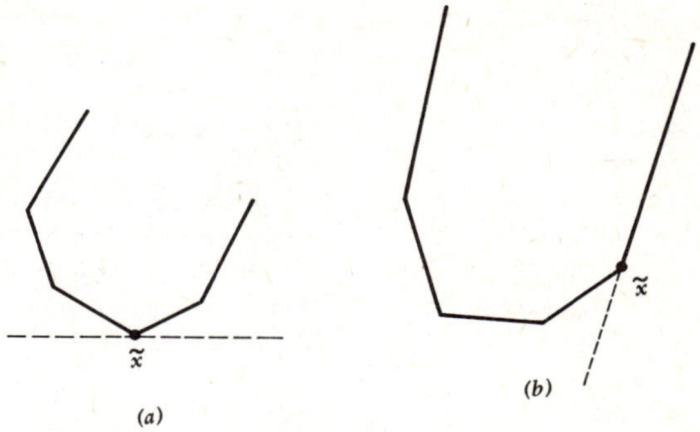

Figure 3.32 (*a*) Face $\{\tilde{x}\}$. This is not a facet. (*b*) Face is the extreme half-line through $\tilde{x}$, which is a facet.

THEOREM 3.9 The set of optimum feasible solutions of an LP is a face of its set of feasible solutions.

Proof: Consider the LP (3.16). Let **K** be the set of feasible solutions of this LP. If $z(x)$ is unbounded below on **K**, there are no optimum feasible solutions; hence, the set of optimum feasible solutions is the empty set; this is a face of **K** by definition. On the other hand, if the LP has an optimum feasible solution, let $\bar{z}$ be the optimum objective value. Then $z(x) = cx \geqq \bar{z}$ for all $x \in \mathbf{K}$, and the set of optimum feasible solutions is the set $\{x: x \in \mathbf{K} \text{ and } cx = \bar{z}\}$; this is a face of **K** by definition. ■

Exercises

3.35 Let **K** be the set of feasible solutions of $Ax = b$, $x \geqq 0$, where A is of order $m \times n$. Let $\mathbf{J} \subset \{1, 2, \ldots, n\}$. Prove that the set of feasible solutions of $Ax = b$, $x \geqq 0$, $x_j = 0$ for all $j \subset \mathbf{J}$, is a face of **K** and conversely that every face of **K** is of this form.

3.36 Let **K** be a convex polyhedron in the two-dimensional plane and suppose that $z(x)$ is a linear objective function defined on **K**. If the problem of minimizing $z(x)$ on **K** has three distinct extreme points of **K** optimal to it, then $z(x)$ is a constant on **K**.

3.11 PIVOT MATRICES

Consider the LP (3.16) where A is a matrix of order $m \times n$ and rank m. The main operations used in the simplex algorithm are the *pivot operations*. The pivot on the input-output coefficient matrix with the sth column as the *pivot column*, the rth row as the *pivot row*, can only be performed if the pivot element a_{rs} is nonzero. It transforms the matrix from $A = (a_{ij})$ into $\bar{A} = (\bar{a}_{ij})$ as in Tableaux 2.2, and 2.3 of Section 2.3.1. Let P be the square matrix of order m that differs from the unit matrix only in its rth column.

$$P = \begin{bmatrix} 1 & \cdots & 0 & -a_{1s}/a_{rs} & 0 & \cdots & 0 \\ \vdots & & \vdots & \vdots & \vdots & & \vdots \\ 0 & \cdots & 1 & -a_{r-1,s}/a_{rs} & 0 & \cdots & 0 \\ 0 & \cdots & 0 & 1/a_{rs} & 0 & \cdots & 0 \\ 0 & \cdots & 0 & -a_{r+1,s}/a_{rs} & 1 & \cdots & 0 \\ \vdots & & \vdots & \vdots & \vdots & & \vdots \\ 0 & \cdots & 0 & -a_{ms}/a_{rs} & 0 & \cdots & 1 \end{bmatrix}$$

It is easily verified that (1) the (r, r)th entry in P is the reciprocal of the pivot element; it is nonzero and hence, P is nonsingular, and (2) $\bar{A} = PA$. For this

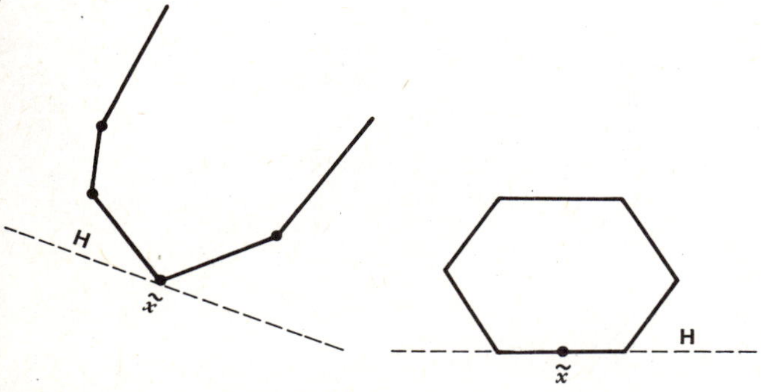

Figure 3.31 Supporting hyperplanes.

reason the matrix P is known as the *pivot matrix* corresponding to this pivot step. The column vector in which P differs from the unit matrix depends on the position of the pivot row in the matrix A. If the pivot row is the rth row, P differs from the unit matrix only in its rth column. Pivot matrices belong to a class of matrices called *elementary matrices* in linear algebra. An elementary matrix is a square nonsingular matrix that is either a permutation matrix or a matrix that differs from the unit matrix in just a row or a column.

In the kth stage of the algorithm, suppose the canonical tableau is:

$$\begin{array}{ccccc|cl}
x_1 \ \cdots & x_s & \cdots \ x_n & & -z & b & \\
\hline
\bar{a}_{11} \ \cdots & \bar{a}_{1s} & \cdots \ \bar{a}_{1n} & & 0 & \bar{b}_1 & \\
\vdots & \vdots & \vdots & & \vdots & \vdots & \\
\bar{a}_{r1} \ \cdots & \textcircled{\bar{a}_{rs}} & \cdots \ \bar{a}_{rn} & & 0 & \bar{b}_r & \textit{Pivot Row} \\
\vdots & \vdots & \vdots & & \vdots & \vdots & \\
\bar{a}_{m1} \ \cdots & \bar{a}_{ms} & \cdots \ \bar{a}_{mn} & & 0 & \bar{b}_m & \\
\bar{c}_1 \ \cdots & \bar{c}_s & \cdots \ \bar{c}_n & & 1 & -\bar{z} &
\end{array}$$

Pivoting on this tableau with the sth column as the pivot column and the rth row as the pivot row is equivalent to multiplying every column in the tableau on the left by the pivot matrix:

$$P^k = \begin{bmatrix}
1 & \cdots & 0 & \overset{r\text{th Column}}{-\bar{a}_{1s}/\bar{a}_{rs}} & 0 & \cdots & 0 \\
\vdots & & \vdots & \vdots & \vdots & & \vdots \\
0 & \cdots & 1 & -\bar{a}_{r-1,s}/\bar{a}_{rs} & 0 & \cdots & 0 \\
0 & \cdots & 0 & 1/\bar{a}_{rs} & 0 & \cdots & 0 \\
0 & \cdots & 0 & -\bar{a}_{r+1,s}/\bar{a}_{rs} & 1 & \cdots & 0 \\
\vdots & & \vdots & \vdots & \vdots & & \vdots \\
0 & \cdots & 0 & -\bar{c}_s/\bar{a}_{rs} & 0 & \cdots & 1
\end{bmatrix} \qquad (3.25)$$

which is of order $m + 1$, since each column in the tableau has $m + 1$ entries, including the cost row. The revised simplex method with the product form of the inverse discussed in Chapter 5 is based on the use of these pivot matrices. The special column vector of a pivot matrix that is different from the corresponding column vector of the unit matrix is known as its *eta column* or *eta vector*, and is sometimes denoted by the symbol η *column*. Notice that the position of the eta column in a pivot matrix is exactly the same as the position of the pivot row in the tableau in the pivot step corresponding to this pivot matrix. Since the pivot matrix differs from the unit matrix of the same size in exactly its eta vector, if we store the eta vector and its position in the pivot matrix, the pivot matrix itself can be constructed using this information. So the special structure of the pivot matrix makes it possible for us to store it compactly using little storage space.

3.12 CANONICAL TABLEAUX IN MATRIX NOTATION

Let B be a feasible basis of the LP (3.16). Let x_B be the basic vector and c_B the corresponding vector of cost coefficients. Suppose x_D is the vector of nonbasic variables and c_D the vector of their cost coefficients. After partitioning into the basic and nonbasic parts, (3.16) is as shown in Tableau 3.2.

Tableau 3.2

x_B	$-z$	x_D	b
B	0	D	b
c_B	1	c_D	0

The canonical tableau with respect to the basis B is obtained by transforming the submatrix under the variables x_B, $-z$ into the unit matrix of order $m + 1$. This is achieved by multiplying each column in the tableau on the left by the matrix

$$T = \left(\begin{array}{c|c} B & 0 \\ \hline c_B & 1 \end{array}\right)^{-1} = \left(\begin{array}{c|c} B^{-1} & 0 \\ \hline -c_B B^{-1} & 1 \end{array}\right) \qquad (3.26)$$

Hence Tableau 3.3 is the canonical tableau with respect to the basis B.

Tableau 3.3

Basic Variables	x_B	$-z$	x_D	*Basic Values*
x_B	I	0	$B^{-1}D$	$B^{-1}b$
$-z$	0	1	$c_D - c_B B^{-1}D$	$-c_B B^{-1}b$

The column vector of x_j in the original tableau is $\begin{pmatrix} A_{.j} \\ \cdots \\ c_j \end{pmatrix}$. The column vector of x_j in the canonical tableau is known as the *updated column vector* of x_j. It is $\begin{pmatrix} \bar{A}_{.j} \\ \cdots \\ \bar{c}_j \end{pmatrix}$, where

$$\bar{A}_{.j} = B^{-1}A_{.j}$$
$$\bar{c}_j = c_j - c_B B^{-1} A_{.j} = c_j - c_B \bar{A}_{.j}$$

Note 3.8: In some linear programming books the quantity $c_B\bar{A}_{.j}$ is denoted by the symbol z_j, and hence in this notation, the relative cost coefficient of x_j is $\bar{c}_j = c_j - z_j$. In these books, the optimality criterion for the minimization problem is stated as $c_j - z_j \geqq 0$ for all j.

The updated right-hand side constants vector is $\begin{pmatrix} \bar{b} \\ \cdots \\ -\bar{z} \end{pmatrix}$ where

$$\bar{b} = B^{-1}b$$
$$\bar{z} = c_B B^{-1} b$$

These are the values of the basic vector and the objective function in the basic solution of (3.16) corresponding to the basis B.

From the results in Section 3.3.2, we know that the updated column vector of x_j, $\begin{pmatrix} \bar{A}_{.j} \\ \cdots \\ \bar{c}_j \end{pmatrix}$ is the representation of the original column of x_j in terms of the columns of basis, that is, $\bar{a}_{ij}, \ldots, \bar{a}_{mj}, \bar{c}_j$ are the coefficients in the expression of $\begin{pmatrix} A_{.j} \\ \cdots \\ c_j \end{pmatrix}$ as a linear combination of the basic columns. The matrix in (3.26) is known as the *inverse matrix* corresponding to this basis. The revised simplex method with the explicit form of the inverse discussed in Chapter 5 is based on the use of inverse matrices.

The canonical tableau is also known as the *updated tableau*. The process of computing it is known as *updating* the tableau. Given the inverse tableau corresponding to the basis, and the original tableau, the columns of the updated tableau can be computed one by one using the formulas given above. As we see in Chapter 5, in order to execute the simplex algorithm it is unnecessary to update the entire tableau. In the revised simplex algorithm discussed in Chapter 5, only the columns that are essential to carry out that particular step of the algorithm are updated.

3.13 WHY IS AN ALGORITHM REQUIRED TO SOLVE LINEAR PROGRAMS?

Let $f(c, A, b)$ denote the optimum objective value in (3.16) as a function of the problem data c, A, and b. Is it possible to obtain $f(c, A, b)$ by a compact formula without computing it by an algorithm for given c, A, b? Consider the simple case where A, c remain fixed, but b may vary. In this case let $g(b) = f(c, A, b)$. The set of $b \in \mathbf{R}^m$ for which (3.16) has a feasible solution is the convex cone $\text{Pos}\{A_{.1}, \ldots, A_{.n}\} = \text{Pos}(A)$. A basis B of (3.16) is a feasible basis iff $B^{-1}b \geqq 0$, that is, iff $b \in \text{Pos}(B)$. B is an optimum basis for (3.16) if it is feasible and if the relative cost coefficients with respect to B are all nonnegative. Any basis B with respect to which the relative cost coefficients are all nonnegative is known as a *dual feasible basis* (discussed in detail in Chapter 4). Hence if B is a dual feasible basis for (3.16) and if $b \in \text{Pos}(B)$, then an optimum solution of (3.16) is

$$x_B = B^{-1}b$$
$$\text{All other } x_j = 0$$
$$\text{Minimum objective value} = g(b) = c_B B^{-1} b$$

Therefore as b varies in the cone $\text{Pos}(B)$, $g(b)$ is a linear function of b. The number of dual feasible bases for (3.16) grows very rapidly with $n - m$. Its growth rate is possibly of the order of $(n - m)!$ Hence, unless $n - m$ is a very small number, the total number of dual feasible bases for (3.16) is likely to be very large.

Even if we are able to tabulate all these cones $\text{Pos}(B)$ corresponding to the various dual feasible bases for (3.16) and the formula for $g(b)$ as a linear function of b in each one of them, the table will be very long because of the large number of these bases. To find out the value of $g(b)$ for any given b, we have to search the table for a cone $\text{Pos}(B)$ in which the vector b lies. This can be a very tedious job unless a systematic algorithm (or an iterative scheme) is developed for this search, because of the length of the table. This is exactly what the simplex algorithm does anyway. It is an iterative scheme for solving (3.16). LPs can only be solved efficiently by using some algorithm that guarantees that the optimal solution will be obtained in a finite number of iterations. The simplex algorithm or one of its modern variants provide such an algorithm.

Why it Is Called the Simplex Algorithm

Let $\{x^1, \ldots, x^{n+1}\}$ be a set of points in $\mathbf{R}^n$ such that $\{x^2 - x^1, \ldots, x^{n+1} - x^1\}$ is a linearly independent set. Then the convex hull of $\{x^1, \ldots, x^{n+1}\}$ is an *n-simplex*. If the set $\{x^1, \ldots, x^n\}$ of points in $\mathbf{R}^n$ is linearly independent, clearly the convex hull of $\{0, x^1, \ldots, x^n\}$ is an n-simplex. The cone $\text{Pos}\{x^1, \ldots, x^n\}$ is a *simplicial cone* in $\mathbf{R}^n$. If $\text{Pos}\{x^1, \ldots, x^n\}$ is a simplicial cone, then for any i, $\text{Pos}\{x^1, \ldots, x^{i-1}, x^{i+1}, \ldots, x^n\}$ is a facet of it. Verify this for $n = 3$ from Figure 3.33.

Consider the LP (3.16) where A is a matrix of order $m \times n$ and rank m. The original column vector including the cost coefficient, associated with the variable x_j in this problem is $\boldsymbol{A}_{.j} = (a_{1j}, \ldots, a_{mj}, c_j)^{\mathrm{T}}$, $j = 1$ to n.

Let $y = (y_1, \ldots, y_m, y_{m+1})$ be the coordinates of a general point in $\mathbf{R}^{m+1}$. Plot each of the points $\boldsymbol{A}_{.j}$, $j = 1$ to n in $\mathbf{R}^{m+1}$. If $x_j \geqq 0$ for all j, the point $\sum_{j=1}^{n} x_j \boldsymbol{A}_{.j}$ will be a point in the cone $\text{Pos}\{\boldsymbol{A}_{.1}, \ldots, \boldsymbol{A}_{.n}\}$. If $\bar{x}$ is a feasible solution of (3.16) with an objective value of $\bar{z}$, then $\sum_{j=1}^{n} \bar{x}_j \boldsymbol{A}_{.j} = (b_1, \ldots, b_m, \bar{z})^{\mathrm{T}}$, and this is a point in $\text{Pos}\{\boldsymbol{A}_{.1}, \ldots, \boldsymbol{A}_{.n}\}$. Conversely, if $(b_1, \ldots, b_m, \alpha)^{\mathrm{T}}$ is a point in $\text{Pos}\{\boldsymbol{A}_{.1}, \ldots, \boldsymbol{A}_{.n}\}$, by the definition of the Pos cone, there must exist nonnegative numbers $\hat{x}_1, \ldots, \hat{x}_n$ such that $(b_1, \ldots, b_m, \alpha)^{\mathrm{T}} = \sum_{j=1}^{n} \hat{x}_j \boldsymbol{A}_{.j}$.

Consider the straight line in $\mathbf{R}^{m+1}$, $\mathbf{L} = \{y = (b_1, \ldots, b_m, z)^{\mathrm{T}}: z \text{ is a real-valued parameter}\}$. This is a line in $\mathbf{R}^{m+1}$, parallel to the y_{m+1} axis, through the point $(b_1, \ldots, b_m, 0)^{\mathrm{T}}$.

From the above argument it is clear that if $(b_1, \ldots, b_m, \alpha)^{\mathrm{T}}$ is a point on $\mathbf{L} \cap \text{Pos}\{\boldsymbol{A}_{.1}, \ldots, \boldsymbol{A}_{.n}\}$, there exists a feasible solution of (3.16) that has an objective value of α (i.e., it makes $cx = \alpha$). Hence the problem of minimizing cx in (3.16) is equivalent to the problem of finding the minimum value of the parameter z such that the point $y \in \mathbf{L}$ is in $\text{Pos}\{\boldsymbol{A}_{.1}, \ldots, \boldsymbol{A}_{.n}\}$. This is the problem of finding the lowest point (in the direction of the y_{m+1} axis) in $\mathbf{L} \cap \text{Pos}\{\boldsymbol{A}_{.1}, \ldots, \boldsymbol{A}_{.n}\}$. Hence the straight line $\mathbf{L}$ is called the *requirement line* and we want to find the lowest point on it that lies in $\text{Pos}\{\boldsymbol{A}_{.1}, \ldots, \boldsymbol{A}_{.n}\}$.

In the course of solving this problem by the simplex algorithm, consider a step in which the feasible basic vector is $(x_1, \ldots, x_m)$, say. Let the objective value of the present BFS be $\bar{z}$. In the present BFS, only the basic variables $x_1, \ldots, x_m$ can take positive values; all the nonbasic variables, $x_{m+1}, \ldots, x_n$ are zero.

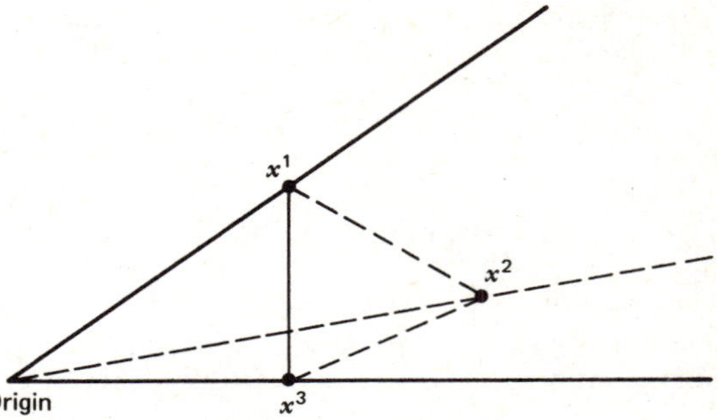

Figure 3.33 A simplical cone in $\mathbf{R}^3$.

These facts imply that the point $(b_1, \ldots, b_m, \bar{z})^{\mathrm{T}}$ is in $\text{Pos}\{\boldsymbol{A}_{.1}, \ldots, \boldsymbol{A}_{.m}\}$.

Let x_{m+1} be the entering variable in this step of the algorithm. Let $\bar{\boldsymbol{A}}_{.m+1} = (\bar{a}_{1,m+1}, \ldots, \bar{a}_{m,m+1}, \bar{c}_{m+1})^{\mathrm{T}}$ be the updated column vector of x_{m+1} with respect to the present basic vector $(x_1, \ldots, x_m)$. By the entering variable choice rule in the algorithm, the relative cost coefficient of x_{m+1}, $\bar{c}_{m+1} < 0$. This implies that $\{\boldsymbol{A}_{.1}, \ldots, \boldsymbol{A}_{.m}, \boldsymbol{A}_{.m+1}\}$ is linearly independent and, hence, that $\text{Pos}\{\boldsymbol{A}_{.1}, \ldots, \boldsymbol{A}_{.m}, \boldsymbol{A}_{.m+1}\}$ is a simplicial cone in $\mathbf{R}^{m+1}$. During this step, the algorithm is at the point $(b_1, \ldots, b_m, \bar{z})^{\mathrm{T}}$ of the requirement line, on the facet $\text{Pos}\{\boldsymbol{A}_{.1}, \ldots, \boldsymbol{A}_{.m}\}$ of this simplicial cone. Two cases can happen now.

Case 1 $\bar{c}_{m+1} < 0$ and $(\bar{a}_{1,m+1}, \ldots, \bar{a}_{m,m+1}) \leqq 0$. In this case, the half-line of the requirement line, obtained by letting the parameter $z \leqq \bar{z}$, lies entirely in the simplicial cone $\text{Pos}\{\boldsymbol{A}_{.1}, \ldots, \boldsymbol{A}_{.m}, \boldsymbol{A}_{.m+1}\}$. Since the simplicial cone $\text{Pos}\{\boldsymbol{A}_{.1}, \ldots, \boldsymbol{A}_{.m+1}\}$ is a subset of $\text{Pos}\{\boldsymbol{A}_{.1}, \ldots, \boldsymbol{A}_{.n}\}$, this implies that the part of the requirement line for $z \leqq \bar{z}$ is completely in $\text{Pos}\{\boldsymbol{A}_{.1}, \ldots, \boldsymbol{A}_{.n}\}$. Hence there is no lowest point in $\mathbf{L} \cap \text{Pos}\{\boldsymbol{A}_{.1}, \ldots, \boldsymbol{A}_{.n}\}$, and the objective function is unbounded below in (3.16).

Case 2 Suppose $\bar{a}_{is} > 0$ for at least one i. In this case, there is a pivot element and when x_{m+1} enters the basic vector, it replaces one of the present basic variables, say x_1, from the basic vector. The new basic vector is $(x_2, \ldots, x_m, x_{m+1})$. Suppose the BFS corresponding to it has an objective value $\hat{z}$. By arguments similar to those used earlier, this implies that the point $(b_1, \ldots, b_m, \hat{z})$ on the requirement line is on $\text{Pos}\{\boldsymbol{A}_{.2}, \ldots, \boldsymbol{A}_{.m}, \boldsymbol{A}_{.m+1}\}$. $\text{Pos}\{\boldsymbol{A}_{.1}, \ldots, \boldsymbol{A}_{.m}\}$ and $\text{Pos}\{\boldsymbol{A}_{.2}, \ldots, \boldsymbol{A}_{.m+1}\}$ are two facets of the simplicial cone Pos$\{\boldsymbol{A}_{.1}, \ldots, \boldsymbol{A}_{.m+1}\}$. Hence the requirement line enters this simplicial cone at the point $(b_1, \ldots, b_m, \bar{z})$ on its facet Pos$\{\boldsymbol{A}_{.1}, \ldots, \boldsymbol{A}_{.m}\}$, and leaves it at the point $(b_1, \ldots, b_m, \hat{z})$ on its facet Pos$\{\boldsymbol{A}_{.2}, \ldots, \boldsymbol{A}_{.m+1}\}$.

This pivot step has the effect of moving from the point $(b_1, \ldots, b_m, \bar{z})$ on the facet Pos$\{\boldsymbol{A}_{.1}, \ldots, \boldsymbol{A}_{.m}\}$, to the point $(b_1, \ldots, b_m, \hat{z})$ lying lower on the requirement line and on the other facet Pos$\{\boldsymbol{A}_{.2}, \ldots, \boldsymbol{A}_{.m+1}\}$ of the simplicial cone Pos$\{\boldsymbol{A}_{.1}, \ldots, \boldsymbol{A}_{.m+1}\}$. Hence each pivot step of the algorithm has the effect of moving down the requirement line from one facet of a simplicial cone to another. Since the algorithm consists of a sequence of such moves across these simplicial cones until the bottom point in $\mathbf{L} \cap$ Pos$\{\boldsymbol{A}_{.1}, \ldots, \boldsymbol{A}_{.n}\}$ is reached, it has been named the simplex algorithm.

Example 3.33

We illustrate with an LP in which $m = 1$, $n = 5$. In this case, the simplicial cones of the earlier discussion will be simplicial cones in $\mathbf{R}^2$, and it is convenient to draw them on the two-dimensional Cartesian plane. The LP is:

x_1	x_2	x_3	x_4	x_5		
1	3	4	1	-1	$= 2$	
3	2	1	-2	-2	$= z$	minimize

$x_j \geqq 0$ for all j.

Since there is only one constraint in this LP, we have to deal with basic vectors of one variable only. The column vectors of x_1 to x_5 (denoted by $\boldsymbol{A}_{.1}$ to $\boldsymbol{A}_{.5}$) are plotted on the y_1, y_2 Cartesian plane as in the earlier discussion. Since the right-hand-side constant $= b_1 = 2$, the requirement line is the line marked in Figure 3.34. The canonical tableau with respect to (x_1) as the basic vector is

Basic Variables	x_1	x_2	x_3	x_4	x_5	$-z$	
x_1	1	3	4	1	-1	0	2
$-z$	0	-7	-11	-5	1	1	-6

The present objective value is 6. Hence the point corresponding to this basic vector is the point P_1 on the requirement line in Figure 3.34. Point P_1 is the intersection of the requirement line and Pos$\{\boldsymbol{A}_{.1}\}$. In this canonical tableau, x_3 has the most negative relative cost coefficient, and we select it as the entering variable. Variable x_3 replaces x_1 as the basic variable. The effect of this change is to move from the point P_1 on the facet Pos$\{\boldsymbol{A}_{.1}\}$ to the point P_2 on the facet Pos$\{\boldsymbol{A}_{.3}\}$ of the simplicial cone Pos$\{\boldsymbol{A}_{.1}$ $\boldsymbol{A}_{.3}\}$. So during this step the algorithm walks across the simplicial cone Pos$\{\boldsymbol{A}_{.1}, \boldsymbol{A}_{.3}\}$. The next canonical tableau is given below.

Basic Variables	x_1	x_2	x_3	x_4	x_5	$-z$	b
x_3	$\frac{1}{4}$	$\frac{3}{4}$	1	$\frac{1}{4}$	$-\frac{1}{4}$	0	$\frac{1}{2}$
$-z$	$\frac{11}{4}$	$\frac{5}{4}$	0	$-\frac{9}{4}$	$-\frac{7}{4}$	1	$-\frac{1}{2}$

Variable x_4 has the most negative relative cost coefficient in this tableau and, hence, it is selected as the next entering variable. It replaces x_3 as the basic variable. The effect is to move from the point P_2 on the facet Pos$\{\boldsymbol{A}_{.3}\}$ and the requirement line to the point P_3 on the facet Pos$\{\boldsymbol{A}_{.4}\}$ of the simplicial cost Pos$\{\boldsymbol{A}_{.3}, \boldsymbol{A}_{.4}\}$. Here is the next canonical tableau.

Basic Variables	x_1	x_2	x_3	x_4	x_5	$-z$	b
x_4	1	3	4	1	-1	0	2
$-z$	5	8	9	0	-4	1	4

Variable x_5 is the only nonbasic variable with a negative relative cost coefficient. The unboundedness criterion of the simplex algorithm is satisfied. We verify that the portion of the requirement line below the point P_3 lies entirely in the simplicial cone Pos$\{\boldsymbol{A}_{.4}, \boldsymbol{A}_{.5}\}$. In Figure 3.34, the various simplicial cones that the algorithm walks through are marked by angle signs.

Note 3.9: Historically, the name "simplex algorithm" originated in the geometry of an LP, which is (3.16) with the additional constraint $\sum x_j = 1$. For this LP, the set to be considered is $\Gamma =$ convex hull of $\{\boldsymbol{A}_{.1}, \ldots, \boldsymbol{A}_{.n}\}$. This LP is the problem of finding the lowest point in $\mathbf{L} \cap \Gamma$. In solving this problem, the simplex algorithm goes through a sequence of moves across simplexes that are subsets of Γ. See Section 7.3 in Dantzig's book, [1.61].

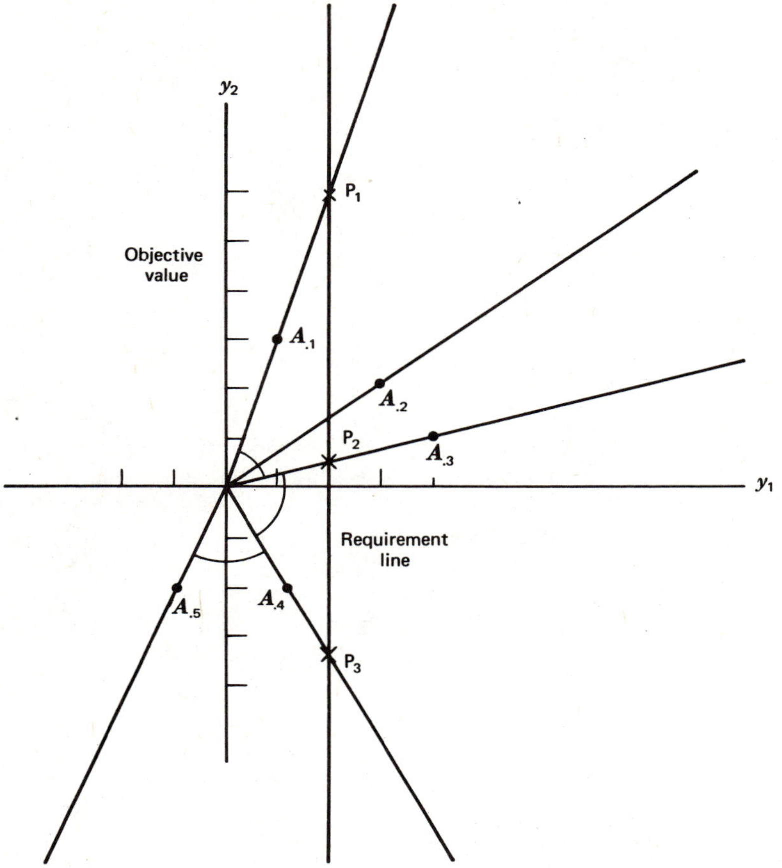

Figure 3.34

3.14 PHASE I OF THE SIMPLEX METHOD USING ONE ARTIFICIAL VARIABLE

Consider the LP (3.16). The aim of Phase I of the simplex method is to find a feasible basis for this problem. The objective row is normally carried in the tableau during Phase I and updated whenever a pivot is made. As in Section 3.3.3, transform the matrix A into row echelon normal form by row operations. If the system $Ax = b$ is inconsistent, this will show up during this step as an equation of the form "$0 =$ some nonzero number," and we terminate. If the system $Ax = b$ is consistent, all the redundant constraints are eliminated in the process of putting the tableau in row echelon normal form. The column vectors of the basic variables in their proper order form a unit matrix in the final transformed tableau. For convenience in referring to them, we will assume that the basic variables are, $x_1, \ldots, x_m$, in that order. The final tableau will be of the same form as Tableau 2.8 of Section 2.5.1. If $\bar{b}_i \geqq 0$ for all i, the present basis is a feasible basis, and we are done. Otherwise $\bar{b}_i < 0$ for some i. In this case let

$$\bar{b}_r = \text{minimum } \{\bar{b}_i : i = 1 \text{ to } m\} \qquad (3.27)$$

Augment the tableau with an artificial variable t whose column vector consists of all -1s. The Phase I tableau is given at the top of page 147.

Basic Variables	$x_1 \ldots x_m$	$x_{m+1} \ldots x_n$	t	$-z$	$-w$	b
x_1	$1 \ldots 0$	$\bar{a}_{1,m+1} \ldots \bar{a}_{1n}$	-1	0	0	$\bar{b}_1$
$\vdots$	$\vdots \quad \vdots$	$\vdots \quad \vdots$	$\vdots$	$\vdots$	$\vdots$	$\vdots$
x_m	$0 \ldots 1$	$\bar{a}_{m,m+1} \ldots \bar{a}_{mn}$	-1	0	0	$\bar{b}_m$
$-z$	$0 \ldots 0$	$\bar{c}_{m+1} \ldots \bar{c}_n$	0	1	0	$-\bar{z}$
$-w$	$0 \ldots 0$	$0 \ldots 0$	1	0	1	0

Replace the present rth basic variable x_r by the artificial variable t. The pivot row for this operation is the rth row and the pivot column is the column vector of t. The new basic vector is $(x_1, \ldots, x_{r-1}, t, x_{r+1}, \ldots, x_m)$, and by the definition of r in equation (3.27), and the column vector of t, this is a feasible basic vector for the Phase I problem.

Example 3.34

The following LP has been transformed into row echelon normal form by selecting (x_1, x_2, x_3) as the basic vector.

Basic Variables	x_1	x_2	x_3	x_4	x_5	x_6	$-z$	b
x_1	1	0	0	0	0	3	0	-3
x_2	0	1	0	-1	1	1	0	-4
x_3	0	0	1	2	1	-3	0	0
$-z$	0	0	0	-7	-14	83	1	-74

$x_j \geqq 0$ for all j; z to be minimized.

In the basic solution corresponding to this basic vector, the most negative basic variable is x_2 with a value of -4, that is, $\bar{b}_2$ is the most negative $\bar{b}_i$. Hence, row 2 will be the pivot row to get a feasible basis for the Phase I problem. Introducing the artificial variable t and the Phase I objective function, w, the Phase I tableau is the first tableau at the bottom. Performing the pivot (the pivot element is circled) leads to the initial canonical tableau for the Phase I problem, the second tableau at the bottom.

Basic Variables	x_1	x_2	x_3	x_4	x_5	x_6	t	$-z$	$-w$	b
x_1	1	0	0	0	0	3	-1	0	0	-3
x_2	0	1	0	-1	1	1	$\boxed{-1}$	0	0	-4
x_3	0	0	1	2	1	-3	-1	0	0	0
$-z$	0	0	0	-7	-14	83	0	1	0	-74
$-w$	0	0	0	0	0	0	1	0	1	0

Basic Variables	x_1	x_2	x_3	x_4	x_5	x_6	t	$-z$	$-w$	b
x_1	1	-1	0	1	-1	2	0	0	0	1
t	0	-1	0	1	-1	-1	1	0	0	4
x_3	0	-1	1	3	0	-4	0	0	0	4
$-z$	0	0	0	-7	-14	83	0	1	0	-74
$-w$	0	1	0	-1	1	1	0	0	1	-4

Starting with this feasible basis, the Phase I problem can be solved in the usual manner.

Exercises

3.37 Complete the solution of this LP.

From a naive point of view, solving a Phase I problem with a single artificial variable, in which only that variable has to be eliminated from the basic vector, appears easier than solving the Phase I problem with a full artificial basis, in which m artificial variables have to be eliminated from the basic vector to get a feasible Phase II basic vector. Even though there is as yet no theoretical justification for this point of view, the empirical work carried out by C. B. Milham [3.33] indicates that the Phase I model using only one artificial variable requires less computational effort on an average than Phase I formulated using a full artificial basis.

3.15 WHY NOT INTRODUCE TWO OR MORE VARIABLES SIMULTANEOUSLY INTO THE BASIC VECTOR?

The main characteristic feature of the simplex algorithm is that it changes the basic vectors by one variable in each step. Efforts were made to generalize the simplex algorithm to bring two or more nonbasic variables into the basic vector simultaneously in each step. The process of bringing two or more nonbasic variables simultaneously into the basic vector is called *block pivoting*. We will use the term *multiple entering variable method* for any method that solves an LP by moving among feasible basic vectors using block pivots. Here we present the main difficulties associated with block pivots.

Consider the LP (3.16) again, where A is a matrix of order $m \times n$ and rank m. Let $\mathbf{K}$ be the set of feasible solutions. Let $\bar{x}$ be a BFS of (3.16) associated with the feasible basic vector $x_B = (x_1, \ldots, x_m)$.

In the usual simplex algorithm, suppose x_{m+1} is the entering variable into x_B in a step. Then the subset of feasible solutions being considered in this step is $\bar{\mathbf{K}} \subset \mathbf{K}$, which is the face of $\mathbf{K}$ in which $x_{m+2}, \ldots, x_n$ (all the nonbasic variables other than the entering variable) are zero. $\bar{\mathbf{K}}$ is an edge of $\mathbf{K}$. If it is bounded, the best feasible solution in $\bar{\mathbf{K}}$ is the extreme point at the other end of this edge, and the simplex algorithm moves to it by walking along the edge. Also, the choice of the entering variable to guarantee that the objective value decreases in this step merely requires that it should satisfy $\bar{c}_j < 0$. Further, once the entering variable is chosen by this simple criterion, the dropping variable that it replaces is easily determined by the minimum ratio step.

In a multiple entering variable method, suppose the nonbasic variables $x_{m+1}, \ldots, x_{m+r}$ are to be brought simultaneously into the current basic vector x_B in a step. The subset of feasible solutions being considered in this step is $\hat{\mathbf{K}} \subset \mathbf{K}$, which is the face of $\mathbf{K}$ in which $x_{m+r+1}, \ldots, x_n$ (those other than the current basic variables or entering variables) are zero. Under general conditions, $\hat{\mathbf{K}}$ will be a face of $\mathbf{K}$ of dimension r. Ideally we would like to move to the best extreme point in $\hat{\mathbf{K}}$ in this step. Even if the current relative costs $\bar{c}_{m+1}, \ldots, \bar{c}_{m+r}$ are all <0, there is no guarantee that the best extreme point in $\hat{\mathbf{K}}$ corresponds to a basic vector in which all $x_{m+1}, \ldots, x_{m+r}$ are basic variables. A more serious difficulty is that it is not easy to determine the subset of current basic variables to be replaced by the entering variables in this block pivot step. Depending on which r basic variables are dropped, the new basic vector may be feasible or infeasible. When $r \geqq 2$, no computationally efficient scheme (such as the minimum ratio test in the usual simplex algorithm) that guarantees that the new basic vector obtained will be feasible is known. Even if the dropping variables for this block pivot step are successfully determined, the only computationally efficient schemes known for obtaining the updated tableau with respect to the new basic vector are those that bring the column vectors of the r entering variables into the basis one at a time. Hence the computational effort of even updating the tableau is the same as the work involved in r steps of the conventional scheme of pivoting one variable at a time. For all these reasons, there exists no successful multiple entering variable method for LPs.

3.16 A MULTISTAGE METHOD FOR SOLVING SYSTEMS OF LINEAR INEQUALITIES

Transform the system of inequalities to be solved into a system of equations in nonnegative variables by the methods of Section 2.2. Suppose this system is

$$\begin{aligned} Ax &= b \\ x &\geqq 0 \end{aligned} \tag{3.28}$$

where A is a matrix of order $m \times n$. In a general stage of this algorithm, we will have a BFS for a system consisting of only a subset of the equality constraints in (3.28). Once an equality constraint in (3.28) is satisfied, all subsequent solutions obtained under the method will also satisfy it. All the solutions obtained under the method will be nonnegative, and hence when a solution satisfying all the equality constraints is obtained, the method terminates. In each stage, either we will determine that (3.28) is infeasible, or obtain a nonnegative solution satisfying one or more equations than the previous solution. Termination occurs in stage m or before.

How the Method Is Initiated

The first equation in (3.28) is $a_{11}x_1 + \cdots + a_{1n}x_n = b_1$. If $b_1 = 0$, select a variable x_j such that $a_{1j} \neq 0$. If $b_1 \neq 0$, select a variable satisfying the properties that $a_{1j} \neq 0$, and b_1, a_{1j} both have the same sign. If no such variable exists, this equation has no nonnegative solution and, hence, (3.28) is infeasible and we terminate. If termination did not occur, for notational convenience assume that the variable selected is x_1.

Let the initial nonnegative solution be $\bar{x}^1 = (b_1/a_{11}, 0, \ldots, 0)^{\mathrm{T}}$. Find out all the equations in (3.28) that $\bar{x}^1$ satisfies. If there are s of these, rearrange the equations in (3.28) so that these are the top s. By performing pivots on these rows select a basic vector for these rows, containing x_1 as a basic variable. In the process of pivoting, if all the coefficients on both the left- and right-hand sides of an equation become zero, then that equation is a redundant equation in (3.28), so eliminate it. If $s = m$, we have a feasible basic vector for (3.28), and we terminate. If $s < m$, every equation other than the top s is violated by the current solution $\bar{x}^1$. With this go to stage 1.

General Stage r Suppose we now have a nonnegative solution $\bar{x}$ that satisfies the top s equations (where $s \geqq r$) and a feasible basic vector corresponding to it for these equations. The present solution $\bar{x}$ violates the ith equation for all $i = s + 1$ to m. If $\sum_{j=1}^{n} a_{s+1,j}\bar{x}_j > b_{s+1}$, leave the $(s + 1)$th equation as it is; otherwise multiply both sides by -1. After this change, we have $\sum_{j=1}^{n} a_{s+1,j}\bar{x}_j > b_{s+1}$. Since (3.28) requires $\sum_{j=1}^{n} a_{s+1,j}x_j = b_{s+1}$, in order to get a solution satisfying this equation, we look at the following LP:

$$\begin{aligned} &\text{Minimize} \quad w(x) = \sum_{j=1}^{n} a_{s+1,j}x_j \\ &\text{Subject to} \quad \sum_{j=1}^{n} a_{ij}x_j = b_i \qquad i = 1 \text{ to } s \\ &\qquad x_j \geqq 0 \qquad \text{for all } j = 1 \text{ to } n \end{aligned} \tag{3.29}$$

The present BFS of (3.29), $\bar{x}$, satisfies $w(\bar{x}) > b_{s+1}$. Beginning with the present basic vector, apply the simplex algorithm on (3.29) until one of the following cases occurs for the first time.

Case 1 An optimum solution for (3.29) is obtained and the optimum objective value in it is strictly greater than b_{s+1}.

Case 2 In the process of solving (3.29), we reach a pivot step in which the BFS of (3.29) is $\tilde{x}$. An entering variable in this pivot step has been selected, but one of the following properties holds. (1) Either $w(\tilde{x}) > b_{s+1}$, and the updated column of the entering variable contains no positive entry. That is, in this step we conclude that $w(x)$ is unbounded in (3.29). (2) Or, the updated column of the entering variable contains at least one positive entry. Let $\hat{x}$ denote the BFS of (3.29) obtained after this pivot step. The property is that there exists an i between $s + 1$ to m, say $i = p$ satisfying: either

$$\sum_{j=1}^{n} a_{pj}\tilde{x}_j < b_p \qquad \text{and} \qquad \sum_{j=1}^{n} a_{pj}\hat{x}_j \geqq b_p$$

or

$$\sum_{j=1}^{n} a_{pj}\tilde{x}_j > b_p \qquad \text{and} \qquad \sum_{j=1}^{n} a_{pj}\hat{x}_j \leqq b_p$$

If Case 1 occurs, no feasible solution of (3.29) satisfies the $(s + 1)$th equation, and hence (3.28) is infeasible, and we terminate.

If Case 2 (2) occurs and $p \neq s + 1$, interchange rows p and $s + 1$. After the rearrangement of the rows, if necessary, it can be verified that if the pivot is performed with the entering column as the pivot column and the $(s + 1)$th row as the pivot row, we get a BFS, say $\hat{x}$, for the system consisting of the top $s + 1$ rows. Find out whether $\hat{x}$ satisfies some of the remaining constraints, namely the $(s + 2)$th to the mth. If so, rearrange the tableau so that the constraints satisfied by $\hat{x}$ are the first to the tth, and the constraints violated by $\hat{x}$ are the $(t+1)$ to the mth. Adjoin the new $(s+2)$th to the tth constraints at the bottom of the current canonical tableau and by performing pivots in these rows, extend the current basic vector into a basic vector for this system. Then go to the next stage.

Exercises

3.38 Discuss whether it is possible to execute this algorithm (by properly defined entering and dropping variable choice rules in each stage) so that the value of $\sum_{j=1}^{n} a_{ij}x_j$ is monotonic (either nondecreasing or nonincreasing throughout the algorithm) for each i.

Note 3.10: In Chapter 5, we discuss how to carry out the simplex algorithm by maintaining the basis inverse instead of computing the canonical tableaux in each step. This leads to a much more efficient implementation of the simplex algorithm. The computations in each stage of the algorithm discussed here can also be carried out by maintaining the basis inverse, and by updating the basis inverse as we move from one stage to the next, increasing the order of the basis as more equations become satisfied (see Exercise 5.5). However, computational experience indicates that the Phase I of the simplex method using artificial variables is more efficient than the method discussed here.

3.17 THE FOURIER ELIMINATION METHOD FOR SOLVING A SYSTEM OF LINEAR INEQUALITIES

This method was first proposed by the mathematician J. B. J. Fourier in a paper "Solution d'une question Particulière du calcul des inégalités," published in 1826. It was used by T. S. Motzkin in his Doctoral thesis in 1936, and is referred to as the *Fourier–Motzkin Elimination Method* in [1.61]. This method is a finite iterative method not based on the simplex algorithm. Let the system of inequalities to be solved be

$$Ax - b \geqq 0 \tag{3.30}$$

where $A = (a_{ij})$ is a matrix of order $m \times n$. If there are any sign restrictions on the variables, they should be included as constraints in (3.30). The method goes through at most n steps before termination. In each step, one variable is eliminated from the system and a new system of linear inequalities in the remaining variables is derived, which is then solved by going to the next step.

Step 1 Select one of the variables in the system, say x_1, for elimination in this step. Look through the constraints in the system (3.30), and partition them into three classes as below. This classification is for eliminating x_1 from the system (3.30). The coefficients a_{i1} of x_1 in the system (3.30) play a role in this classification.

Class 1: The ith constraint in (3.30) belongs to this class if $a_{i1} > 0$. Let $\mathbf{I}_1 = \{i\colon i, \text{ such that } a_{i1} > 0\}$.

Class 2: The ith constraint in (3.30) belongs to this class if $a_{i1} < 0$. Let $\mathbf{I}_2 = \{i\colon i, \text{ such that } a_{i1} < 0\}$.

Class 3: The ith constraint in (3.30) belongs to this class if $a_{i1} = 0$. Let $\mathbf{I}_3 = \{i\colon i, \text{ such that } a_{i1} = 0\}$.

The Class 1 Constraints that x_1 Has to Satisfy If $i \in \mathbf{I}_1$, the ith constraint in (3.30) is equivalent to $x_1 \geqq (b_i/a_{i1}) - \sum_{j=2}^{n} (a_{ij}/a_{i1})x_j$. Inequalities of this type are known as the *Class 1 constraints that x_1 has to satisfy*.

Class 2 Constraints that x_1 Has to Satisfy If $i \in \mathbf{I}_2$, the ith constraint in (3.30) is equivalent to $(b_i/a_{i1}) - \sum_{j=2}^{n} (a_{ij}/a_{i1})\ x_j \geqq x_1$. Inequalities of this type are known as the *Class 2 constraints that x_1 has to satisfy*.

For convenience of notation, we will denote the general Class 1 constraint that x_1 has to satisfy by

$$x_1 \geqq \alpha_{p0} + \sum_{j=2}^{n} \alpha_{pj}x_j \qquad \text{for } p \in \mathbf{I}_1 \tag{3.31}$$

and the general Class 2 constraint that x_1 has to satisfy by

$$\beta_{t0} + \sum_{j=2}^{n} \beta_{tj}x_j \geqq x_1 \qquad \text{for } t \in \mathbf{I}_2 \tag{3.32}$$

Also given specific values for the remaining variables—say $x_j = \bar{x}_j$ for $j = 2$ to n—define

$$\begin{aligned}\theta_{\mathrm{L}}(\bar{x}_2, \ldots, \bar{x}_n) &= \text{maximum}\left\{\alpha_{p0} + \sum_{j=2}^{n} \alpha_{pj}\bar{x}_j\colon p \in \mathbf{I}_1\right\} \quad \text{if } \mathbf{I}_1 \neq \emptyset \\ &= -\infty \quad \text{if } \mathbf{I}_1 = \emptyset.\end{aligned}$$

$$\theta_U(\bar{x}_2, \ldots, \bar{x}_n) = \text{minimum}\left\{\beta_{t0} + \sum_{j=2}^{n} \beta_{tj}\bar{x}_j : t \in \mathbf{I}_2\right\} \quad \text{if } \mathbf{I}_2 \neq \emptyset$$
$$= +\infty \quad \text{if } \mathbf{I}_2 = \emptyset. \tag{3.33}$$

We will now consider several cases separately.

Case 1 Suppose $\mathbf{I}_2 = \mathbf{I}_3 = \emptyset$. In this case, give the remaining variables $x_2, \ldots, x_n$ arbitrary real values, say $x_j = \bar{x}_j$, for $j = 2$ to n. Then $(x_1, \bar{x}_2, \ldots, \bar{x}_n)^T$ is a feasible solution of (3.30) for any $x_1 \geqq \theta_L(\bar{x}_2, \ldots, \bar{x}_n)$ in this case. Terminate.

Case 2 Suppose $\mathbf{I}_1 = \mathbf{I}_3 = \emptyset$. Give the remaining variables arbitrary real values, say $x_j = \bar{x}_j$ for $j = 2$ to n. Then $(x_1, \bar{x}_2, \ldots, \bar{x}_n)^T$ is a feasible solution of (3.30) for any $x_1 \leqq \theta_U(\bar{x}_2, \ldots, \bar{x}_n)$ in this case. Terminate.

Case 3 Suppose $\mathbf{I}_1 = \emptyset$, but $\mathbf{I}_2 \neq \emptyset$, $\mathbf{I}_3 \neq \emptyset$. Store all the Class 2 constraints (3.32) that x_1 has to satisfy in some location. Go to Step 2 to find a feasible solution of the system consisting of the ith constraint in (3.30) only for $i \in \mathbf{I}_3$. All these constraints involve only the remaining variables $x_2, \ldots, x_n$. If $(\bar{x}_2, \ldots, \bar{x}_n)^T$ is a feasible solution of this system, then $(x_1, \bar{x}_2, \ldots, \bar{x}_n)$ is a feasible solution of (3.30) for any $x_1 \leqq \theta_U(\bar{x}_2, \ldots, \bar{x}_n)$ in this case.

Case 4 Suppose $\mathbf{I}_2 = \emptyset$, but $\mathbf{I}_1 \neq \emptyset$, $\mathbf{I}_3 \neq \emptyset$. Store all the Class 1 constraints (3.31) that x_1 has to satisfy in some location. Go to Step 2 to find a feasible solution of the system consisting of the ith constraint in (3.30) only for $i \in \mathbf{I}_3$. If $(\bar{x}_2, \ldots, \bar{x}_n)^T$ is a feasible solution of this system, $(x_1, \bar{x}_2, \ldots, \bar{x}_n)^T$ is a feasible solution of (3.30) for any $x_1 \geqq \theta_L(\bar{x}_2, \ldots, \bar{x}_n)$ in this case.

Case 5 Suppose $\mathbf{I}_1 \neq \emptyset, \mathbf{I}_2 \neq \emptyset$. If $\bar{x} = (\bar{x}_1, \ldots, \bar{x}_n)^T$ is feasible to (3.30), we must clearly have $\beta_{t0} + \sum_{j=2}^{n} \beta_{tj}\bar{x}_j \geqq \bar{x}_1 \geqq \alpha_{p0} + \sum_{j=2}^{n} \alpha_{pj}\bar{x}_j$ for $p \in \mathbf{I}_1$, $t \in \mathbf{I}_2$. This implies that the remaining variables $x_2, \ldots, x_n$ must satisfy

$$\left(\beta_{t0} + \sum_{j=2}^{n} \beta_{tj}x_j\right) - \left(\alpha_{p0} + \sum_{j=2}^{n} \alpha_{pj}x_j\right) \geqq 0 \tag{3.34}$$
$$\text{for } p \in \mathbf{I}_1, \quad t \in \mathbf{I}_2$$

For some $p \in \mathbf{I}_1$, $t \in \mathbf{I}_2$, if $\beta_{tj} = 0 = \alpha_{pj}$ for each $j = 2$ to n, and $\beta_{t0} < \alpha_{p0}$, (3.34) is violated, and this implies that the original system (3.30) has no feasible solution. Terminate. On the other hand, if $\beta_{tj} = 0 = \alpha_{pj}$ for each $j = 2$ to n, and $\beta_{t0} \geqq \alpha_{p0}$, that constraint in (3.34) is a redundant constraint, so delete it. If at least one of α_{pj}, β_{tj} is nonzero for some j, that constraint in (3.34) is one that the remaining variables $x_2, \ldots, x_n$ must satisfy. These constraints are known as the *additional constraints* in the remaining variables, obtained when x_1 is eliminated from the system (3.30). The set of all the constraints that the remaining variables $x_2, \ldots, x_n$ must satisfy includes all these additional constraints of the form (3.34), and the ith constraints in the original system for each $i \in \mathbf{I}_3$.

Store all the Class 1 and 2 constraints (3.31), (3.32) that x_1 must satisfy in some location, and go to the next step to solve the system of constraints (consisting of those in the original system for $i \in \mathbf{I}_3$ and the additional constraints generated above) in the remaining variables $x_2, \ldots, x_n$. Let r_1, r_2, r_3 be the cardinalities of $\mathbf{I}_1, \mathbf{I}_2, \mathbf{I}_3$, respectively. The number of additional inequalities of the type (3.34) that the remaining variables must satisfy is at most r_1r_2. So, including the Class 3 constraints, the system of constraints that the remaining variables must satisfy may consist of up to $r_3 + r_1r_2$ inequalities. If a feasible solution $(\bar{x}_2, \ldots, \bar{x}_n)^T$ for this system is found, determine $\theta_L(\bar{x}_2, \ldots, \bar{x}_n)$ and $\theta_U(\bar{x}_2, \ldots, \bar{x}_n)$ as in (3.33); the range for x_1 is then $\theta_L(\bar{x}_2, \ldots, \bar{x}_n)$ to $\theta_U(\bar{x}_2, \ldots, \bar{x}_n)$. The fact that $(\bar{x}_2, \ldots, \bar{x}_n)^T$ is feasible to this system of constraints implies that $\theta_L(\bar{x}_2, \ldots, \bar{x}_n) \leqq \theta_U(\bar{x}_2, \ldots, \bar{x}_n)$ (this follows from the additional constraints (3.34) that $(\bar{x}_2, \ldots, \bar{x}_n)^T$ satisfies), and hence the range for x_1 is nonempty. When $\bar{x}_1$ is any value in this range for x_1, $(\bar{x}_1, \bar{x}_2, \ldots, \bar{x}_n)^T$ is a feasible solution of the original system of constraints (3.30).

Thus, once the values for $x_2, \ldots, x_n$ are specified, the Class 1 and 2 constraints for x_1 specify a range for x_1 within which it must lie in order to lead to a feasible solution of the original system.

If the method did not terminate in Step 1, we would have generated a system of linear inequalities in the remaining variables with which to go to the next step. In that step we select another variable, eliminate it from that system in the same manner, and continue in the same way.

The method has to terminate in step n or before. If the method goes to step n, the system of inequality constraints in that final step contains inequalities involving only one variable, say x_n, and hence these must be lower and upper bound constraints on that variable. By dividing both sides of each of these constraints by the coefficient of x_n in it, these constraints can be classified into two classes. *The Class 1 constraints for x_n* are the constraints among these of the form $x_n \geqq \alpha_p$, $p = 1$ to r. The *Class 2 constraints for x_n* are the constraints among these of the form $\beta_t \geqq x_n$, $t = 1$ to s. Define

$$\begin{aligned} \theta_U &= +\infty, \text{ if there are no Class 2 constraints for } x_n \\ &= \text{minimum } \{\beta_t : t = 1 \text{ to } s\}, \text{ otherwise} \\ \theta_L &= -\infty, \text{ if there are no Class 1 constraints} \\ &= \text{maximum } \{\alpha_p : p = 1 \text{ to } r\}, \text{ otherwise} \end{aligned}$$

If $\theta_L > \theta_U$, the original system of constraints (3.30) is infeasible. If $\theta_L \leqq \theta_U$, the range θ_L to θ_U is the range of values of x_n among the feasible solutions of the original system of constraints (3.30). Select a value, say $\bar{x}_n$, from this range. Suppose the variables eliminated in the various steps are $x_1, x_2, \ldots, x_{n-1}$, in that order. Then by substituting $x_n = \bar{x}_n$ in the Class 1 and 2 constraints on the variable x_{n-1} determined in step $n-1$, obtain the range $\theta_L(\bar{x}_n)$ to $\theta_U(\bar{x}_n)$ for x_{n-1}. Select any value for x_{n-1}, say $\bar{x}_{n-1}$, from this range. Now substitute $x_{n-1} = \bar{x}_{n-1}$, $x_n = \bar{x}_n$ in the Class 1 and 2 constraints determined for x_{n-2} in step $n-2$, obtain the range $\theta_L(\bar{x}_{n-1}, \bar{x}_n)$ to $\theta_U(\bar{x}_{n-1}, \bar{x}_n)$ for x_{n-2}, and choose a value for x_{n-2} in this range. Continuing in this manner, we generate a feasible solution for the original system of constraints (3.30). This is known as the *backward process* of generating a feasible solution. In this backward process, the values of the variables in a feasible solution are determined in the reverse order to which the variables are eliminated. Also, by selecting a suitable value for each variable in its range in this backward process, all feasible solutions of the original system can be obtained.

So the method terminates in at most n steps. However, as we move from one step to the next, the number of inequality constraints in the system that the remaining variables have to satisfy tends to grow very rapidly. For example, if x_1 appears in all the constraints in (3.30) with a nonzero coefficient, and if this coefficient is positive or negative in approximately half of the constraints each, then after eliminating x_1 we are left with a system consisting of about $(m/2)^2$ inequality constraints in the remaining variables. If the same thing repeats, after only a few steps we are left with a system consisting of an enormous number of inequalities in the remaining variables. Thus, even though this method is very simple and easy to implement, it is not practically useful to solve linear inequalities, unless the number of variables in the system is very very small.

Example 3.35

Consider the following system:

$$\begin{aligned} -x_1 + 2x_2 - x_3 - 4 &\geqq 0 \\ x_1 - 2x_2 - 2x_3 + 5 &\geqq 0 \\ x_2 + x_3 - 1 &\geqq 0 \\ x_1 - 2 &\geqq 0 \\ -x_1 + 10 &\geqq 0 \\ x_2 - 1 &\geqq 0 \\ -x_2 + 15 &\geqq 0 \\ x_3 &\geqq 0 \\ -x_3 + 20 &\geqq 0 \end{aligned}$$

Step 1 We eliminate x_1 in this step. x_1 appears with a positive coefficient in inequalities 2 and 4, and with a negative coefficient in inequalities 1 and 5. So the Class 1 and Class 2 constraints on x_1 are

Class 1	Class 2
$x_1 \geqq 2x_2 + 2x_3 - 5$	$2x_2 - x_3 - 4 \geqq x_1$
$x_1 \geqq 2$	$10 \geqq x_1$

So the system of constraints that the remaining variables have to satisfy is

$$\begin{aligned} x_2 + x_3 - 1 &\geqq 0 \\ x_2 - 1 &\geqq 0 \\ -x_2 + 15 &\geqq 0 \\ x_3 &\geqq 0 \\ -x_3 + 20 &\geqq 0 \\ (2x_2 - x_3 - 4) - (2x_2 + 2x_3 - 5) &\geqq 0 \\ (2x_2 - x_3 - 4) - 2 &\geqq 0 \\ 10 - (2x_2 + 2x_3 - 5) &\geqq 0 \\ 10 - 2 &\geqq 0 \end{aligned} \tag{3.35}$$

The bottommost constraint here is redundant, and can be eliminated.

Step 2 In this step we eliminate x_2 from (3.35). The Class 1 and Class 2 constraints on x_2 are

$x_2 \geqq -x_3 + 1$	$15 \geqq x_2$
$x_2 \geqq 1$	$-x_3 + \frac{15}{2} \geqq x_2$
$x_2 \geqq \frac{x_3}{2} + 3$	

So the system of constraints that the remaining variable x_3 has to satisfy is

$$x_3 \geqq 0$$
$$-x_3 + 20 \geqq 0$$
$$-3x_3 + 1 \geqq 0$$
$$15 - (-x_3 + 1) \geqq 0$$
$$15 - \left(\frac{x_3}{2} + 3\right) \geqq 0$$
$$-x_3 + \frac{15}{2} - (-x_3 + 1) \geqq 0$$
$$-x_3 + \frac{15}{2} - 1 \geqq 0$$
$$-x_3 + \frac{15}{2} - \left(\frac{1}{2}x_3 + 3\right) \geqq 0$$

which on simplification leads to $(1/3) \geqq x_3 \geqq 0$, and hence any value of x_3 in the interval 0 to (1/3) leads to a feasible solution of the original system. Suppose we select $x_3 = \bar{x}_3 = 1/3$. Substituting $\bar{x}_3$ for x_3 in the Class 1 and Class 2 constraints for x_2 yields $(43/6) \geqq x_2 \geqq (19/6)$. Suppose we select $x_2 = \bar{x}_2 = 19/6$. Substituting these values for x_2, x_3 in the Class 1 and Class 2 constraints for x_1 leads to $2 \geqq x_1 \geqq 2$. So $\bar{x} = (2, 19/6, 1/3)^T$ is a feasible solution for the original system.

How to Solve an LP Using the Fourier Elimination Method

Put the LP in the form: minimize $z(x) = cx$ subject to $Ax \geqq b$, where A is of order $m \times n$. Consider the system of constraints $Ax - b \geqq 0$, $z - cx \geqq 0$ in $(n + 1)$ variables. In this system eliminate the variables $x_1, \ldots, x_n$ in some order by Fourier elimination. If infeasibility termination does not occur in some intermediate step, we will end up with bounds on the remaining variable z. If there is no lower bound on z then, z is unbounded below in the problem. Otherwise, make z equal to its lower bound (which is the optimum objective value in the problem) and determine the optimum solution by going through the backward process.

3.18 EQUIVALENT LINEAR PROGRAMS

3.18.1 Equivalent Systems of Simultaneous Linear Equations

The concept of *equivalence* between two systems of simultaneous linear equations is a well-established concept in linear algebra, and we briefly discuss this before an analogous equivalence for linear programs. Consider two systems of linear equations, say $Ax = b$, and $Dy = d$. For these systems to be equivalent, the number of variables in the two systems must be equal. Also it should be possible to pass from one of these two systems to the other by performing a sequence of operations of the following types: (1) Transform the variables by a nonsingular linear transformation; (2) in the resulting system, multiply both sides of an equality constraint by the same nonzero number; (3) in the resulting system, add a nonzero multiple of one of the equality constraints to another; and (4) if, after performing operations (2) and (3) several times, all the coefficients of the variables and the right-hand side constant in an equality constraint all become equal to zero, it is a redundant constraint, discard it. In general, delete any equality constraint from the system that can be verified to be a redundant constraint, as mentioned in Note 3.3 of Section 3.3.5. Alternatively generate a linear combination of the existing equality constraints in the system as an additional equality constraint (all such additional equality constraints are obviously redundant equality constraints). Suppose **L** and **H** are the sets of solutions of these systems, respectively. If these systems are equivalent, and the set of solutions of one of them, say **L**, is nonempty, then clearly $\mathbf{H} \neq \emptyset$ also, and **H** can be obtained from **L** by a nonsingular linear transformation on the variables.

3.18.2 Equivalent Linear Programs

In linear programming we deal with equality and/or inequality constraints, and unrestricted and/or sign restricted variables. In addition to these, we are required to minimize or maximize a linear objective function. We use transformations in which an unrestricted variable is transformed into the difference of two nonnegative variables (see Section 2.2). Also we transform inequality constraints into equality constraints by introducing appropriate slack variables. Hence the number of variables in two equivalent linear programs need not be the same. Also minimizing a function $f(x)$, subject to some constraints, is equivalent to minimizing $vf(x) + \alpha$, or maximizing $-vf(x) + \beta$ subject to the same constraints, where α and β are arbitrary real numbers, and v is a strictly positive number. Hence the optimum objective values in two equivalent linear programs need not be the same, and a minimization problem may be equivalent to a maximization problem.

Some definitions first. Consider the following LP

$$\begin{aligned} \text{Minimize} \quad & cx + fy \\ \text{Subject to} \quad & Ax + Dy = b \\ & Ex + Fy \geqq d \\ & x \geqq 0 \qquad y \text{ unrestricted in sign} \end{aligned} \tag{3.36}$$

Redundant Equality Constraint
An equality constraint in (3.36) is said to be a *redundant equality constraint* in it, if it can be obtained as a linear combination of the other equality constraints in it. See Note 3.3 in Section 3.3.5. In general it is easy to see that any change in the right-hand-side constant of a redundant equality constraint would make the system automatically inconsistent.

Redundant Inequality Constraints
An inequality constraint in (3.36) is said to be a *redundant inequality constraint*, if the set of feasible solutions of the system is unchanged by deleting that constraint. Let A have m_1 rows and E have m_2 rows. Then the constraints in (3.36) can be written as $A_{i.}x + D_{i.}y = b_i$, $i = 1$ to m_1, $E_{p.}x + F_{p.}y \geqq d_p$, $p = 1$ to m_2, $x \geqq 0$, y unrestricted. The inequality constraint $E_{t.}x + F_{t.}y \geqq d_t$ is a redundant inequality constraint in (3.36) iff $\hat{d}_t \geqq d_t$, where $\hat{d}_t$ equals the minimum value of $E_{t.}x + F_{t.}y$, subject to $A_{i.}x + D_{i.}y = b_i$, $i = 1$ to m_1, $E_{p.}x + F_{p.}y \geqq d_p$, $p = 1$ to m_2, $p \neq t$, $x \geqq 0$, y unrestricted.

Example 3.36

Consider the following system of constraints:

$$\begin{aligned} x_1 \qquad & \geqq 6 \\ x_2 & \geqq 2 \\ 2x_1 + x_2 & \geqq 11 \end{aligned}$$

Obviously the third inequality constraint is a redundant inequality constraint in the system. Neither of the first two constraints are redundant. The set of feasible solutions of this system is plotted in Figure 3.35. If the right-hand-side constant in the third constraint in the system is changed from 11 to $14 + \varepsilon$, where ε is any positive number, the third inequality is no longer redundant.

Thus one way to find out whether an inequality constraint is redundant in a system of constraints is to solve an LP. Also the redundant inequality constraint might become a nonredundant or effective inequality constraint, when the right-hand-side constant in it is changed appropriately. Notice the difference in the behavior of a redundant equality constraint and a redundant inequality constraint, under changes in the right-hand-side constants in them.

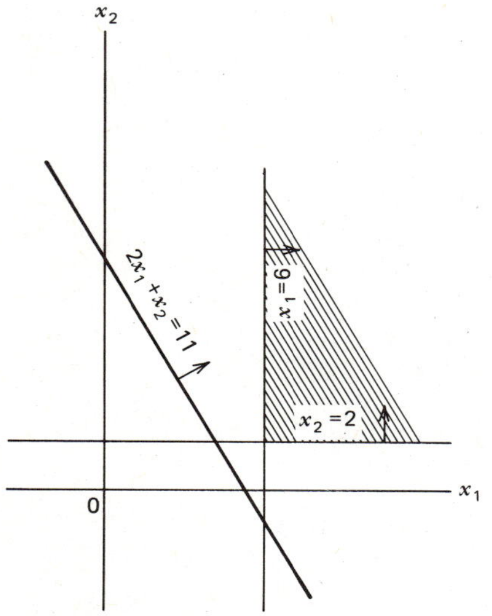

Figure 3.35

Binding Inequality Constraint
An inequality in (3.36) is said to be a *binding inequality constraint* in it if the set of feasible solutions of (3.36) is unaltered by treating it as an equation.

Hence, $E_{t.}x + F_{t.}y \geqq d_t$ is binding if the set of feasible solutions is unaltered when it is replaced by $E_{t.}x + F_{t.}y = d_t$. This happens iff $\tilde{d}_t = d_t$, where $\tilde{d}_t$ equals the maximum value of $E_{t.}x + F_{t.}y$, subject to $A_{i.}x + D_{i.}y = b_i$, $i = 1$ to m_1; $E_{p.}x + F_{p.}y \geqq d_p$, $p = 1$ to m_2, $p \neq t$; $x \geqq 0$, y unrestricted.

Example 3.37

Consider the following system of constraints:

$$\begin{aligned} x_1 \quad\quad &\geqq 6 \\ x_2 &\geqq 2 \\ -2x_1 - x_2 &\geqq -14 \end{aligned}$$

Each inequality constraint in this system is a binding inequality constraint. The set of feasible solutions of this system is the singleton point $(x_1, x_2)^{\mathrm{T}} = (6, 2)^{\mathrm{T}}$, and it is plotted in Figure 3.36. If the right-hand-side constant in the third constraint in this system is changed from -14 to $-14 - \varepsilon$, where ε is any positive number, the inequality constraints in this system will no longer be binding.

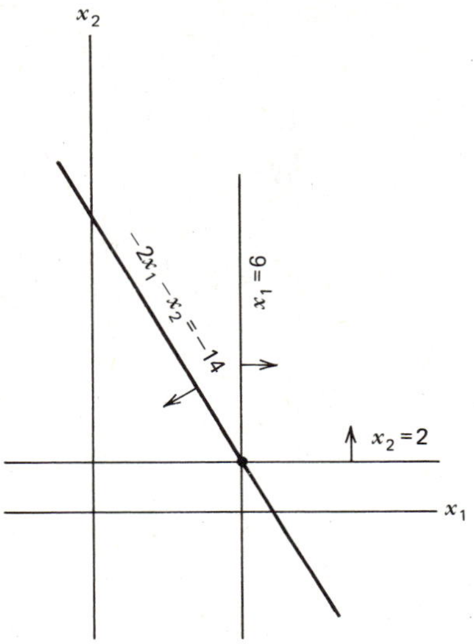

Figure 3.36

3.18.3 Elimination of Redundant Constraints

Redundant equality constraints are identified in the process of transforming the system of equations into row echelon normal form, with no effort beyond that. On the other hand, to identify a redundant or a binding inequality constraint in the system is much more difficult, and it usually requires the solving of a separate LP. Mathematically, it is a very appealing idea to think of identifying all the redundant inequality constraints in (3.36) and eliminating them, and also to identify all the binding inequality constraints and replace each of them by the corresponding equation. This leads to a representation of the set of feasible solutions of (3.36) in terms of the least number of constraints. But the enormous and costly effort required to identify the redundant inequality constraints and the binding inequality constraints makes this impractical in most cases.

For a complete and mathematically rigorous discussion on the representation of the set of feasible solutions in terms of a minimal set of constraints, see references [3.16, 3.38, and Chapter 5 in 1.88].

In practical applications, a linear programming model is used not only to compute an optimum solution, but also to investigate how the optimum solution changes when the right-hand-side constants change (see Sections 1.1.4, 4.6.3, and 8.15). Marginal analysis and sensitivity analysis with the right-hand-side constants provide very useful economic information for planning purposes. Any change in the value of the right-hand-side constant of a reduntant equality constraint makes the system of constraints infeasible, and hence no useful information can be gathered by performing marginal or sensitivity analysis on the right-hand-side constant of a redundant equality constraint. Thus, the resulting model when this redundant equality constraint is eliminated will be considered equivalent to the original model.

However, when the right-hand-side constant in a redundant or binding inequality constraint is changed appropriately, the system of constraints might remain feasible. Each inequality constraint in a linear programming model is the material balance inequality of some *item*, and the right-hand-side constant in it represents either the availability or the requirement for this item. Marginal and sensitivity analysis on this right-hand-side constant provides very useful economic information on the behavior of the optimum solution when the availability or requirement for this item changes. Even if the inequality constraint is redundant, if it is eliminated from the model, we will be unable to derive all this useful economic information about the item. Hence, the resulting model, when this redundant inequality constraint is eliminated, has the same set of feasible solutions (and, in fact, the same optimum solutions) as the original model, but it will not be considered equivalent to the original model. In a similar manner, the model resulting when a binding inequality constraint is replaced by the corresponding equation, will not be considered equivalent to the original model.

When formulating a practical problem as an LP, *we do not* first identify the set of feasible solutions as a geometric object, and try to represent it mathematically using the least number of constraints. The constraints in LP models of practical problems arise from practical considerations, and each constraint corresponds to an item of which there is either a limited supply available, or for which there is a specific requirement. In building the model, each constraint is constructed one by one by examining one item at a time from the set of items in the problem that the decision maker is aware of. After the formulation is completed and the whole model put together, it may happen that some of the inequality constraints in the model are mathematically redundant. If we eliminate those redundant inequality constraints, the output from the remaining problem cannot provide any clues on how the optimum objective value or the optimum solution might vary when changes take place in the availability or the requirements of the items associated with the eliminated inequality constraints. In other words, even though an inequality constraint in an LP model may be mathematically redundant, we cannot eliminate it from the model, since the constraint has a practical meaning.

Dependent Unrestricted Variables

Let the unrestricted variables $y_1, y_2, \ldots, y_r$ in (3.36) be such that in each of the constraints and in the objective function, the (coefficient of y_1) $= \sum_{s=2}^{r} \alpha_s$ (coefficient of y_s), where $\alpha_2, \ldots, \alpha_r$, are nonzero real numbers. Then the set of unrestricted variables $\{y_1, y_2, \ldots, y_r\}$ is said to be a *cycle of dependent unrestricted variables* in (3.36). Each unrestricted variable in the cycle is dependent on the other variables in the cycle. In this case, let $y'_1 = 0, y'_s = y_s + \alpha_s y_1$, for $s = 2$ to r. The system (3.36) can be transformed by replacing y_s by y'_s for each $s = 1$ to r, and leaving all the other variables and their coefficients unchanged. Then, in each constraint and in the objective function, the contribution from the variables $y_1, \ldots, y_r$ in (3.36) is identical to the contribution from the variables $y'_1, \ldots, y'_r$ of the transformed problem. Hence we can consider the transformed problem to be equivalent to (3.36). Since $y_1, \ldots, y_r$ are unrestricted variables in (3.36), the new variables $y'_1, \ldots, y'_r$ are unrestricted variables in the transformed problem. In effect, the transformed problem is obtained by eliminating the dependent unrestricted variable y_1 from the model (i.e., setting it equal to zero), and changing the name of the unrestricted variables y_s to y'_s for $s = 2$ to r. Without any loss of generality we might as well retain the same name y_s to denote the new variables for $s = 2$ to r. Thus one of the variables in any cycle of dependent unrestricted variables can be eliminated from the problem by setting it equal to zero. If the remaining problem contains another cycle of dependent unrestricted variables again, one variable from that new cycle can again be eliminated, and the same procedure repeated until finally the problem reduces to one in which there exists no cycle of dependent unrestricted variables.

3.18.4 Transformations that Lead to Equivalent Linear Programs

Based on this discussion, and considering the usual transformations of LPs from one form to another as in Chapter 2 and the earlier portions of this chapter, we consider two linear programming models as being *equivalent* if any one of them can be derived from the other by making a sequence of transformations of the following types.

1. Transform an inequality constraint into the corresponding equality constraint by introducing the appropriate slack variable.
2. If there is an equality constraint of the form $\sum_{j=1}^{n} a_j x_j = h$, where $a_1 \neq 0$, and x_1 is a sign restricted nonnegative variable, then derive an expression for x_1 in terms of the other variables using this equality constraint $x_1 = (h/a_1) - \sum_{j=1}^{n} (a_j/a_1)x_j$ and eliminate x_1 from all the other constraints and the objective function by substituting this expression for it. Then replace the original constraints $\sum_{j=1}^{n} a_j x_j = h$, $x_1 \geqq 0$, by $(h/a_1) - \sum_{j=2}^{n} (a_j/a_1)x_j \geqq 0$. The original nonnegative variable x_1, plays the role of the slack variable corresponding to this new inequality constraint in the transformed model.
3. Replace an equality constraint $\sum a_j x_j = h$ by the equivalent pair of opposing inequality constraints; $\sum a_j x_j \geqq h$, $-\sum a_j x_j \geqq -h$.

4 If there is an opposing pair of inequality constraints of the form $\sum a_j x_j \geqq h$, $-\sum a_j x_j \geqq -h$, replace them by the equivalent equality constraint $\sum a_j x_j = h$.

5 Replace any unrestricted variable by a difference of two nonnegative variables, as in Section 2.2, in all the constraints and in the objective function.

6 If there are two sign restricted nonnegative variables, say, x_1, x_2 with the property that the coefficients of x_1 and x_2 sum to zero in each constraint and the objective function, eliminate x_1 and x_2 from the model by substituting y_1 for $x_1 - x_2$ and treating y_1 as an unrestricted variable in the resulting model.

7 Multiply both sides of any inequality constraint by the same negative number, and reverse the inequality sign.

8 Multiply both sides of any inequality constraint by the same positive number.

9 Multiply both sides of any equality constraint by the same nonzero number.

10 Add a nonzero multiple of any of the equality constraints to another constraint or to the objective function.

11 Delete any equality constraint from the system that can be verified to be a redundant equality constraint. Or generate a linear combination of the existing equality constraints in the system, and introduce it into the system as an additional equality constraint.

12 Eliminate a dependent unrestricted variable from the model. Or introduce a new dependent unrestricted variable into the model, with its coefficient equal to the same linear combination of the coefficients of the existing unrestricted variables in each of the constraints and the objective function.

13 Transform the variables by using a nonsingular linear transformation and substituting the transformed expressions into all the constraints, sign restrictions, and the objective function.

14 Add an arbitrary constant to the objective function, multiply the objective function by any positive number, or multiply the objective function by a negative number and replace "minimization" by "maximization" or vice versa. Eliminate any constant terms from the objective function.

Equivalence and Stability in Linear Programming

Stability deals with the study of changes in the optimum solutions of a linear programming problem when small perturbations occur in its data. Stability in linear programming is discussed in detail in Section 4.8. It should be pointed out that even though two LPs are equivalent in the sense discussed here, their stability behavior may be entirely different. Whenever a transformation of type 3 above is carried out on an LP, the result is an equivalent LP, but it is unstable. Likewise including a redundant equality constraint in an LP leads to an equivalent LP that is unstable.

Note 3.11: The concept of equivalence of LPs defined above is based on preserving not only the sets of feasible solutions and optimum solutions, but also the marginal analysis information associated with the problem. After the dual of an LP is defined in Chapter 4, the reader can verify that this concept of equivalence between LPs is motivated to keep the theorem "the duals of equivalent LPs are equivalent" valid (see Section 4.5).

3.19 EXTREME POINT RANKING

Consider the LP (3.16) again, where A is of order $m \times n$ and rank m. Let **K** be the set of feasible solutions. Assume that it has a finite optimum solution. When the simplex method is applied on this LP, it finds an optimum BFS, x^1, which is an extreme point of **K** at which $z(x)$ achieves its minimum value in **K**. Since **K** has a finite number of extreme points, the set $\{x: x \text{ an extreme point of } \mathbf{K}, \text{ excluding } x^1\}$ is a well-defined set containing a finite number of elements. Let x^2 be an extreme point of **K** satisfying $z(x^2) = \text{minimum } \{z(x): x \text{ an extreme point of } \mathbf{K}, \text{ excluding } x^1\}$, then x^2 is known as a *second best extreme point solution for* (*3.16*), or a *second best BFS for* (*3.16*). It is possible that $z(x^2) = z(x^1)$ and in that case, x^2 is an alternate optimum BFS of (3.16). In some applications it may be very useful to obtain the second best extreme point solution of (3.16). In fact, it may be useful to rank all the extreme points of **K** (i.e., BFSs of (3.16)) in *increasing* (actually, *nondecreasing*) order of the value of $z(x)$; that is, to order the extreme points of **K** as a sequence x^1, x^2,

$x^3, \ldots,$ with the property that

$$z(x^{r+1}) = \text{minimum } \{z(x): x \text{ an extreme point of } \mathbf{K}, \text{ excluding } x^1, \ldots, x^r\} \quad (3.37)$$

for each $r \geqq 1$. When determined in this manner, x^{r+1} is known as an *$(r+1)$th best BFS for (3.16)* or an *$(r+1)$th best extreme point solution for (3.16)*. Here we discuss an algorithm for ranking the extreme points of **K** in this manner, starting with an optimum extreme point x^1, obtained by the simplex method. We first prove some theorems that are used in developing this ranking algorithm.

THEOREM 3.10 Let $\hat{x}$ and $\tilde{x}$ be a pair of extreme points of **K**. Then there exists an edge path on **K** between $\hat{x}$ and $\tilde{x}$.

Proof: $\hat{x}$ is a BFS of (3.16). Let $\mathbf{J} = \{j: j, \text{ such that } \hat{x}_j > 0\} = \{j_1, \ldots, j_r\}$, where $r \leqq m$. Define $d_j = 0$, if $j \in \mathbf{J}$, 1 otherwise, and let $v(x) = \sum_{j=1}^{n} d_j x_j$. Then $v(\hat{x}) = 0$, and for every $x \in \mathbf{K}$, $x \neq \hat{x}$, we have $v(x) > 0$. So $\hat{x}$ is the unique optimum solution of the LP: minimize $v(x)$, subject to $Ax = b$, $x \geqq 0$. Apply the simplex algorithm to solve this LP, starting with $\tilde{x}$ as the initial BFS. Since $\hat{x}$ is the unique optimum solution of this LP, the simplex algorithm walks along an edge path of **K**, which must terminate with $\hat{x}$. ■

THEOREM 3.11 Let x^1 be an optimum BFS of (3.16) and let $\tilde{x}$ be another BFS of (3.16). There exists an edge path of **K** from $\tilde{x}$ to x^1 with the property that as we walk along this edge path from $\tilde{x}$ to x^1, the value of $z(x)$ is decreasing (actually nonincreasing).

Proof: If $c = 0$, then all the feasible solutions of (3.16) have the same objective value of zero, and the result follows from Theorem 3.10. So assume $c \neq 0$. Apply the simplex algorithm to solve (3.16), starting from the BFS $\tilde{x}$. If x^1 is the unique optimum solution for (3.16), the simplex algorithm traces an edge path of **K** along which $z(x)$ is decreasing, which must terminate with x^1. If x^1 is not the unique optimum, it may terminate with an alternate optimum BFS $\bar{x}$ for (3.16). In this case the set of all optimum solutions of (3.16) is the face of **K**, $\mathbf{F} = \{x: Ax = b, cx = z(x^1), x \geqq 0\}$. Since **F** is a face of **K**, all extreme points and edges of **F** are actually extreme points and edges of **K**. Now x^1 and $\bar{x}$ are two extreme points of **F**. From Theorem 3.10, there exists an edge path in **F** from $\bar{x}$ to x^1, and so every point on this path satisfies $z(x) = \bar{z} = z(x^1)$. Combining the edge path from $\tilde{x}$ to $\bar{x}$ obtained earlier during the simplex algorithm with the edge path from $\bar{x}$ to x^1 in **F**, we have the required path. ■

THEOREM 3.12 Suppose $x^1, \ldots, x^r$, the r best extreme points of **K** in the ranked sequence, have already been determined. Then the $(r+1)$th best extreme point of **K**, x^{r+1}, can be taken to be an adjacent extreme point of x^1, or $x^2, \ldots,$ or x^r on **K**, which is distinct from $x^1, \ldots, x^r$ and minimizes $z(x)$ among these points.

Proof: Let $\{y^1, \ldots, y^t\}$ be the set of all adjacent extreme points of x^1, or $x^2, \ldots,$ or x^r on **K**, excluding $x^1, \ldots, x^r$. Let x^{r+1} be the point in $\{y^1, \ldots, y^t\}$ that minimizes $z(x)$ among them. If $\bar{x}$ is any extreme point of **K** distinct from $x^1, \ldots, x^r, y^1, \ldots, y^t$, any edge path in **K** from x^1 to $\bar{x}$ must contain one of the points $y^1, \ldots, y^t$ as an intermediate point. So, by Theorem 3.11, we must have $z(\bar{x}) \geqq z(x^{r+1})$. So x^{r+1} determined in this way satisfies (3.37). ■

Ranking Algorithm

Theorem 3.12 provides the main result for ranking the extreme points of **K** in increasing order of the value of $z(x)$. This algorithm is from [3.34]. The algorithm begins with an optimum BFS x^1 of (3.16).

Step 1 x^1 is the first element in the ranked sequence. Obtain all the adjacent extreme points of x^1 in **K** and store these in a *list* in increasing order of the objective value $z(x)$ from top to bottom. The algorithm discussed in Section 3.7.1 can be used to generate the adjacent extreme points of x^1 in **K**. While storing an extreme point of **K** in the list, it is convenient to store a basic vector corresponding to it, the actual values of the variables in this extreme point, and the corresponding objective value. Then go to the next step.

General Step Suppose the extreme points $x^1, \ldots, x^r$ in the ranked sequence have already been obtained. The list at this stage contains all the adjacent extreme points of x^1, or $x^2, \ldots,$ or x^r, excluding $x^1, \ldots, x^r$, arranged in increasing order of their objective value from top to bottom. Select the extreme point at the top of the current list and make it x^{r+1}, the next extreme point in the ranked sequence, and remove it from the list. Obtain all the adjacent extreme points of x^{r+1} in **K**, which satisfy $z(x) \geqq z(x^{r+1})$ by using the algorithm discussed in Section

3.7.1, and if any of them is not in the current list and not equal to any of the extreme points in the ranked sequence obtained so far, insert it in the appropriate slot in the list according to its objective value. With this new list, go to the next step and continue until as many extreme points in the ranked sequence as are required, are obtained.

Example 3.38

Consider the LP:

x_1	x_2	x_3	x_4	x_5	x_6	$-z$	b
1	0	0	1	−1	2	0	2
0	1	0	1	2	1	0	3
0	0	1	−1	1	−1	0	6
0	0	0	3	4	5	1	0

$x_j \geqq 0$ for all j; z to be minimized.

where x^1, the first extreme point solution in the ranked sequence, is obviously $(2, 3, 6, 0, 0, 0)^T$ with an objective value of zero. The list at the end of step 1 is:

List

Extreme Point	*Objective Value*	*Associated Basic Vector*
$(0, 2, 7, 0, 0, 1)^T$	5	(x_6, x_2, x_3)
$(0, 1, 8, 2, 0, 0)^T$	6	(x_4, x_2, x_3)
$\left(\frac{7}{2}, 0, \frac{9}{2}, 0, \frac{3}{2}, 0\right)^T$	6	(x_1, x_5, x_2)

Going to step 2, we take $(0, 2, 7, 0, 0, 1)^T$ with an objective value of 5, at the top of the current list, as the second extreme point solution x^2 in the ranked sequence. The canonical tableau with respect to the basic vector (x_6, x_2, x_3) is obtained, and the adjacent extreme points of x^2 with objective value $\geqq 5$ are generated and the list updated. The new list is:

Extreme Point	*Objective Value*	*Associated Basic Vector*
$(0, 1, 8, 2, 0, 0)^T$	6	(x_4, x_2, x_3)
$\left(\frac{7}{2}, 0, \frac{9}{2}, 0, \frac{3}{2}, 0\right)^T$	6	(x_1, x_5, x_3)
$\left(0, 0, \frac{33}{5}, 0, \frac{4}{5}, \frac{7}{5}\right)^T$	$\frac{51}{5}$	(x_6, x_5, x_3)

Going to the next step, the next extreme point solution in the ranked sequence is $x^3 = (0, 1, 8, 2, 0, 0)^T$, with an objective value of 6. Adjacent extreme points of x^3 whose objective value is $\geqq 6$ are generated and the list updated. The new list is:

Extreme Point	*Objective Value*	*Associated Basic Vector*
$\left(\frac{7}{2}, 0, \frac{9}{2}, 0, \frac{3}{2}, 0\right)^T$	6	(x_1, x_5, x_3)
$\left(0, 0, 8, \frac{7}{3}, \frac{1}{3}, 0\right)^T$	$\frac{25}{3}$	(x_4, x_5, x_3)
$\left(0, 0, \frac{33}{5}, 0, \frac{4}{5}, \frac{7}{5}\right)^T$	$\frac{51}{5}$	(x_6, x_5, x_3)

So the next extreme point solution in the ranked sequence is $x^4 = (7/2, 0, 9/2, 0, 3/2, 0)^T$ with an objective value of 6. The algorithm can be continued in a similar manner until enough extreme point solutions in the ranked sequence have been obtained.

General Comments If only the first k extreme point solutions in the ranked sequence are desired (for k specified in advance), keep only the top most k extreme points in the list at each stage, and discard all the remaining bottom ones. If only the extreme point solutions of the ranked sequence whose objective value is less than or equal to some specified value α are desired, discard all the adjacent extreme points generated that correspond to an objective value greater than α, and do not put them in the list.

Problems Posed by Degeneracy In the course of the algorithm, if a degenerate BFS appears in the ranked sequence, then finding all its adjacent extreme points might be a computationally difficult job, as discussed in Section 3.7.1.

How to Enumerate All the Extreme Points of a Convex Polyhedron

In some applications, it may be required to enumerate all the extreme points of a convex polyhedron specified by linear constraints. Consider the LP (3.16). Suppose we wish to enumerate all the extreme point feasible solutions of (3.16). Since the number of these extreme points tends to increase very rapidly with $n - m$, this might be a difficult job if $n - m$ is large. So a task such as this can only be attempted if $n - m$ is small.

We can pick a dummy linear objective function cx by choosing the vector c appropriately, and then use the algorithm discussed above to generate the extreme points in order of increasing value of cx. When this method terminates, we have all the extreme points. However, this method is inefficient because of the computational effort involved in ranking, which is unnecessary for this problem.

Another approach is to generate an extreme point, say x^1, by some method such as Phase I of the simplex method. The singleton set $\{x^1\}$ is the level 0 set of extreme points. Now use the algorithm discussed in Section 3.7.1 to generate the set of all adjacent extreme points of x^1, say $\{x^2, \ldots, x^r\}$, which is the level 1 set of extreme points. Using the algorithm of Section 3.7.1 again repeatedly on each of the extreme points in the level 1 set, obtain the set of all extreme points that are adjacent to at least one extreme point in the level 1 set, but not contained in the level 0 or 1 sets. Call this the level 2 set of extreme points. Repeat the same process with the level 2 set and continue in the same manner. In general, the level $t + 1$ set contains all the extreme points that are adjacent to at least one extreme point in the level t set, but not contained in any level p set for $p \leqq t$. The procedure is continued until at some stage, the new level set turns out to be empty. At that stage, the union of all the level sets generated so far is the set of all extreme points. The repeated checking at each stage whether an adjacent extreme point is already contained in one of the level sets generated so far makes this approach tedious. Bookkeeping methods have been developed to minimize this effort and to make the enumeration efficient. For these, and for other approaches to enumerate all the extreme points of a convex polyhedron, see references [3.17–3.20, 3.23, 3.30, 3.31, 3.36, 3.37].

One final remark. Let A, b, be given matrices of orders $m \times n$ and $m \times 1$, respectively. Let $\mathbf{K} = \{x: Ax \geqq b\}$. Suppose it is known that $\mathbf{K}$ is nonempty and bounded. Even though the number m of constraints in the system $Ax \geqq b$, is small, the number of extreme points of $\mathbf{K}$ usually grows exponentially with m. For example, consider the system of $2n$ constraints in $\mathbf{R}^n$

$$-1 \leqq x_j \leqq 1 \qquad j = 1 \text{ to } n$$

The set of feasible solutions of this system is a cube in $\mathbf{R}^n$ and it has 2^n extreme points. These are all the points of the form $(\alpha_1, \ldots, \alpha_n)^{\mathrm{T}}$ where each α_j is either -1 or $+1$.

Let us consider the reverse problem. If $\{x^1, \ldots, x^r\}$ is a given finite set of points in $\mathbf{R}^n$, its convex hull can be represented as the set of feasible solutions of a system of linear inequalities. Ironically, it usually happens that, even though r is small, the number of linear inequalities required to represent the convex hull of $\{x^1, \ldots, x^r\}$ grows exponentially with r. As an example, consider the set of $2n$ points in $\mathbf{R}^n$, $\Gamma = \{I_{.j}, -I_{.j}: j = 1 \text{ to } n\}$, where I is the unit matrix of order n. The minimal representation for the convex hull of Γ as the set of feasible solutions of a system of linear constraints requires 2^n inequality constraints. These are all the constraints of the form: $\alpha_1 x_1 + \cdots + \alpha_n x_n \leqq 1$, where each of the α_j is either $+1$ or -1. See reference [4.6] of B. C. Eaves and R. M. Freund.

3.20 FRACTIONAL PROGRAMMING

Consider the following optimization problem

$$\begin{aligned} \text{Minimize} \quad & f(x) = (\alpha + cx)/(\beta + dx) \\ \text{Subject to} \quad & Ax = b \\ & x \geqq 0 \end{aligned} \tag{3.38}$$

where α and β are given real numbers, c and d are given row vectors in $\mathbf{R}^n$, A is a given $m \times n$ matrix, b is a given column vector in $\mathbf{R}^m$, and $x \in \mathbf{R}^n$ is the column vector of decision variables. Let $g(x) = \alpha + cx$, $h(x) = \beta + dx$. A problem of this type is known as a *fractional program* or a *programming problem with linear fractional functionals*. Let $\mathbf{K}$ be the set of feasible solutions of (3.38). Without any loss of generality we assume that $\mathbf{K} \neq \emptyset$. In the real number system, the operation of division by the number zero is not defined. Hence, if the denominator $h(x)$ assumes the value zero at some point $x \in \mathbf{K}$, (3.38) is not well defined. If there exists feasible solutions $x^1 = (x_1^1, \ldots, x_n^1)^{\mathrm{T}}$ and $x^2 = (x_1^2, \ldots, x_n^2)^{\mathrm{T}}$, such that $h(x^1) < 0$ and $h(x^2) > 0$, let $\lambda = (h(x^2))/(h(x^2) - h(x^1))$. Then clearly since $0 < \lambda < 1$, $\hat{x} = \lambda x^1 + (1 - \lambda)x^2$ is also a feasible solution, and it can be verified that $h(\hat{x}) = \lambda h(x^1) + (1 - \lambda)h(x^2) = 0$. Hence (3.38) is well defined iff $h(x)$ is either strictly positive for all $x \in \mathbf{K}$, or strictly negative for all $x \in \mathbf{K}$.

Since $f(x)$ is nonlinear when $\mathbf{K}$ is not bounded, it is possible that $f(x)$ is bounded below on $\mathbf{K}$, and yet (3.38) may not have a finite optimum solution. This happens when $\delta = \text{infimum } \{f(x): x \in \mathbf{K}\}$ is finite and $f(x)$ approaches the value δ only asymptotically as the point x travels along a half-line in $\mathbf{K}$ to infinity.

As an example, consider the problem: minimize $f(x_1, x_2) = 1/(x_1 + 1)$, subject to $x_1, x_2 \geqq 0$; a special case of (3.38). Here, the infimum, 0, of $f(x_1, x_2)$ in the problem, is not attained at any finite feasible joint. $f(x_1, x_2)$ approaches its infimum asymptotically as x_1 tends to $+\infty$ along the x_1-axis. See Figure 3.37.

It is possible to process (3.38) by solving an LP, but the LP to be used depends on whether $h(x)$ is strictly positive or strictly negative on **K**. There are several cases to consider.

Case 1 Suppose it can be determined directly from the data that $h(x) > 0$ on **K**. This occurs, for example, if $\beta > 0$ and $d \geqq 0$, or if $\beta \geqq 0$, $b \neq 0$ and $d > 0$. Consider the following LP:

$$\begin{aligned} \text{Minimize} \quad & z(y, t) = cy + \alpha t \\ \text{Subject to} \quad & Ay - bt = 0 \\ & dy + \beta t = 1 \\ y = (y_1, \ldots, y_n)^{\mathrm{T}} \geqq 0 \quad & t \geqq 0 \end{aligned} \tag{3.39}$$

The LP (3.39) contains the additional nonnegative variable t. We have the following theorem.

THEOREM 3.13 (1) If **K** is bounded and $h(x) > 0$ on **K**, the following results hold: (a) If (y, t) is a feasible solution for (3.39), then $t > 0$; (b) (3.38) has a feasible solution iff (3.39) has a feasible solution; (c) if $\tilde{x}$ is an optimum feasible solution of (3.38), let $\tilde{t} = 1/(d\tilde{x} + \beta)$ and $\tilde{y} = \tilde{t}\tilde{x}$. Then $(\tilde{y}, \tilde{t})$ is an optimum feasible solution of (3.39); (d) If $(\tilde{y}, \tilde{t})$ is an optimum feasible solution of (3.39), then $\tilde{x} = (1/\tilde{t})\tilde{y}$ is an optimum feasible solution of (3.38).

(2) If $h(x) > 0$ on **K** and **K** is unbounded, the following results hold: (e) If the objective value in the LP (3.39) is unbounded below, then $f(x)$ is unbounded below on **K** and vice versa. In this case, from the extreme half-line of (3.39) along which $z(y, t)$ diverges to $-\infty$, we can construct a half-line of **K** along which $f(x)$ diverges to $-\infty$. (f) If (3.38) has a finite optimum feasible solution, say $\bar{x}$, let $\bar{t} = 1/(d\bar{x} + \beta)$, $\bar{y} = \bar{t}\bar{x}$. Then $(\bar{y}, \bar{t})$ is an optimum feasible solution of the LP (3.39). Conversely if $(\bar{y}, \bar{t})$ is an optimum feasible solution of (3.39) satisfying $\bar{t} > 0$, let $\bar{x} = (1/\bar{t})\bar{y}$. Then $\bar{x}$ is an optimum feasible solution of (3.38). System (3.38) has a finite optimum feasible solution iff the LP (3.39) has an optimum feasible solution (y, t) that satisfies $t > 0$. (g) Let $\delta =$ infimum $\{f(x) : x \in \mathbf{K}\}$. If δ is finite, the LP (3.39) has an optimum feasible solution, and δ is the optimum

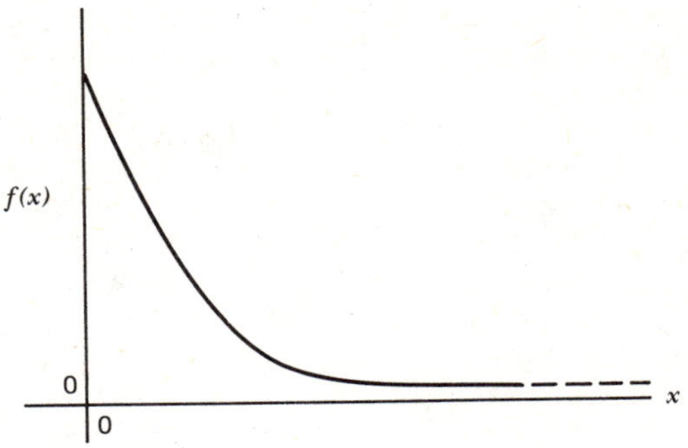

Figure 3.37 The infimum of $f(x)$ over $x \geqq 0$ is 0, but $f(x)$ approaches 0 only asymptotically as x tends to $+\infty$.

objective value in it. If every optimum feasible solution (y, t) for (3.39) satisfies $t = 0$, (3.38) has no finite optimum feasible solution, but δ is finite and it is equal to the optimum objective value in (3.39). In this case, we can construct a half-line of **K** along which $f(x)$ asymptotically approaches its infimum δ in **K**.

Proof: (1) Here we are assuming that **K** is bounded. So, by the results in Section 3.7, the system $Ay = 0$, $y \geqq 0$ has the unique solution $y = 0$. Hence the system $Ay = 0$, $dy = 1$, $y \geqq 0$ has no feasible solution. This implies (1)(a).

If $\bar{x} \in \mathbf{K}$, by the hypothesis $h(\bar{x}) > 0$ it can be verified that $(\bar{y}, \bar{t})$ is a feasible solution of (3.39), where $\bar{t} = 1/h(\bar{x})$, and $\bar{y} = \bar{t}\bar{x}$. Conversely, if $(\bar{y}, \bar{t})$ is a feasible solution of (3.39), $\bar{t} > 0$ by (a), and it can be verified that $\bar{x} = (1/\bar{t})\bar{y}$ is a feasible solution of (3.38). This proves (b).

Let $\bar{x}$ be a feasible solution of (3.38). Let $\bar{t} = 1/h(\bar{x})$ and $\bar{y} = \bar{t}\bar{x}$. Then $(\bar{y}, \bar{t})$ is a feasible solution of (3.39), and it can be verified that $f(\bar{x}) = z(\bar{y}, \bar{t})$. Conversely if $(\bar{y}, \bar{t})$ is a feasible solution of (3.39), let $\bar{x} = (1/\bar{t})\bar{y}$. Then $\bar{x}$ is a feasible solution of (3.38), and it satisfies $f(\bar{x}) = z(\bar{y}, \bar{t})$. These facts together imply (c) and (d).

(2) (e) Here we are assuming that **K** is unbounded. If $f(x)$ is unbounded below on **K**, we can find a sequence of points $\{x^1, x^2, \ldots, x^r, \ldots\}$ all lying in **K** such that $f(x^r)$ diverges to $-\infty$ as r goes to ∞. By the hypothesis $\beta + dx^r > 0$ for all r. Define $t^r = 1/(\beta + dx^r)$ and $y^r = t^r x^r$. Clearly (y^r, t^r) is feasible to (3.39), and $z(y^r, t^r) = f(x^r)$ for each $r = 1, 2, \ldots$. This clearly implies that if $f(x)$ is unbounded below on **K**, so is $z(y, t)$ in the LP (3.39). Conversely suppose the objective value in the LP (3.39) is unbounded below. Then by the results in Section 3.7 there exists

a homogeneous solution $(\hat{y}, \hat{t})$ to (3.39) satisfying $A\hat{y} - b\hat{t} = 0$, $d\hat{y} + \beta\hat{t} = 0$, $c\hat{y} + \alpha\hat{t} < 0$, $(\hat{y}, \hat{t}) \geq 0$. In this system, if $\hat{t} > 0$, define $\hat{x} = \hat{y}/\hat{t}$, and we conclude that $\hat{x} \in \mathbf{K}$ and $d\hat{x} + \beta = 0$, which is a contradiction to the hypothesis that $h(x) > 0$ over $\mathbf{K}$. So we must have $\hat{t} = 0$. Let $\hat{x}$ be any feasible solution for (3.38). Since $\hat{t} = 0$ in the above system, the half-line $\{\hat{x} + \lambda\hat{y}: \lambda \geqq 0\}$ is completely contained in $\mathbf{K}$. Also, since $\hat{t} = 0$ in the above system, we have $d\hat{y} = 0$ and $c\hat{y} < 0$, and from this we have $f(\hat{x} + \lambda\hat{y}) = (\alpha + c\hat{x} + \lambda c\hat{y})/(\beta + d\hat{x} + \lambda d\hat{y}) = (\alpha + c\hat{x} + \lambda c\hat{y})/(\beta + d\hat{x}) = (\alpha + c\hat{x})/(\beta + d\hat{x}) + \lambda(c\hat{y}/(\beta + d\hat{x}))$, and since $c\hat{y} < 0$, $\beta + d\hat{x} = h(\hat{x}) > 0$ (since $\hat{x} \in \mathbf{K}$), we conclude that $f(\hat{x} + \lambda\hat{y})$ diverges to $-\infty$ as λ goes to $+\infty$. Hence, on the half-line $\{\hat{x} + \lambda\hat{y}: \lambda \geqq 0\}$, $f(x)$ diverges to $-\infty$ as λ goes to $+\infty$. Suppose (y^1, t^1) is the BFS of (3.39) at the stage where the unboundedness of $z(y, t)$ is recognized in the process of solving the LP (3.39) by the simplex method. From the column in which the unboundedness criterion is satisfied, we can construct the homogeneous solution $(\hat{y}, \hat{t} = 0)$ discussed above, using the arguments presented in Section 3.7; and, in fact, the extreme half-line of (3.39) along which $z(y, t)$ diverges to $-\infty$, constructed as discussed in Chapter 2, will be $\{(y^1, t^1) + \lambda(\hat{y}, \hat{t}): \lambda \geqq 0\}$.

(2) (f) Follows from the arguments used in the proof of (1).

(2) (g) For proving these, we assume that the infimum δ of $f(x)$ on $\mathbf{K}$ is finite. For any $x \in \mathbf{K}$, define $t = 1/(\beta + dx)$ (which is possible, since by hypothesis, $h(x) > 0$ over $\mathbf{K}$). Then $(y = tx, t)$ can be verified to provide a feasible solution for (3.39), which satisfies $z(y, t) = f(x)$. This implies that the optimum objective value in (3.39) is $\leqq \delta$. Now suppose the optimum objective value in (3.39) is $\gamma < \delta$. Let (y', t') be an optimum feasible solution for (3.39). So $z(y', t') = \gamma$. If $t' > 0$, then define $x' = y'/t'$, and it can be verified that x' is feasible to (3.38) and satisfies $f(x') = z(y', t') = \gamma < \delta$, a contradiction, since δ is the infimum of $f(x)$ over $\mathbf{K}$. So we must have $t' = 0$. Let $\varepsilon = (\delta - \gamma)/4$. Since δ is the infimum of $f(x)$ over $\mathbf{K}$, we can find a point $x'' \in \mathbf{K}$ satisfying $\delta \leqq f(x'') < \delta + \varepsilon$. Define $t'' = 1/(\beta + dx'')$, $y'' = t''x''$. By hypothesis, $t'' > 0$. So, (y'', t'') is feasible to (3.39) and satisfies $z(y'', t'') = f(x'')$. Let $(y''', t''') = (1/2)(y' + y'', t' + t'')$. So (y''', t''') is feasible to (3.39), $t''' > 0$ and $z(y''', t''') = (1/2)(z(y', t') + z(y'', t'')) \leqq (1/2)(\gamma + \delta + \varepsilon) < \delta$. Since $t''' > 0$, define $x''' = y'''/t'''$. Clearly x''' is feasible to (3.38) and $f(x''') = z(y''', t''') < \delta$, a contradiction to the fact that δ is the infimum of $f(x)$ over $\mathbf{K}$. So, in this case, the optimum objective value in the LP (3.39) must be equal to δ.

Let $(y^1, t^1 = 0)$ be an optimum solution of (3.39), so $z(y^1, t^1) = cy^1 = \delta$. Let x^1 be any feasible solution of (3.38). Since $(y^1, t^1 = 0)$ is feasible to (3.39), we have $Ay^1 = 0$, $dy^1 = 1$, $y^1 \geq 0$. So $x^1 + \lambda y^1$ is feasible to (3.38) for all $\lambda \geqq 0$, and $f(x^1 + \lambda y^1) = (\alpha + cx^1 + \lambda cy^1)/(\beta + dx^1 + \lambda dy^1) = (\alpha + cx^1 + \lambda\delta)/(\beta + dx^1 + \lambda)$ and, as λ tends to $+\infty$, this $f(x^1 + \lambda y^1)$ approaches δ. Thus, in this case, as one travels along the half-line $\{x^1 + \lambda y^1: \lambda > 0\}$ to infinity, the objective function $f(x)$ approaches its infimum δ in $\mathbf{K}$ asymptotically. ■

Hence, in this case, the fractional program (3.38) can be solved by solving the LP (3.39). When (3.39) is solved, if optimum solution $(\bar{y}, \bar{t})$ is obtained for it, and $\bar{t} > 0$, $\bar{x} = \bar{y}/\bar{t}$ is an optimum solution for (3.38). On the other hand, if $\bar{t} = 0$, solve the LP

$$\begin{array}{lll} \text{Maximize} & t & \\ \text{Subject to:} & Ay - bt = 0 & \\ & dy + \beta t = 1 & \\ & cy + \alpha t = c\bar{y} & \\ y \geqq 0 & t \geqq 0. & \end{array}$$

If the optimum objective value in this LP is >0, it provides an optimum feasible solution (y, t) for (3.39) in which $t > 0$, and from it an optimum feasible solution for (3.38) can be obtained. On the other hand, if the optimum objective value in this LP is zero, the situation described in (2)(g) of Theorem 3.13 occurs, and (3.38) has an infimum but no finite optimum solution.

Case 2 Suppose it can be determined directly from the data that $h(x) < 0$ on $\mathbf{K}$. This occurs, for example, if $\beta < 0$ and $d \leqq 0$, or if $\beta \leqq 0$, $b \neq 0$ and $d < 0$. All the results stated in Theorem 3.13 continue to hold if the LP (3.39) is replaced by

$$\begin{array}{lll} \text{Minimize} & z(y, t) = cy + \alpha t & \\ \text{Subject to} & Ay - bt = 0 & \\ & dy + \beta t = -1 & \\ y \geqq 0 & t \geqq 0. & \end{array} \qquad (3.40)$$

Hence, in this case, (3.38) can be processed by solving the LP (3.40).

Case 3 Suppose it is not known whether $h(x)$ is >0 or <0 on $\mathbf{K}$. Find a feasible solution, say x^1. If $h(x^1) < 0$, replace both $h(x)$ and $g(x)$ by their negatives. After this change, we can assume that $h(x^1) > 0$. Now, solve the LP: Minimize $h(x)$ over $x \in \mathbf{K}$. If the minimum objective value in this LP is >0, we are done, go to Case 1. In the process of solving this LP, if $h(x)$ ever assumes a value $\leqq 0$, (3.38) is not well defined, terminate.

We will now discuss some results about the optimum solutions of the linear fractional program (3.38). Without any loss of generality we assume that the rank of A is m. If the rank of the system of equality constraints in (3.39) is also m, $\beta + dx$ must be a constant on $\mathbf{K}$, and (3.38) is actually an LP. So in proving the following results, we assume that the rank of the system of equality constraints in (3.39) is $m + 1$. We also assume that $h(x) > 0$ for all $x \in \mathbf{K}$.

THEOREM 3.14 Suppose $(\bar{y}, \bar{t})$ is an optimum BFS of (3.39) in which $\bar{t} > 0$. Then $\bar{x} = \bar{y}/\bar{t}$ is an optimum solution of (3.38), which is a BFS of (3.38).

Proof: Suppose $(\bar{y}, \bar{t})$ is the BFS associated with the basic vector $(y_1, \ldots, y_m, t)$ for (3.39). So the right-hand-side column vector in (3.39), which is $I_{.m+1}$ (the $(m+1)$th column vector of the unit matrix of order $m+1$), has a unique expression as a linear combination of the columns of $y_1, \ldots, y_m, t$ in (3.39). This implies that the column of t in (3.39) has a unique expression as a linear combination of $I_{.m+1}$ and the columns of $y_1, \ldots, y_m$ in (3.39). This implies that the matrix $\boldsymbol{B}$ of order $(m+1)$ consisting of the columns of $y_1, \ldots, y_m$ in (3.39) and $I_{.m+1}$ is nonsingular. Expanding the determinant of $\boldsymbol{B}$ through the last column $I_{.m+1}$, we conclude that its submatrix $B = (A_{.1} \vdots \cdots \vdots A_{.m})$ is nonsingular. So B is a basis for (3.38), and clearly $\bar{x} = (1/\bar{t})\bar{y}$ is the BFS of (3.38) corresponding to it; $\bar{x}$ is also optimal to (3.38) by Theorem 3.13. ■

THEOREM 3.15 If the linear fractional program (3.38) has an optimum solution, then it has an optimum solution that is a BFS of (3.38)—that is—an extreme point of $\mathbf{K}$, and the procedure discussed above finds such an optimum solution.

Proof: Follows from Theorems 3.13 and 3.14. ■

THEOREM 3.16 Assume that $\beta + dx > 0$ for all $x \in \mathbf{K}$. Let $x^1 \in \mathbf{K}$, $\lambda^1 = f(x^1)$. Consider the LP minimize $(c - \lambda^1 d)x$, subject to $Ax = b$, $x \geqq 0$. If x^1 is an optimum solution for this LP, then x^1 is an optimum solution for (3.38).

Proof: Suppose x^1 is optimal for this LP. So we have $(c - \lambda^1 d)x \geqq (c - \lambda^1 d)x^1$ for all $x \in \mathbf{K}$. This obviously implies that $(cx + \alpha) - \lambda^1(dx + \beta) \geqq (cx^1 + \alpha) - \lambda^1(dx^1 + \beta) = 0$, for all $x \in \mathbf{K}$ (since $\lambda^1 = f(x^1)$). Also, since $dx + \beta > 0$ for all $x \in \mathbf{K}$, this implies that $(cx + \alpha)/(dx + \beta) \geqq \lambda^1$ for all $x \in \mathbf{K}$, which in turns implies that x^1 is optimal to (3.38). ■

THEOREM 3.17 Assume that $\beta + dx > 0$ for all $x \in \mathbf{K}$. If $x^1, x^2 \in \mathbf{K}$ satisfy $(c - \lambda^1 d)x^2 < (c - \lambda^1 d)x^1$, where $\lambda^1 = f(x^1)$, then $f(x^2) < f(x^1)$.

Proof: Since $\lambda^1 = f(x^1)$, we have $(cx^1 + \alpha) - \lambda^1(dx^1 + \beta) = 0$. The hypothesis $(c - \lambda^1 d)x^2 < (c - \lambda^1 d)x^1$ clearly implies that

$$(cx^2 + \alpha) - \lambda^1(dx^2 + \beta) < (cx^1 + \alpha) - \lambda^1(dx^1 + \beta) = 0.$$

Since $dx^2 + \beta > 0$, this implies that $f(x^2) = (cx^2 + \alpha)/(dx^2 + \beta) < \lambda^1 = f(x^1)$. ■

The method discussed above for solving (3.38) through (3.39) is a variant of a method due to A. Charnes and W. W. Cooper [3.3]. It can process (3.38) without any boundedness conditions on $\mathbf{K}$ or any other conditions, and requiring the solution of at most three LPs. It is quite efficient, and when optimum solutions exist, it does produce an extreme-point optimum solution. However, it has one minor drawback. If the system of constraints in (3.38) has special structure (for example, the constraints may be constraints of the transportation, or assignment, or general network flow type), this special structure is lost in going over to (3.39), because of the last equality constraint in (3.39). Linear programs with special structure can usually be solved much more efficiently by special algorithms that take advantage of the structure than by the general simplex method. When the constraints in (3.38) have some special structure, it may be worthwhile to look for ways of solving it using the system of constraints as it is, that is, without adding any new constraints. We will now provide such an approach based on methods for solving (3.38) proposed in [3.7 and 3.8]. It turns out that the linear fractional program can be solved by an *adjacent vertex method*, which is similar to the simplex algorithm, moving along an edge path in $\mathbf{K}$, making a strict improvement in the objective value after each move. Although this method can also process (3.38) in general, whether **K** is bounded or not, for the sake of simplicity, we will assume that **K** is bounded in the following discussion (the results and the method discussed below can be extended quite easily to the case where **K** is not bounded: see references [3.7 and 3.8]). In most practical applications, there are usually finite bounds imposed on the variables, which make **K** bounded automatically, and hence the method discussed below can be applied on them directly. Since we are assuming that **K** is nonempty and bounded, (3.38) does have an optimum solution, and the unboundedness and the asymptotic infimum cases do not arise.

THEOREM 3.18 Let **K** be bounded and x^1 be an extreme point of **K**. x^1 is not an optimum solution of (3.38) iff there exists an adjacent extreme point x^2 of x^1 on **K**, such that as one travels from x^1 to x^2 along the edge of **K** joining them, $f(x)$ decreases strictly.

Proof: By the results in Sections 3.6, 3.7, and 3.8, we know that the corresponding results hold for linear programs. Let $\lambda^1 = f(x^1)$. By Theorem 3.16, x^1 is not an optimum solution of (3.38) iff x^1 is not an optimum solution for the LP: minimize $(c - \lambda^1 d)x$ over $x \in \mathbf{K}$. If x^1 is not an optimum solution for this LP, we know by the results in Section 3.8 and the fact that **K** is bounded here that there exists an adjacent extreme point x^2 of x^1 on **K**, such that $(c - \lambda^1 d)x^2 < (c - \lambda^1 d)x^1$. Theorem 3.17 now implies that as you move from x^1 to x^2 along the edge joining them, $f(x)$ strictly decreases. ■

Adjacent Vertex Method for (3.38) when K is Bounded and $(\beta + dx) > 0$ for all $x \in \mathbf{K}$

The method begins with an initial extreme point of **K** obtained by Phase I of the simplex method. Let x^1 be the current extreme point of **K**, associated with the present basic vector x_{B_1} for (3.38) and let $\lambda^1 = f(x^1)$. Consider the LP:

$$\begin{array}{ll} \text{Minimize} & (c - \lambda^1 d)x \\ \text{Subject to} & Ax = b \\ & x \geqq 0 \end{array}$$

Begin solving this LP by the simplex algorithm initiated with the present feasible basic vector x_{B_1}, but proceed only until the first nondegenerate pivot step occurs. If x^1 is confirmed as an optimum solution of this LP, x^1 is also optimal to (3.38), terminate. If a nondegenerate pivot step occurs, let x^2 be the adjacent BFS of x^1 after this pivot step, and x_{B_2} the associated basic vector. Go to the next step of this method with x^2, x_{B_2}, and continue in the same way.

It can be verified that the gradient vector of $f(x)$ at $x^1 = \nabla f(x^1) = (c - \lambda^1 d)/(\beta + dx^1)$, and since $\beta + dx^1 > 0$, $(c - \lambda^1 d)$ is a positive multiple of the gradient vector of $f(x)$ at x^1. From this one can easily verify that the above method is a *reduced gradient method* for solving (3.38).

Fractional programming finds many applications in problems in which the aim is to optimize productivity. If $g(x)$ is the net profit, and $h(x)$ is the total number of man hours required when the feasible solution x is adopted; then $f(x) = g(x)/h(x)$ is the net return per man hour employed. The problem of maximizing the net return per man hour employed is an optimization problem with a fractional objective function, and it can be solved by the methods discussed above (see Exercise 3.65).

3.21 PRECISION IN COMPUTATION

In the algorithms, discussed in this chapter, for testing the linear independence of the row vectors of a matrix, or for computing the rank of a matrix, etc., it is necessary to check whether there exists a nonzero entry in a row of the current tableau. For this we have to look at each entry in that row and decide whether it is zero or not. This sounds like a trivial task, but sometimes it can be very hard to do correctly. Surely, this task is easy while solving a small problem by hand using integer arithmetic. However, to solve a large problem, we have to implement the algorithm on a computer by writing a computer program for it. Digital computers are finite precision machines, and usually any fraction encountered in the computation is converted into decimal form and rounded off to a fixed number of significant digits. This introduces a small round-off error at this stage, and as computations continue, these errors accumulate in every entry of the current tableau. If an entry in the current tableau is nonzero, but has a very small absolute value, it is very hard to decide whether that entry would have been zero or a nonzero number of small absolute value if all computations are performed exactly. Also, because of its finite precision feature, computers cannot distinguish between zero and a number whose absolute value is less than the machines precision limit. For this reason, algorithms that require checking to determine whether an entry in the current tableau is zero or nonzero could be very hard to carry out correctly, particularly when dealing with large ill-conditioned matrices (see reference [7.27] for definitions of these terms). Normally, in computer programs a control parameter known as a *tolerance* (which is a small positive number such as 10^{-20} or so, depending on the machine's precision) is specified, and any entry in the current tableau whose absolute value is less than the tolerance is treated as being equal to zero. Because of this, when solving large problems, the final answer produced by the computer may not be the true answer to the problem; it may only be approximately correct to within the limits of precision that we can achieve.

The branch of mathematics known as *numerical analysis* deals with this issue of precision in computation and how to improve it. Numerical analysts have developed several techniques for improving precision and achieving a numerically stable implementation, when implementing algorithms similar to those discussed in this chapter. Some of these techniques are discussed in Section 3.3.8. Another technique that is commonly used is not to compute a basis inverse explicitly, but to keep it in product form (that is, as a product of an ordered sequence of pivot matrices, each of which is stored separately). The implementation of the simplex algorithm using the product form of the basis inverse is discussed in Chapter 5. A further improvement on this is to actually carry out all the computations for which the basis inverse is needed, using instead a factorization or a triangular decomposition of the basis (e.g., the *LU* decomposition of the basis). The implementation of the simplex algorithm using the *LU* decomposition of the basis is discussed in Chapter 7. For a detailed discussion of numerical analysis methods for improving precision in implementations of pivotal algorithms, see references [3.10, 3.12, 3.13, 3.15] and those listed at the end of Chapter 7.

Comment 3.1: In a paper written in 1824, the famous mathematician J. B. J. Fourier studied the problem of finding an approximate solution to a (possibly inconsistent) system of linear equations that minimizes the maximum absolute error in any equation of the system (the Chebyshev approximation). He states that this is equivalent to the problem of finding the lowest point of a convex polyhedron. In geometrical terms he describes that this problem can be solved by starting at an extreme point of the convex polyhedron, descending to another adjacent extreme point by moving along an edge of the convex polyhedron (always following the edge that descends most steeply from the extreme point), and continuing in the same manner until the lowest point of the convex polyhedron is reached. He states that there is a sequence of numerical operations that corresponds to these geometrical operations, but he does not describe them (See page 21 in the book by G. B. Dantzig, [1.61] and the paper by D. A. Kohler, [3.29]). From the results in this chapter we see that the *extreme point to adjacent extreme-point descent* is the principle behind the simplex method developed by G. B. Dantzig in 1947.

Exercises

3.39 Let $A = \{A_{.1}, \dots, A_{.n}\}$ be a basis for $\mathbf{R}^n$. Let $B = \{B_{.1}, \dots, B_{.k}\}$ be a linearly independent set of vectors in $\mathbf{R}^n$. Prove that a subset of $n - k$ vectors in A can be found that together with the vectors in B forms a basis for $\mathbf{R}^n$.

3.40 (a) Let $B_{.1} = (1, 1, 3)^T$, $B_{.2} = (1, -2, 1)^T$, $B_{.3} = (1, -1, 4)^T$, $b = (1, -4, 5)^T$. Let $I_{.j}$ be the jth column vector of the unit matrix of order 3. Find a representation of b as a linear combination of (1) $\{B_{.1}, I_{.2}, I_{.3}\}$, (2) $\{B_{.1}, I_{.2}, B_{.3}\}$, and (3) $\{B_{.1}, B_{.2}, B_{.3}\}$. In each case check whether the representation is unique.

(b) Consider the following system of equations. Find the set of all α for which this system is (1) inconsistent, and (2) has at least one solution. In this case find all the solutions of the system.

$$\begin{aligned} 2x_1 - 4x_2 - 2x_3 &= -2 \\ 3x_1 - 2x_2 + x_3 &= 2 \\ -x_1 + 3x_2 + 2x_3 &= \alpha \end{aligned}$$

3.41 For the following system find the basic solution associated with the basic vector (x_1, x_3, x_4).

$$\begin{aligned} -x_3 + x_4 &= 0 \\ x_2 + x_3 + x_4 &= 3 \\ -x_1 + x_2 + x_3 + x_4 &= 2 \\ x_j &\geqq 0 \quad \text{for all } j \end{aligned}$$

3.42 Let $\{A_{1.}, \dots, A_{r.}\} \subset \mathbf{R}^n$ be a given set of row vectors. Suppose $A_{r+1.}$ is another row vector in $\mathbf{R}^n$, which is not in the linear hull of $\{A_{1.}, \dots, A_{r.}\}$. Prove that the system of equations $A_{i.}x = 0$, for $i = 1$ to r, $A_{i.}x = 1$, for $i = r + 1$, has a solution x.

3.43 Let $\mathbf{F} = \{F_{.1}, \ldots, F_{.n}\}$ be a set of n column vectors in $\mathbf{R}^m$. Suppose the rank of $\mathbf{F}$ is known to be r. Let $y = \sum_{j=1}^{n} F_{.j}x_j$ be a given linear combination of the column vectors in $\mathbf{F}$. Prove that it is possible to express y as a linear combination of the column vectors in $\mathbf{F}$ as $y = \sum_{j=1}^{n} F_{.j}\bar{x}_j$, where at most r of the $\bar{x}_j$, $j = 1$ to n, are nonzero.

3.44 Consider the LP in standard form (3.16), where A is a matrix of order $m \times n$ and rank m. Suppose it is known that (3.16) is nondegenerate. Prove that if $\bar{x}$ is a solution of $Ax = b$, then the rank of the set of column vectors $\{A_{.j}: j$, such that $\bar{x}_j \neq 0\}$ is m. Also prove that every feasible solution of (3.16) in which exactly m of the x_j are strictly positive must be a BFS, and conversely. Construct examples of degenerate LPs in standard form in which the above conclusions are false.

3.45 For the following system, is $(3, 2, 1, 2, 2, 0, 0, 0, 0)^T$ a BFS? If not, construct a BFS from it.

x_1	x_2	x_3	x_4	x_5	x_6	x_7	x_8	x_9	b
1	1	−2	0	3	1	0	0	0	9
0	−1	0	−1	−1	0	1	0	0	−6
0	0	1	1	−1	0	0	1	0	1
1	−1	2	2	−3	0	0	0	1	1
2	−1	1	2	−2	1	1	1	1	5

$x_j \geqq 0$ for all j.

3.46 For the following system, is $(1, 2, 1, 1, 0, 0, 0)^T$ a BFS? If not, construct a BFS from it.

x_1	x_2	x_3	x_4	x_5	x_6	x_7	b
1	−3	0	−8	1	1	0	−13
0	0	−2	−2	0	1	3	−4
−1	−4	0	−10	0	0	2	−19
0	0	3	3	0	1	0	6

$x_j \geqq 0$ for all j.

3.47 For the following system, are 0, $(1, 2, 3, 0, 0, 0)^T$ and $(1, 2, 3, 1, 0, 0)^T$ BFSs?

x_1	x_2	x_3	x_4	x_5	x_6
1	−1	2	2	2	$-1 \leqq 7$
−1	2	−1	0	2	$-1 = 0$
1	2	2	1	2	$-1 \leqq 12$
−1	2	2	2	−1	$-1 \leqq 11$

$x_j \geqq 0$ for all j.

3.48 Find all the BFSs of the system $-6 \leqq x_1 \leqq 2$, $-5 \leqq x_2 \leqq 3$.

3.49 Consider the system of constraints (3.15) for a bounded variable LP. Let $\bar{x}$ be a feasible solution of (3.15). Using arguments similar to those in the proof of Theorem 3.2, prove that $\bar{x}$ can be expressed as $\alpha x^1 + (1 - \alpha)x^2$ for some $0 < \alpha < 1$, where x^1 and x^2 are two distinct feasible solutions of (3.15) iff the set of column vectors $\{A_{.j}: j$, such that $\ell_j < \bar{x}_j < U_j\}$ is linearly dependent. From this, provide a direct proof of the statement that $\bar{x}$ is an extreme point of the set of feasible solutions of (3.15) iff the set $\{A_{.j}: j$, such that $\ell_j < \bar{x}_j < U_j\}$ is linearly independent, using the geometric definition of extreme points given in Section 3.4.1.

3.50 Find all the adjacent extreme points of the BFS associated with the basic vector (x_1, x_2, x_3) in the following system.

x_1	x_2	x_3	x_4	x_5	x_6	b
1	0	0	1	0	−1	0
0	1	0	−1	1	2	3
0	0	1	2	−1	1	2

$x_j \geqq 0$ for all j.

3.51 For the following system, is $(4, 9, 0, 3, 0, 0)^T$ an extreme point? Why? If not, is it on an edge? If it is, is the edge bounded or unbounded?

$$\begin{aligned} x_1 + x_2 \quad\quad - 3x_4 + 3x_5 + x_6 &= 4 \\ x_1 + 2x_2 \quad\quad - 5x_4 + 5x_5 + 3x_6 &= 7 \\ - x_2 + x_3 + 2x_4 - 5x_5 + x_6 &= -3 \\ x_j \geqq 0 \quad \text{for all } j \end{aligned}$$

3.52 For the system

$$\begin{aligned} 4x_1 + x_2 &\geqq 4 \\ x_1 - x_2 &\leqq 2 \\ -2x_1 + x_2 &\leqq 2 \\ x_1 + x_2 &\geqq 2 \\ x_1 + 3x_2 &\geqq 3 \end{aligned}$$

find all the extreme points, extreme homogeneous solutions, and extreme half-lines. Derive an expression for a general feasible solution of this system as in Resolution Theorem II.

3.53 Solve the following LPs using only one artificial variable.

(a)
$$\begin{array}{ll} \text{Minimize} & -x_1 + 2x_2 - 3x_3 + x_4 + 3x_5 - x_6 \\ \text{Subject to} & x_1 - 2x_4 + x_5 + x_6 = 3 \\ & x_2 - 2x_4 - 2x_5 + x_6 = -6 \\ & x_3 + x_4 - 2x_5 - 2x_6 = -12 \\ & x_j \geqq 0 \quad \text{for all } j \end{array}$$

(b)
$$\begin{array}{ll} \text{Maximize} & x_8 \\ \text{Subject to} & -x_1 - x_3 + x_4 + x_5 = -2 \\ & x_1 - 2x_3 - 3x_4 + x_6 - x_8 = 4 \\ & x_1 - 2x_4 + x_7 = 3 \\ & 2x_1 + x_2 - 5x_4 - x_8 = 6 \\ & x_j \geqq 0 \quad \text{for all } j \end{array}$$

3.54 Solve the following LPs. Determine the set of alternate optima in each case.

(a)
$$\begin{array}{ll} \text{Maximize} & 12x_1 + x_2 + 2x_3 + 5x_4 + 4x_5 - 6x_6 \\ \text{Subject to} & 8x_1 + x_2 + 3x_3 + 2x_4 + 3x_5 - 3x_6 = 17 \\ & -5x_1 - x_2 - x_3 - x_4 - 2x_5 + 2x_6 = -12 \\ & -5x_1 - x_3 - 2x_4 - x_5 + 4x_6 \leqq -8 \\ & x_j \geqq 0 \quad \text{for all } j \end{array}$$

(b)
$$\begin{array}{ll} \text{Minimize} & -2x_1 + x_2 + 18x_3 + 6x_4 - 2x_8 \\ \text{Subject to} & x_1 + x_3 - x_4 - 2x_6 + x_7 - x_8 = 6 \\ & x_2 - x_3 + x_4 - 2x_5 + x_6 + 2x_7 + x_8 = 3 \\ & x_3 + 2x_4 - 3x_5 + x_6 - 2x_7 + x_8 = 3 \\ & x_j \geqq 0 \quad \text{for all } j \end{array}$$

3.55 For the following system, let $\bar{x}$ be the BFS associated with the basic vector (x_1, x_2, x_3). Find $\bar{x}$ and all the adjacent extreme points of $\bar{x}$. Use this information to check whether $\bar{x}$ minimizes $z(x) = 3x_1 - 2x_2 - 10x_3 - 11x_4 + 18x_5 - 10x_6$ on the set of feasible solutions of this system.

x_1	x_2	x_3	x_4	x_5	x_6	b
1	0	0	1	−1	1	3
0	1	0	2	2	−1	4
0	0	1	−1	1	2	6

$x_j \geqq 0$ for all j.

3.56 Let **K** be a convex polyhedron and let $\bar{x}$ be a point in it. Prove that $\bar{x}$ is an extreme point of **K** iff $\mathbf{K}\backslash\{\bar{x}\}$ is also a convex set. Develop and prove a corresponding result for higher dimensional faces.

3.57 Solve the following LP using (x_5, x_6, x_7) as the initial basic vector. Also show that the set of all optimum solutions of this problem is an unbounded set.

x_1	x_2	x_3	x_4	x_5	x_6	x_7	x_8	$-z$	$= b$
1	1	0	0	−2	0	3	−3	0	7
−2	1	0	−3	1	3	−3	0	0	−5
0	1	1	−2	0	0	3	4	0	9
1	1	1	3	4	2	7	−5	1	0

$x_j \geqq 0$ for all j; z to be minimized.

3.58 Consider the LP (3.16). If c_j is decreased, keeping all the other cost coefficients unaltered, prove that the value of x_j in an optimum solution increases.

3.59 Consider the following LP. Verify that the pivot step of bringing either the nonbasic variable

x_1 or the nonbasic variable x_2 into the feasible basic vector (x_3, x_4, x_5) are degenerate pivot steps. Hence in this problem, it is impossible to make any improvement in the objective value by performing only one pivot step in the present basic vector (x_3, x_4, x_5). However, verify that the present BFS is not optimal to this problem.

x_1	x_2	x_3	x_4	x_5	$-z$	b
-2	1	1	0	0	0	0
1	-2	0	1	0	0	0
1	1	0	0	1	0	6
-2	-1	0	0	0	1	0

$x_j \geqq 0$ for all j; minimize z.

—(*R. E. Rosenthal*)

3.60 Consider the LP (3.16) where A is a matrix of order $m \times n$ and rank m. Let **K** be the set of feasible solutions of this LP. Assume that $\mathbf{K} \neq \varnothing$ and that $z(x)$ is unbounded below on **K**. In this case, prove that there exists a feasible basic vector x_B for (3.16) that satisfies the following properties. (1) Let $\bar{c} = (\bar{c}_j)$ be the vector of relative cost coefficients and $\bar{A} = (\bar{a}_{ij})$ the updated input–output coefficient matrix with respect to the basic vector x_B. At least one of the $\bar{c}_j$ is strictly negative. (2) For each j such that $\bar{c}_j < 0$, either we have $\bar{a}_{ij} \leqq 0$ for all i, or $\bar{a}_{ij} > 0$ for any i implies $\bar{b}_i = 0$, where $\bar{b} = (\bar{b}_i)$ is the current updated right-hand-side constants vector. Give a geometric interpretation.

3.61 Let **K** be a convex polyhedral subset of $\mathbf{R}^n$, and **H** a hyperplane of dimension $n-1$ in $\mathbf{R}^n$. Suppose that $\mathbf{K} \cap \mathbf{H} \neq \varnothing$. (1) If $x^1 \in \mathbf{K} \cap \mathbf{H}$, prove that x^1 is an extreme point of $\mathbf{K} \cap \mathbf{H}$ iff either x^1 is an extreme point of **K** in **H**; or else the intersection of **H** with an edge of **K** that does not completely lie in **H**. (2) x^1 and x^2 be two distinct points in $\mathbf{K} \cap \mathbf{H}$. (a) If x^1 and x^2 are extreme points of **K** in **H**, prove that they are adjacent on $\mathbf{K} \cap \mathbf{H}$ iff they are adjacent on **K**. (b) If x^1 and x^2 are extreme points of $\mathbf{K} \cap \mathbf{H}$, at least one of which is not an extreme point of **K**, prove that x^1 and x^2 are adjacent on $\mathbf{K} \cap \mathbf{H}$ iff both x^1 and x^2 are points on the edges of a two-dimensional face of **K** [3.35].

3.62 Let **K** be a nonempty convex polyhedral subset of $\mathbf{R}^n$ satisfying the property that it does not contain any entire straight line as a subset. Prove that **K** has at least one extreme point.

3.63 Let $\mathbf{K} \subset \mathbf{R}^n$ be the set of feasible solutions of the system of linear inequalities: $A_{i.}x \geqq b_i$, $i = 1$ to m. If $\mathbf{K} \neq \varnothing$, prove that there exists a nonempty subset $\mathbf{I} = \{i_1, \ldots, i_r\} \subset \{1, \ldots, m\}$, such that every solution of the system of equations $A_{i.}x = b_i$, $i \in \mathbf{I}$, is in **K**.

3.64 Let **K** be a convex polyhedron defined by the general system of linear constraints: $Ax + Dy = b$, $Ex + Fy \geqq d$, $x \geqq 0$, y unrestricted. If **F** is a face of **K**, prove that **F** is the set of feasible solutions of the system obtained after modifying a subset of the inequality constraints among $Ex + Fy \geqq d$ and $x \geqq 0$ into the corresponding equations, and leaving everything else unchanged.

3.65 Suppose a company has n economic acitivites that it can pursue. Let x_j be the level at which activity j is carried out by the company. Let d_j be the units of fuel (or some form of energy) required per unit level of carrying out activity j, $j = 1$ to n. Let c_j be the profit in monetary units obtained per unit level of carrying out activity j, $j = 1$ to n. The commitments of the company to meet its contractual obligations and the resources available to the company for carrying out the work lead to the following constraints on $x = (x_1, \ldots, x_n)^{\mathrm{T}}$: $Ax = b$, $Dx \leqq p$, $x \geqq 0$. It is required to determine the optimal levels for the various activities, which maximizes the net profit per unit consumption of fuel, subject to the constraints given above. Formulate this problem and show how to transform it into an LP.

3.66 Let $\varnothing \neq \mathbf{K}_1 = \{x\colon Ax \geqq b\}$, $\varnothing \neq \mathbf{K}_2 = \{x\colon Dx \geqq d\}$, where A, D, b, d, are given. Discuss an efficient procedure for checking whether $\mathbf{K}_1 \subset \mathbf{K}_2$.

3.67 Let **K** be the set of feasible solutions of $Ax = b$, $x \geqq 0$, where A, b are given matrices of orders $m \times n$ and $m \times 1$, respectively. Let I be the unit matrix of order n. If $\mathbf{K} \neq \varnothing$, prove that the dimension of **K** is $n - \alpha$, where $\alpha = \text{rank}\ (\{A_{i.}, i = 1 \text{ to } m\} \cup \{I_{t.}\colon t \text{ such that } x_t = 0 \text{ for all } x \in \mathbf{K}\})$.

3.68 Let A be a matrix of order $m \times n$, and b, a column vector in $\mathbf{R}^m$. Let $\mathbf{K}$ be the set of feasible solutions of $Ax \geqq b$. Prove that if $\mathbf{K} \neq \emptyset$, its dimension is $n - \alpha$ where $\alpha = \text{rank}\,\{A_{i.} : i \text{ such that } A_{i.}x = b_i, \text{ for all } x \in \mathbf{K}\}$.

3.69 Let $\mathbf{K}$ be the set of feasible solutions of $Ax \leqq b$, where A is a given matrix of order $m \times n$, and b is a given column vector in $\mathbf{R}^m$. It is known that $\mathbf{K} \neq \emptyset$, and its dimension is n. It is required to find an n-dimensional sphere of maximum radius that is completely contained in $\mathbf{K}$. Formulate this as a linear programming problem. Can the problem of finding the smallest diameter n-dimensional sphere containing $\mathbf{K}$ be posed as a linear programming problem also? Why?

3.70 Consider the following systems:

(3.41)	(3.42)	(3.43)
$Ax = b$	$Ay = 0$	$Ay = 0$
$Dx \geqq d$	$Dy \geqq 0$	$Dy \geqq 0$
$x \geqq 0$	$y \geqq 0$	$\sum_i D_{i.}y + \sum_j y_j = 1$
		$y \geqq 0$

If (3.41) has a feasible solution, prove that it must have a BFS. If 0 is the only feasible solution for (3.42), prove that every feasible solution for (3.41) can be expressed as a convex combination of BFSs of (3.41). Also prove that if (3.41) is feasible, its set of feasible solutions is bounded iff (3.42) has no nonzero feasible solution. If (3.42) has a nonzero feasible solution, prove that every feasible solution of (3.41) can be expressed as the sum of (1) a convex combination of BFSs of (3.41) and (2) a nonnegative combination of BFSs of (3.43).

3.71 Consider the systems given at the bottom, where $x \in \mathbf{R}^{n_1}$, $y \in \mathbf{R}^{n_2}$, A is of order $m_1 \times n_1$, E is of order $m_2 \times n_1$, and the other matrices and vectors are of appropriate orders, respectively, and for any r, e_r is the column vector of all 1s in $\mathbf{R}^r$. We assume that there is at least one unrestricted variable such as y in the system, and at least one inequality constraint or a sign restricted variable such as x (i.e., we assume $n_2 \geqq 1$, either $m_2 > 0$ or $n_1 > 0$). Let $P = \begin{pmatrix} D \\ \cdots \\ F \end{pmatrix}$.

Let (x^1, y^1), (x^2, y^2) be two distinct BFSs of (3.44). Develop a criterion for checking whether (x^1, y^1) and (x^2, y^2) are adjacent extreme points, in terms of the rank of a set of column vectors, as in Section 3.6.2. Also if $(\bar{x}, \bar{y})$ is a feasible solution for (3.44), develop an algorithm for checking whether it is on an edge.

Let $\mathbf{K}$ be the set of feasible solutions of (3.44). Assuming that $\mathbf{K} \neq \emptyset$, prove that (3.44) has a BFS iff the column vectors of P form a linearly independent set.

If $\mathbf{K} \neq \emptyset$, prove that $\mathbf{K}$ is bounded iff (3.45) has no nonzero feasible solution. Also if this holds, prove that the column vectors of P must form a linearly independent set. In this case, prove that every point in $\mathbf{K}$ can be expressed as a convex combination of BFSs of (3.44).

If $\mathbf{K} \neq \emptyset$, the column vectors of P form a linearly independent set, and (3.45) has at least one nonzero feasible solution, prove that every feasible solution of (3.44) can be expressed as a sum of (1) a convex combination of BFSs of (3.44) and (2) a nonnegative combination of BFSs of (3.46).

Now consider the case where $\mathbf{K} \neq \emptyset$, but the set of column vectors of P is a linearly dependent set. Let $\mathbf{L}$ be the subspace of the set of feasible solutions of (3.47) given in page 170. Let $\mathbf{J} = \{j_1, \ldots, j_t\}$ be such that the columns of P for $j \in \mathbf{J}$ form a maximal linearly independent subset of columns of P. So, for each $j \in \{1, \ldots, n_2\} \backslash \mathbf{J}$, y_j is a dependent

(3.44)	(3.45)	(3.46)
$Ax + Dy = b$	$A\xi + D\eta = 0$	$A\xi + D\eta = 0$
$Ex + Fy \geqq d$	$E\xi + F\eta \geqq 0$	$E\xi + F\eta \geqq 0$
$x \geqq 0$	$\xi \geqq 0$	$e_{n_1}^{\mathrm{T}}\xi + e_{m_1}^{\mathrm{T}}(E\xi + F\eta) = 1$
		$\xi \geqq 0$

(3.47)	(3.48)	(3.49)	(3.50)
$A\xi + D\eta = 0$ $E\xi + F\eta = 0$ $\xi = 0$	$Ax + Dy = b$ $Ex + Fy \geqq d$ $y_j = \alpha_j$ for $j \notin \mathbf{J}$ $x \geqq 0$	$A\xi + D\eta = 0$ $E\xi + F\eta \geqq 0$ $\xi \geqq 0$ $\eta_j = 0$ for $j \notin \mathbf{J}$	$A\xi + D\eta = 0$ $E\xi + F\eta \geqq 0$ $\eta_j = 0$ for $j \notin \mathbf{J}$ $e_{n_1}^{\mathrm{T}}\xi + e_{m_1}^{\mathrm{T}}(E\xi + F\eta) = 1$ $\xi \geqq 0$

unrestricted variable in (3.44). In system (3.48), for each $j \notin \mathbf{J}$, α_j is some arbitrarily selected but fixed real number (for example, $\alpha_j = 0$ for all $j \notin \mathbf{J}$). Prove that if (3.49) has no nonzero feasible solution, then every feasible solution of (3.44) can be expressed as a sum of (1) a convex combination of BFSs of (3.48) and (2) a point in the subspace **L**. Also prove that if (3.49) has a nonzero feasible solution, then every feasible solution of (3.44) can be expressed as a sum of (1) a convex combination of BFSs of (3.48), (2) a nonnegative combination of BFSs of (3.50), and (3) a point in the subspace **L**. In general, if **P** denotes the convex hull of all BFSs of (3.48) and **C** denotes the cone of feasible solutions of (3.49), then these results state that in this case $\mathbf{K} = \mathbf{P} + \mathbf{C} + \mathbf{L}$. These results constitute the resolution theorem (also called the *decomposition theorem*) for general convex polyhedra.

3.72 *MULTIOBJECTIVE PROGRAMMING.* Let **K** be the set of feasible solutions of $Ax = b$, $x \geqq 0$, where A is a matrix of order $m \times n$ and rank m. $z_r(x) = c^r x$, $r = 1$ to t, are t different linear objective functions. The objective functions are required to be minimized in the priority order $r = 1$ to t by the sequential procedure discussed in Section 1.4.2. Discuss an efficient method for implementing this procedure so that in each stage we deal only with bases of order m. Apply the method on the following numerical problem: $z_1(x) = x_1 + x_2 + x_3 - x_4 + 2x_5 + 2x_6 + 2x_7 + 9x_8 + 12x_9$, $z_2(x) = -x_1 + 2x_2 + x_3 + x_4 - 2x_5 + 10x_7 - 4x_8$, $z_3(x) = -x_1 + 3x_2 - x_3 + 3x_5 - 2x_6 - 4x_7 - 3x_8$.

x_1	x_2	x_3	x_4	x_5	x_6	x_7	x_8	x_9	b
1	0	0	−1	2	1	−1	2	2	3
0	1	0	1	−1	2	1	−1	2	2
0	0	1	−1	1	−1	2	2	−1	5

$x_j \geqq 0$ for all j.

3.73 *UNRESTRICTED VARIABLES AS PERMANENT BASIC VARIABLES.* Consider the general LP

$$\begin{aligned} \text{Minimize} \quad & z = dy + cx \\ \text{Subject to} \quad & Dy + Ax = b \\ & y \text{ unrestricted} \quad x \geqq 0 \end{aligned} \tag{3.51}$$

where D and A are matrices of order $m \times k$ and $m \times n$, respectively. Without any loss of generality, we assume that there are no redundant constraints in (3.51), that is, that the rank of $(D \vdots A)$ is m. (i) Prove that the number of unrestricted variables that can be eliminated from (3.51), as discussed in Chapter 2, is equal to the rank of D. If the rank of D is k (which happens iff the set of column vectors of D is linearly independent), all the unrestricted variables y can be eliminated and (3.51) can be transformed into an LP involving the nonnegative variables x only. (ii) Assume that the rank of D is r where $r \leqq k$. Transform the system of constraints in (3.51) into row echelon normal form by performing Gauss–Jordan pivot steps on it, choosing an unrestricted variable y_i for some i as a basic variable as far as possible. Show that this leads to a basic vector containing exactly r of the ys as basic variables.

For convenience in notation assume that the basic vector obtained is $(y_1, \ldots, y_r; x_1, \ldots, x_{m-r})$. Rearrange the rows of the row echelon normal form so that the first r rows are the rows in which the basic variables are $y_1, \ldots, y_r$, in that order, and the bottom $(m - r)$ rows are the rows in which the basic variables are $x_1, \ldots, x_{m-r}$. If $r < k$, prove that all the entries in the bottom $(m - r)$ rows in the updated column vector of any nonbasic y-variable are zero.

Suppose $r < k$. Update the cost row (i.e., price out all the basic variables in it) and suppose the canonical tableau with respect to the present basic vector is Tableau 3.4

Tableau 3.4

Basic Variables	$y_1 \cdots y_r$	$y_{r+1} \cdots y_k$	$x_1 \cdots x_{m-r}$	$x_{m-r+1} \cdots x_n$	$-z$	
$y_1 \vdots y_r$	I_r	$\bar{E}$	0	$\bar{P}$	0	$\bar{a}$
$x_1 \vdots x_{m-r}$	0	0	I_{m-r}	$\bar{F}$	0	$\bar{f}$
$-z$	0	$\bar{p}$	0	$\bar{q}$	1	$-\bar{v}$

where I_r and I_{m-r} are unit matrices of orders r and $m-r$, respectively. In this case, if $\bar{p} \neq 0$, prove that z is unbounded below if (3.51) is feasible. Also prove the following: (1) the system (3.51) is feasible iff there is a feasible solution for it in which all the variables $y_{r+1}, \ldots, y_k$ are zero; (2) if (3.51) has an optimum solution, $\bar{p} = 0$ and (3.51) has an optimum solution in which $y_{r+1}, \ldots, y_k$ are all zero; and (3) if $\bar{p} \neq 0$, (3.51) is feasible, and z is unbounded below, there exists a half-line in the set of feasible solutions of (3.51) along which z diverges to $-\infty$, and every point on this half-line satisfies $y_{r+1} = \cdots = y_k = 0$.

So, if $\bar{p} = 0$, we can substitute $y_{r+1} = \cdots = y_k = 0$ and eliminate these variables from the problem. If $\bar{p} \neq 0$, (3.51) is either infeasible or it is feasible and z is unbounded below in it. To determine which of these two possibilities occurs when $\bar{p} \neq 0$, eliminate the objective function and do a Phase I formulation on the remaining problem after setting $y_{r+1} = \cdots = y_k = 0$.

(iii) To solve the remaining problem, proceeding as in Chapter 2, we would solve the LP

Minimize

$$\bar{q}(x_{m-r+1}, \ldots, x_n)^{\mathrm{T}}$$

Subject to

$$(x_1, \ldots, x_{m-r})^{\mathrm{T}} + \bar{F}(x_{m-r+1}, \ldots, x_n)^{\mathrm{T}} = \bar{f}$$
$$x_j \geqq 0 \quad \text{for all } j \qquad (3.52)$$

Prove that solving (3.52) by the simplex method is the same as solving the LP in Tableau 3.4 by the simplex method leaving $y_1, \ldots, y_r$ as permanent basic variables in every basic vector, letting them take any real value in the basic solution. If Phase I is carried out, artificial variables are introduced only in the bottom $m-r$ rows. In carrying out the simplex method, the pivot row would always be one of rows $r+1$ to m in which a nonnegative x-variable or an artificial variable is basic. For carrying out the minimum ratio test, the ratios are only computed in the bottom $(m-r)$ rows.

3.74 Let **K** denote the set of feasible solutions of the general system of linear constraints

$$\begin{aligned} A_{i.}x &= b_i, \qquad i = 1 \text{ to } m \\ &\geqq b_i, \qquad i = m+1 \text{ to } m+p \end{aligned} \qquad (3.53)$$

Any sign restrictions or lower or upper bound constraints on the variables are included among the inequality constraints in (3.53). In (3.53), the matrix of input-output coefficients, A, in of order $(m+p) \times n$. Let $\bar{x} \in \mathbf{K}$. Define $\mathbf{I}(\bar{x}) = \{i: 1 \leqq i \leqq m+p$, and i such that $A_{i.}\bar{x} = b_i\}$. The ith constraint in (3.53) is said to be *active* at $\bar{x}$ iff $i \in \mathbf{I}(\bar{x})$. Thus the active constraints at $\bar{x}$ are all the equality constraints, and all the inequality constraints that hold as equations at $\bar{x}$.

(i) Prove that $\bar{x}$ is an extreme point of **K** or, equivalently, that $\bar{x}$ is a BFS of (3.53), iff $\bar{x}$ is the unique solution for the system of equality constraints:

$$A_{i.}x = b_i, \qquad i \in \mathbf{I}(\bar{x})\}$$

Equivalently, this holds iff the rank of the set $\{A_{i.}: i \in \mathbf{I}(\bar{x})\}$ is n.

(ii) Assume that $\bar{x}$ is an extreme point of **K**. Prove that every edge of **K** containing $\bar{x}$ is of the form $\{\bar{x} + \lambda y: 0 \leqq \lambda \leqq \bar{\lambda}\}$ where

$y \in \mathbf{R}^n$ and $\tilde{\lambda}$ satisfy the following properties. There exists a nonempty subset $\mathbf{J} \subset \mathbf{I}(\bar{x}) \cap \{m+1, \ldots, m+p\}$ such that the rank of $\{A_{i.} : i \in \mathbf{I}(\bar{x}) \backslash \mathbf{J}\}$ is $n-1$, and y satisfies

$$\begin{aligned} A_{i.}y &> 0, \qquad \text{for all } i \in \mathbf{J} \\ &= 0, \qquad \text{for all } i \in \mathbf{I}(\bar{x}) \backslash \mathbf{J} \end{aligned} \tag{3.54}$$

and $\tilde{\lambda} = +\infty$ if $A_{i.}y \geqq 0$ for all $i \in \{m+1, \ldots, m+p\} \backslash \mathbf{I}(\bar{x})$, or $=$ minimum $\{(A_{i.}\bar{x} - b_i)/(-A_{i.}y) : i \in \{m+1, \ldots, m+p\} \backslash \mathbf{I}(\bar{x})$ and satisfies $A_{i.}y < 0\}$ otherwise. (It can be verified that the conditions satisfied by $\mathbf{J}$ imply that once $\mathbf{J}$ is given, the solution y of (3.54) is unique, except for a positive scalar multiplier. It can also be verified that this edge is the set of feasible solutions of: $A_{i.}x = b_i$, for all $i \in \mathbf{I}(\bar{x}) \backslash \mathbf{J}$ and $\geqq b_i$ for all other i.)

(iii) Again, assuming that $\bar{x}$ is an extreme point of $\mathbf{K}$, provide a similar characterization for an r-dimensional face of $\mathbf{K}$ containing $\bar{x}$ for $r \geqq 2$.

(iv) Given $\bar{x} \in \mathbf{K}$, develop the necessary and sufficient conditions for checking whether $\mathbf{K} = \{\bar{x}\}$ and an efficient procedure for verifying whether these conditions hold.

(v) Assuming that $\mathbf{K} \neq \emptyset$, discuss an efficient procedure based on the simplex algorithm for checking whether $\mathbf{K}$ is bounded or not.

(vi) Let $\mathbf{C}$ denote the cone $\{\alpha x : \alpha \geqq 0, x \in \mathbf{K}\}$. Assume that $\mathbf{K} \neq \emptyset$. $\mathbf{C}$ is clearly a convex polyhedral cone and it can be represented as the set of feasible solutions of a homogeneous system of linear inequalities. It is required to find this system. Discuss an efficient procedure for determining it.

(vii) Let $\xi = (\xi_1, \ldots, \xi_q) \in \mathbf{R}^q$ and let $\mathbf{\Gamma}$ be the set of feasible solutions of the system

$$D\xi \geqq d$$

A well-known result in linear algebra states that $\mathbf{\Gamma}$ has dimension q iff there exists a $\bar{\xi} \in \mathbf{\Gamma}$ satisfying $D\bar{\xi} > d$. This result can be used to determine the dimension of a convex polyhedron represented by a system of linear inequalities. For example, consider the convex polyhedron $\mathbf{K}$, defined above. If a linear equality constraint satisfied by all the points in $\mathbf{K}$ is identified, we know that it can be used to eliminate a variable from the system defining $\mathbf{K}$. Using this technique and the result discussed above, develop an efficient procedure based on the simplex algorithm for determining the dimension of $\mathbf{K}$.

3.75 Read each of the following statements carefully to see whether it is *true* or *false*. Justify your answer by constructing a simple illustrative example, if possible, or by a simple proof. In all these statements problem (1) is the LP.

$$\begin{aligned} \text{Minimize} \quad & z(x) = cx \\ \text{Subject to} \quad & Ax = b \\ & x \geqq 0 \end{aligned} \tag{1}$$

where A is a matrix of order $m \times n$, $\mathbf{K}$ is the set of feasible solutions of problem (1), and $f(b)$ is the minimum objective value in this problem, as a function of the right-hand-side constant vector b.

1. The number of positive x_j in any BFS of (1) can never exceed the rank of A.
2. If rank of A is m, and (1) is nondegenerate, every feasible solution in which exactly m variables are positive, must be a BFS.
3. The unboundedness criterion will never be satisfied while solving the Phase I problem.
4. While solving the Phase I problem associated with (1), if the unboundedness criterion is satisfied, (1) must be infeasible.
5. If (1) is feasible, termination occurs in solving the Phase I problem associated with (1) only when all the artificial variables leave the basic vector.
6. In solving (1) by the primal simplex method, the pivot element in each pivot step must be a positive number.
7. If a tie occurs in the pivot row choice during a pivot step while solving (1) by

the simplex algorithm, the BFS obtained after this pivot step is degenerate.

8 While solving (1) by the simplex algorithm, if the BFS in the beginning of a pivot step is degenerate, the objective value remains unchanged in this pivot step.

9 The simplex algorithm moves away from an extreme point of **K** only in a nondegenerate pivot step.

10 In solving an LP by the simplex algorithm, a new feasible solution is generated after every pivot step.

11 A feasible basis for (1) can be found by transforming the system into row echelon normal form.

12 Let the rank of A in (1) be m. If (1) has an optimum solution, there must exist an optimum basis. Every basis consists of m column vectors and a pivot step changes one column vector in a basis. Hence starting from a feasible basis, the simplex algorithm finds an optimum basis in at most m pivot steps.

13 Every feasible solution of (1) in which m variables are positive is a BFS.

14 If $\bar{x}$ and $\tilde{x}$ are two adjacent BFSs of (1), the total number of variables that are positive in either $\bar{x}$ or $\tilde{x}$, or both, is at most $m + 1$.

15 If $\bar{x}$ and $\tilde{x}$ are two BFSs for (1) such that the total number of variables that are positive in either $\bar{x}$ or $\tilde{x}$ or both is at most $m + 1$, then $\bar{x}$ and $\tilde{x}$ must be adjacent on **K**.

16 If $\bar{x}$ is a feasible solution of (1) and the rank of $\{A_{.j}: j, \text{ such that } \bar{x}_j > 0\}$ is one less than its cardinality, $\bar{x}$ lies on an edge of **K**.

17 No feasible solution of (1) in which $m + 2$ or more variables are positive can lie on an edge of **K**.

18 Every nondegenerate BFS of (1) lies on exactly $n - m$ edges of **K**.

19 Every convex set has an extreme point.

20 Every nonempty convex polytope has an extreme point.

21 The number of extreme points of any convex set is always finite.

22 If (1) has an optimum solution, **K** must be a bounded set.

23 In every optimum solution of (1), no more than m variables can be positive.

24 An LP can have more than one optimum solution iff it is degenerate.

25 The total number of optimum feasible solutions of an LP is always finite.

26 The total number of BFSs of any LP is always finite.

27 If an LP has more than one optimum solution, it must have an uncountable number of optimum solutions.

28 The set of optimum solutions of any LP is always a bounded set.

29 If (1) has an optimum solution in which $m + 1$ or more variables are positive, it must have an uncountable number of optimum solutions.

30 If (1) is feasible and there is no semipositive linear combination of columns of A that equals the 0-vector, $f(b)$ is finite.

31 Any consistent system of linear equations is equivalent to a system in which the number of constraints is less than or equal to the number of variables.

32 A system of linear equations has a unique solution iff the number of constraints in the system is equal to the number of variables.

33 A system of n independent linear equations in n unknowns has a unique solution. Likewise a system of n linear inequalities in n unknowns has a unique solution.

34 A straight line in $\mathbf{R}^2$ is the set of solutions of a single linear equation. Likewise a straight line in $\mathbf{R}^n$ is the set of solutions of a single linear equation.

35 If all the points on a straight line in $\mathbf{R}^n$ lie in a convex polyhedron Γ, that straight line contains no extreme points of Γ.

36 Let Γ be the convex hull of $\{x^1, \ldots, x^r\}$. The set of extreme points of Γ is a subset of $\{x^1, \ldots, x^r\}$.

37 If $\mathbf{K} \neq \emptyset$, its dimension is $n - r$ or less, where r is the rank of A.

38 If every edge of **K** contains exactly two extreme points, **K** must be bounded.

39 If $n = m + 1$ (i.e., (1) is a system of m equations in $m + 1$ nonnegative variables), then the number of BFS's of (1) is at most 2.

40 The number of edges of **K** through an extreme point of **K** is always greater than or equal to the dimension of **K**.

41 If b is nondegenerate, the number of edges of **K** through any extreme point of **K** is exactly equal to the dimension of **K**.

42 If **K** is unbounded there must exist a semipositive homogeneous solution corresponding to (1).

43 The set of homogeneous solutions corresponding to (1) is a subset of **K**.

44 In a pivot step of the simplex algorithm, if the minimum ratio is zero, there is no change in the objective value. In general, the minimum ratio in a pivot step is the amount by which the objective value changes in that step.

45 Consider a step in solving (1) by the simplex algorithm in which x_j is the entering variable. The minimum ratio in this step is the value of x_j in the solution that will be obtained after this pivot step.

46 Let the rank of A be m. In the course of solving (1) by the simplex method, let $\bar{A}_{.j}$ be the updated column vector of x_j in some stage. Then $\bar{A}_{.j}$ is the column vector of the coefficients in an expression of $A_{.j}$ as a linear combination of the columns of the basis in that stage.

47 In solving (1) by the simplex algorithm, if x_j is the entering variable during a pivot step, then the pivot matrix corresponding to the pivot in that step differs from the unit matrix only in the jth column.

48 A company's product mix model is an LP with a single constraint in nonnegative variables. If a maximum profit solution exists, there is a maximum profit solution in which the company manufactures only one product.

49 A company's product mix model turns out to be an LP model involving 1000 activities and 20 items. If there is an optimum solution for this problem, there is an optimum solution in which the company performs at most 20 of the 1000 possible activities at positive levels.

50 A problem in which a separable piecewise linear concave function has to be minimized subject to linear constraints can be transformed into an LP.

51 The problem of minimizing the pointwise maximum of a pair of linear functions on a convex polyhedral set can always be posed as an LP.

52 Consider the problem of minimizing the maximum $\{c^1x, c^2x\}$ subject to $Ax = b$, $x \geqq 0$. This can be transformed into an LP. So if this problem has an optimum feasible solution, it must have an optimum feasible solution that is a BFS of $Ax = b$, $x \geqq 0$.

53 As long as the parameters to be estimated in a curve-fitting problem appear linearly in the formula for the curve, and the constraints, if any, on these parameters are linear, the values of these parameters that give the best fit can be obtained by an LP formulation for minimizing the L_1 or L_∞ measures of deviation.

54 The purpose of Phase I in the simplex method is to find a BFS.

55 The big-***M*** method for solving (1) combines the problem of finding an initial BFS and then an optimum BFS into a single application of the simplex algorithm. The usual simplex method using Phase I, II applies the simplex algorithm twice, once on Phase I and then on Phase II. Hence the Big ***M*** method solves (1) with half the computational effort required by the Phase I, II simplex method for solving the same problem.

56 If (1) is solved by the Big-***M*** approach, where ***M*** is given a fixed positive value at the beginning of the algorithm (instead of being treated as a parameter having an arbitrarily large positive value), it is possible that some artificial variables will have positive values in the final solution obtained.

57 If the unboundedness criterion is satisfied when (1) is being solved by the big-M method, then (1) cannot have an optimum solution.

58 If (1) is solved by the big-M approach (where M is treated as an arbitrary large positive parameter), using some pivot column choice rule, the basic vectors obtained will be the same as those obtained when (1) is solved by the usual Phase I, II approach, using the same pivot column choice rule.

59 If the objective function in the Big-M problem corresponding to an LP is unbounded below, the objective function in the original LP must itself be unbounded below.

60 It is always possible to set up the Phase I problem corresponding to an LP using only one artificial variable.

61 Since the aim in Phase I is to get rid of the artificial variables from the basic vector, a Phase I formulation for an LP using only one artificial variable can be solved with much less computational effort than a Phase I formulation for the same LP using many artificial variables.

62 If A contains the unit matrix of order m as a submatrix, and $b \ngeq 0$, it is possible to set up a Phase I problem corresponding to (1) using one artificial variable only.

63 If a subset $\mathbf{S} \subset \mathbf{R}^n$ is nonconvex, it is possible to find a point outside of $\mathbf{S}$ that can be expressed as a convex combination of a pair of points in $\mathbf{S}$.

64 The computational effort required by the simplex algorithm for solving the Phase I problem is directly proportional to the number of artificial variables introduced.

65 To set up the Phase I problem corresponding to (1), m artificial variables are always needed.

66 During Phase I, an artificial variable is never chosen as the entering variable.

67 When (1) is solved by the two-phase simplex method, Phase I finds an initial feasible solution of (1), and starting from that, Phase II walks to the optimum solution. Since it is easier to find an arbitrary feasible solution than an optimum solution for (1), the number of pivot steps to be performed in Phase I will always be less than the number of pivot steps required in Phase II.

68 In carrying out Phase I with a full artificial basis on (1), in each pivot step one artificial variable drops out of the basic vector.

69 In order to apply the simplex algorithm on a linear programming problem, we must first have a basic feasible solution of the problem.

70 A feasible solution for (1) can be found by the Gauss–Jordan or the Gaussian elimination method.

71 Any system of linear inequalities can always be transformed into a system of linear equations by introducing the appropriate slack variables, and then solved by a straightforward application of Gaussian elimination method.

72 Whenever there is an equality constraint in an LP model, we can use it to eliminate one variable from the model and also reduce the number of constraints by one.

73 Whenever there is an unrestricted variable in an LP model, by eliminating it, we can reduce this LP to another of smaller order. On the other hand, eliminating a sign-restricted variable from an LP does not, in general, reduce the order of the problem.

74 Unrestricted variables can always be eliminated from a linear programming model, but any variable restricted to be nonnegative cannot be eliminated in a similar manner.

75 If the system of linear equations $Ax = b$ has a solution, it must have a nonnegative solution.

76 If rank of A is m, (1) always has a solution (though not necessarily a feasible solution).

77 Pos(A) is a cone in $\mathbf{R}^m$, consisting of the set of all vectors b for which (1) has a feasible solution.

78 The constraints in (1) can be posed as a system of $(r + 1)$ linear inequalities, where $r = \text{rank}(A)$.

79 If (1) has at least one solution (not necessarily feasible), then the set of solutions for (1) is a translate of a subspace of $\mathbf{R}^n$ of dimension $n - r$ where $r = \text{rank}(A)$.

80 Suppose an LP is put into standard form by using the transformations discussed in Chapter 2. Every extreme point of the standard form, corresponds to a unique extreme point of the original LP and vice versa.

81 The set of solutions of a system of n linearly independent linear equations in $n + 1$ unknowns is a straight line.

82 The system of equations $Ax = b$ has a solution x iff the inclusion of b as an additional column vector in A would not increase its rank.

83 If the set of row vectors of a square matrix form a linearly independent set, the set of column vectors of this matrix must also be linearly independent.

84 If the set of row vectors of an arbitrary real matrix forms a linearly independent set, the set of column vectors of this matrix must also be linearly independent.

85 The set of solutions of a system of linear equations remains unchanged when an equation is multiplied by a real number and added to another equation.

86 The set of solutions of a system of linear inequalities of $\geqq$ type remains unchanged if one inequality is multiplied by a positive number and added to another.

87 A straight line is the convex hull of any pair of distinct points on it.

88 The affine hull of a set of vectors, Δ, is a subset of the convex hull of Δ.

89 A straight line in $\mathbf{R}^n$ is the linear hull of any pair of distinct points on it.

90 If $\{x^1, \ldots, x^r\}$ is a linearly independent set of vectors in $\mathbf{R}^n$, the linear hull of $\{x^1, \ldots, x^r\}$ has dimension r, and the affine hull of $\{x^1, \ldots, x^r\}$ has dimension $r - 1$.

91 Let $\{x^1, \ldots, x^r\}$ be a subset of points in $\mathbf{R}^n$. The affine hull of $\{x^1, \ldots, x^r\}$ and the convex hull of $\{x^1, \ldots, x^r\}$ have the same dimension.

92 If $\{x^1, \ldots, x^n\}$ is a basis for $\mathbf{R}^n$ and y is another vector in $\mathbf{R}^n$, y can replace any of the x^j, to yield another basis for $\mathbf{R}^n$.

93 The rank of a matrix and the rank of its transpose are always equal.

94 The intersection of any pair of convex subsets of $\mathbf{R}^n$ is a convex set.

95 The union of any pair of convex subsets of $\mathbf{R}^n$ is a convex set.

96 A convex polytope is the convex hull of its extreme points.

97 If an additional constraint is imposed on an LP model, the effect is to make the set of feasible solutions smaller.

98 A BFS of a system of linear equations and inequalities always corresponds to an extreme point of the set of feasible solutions.

99 It is possible for $\mathbf{K}$ to be nonempty, and yet not have any extreme points.

100 The number of positive variables in any nondegenerate BFS of (1) is always equal to the rank of $\mathbf{A}$.

101 Every feasible solution for (1) can be expressed as a convex combination of BFSs of (1).

102 Every optimum solution for (1) can be expressed as a sum of (i) a convex combination of optimum BFSs of (1), and (ii) a nonnegative combination of extreme homogeneous solutions associated with (1) that satisfy $cy = 0$.

103 Let $\mathbf{S}$ be a set of vectors from $\mathbf{R}^n$. If the rank of $\mathbf{S}$ is r, any subset of $r + 1$ vectors from $\mathbf{S}$ is linearly dependent.

104 Every edge of a convex polyhedron contains two extreme points of the convex polyhedron.

105 A BFS of $Ax \geqq b$, $x \geqq 0$ is a feasible solution $\bar{x}$ satisfying the property that the set $\{A_{.j}$: j, such that $\bar{x}_j > 0\}$ is linearly independent.

106 If a system of linear equations and/or inequalities has a feasible solution, it has a BFS.

107 If $\mathbf{K} \neq \emptyset$, $\mathbf{K}$ is bounded iff $Ax = 0$, $x \geqq 0$ has no nonzero solution.

108 If (1) has at least one feasible solution, the dimension of the set of feasible solutions for (1) is $n - r$, where $r =$ rank (A).

109 Extreme points of (1) are faces of $\mathbf{K}$ of dimension 0.

110 A convex cone that contains a point other than the origin cannot be a convex polytope.

111 Suppose rank (A) is m. Then every primal feasible basis for (1) is associated with a unique BFS for (1) and vice versa.

112 $\mathbf{K}$ is bounded iff it has no extreme half-lines.

113 Suppose rank (A) is m. Then every BFS of (1) has exactly $n - m$ adjacent extreme points on $\mathbf{K}$.

114 Let $\bar{x}$ be a feasible solution for (1) and let $\mathbf{J} = \{j: \bar{x}_j > 0\}$. $\bar{x}$ is on an edge of $\mathbf{K}$ iff the system $\sum_{j \in \mathbf{J}} \alpha_j A_{.j} = 0$ has a unique nonzero solution $(\alpha_j: j \in \mathbf{J})$ (unique up to a scalar multiple).

115 Either (1) has an optimum feasible solution for all $b \in \text{Pos}(A)$, or $z(x)$ is unbounded below in (1) whenever $b \in \text{Pos}(A)$.

116 If the cost vector c is a linear combination of the row vectors of A, every feasible solution for (1) is an optimum feasible solution for (1).

117 Problem (1) has an optimum feasible solution iff $\mathbf{K} \neq \emptyset$ and $z(x)$ is bounded below on $\mathbf{K}$. Hence the set of optimum feasible solutions for (1) is always a bounded set.

118 Suppose $\bar{z}$ is the finite optimum objective value in (1). Consider the system

$$\begin{aligned} Ax &= b \\ cx &= \bar{z} \qquad (2) \\ x &\geqq 0 \end{aligned}$$

which consists of one more constraints over those in (1). Every extreme point for (2) is also an extreme point for (1).

119 Let $\{x^1, \ldots, x^k\}$ be a set of points in $\mathbf{R}^n$. The convex hull of $\{x^1, \ldots, x^k\}$ is always a subset of the affine hull of $\{x^1, \ldots, x^k\}$.

120 The convex hull of a set consisting of a finite number of points in $\mathbf{R}^n$ is always a bounded set.

121 A convex polyhedron always has a finite number of extreme points.

122 A degenerate BFS of (1) is usually associated with many alternate feasible bases for (1).

123 The minimum value of a linear function on a convex polyhedral set always occurs at an extreme point.

124 Among all the pivot column choice rules that can be used in the simplex algorithm, the steepest descent rule always leads to a solution of the problem with the least amount of computational effort.

125 If the *maximum improvement criterion* is being used to select the entering variable in the simplex algorithm, and the actual improvement in a step happens to be zero, then the BFS in that step must be an optimum solution.

126 If (1) is degenerate, all the BFSs obtained while solving (1) by the simplex algorithm will be degenerate.

127 If (1) is nondegenerate during Phase II of the simplex method for solving (1), the pivot row is uniquely identified by the minimum ratio test in every pivot step.

128 If (1) is degenerate, it is possible that the simplex algorithm fails to solve (1), unless special precautions are taken to resolve degeneracy.

129 If (1) has alternate optimum feasible solutions, it has at least two BFSs that are optimal.

130 If the BFS at the beginning of a pivot step in the simplex algorithm is nondegenerate, that pivot step will be a nondegenerate pivot step.

131 If there is no change in the objective value during a pivot step in the simplex algorithm, the current feasible solution remains unchanged during that pivot step.

132 If there is no change in the objective value during several consecutive pivot steps in the simplex algorithm, the current feasible solution must be very close to the optimum.

133 In the LP (1), degeneracy arises iff there is a redundant constraint in (1).

134 If $c \geqq 0$, while solving (1), the unboundedness criterion will never be satisfied.

135 If $c_1 < 0$ and $a_{11}, \ldots, a_{1n}$ are all $\leqq 0$, the objective function is unbounded below in the LP (1).

136 If $c_j < 0$ for all j, the objective function must be unbounded below in the LP (1).

137 In an LP model, there are finite lower- and upper-bound constraints on each decision variable. The unboundedness criterion will not be satisfied while solving this LP.

138 The unboundedness criterion is satisfied in a step of the simplex algorithm if the canonical tableau in that step contains a column that is nonpositive.

139 While solving (1) by the simplex algorithm, the unboundedness criterion will only be satisfied when the algorithm has identified a homogeneous solution corresponding to (1) at which the objective function assumes a negative value, and vice versa.

140 Let x_B be a basic vector for an LP in standard form, and let x_D be the corresponding vector of nonbasic variables. It is possible to eliminate the variables x_B from the problem and represent this LP purely in terms of the nonbasic variables x_D only. The gradient vector of the objective function in this representation is the vector of relative cost coefficients of the nonbasic variables with x_B as the basic vector.

141 While solving (1) by the simplex algorithm, if a variable drops out of the basic vector at some stage, we can erase that variable from the tableau and conclude that its value is zero in an optimum solution of (1).

142 While solving (1) by the simplex algorithm, if x_1 is the dropping variable in some pivot step, just after this pivot step the relative cost coefficient of x_1 will be strictly positive.

143 It is possible to operate the simplex algorithm for solving (1) so that once the relative cost coefficient of a variable becomes nonnegative in the algorithm, it remains nonnegative in all subsequent steps.

144 The steepest descent entering variable choice rule in the simplex algorithm guarantees that in each step of the algorithm, the objective value decreases by the maximum possible amount.

145 If a variable is the entering variable in some step of the simplex algorithm for solving (1), that variable must be in the final optimum basic vector.

146 A variable that drops from the basic vector in some step of the simplex algorithm for solving (1) may be chosen as the entering variable into the basic vector at a later step.

147 Let $x_{B_1} = (x_1, \ldots, x_m)$ be a feasible basic vector for (1) and suppose we know that when any of the nonbasic variables $x_{m+1}, \ldots, x_n$ is brought into x_{B_1}, we do not decrease the objective value. Then the BFS corresponding to x_{B_1} must be optimal to (1).

148 An eligible variable in a step of the simplex algorithm is any variable x_j, whose relative cost coefficient is $\leqq 0$ in that step.

149 When solving an LP by the simplex algorithm, the objective value strictly decreases in every pivot step.

150 The simplex algorithm chooses the entering variable in a pivot step and makes its value positive in the solution only because doing so leads to a solution with a better objective value. Hence, once a variable is chosen as the entering variable in a step, it stays in the basic vector in all subsequent steps.

151 During the simplex algorithm, a variable drops out of the basic vector in a nondegenerate pivot step if its value drops to zero from its previous positive value. Once this happens, the value of this variable remains equal to zero in all subsequent steps of the algorithm.

152 A feasible solution $\bar{x}$ of (1) is on an edge of **K** iff the rank of $\{A_{.j}\, j$, such that $\bar{x}_j > 0\}$ is either equal to its cardinality or one less than the cardinality.

153 If all the extreme points of **K** on a face of **K** are optimal to (1), then every point in that face must be an alternate optimal solution of the LP.

154 If there exists one nonzero homogeneous solution corresponding to (1), then (1) has at least one feasible solution.

155 The dimension of a convex polyhedron is always less than or equal to the number of its edges through any of its extreme points.

156 The dimension of **K** is $n - m$.

157 A system of n linear equations in n unknowns always has a unique solution.

158 A system of n linear equations in $n + 1$ nonnegative variables has at most two BFSs.

159 The set of optimum feasible solutions of any LP is always a face of the set of feasible solutions of this LP.

160 If (1) has alternate optimum solutions, it must be degenerate.

161 Every convex combination of any pair of optimum solutions of an LP is also optimal.

162 If a point in the relative interior of an edge of **K** is optimal for (1), then every point on that edge is an alternate optimum solution for (1).

163 If there exists no nonzero homogeneous solution corresponding to (1), the unboundedness criterion will not be satisfied when (1) is solved.

REFERENCES

References On Fractional Programming

3.1 C. R. Bector, "Programming Problems with Convex Fractional Functions," *Operations Research 16*, 1968, pp. 383–391.

3.2 A. Charnes and W. W. Cooper, "Nonlinear Power of Adjacent Extreme Point Methods in Linear Programming," *Econometrica 25*, 1956, pp. 132–153.

3.3 A. Charnes and W. W. Cooper, "Programming with Linear Fractional Functionals," *Naval Research Logistics Quarterly 9*, 1962, pp. 181–186.

3.4 W. Dinkelbach, "On Nonlinear Fractional Programming," *Management Science 12*, 1966, pp. 609–615.

3.5 R. Jagannathan, "On Some Properties of Programming Problems in Parametric Form Pertaining to Fractional Programming," *Management Science 12*, 1966, pp. 609–615.

3.6 H. C. Joksch, "Programming with Fractional Linear Objective Functions," *Naval Research Logistics Quarterly 11*, 1964, pp. 197–204.

3.7 B. Martos, "Hyperbolic Programming," *Naval Research Logistics Quarterly 11*, 1964, pp. 135–155.

3.8 B. Martos, "The Direct Power of Adjacent Vertex Programming Methods," *Management Science 12*, 1965, pp. 241–252.

Selected References on Linear Algebra

3.9 D. Gale, *The Theory of Linear Economic Models*, McGraw-Hill, New York, 1960.

3.10 D. K. Faddeev and V. N. Faddeeva, *Computational Methods of Linear Algebra*, Freeman San Francisco, Calif. 1963.

3.11 P. R. Halmos, *Finite Dimensional Vector Spaces*, D. Van Nostrand, New York, 1969.

3.12 B. Noble and J. W. Daniel, *Applied Linear Algebra*, 2nd Ed., Prentice-Hall, Englewood Cliffs, N.J., 1977.

3.13 D. I. Steinberg, *Computational Matrix Algebra*, McGraw-Hill, New York, 1974.

3.14 R. M. Thrall and L. Tornheim, *Vector Spaces and Matrices,* Wiley, New York, 1957.

3.15 R. S. Varga, *Matrix Iterative Analysis*, Prentice-Hall, Englewood Cliffs, N.J., 1962.

Other References

3.16 I. Adler, "Equivalent Linear Programs," Dept. of IE & OR, University of California, Berkeley, 1976.

3.17 M. L. Balinski, "An Algorithm for Finding all Vertices of Convex Polyhedral Sets," *SIAM Journal on Applied Mathematics IX*, 1961, pp. 72–88.

3.18 C. A. Burdet, "Generating all the Faces of a Polyhedron," *SIAM Journal on Applied Mathematics XXVI*, 1974, pp. 479–489.

3.19 N. V. Chernikova, "Algorithm for Finding a General Formula for the Nonnegative Solutions of a System of Linear Equations," *U.S.S.R. Computational Mathematics and Mathematical Physics IV*, no. 4, 1964, pp. 151–158.

3.20 N. V. Chernikova, "Algorithm for Finding a General Formula for the Nonnegative Solutions of a System of Linear Inequalities," *U.S.S.R. Computational Mathematics and Mathematical Physics V*, no. 2, 1965, pp. 228–233.

3.21 G. B. Dantzig and B. C. Eaves, "Fourier-Motzkin Elimination and its Dual," *Journal of Combinatorial Theory, Series A XIV* 1973, pp. 228–237.

3.22 R. J. Duffin, "On Fourier's Analysis of Linear Inequality Systems," *Mathematical Programming Study 1*, Elsevier, New York, 1974.

3.23 M. E. Dyer and L. G. Proll, "An Algorithm for Determining all Extreme Points of a Convex Polytope," *Mathematical Programming 12*, 1977, pp. 81–96.

3.24 U. Eckhardt, "Theorems on the Dimension of Convex Sets," *Linear Algebra and Its Applications 12*, 1975, pp. 63–76.

3.25 J. B. J. Fourier, *Mémoires de l'Academie Royale des Sciences de l'institute de France 7*, 1824, pp. XIVII–IV; also, *Second Extrait* in G. Darboux (Ed.) *Oeuvres de Fourier 2*, 1890, pp. 324–328.

3.26 A. J. Goldman, "Resolution and Separation Theorems for Polyhedral Convex Sets" in *Linear Inequalities and Related Systems*, H. W. Kuhn and A. W. Tucker (Eds.), Princeton University Press, Princeton, N.J., 1956.

3.27 I. Grattan-Guinness, "Joseph Fourier's Anticipation of Linear Programming," *Operational Research Quarterly 21*, 3, 1970, pp. 361–364.

3.28 B. Grunbaum, *Convex Polytopes*, Wiley, New York, 1967.

3.29 D. A. Kohler, "Translation of a Report by Fourier on His Work on Linear Inequalities," *OPSEARCH 10*, 1973, pp. 38–42.

3.30 M. Manas and J. Nedoma, "Finding all Vertices of a Convex Polyhedron," *Numerische Mathematic 12*. 1968, pp. 226–229.

3.31 T. H. Matheiss and D. S. Rubin, "A Survey and Comparison of Methods for Finding all Vertices of Convex Polyhedral Sets," *Mathematics of Operations Research 5*, 2 May 1980, 167–185.

3.32 P. G. McKeown, "A Vertex Ranking Procedure for Solving the Linear Fixed Charge Problem," *Operations Research 23*, 6, 1975, pp. 1183–1191.

3.33 C. B. Millham, "Fast Feasibility Methods for Linear Programming," *OPSEARCH 13*, 3–4, 1976, pp. 198–204.

3.34 K. G. Murty, "Solving the Fixed Charge Problem by Ranking the Extreme Points," *Operations Research 16*, 1968, pp. 268–279.

3.35 K. G. Murty, "Adjacency on Convex Polyhedra," *SIAM Review 13*, 3 (July 1971), 377–386.

3.36 H. Raiffa, G. L. Thompson, and R. M. Thrall, "The Double Description Method," in H. W. Kuhn and A. W. Tucker (Eds.), *Contributions to the Theory of Games II*, *Annals of Mathematics Study No. 28*, Princeton University Press, Princeton, N.J., 1953.

3.37 D. S. Rubin, "Neighboring Vertices on Convex Polyhedral Sets," Graduate School of Business Administration, University of North Carolina at Chapel Hill, 1972.

3.38 A. Shefi, "Reduction of Linear Inequality Constraints and Determination of all Feasible Extreme Points," Ph.D. dissertation, Department of Engineering-Economic Systems, Stanford University, Stanford, Calif., October 1969.

3.39 D. W. Walkup and R. J.-B. Wets, "Lifting Projections of Convex Polyhedra," *Pacific Journal of Mathematics* *28*, 2, 1969, pp. 465–475.

3.40 R. J.-B. Wets and C. Witzgall, "Towards an Algebraic Characterization of Convex Polyhedral Cones, *Numerische Mathematik* *12*, 1968, pp. 134–138.

chapter 4

Duality in Linear Programming

4.1 INTRODUCTION

When an optimum solution of an LP is obtained by the simplex algorithm, the updated cost row in the final canonical tableau is nonnegative. This updated cost row is obtained by subtracting suitable multiples of the constraint rows from the original cost row. For the LP (3.16), if B is an optimum basis and $\bar{c}$ is the updated cost row with respect to it, it was shown in Section 3.12 that $\bar{c} = c - (c_B B^{-1})A$. By treating these multiplying coefficients as variables, another LP known as the *dual problem* associated with the original problem (which in this context is called the *primal problem*) can be constructed. The variables in the dual problem are multipliers associated with the constraints in the primal problem.

Principles of duality appear in various branches of mathematics, physics, and statistics. In linear programming, duality theory turns out to be of great practical use. Also, duality in linear programming admits an elegant economic interpretation.

4.2 EXAMPLES OF DUAL PROBLEMS

The fact that the dual problem arises from economic considerations is illustrated by the following diet problem of the *cost minimizing family* and the *profit maximizing vitamin pill manufacturer.* A family is trying to make a *minimal cost diet* from six available primary foods (called 1, 2, 3, 4, 5, 6) so that the diet contains at least 9 units of vitamin A and 19 units of vitamin C. Here are the data on the foods.

Nutrient	*Number of Units of Nutrients per kilogram of Food* 1	2	3	4	5	6	*Minimum Daily Requirement (MDR) of Nutrient (in Units)*
Vitamin A	1	0	2	2	1	2	9
Vitamin C	0	1	3	1	3	2	19
Cost of food (cents/kg)	35	30	60	50	27	22	

To formulate the family's diet problem suppose the diet consists of x_j kg of food j, $j = 1$ to 6. The constraints are: vitamin A content $= x_1 + 2x_3 + 2x_4 + x_5 + 2x_6 \geqq 9$, vitamin C content $= x_2 + 3x_3 + x_4 + 3x_5 + 2x_6 \geqq 19$, and $x_j \geqq 0$ for all j. The objective function to be minimized is the cost of the diet $z(x) = 35x_1 + 30x_2 + 60x_3 + 50x_4 + 27x_5 + 22x_6$. In tabular form here is the family's diet problem.

The Family's Diet Problem

x_1	x_2	x_3	x_4	x_5	x_6		
1	0	2	2	1	2	$\geqq 9$	
0	1	3	1	3	2	$\geqq 19$	
35	30	60	50	27	22	$= z(x)$	minimize

And $x_j \geqq 0$ for all j.

(4.1)

A manufacturer visualizes the scope for starting a new business in this situation. The manufacturer proposes to make synthetic pills of each nutrient and to sell them to the family. For the business to thrive the manufacturer has to persuade the family to meet all the nutrient requirements by using the pills instead of the primary foods. However, since the family is cost conscious, they will not use the pills unless the manufacturer can convince them that the prices of the pills are competitive in comparison with each of the available primary foods. This imposes several constraints on the prices the manufacturer can charge for the pills. Let π_1, π_2 be the prices of vitamin

A and C respectively in pill form (cents per unit). Consider one primary food, say food 5. One kilogram of it contains one unit of vitamin A and three units of vitamin C. In terms of the manufacturer's prices, its intrinsic worth is therefore $\pi_1 + 3\pi_2$. One kilogram of this food costs 27 cents. The family will obviously compare these two quantities and unless $\pi_1 + 3\pi_2 \leqq 27$, they will conclude that the manufacturer's prices are not competitive. Hence, to win their business, the manufacturer's price vector must satisfy the above constraint. A similar constraint on (π_1, π_2) arises from considering each primary food.

Since the manufacturer is trying to make money and not give it away, both π_1 and π_2 should be nonnegative. Also, since the family is cost conscious, if they decide to use the pills instead of the primary foods, they will buy just as many pills as are required to satisfy the minimal nutrient requirements exactly (buying more would cost extra money since the prices are nonnegative). Hence, the manufacturer's sales revenue will be $v(\pi) = 9\pi_1 + 19\pi_2$; and the manufacturer would like to see this maximized. Thus the prices that the manufacturer can charge for the nutrient pills are obtained by solving the problem:

The Pillmaker's Problem

π_1	π_2		
1	0	$\leqq 35$	
0	1	$\leqq 30$	
2	3	$\leqq 60$	
2	1	$\leqq 50$	(4.2)
1	3	$\leqq 27$	
2	2	$\leqq 22$	
9	19	$= v(\pi)$, maximize	

And $\pi_i \geqq 0$ for all i.

The following facts are clear:

1 The input-output coefficient tableau corresponding to the manufacturer's problem in (4.2) is just the transpose of the input-output tableau of the family's diet problem in (4.1) and vice versa.

2 The right-hand-side constants in (4.1) are the coefficients of the objective function in (4.2) and vice versa.

3 Each column vector of (4.1) leads to a constraint in (4.2) and vice versa.

4 There is a variable in (4.2) corresponding to each constraint in (4.1), and (4.2) contains a constraint corresponding to each variable in (4.1).

5 (4.1) is a minimization problem in which the constraints are $\geqq$ type, and (4.2) is a maximization problem, in which the constraints are $\leqq$ type.

6 From the arguments of the family it is clear that they will not buy the pills unless they provide the required nutrients just as cheaply as the primary foods. This seems to suggest that

$$\text{The maximum value of } v(\pi) \leqq \text{the minimum value of } z(x) \tag{4.3}$$

This is in fact true. This follows from the weak duality theorem to be proved later. The LPs (4.1) and (4.2) are the *duals of each other*. The inequality (4.3) is one relationship between them.

In the pill manufacturer's problem, the nonnegative vitamin A price π_1 is associated with the nonnegative primal slack variable $x_1 + 2x_3 + 2x_4 + x_5 + 2x_6 - 9$, and together they form a pair $(x_1 + 2x_3 + 2x_4 + x_5 + 2x_6 - 9, \pi_1)$ that is known as a *complementary pair* in these LPs, which are in duality relationship. Similarly, the pair $(x_2 + 3x_3 + x_4 + 3x_5 + 2x_6 - 19, \pi_2)$ is another complementary pair. Likewise the nonnegative variable x_1 in the family's diet problem, is associated with the nonnegative slack variable $35 - \pi_1$ in the pill manufacturer's problem, and $(x_1, 35 - \pi_1)$ form another complementary pair. In a similar way $(x_2, 30 - \pi_2)$, $(x_3, 60 - 2\pi_1 - 3\pi_2)$, $(x_4, 50 - 2\pi_1 - \pi_2)$, $(x_5, 27 - \pi_1 - 3\pi_2)$, and $(x_6, 22 - 2\pi_1 - 2\pi_2)$ form other complementary pairs.

If π_1, the price of vitamin A, is strictly positive in an optimum solution of the pill manufacturer's problem, it seems intuitively reasonable to expect that in a minimum-cost diet for the family the requirements of Vitamin A will be met exactly, that is the vitamin A slack will be zero at an optimum solution of the family's diet problem. Similar reasoning suggests that at optimum solutions of the family's and the pill manufacturer's problems, at least one quantity in each complementary pair will be zero. This result is in fact true. Similar results hold in any pair of LPs in duality relationship (see Section 4.5.8).

In an LP model, the rate of change in the optimum objective value per unit change in the right-hand-side constant in one of the constraints from its present value (while all the other data in the problem remain unchanged) is known as the *marginal value or marginal rate of the item corresponding to that constraint,* when these rates exist and are uniquely determined. For example, in the family's diet problem (4.1), the marginal value of a vitamin represents the amount of extra money the family has to spend by using an optimum diet, per unit increase in the requirement of that vitamin, and this marginal value therefore represents the cost of one unit of that vitamin in the family's diet budget under the present conditions. Hence the price vector charged by the pillmaker per unit of the various vitamins, will be acceptable to the family if the price of each vitamin is less than or equal to the marginal value of that vitamin in (4.1). Also, since the pillmaker would like to make the price per unit of each vitamin as large as possible, that price will be exactly equal to that marginal value. Thus in an optimum solution of the dual problem (4.2), the π_1 and π_2 correspond to marginal values of vitamins A and C respectively. In the same manner, in any LP, the dual variables are the marginal values of the items associated with the constraints in the primal problem. These marginal values depend, of course, on the data, and may change if the data change.

The Dual of the Fertilizer Manufacturer's Problem

Consider the fertilizer manufacturer's problem discussed in Section 1.1.2. The fertilizer manufacturer gets a supply of 1500 tons of raw material 1, 1200 tons of raw material 2, and 500 tons of raw material 3 per month from the company's quarries. The fertilizer manufacturer can use these raw materials to manufacture Hi-phosphate, or Lo-phosphate fertilizers, to make a profit. The data for the problem are given in Section 1.1.2.

There is a big *detergent manufacturer* in the area that needs raw materials 1, 2, 3 for the detergent company. The detergent manufacturer wants to persuade the fertilizer manufacturer to give up the fertilizer making business, and instead sell the supply of raw materials to the detergent company. Being very profit conscious, the fertilizer manufacturer will not agree to this deal unless the prices offered by the detergent manufacturer for the raw materials fetch as much income as each of the options in the fertilizer manufacturing business. Let π_i = price offered by the detergent manufacturer per ton of raw material i (\$/ton), for $i = 1, 2, 3$. Obviously these prices have to be nonnegative. Now consider the Hi-phosphate fertilizer. One ton of it yields a profit of \$15 for the fertilizer manufacturer, and uses up 2 tons of raw material 1, 1 ton of raw material 2, and 1 ton of raw material 3. The same basket of raw materials fetches a price of $2\pi_1 + \pi_2 + \pi_3$ from the detergent manufacturer. Being profit conscious, the fertilizer manufacturer will not find the price vector $\pi = (\pi_1, \pi_2, \pi_3)$ acceptable unless $2\pi_1 + \pi_2 + \pi_3 \geqq 15$. Similar economic analysis with the Lo-phosphate fertilizer leads to the constraint $\pi_1 + \pi_2 \geqq 10$. At the price vector π, the cost to the detergent company, of acquiring the raw material supply available monthly is $1500\pi_1 + 1200\pi_2 + 500\pi_3$, and the detergent manufacturer would like to see this minimized. Thus the prices of the detergent manufacturer are obtained by solving the following LP.

Detergent Manufacturer's Problem

π_1	π_2	π_3	
2	1	1	$\geqq 15$
1	1	0	$\geqq 10$
1500	1200	500	$= v(\pi)$, minimize

And $\pi_i \geqq 0$ for all i.

The fertilizer manufacturer's problem and the detergent manufacturer's problem are the duals of each other. In this pair of problems, if the fertilizer manufacturer's problem is called the *primal problem,* the detergent manufacturer's problem is the *dual problem,* and vice versa. This pair of problems is a *primal-dual pair* of LPs. It can be verified that the complementary pairs in this primal, dual pair of LPs are $(\pi_1, 1500 - 2x_1 - x_2)$, $(\pi_2, 1200 - x_1 - x_2)$, $(\pi_3, 500 - x_1)$, $(2\pi_1 + \pi_2 + \pi_3 - 15, x_1)$, $(\pi_1 + \pi_2 - 10, x_2)$.

The marginal value of raw material i in the fertilizer manufacturer's problem is the rate of change in the maximum profit per unit change in the availability of raw material i from its present value; thus it is the net worth of one additional unit of supply of raw material i (over the present supply), for $i = 1, 2, 3$, to the fertilizer manufacturer. Hence if the detergent manufacturer offered to buy raw material i at a price greater than or equal to its marginal value, for $i = 1, 2, 3$, the fertilizer manufacturer would find the deal acceptable. Being cost conscious, the detergent manufacturer wants to make the price offered for any raw material be the smallest price for it that will be acceptable to the fertilizer manufacturer. Hence an optimum solution of the detergent manufacturer's problem, the π_i will be the marginal value of raw material i, for $i = 1, 2, 3$, in the fertilizer manufacter's problem. So here again the dual variables are the marginal values of the items in the primal problem.

The Dual of the Ore Shipping Problem

Consider the ore shipping problem discussed in Example 1.4 in Section 1.1.4. Transforming all the constraints into inequality constraints of the $\geqq$ type, it can be written in an equivalent form as below:

Steel Company's Problem

x_{11}	x_{12}	x_{13}	x_{21}	x_{22}	x_{23}		
-1	-1	-1				$\geqq -103$	
			-1	-1	-1	$\geqq -197$	
1			1			$\geqq 71$	
	1			1		$\geqq 133$	
		1			1	$\geqq 96$	
9	16	28	14	29	19	$= z(x)$,	minimize

And $x_{ij} \geqq 0$ for all i, j.

There is a big *Iron Ore Trading Company* that has its own iron ore dumps at all five locations (the two mines and the three steel plants). They want to explore the possibility of buying the ore from the steel company at its mine locations and delivering the required ore to its steel plants. Let u_i, v_j be the price (cents per ton) the Iron Ore Trading Company offers to pay for ore at mine i and offers to sell it at steel plant j location, respectively, for $i = 1, 2$, and $j = 1, 2, 3$.

Obviously, u_i, v_j will all be nonnegative, and since the steel company would like to have its steel plants supplied with the necessary ore at minimal cost, it would like to sell all the available ore at each mine and buy only the minimum required ore at each steel plant. Hence, at these prices, the net return to the Iron Ore Trading Company will be $-103u_1 - 197u_2 + 71v_1 + 133v_2 + 96v_3$ cents, and they would like to see this maximized.

The steel company can ship 1 ton of ore from mine 1 to steel plant 1 at a cost of 9 cents. They receive u_1 cents/ton of ore at mine 1, and have to pay at v_1 cents/ton of ore at steel plant 1. Thus the difference ($v_1 - u_1$ cents) is what the Iron Ore Trading Company is charging for the transfer of ore from mine 1 to steel plant 1, and the steel company can itself effect this transfer at a cost of 9 cents. Since the steel company is cost conscious, it will not accept the Iron Ore Trading Company's deal unless $v_1 - u_1 \leqq 9$. Other dual constraints are obtained by similar economic arguments on each mine-to-plant route. Thus the prices that the Iron Ore Trading Company can charge are obtained by solving the following LP.

Iron Ore Trading Company's Problem

u_1	u_2	v_1	v_2	v_3		
-1		1			$\leqq 9$	
-1			1		$\leqq 16$	
-1				1	$\leqq 28$	
	-1	1			$\leqq 14$	
	-1		1		$\leqq 29$	
	-1			1	$\leqq 19$	
-103	-197	71	133	96	$= w(u, v)$	maximize

And u_i, $v_j \geqq$ for all i, j.

The various complementary pairs in this primal dual pair can be verified as $(u_1, 103 - x_{11} - x_{12} - x_{13})$, $(u_2, 197 - x_{21} - x_{22} - x_{23})$, $(v_1, x_{11} + x_{21} - 71)$, $(v_2, x_{12} + x_{22} - 133)$, $(v_3, x_{13} + x_{23} - 96)$, $(9 + u_1 - v_1, x_{11})$, $(16 + u_1 - v_2, x_{12})$, $(28 + u_1 - v_3, x_{13})$, $(14 + u_2 - v_1, x_{21})$, $(29 + u_2 - v_2, x_{22})$, $(19 + u_2 - v_3, x_{23})$. Using arguments similar to those in the above examples, it can be verified that u_i, v_j are the marginal values (in terms of cents in shipping budget) of 1 ton of ore available at mine i and required at plant j, respectively, for $i = 1, 2$, $j = 1, 2, 3$.

4.3 DUAL VARIABLES ARE THE MARGINAL VALUES ASSOCIATED WITH THE ITEMS, WHEN THE DUAL PROBLEM HAS A UNIQUE OPTIMUM SOLUTION

Given a general LP, the dual problem associated with it can be constructed using arguments similar to those in Section 4.2. In this context, the original LP is known as the *primal problem*. The primal problem and its dual constitute a pair of LPs known as the *primal-dual pair* of LPs. It will be shown that the dual of the dual problem is an LP that is equivalent to the primal problem. Hence each problem in a primal-dual pair of LPs is the dual of the other.

In LPs, each constraint usually comes from the requirement that the total amount of some item utilized should be less than or equal to the total amount of this item available, or that the total number of units of some items produced should be greater than or equal to the known requirement for this item. In the dual problem there will be a dual variable associated with this primal constraint, and it can be interpreted as the *price* of the item corresponding to that constraint, or the *marginal value* of that item in the primal problem, provided the dual problem has a unique optimum solution. When the dual problem has a unique optimum solution, the dual variables are also known as *shadow prices*. The dual variables are also called *simplex multipliers* or *Lagrange multipliers*.

A Word of Caution

Let b represent the right-hand-side constants vector in the system of constraints in an LP model and let $f(b)$ denote the optimum objective value function in this LP as a function of the b-vector, when b may vary, but all the other data remain fixed at their present values. It is shown in Chapter 8 that $f(b)$ is a piecewise linear function. It is differentiable almost everywhere in a statistical sense, except when b lies in a set of points of measure zero. When b is such that $f(b)$ is differentiable at b, the dual problem has a unique optimum solution that is equal to the vector of partial derivatives $\partial f(b)/\partial b$. When the point b is such that $f(b)$ is not differentiable there, the dual problem will have alternate optimum solutions. At such points b, the definition of marginal values, is mathematically technical, so we deter further discussion to Section 8.15.

If the primal problem is to minimize the cost, the dual problem may be interpreted as that of determining the prices of the items to maximize the profit. In the dual problem the dual variables associated with inequality constraints of the primal problem will be restricted in sign.

In a general LP there may also be some equality constraints. In writing the corresponding dual problem, a dual variable is also associated with each primal equality constraint. However, such dual variables will be unrestricted in sign in the dual problem. The justification for this is provided in Exercise 4.4. A negative-valued dual variable can also be interpreted as a price by treating negative prices as subsidies; for example, many cities would be happy to sell their garbage at $-\$5$ per ton, which is another way of saying that the city would be happy to pay \$5 per ton to someone who offers to remove the garbage.

4.4 HOW TO WRITE THE DUAL OF A GENERAL LINEAR PROGRAM

Suppose the primal problem consists of some equality constraints, some inequality constraints, and sign restrictions on the variables. As in Section 2.2 the word *constraint* is used here to refer to the equality or inequality conditions in the problem (i.e., the conditions that lead to a row in the original input-output tableau); the sign restrictions on the variables are considered separately.

By transforming the variables, if necessary, make all the sign restrictions in the problem into nonnegativity restrictions on the variables. If there are any upper-bound constraints on some of the variables, they should be treated as constraints in writing the dual problem. The nonnegativity restriction on a variable is actually a lower-bound restriction on it. Hence, if there is a nonnegativity restriction on a variable and a lower-bound constraint, the implicit sign restriction becomes redundant and can be deleted. Thus any variable on which there is a lower-bound constraint is treated as an unrestricted variable. All lower-bound constraints on the variables (excepting nonnegativity restrictions) are treated as constraints in writing the dual problem. *It is not necessary to transform the problem into standard form before writing its dual.*

Make sure that all the inequality constraints in the problem are of the *right type* before writing the dual. If the problem is a minimization problem, all the inequalities should be of the $\geqq$ type (i.e., of the form $\sum a_i x_i \geqq b$). If the problem is a maximization problem, all the inequalities should be of the $\leqq$ type (i.e., of the form $\sum a_i x_i \leqq b$). Any inequality of the wrong type can be converted into an inequality of the right type by multiplying both sides of it by -1 and reversing the inequality sign. Also, all sign restrictions on the variables should be nonnegativity restrictions.

Mnemonic for the Right Type of Inequalities

Imagine the problem of minimizing on the real line. Let $x \in \mathbf{R}^1$. Suppose the problem is to minimize x subject to a constraint of the form $x \leqq b$, for some real number b. This minimization problem has no finite optimum solution, because the value of x can be decreased arbitrarily on the set of feasible solutions for this constraint. On the other hand, if the problem is to minimize x, subject to $x \geqq b$, there is an optimum solution, namely, $x = b$. This suggests that for a minimization problem, inequality constraints should be expressed in the $\geqq$ form to be of the right type. In a similar manner, for a maximization problem, inequality constraints should be expressed in the $\leqq$ form to be of the right type.

Dual Problem

There will be one dual variable associated with each constraint in the primal (excluding nonnegativity restrictions on individual variables, if any). If the primal constraint is an inequality constraint of the right type, the associated dual variable in the dual problem is restricted to being nonnegative. If the primal constraint is an equality constraint, the associated dual variable is unrestricted in sign.

There is one dual constraint corresponding to each primal variable. Thus each column vector of the primal tableau leads to a constraint in the dual problem. Let π be the row vector of dual variables. Let $A_{.j}$ be the column vector of the coefficients of the primal variable x_j in the primal constraints, and c_j, the coefficient of x_j in the primal objective function. Then the dual constraint corresponding to x_j is $\pi A_{.j} = c_j$, if x_j is unrestricted in sign in the primal problem; or either the inequality constraint $\pi A_{.j} \leqq c_j$ or $\pi A_{.j} \geqq c_j$, whichever is of the right type for the dual problem, if x_j is restricted to being nonnegative in the primal problem.

If the primal is a minimization problem, the dual is a maximization problem and vice versa. The right-hand-side constants of the primal problem are the coefficients of the dual objective function and vice versa.

Example 4.1

Let the primal problem be

Minimize

$$z(x) = 5x_1 - 6x_2 + 7x_3 + x_4$$

Subject to

$$\begin{aligned} x_1 + 2x_2 - x_3 - x_4 &= -7 \\ 6x_1 - 3x_2 + x_3 + 7x_4 &\geqq 14 \\ -2.8x_1 - 17x_2 + 4x_3 + 2x_4 &\leqq -3 \\ x_1 \geqq 0 \quad x_2 &\geqq 0 \end{aligned}$$

x_3 and x_4 unrestricted in sign

By converting the inequalities to the right type, the problem becomes (in detached coefficient form)

Associated Dual Variable	Primal Problem					
	x_1	x_2	x_3	x_4		
π_1	1	2	−1	−1 =	−7	
π_2	6	−3	1	−7 ≧	14	
π_3	2.8	17	−4	−2 ≧	3	
	5	−6	7	1 =	$z(x)$	minimize

$x_1 \geqq 0$, $x_2 \geqq 0$, x_3 and x_4 unrestricted.

Here is the dual problem in detached coefficient form. Also verify that the dual of the dual problem is the primal.

Associated Primal Variable	π_1	π_2	π_3		
x_1	1	6	2.8 ≦	5	
x_2	2	−3	17 ≦	−6	
x_3	−1	1	−4 =	7	
x_4	−1	−7	−2 =	1	
	−7	14	3 =	$v(\pi)$	maximize

π_1 unrestricted, $\pi_2 \geqq 0$, $\pi_3 \geqq 0$.

Example 4.2

Consider the primal problem:

Maximize

$$z'(x) = 3x_1 - 2x_2 - 5x_3 + 7x_4 + 8x_5$$

Subject to

$$\begin{aligned} x_2 - x_3 + 3x_4 - 4x_5 &= -6 \\ 2x_1 + 3x_2 - 3x_3 - x_4 &\geqq 2 \\ -x_1 + 2x_3 - 2x_4 &\leqq -5 \\ -2 \leqq x_1 &\leqq 10 \\ 5 \leqq x_2 &\leqq 25 \\ x_3, x_4 &\geqq 0 \end{aligned}$$

x_5 unrestricted

Transforming all the inequality constraints into the right type, here is the primal in detached coefficient form:

Associated Dual Variable	x_1	x_2	x_3	x_4	x_5			
π_1		1	−1	3	−4	=	−6	
π_2	−2	−3	3	1		≦	−2	
π_3	−1		2	−2		≦	−5	
π_4	1					≦	10	
π_5	−1					≦	2	
π_6		1				≦	25	
π_7		−1				≦	−5	
	3	−2	−5	7	8	=	$z'(x)$	maximize

x_1, x_2, x_5 unrestricted; $x_3, x_4 \geqq 0$.

By the constraints here, x_2 is obviously a nonnegative variable, but since $x_2 \geqq 5$ is already included in the system of constraints, it is unnecessary to include an additional sign restriction on x_2. So we treat x_2 as an unrestricted variable in this problem. The dual problem in detached coefficient form is given at the bottom.

Example 4.3

Let A be a $m \times n$ matrix, b an $m \times 1$ column vector, c a $1 \times n$ row vector. If the primal is (4.4), the dual is (4.5), where π is the $1 \times m$ row vector of dual variables and x, the $n \times 1$ column vector of primal variables. In (4.4) all the constraints are inequalities and all the variables are restricted to being nonnegative. Its dual (4.5) is also of the same type. That's why an LP of the type (4.4) is said to be in *symmetric form*.

If the primal is

$$\begin{aligned} \text{Minimize} \quad & z(x) = cx \\ \text{Subject to} \quad & Ax \geqq b \\ & x \geqq 0 \end{aligned} \qquad (4.4)$$

The dual is

$$\begin{aligned} \text{Maximize} \quad & v(\pi) = \pi b \\ \text{Subject to} \quad & \pi A \leqq c \\ & \pi \geqq 0 \end{aligned} \qquad (4.5)$$

Example 4.4

If the primal is

$$\begin{aligned} \text{Minimize} \quad & z(x) = cx \\ \text{Subject to} \quad & Ax = b \\ & x \geqq 0 \end{aligned} \qquad (4.6)$$

The dual is

$$\begin{aligned} \text{Maximize} \quad & v(\pi) = \pi b \\ \text{Subject to} \quad & \pi A \leqq c \\ & \pi \text{ unrestricted} \end{aligned} \qquad (4.7)$$

Example 4.5

In addition, let E be an $r \times n$ matrix, and d an $r \times 1$ column vector.

Associated Primal Variable	π_1	π_2	π_3	π_4	π_5	π_6	π_7			
x_1		−2	−1	1	−1			=	3	
x_2	1	−3				1	−1	=	−2	
x_3	−1	3	2					≧	−5	
x_4	3	1	−2					≧	7	
x_5	−4							=	8	
	−6	−2	−5	10	2	25	−5	=	$v(\pi)$	minimize

π_1 unrestricted; π_2 to $\pi_7 \geqq 0$.

If the primal is:

$$\begin{aligned} \text{Minimize} \quad & z(x) = cx \\ \text{Subject to} \quad & Ax = b \\ & Ex \geqq d \\ & x \geqq 0 \end{aligned}$$

The dual is

$$\begin{aligned} \text{Maximize} \quad & v(\pi, \mu) = \pi b + \mu d \\ \text{Subject to} \quad & \pi A + \mu E \leqq c \\ \pi \text{ unrestricted} \quad & \mu \geqq 0 \end{aligned}$$

where π is a row vector of order $1 \times m$, and μ is a row vector of order $1 \times r$.

Example 4.6

Consider the general LP

$$\begin{aligned} \text{Minimize} \quad & z(x, y) = cx + dy \\ \text{Subject to} \quad & Ax + By = b \\ & Ex + Fy \geqq g \qquad (4.8) \\ x \geqq 0 \quad & y \text{ unrestricted in sign} \end{aligned}$$

where $A(m_1 \times n_1)$, $B(m_1 \times n_2)$, $b(m_1 \times 1)$, $E(m_2 \times n_1)$, $F(m_2 \times n_2)$, $g(m_2 \times 1)$, $c(1 \times n_1)$, and $d(1 \times n_2)$ are given matrices, and $x(n_1 \times 1)$ and $y(n_2 \times 1)$ are the variables. Let $\pi(1 \times m_1)$ be the vector of dual variables associated with the equality constraints in (4.8) and $\mu(1 \times m_2)$ be the vector of dual variables associated with the inequality constraints. Then the dual problem is

$$\begin{aligned} \text{Maximize} \quad & v(\pi, \mu) = \pi b + \mu g \\ \text{Subject to} \quad & \pi A + \mu E \leqq c \\ & \pi B + \mu F = d \qquad (4.9) \\ \pi \text{ unrestricted} \quad & \mu \geqq 0 \end{aligned}$$

Exercises

4.1 Consider the transportation problem in which the constraints are all stated as equations. A transportation problem in this form will arise if b_j is the exact amount required in plant j and a_i is the exact amount to be shipped out of mine i. In this form the transportation problem has a feasible solution iff $\sum a_i = \sum b_j$. That's why a transportation problem in this form is known as a *balanced transportation problem.* Write down the dual of this problem. Prove that the dual problem always has a feasible solution.

Minimize

$$z(x) = \sum\sum c_{ij} x_{ij}$$

Subject to

$$\begin{aligned} \sum_j x_{ij} &= a_i \qquad i = 1 \text{ to } m \\ \sum_i x_{ij} &= b_j \qquad j = 1 \text{ to } n \qquad (4.10) \\ x_{ij} &\geqq 0 \qquad \text{for all } i, j \end{aligned}$$

4.5 DUALITY THEORY FOR LINEAR PROGRAMMING

The primal and the dual problems are generally two distinct LPs. (There is a small class of LPs called *self-dual linear programs*. Any LP in this class is equivalent to its dual; see Exercise 4.37.) The dual variables do not appear in the statement of the primal problem and vice versa.

THEOREM 4.1 Dual of the dual is primal.

Proof: Let us obtain the dual problem of the dual LP, (4.5) in Example 4.3. The problem (4.5) can be rewritten as

$$\begin{aligned} \text{Maximize} \quad & v(\pi) = b^T \pi^T \\ \text{Subject to} \quad & A^T \pi^T \leqq c^T \\ & \pi^T \geqq 0 \end{aligned}$$

Therefore, there are n constraints in this problem. Let x_j denote the dual variable associated with the jth constraint. Let $x^T = (x_1, \ldots, x_n)$ be the dual vector. Then it can be verified that the dual problem of (4.5) is (4.4) of Example 4.3. In the same manner, it can be verified that in any primal-dual pair of LPs, each is the dual of the other. ■

THEOREM 4.2 Duals of equivalent problems are equivalent. Let (P) refer to an LP and let (D) be its dual. Let ($\tilde{P}$) be an LP that is equivalent to (P). Let ($\tilde{D}$) be the dual of ($\tilde{P}$). Then ($\tilde{D}$) is equivalent to (D).

Proof: For example, consider the primal (4.4) in Example 4.3. Including the slack vector s (a column vector of order $m \times 1$), the problem may be written in the equivalent form:

$$\begin{aligned} \text{Minimize} \quad & z(x, s) = cx + 0s \\ \text{Subject to} \quad & Ax - Is = b \\ & x \geqq 0 \quad s \geqq 0 \end{aligned}$$

This is a problem of the form discussed in Example 4.4. Denoting the dual vector by π (a row vector of order $1 \times m$), we find that the dual is

$$\begin{aligned} \text{Maximize} \quad & v(\pi) = \pi b \\ \text{Subject to} \quad & \pi A \leqq c \\ & \pi(-I) \leqq 0 \\ & \pi \text{ unrestricted} \end{aligned}$$

The constraint $\pi(-I) \leqq 0$ implies $\pi \geqq 0$. So this dual problem is equivalent to (4.5).

It can be verified that the dual variable corresponding to a redundant equality constraint in an LP will be a dependent, unrestricted variable (see Section 3.18 for definitions of these terms) in its dual and vice versa. Hence the operation of deleting a redundant equality constraint from an LP model has the effect of eliminating the corresponding dependent, unrestricted variable from the dual problem and vice versa.

In a similar manner, it can be verified that if $(\bar{P})$ is an LP obtained from the LP (P) by making any one of the transformations from (1) to (14) discussed in Section 3.18, and if (D), $(\bar{D})$ are the duals of (P), $(\bar{P})$, respectively, then (D) and $(\bar{D})$ are equivalent LPs as defined in Section 3.18. If $(\tilde{P})$ is obtained by making the transformations of the type (1) to (14) of Section 3.18 on (P) one or more times, using this argument repeatedly, we conclude that $(\tilde{D})$, the dual of $(\tilde{P})$, is equivalent to (D). Hence the duals of the equivalent LPs are equivalent. ■

Exercises

4.2 Verify that the dual of the dual is the primal for the problems in Examples 4.4, 4.5, and 4.6.

4.3 The LP (4.8) in Example 4.6 can be written in an equivalent form by expressing each unrestricted variable y_j as the difference of two nonnegative variables, $y_j^+ - y_j^-$. Show that the dual of this equivalent problem is equivalent to (4.9). In general, for each transformation of the type (1) to (14) of Section 3.18 on the primal, write down the corresponding transformation on the dual.

4.4 An equivalent LP to the one in Example 4.4 is obtained by replacing the constraints "$Ax = b$" by the equivalent system of inequality constraints: $Ax \geqq b$, $Ax \leqq b$. Show that the dual of this equivalent problem is equivalent to the dual found in Example 4.4.

This explains why the dual variables associated with primal equality constraints are unrestricted in sign.

4.5 Express the primal in Example 4.5 in an equivalent form by introducing slack variables corresponding to the inequality constraints. Show that the dual of this equivalent problem is equivalent to the dual obtained in Example 4.5.

4.5.1 The Weak Duality Theorem and Its Corollaries

THEOREM 4.3 WEAK DUALITY THEOREM In a primal-dual pair of LPs, let x be a primal feasible solution and $z(x)$ the corresponding value of the primal objective function that is to be *minimized*. Let π be a dual feasible solution and $v(\pi)$ the corresponding dual objective function that is to be *maximized*. Then, $z(x) \geqq v(\pi)$.

Proof: We prove it for the case where the primal and the dual are stated as in Example 4.3. The primal is (4.4) and the dual is (4.5) Then,

$$Ax \geqq b \text{ (because } x \text{ is primal feasible)}$$

$$\pi Ax \geqq \pi b \text{ (because } \pi \geqq 0) \tag{4.11}$$

$$\pi A \leqq c \text{ (because } \pi \text{ is dual feasible)}$$

$$\pi Ax \leqq cx \text{ (because } x \geqq 0) \tag{4.12}$$

Combining (4.11) and (4.12) we get $cx \geqq \pi Ax \geqq \pi b$; that is, $z(x) \geqq v(\pi)$, which proves the theorem when the primal and dual are stated in this form.

In general, every LP can be transformed into an equivalent problem in standard form by the transformations discussed in Chapter 2. Suppose the problem in standard form is: minimize cx, subject to $H(x) = h$, $x \geqq 0$. This same problem can be restated as

$$\begin{aligned} \text{Minimize} \quad & cx \\ \text{Subject to} \quad & Hx \geqq h \\ & -Hx \geqq -h \\ & x \geqq 0 \end{aligned}$$

which is in the same form as (4.4). So the proof given above applies to it. However, by Theorem 4.2, the dual of this problem is equivalent to the dual of the original problem. Therefore the weak duality theorem must hold for any primal-dual pair. ■

Note 4.1: One may be tempted to summarize the statement of the weak duality theorem as: "In any primal dual pair of LPs the primal objective value of any primal feasible solution is always greater than or equal to the dual objective value of any dual feasible solution." The weak duality theorem states that the above statement is true *only when* the minimization problem in the pair is called the primal and the other is called the dual.

Exercises

4.6 Prove the weak duality theorem from first principles (i.e., using arguments similar to those used here) for the cases when the primal and dual are stated as in Examples 4.4, 4.5, and 4.6.

Corollaries of the Weak Duality Theorem

Considering any primal-dual pair of LPs, let the *primal* refer to the minimization problem in the pair and the *dual* refer to the maximization problem in the pair.

1 The primal objective value of any primal feasible solution is an upper bound to the maximum value of the dual objective in the dual problem.

2 The dual objective value of any dual feasible solution is a lower bound to the minimum value of the primal objective in the primal problem.

3 If the primal problem is feasible and its objective function is unbounded below on the primal feasible solution set, the dual problem cannot have a feasible solution.

4 If the dual problem is feasible and its objective function is unbounded above on the dual feasible solution set, the primal problem cannot have a feasible solution.

5 The converse of (3) is the following: "If the dual problem is infeasible and the primal problem is feasible, the primal objective function is unbounded below on the primal feasible solution set."
Similarly, the converse of (4) is: "If the primal problem is infeasible and the dual problem is feasible, the dual objective function is unbounded above on the dual feasible solution set."
Both these results are true and will be proved after a discussion of the fundamental duality theorem.

It is possible that both the primal and the dual problems in a primal-dual pair have no feasible solutions. For example, consider the following:

$$\begin{array}{lrl} \text{Minimize} & z(x) = & 2x_1 - 4x_2 \\ \text{Subject to} & & x_1 - x_2 = 1 \\ & & -x_1 + x_2 = 2 \\ & x_1 \geqq 0 & x_2 \geqq 0 \end{array}$$

$$\begin{array}{lrl} \text{Maximize} & v(\pi) = & \pi_1 + 2\pi_2 \\ \text{Subject to} & & \pi_1 - \pi_2 \leqq 2 \\ & & -\pi_1 + \pi_2 \leqq -4 \\ & & \pi_1, \pi_2 \text{ unrestricted} \end{array}$$

Clearly both problems in the pair are infeasible. Thus, even though the result in (5) is true, the fact that the dual problem is infeasible in a primal-dual pair of LPs does not imply that the primal objective function is unbounded on the primal feasible solution set, unless it is known that the primal is feasible.

THEOREM 4.4 SUFFICIENT OPTIMALITY CRITERION IN LINEAR PROGRAMMING In a primal-dual pair of LPs, let $z(x)$ be the primal objective function and $v(\pi)$, the dual objective function. If $\bar{x}$, $\bar{\pi}$ are a pair of primal and dual feasible solutions satisfying $z(\bar{x}) = v(\bar{\pi})$, then $\bar{x}$ is an optimal feasible solution of the primal and $\bar{\pi}$ is an optimum feasible solution of the dual.

Proof: Suppose the primal denotes the minimization problem in the primal-dual pair. Let x be any primal feasible solution. By the weak duality theorem, we have $z(x) \geqq v(\bar{\pi})$ because $\bar{\pi}$ is dual feasible. But $z(\bar{x}) = v(\bar{\pi})$ by hypothesis. So $z(x) \geqq z(\bar{x})$ for all x primal feasible. Thus $\bar{x}$ is optimal to the primal problem. Similarly, $\bar{\pi}$ is optimal to the dual problem. ■

Example 4.7

Considering the problems of the family and the pillmaker, let (4.1) be the primal and (4.2) the dual. We verify that $\bar{x} = (\bar{x}_1, \bar{x}_2, \bar{x}_3, \bar{x}_4, x_5, x_6)^{\mathrm{T}} = (0, 0, 0, 0, 5, 2)^{\mathrm{T}}$ is a primal feasible solution, and that $\bar{\pi} = (\bar{\pi}_1, \bar{\pi}_2) = (3, 8)$ is a dual feasible solution. Also $z(\bar{x}) = v(\bar{\pi}) =$ 179 cents. Therefore, by the sufficient optimality, criterion $\bar{x}$ represents the optimal diet for the family and $\bar{\pi}$ represents the optimal prices that the pill-maker can adopt to maximize the profit while keeping the nutrient pills economically competitive.

The converse of this sufficient optimality criterion is the fundamental duality theorem that is discussed in Section 4.5.3.

The book [1.69] by David Gale has an excellent discussion of the sufficient optimality criterion in linear programming.

4.5.2 The Supervisor Principle

Suppose your supervisor asks you to solve the primal-dual pair of linear programs (4.4) and (4.5). You are free to solve the problems by hand or computer or whatever; use any algorithm or method, or even guesswork. However, the supervisor happens to be an extremely suspicious man when you give him the pair $\bar{x}$, $\bar{\pi}$ with the claim that they are optimal, respectively, to (4.4), (4.5); he does not accept your word for it unless he can verify the claim personally. The sufficient optimality condition proved above makes it possible for the supervisor to verify the optimality of $\bar{x}$, $\bar{\pi}$ to (4.4), (4.5), respectively, very efficiently. All that the supervisor has to do is to verify that $\bar{x}$ is feasible to (4.4), that $\bar{\pi}$ is feasible to (4.5), and that $c\bar{x} = \bar{\pi}b$. If these three easily verified conditions hold, the weak duality theorem guarantees that $\bar{x}$ is indeed optimal to (4.4) and that $\bar{\pi}$ is optimal to (4.5). This is called the *supervisor principle* for the primal-dual pair of LPs (4.4) and (4.5).

A *supervisor principle* for an optimization problem is a fast method to check whether a given solution is optimal for it. The weak duality theorem provides a very convenient supervisor principle for a pair of primal-dual LPs. Contrast this with the situation for many integer programming and nonconvex optimization problems, for which good optimality conditions are not known for us to be able to construct a supervisor principle for them. One major contribution of the duality theory of linear programming is to provide a very simple and fast supervisor principle for a pair of primal-dual LPs.

4.5.3 The Fundamental Duality Theorem and Its Corollaries

THEOREM 4.5 THE FUNDAMENTAL DUALITY THEOREM In a primal-dual pair of LPs, if either the primal or the dual problem has an optimal feasible solution, then the other does also, and the two optimal objective values are equal.

Proof: We will prove the fundamental duality theorem for the case where the primal and dual problems are stated as in Example 4.4. Let the primal problem be (4.6), where A is of order $m \times n$ and rank m. Suppose the primal problem has an optimal feasible solution. By the results in Section 3.5.6, the simplex method applied to (4.6) terminates after a finite number of pivot steps, by obtaining a feasible basic vector for it, say $x_B = (x_1, \ldots, x_m)$, in which the optimality criterion of Section 2.5.3 is satisfied. Let B be the basis associated with the basic vector x_B in (4.6).

Let D be the $m \times (n - m)$ matrix of the column vectors in A associated with the nonbasic variables. After partitioning into the basic and nonbasic parts, the problem can be represented in tableau form as in Tableau 3.2 of Section 3.12. By the results in Section 3.12, the canonical tableau with respect to the optimum basis B is Tableau 3.3. The last row in this optimum canonical tableau is $\bar{c} = c - c_B B^{-1} A$. Since the optimality criterion is satisfied in this tableau, we have $\bar{c} = c - c_B B^{-1} A \geqq 0$, that is

$$(c_B B^{-1})A \leqq c \tag{4.13}$$

From Tableau 3.3 it is clear that the optimal objective value is $(c_B B^{-1})b$. If we define $\pi = c_B B^{-1}$, by (4.13), we have $\pi A \leqq c$ and $\pi b = (c_B B^{-1})b =$ optimal primal objective value. By the sufficient optimality criterion, Theorem 4.4, π is a dual optimal solution. This completes the proof when the primal and dual are as stated here.

In general, every LP can be transformed into an equivalent problem in standard form by the transformations discussed in Chapter 2. The equivalent problem in standard form is of the same type as the primal problem in Example 4.4; hence, our proof of the fundamental duality theorem applies to it. However, by Theorem 4.2, the dual of the equivalent problem in standard form is equivalent to the dual of the original problem. Thus, the fundamental duality theorem must hold for it too. This completes the general proof of the fundamental duality theorem. ∎

Example 4.8

From Example 4.7 verify that the fundamental duality theorem holds for the problems of the family and the pill manufacturer.

Comment 4.1: In some textbooks, the fundamental duality theorem is referred to as the *strong duality theorem*. Consider the primal and dual problems as stated in Theorem 4.3 of Section 4.5.1. In that theorem we proved that the minimum value of $z(x)$ is greater than or equal to the maximum value of $v(\pi)$. In Theorem 4.5, we proved the stronger statement that actually the minimum value of $z(x)$ is equal to the maximum value of $v(\pi)$. This explains the names weak, strong duality theorems for Theorems 4.3 and 4.5 respectively.

Corollaries of the Fundamental Duality Theorem

Alternative Statement of the Duality Theorem An alternate statement of the fundamental duality theorem is: "If both the problems in a primal-dual pair of LPs have feasible solutions, then both have optimal feasible solutions, and the optimal objective values of the two problems are equal." This is easily proved by using the weak duality theorem and the fundamental duality theorem.

Separation Property of Objective Values Consider a primal-dual pair of LPs and suppose the minimization problem in the pair is the primal with the objective function $z(x)$. Suppose the dual objective function is $v(\pi)$. If both the problems have feasible solutions, then the values assumed by the two objective functions at feasible solutions of the respective problems are separated on the real line as in Figure 4.1.

Primal Objective Unbounded If the primal is the minimization problem in a primal-dual pair, and if the primal is feasible and the dual is infeasible, then the primal cannot have an optimal feasible solution, that is, the primal objective function is unbounded below.

Dual Objective Unbounded If the dual is the maximization problem in a primal-dual pair, and if the dual is feasible and the primal infeasible, then the dual cannot have an optimal feasible solution, that is, the dual objective function is unbounded above.

Note 4.2: The latter two sections are the converses of parts (3) and (4) of Section 4.5.1 respectively, discussed in part (5) of Section 4.5.1.

THEOREM 4.6 NECESSARY AND SUFFICIENT OPTIMALITY CONDITIONS FOR LP Consider a primal-dual pair of LPs. Let x, π be the vectors of primal and dual variables, respectively, and let $z(x)$, $v(\pi)$ be the primal and dual objective functions, respectively. If $\bar{x}$ is a primal feasible solution, it is an optimum solution of the primal problem iff there exists a dual feasible solution $\bar{\pi}$ satisfying $z(\bar{x}) = v(\bar{\pi})$.

Proof: Follows directly from Theorems 4.5 and 4.4. ■

Exercises

4.7 For each of the LPs discussed in Examples 4.3, 4.4, 4.5, and 4.6, write out the necessary and sufficient conditions for the optimality of a given primal feasible solution, developed in the above theorem.

Another set of necessary and sufficient conditions for the optimality of a feasible solution of an LP is derived in Section 4.5.8.

4.5.4 Lower Bounds for the Optimum Object Value Using the Dual Problem

Consider the LP (4.4). Let $\hat{z}$ denote the unknown optimum objective value in it. If $\bar{x}$ is a feasible solution for (4.4), $z(\bar{x})$ is an upper bound for $\hat{z}$. This

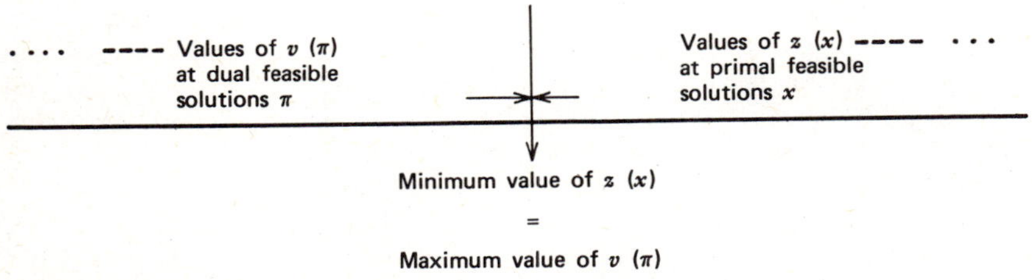

Figure 4.1

information is not useful for checking whether $\bar{x}$ is optimal for (4.4). On the other hand, if we are given a lower bound α for the optimum objective value in (4.4) (i.e., this α is a real number satisfying $z(x) \geqq \alpha$ for all feasible solutions x of (4.4)), and if it so happens that $\alpha = z(\bar{x})$, we can immediately conclude that $\bar{x}$ is optimal to (4.4). Even if $\alpha < z(\bar{x})$, we know that $z(\bar{x})$ is no greater than $(z(\bar{x}) - \alpha)$ from the optimum objective value in (4.4); and if $z(\bar{x}) - \alpha$ is small we can conclude that $\bar{x}$ is near optimal for (4.4). For these reasons, when faced with a minimization problem, it is always very useful to construct a lower bound for the minimum objective value in the problem. Mathematically, the dual of (4.4) constructed so that the dual objective value at any dual feasible solution is a lower bound for the minimum objective value in (4.4). The quality of a lower bound depends on how close it is to the quantity it is bounding. Since the dual objective value at any dual feasible solution is a lower bound for $\hat{z}$, we can get the best lower bound for $\hat{z}$ using the dual problem by maximizing the dual objective value in it. That's why the dual of (4.4) is a maximization problem. The fundamental duality theorem states that if a lower bound exists for $\hat{z}$, this best lower bound for $\hat{z}$ obtained using the dual problem is, in fact, equal to $\hat{z}$, that is, this bound is tight. One of the main goals of optimization theory is to construct such efficient, tight lower-bounding schemes for minimization problems. The duality theory of LP provides an efficient, tight lower-bounding scheme for LP. Unfortunately, for many of the nonconvex programming problems (which include discrete optimization, integer and combinatorial programming, and nonconvex continuous nonlinear programming) efficient, tight lower-bounding schemes are not known.

4.5.5 Dual Feasibility of a Basis

Consider the primal LP in standard form, (4.6), in Example 4.4, where A is a matrix of order $m \times n$ and rank m. Let B be a basis for (4.6), and let x_B denote the corresponding basic vector. Let c_B be the row vector of the basic cost coefficients. From the arguments in Theorem 4.5 the dual solution corresponding to the basis B is given by $\pi = c_B B^{-1}$. Partitioning the variables, column vectors, etc., into basic and nonbasic parts as in Theorem 4.5, the primal and dual can be written as:

Primal

Minimize	$z(x) = c_B x_B + c_D x_D$	
Subject to	$Bx_B + Dx_D = b$	
	$x_B \geqq 0$	$x_D \geqq 0$

Dual

Maximize	$v(\pi) = \pi b$
Subject to	$\pi B \leqq c_B$
	$\pi D \leqq c_D$
	π unrestricted

The primal basic solution of (4.6) corresponding to the basis B is obtained by solving the system

$$x_D = 0$$

and

$$Bx_B = b$$

and the basis is *primal feasible* iff the solution satisfies the remaining nonnegativity constraints on the primal variables, that is, iff $x_B = B^{-1}b \geqq 0$. The dual solution corresponding to the basis B is obtained by solving

$$\pi B = c_B \tag{4.14}$$

The dual solution obtained from (4.14) may not be dual feasible if it does not satisfy all the constraints in the dual problem. This solution is a dual feasible solution and the basis B is *dual feasible* if this dual solution satisfies the remaining dual constraints, namely, $\pi D \leqq c_D$. Hence, if B is a dual feasible basis, the dual solution corresponding to it is $\pi = c_B B^{-1}$, and this will satisfy all the dual constraints if $\pi A \leqq c$. That is,

$$[c - (c_B B^{-1})A] \geqq 0 \tag{4.15}$$

By the results in Section 3.12, $c - (c_B B^{-1})A = \bar{c}$ is just the vector of relative cost coefficients in this LP, with respect to the basis B. Hence a basis for the LP (4.6) is dual feasible iff the vector of relative cost coefficients with respect to it is nonnegative. *The optimality criterion in the primal simplex algorithm* (see Section 2.5.3) *is therefore the dual feasibility criterion for the current basis.* A basis B for (4.6) is said to be an *optimum basis,* and the basic vector x_B an *optimum basic vector* iff they are both primal and dual feasible.

The simplex algorithm always deals with primal feasible bases. But until (4.7) is satisfied, it does not reach an optimal basis. Thus, all but the final optimal basis encountered in the simplex algorithm are dual infeasible. When dual feasibility is attained, the simplex algorithm terminates. The simplex algorithm starts with a primal feasible (but dual infeasible) basis, and it tries to attain dual feasibility, keeping primal feasibility throughout the algorithm. On the other hand, it is possible to develop an algorithm that starts with a dual feasible but primal infeasible basis and tries to attain primal feasibility, keeping dual feasibility throughout the algorithm. Such an algorithm is appropriately called the *dual simplex algorithm,* and it is discussed in Chapter 6.

In (4.6), let A be of order $m \times n$ and rank m. If the basic vector $x_B = (x_1, \ldots, x_m)$, these facts are summarized in Tableau 4.1.

Comment 4.2: A tableau such as Tableau 4.1, in which both the primal and dual problems in an LP pair are displayed simultaneously, with the primal variables acting on the columns of the tableau (hence the primal problem is obtained by reading the tableau *across*, with each row of the tableau leading to a primal constraint) and the dual variables acting on the rows of the tableau (hence the dual problem is obtained by reading the tableau *down*, with each column of the tableau leading to a dual constraint) is known as the *Tucker tableau,* or the *Tucker diagram*, in honor of A. W. Tucker.

Example 4.9

Consider the LP in standard form given in Example 4.10. Let B be the basis corresponding to the basic vector $x_B = (x_6, x_2, x_3)$. The primal-dual basic solutions corresponding to this basis are obtained by solving.

x_6	x_2	x_3	
4	0	0	7
6	1	0	19
-7	0	1	21

$x_1 = x_4 = x_5 = 0.$

π_1	π_2	π_3	
4	6	-7	19
0	1	0	-4
0	0	1	-8

and they are $\bar{x} = (0, \frac{17}{2}, \frac{133}{4}, 0, 0, \frac{7}{4})^{\mathrm{T}}$, $\bar{\pi} = (-\frac{13}{4}, -4, -8)$. Since $\bar{x}$ is nonnegative, B is primal feasible for this LP; $\bar{\pi}$ satisfies the remaining dual constraints, namely $\pi_1 \leqq 3$, $-\pi_1 + \pi_2 + 2\pi_3 \leqq 16$, $2\pi_1 - 2\pi_2 + 3\pi_3 \leqq -17$ and hence B is dual feasible too. Since B

Tableau 4.1

The primal problem operates on the columns of A. The primal variables are the coefficients in an expression of b as a linear combination of columns of A. The primal solution is feasible if all primal variables $\geqq 0$.

The dual problem operates on the rows of A. The dual variables are the coefficients in a linear combination of rows of A. The dual solution is feasible if the linear combination of rows of A is $\leqq c$.

Dual variables	Primal basic variables associated with columns of the basis B	Primal variables not in basic vector x_B, are equal to 0 in the primal basic solution corresponding to the basis B	
	$x_1 \;\ldots\; x_m$	$x_{m+1} \;\ldots\; x_j \;\ldots\; x_n$	
π_1	$a_{11} \ldots a_{1m}$	$a_{1,m+1} \ldots a_{1j} \ldots a_{1n}$	$= b_1$
$\vdots$	$\vdots \quad \vdots$	$\vdots \quad \vdots \quad \vdots$	$\vdots$
π_m	$a_{m1} \ldots a_{mm}$	$a_{m,m+1} \ldots a_{mj} \ldots a_{mn}$	$= b_m$
	$\leqq \quad \leqq$	$\leqq \;\ldots \leqq \ldots \leqq$	$= v(\pi)$ maximize
	$c_1 \;\ldots\; c_m$	$c_{m+1} \;\ldots\; c_j \;\ldots\; c_n$	$= z(x)$ minimize
	These dual constraints are treated as equations to obtain the dual solution corresponding to the basis B	Dual basic solution corresponding to the basis B is dual feasible if it satisfies these remaining dual constraints	

is both primal and dual feasible, it is an optimum basis for this problem, $\bar{x}$ is an optimum solution, and $\bar{\pi}$ is an optimum dual solution associated with this problem.

4.5.6 Relative Cost Coefficients Are Dual Slack Variables

Consider the LP (4.6) in standard form again. Suppose the basis B, associated with the basic vector x_B and cost vector c_B, is an optimum basis for this problem. Then $\bar{\pi} = c_B B^{-1}$ is an optimal feasible solution of the dual of (4.6). The dual constraints are $\pi A_{.j} \leqq c_j$ for $j = 1$ to n. Hence, the dual slacks corresponding to the optimum dual feasible solution π are $\bar{c}_j = (c_j - \bar{\pi} A_{.j}) = (c_j - c_B B^{-1} A_{.j})$. Thus, the relative cost coefficients of the simplex algorithm are the dual slack variables. If any of the relative cost coefficients with respect to the basis B are negative, since that slack variable is negative, the associated dual solution is dual infeasible and, hence, the basis B is dual infeasible. If all the relative coefficients are nonnegative, the associated dual solution is dual feasible and, hence, B is a dual feasible basis. *Therefore, the optimality criterion of the primal simplex algorithm is just the dual feasibility criterion.* The relative cost coefficients from an optimum canonical tableau are also known as the *shadow costs* associated with the activities in the original LP model.

Example 4.10

Consider the LP:

x_1	x_2	x_3	x_4	x_5	x_6	$-z$	b
1	0	0	−1	2	4	0	7
0	1	0	1	−2	6	0	19
0	0	1	2	3	−7	0	21
3	−4	−8	16	−17	19	1	0

$x_j \geqq 0$ for all j; z to be minimized.

The dual constraints can be written in an equivalent manner as

$$\begin{aligned} 3 - \pi_1 \qquad\qquad\qquad &\geqq 0 \\ -4 \qquad - \pi_2 \qquad\quad &\geqq 0 \\ -8 \qquad\qquad - \pi_3 &\geqq 0 \\ 16 + \pi_1 - \pi_2 - 2\pi_3 &\geqq 0 \\ -17 - 2\pi_1 + 2\pi_2 - 3\pi_3 &\geqq 0 \\ 19 - 4\pi_1 - 6\pi_2 + 7\pi_3 &\geqq 0 \end{aligned}$$

and the expressions on the left-hand side in this system are obviously the slacks corresponding to the constraints in the dual problem. Consider the basic vector (x_1, x_2, x_3). Here is the canonical tableau with respect to this basic vector.

x_1	x_2	x_3	x_4	x_5	x_6	$-z$	b
1	0	0	−1	2	4	0	7
0	1	0	1	−2	6	0	19
0	0	1	2	3	−7	0	21
0	0	0	39	−7	−25	1	223

Hence the relative cost coefficient vector with respect to the basic vector (x_1, x_2, x_3) is $\bar{c} = (0, 0, 0, 39, -7, -25)$. The dual solution associated with the basis corresponding to the basic vector (x_1, x_2, x_3) is derived by solving the system obtained by treating the first three constraints in the above system as equations. It is $\pi = (3, -4, -8)$. Substituting this solution, we verify that the dual slack vector is equal to the relative cost vector $\bar{c}$ obtained from the canonical tableau.

The most commonly used method for solving linear programming problems is the revised simplex method (see Chapter 5). It obtains and uses the dual solution at each stage, to compute the relative cost coefficients.

4.5.7 Unboundedness Criterion and Dual Infeasibility

Again consider the LP (4.6) in standard form. When this problem is solved by the simplex algorithm, suppose the unboundedness criterion is satisfied. Let the terminal basis obtained be B, and for convenience in referring to it, let us assume that the associated basic vector is $(x_1, \ldots, x_m)$. Since the unboundedness criterion is satisfied at this stage, there must exist a nonbasic variable, say x_{m+1}, such that the column vector of x_{m+1} in the canonical tableau at this stage satisfies the unboundedness criterion. Let Tableau 2.8 of Section 2.5.1 be the canonical tableau with respect to this basis.

Our assumptions are that

$$\bar{c}_{m+1} < 0, \qquad \bar{a}_{i,m+1} \leqq 0 \qquad \text{for } i = 1 \text{ to } m \qquad (4.16)$$

From the canonical tableau with respect to the basis

B, the LP (4.6) is equivalent to

$$\text{Minimize} \quad z(x) = \left(\sum_{j=m+1}^{n} \bar{c}_j x_j \right) + \bar{z}$$

$$\text{Subject to} \quad x_i + \sum_{j=m+1}^{n} \bar{a}_{ij} x_j = \bar{b}_i \qquad i = 1 \text{ to } m$$

$$x_j \geqq 0 \qquad \text{for all } j = 1 \text{ to } n$$

(4.17)

Since $\bar{z}$ is a known constant, it can be ignored in the minimization process. The dual of (4.17) is

$$\text{Maximize} \quad v(\pi) = \left(\sum_{i=1}^{m} \bar{b}_i \pi_i \right) + \bar{z}$$

$$\text{Subject to} \quad \pi_i \leqq 0 \qquad \text{for } i = 1 \text{ to } m$$

$$\sum_{i=1}^{m} \bar{a}_{ij} \pi_i \leqq \bar{c}_j \qquad \text{for } j = m+1 \text{ to } n$$

(4.18)

However, the constraints $\pi_i \leqq 0$, for $i = 1$ to m, and $\sum_{i=1}^{m} \bar{a}_{i,m+1} \pi_i \leqq \bar{c}_{m+1}$, are inconsistent because of (4.16). Thus, the dual constraints (4.18) are inconsistent, and hence, the dual of (4.17) is infeasible. But (4.17) is equivalent to (4.6), and so the dual of (4.17) is equivalent to the dual of (4.6). So the dual of (4.6) must be infeasible too. *This shows that the unboundedness criterion in the primal simplex algorithm just verifies dual inconsistency.*

Unboundedness Under Variable Right-Hand-Side Constants Vector

Consider the LP (4.6) in standard form again. If (4.6) is feasible and unbounded below, that is, $z(x)$ can be made to diverge to $-\infty$ on the set of feasible solutions of (4.6), then the dual constraints in (4.7) must be inconsistent. But the dual constraints are independent of the primal right-hand-side constants vector b, and hence, if they are inconsistent for some b, they remain inconsistent for any b. We therefore conclude that (4.6) remains unbounded, even if we change b, as long as it remains feasible. Thus,

1 If the LP (4.6) is unbounded below for some b, then it is unbounded below for every b for which it is feasible.

2 If the LP (4.6) has an optimal feasible solution, then it will have an optimal feasible solution even after changing b, as long as it remains feasible.

This provides another argument to justify Corollary 3.3 of Section 3.7. These results will be useful when we consider sensitivity analysis and parametric linear programming.

Unboundedness in the Big-M Method

Again consider the LP in standard form (4.6), where A is a matrix of order $m \times n$. Its dual is (4.7). To solve (4.6) by the big-$\boldsymbol{M}$ method (see Section 2.7) the artificial variables $t_1, \ldots, t_m$ are introduced, and the problem modified to (2.14) of Section 2.7, where M is an arbitrarily large positive number. Its dual is

$$\begin{aligned} &\text{Maximize} && \pi b \\ &\text{Subject to} && \pi A \leqq c \qquad\qquad (4.19)\\ & && \pi_i \leqq \boldsymbol{M} \qquad \text{for } i = 1 \text{ to } m \\ & && \pi \text{ unrestricted in sign} \end{aligned}$$

The only constraints in (4.19) that (4.7) does not include are the restrictions $\pi_i \leqq \boldsymbol{M}$, for $i = 1$ to m. Clearly, if (4.19) is infeasible for arbitrarily large $\boldsymbol{M}$, (4.7) must be infeasible too.

In the course of solving (2.14) by the primal simplex algorithm, if the unboundedness criterion is satisfied independently of how large a value $\boldsymbol{M}$ has, then its dual, (4.19) must be infeasible. So (4.7) must be infeasible too. Hence, in this case $z(x)$ is unbounded below if (4.6) is feasible. This proves the results discussed in Section 2.7.

4.5.8 Complementary Slackness Property

Another important corollary of the duality theorem is known as the *complementary slackness theorem*. When problems in mathematical economics are modeled as LPs, this theorem can be interpreted as indicating an *equilibrium* or *economic stability* when optimality conditions are satisfied. That is why this theorem is also called the *equilibrium theorem* or the *price equilibrium theorem*. In this context any optimum feasible dual vector is called an *equilibrium vector*. Consider a pair of primal and dual LPs. Each complementary pair for these consists of a sign restricted variable in one of the LPs, and the associated slack variable in the other problem.

THEOREM 4.7 COMPLEMENTARY SLACKNESS THEOREM (EQUILIBRIUM THEOREM) A pair of primal and dual feasible solutions are optimal to the respective problems in a primal-dual pair of LPs, iff, whenever these feasible solutions make a slack variable in one problem strictly positive, the value (in these feasible solutions) of the associated nonnegative variable of the other problem is zero.

Proof: We prove this theorem for the case where the primal-dual pair of problems is stated as in (4.4) and (4.5) in Example 4.3 where A is of order $m \times n$. The dual variable π_i is associated with the primal constraint $A_{i.}x \geqq b_i$ and hence, its associated primal slack variable is $A_{i.}x - b_i = v_i$ for $i = 1$ to m. The primal variable x_j is associated with the dual constraint $\pi A_{.j} \leqq c_j$ and, hence, its associated dual slack variable is $c_j - \pi A_{.j} = u_j$ for $j = 1$ to n. In this example, each primal and dual variable is restricted to be nonnegative; hence, each of them has an associated slack variable of the other problem. Let $\bar{x}$, $\bar{\pi}$ be a pair of primal and dual feasible solutions, respectively. Then $\bar{v} = A\bar{x} - b$, and $\bar{u} = c - \bar{\pi}A$ are the vectors of the slack variables corresponding to these solutions. The theorem states that $\bar{x}$, $\bar{\pi}$ are optimal to the respective problems iff

Whenever

$$\bar{v}_i = A_{i.}\bar{x} - b_i > 0 \qquad \text{we have } \bar{\pi}_i = 0 \qquad (4.20)$$

And whenever

$$\bar{u}_j = c_j - \bar{\pi}A_{.j} > 0 \qquad \text{we have } \bar{x}_j = 0 \qquad (4.21)$$

Another way of writing (4.20) and (4.21) is

$$\bar{v}_i\bar{\pi}_i = (A_{i.}\bar{x} - b_i)\bar{\pi}_i = 0 \qquad \text{for all } i = 1 \text{ to } m \qquad (4.22)$$

$$\bar{u}_j\bar{x}_j = (c_j - \bar{\pi}A_{.j})\bar{x}_j = 0 \qquad \text{for all } j = 1 \text{ to } n \qquad (4.23)$$

We prove the "if" portion of the theorem first. By primal feasibility $A_{i.}\bar{x} \geqq b_i$ for all $i = 1$ to m. Multiplying both sides by $\bar{\pi}_i \geqq 0$ (by dual feasibility), summing over i, and using (4.22)

$$\sum \bar{\pi}_i(A_{i.}\bar{x}) = \sum \bar{\pi}_i b_i \qquad (4.24)$$

Similarly, from dual feasibility, $\bar{\pi}A_{.j} \leqq c_j$ for all $j = 1$ to n. Multiplying both sides of this by $\bar{x}_j$ and summing over j and using (4.23) leads to

$$\sum \bar{x}_j(\bar{\pi}A_{.j}) = \sum \bar{x}_j c_j \qquad (4.25)$$

However, $\sum \bar{\pi}_i(A_{i.}\bar{x}) = \sum \bar{x}_j(\bar{\pi}A_{.j}) = \bar{\pi}A\bar{x}$. From (4.24) and (4.25) $\sum \bar{x}_j c_j = \sum \bar{\pi}_i b_i$. Hence, from the sufficient optimality criterion $\bar{x}$, $\bar{\pi}$ are optimal feasible solutions of the respective problems.

To prove the "only if" portion of the theorem suppose $\bar{x}$ and $\bar{\pi}$ are a pair of optimal feasible solutions. We have to show that they satisfy the complementary slackness conditions (4.20) and (4.21). From the weak duality theorem we know that $c\bar{x} \geqq \bar{\pi}A\bar{x} \geqq \bar{\pi}b$. But from the duality theorem $c\bar{x} = \bar{\pi}b$. So $c\bar{x} = \bar{\pi}A\bar{x} = \bar{\pi}b$. So $c\bar{x} - \bar{\pi}A\bar{x} = 0$. This implies that

$$\sum (c_j - \bar{\pi}A_{.j})\bar{x}_j = 0 \qquad (4.26)$$

However, $\bar{x}_j \geqq 0$ and $c_j - \sum_i \bar{\pi}_i a_{ij} \geqq 0$ by primal and dual feasibility. Therefore, the left side of (4.26) is a sum of nonnegative quantities. It is zero only if each term in the sum is zero. Thus (4.23) must hold. Similarly, $\bar{\pi}(A\bar{x} - b) = 0$ implies that (4.22) must hold. This completes the proof for the case where the primal and dual are as stated in Example 4.3. Proof of complementary slackness theorem, when the primal and dual problems are stated in a different form, is similar. ■

Note 4.3: (1) Conditions (4.20) or (4.22) only require that if $\bar{v}_i > 0$, then $\bar{\pi}_i = 0$. They *do not require* that if $\bar{v}_i = 0$, then $\bar{\pi}_i$ be > 0; that is, both $\bar{v}_i$ and $\bar{\pi}_i$ could be equal to zero, and the conditions of the complementary slackness theorem would be satisfied. (2) Conditions (4.20) and (4.21) automatically imply that if $\bar{\pi}_i > 0$, then $\bar{v}_i = 0$, and that if $\bar{x}_j > 0$, then $\bar{u}_j = 0$.

Note 4.4: The complementary slackness theorem does not say anything about the values of unrestricted variables (corresponding to equality constraints in the other problem) in a pair of optimal feasible solutions of the primal and dual problems, respectively. It is concerned only with nonnegative variables of one problem and the slack variables corresponding to the associated inequality constraints in the other problem of the pair of primal and dual LPs.

COROLLARY 4.1 Consider a primal, dual pair of LPs. Let $\bar{x}$ be an optimal feasible solution of the primal LP. Then the following statements can be made about every dual optimum feasible solution.

1 If x_j is a variable restricted to be nonnegative in the primal problem and if $\bar{x}_j > 0$, then the dual inequality constraint associated with the primal variable x_j is satisfied as an equation by *every* dual optimum feasible solution.

2 If the primal problem consists of any inequality constraints, let $\bar{v}$ represent the values of the corresponding slack variables, at the primal feasible solution $\bar{x}$. Then, if a slack variable $\bar{v}_i > 0$, the dual variable associated with it is equal to zero in *every* dual optimum feasible solution.

Corresponding symmetric statements can be made about the positive value in $\bar{\pi}$ of sign restricted dual variables, and primal inequality constraints that must hold as equations in every primal optimum solution.

Example 4.11

Consider the LP (4.6) in standard form and its dual (4.7). The complementary pairs in these problems are $(x_j, c_j - \pi A_{.j})$, $j = 1$ to n. The complementary slackness conditions for optimality in this primal-dual pair of LPs are: $x_j(c_j - \pi A_{.j}) = 0$ for $j = 1$ to n. Denoting $c_j - \pi A_{.j}$ by $\bar{c}_j$, these conditions are $x_j\bar{c}_j = 0$ for $j = 1$ to n.

THEOREM 4.8 NECESSARY AND SUFFICIENT CONDITIONS FOR OPTIMALITY IN LP Let x and π be the vectors of variables in an LP and its dual, respectively. If $\bar{x}$ is a feasible solution of the LP, it is an optimum solution iff there exists a dual feasible solution $\bar{\pi}$, such that $\bar{x}$ and $\bar{\pi}$ together satisfy the complementary slackness conditions for optimality in this primal-dual pair.

Proof: Follows directly from Theorems 4.5 and 4.7. ■

Given an optimal feasible solution of one of the problems in the primal-dual pair, the above results can be used to characterize the set of all optimal feasible solutions of the other problem.

Example 4.12

Consider the problems of the family and the pill manufacturer. The family's diet problem is (4.1). The slack variables in this problem are $v_1 = (x_1 + 2x_3 + 2x_4 + x_5 + 2x_6 - 9)$, $v_2 = (x_2 + 3x_3 + x_4 + 3x_5 + 2x_6 - 19)$. Suppose we are told that $\bar{x} = (0, 0, 0, 0, 5, 2)^T$ is an optimal feasible solution to the family's problem. The dual problem is (4.2). Since x_5 and x_6 are positive, every dual optimum feasible solution must satisfy

$$\pi_1 + 3\pi_2 = 27$$
$$2\pi_1 + 2\pi_2 = 22$$

But this system of equations has a unique solution, $\bar{\pi} = (3, 8)$, and we verify that this is dual feasible. This also verifies the assertion that $\bar{x}$ is indeed optimal to the primal problem. Also, since $\bar{\pi}_1 > 0$ and $\bar{\pi}_2 > 0$, (v_1, v_2) must be equal to 0 at every primal optimum feasible solution. We verify that $\bar{v} = 0$. Thus, the dual optimum feasible solution is unique, and it is $\bar{\pi}$.

Example 4.13

Consider the following LP:

x_1	x_2	x_3	x_4	x_5	x_6	
1	0	1	1	−2	−1	= 2
1	1	0	−2	1	−2	= −4
0	1	1	1	−1	−1	= 2
3	−6	−2	−4	0	−7	= $z(x)$ minimize

$x_j \geqq 0$ for all j.

Let $\bar{x} = (0, 0, 0, 2, 0, 0)^T$. Suppose we are told that $\bar{x}$ is an optimum feasible solution of this LP. Since $\bar{x}_4 = 2 > 0$, by the complementary slackness theorem the dual constraint corresponding to x_4, namely, $\pi_1 - 2\pi_2 + \pi_3 \leqq -4$, must hold as an equation at every dual optimum solution. This is the only information we can get about dual optimum solutions by applying the complementary slackness theorem. From this we conclude that the set of optimum solutions of the dual problem is the set of feasible solutions of

$$\begin{aligned}
\pi_1 + \pi_2 \qquad &\leqq 3 \\
\pi_2 + \pi_3 &\leqq -6 \\
\pi_1 \qquad + \pi_3 &\leqq -2 \\
\pi_1 - 2\pi_2 + \pi_3 &= -4 \\
-2\pi_1 + \pi_2 - \pi_3 &\leqq 0 \\
-\pi_1 - 2\pi_2 - \pi_3 &\leqq -7
\end{aligned}$$

Here, the complementary slackness theorem was only able to determine that one dual constraint must hold as an equation at every dual optimum solution. This single equation is not sufficient to compute a dual optimum solution. But one can find a dual optimum solution by solving the above system using a Phase I approach.

Exercises

4.8 Prove from first principles, as previously, that $\bar{x}$ and $\bar{\pi}$ are primal and dual optimal, respectively, to (4.6) iff they are feasible to the respective problems and satisfy the conditions in in Example 4.11.

4.9 Write down the complementary slackness conditions when the primal and dual are as stated in Examples 4.5 and 4.6 and prove the complementary slackness theorem for each case from first principles.

4.10 Write down the complementary slackness conditions for the transportation problems discussed in Sections 4.2 and 4.4.

Suppose B is an optimum basis for the LP (4.6) in standard form. Let $\bar{c}$ be the vector of relative cost coefficients with respect to the basis B. If $\bar{c}_j > 0$, then $x_j = 0$ in every optimum feasible solution of (4.6), from the complementary slackness conditions for optimality. Also any feasible solution of (4.6) in which $x_j = 0$, for all j such that $\bar{c}_j > 0$, is an optimum feasible solution of (4.6). This shows that the set of optimum feasible solutions of (4.6) is characterized as in Section 3.10.

Economic Interpretation

Consider a diet problem. An example of a diet problem is the problem of the family (4.1) and its dual is the nutrient pill manufacturer's problem (4.2). The pill manufacturer has to set the prices of the nutrient pills in such a way that the total value (in terms of these pill prices) of the nutrients in a unit of each food available to the family is less than or equal to the cost per unit of this food in the market.

In a general diet problem, suppose there are m nutrients, and n foods, and let a_{ij} be the number of units of nutrient i per unit of food j. Let c_j be the cost in the market per unit of food j. Let b_i be the minimal requirement for nutrient i. Let π_i denote the price charged per unit of nutrient i (in the form of a pill) by the pill manufacturer. The constraints on the prices are $\sum_{i=1}^{m} \pi_i a_{ij} \leqq c_j$, for all $j = 1$ to n. Given these prices, the family checks whether it can meet its nutrient requirements cheaper by buying the foods rather than the nutrient pills. Let x_j denote the number of units of food j that it buys. Since the cost per unit of food j is c_j, and it can get all the nutrition contained in one unit of food j from the pill manufacturer at a cost of $\sum_i \pi_i a_{ij}$, the family would definitely not like to buy food j if $\sum_i \pi_i a_{ij} < c_j$. Thus it would make $x_j = 0$ whenever $\sum_i \pi_i a_{ij} < c_j$ holds. Similarly, if the pill manufacturer associates a positive price per unit of nutrient i (i.e., $\pi_i > 0$), then the family would try to meet the bare minimal requirements for this nutrient (since the restriction is only that the amount of nutrient i in the diet should be greater than or equal to the minimal requirement, and getting any more than the minimal requirement costs money under this setup). Thus, if $\pi_i > 0$, the family will try to satisfy $\sum_j a_{ij}x_j = b_i$. A similar interpretation can be given about the prices that the pill manufacturer will adopt knowing the amounts of foods that the family is buying.

These are precisely the complementary slackness conditions for this primal-dual pair. When they are satisfied, there is no incentive for the family to change the levels of its food purchases or for the pill manufacturer to change its prices. The minimum cost incurred by the family is precisely the maximum revenue that the pill manufacturer can make. Therefore, these conditions may be considered *economic stability conditions* or *equilibrium conditions*.

4.5.9 How to Check Whether a Given Feasible Solution of a Linear Program Is Optimal

In some practical applications, a feasible solution of the LP model can be constructed either from the current decisions under use or by clever guesswork. The methods discussed in this section are useful in such applications. If the LP under discussion is in standard form, and if the feasible solution available is a BFS for which an associated basis is known, then we can check whether this BFS is optimal by checking whether the basis satisfies the optimality criterion. The methods discussed in this section rely on the complementary slackness theorem, and they can be used to check whether any given feasible solution of an LP (whether it is a BFS or not) is optimal to it.

Consider the LP (4.8) and its dual (4.9). Suppose we want to check whether the known feasible solution of (4.8), $(\bar{x}, \bar{y})$ is optimal. Let $\mathbf{J} = \{j: j,$ such that $1 \leqq j \leqq n_1$ and $\bar{x}_j > 0\}$, $\mathbf{P} = \{p: p,$ such that $1 \leqq p \leqq m_2$ and $E_{p.}x + F_{p.}y - g_p > 0\}$. Consider the following system of equality constraints in the dual variables.

$$\begin{aligned} \pi B_{.t} + \mu F_{.t} &= d_t && \text{for } t = 1 \text{ to } n_2 \\ \pi A_{.j} + \mu E_{.j} &= c_j && \text{for all } j \in \mathbf{J} \\ \mu_p &= 0 && \text{for all } p \in \mathbf{P} \end{aligned} \qquad (4.27)$$

The system (4.27) consists of all the equality constraints in the dual problem, and all the constraints that must hold as equations if $(\bar{x}, \bar{y})$ is optimal to (4.8), as determined by the complementary slackness theorem. If (4.27) is inconsistent, $(\bar{x}, \bar{y})$ is not optimal to (4.8). If (4.27) has a unique solution $(\bar{\pi}, \bar{\mu})$; $(\bar{x}, \bar{y})$ is an optimum solution of (4.8) iff $(\bar{\pi}, \bar{\mu})$ satisfies the remaining dual constraints, namely,

$$\begin{aligned} \pi A_{.j} + \mu E_{.j} &\leqq c_j \quad \text{for all } j = 1 \text{ to } n_1 \quad j \notin \mathbf{J}, \\ \mu_p &\geqq 0 \quad \text{for all } p = 1 \text{ to } m_2 \quad p \notin \mathbf{P} \end{aligned} \tag{4.28}$$

If (4.27) is consistent but does not have a unique solution, the complementary slackness conditions have not led to enough equality constraints to compute the dual solution uniquely. We can only conclude that $(\bar{x}, \bar{y})$ is optimal to (4.8) iff the system of constraints (4.28) and (4.27) together have a feasible solution.

4.5.10 Discussion on How Various Algorithms Solve the Problem

There are three sets of conditions to be satisfied by the optimal solutions of a primal-dual pair of LPs: (1) primal feasibility, (2) dual feasibility, and (3) complementary slackness property.

The primal simplex algorithm discussed in Chapter 2 obtains a sequence of solutions satisfying primal feasibility and complementary slackness properties. When dual feasibility is attained, it terminates. The dual simplex algorithm (see Chapter 6) obtains a sequence of solutions satisfying complementary slackness and dual feasibility properties. Hence, that algorithm terminates when it attains primal feasibility. There are other algorithms for solving LPs, known as *out-of-kilter algorithm*, *complementary pivot algorithm*, etc. These algorithms generate a sequence of solutions of the primal-dual pair of LPs, satisfying primal and dual feasibility properties. These algorithms terminate when the complementary slackness property is satisfied.

4.5.11 Which Problem in a Primal-Dual Pair Is Easier to Solve?

When you solve an LP in standard form by the simplex algorithm, you simultaneously solve the dual as well. The primal and the dual solutions corresponding to any basis for this problem can be obtained as in Section 4.5.5. Actually, the revised simplex algorithm, which is a computationally efficient method of carrying out the simplex algorithm, obtains both the primal and dual solutions corresponding to each basis, and at termination it provides the solutions of both the primal and dual problems. Thus, solving either of the problems of a primal-dual pair of LPs poses the same order of difficulty. However, one of the problems in the pair might have some special structure for which there is an efficient special algorithm. In that case, it may be advantageous to solve that problem in the pair using the efficient special algorithm. When the optimum solution of that problem is obtained, an optimum solution of the other problem in the pair can be obtained directly by using the complementary slackness conditions. An example of this is the problem of computing the project cost curve in a project network, discussed in reference [1.68]. Its dual turns out to be a minimum cost flow problem on a directed network for which there are several efficient special algorithms. Another example of this is a linear programming problem of the following form: maximize cx, subject to $Ax \leqq b$, where A is a given matrix of order $m \times n$. An LP of this type appears in a method for solving linear complementarity problems discussed in reference [4.29]. We can, of course, solve this problem by the primal simplex method by transforming it into standard form first. The transformation of this LP into standard form involves the elimination of the unrestricted variables x after slack variables are introduced to make $Ax \leqq b$ into equality constraints, a lot of cumbersome work. However, it can be verified that the dual of this problem is already in standard form. So, for LPs with this special structure, it is much more convenient to apply the primal simplex method on its dual.

Thus, for solving a specially structured LP, it may be worthwhile to write its dual and solve that problem in the primal-dual pair, for which an efficient special algorithm exploiting the structure of that problem is available.

4.6 OTHER INTERPRETATIONS AND APPLICATIONS OF DUALITY

4.6.1 Dual Variables as Lagrange Multipliers

Problems in which an objective function has to be optimized subject to equality constraints can be handled by the Lagrange multiplier technique of classicial calculus. However, when the constraints in the problem include some inequality constraints, the Lagrange multipliers associated with them have to satisfy sign restrictions and complementary slackness conditions. In this context, the Lagrange multipliers are known as the Karush–Kuhn–Tucker–Lagrange multipliers. For LPs, it turns out that the dual variables are the Karush–Kuhn–Tucker–Lagrange multipliers. Also the complementary slackness conditions in the *Karush–Kuhn–Tucker necessary conditions for optimality* (see references [4.12, 4.13, 4.14]) in an LP are precisely the complementary slackness conditions for optimality in a primal-dual pair of LPs. Consider the LP (4.4). Associate the multiplier vector π to the constraints and the multiplier vector μ to the sign restrictions. The *Karush–Kuhn–Tucker–Lagrangian* corresponding to (4.4) is $L(x; \pi, \mu) = cx - \pi(Ax - b) - \mu x$. The Karush–Kuhn–Tucker necessary conditions for optimality in (4.4) are

$$\frac{\partial L}{\partial x} = c - \pi A - \mu = 0$$

$$Ax \geqq b$$

$$x \geqq 0$$

$$\pi(Ax - b) = 0$$

$$\mu x = 0$$

$$\pi \geqq 0 \qquad \mu \geqq 0$$

Equivalently, these are

$$\pi A \leqq c, \qquad \pi \geqq 0$$

$$Ax \geqq b, \qquad x \geqq 0$$

$$\pi(Ax - b) = (c - \pi A)x = 0$$

These are precisely the dual and primal feasibility and the complementary slackness conditions for optimality.

4.6.2 The Saddle Point Property

Consider the primal LP (4.4) and its dual, (4.5). The Karush–Kuhn–Tucker–Lagrangian associated with (4.4) is $L(x: \pi, \mu) = cx - \pi(Ax - b) - \mu x$. The point $(\bar{x}, \bar{\pi}, \bar{\mu})$ is said to be a *saddle point* of $L(x; \pi, \mu)$ if $\bar{\pi} \geqq 0$, $\bar{\mu} \geqq 0$, and

$$L(x, \bar{\pi}, \bar{\mu}) \geqq L(x, \bar{\pi}, \bar{\mu}) \geqq L(\bar{x}; \pi, \mu)$$

$$\text{for all } \pi \geqq 0,\ \mu \geqq 0, \text{ and all } x$$

Exercises

4.11 Prove that $(\bar{x}, \bar{\pi}, \bar{\mu})$ is a saddle point of $L(x; \pi, \mu)$ iff $\bar{x}$ is an optimum feasible solution of (4.4), $\bar{\pi}$ is an optimum feasible solution of (4.5), and $\bar{\mu} = c - \bar{\pi}A$.

4.6.3 Dual Variables as Partial Derivatives of the Optimal Objective Value

Consider the LP in standard form (4.6). Assume that A, c are fixed, and for a given b let $f(b)$ be the minimum value of $z(x)$ in the problem. The dual problem is (4.7). We wish to investigate how $f(b)$ behaves as b changes, but A, c remain fixed. Let $\bar{x}$, $\bar{\pi}$ be a pair of primal and dual primal feasible solutions for fixed b and assume that $\bar{\pi}$ is the unique dual optimum solution. By the duality theorem, $f(b) = \bar{\pi}b$. So in a crude sense one could conclude that $\partial f/\partial b = \bar{\pi}$; that is, if the optimum dual feasible solution is unique, it is the partial derivative vector of the optimal objective value of the LP with respect to the right-hand-side constants. When the dual optimum feasible solution $\bar{\pi}$ is unique, the $\bar{\pi}_i$ are the *marginal rates of change of* $f(b)$ for small perturbations in b_i. Thus $\bar{\pi}_i$ is the *marginal value* of the item associated with the ith constraint in (4.6). Mathematical proofs of the differentiability properties of the function $f(b)$ are given in Section 8.14. There it is shown that $f(b)$ is a piecewise linear function that is differentiable at a point b if the optimum dual feasible solution at that point is unique. At points b where (4.6) has alternate optimum dual feasible solutions, $f(b)$ is not differentiable, and for the determination of marginal values at such points b see Section 8.15.

Thus when it is unique, the optimal dual feasible solution can be used in sensitivity analysis when A, c remain fixed, but when small perturbations occur in the right-hand-side constants, $f(b)$ is not known explicitly as a function of b. However, for any given b vector, the value of $f(b)$ can be computed by solving (4.6), and the vector of partial derivatives of $f(b)$, that is, $\nabla f(b)$, is the optimum dual solution. If it is required to explore for an optimum b vector in some specified feasible region, this provides the necessary information for finding it by using some subgradient optimization method. Methods for solving parametric LPs discussed in Chapter 8 provide very useful tools to explore for an optimum b vector.

Similarly, suppose A, b remain fixed, but the cost vector c is likely to vary. Let $g(c)$ be the minimum objective value in the problem as a function of c. If $\bar{x}$ is the unique optimum feasible solution of this LP, then $g(c) = c\bar{x}$. Hence, in a crude sense, we could conclude that $\partial g/\partial c = \bar{x}$ and that $\bar{x}_j$ are the marginal rates of change in $g(c)$ when A, b remain fixed but when c_j is subject to small perturbations.

The Usefulness of the Optimum Dual Solution in Practical Applications

Consider the family's diet problem (4.1) and its dual (4.2). From Example 4.7 we know that the optimum dual solution is $\bar{\pi} = (3, 8)$. From the above discussion, we conclude that under the present conditions, the family can expect to spend 3 cents per each additional unit of vitamin A required and 8 cents per each additional unit of vitamin C required. This kind of analysis, in which the values of the dual variables in the optimum dual solution are interpreted as the rates of change in the optimum objective value per unit change in the values of the primal right-hand-side constants, is known as *marginal analysis.* One should, of course, remember that these rates 3 and 8 are only valid for small changes in the requirements from their present levels and if large changes occur in the requirement levels, the values of these marginal values might change.

Again consider the product mix problem of a company that can manufacture n different products using m different resources (raw materials, etc.). Let b_i units be the amount of resource i available, $i = 1$ to m. Let c_j \$/unit be the profit per unit of product j manufactured. Suppose the company wants to maximize its total profit. The product mix model for this company is a profit maximization LP, in n nonnegative variables, subject to m resource availability constraints. Let π_i be the dual variable associated with the constraint imposed by the ith resource availability. Let $\bar{\pi} = (\bar{\pi}_1, \ldots, \bar{\pi}_m)$ be the optimum dual solution. Then $\bar{\pi}_i$ \$ is the amount by which the company's profits go up, if the availability of resource i goes up by one unit from its present level of b_i units. If $\bar{\pi}_i > 0$, this implies that the company can increase its profits by increasing its supply of resource i. In most practical situations, the supplies of resources can be increased by spending some money. Suppose the company can acquire more units of resource i at the cost of α_i \$/unit. If $\bar{\pi}_i > \alpha_i$, the company can expect to increase its total net profit at the marginal value of $(\bar{\pi}_i - \alpha_i)$\$ per each additional unit of resource i acquired at this cost. By comparing the marginal values corresponding to the various resources, the company can, for example, determine which resource is its most critical resource.

Example 4.14

Consider the fertilizer problem (1.1) discussed in Section 1.1.3. Let π_i be the dual variable associated with the material balance inequality of raw material i in this problem, $i = 1$ to 3. The resources of this company are raw materials 1, 2, 3. The optimum dual solution corresponding to this problem can be verified to be $\bar{\pi} = (5, 5, 0)$. This implies that the company's total net profit is not expected to change when small changes occur in the availability of raw material 3 (from its present availability level of 500 tons per month), because $\bar{\pi}_3 = 0$.

Also, availability of additional quantities (from present levels) of either raw material 1 or 2 in small amounts is expected to increase the company's total net profit by \$5/ton, because $\bar{\pi}_1 = \bar{\pi}_2 = 5$. We illustrate this. With the present data we know that the optimum solution of this problem is $(x_1, x_2) = (300, 900)$ with a maximum profit of \$13,500. Now suppose the availability of raw material 1 changes from 1500 tons to 1501 tons, while all the other data in the problem remain unchanged. It can be verified that the optimum solution of the modified problem is $(x_1, x_2) = (301, 899)$, leading to a maximum net profit of \$13,505, which is \$5 $= \bar{\pi}_1$ more than that in the original problem. Similarly, suppose that all the data in the original problem (1.1) remain unchanged, except that the availability of raw material 2 is changed from 1200 tons to 1201 tons. It can be verified that the optimum solution of this modified problem is $(x_1, x_2) = (299, 902)$, leading to a maximum net profit of \$13,505, which is \$5 $= \bar{\pi}_2$ more than that in the original problem. These facts imply that the fertilizer manufacturing company can improve its profit by acquiring additional amounts of raw material 1 or 2 at a cost of at most \$5/ton for either raw material.

In practical linear programming models, the right-hand-side constants may be either the limits on resource availability or targets for production levels of the various products. By comparing the marginal values corresponding to the various resource availabilities and production targets, the company can determine its most critical resource, or its most critical production target, etc. These marginal values are the weights for assessing the relative importance (under present conditions) of the various resources in terms of their contribution to the objective function.

They can be used to set up priorities for additional resource acquisition and other such planning activities. A linear programming model can be used to determine not only the present optimum solution, but, by utilizing the optimum dual solution, it will indicate changes that can be made in resource availabilities, etc., that will lead to further improvements in the objective function. When coupled with the techniques of sensitivity analysis (see Chapters 8 and 9), this provides very useful quantitative information for planning to change the existing constraints to enable better performance.

Note 4.5: *A word of caution is necessary.* Let $b = (b_1, \ldots, b_m)^{\mathrm{T}}$ be the vector of right-hand-side constants in an LP, and $\bar{\pi} = (\bar{\pi}_1, \ldots, \bar{\pi}_m)$ the associated optimum dual solution. Suppose all the data in this LP remain unchanged except for one b_i. In Chapters 8 and 9, it is shown that the set of all values of b_i for which $\bar{\pi}$ remains an optimum dual solution is a closed interval, say $b_i^1 \leqq b_i \leqq b_i^2$. Methods for computing this interval are discussed in Chapter 9. As long as all the data in the LP other than b_i remains unchanged, and the value of b_i satisfies $b_i^1 \leqq b_i \leqq b_i^2$, $\bar{\pi}$ remains an optimum dual solution, and $\bar{\pi}_i$ is the marginal value. Once the value of b_i leaves this interval, the marginal values may change.

4.6.4 Solving a Linear Program Is Equivalent to Solving a System of Equations in Nonnegative Variables

From the results in Chapters 2 and 3, every LP can be expressed in the form (4.4), whose dual is (4.5). From the sufficient optimality criterion of Theorem 4.4, if x, π are, respectively, primal and dual feasible and satisfy $cx = \pi b$, then x, π must be optimal to the respective problems. Rewriting all these constraints together, we have a system of inequalities. Introducing $y = \pi^{\mathrm{T}}$, and introducing the slack vectors u, v, this system becomes

$$\begin{aligned} Ax \qquad\qquad - v \quad &= b \\ A^{\mathrm{T}}y \quad + u \quad &= c^{\mathrm{T}} \\ cx - b^{\mathrm{T}}y \qquad\qquad &= 0 \\ x \geqq 0 \quad y \geqq 0 \quad u \geqq 0 \quad & v \geqq 0 \end{aligned} \tag{4.29}$$

Thus, if $(\bar{x}, \bar{y}, \bar{u}, \bar{v})$ is a feasible solution of (4.29), then $\bar{x}$ is an optimum feasible solution of (4.4) and $\bar{\pi} = \bar{y}^{\mathrm{T}}$ is an optimum feasible solution of the dual problem.

Conversely, if $\bar{x}$ is any optimum feasible solution of (4.4), and $\bar{y} = \bar{\pi}^{\mathrm{T}}$ is any optimum feasible solution of the dual, and $\bar{v} = A\bar{x} - b$, $\bar{u} = c^{\mathrm{T}} - A^{\mathrm{T}}\bar{y}$, then $(\bar{x}, \bar{y}, \bar{u}, \bar{v})$ is feasible to (4.29).

Thus, solving the LP (4.4) is equivalent to solving the system of equations (4.29) in nonnegative variables. Hence any algorithm for solving systems of simultaneous linear equations in nonnegative variables can be used for solving LPs directly without a need for optimization. Conversely, any algorithm for solving LPs can be used to solve a system of simultaneous linear equations in nonnegative variables, by solving a Phase-I-type problem.

Numerous algorithms for solving systems of simultaneous linear equations in unrestricted variables (based on pivoting, elimination, reduction, triangularization, etc.) are discussed in classical linear algebra. However, the problem of solving a system of simultaneous linear equations in *nonnegative variables* is much harder, and had to wait until the linear programming era.

4.6.5 Formulation of a Linear Program as a Linear Complementarity Problem

Consider the LP (4.4) and its dual (4.5). Introducing slack vectors, these systems can be written together as $u = -A^{\mathrm{T}}y + c^{\mathrm{T}}$, $v = Ax - b$; $u, v, x, y \geqq 0$. From the complementary slackness theorem, if the solution u, v, x, y of this satisfies $u^{\mathrm{T}}x + v^{\mathrm{T}}y = 0$, then x is optimal to (4.4) and $\pi = y^{\mathrm{T}}$ is optimal to its dual (4.5). Let

$$\xi = \begin{pmatrix} u \\ \cdots \\ v \end{pmatrix} \quad \eta = \begin{pmatrix} x \\ \cdots \\ y \end{pmatrix} \quad M = \begin{pmatrix} 0 & \vdots & -A^{\mathrm{T}} \\ \cdots & \cdots & \cdots \\ A & \vdots & 0 \end{pmatrix} \quad \nu = \begin{pmatrix} c^{\mathrm{T}} \\ \cdots \\ -b \end{pmatrix}$$

Solving the LP (4.4) is equivalent to solving the system:

$$\begin{gathered} \xi - M\eta = \nu \\ \xi \geqq 0 \qquad \eta \geqq 0 \\ \xi^{\mathrm{T}}\eta = 0 \end{gathered} \tag{4.30}$$

The constraints $\xi \geqq 0$, $\eta \geqq 0$, $\xi^{\mathrm{T}}\eta = 0$ imply that $\xi_r \eta_r = 0$ for all r. Hence at most one variable in each pair (ξ_r, η_r) can take a positive value in any solution of (4.30). Hence each pair of variables (ξ_r, η_r) is known as a *complementary pair of variables* in (4.30). Problems similar to (4.30) are known as *linear complementarity problems*. See references [1.92, 1.93] and the papers by R. W. Cottle, G. B. Dantzig, C. E. Lemke, and J. T. Howson [4.4, 4.15, 4.16] for algorithms to solve them.

4.6.6 Applications in Two-Person Zero-Sum Matrix Games

A *zero-sum two-person game consists* of two players, each with a fixed finite set of choices in each play of the game. Suppose player I has m choices, namely, 1, 2, . . . , m. Suppose player II has n choices that are called 1, 2, . . . , n. If in a play of the game, player I chooses i and player II chooses j, player I gains a_{ij} dollars and player II gains $-a_{ij}$ dollars. In this case, a negative gain is considered as a loss and vice versa. Thus, the matrix $A = (a_{ij})$ is known as player I's payoff matrix and $-A$ is player II's payoff matrix.

In playing a game of this type, the players might pick their choices in a probabilistic manner just before each play. Let x_i be the probability with which player I picks his choice i, for $i = 1$ to m, and let y_j be the probability with which player II picks her choice j, for $j = 1$ to n.

The probability vector $x = (x_1, \ldots, x_m)^{\mathrm{T}}$ is known as player I's *mixed strategy*, and correspondingly the vector $y = (y_1, \ldots, y_n)^{\mathrm{T}}$ is player II's mixed strategy.

Suppose player I is interested in finding a mixed strategy x, which gives an expected payoff of at least α. The expected payoff of player I, if he uses mixed strategy x and player II chooses j, is $x_1 a_{1j} + \cdots + x_m a_{mj}$. Since α is the guaranteed minimum expected payoff to player I under mixed strategy x, irrespective of what player II does, it must satisfy $x_1 a_{1j} + \cdots + x_m a_{mj} \geqq \alpha$ for all $j = 1$ to n. Hence, player I's problem is the LP

$$\begin{array}{lrl}
\text{Maximize} & 0x_1 + \cdots + 0x_m + \alpha & \\
\text{Subject to} & a_{11}x_1 + \cdots + a_{m1}x_m - \alpha & \geqq 0 \\
 & \vdots \qquad \vdots \qquad \vdots & \vdots \\
 & a_{1n}x_1 + \cdots + a_{mn}x_m - \alpha & \geqq 0 \\
 & x_1 + \cdots + x_m & = 1 \\
 & x_i \geqq 0 \quad \text{for all } i & \\
 & \alpha \text{ unrestricted in sign} &
\end{array} \qquad (4.31)$$

In a similar manner, let β denote the maximum expected payoff of player II, irrespective of what player I does. The problem of finding an optimal strategy y and payoff β of player II is the LP:

$$\begin{array}{lrl}
\text{Maximize} & 0y_1 + \cdots + 0y_n + \beta & \\
\text{Subject to} & -a_{11}y_1 - \cdots - a_{1n}y_n - \beta & \geqq 0 \\
 & \vdots \qquad \vdots \qquad \vdots & \vdots \\
 & -a_{m1}y_1 - \cdots - a_{mn}y_n - \beta & \geqq 0 \\
 & y_1 + \cdots + y_n & = 1 \\
 & y_j \geqq 0 \quad \text{for all } j & \\
 & \beta \text{ unrestricted in sign} &
\end{array} \qquad (4.32)$$

Exercises

4.12 Construct one feasible solution (x, α) to (4.31) in terms of (a_{ij}), thus proving that it is always feasible. Do the same for (4.32).

It is easily seen that problems (4.31) and (4.32) are a primal-dual pair of LPs. Since both problems are feasible, we conclude from the fundamental duality theorem that the maximum expected payoff of player I is equal to (−minimum expected loss of player II). See the book by David Gale [1.69] for a detailed treatment of the theory of zero-sum games and its relationship to linear programming.

4.6.7 Applications in Proving Theorems of Alternatives for Linear Systems

An important theorem in the theory of convex sets is the *separating hyperplane theorem*. A consequence of this theorem is the following. Let $\mathbf{K}$ be a closed convex subset of $\mathbf{R}^m$ and $b \in \mathbf{R}^m$. If $b \notin \mathbf{K}$, there exists a hyperplane in $\mathbf{R}^m$ that separates the point b from the subset $\mathbf{K}$; that is, there exists $(\pi_1, \ldots, \pi_m) \neq 0$ and real number α, such that

$$\begin{aligned} \pi b &> \alpha \\ \pi y &\leqq \alpha \qquad \text{for all } y \in \mathbf{K} \end{aligned} \qquad (4.33)$$

In this case, the hyperplane $\pi y = \alpha$ is known as a *separating hyperplane* separating the point b from the convex set $\mathbf{K}$ (Figure 4.2). Consider the special case of this theorem when $\mathbf{K}$ is a convex polyhedral cone. If $\mathbf{K}$ is a convex polyhedral cone, every point in it is a nonnegative linear combination of the extreme rays of the cone. Hence there exists a finite set of points, say, $\{A_{.1}, \ldots, A_{.n}\}$ in $\mathbf{R}^m$ such that $\mathbf{K} = \{y : y = x_1 A_{.1} + \cdots + x_n A_{.n},\ x_j \geqq 0 \text{ for all } j\}$. Applying the separating hyperplane theorem to this case, we conclude that if b is not in the convex

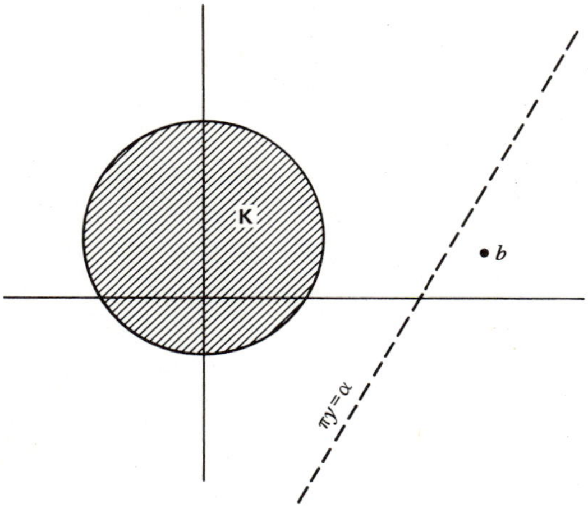

Figure 4.2 **K** is the set of all points inside or on the circle. The dashed hyperplane separates the point b from **K**.

polyhedral cone **K**, there exists $\pi = (\pi_1, \ldots, \pi_m) \neq 0$, and a real number α satisfying (4.33). Since $0 \in \mathbf{K}$, we must have $\alpha \geqq 0$. Clearly $\pi y \leqq \alpha$ for all $y \in \mathbf{K}$ iff $\pi A_{.j} \leqq \alpha$ for all $j = 1$ to n, and $\alpha \geqq 0$. If there exists a $y \in \mathbf{K}$ such that $\pi y > 0$, since **K** is a cone, πy is unbounded above on **K**. These facts imply that α must be equal to zero.

Combining these, we get the following result for this case. Either $b \in \mathbf{K}$ or there exists a $\pi = (\pi_1, \ldots, \pi_m) \neq 0$, satisfying $\pi b > 0$ and $\pi A_{.j} \leqq 0$ for all $j = 1$ to n. Equivalently, either the following system (I) has a solution x, or system (II) has a solution π, but not both.

$Ax = b$	(I)	$\pi A \leqq 0$	(II)
$x \geqq 0$		$\pi b > 0$	

This result guarantees that exactly one of the two systems (I) or (II) has a feasible solution, and the other system is inconsistent. In mathematical programming literature results of this type are known as *theorems of the alternatives*. The theorem of the alternative that is deduced here from the separating hyperplane theorem of convex sets is known as *Farkas' lemma*. It is a fundamental result that has found numerous applications in linear, nonlinear, and integer programming. We now show how Farkas' lemma can be proved using the duality theorem.

THEOREM 4.9 FARKAS' LEMMA Let A be a given $m \times n$ matrix and b a column vector of order $m \times 1$, then exactly one of the systems (I), (II) above has a feasible solution, and the other system is inconsistent.

Proof: Consider the primal LP

$$\begin{aligned} \text{Minimize} \quad & z(x) = 0x \\ \text{Subject to} \quad & Ax = b \\ & x \geqq 0 \end{aligned}$$

Its dual is

$$\begin{aligned} \text{Maximize} \quad & v(\pi) = \pi b \\ \text{Subject to} \quad & \pi A \leqq 0 \end{aligned}$$

If system (I) has a feasible solution, it is optimal to the primal problem, since $z(x) = 0$ for every x. So by the duality theorem the maximum value of $v(\pi)$ is zero in the dual problem. Hence, there cannot exist any dual feasible solution that makes $v(\pi) > 0$. Thus (II) is infeasible.

The dual problem here is always feasible, since $\pi = 0$ is a dual feasible solution. If system (I) is infeasible, the primal problem is infeasible, and by the fundamental duality theorem $v(\pi)$ is unbounded above on the dual feasible solution set. Thus, (II) must have a feasible solution. ■

Example 4.15

We will now provide a geometric illustration of Farkas' lemma. Let

$$A = \begin{pmatrix} 1 & 3 & 3 \\ 3 & 2 & 1 \end{pmatrix} \qquad b = \begin{pmatrix} 4 \\ 0 \end{pmatrix}$$

Each of the column vectors b, $A_{.j}$ for $j = 1, 2, 3$, and the row vector $\pi = (\pi_1, \pi_2)$ can be represented as points in $\mathbf{R}^2$. Let $\mathbf{H}_j$ be the hyperplane in $\mathbf{R}^2$ through the origin (hence it is a straight line through the origin in $\mathbf{R}^2$) for which the ray of $A_{.j}$ is the normal at the origin, $j = 1, 2, 3$. The inequality $\pi A_{.j} \leqq 0$ implies that either $\pi = 0$ or that the angle between the rays of π and $A_{.j}$ is not less than 90°. So $\pi A_{.j} \leqq 0$ holds iff π lies in the closed half-space $\mathbf{H}_j^{\leqq}$, on the side of $\mathbf{H}_j$ not containing the point $A_{.j}$, indicated by an arrow in Figure 4.3. So $\pi A \leqq 0$ holds iff $\pi \in \bigcap_{j=1}^{3} \mathbf{H}_j^{\leqq}$. Let the hyperplane $\mathbf{H}_b$ be the hyperplane through the origin to which the ray of b is the normal at the origin. Here it is the vertical axis in Figure 4.3. The inequality $\pi b > 0$ will hold iff π lies in the open half-space $\mathbf{H}_b^{>}$, marked by an arrow on $\mathbf{H}_b$ in Figure 4.3. So the vector $\pi = (\pi_1, \pi_2)$ satisfies (II) in

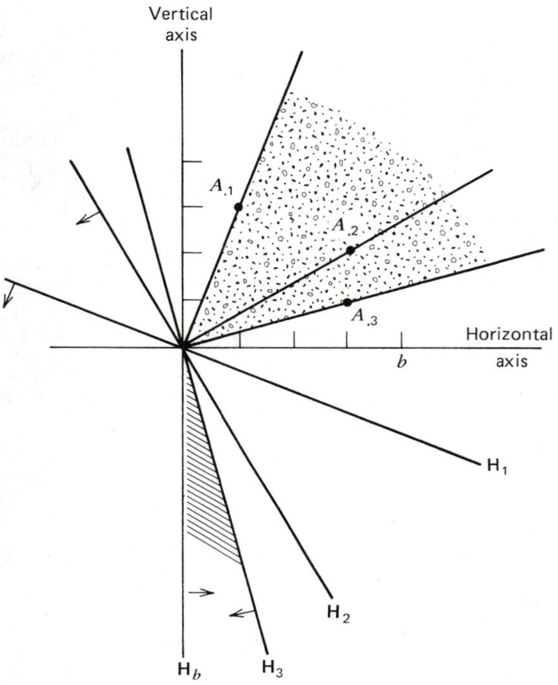

Figure 4.3

Farkas' lemma with the data given above, iff $\pi \in \mathbf{H}_b^> \cap (\bigcap_{j=1}^3 \mathbf{H}_j^{\leqq})$. Also, there exists a solution for system (I) in Farkas' lemma with the data given above iff $b \in \text{Pos}(A) = \{y: y = Ax, \text{ for some } x \geqq 0\}$, marked by dots in Figure 4.3. So, in this case, Farkas' lemma states that either $b \in \text{Pos}(A)$ or that $\mathbf{H}_b^> \cap (\bigcap_{j=1}^3 \mathbf{H}_j^{\leqq}) \neq \varnothing$, but not both. From Figure 4.3, we see that $b \notin \text{Pos}(A)$ and that $\mathbf{H}_b^> \cap (\bigcap_{j=1}^3 \mathbf{H}_j^{\leqq}) \neq \varnothing$. In fact, $\mathbf{H}_b^> \cap (\bigcap_{j=1}^3 \mathbf{H}_j^{\leqq})$ is the conical region shaded by lines in Figure 4.3.

It can also be verified that as the point b moves from its present position into the dotted cone Pos(A) (after such a change, system (I) will have a feasible solution), the hyperplane $\mathbf{H}_b$ moves correspondingly in such a way that the modified open half-space $\mathbf{H}_b^>$ has an empty intersection with $\bigcap_{j=1}^3 \mathbf{H}_j^{\leqq}$ (so, after such a change, system (II) will have no solution).

Exercises

4.13 Construct a proof of the fundamental duality theorem using Farkas' lemma.

4.14 Using Farkas' lemma, prove the following theorem of the alternative. If A is an $m \times n$ matrix and $b \in \mathbf{R}^m$, exactly one of the following two systems has a solution.

$$\text{Either} \quad Ax = b \qquad \text{(I)}$$

$$\text{or} \quad \pi A = 0 \quad \pi b = 1 \qquad \text{(II)}$$

Theorems of the alternatives have numerous applications in mathematical programming. We state two important theorems of alternatives.

THEOREM 4.10 MOTZKIN'S THEOREM OF THE ALTERNATIVES Let A, C, D be given matrices of order $m_1 \times n$, $m_2 \times n$, $m_3 \times n$, respectively, where $m_1 \geqq 1$. Exactly one of the following systems has a feasible solution, and the other system is inconsistent.

Either		Or
$Ax > 0$		$\pi A + \mu C + \gamma D = 0$
$Cx \geqq 0$	(4.34)	$\pi \geq 0, \quad \mu \geqq 0$
$Dx = 0$		(4.35)

Proof: It is clear that if (π, μ, γ) is a feasible solution of (4.35), then $(\alpha\pi, \alpha\mu, \alpha\gamma)$ is also a solution of (4.35) for any $\alpha > 0$. Using this, we can replace the constraints $\pi \geq 0$ in (4.35) by $\sum_{i=1}^m \pi_i = 1$, $\pi \geqq 0$. Also replace γ by $\gamma^+ - \gamma^-$, where $\gamma^+ \geqq 0$, $\gamma^- \geqq 0$. Write down the resulting system in the form of system (I) discussed under Farkas' lemma (Theorem 4.9). Now verify that the associated system (II) under Farkas' lemma corresponding to this is just equivalent to (4.34). This, by Farkas' lemma, proves this theorem. ■

THEOREM 4.11 TUCKER'S THEOREM OF THE ALTERNATIVES Let A, C, D be given matrices of orders $m_1 \times n$, $m_2 \times n$, $m_3 \times n$, respectively, where $m_1 \geqq 1$. Exactly one of the following systems has a feasible solution, and the other system has no feasible solution.

Either	Or
$Ax \geq 0$	$\pi A + \mu C + \nu D = 0$
$Cx \geqq 0$	$\pi > 0 \quad \mu \geqq 0$
$Dx = 0$	

Proof: This theorem of the alternatives can be proved using Farkas' lemma in a manner very similar to Motzkin's theorem of the alternatives. The proof is left to the reader. ■

For a complete discussion of theorems of alternatives and direct proofs for them without relying on the fundamental duality theorem of linear programming, see Chapter 2 of O. L. Mangasarian's book [2.16].

4.7 NECESSARY AND SUFFICIENT CONDITIONS FOR A LINEAR PROGRAM TO HAVE A UNIQUE OPTIMUM SOLUTION

In Section 3.10, we developed conditions under which an LP in standard form has a unique optimum solution, and efficient methods for checking this uniqueness. Here we discuss necessary and sufficient conditions for a given optimum feasible solution of an LP in general form to be the unique optimum feasible solution, and provide computationally efficient methods to check this uniqueness. These results are from O. L. Mangasarian [4.17]. For convenience in dealing with this general model we assume that any sign restrictions or upper- and lower-bound restrictions on the variables are included among the set of inequality constraints in the model. Once this is done, the LP assumes the form:

$$\begin{aligned} \text{Minimize} \quad & z(x) = cx \\ \text{Subject to} \quad & Ax = b \\ & Dx \geqq d \end{aligned} \tag{4.36}$$

where A and D are given matrices of orders $m_1 \times n$, $m_2 \times n$, respectively. Let $\pi = (\pi_1, \ldots, \pi_{m_1})$, $\mu = (\mu_1, \ldots, \mu_{m_2})$ be the row vectors of dual variables associated with the equality and inequality constraints in (4.36), respectively. The dual of (4.36) is:

$$\begin{aligned} \text{Maximize} \quad & \pi b + \mu d \\ \text{Subject to} \quad & \pi A + \mu D = c \\ & \mu \geqq 0 \end{aligned} \tag{4.37}$$

THEOREM 4.12 Let $\bar{x}$ be a known optimum feasible solution for (4.36). This $\bar{x}$ is the unique optimum feasible solution for (4.36) iff it remains an optimum feasible solution to a perturbed LP obtained by introducing an arbitrary but sufficiently small perturbation in the cost vector in (4.36). That is, for each $c^* \in \mathbf{R}^n$, there exists a positive number α such that $\bar{x}$ remains an optimum solution to the perturbed LP (4.38) for all $0 \leqq \lambda \leqq \alpha$.

$$\begin{aligned} \text{Minimize} \quad & (c + \lambda c^*)x \\ \text{Subject to} \quad & Ax = b \\ & Dx \geqq d \end{aligned} \tag{4.38}$$

Outline of Proof: Let $\mathbf{K}$ denote the set of feasible solutions of (4.36). If $\bar{x}$ is the unique optimum solution for (4.36), by the results in Chapter 3 it must be an extreme point of $\mathbf{K}$. Let $\mathbf{H} = \{x: cx = c\bar{x}\} \subset \mathbf{R}^n$. If $c = 0$, $\bar{x}$ is the unique optimum solution of (4.36), iff $\mathbf{K} = \{\bar{x}\}$, and in this case the theorem is verified to be trivially true. Thus we assume that $c \neq 0$. So $\mathbf{H}$ is a hyperplane in $\mathbf{R}^n$, and $\bar{x}$ is the unique optimum solution for (4.36) iff $\mathbf{K} \cap \mathbf{H} = \{\bar{x}\}$ and $\mathbf{K} \subset \{x: cx \geqq c\bar{x}\}$ (see Figure 4.4). In this case, as we move from $\bar{x}$ along any of the edges of $\mathbf{K}$ containing $\bar{x}$, the value of cx strictly increases. By introducing a sufficiently small perturbation, suppose c is changed to $\hat{c}$, $\hat{c} \neq 0$. Let $\hat{\mathbf{H}} = \{x: \hat{c}x = \hat{c}\bar{x}\}$. When $\|c - \hat{c}\|$ is sufficiently small, clearly the conditions $\mathbf{K} \cap \mathbf{H} = \{\bar{x}\}$, $\mathbf{K} \subset \{x: cx \geqq c\bar{x}\}$ hold iff $\mathbf{K} \cap \hat{\mathbf{H}} = \{\bar{x}\}$, $\mathbf{K} \subset \{x: \hat{c}x \geqq \hat{c}\bar{x}\}$. This provides an intuitive justification for the theorem. For a rigorous proof of this theorem, see [4.17]. ■

Theorem 4.12 can also be proved directly, using the result in Exercise 3.34. See Exercise 4.47.

THEOREM 4.13 Let $(\bar{\pi}, \bar{\mu})$ be an optimum dual solution associated with (4.36). This $(\bar{\pi}, \bar{\mu})$ is the unique optimum dual solution associated with (4.36) iff it remains an optimum dual solution for the perturbed problem obtained by introducing an arbitrary but sufficiently small perturbation in the right-hand-side constants vectors in (4.36); that is, given column vectors $g \in \mathbf{R}^{m_1}$ and $h \in \mathbf{R}^{m_2}$, there

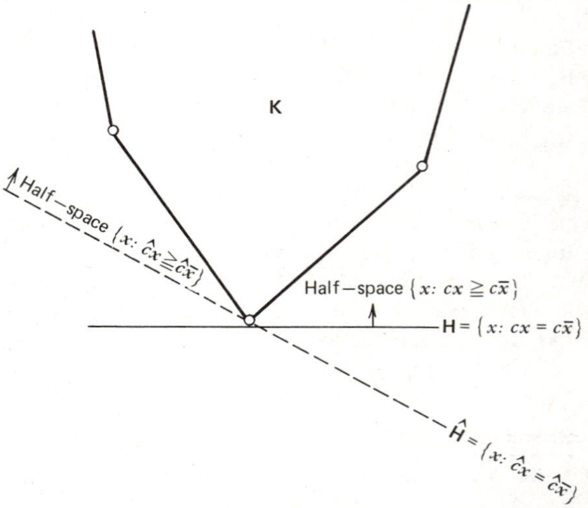

Figure 4.4 If $\mathbf{K} \cap \mathbf{H} = \{\bar{x}\}$ and $\mathbf{K} \subset \{x: cx > c\bar{x}\}$, these conditions continue to hold when $\mathbf{H}$ is tilted slightly around $\bar{x}$ by perturbing c.

exists a positive number α such that $(\bar{\pi}, \bar{\mu})$ remains an optimum dual solution for the perturbed LP:

$$\begin{aligned} &\text{Minimize} && cx \\ &\text{Subject to} && Ax = b + \lambda g \qquad (4.39) \\ &&& Dx \geqq d + \lambda h \end{aligned}$$

for all $0 \leqq \lambda \leqq \alpha$.

Proof: This follows by applying the result in Theorem 4.12 to the dual problem (4.37). ■

We now state two theorems that provide efficient computational procedures for checking the uniqueness of the optimum solutions of (4.36) and its dual (4.37).

THEOREM 4.14 Let $\bar{x}$ be a given optimum solution of (4.36) and let $(\bar{\pi}, \bar{\mu})$ be an optimum dual solution associated with (4.36). Define

$$\mathbf{I} = \{t: D_{t.}\bar{x} = d_t\} \qquad \mathbf{J} = \{t: \bar{\mu}_t > 0\}$$
$$\mathbf{L} = \{t: t \in \mathbf{I} \text{ and } \bar{\mu}_t = 0\}$$

$D_{\mathbf{I}}$, $D_{\mathbf{J}}$, $D_{\mathbf{L}}$ are matrices whose rows are $D_{t.}$ for t in sets **I**, **J**, **L**, respectively.

$d_{\mathbf{I}}$, $d_{\mathbf{J}}$, $d_{\mathbf{L}}$ are column vectors of d_t for t in sets **I**, **J**, **L**, respectively.

$\mu_{\mathbf{J}}$, $\mu_{\mathbf{L}}$ are row vectors of μ_t for t in **J**, **L**, respectively.

$\bar{x}$ is the unique optimum solution of (4.36) iff the set of column vectors of $\begin{pmatrix} A \\ \cdots \\ D_{\mathbf{I}} \end{pmatrix}$ is linearly independent, and either $\mathbf{L} = \emptyset$, or $\mathbf{L} \neq \emptyset$, and the optimum objective value in the LP

$$\begin{aligned} &\text{Maximize} && \sum_{t \in \mathbf{L}} D_{t.}x \\ &\text{Subject to} && Ax = 0 \qquad D_{\mathbf{J}}x = 0 \qquad D_{\mathbf{L}}x \geqq 0 \end{aligned}$$

is zero. Equivalently, the requirement when $\mathbf{L} \neq \emptyset$ is that the system

$$Ax = 0 \qquad D_{\mathbf{J}}x = 0 \qquad D_{\mathbf{L}}x \geqq 0 \qquad \sum_{t \in \mathbf{L}} D_{t.}x = 1$$

has no feasible solution. Another equivalent requirement when $\mathbf{L} \neq \emptyset$, is that the system

$$\pi A + \mu_{\mathbf{J}} D_{\mathbf{J}} + \mu_{\mathbf{L}} D_{\mathbf{L}} = 0$$
$$\mu_t \geqq 1 \qquad \text{for each } t \in \mathbf{L}$$

has at least one feasible solution.

THEOREM 4.15 In the notation of Theorem 4.14, $(\bar{\pi}, \bar{\mu})$ is the unique optimum solution of the dual (4.37) iff the set of row vectors of $\begin{pmatrix} A \\ \cdots \\ D_{\mathbf{J}} \end{pmatrix}$ is linearly independent, and either $\mathbf{L} = \emptyset$ or $\mathbf{L} \neq \emptyset$, and the optimum objective value in the LP

$$\begin{aligned} &\text{Maximize} && \sum_{t \in \mathbf{L}} \mu_t \\ &\text{Subject to} && \pi A + \mu_{\mathbf{J}} D_{\mathbf{J}} + \mu_{\mathbf{L}} D_{\mathbf{L}} = 0 \\ &&& \mu_{\mathbf{L}} \geqq 0 \end{aligned}$$

is zero. Equivalently, the requirement when $\mathbf{L} \neq \emptyset$ is that the system

$$\pi A + \mu_{\mathbf{J}} D_{\mathbf{J}} + \mu_{\mathbf{L}} D_{\mathbf{L}} = 0 \qquad \sum_{t \in \mathbf{L}} \mu_t = 1 \qquad \mu_{\mathbf{L}} \geqq 0$$

has no feasible solution. Another equivalent requirement when $\mathbf{L} \neq \emptyset$ is that the following system have at least one feasible solution x.

$$Ax = 0 \qquad D_{\mathbf{J}}x = 0 \qquad D_{t.}x \geqq 1 \qquad \text{for each } t \in \mathbf{L}$$

For proofs of Theorems 4.14 and 4.15, see reference [4.17] of O. L. Mangasarian. Also, see Exercise 4.47.

4.8 STABILITY OF THE LINEAR PROGRAMMING MODEL

The results presented in this section need some background knowledge about the convergence properties of infinite sequences of points in $\mathbf{R}^n$. We provide a summary of these properties first.

An infinite sequence of points $\{x^r: r = 1, 2, \ldots\}$ in $\mathbf{R}^n$ is said to converge in the *limit* to the given point x^* if, for each $\varepsilon > 0$, there exists a positive integer N such that $\|x^r - x^*\| < \varepsilon$ for all $r \geqq$ N. As an example the sequence in $\mathbf{R}^1$, $\{x^r: x^r = (1/r), r \geqq 1$ and integer$\}$ converges to zero. However, the sequence $\{x^r$: where $x^r = (1/r)$ if $r = 2s$ for some positive integer s, and $x^r = 1$ if $r = 2s + 1$ for some positive integer $s\}$ does not converge. A point $x^* \in \mathbf{R}^n$, is said to be a *limit point* or an *accumulation point* for the infinite sequence $\{x^r: r = 1, 2, \ldots\}$ of points in $\mathbf{R}^n$ if for every $\varepsilon > 0$ and positive integer N, there exists a positive integer $r >$ N such that $\|x^r - x^*\| < \varepsilon$. If x^* is a limit point of the sequence $\{x^r: r = 1, 2, \ldots\}$, then there exists a subsequence of this sequence, say $\{x^{r_k}: k = 1, 2, \ldots\}$, where $\{r_k: k = 1, 2, \ldots\}$ is a monotonic increasing sequence of positive integers, which converges in the limit to x^*. If the sequence $\{x^r: r = 1, 2, \ldots\}$ converges in the limit to x^*, then x^* is the only limit point for this sequence. A sequence that does not converge may have no limit point (for example, the sequence of positive integers in $\mathbf{R}^1$ has no limit point) or may have any number of limit points. As an example, consider the sequence of numbers in $\mathbf{R}^1$, $\{x^r$, where $x^r = 1/r$, if $r = 2s$ for some positive integer s, otherwise $x^r = 1 + (1/r)$, if $r = 2s + 1$ for some nonnegative integer $s\}$. This sequence has two limit points, namely 0 and 1. The subsequence $\{x^{2s}: s = 1, 2, \ldots\}$ of this sequence converges to the limit point 0, while the subsequence $\{x^{2s+1}: s = 1, 2, \ldots\}$ converges to the limit point 1.

The discussion in this section also needs knowledge of some of the basic properties of compact subsets of $\mathbf{R}^n$. See references [4.9, 4.23, 4.27, 4.28] for this background information.

When using an optimization model, there are two aspects of stability involved. One is the numerical stability of the algorithm used to obtain an optimum solution of the model. During the computational process of applying the algorithm, round-off errors creep in whenever multiplications or divisions are performed, and these errors accumulate. For this reason, the optimum solution obtained at the end of the algorithm may be quite different from the true optimum solution of the original model. An algorithm, or its implementation, is said to be *numerically stable*, if the net effect of the round-off error accumulation during the computation is less than a specified error bound. Numerical stability of algorithms for LPs is discussed in Sections 5.6, 5.7, and Chapter 7.

Another aspect of stability is the stability of the model itself under small perturbations in the values of the data elements in the model. In most practical applications of optimization, the values of the data elements are usually estimated from practical considerations, and they are liable to include unknown error terms. Even when they are known precisely, their values are likely to change over time. If the model yields an optimum solution with the data fixed at some values, but becomes infeasible when small but arbitrary perturbations occur in these values, or if small changes in the values of the data elements lead to large changes in the optimum objective value, the *model is unstable*. An optimization model is considered to be *stable* if small but arbitrary changes in the values of the data elements in the model lead only to correspondingly small changes in the optimum objective value and the optimum solution. In this section, we study only this aspect of *stability in linear programming models*. We now present some illustrative examples.

Example 4.16

We first consider the stability of systems of linear equations with no optimization involved. Consider the following two systems of linear equations.

$$x_1 + x_2 + x_3 + x_4 = 1 \tag{4.40}$$

and the system,

$$\begin{aligned} x_1 + x_2 + x_3 + x_4 &= 1 \\ 2x_1 + 2x_2 + 2x_3 + 2x_4 &= 2 \end{aligned} \tag{4.41}$$

One of the constraints in (4.41) in redundant, and actually (4.40) and (4.41) are *equivalent*, as discussed in Section 3.18. However, when a small change occurs in one of the right-hand-side constants in (4.41), while all the other data in it remain unchanged, (4.41) becomes infeasible. For example, if the first right-hand-side constant in (4.41) is changed to $1 + \varepsilon$, we led to the perturbed system

$$\begin{aligned} x_1 + x_2 + x_3 + x_4 &= 1 + \varepsilon \\ 2x_1 + 2x_2 + 2x_3 + 2x_4 &= 2 \end{aligned}$$

which is inconsistent for every $\varepsilon \neq 0$, however small. So the system (4.41) is unstable, while (4.40) is stable, even though both these systems are equivalent in a mathematical sense, as discussed in Section 3.18.

In general, any feasible system of linear equations that contains a redundant equation is unstable, because it can be made infeasible by changing only one right-hand-side constant in it slightly.

Example 4.17

Consider the following LP in one variable $x \in \mathbf{R}^1$, due to D. Gale [1.69, Exercise 13, pp. 94–95].

$$\begin{aligned} f(\alpha) = \text{Maximum value of} \quad & x \in \mathbf{R}^1 \\ \text{Subject to} \quad & x \leqq 1 \\ & \alpha x \leqq 0 \\ & x \geqq 0 \end{aligned}$$

where α is a nonnegative real number. The optimum

objective value in this LP $f(\alpha)$ exists for all $\alpha \geqq 0$. It can be verified that $f(0) = 1$ and $f(\alpha) = 0$, for $\alpha > 0$. But in the neighborhood of $\alpha = 0$, a change in the value of α, however small, leads to a change in the value of $f(\alpha)$ from 0 to 1. Hence this LP is unstable in the neighborhood of $\alpha = 0$.

Example 4.18

Consider the following LP due to S. M. Robinson [4.26]:

$$\begin{array}{lrl} \text{Minimize} & 3x_1 + & x_2 + x_3 + 3x_4 \\ \text{Subject to} & x_1 + (4/3)\, x_2 + 2x_3 & = 3/2 \\ & x_2 + 3x_3 & = 3/2 \\ & x_1 + x_2 + x_3 + x_4 & = 1 \\ & x_j > 0 & \text{for all } j \end{array}$$

$\bar{x} = (0, 3/4, 1/4, 0)^{\mathrm{T}}$ is an optimum solution of this problem, and $\bar{\pi} = (1/2, -1/6, 1/2)$ is an optimum dual solution. The optimum objective is 1. Changing the input-output coefficient 4/3 to $((4/3) - \varepsilon)$ for $\varepsilon > 0$, however small, leads to a perturbed problem, for which $\hat{x} = (1/2, 0, 1/2, 0)^{\mathrm{T}}$ is an optimum solution, with $\hat{\pi} = (0, -2/3, 3) + (1/\varepsilon)(4/3, -4/9, -4/3)$ is an optimum dual solution, and 2 is the optimum objective value. So this LP is unstable.

When studying the stability of an LP model that belongs to a special class of models, it may be necessary to restrict the perturbations in the values of the data elements in the model to satisfy certain properties, so that the perturbed model also belongs to the same special class. As an example, consider the balanced transportation model (13.1) discussed in Chapter 13. Since this model contains a redundant equality constraint, it is unstable when viewed as a general LP model. However, in the class of balanced transportation models, the input-output coefficients in (13.1) never change, and the right-hand-side constants a_i, b_j always satisfy $a_i, b_j > 0$ for all i, j; and $\sum_i a_i = \sum_j b_j$. When the stability of the balanced transportation model (13.1) is examined under perturbation in which the c_{ij} may change, and the a_i, b_j may change subject to the above conditions, it turns out to be quite stable. So in studying the stability of a particular LP model with a special character, it is always necessary to take the conditions satisfied by the data elements in the model into account, and to allow only perturbations that retain the special character of the model.

In this section we discuss the stability of two classes of LP models. One of them is the LP model in symmetric form, the other is the general LP model in which there may be some equality and some inequality constraints, some unrestricted and some sign restricted variables.

Stability in the Linear Programming Model in Symmetric Form

Consider the LP (4.4) in Example 4.3 in symmetric form, where A is a matrix of order $m \times n$. We now study the stability of this LP model under perturbations in the data elements in A, b, c. In this discussion we assume that the nonnegativity restrictions on the variables remain unchanged under these perturbations. All the data can be conveniently represented in matrix form as

$$\boldsymbol{A} = \begin{pmatrix} A & \vdots & b \\ \cdots & \cdots & \cdots \\ c & \vdots & 0 \end{pmatrix}$$

$\boldsymbol{A}$ is a matrix of order $(m+1) \times (n+1)$, the $(m+1, n+1)$ element is always zero, and except for this element, all the remaining elements correspond to data elements. Now define $\psi(\boldsymbol{A}, x, \pi) = cx - \pi(Ax - b) = cx + \pi b - \pi Ax$. An $\bar{x} \in \mathbf{R}^n$, $\bar{\pi} \in \mathbf{R}^m$ together satisfy the *saddle point conditions* corresponding to the primal-dual pair of LPs (4.4), (4.5) if

$$\bar{x} \geqq 0 \qquad \bar{\pi} \geqq 0$$

and

$$\psi(\boldsymbol{A}, \bar{x}, \pi) \leqq \psi(\boldsymbol{A}, \bar{x}, \bar{\pi}) \leqq \psi(\boldsymbol{A}, x, \bar{\pi}) \qquad (4.42)$$

for all

$$x \geqq 0 \qquad \pi \geqq 0$$

NOTE 4.6: In writing the Lagrangian in Section 4.6.2, the nonnegativity restrictions on x are also considered as constraints and they are multiplied by the multiplier vector μ and included in the Lagrangian. Hence the inequality in the saddle point condition mentioned in Section 4.6.2 must hold for all $x \in \mathbf{R}^n$. Here, the nonnegativity restrictions on x are not included in writing the Lagrangian $\psi(\boldsymbol{A}, x, \pi)$. Hence, the inequality in the saddle point condition stated here has to hold over $x \geqq 0$.

We now prove a theorem relating optimum solutions for (4.4) and (4.5) with saddle points for $\psi(\boldsymbol{A}, x, \pi)$.

This theorem is used later to establish results on stability in the model (4.4).

THEOREM 4.16 Given $\bar{x} \in \mathbf{R}^n$, $\bar{\pi} \in \mathbf{R}^m$, $\bar{x}$ is optimal to (4.4) and $\bar{\pi}$ is optimal to (4.5) iff $(\bar{x}, \bar{\pi})$ together satisfy the saddle point conditions (4.42).

Proof: We first prove that if $\bar{x}$ is optimal to (4.4), and $\bar{\pi}$ is optimal to (4.5), then $(\bar{x}, \bar{\pi})$ is a saddle point of $\psi(\boldsymbol{A}, x, \pi)$. By the complementary slackness theorem and the fundamental duality theorem, we have

$$\bar{\pi}(A\bar{x} - b) = 0 \tag{4.43}$$

$$c\bar{x} = \bar{\pi}b \tag{4.44}$$

Now, for all $\pi \geqq 0$, we have $\psi(\boldsymbol{A}, \bar{x}, \pi) = c\bar{x} - \pi(A\bar{x} - b) \leqq c\bar{x}$ (because $A\bar{x} - b \geqq 0$ by primal feasibility of $\bar{x}$) $= c\bar{x} - \bar{\pi}(A\bar{x} - b)$ (by (4.43)) $= \psi(\boldsymbol{A}, \bar{x}, \bar{\pi})$. And for all $x \geqq 0$, we have $\psi(\boldsymbol{A}, x, \bar{\pi}) = cx - \bar{\pi}(Ax - b) = (c - \bar{\pi}A)x + \bar{\pi}b \geqq \bar{\pi}b$ (since $c - \bar{\pi}A \geqq 0$ by the dual feasibility of $\bar{\pi}$) $= c\bar{x}$ (by (4.44)) $= c\bar{x} - \bar{\pi}(A\bar{x} - b) = \psi(\boldsymbol{A}, \bar{x}, \bar{\pi})$. These facts together imply that $\bar{x}$, $\bar{\pi}$ together satisfy (4.42).

We will now prove that if $\bar{x}$, $\bar{\pi}$ satisfy (4.42), they are optimal to (4.4) and (4.5), respectively. From (4.42) we have

$$\pi(A\bar{x} - b) \geqq \bar{\pi}(A\bar{x} - b) \qquad \text{for all } \pi \geqq 0 \tag{4.45}$$

The right-hand side of (4.45) is a constant. If $A_{i.}\bar{x} - b_i < 0$ for any i, by choosing a nonnegative π in which π_i is arbitrarily large, the left-hand side of (4.45) can be made into a negative number with an arbitrarily large absolute value, violating (4.45). So (4.45) implies that $A_{i.}\bar{x} - b_i \geqq 0$ for all i, that is, $A\bar{x} \geqq b$. Together with the nonnegativity conditions on $\bar{x}$ in (4.42), this implies that $\bar{x}$ is feasible to (4.4). Again, from (4.42) we have

$$(c - \bar{\pi}A)\bar{x} \leqq (c - \bar{\pi}A)x \qquad \text{for all } x \geqq 0 \tag{4.46}$$

The left-hand side of (4.46) is a constant. Since (4.46) holds for all $x \geqq 0$ by an argument similar to the above, it implies that $(c - \bar{\pi}A) \geqq 0$. This, together with the nonnegativity restrictions on $\bar{\pi}$ in (4.42), implies that $\bar{\pi}$ is feasible to (4.5).

Substituting $\pi = 0$ in (4.42) we get $c\bar{x} \leqq c\bar{x} - \bar{\pi}(A\bar{x} - b)$, that is $\bar{\pi}(A\bar{x} - b) \leqq 0$. Since $\bar{\pi} \geqq 0$, $(A\bar{x} - b) \geqq 0$, we also have $\bar{\pi}(A\bar{x} - b) \geqq 0$. These together imply $\bar{\pi}(A\bar{x} - b) = 0$. Substituting $x = 0$ in (4.42), we get $c\bar{x} - \bar{\pi}(A\bar{x} - b) \leqq \bar{\pi}b$, that is $(c - \bar{\pi}A)\bar{x} \leqq 0$. Since $\bar{x} \geqq 0$, $(c - \bar{\pi}A) \geqq 0$, we also have $(c - \bar{\pi}A)\bar{x} \geqq 0$. These together imply $(c - \bar{\pi}A)\bar{x} = 0$. So, $\bar{x}$, $\bar{\pi}$ are feasible to the respective problems and satisfy the complementary slackness conditions, and hence $\bar{x}$ is optimal to (4.4), and $\bar{\pi}$ is optimal to (4.5). ■

Let $\mathbf{K}(\boldsymbol{A})$, $\boldsymbol{\Gamma}(\boldsymbol{A})$ denote the set of feasible solutions of (4.4) and (4.5), respectively. Let $\mathbf{K}^0(\boldsymbol{A})$, $\boldsymbol{\Gamma}^0(\boldsymbol{A})$ denote the set of optimum feasible solutions of (4.4) and (4.5), respectively. We now prove a theorem on the necessary and sufficient conditions for $\mathbf{K}^0(\boldsymbol{A})$, $\boldsymbol{\Gamma}^0(\boldsymbol{A})$, to be bounded.

THEOREM 4.17 (1) $\mathbf{K}^0(\boldsymbol{A})$ is nonempty and bounded iff $\mathbf{K}(\boldsymbol{A})$ is nonempty and

$$\begin{gathered}\text{there exists no } y \text{ satisfying}\\ Ay \geqq 0 \qquad cy \leqq 0 \qquad y \geq 0\end{gathered} \tag{4.47}$$

(2) $\boldsymbol{\Gamma}^0(\boldsymbol{A})$ is nonempty and bounded iff $\boldsymbol{\Gamma}(\boldsymbol{A})$ is nonempty and

$$\begin{gathered}\text{there exists no } \mu \text{ satisfying}\\ \mu A \leqq 0 \qquad \mu b \geqq 0 \qquad \mu \geq 0\end{gathered} \tag{4.48}$$

(3) $\mathbf{K}^0(\boldsymbol{A})$, $\boldsymbol{\Gamma}^0(\boldsymbol{A})$ are both nonempty and bounded iff both (4.47) and (4.48) hold.

Proof of (1): Suppose $\mathbf{K}^0(\boldsymbol{A})$ is nonempty and bounded. Let I_r denote the identity matrix of order r. By the fundamental duality theorem of linear programming, both (4.4) and (4.5) must be feasible. So $A^{\mathrm{T}}\pi^{\mathrm{T}} + I_n u^{\mathrm{T}} = c^{\mathrm{T}}$, π^{T}, $u^{\mathrm{T}} \geqq 0$ has a feasible solution $(\pi^{\mathrm{T}}, u^{\mathrm{T}})$. So by Farkas' lemma (Theorem 4.9)

$$\begin{gathered}\text{there exists no } y \text{ satisfying}\\ Ay \geqq 0 \qquad cy < 0 \qquad y \geq 0\end{gathered} \tag{4.49}$$

Now, let $\bar{x}$ be an optimum feasible solution of (4.4). So, $\mathbf{K}^0(\boldsymbol{A})$ is the set of feasible solutions of $Ax \geqq b$, $cx = c\bar{x}$, $x \geqq 0$, and it is bounded. Equivalently, the set of feasible solutions of

$$\begin{aligned} Ax - I_m s &= b \\ cx \qquad &= c\bar{x} \\ x \geqq 0 \qquad s &\geqq 0 \end{aligned} \tag{4.50}$$

is nonempty and bounded. By the results of Section 3.7, this implies that $y = 0$, $t = 0$ is the only feasible solution of the homogeneous system $Ay - I_m t = 0$, $cy = 0$, $y \geqq 0$, $t \geqq 0$. Equivalently, this implies that

$$\begin{gathered}\text{there exists no } y \text{ satisfying}\\ Ay \geqq 0 \qquad cy = 0 \qquad y \geq 0\end{gathered} \tag{4.51}$$

(4.49) and (4.51) together imply (4.47).

Conversely, suppose (4.4) is feasible and (4.47) holds. So (4.49) holds. By Farkas' lemma, this implies that (4.5) is feasible. Since (4.4) and (4.5) are both feasible, by the fundamental duality theorem, $\mathbf{K}^0(\boldsymbol{A}) \neq \emptyset$. Let $\bar{x} \in \mathbf{K}^0(\boldsymbol{A})$. Condition (4.47) implies (4.51) and by the results in Section 3.7, this implies that the set of feasible solutions of (4.50) is bounded, which in turn implies that $\mathbf{K}^0(\boldsymbol{A})$ is bounded. ■

Proof of (2): This is similar to the proof of (1). ■

Proof of (3): If $\mathbf{K}^0(\boldsymbol{A})$, $\boldsymbol{\Gamma}^0(\boldsymbol{A})$ are both nonempty and bounded, by (1) and (2) both (4.47) and (4.48) hold.

Conversely, suppose (4.47) and (4.48) hold. Condition (4.47) implies (4.49), and this implies that (4.5) is feasible, by the arguments used above. Similarly (4.48) implies that (4.4) is feasible, since both (4.4) and (4.5) are feasible, $\mathbf{K}^0(\boldsymbol{A})$, $\boldsymbol{\Gamma}^0(\boldsymbol{A})$ are both nonempty by the fundamental duality theorem. Also when (4.47) and (4.48) hold, by the arguments used above, both $\mathbf{K}^0(\boldsymbol{A})$ and $\boldsymbol{\Gamma}^0(\boldsymbol{A})$ are bounded. Hence $\mathbf{K}^0(\boldsymbol{A})$, $\boldsymbol{\Gamma}^0(\boldsymbol{A})$ are both nonempty and bounded if (4.47) and (4.48) hold. ■

The conditions (4.47) and (4.48) are known as the *regularity conditions* for the constraints in the primal-dual pair of LPs (4.4) and (4.5). Conditions (4.47) are the conditions for the regularity of the constraints in the dual problem (4.5), conditions (4.48) are the conditions for the regularity of the constraints in the primal problem (4.4). An equivalent way of stating these conditions is the following:

$$\text{for all } y \text{ such that } Ay \geqq 0,\ y \geq 0, \text{ we have } cy > 0 \tag{4.52}$$

$$\text{for all } \mu \text{ such that } \mu A \leqq 0,\ \mu \geq 0, \text{ we have } \mu b < 0 \tag{4.53}$$

Clearly (4.47) is equivalent to (4.52), and (4.48) is equivalent to (4.53).

Let A^* be an $m \times n$ matrix, b^* an $m \times 1$ vector, and c^* a $1 \times n$ vector. Consider the perturbed problem:

$$\begin{aligned} \text{Minimize} \quad & (c + \lambda c^*)x \\ \text{Subject to} \quad & (A + \lambda A^*)x \geqq (b + \lambda b^*) \\ & x \geqq 0 \end{aligned} \tag{4.54}$$

and its dual

$$\begin{aligned} \text{Maximize} \quad & \pi(b + \lambda b^*) \\ \text{Subject to} \quad & \pi(A + \lambda A^*) \leqq c + \lambda c^* \\ & \pi \geqq 0 \end{aligned} \tag{4.55}$$

The matrix:

$$\boldsymbol{A}^* = \begin{pmatrix} A^* & \vdots & b^* \\ \cdots & & \cdots \\ c^* & \vdots & 0 \end{pmatrix}$$

is known as the *perturbation matrix* corresponding to this perturbed problem. We denote the optimum objective value in the perturbed problem (4.54) by $f(\lambda, \boldsymbol{A}^*)$. We establish conditions under which there exists, for every possible choice of the perturbation matrix $\boldsymbol{A}^*$, an $\alpha > 0$, such that for all $0 \leqq \lambda \leqq \alpha$, (4.54) has an optimum feasible solution (i.e., $f(\lambda, \boldsymbol{A}^*)$ exists and is finite for each $0 \leqq \lambda \leqq \alpha$); $f(\lambda, \boldsymbol{A}^*)$ is continuous to the right of $\lambda = 0$; and the limit

$$f'(0, \boldsymbol{A}^*) = \text{limit of } (f(\lambda, \boldsymbol{A}^*) - f(0, \boldsymbol{A}^*))/\lambda \text{ over } \lambda \text{ tending to zero through positive values.} \tag{4.56}$$

exists. This limit $f'(0, \boldsymbol{A}^*)$ is known as the *marginal value of the LP* (4.4) with respect to the perturbation matrix $\boldsymbol{A}^*$.

Let $\mathbf{K}(\boldsymbol{A} + \lambda\boldsymbol{A}^*)$, $\mathbf{K}^0(\boldsymbol{A} + \lambda\boldsymbol{A}^*)$ denote the set of feasible solutions and the set of optimum feasible solutions, respectively, of (4.54). Let $\boldsymbol{\Gamma}(\boldsymbol{A} + \lambda\boldsymbol{A}^*)$, $\boldsymbol{\Gamma}^0(\boldsymbol{A} + \lambda\boldsymbol{A}^*)$ denote set of feasible solutions and the set of optimum feasible solutions, respectively, of the dual problem (4.55).

THEOREM 4.18 For each possible choice of the perturbation matrix $\boldsymbol{A}^*$, if there exists an $\alpha > 0$ such that for all $0 \leqq \lambda \leqq \alpha$, the perturbed problems (4.54) and (4.55) have optimum feasible solutions, then the regularity conditions (4.47) and (4.48) hold.

Proof: Let e_r denote the column vector in $\mathbf{R}^r$, all of whose entries are 1. Consider the perturbed problem obtained by selecting the perturbation matrix as $A^* = 0$, $b^* = 0$, $c^* = -e_n^{\mathrm{T}}$. For this choice of perturbation matrix, by the hypothesis of the theorem, there exists a positive number $\tilde{\alpha}$ such that (4.54) and (4.55) have optimum solutions $\tilde{x}(\lambda)$, $\tilde{\pi}(\lambda)$, respectively, for all $0 \leqq \lambda \leqq \tilde{\alpha}$. So by Theorem 4.16 we have

$$\begin{aligned} &(c - \tilde{\alpha}e_n^{\mathrm{T}})\tilde{x}(\tilde{\alpha}) - \tilde{\pi}(\tilde{\alpha})(A\tilde{x}(\tilde{\alpha}) - b) \\ &\quad \leqq (c - \tilde{\alpha}e_n^{\mathrm{T}})x - \tilde{\pi}(\tilde{\alpha})(Ax - b) \quad \text{for all } x \geqq 0 \end{aligned} \tag{4.57}$$

If there exists a y satisfying $Ay \geqq 0$, $y \geq 0$, substitute $x = \tilde{x}(\tilde{\alpha}) + y$ in (4.57). This leads to

$$0 \leqq (c - \tilde{\alpha}e_n^{\mathrm{T}})y - \tilde{\pi}(\tilde{\alpha})Ay \tag{4.58}$$

Since $y \geq 0$ and $\tilde{\alpha} > 0$, $\tilde{\alpha}e_n^{\mathrm{T}}y > 0$. Also, since $\tilde{\pi}(\tilde{\alpha}) \geqq 0$ and $Ay \geqq 0$, (4.58) implies that $cy > 0$. This implies that (4.52) holds, and hence the equivalent (4.47) holds.

By taking the perturbation matrix as $A^* = 0$, $b^* = e_m$, $c^* = 0$, and using the first inequality in the saddle point theorem corresponding to this perturbed problem, we can establish from similar arguments that (4.53) also holds. Hence the equivalent (4.48) holds. ■

THEOREM 4.19 If the regularity conditions (4.47) and (4.48) hold, then for each choice of the perturbation matrix $\boldsymbol{A}^*$ there exists an $\alpha > 0$ such that for all $0 \leqq \lambda \leqq \alpha$, the perturbed problems (4.54) and (4.55) have optimum feasible solutions.

Proof: Suppose not. Then there exists a perturbation matrix $\boldsymbol{A}^*$ such that for any positive number α_1, however small, there exists another positive number $\alpha_h < \alpha_1$, for which at least one of the problems among the perturbed primal or dual is infeasible when $\lambda = \alpha_h$. Make the perturbation matrix $\boldsymbol{A}^*$ as that satisfying this condition, and keep it fixed. Suppose it is the perturbed primal that is infeasible when $\lambda = \alpha_h$. So $(A + \alpha_h A^*)x - I_m s = b + \alpha_h b^*$, $x \geqq 0$, $s \geqq 0$, has no feasible solution (x, s). By Farkas' lemma this implies that there exists a vector $\mu(\alpha_h)$ satisfying

$$\begin{aligned} &\mu(\alpha_h)(A + \alpha_h A^*) \leqq 0 \\ &\mu(\alpha_h)(b + \alpha_h b^*) > 0 \qquad \mu(\alpha_h) \geqq 0 \end{aligned} \tag{4.59}$$

Clearly $\mu(\alpha_h) \neq 0$, and dividing it by its Euclidean norm if necessary, we assume that (4.59) has a solution $\mu(\alpha_h)$ satisfying $\|\mu(\alpha_h)\| = 1$. Instead of the perturbed primal, if it is the perturbed dual that is infeasible when $\lambda = \alpha_h$, using similar arguments we conclude that there exists a vector $y(\alpha_h)$ satisfying

$$\begin{aligned} (A + \alpha_h A^*)y(\alpha_h) &\geqq 0 \\ \|y(\alpha_h)\| = 1, \qquad y(\alpha_h) &\geqq 0 \\ (c + \alpha_h c^*)y(\alpha_h) &< 0 \end{aligned} \tag{4.60}$$

By using this argument over and over again, we generate an infinite monotonic decreasing sequence of positive numbers $\alpha_1, \alpha_2, \ldots$, converging to zero, satisfying the property that for each $h = 1, 2, \ldots$, either the system (4.59) has a solution $\mu(\alpha_h)$ satisfying $\|\mu(\alpha_h)\| = 1$, or (4.60) has a solution $y(\alpha_h)$, or both. Since for each $h = 1, 2, \ldots$, either (4.59) or (4.60) or both have solutions, it is possible to select from the sequence of numbers $\{\alpha_1, \alpha_2, \ldots\}$ an infinite subsequence $\{\alpha_{g_1}, \alpha_{g_2}, \ldots\}$ such that whenever $h = g_1, g_2, \ldots$, one of the two systems (4.59) or (4.60), say (4.59), always has a solution satisfying $\|\mu(\alpha_h)\| = 1$. Now, the infinite sequence of vectors $\mu(\alpha_{g_1})$, $\mu(\alpha_{g_2}), \ldots$ satisfies $\|\mu(\alpha_{g_r})\| = 1$, for all $r = 1, 2, \ldots$ and hence all these vectors lie on the boundary of the unit sphere in $\mathbf{R}^m$ with the origin as the center. Since the unit sphere in $\mathbf{R}^m$ is a compact set, given an infinite sequence of points in it, a subsequence of the sequence can be selected that converges to a limit point of the sequence (see references [4.9, 4.23]). Using this property select a subsequence of $\mu(\alpha_{g_1})$, $\mu(\alpha_{g_2}), \ldots$, which we will denote by $\mu(\alpha^1)$, $\mu(\alpha^2), \ldots$ such that $\mu(\alpha^r)$ tends to a limit $\hat{\mu}$ on the boundary of the unit sphere in $\mathbf{R}^m$. Since the sequence $\alpha_{g_1}, \alpha_{g_2}, \ldots$ is a monotonic decreasing sequence converging to zero, its subsequence $\alpha^1, \alpha^2, \ldots$ also converges to zero. By letting r tend to infinity in $\mu(\alpha^r)$, these facts imply that $\hat{\mu}$ satisfies: $\hat{\mu}A \leqq 0$, $\hat{\mu} \geqq 0$, $\hat{\mu}b \geqq 0$, $\|\hat{\mu}\| = 1$. So $\hat{\mu}$ satisfies

$$\hat{\mu}A \leqq 0 \qquad \hat{\mu}b \geqq 0 \qquad \hat{\mu} \geq 0$$

which contradicts (4.48). If the initial subsequence $\alpha_{g_1}, \alpha_{g_2}, \ldots$ is such that whenever $h = g_1, g_2, \ldots$, the system (4.60) has a solution. Using similar arguments, we obtain a vector $\hat{y}$ that contradicts (4.47). So, if (4.47) and (4.48) hold, for any $\boldsymbol{A}^*$ there always exists a positive number α such that $\mathbf{K}(\boldsymbol{A} + \lambda\boldsymbol{A}^*)$ and $\boldsymbol{\Gamma}(\boldsymbol{A} + \lambda\boldsymbol{A}^*)$ are both nonempty for all $0 \leqq \lambda \leqq \alpha$, and by the duality theorem this implies that (4.54), (4.55) both have optimum feasible solutions for all $0 \leqq \lambda \leqq \alpha$. ■

THEOREM 4.20 If the regularity conditions (4.47) and (4.48) hold, then for each choice of the perturbation matrix $\boldsymbol{A}^*$ there exists an $\alpha' > 0$ such that for all $0 \leqq \lambda \leqq \alpha'$, the perturbed problems (4.54) and (4.55) both have optimum feasible solutions, and $f(\lambda, \boldsymbol{A}^*)$ is bounded as λ varies in this interval.

Proof: Select the perturbation matrix $\boldsymbol{A}^*$, and keep it fixed. By Theorem 4.19 there exists a positive number α, such that for all $0 \leqq \lambda \leqq \alpha$, both the perturbed problems (4.54) and (4.55) have optimum feasible solutions, and hence $f(\lambda, \boldsymbol{A}^*)$ exists and is finite in the interval $0 \leqq \lambda \leqq \alpha$. We now prove that there exists a positive number $\alpha' \leqq \alpha$, such that for all λ in the interval $0 \leqq \lambda \leqq \alpha'$, $f(\lambda, \boldsymbol{A}^*)$ is bounded. Using the same argument as in Theorem 4.19 this result follows, if we can show that whenever α_1, $\alpha_2, \ldots$ is a monotonic decreasing sequence of positive numbers, all of them less than or equal to α, converging to zero, then the values of $f(\alpha_h, \boldsymbol{A}^*)$ are bounded for all $h = 1, 2, \ldots$. Suppose not. Then there must exist a monotonic decreasing sequence of positive numbers $\alpha_1, \alpha_2, \ldots$ converging to zero, such that $\alpha_h < \alpha$ for all $h = 1, 2, \ldots$, and $\{f(\alpha_h, \mathbf{A}^*): h = 1, 2, \ldots\}$ is unbounded.

Let $\bar{x}(\alpha_h)$ be an optimum feasible solution of (4.54) when $\lambda = \alpha_h$, $h = 1, 2, \ldots$. $\{f(\alpha_h, \mathbf{A}^*): h = 1, 2, \ldots\}$ is unbounded only if $\{\bar{x}(\alpha_h): h = 1, 2, \ldots\}$ is unbounded. In this case, it is possible to select a subsequence $\{\bar{x}(\alpha_h): h = g_1, g_2, \ldots\}$ such that $\{\|\bar{x}(\alpha_{g_r})\|: r = 1, 2, \ldots\}$ is a strictly monotonic increasing sequence of numbers diverging to $+\infty$. Define $y(\alpha_{g_r}) = \bar{x}(\alpha_{g_r})/(\|\bar{x}(\alpha_{g_r})\|)$, for $r = 1, 2, \ldots$. From the definitions, we have

$$\begin{aligned} (A + \alpha_h A^*)\bar{x}(\alpha_h) &\geqq b + \alpha_h b^* && h = 1, 2, \ldots \\ (c + \alpha_h c^*)\bar{x}(\alpha_h) &= f(\alpha_h, \mathbf{A}^*) && h = 1, 2, \ldots \end{aligned} \qquad (4.61)$$

So

$$(A + \alpha_{g_r} A^*) y(\alpha_{g_r}) \geqq (b + \alpha_{g_r} b^*)/\|\bar{x}(\alpha_{g_r})\| \quad r = 1, 2, \ldots \qquad (4.62)$$

We also have $y(\alpha_{g_r}) \geqq 0$ and $\|y(\alpha_{g_r})\| = 1$ for all $r = 1, 2, \ldots$. So $y(\alpha_{g_r}) \geq 0$ for all $r = 1, 2, \ldots$. All the points $y(\alpha_{g_r})$, $r = 1, 2, \ldots$ lie on the boundary of the unit sphere in $\mathbf{R}^n$, which is a compact set. So it is possible to select a subsequence of $y(\alpha_{g_r})$: $r = 1, 2, \ldots$, which we denote by $y(\alpha^1), y(\alpha^2), \ldots$, which converges to a limit point $\bar{y}$ of the sequence. The limit point $\bar{y}$ is also on the boundary of the unit sphere in $\mathbf{R}^n$. These facts imply that $\bar{y} \geq 0$. From (4.62), we have

$$(A + \alpha^p A^*) y(\alpha^p) \geqq (b + \alpha^p b^*)/\|\bar{x}(\alpha^p)\| \qquad (4.63)$$

Since $\{\alpha^p, p = 1, 2, \ldots\}$ is a subsequence of $\alpha_1, \alpha_2, \ldots$, and we know that α^p converges to zero as p tends to ∞ in (4.62), we have $A\bar{y} \geqq 0$, since $\|\bar{x}(\alpha^p)\|$ tends to $+\infty$ as p tends to ∞. Since $\bar{y} \geq 0$, $A\bar{y} \geqq 0$, by (4.47) we must have $c\bar{y} > 0$. But the limit of $f(\alpha^p, \mathbf{A}^*)/(\|\bar{x}(\alpha^p)\|)$ as p tends to ∞ is $c\bar{y}$ from (4.61), and hence this limit is strictly positive. So

$$f(\alpha^p, \mathbf{A}^*) > 0 \qquad \text{for all } p \text{ very large} \qquad (4.64)$$

Now let $\bar{\pi}(\alpha^p)$ be an optimum feasible solution of (4.55) when $\lambda = \alpha^p$, $p = 1, 2, \ldots$. So

$$\bar{\pi}(\alpha^p)(A + \alpha^p \mathbf{A}^*) \leqq c + \alpha^p c^* \qquad (4.65)$$

$$\bar{\pi}(\alpha^p)(b + \alpha^p b^*) = f(\alpha^p, \mathbf{A}^*) \qquad (4.66)$$

Since $\{f(\alpha^p, \mathbf{A}^*): p = 1, 2, \ldots\}$ is unbounded, from (4.66), $\{\bar{\pi}(\alpha^p): p = 1, 2, \ldots\}$ is unbounded, too. So it is possible to select a subsequence of this, which we will denote by $\{\bar{\pi}(\alpha_{q_r}): r = 1, 2, \ldots\}$ such that $\|\bar{\pi}(\alpha_{q_r})\| > 0$ for all r, and is monotonically increasing with r and diverging to $+\infty$. Let $\mu(\alpha_{q_r}) = \bar{\pi}(\alpha_{q_r})/\|\bar{\pi}(\alpha_{q_r})\|$. Using the same arguments as before, it is possible to select a subsequence of $\{\mu(\alpha_{q_r}): r = 1, 2, \ldots\}$, which converges to a limit point $\bar{\mu}$, where $\bar{\mu}$ satisfies $\bar{\mu}A \leqq 0$, $\bar{\mu} \geq 0$, and $\bar{\mu}b =$ limit of $f(\alpha_t, \mathbf{A}^*)/\|\bar{\pi}(\alpha_t)\|$ for α_t in this subsequence, and t tending to ∞. From (4.48) we have $\bar{\mu}b < 0$. From the above this implies that for all t very large such that α_t is in the above sequence, we have

$$f(\alpha_t, \mathbf{A}^*) < 0 \qquad (4.67)$$

The subsequence $\{\alpha_t\}$ above is itself a subsequence of α^p, $p = 1, 2, \ldots$, and therefore (4.67) contradicts (4.64). Hence $\{f(\alpha_h, \mathbf{A}^*): h = 1, 2, \ldots$ in the original sequence$\}$ is bounded. ■

THEOREM 4.21 Suppose the regularity conditions (4.47) and (4.48) hold. Choose the perturbation matrix $\mathbf{A}^*$ and fix it. Determine $\alpha' > 0$ as in Theorem 4.20. Let α_h, $h = 1, 2, \ldots$ be a monotonic decreasing sequence of positive numbers converging to zero, where $\alpha_h \leqq \alpha'$. For $h = 1, 2, \ldots$, let $\bar{x}(\alpha_h)$, $\bar{\pi}(\alpha_h)$ be optimum feasible solutions of (4.54) and (4.55), respectively, when $\lambda = \alpha_h$. Then the sequences $\bar{x}(\alpha_h)$, $h = 1, 2, \ldots$ and $\bar{\pi}(\alpha_h)$, $h = 1, 2, \ldots$ are both bounded.

Proof: Suppose on the contrary that one of these sequences, say $\bar{x}(\alpha_h)$, $h = 1, 2, \ldots$, is not bounded. In this case, it is possible to select a subsequence of this, $\{\bar{x}(\alpha_h): h = g_1, g_2, \ldots\}$, such that $\|\bar{x}(\alpha_{g_r})\|: r = 1, 2, \ldots$ is a strictly monotonic increasing sequence of numbers diverging to $+\infty$. Define $y(\alpha_{g_r}) = \bar{x}(\alpha_{g_r})/(\|\bar{x}(\alpha_{g_r})\|$ for $r = 1, 2, \ldots$, and from the sequence $y(\alpha_{g_r})$, $r = 1, 2, \ldots$ select a subsequence that converges to a limit point of this sequence, say $\bar{y}$, as in the proof of Theorem 4.20. From the arguments used in the proof of Theorem 4.20, we conclude that this $\bar{y}$ satisfies $\bar{y} \geq 0$, $A\bar{y} \geqq 0$. However, from (4.61) we have $(c + \alpha_{g_r} c^*)\bar{x}(\alpha_{g_r}) = f(\alpha_{g_r}, \mathbf{A}^*)$, that is

$$(c + \alpha_{g_r} c^*) y(\alpha_{g_r}) = f(\alpha_{g_r}, \mathbf{A}^*)/\|\bar{x}(\alpha_{g_r})\| \qquad (4.68)$$

By Theorem 4.20, $f(\alpha_{g_r}, \mathbf{A}^*)$ is bounded for all $r = 1, 2, \ldots$. So letting r tend to ∞ in (4.68) we conclude that $c\bar{y} = 0$. So $\bar{y}$ satisfies $\bar{y} \geq 0$, $A\bar{y} \geqq 0$, $c\bar{y} = 0$, a contradiction to (4.47). So the sequence $\{\bar{x}(\alpha_h), h = 1, 2, \ldots\}$ must be bounded. A similar argument involving (4.48) establishes that the sequence $\{\bar{\pi}(\alpha_h), h = 1, 2, \ldots\}$ must be bounded. ■

THEOREM 4.22 Let α be a positive number such that for all $0 \leqq \lambda \leqq \alpha$, the perturbed problems (4.54) and (4.55) both have optimum feasible solutions. Let $\bar{x} \in \mathbf{K}^0(\boldsymbol{A})$, $\bar{\pi} \in \boldsymbol{\Gamma}^0(\boldsymbol{A})$, $\bar{x}(\lambda) \in \mathbf{K}^0(\boldsymbol{A} + \lambda \boldsymbol{A}^*)$, $\bar{\pi}(\lambda) \in \boldsymbol{\Gamma}^0(\boldsymbol{A} + \lambda \boldsymbol{A}^*)$ for $0 \leqq \lambda \leqq \alpha$. Then for all $0 \leqq \lambda \leqq \alpha$,

$$\lambda \psi(\boldsymbol{A}^*, \bar{x}(\lambda), \bar{\pi}) \leqq f(\lambda, \boldsymbol{A}^*) - f(0, \boldsymbol{A}^*) \leqq \lambda \psi(\boldsymbol{A}^*, \bar{x}, \bar{\pi}(\lambda)) \tag{4.69}$$

Proof: Applying Theorem 4.16 to the perturbed LP (4.54) we get

$$\psi(\boldsymbol{A} + \lambda \boldsymbol{A}^*, \bar{x}(\lambda), \pi) \leqq \psi(\boldsymbol{A} + \lambda \boldsymbol{A}^*, \bar{x}(\lambda), \bar{\pi}(\lambda)) \leqq \psi(\boldsymbol{A} + \lambda \boldsymbol{A}^*, x, \bar{\pi}(\lambda)) \tag{4.70}$$

for all $x \geqq 0$, $\pi \geqq 0$. Also

$$-\psi(\boldsymbol{A}, x, \bar{\pi}) \leqq -\psi(\boldsymbol{A}, \bar{x}, \bar{\pi}) \leqq -\psi(\boldsymbol{A}, \bar{x}, \pi) \tag{4.71}$$

for all $x \geqq 0$, $\pi \geqq 0$. Notice that $f(0, \boldsymbol{A}^*)$ is independent of the perturbation matrix $\boldsymbol{A}^*$, and it is the optimum objective value in the original LP (4.4). Now substitute $x = \bar{x}$, $\pi = \bar{\pi}$ in (4.70), and substitute $x = \bar{x}(\lambda)$, $\pi = \bar{\pi}(\lambda)$ in (4.71), and add the resulting inequalities. Using the facts that $\bar{\pi}(A\bar{x} - b) = 0$, $\bar{\pi}(\lambda)((A + \lambda A^*)\bar{x}(\lambda) - (b + \lambda b^*)) = 0$ (these follow from the complementary slackness theorem), this leads to the result in this theorem. ∎

THEOREM 4.23 Suppose the regularity conditions (4.47) and (4.48) hold. Let $\boldsymbol{A}^*$ be a perturbation matrix. Then $f(\lambda, \boldsymbol{A}^*)$ is continuous to the right at $\lambda = 0$.

Proof: In this case, by Theorem 4.20, there exists a positive number α' satisfying the results in Theorems 4.20 and 4.21. Let $\alpha_1, \alpha_2, \ldots$ be a monotonic decreasing sequence of positive numbers converging to zero, such that $\alpha_1 \leqq \alpha'$. Let $\bar{x}(\alpha_h) \in \mathbf{K}^0(\boldsymbol{A} + \alpha_h \boldsymbol{A}^*)$, $\bar{\pi}(\alpha_h) \in \boldsymbol{\Gamma}^0(\boldsymbol{A} + \alpha_h \boldsymbol{A}^*)$. Then the sequences $\bar{x}(\alpha_h)$, $h = 1, 2, \ldots$ and $\bar{\pi}(\alpha_h)$, $h = 1, 2, \ldots$ are both bounded by Theorem 4.21. So there exist numbers γ_1, γ_2 such that, $\gamma_1 \leqq \psi(\boldsymbol{A}^*, \bar{x}(\alpha_h), \bar{\pi})$, $\gamma_2 \geqq \psi(\boldsymbol{A}^*, \bar{x}, \bar{\pi}(\alpha_h))$. Using this in the result of Theorem 4.22, we conclude that $\alpha_h \gamma_1 \leqq f(\alpha_h, \boldsymbol{A}^*) - f(0, \boldsymbol{A}^*) \leqq \alpha_h \gamma_2$ for all $h = 1, 2, \ldots$. Since α_h converges to zero as h tends to ∞, by letting h be arbitrarily large, the difference $|f(\alpha_h, \boldsymbol{A}^*) - f(0, \boldsymbol{A}^*)|$ can be made arbitrarily small. Hence $f(\lambda, \boldsymbol{A}^*)$ is continuous to the right at $\lambda = 0$. ∎

Suppose the regularity conditions (4.47) and (4.48) hold. Let $\boldsymbol{A}^*$ be a perturbation matrix, which has been selected and fixed. Let α' be a positive number satisfying the result in Theorem 4.20. Consider all possible monotonic decreasing sequences of positive numbers $\alpha_1, \alpha_2, \ldots$ converging to zero, with $\alpha_1 \leqq \alpha'$. Let $\bar{x}(\alpha_h) \in \mathbf{K}^0(\boldsymbol{A} + \alpha_h \boldsymbol{A}^*)$, $\bar{\pi}(\alpha_h) \in \boldsymbol{\Gamma}^0(\boldsymbol{A} + \alpha_h \boldsymbol{A}^*)$, for $h = 1, 2, \ldots$. By Theorem 4.21, each of the sequences $\bar{x}(\alpha_1), \bar{x}(\alpha_2), \ldots$ and $\bar{\pi}(\alpha_1), \bar{\pi}(\alpha_2), \ldots$ is always bounded. So each of these sequences lies inside some compact subset of $\mathbf{R}^n$ or $\mathbf{R}^m$, respectively. So these sequences have limit points. Let $\mathbf{X}$ denote the set of all limit points of all possible sequences such as $\bar{x}(\alpha_1), \bar{x}(\alpha_2), \ldots$. Let $\boldsymbol{\Pi}$ be the set of all limit points of all possible sequences such as $\bar{\pi}(\alpha_1), \bar{\pi}(\alpha_2), \ldots$. It can be verified that $\mathbf{X}$, and $\boldsymbol{\Pi}$ are both nonempty closed and bounded sets.

THEOREM 4.24 Suppose the regularity conditions (4.47) and (4.48) hold. Select a perturbation matrix $\boldsymbol{A}^*$ and fix it. Let $\mathbf{X}$, $\boldsymbol{\Pi}$ be sets defined as above. Then $\mathbf{X} \subset \mathbf{K}^0(\boldsymbol{A})$, $\boldsymbol{\Pi} \subset \boldsymbol{\Gamma}^0(\boldsymbol{A})$, and

$$\begin{aligned}\underset{\pi \in \boldsymbol{\Gamma}^0(\boldsymbol{A})}{\text{maximum}} \left(\underset{x \in \mathbf{X}}{\text{minimum}}\ \psi(\boldsymbol{A}^*, x, \pi) \right) &\geqq \underset{\pi \in \boldsymbol{\Gamma}^0(\boldsymbol{A})}{\text{maximum}} \left(\underset{x \in \mathbf{K}^0(\boldsymbol{A})}{\text{minimum}}\ \psi(\boldsymbol{A}^*, x, \pi) \right) \\ &\geqq \underset{\pi \in \boldsymbol{\Pi}}{\text{maximum}} \left(\underset{x \in \mathbf{K}^0(\boldsymbol{A})}{\text{minimum}}\ \psi(\boldsymbol{A}^*, x, \pi) \right)\end{aligned} \tag{4.72}$$

Proof: Let α' be a positive number satisfying the result in Theorem 4.20. Let $\tilde{x} \in \mathbf{X}$. Then there exists a monotonic decreasing sequence of positive numbers $\alpha_1, \alpha_2, \ldots$ converging to zero with $\alpha_1 \leqq \alpha'$, and $\bar{x}(\alpha_h) \in \mathbf{K}^0(\boldsymbol{A} + \alpha_h \boldsymbol{A}^*)$, for $h = 1, 2, \ldots$ such that $\tilde{x}$ is a limit point of the sequence $\{\bar{x}(\alpha_1), \bar{x}(\alpha_2), \ldots\}$. Let $\bar{x}(\alpha_{g_1}), \bar{x}(\alpha_{g_2}), \ldots$ be a subsequence that converges to $\tilde{x}$. Since

$$(A + \alpha_{g_r} A^*)\bar{x}(\alpha_{g_r}) \geqq b + \alpha_{g_r} b^* \tag{4.73}$$

letting r tend to ∞ in (4.73) we get $A\tilde{x} \geqq b$. So $\tilde{x} \in \mathbf{K}(\boldsymbol{A})$. Now let $\bar{x} \in \mathbf{K}^0(\boldsymbol{A})$. We have

$$f(\alpha_{g_r}, \boldsymbol{A}^*) = (c + \alpha_{g_r} c^*)\bar{x}(\alpha_{g_r}) \qquad f(0, \boldsymbol{A}^*) = c\bar{x}$$

$$\begin{aligned}\therefore f(\alpha_{g_r}, \boldsymbol{A}^*) - f(0, \boldsymbol{A}^*) \\ = c\bar{x}(\alpha_{g_r}) - c\bar{x} + \alpha_{g_r} c^* \bar{x}(\alpha_{g_r})\end{aligned} \tag{4.74}$$

In (4.74) let r tend to ∞. Since $f(\lambda, \boldsymbol{A}^*)$ is continuous to the right at $\lambda = 0$ by Theorem 4.23, the left-hand side of (4.74) converges to zero as r tends to ∞. The right-hand side of (4.74) converges to $c\tilde{x} - c\bar{x}$. So

$$c\tilde{x} - c\bar{x} = 0 \tag{4.75}$$

Since $\tilde{x} \in \mathbf{K}(\boldsymbol{A})$ and $\bar{x} \in \mathbf{K}^0(\boldsymbol{A})$, (4.75) implies that $\tilde{x} \in \mathbf{K}^0(\boldsymbol{A})$ too. So whenever $\tilde{x} \in \mathbf{X}$, we have $\tilde{x} \in \mathbf{K}^0(\boldsymbol{A})$. So $\mathbf{X} \subset \mathbf{K}^0(\boldsymbol{A})$. Using exactly similar arguments, it can be proved that $\mathbf{\Pi} \subset \mathbf{\Gamma}^0(\boldsymbol{A})$.

$\mathbf{X}, \mathbf{\Pi}, \mathbf{K}^0(\boldsymbol{A}), \mathbf{\Gamma}^0(\boldsymbol{A})$ are all closed bounded sets. Also whenever x or π is fixed, $\psi(\boldsymbol{A}^*, x, \pi)$ is an affine function in the remaining variables. These facts and the fact that $\mathbf{X} \subset \mathbf{K}^0(\boldsymbol{A})$ implies that for any $\pi \in \mathbf{\Gamma}^0(\boldsymbol{A})$.

$$\underset{x \in \mathbf{X}}{\text{minimum}}\ \psi(\boldsymbol{A}^*, x, \pi) \geqq \underset{x \in \mathbf{K}^0(\boldsymbol{A})}{\text{minimum}}\ \psi(\boldsymbol{A}^*, x, \pi)$$

So

$$\underset{\pi \in \mathbf{\Gamma}^0(\boldsymbol{A})}{\text{maximum}} \left(\underset{x \in \mathbf{X}}{\text{minimum}}\ \psi(\boldsymbol{A}^*, x, \pi) \right) \geqq \underset{\pi \in \mathbf{\Gamma}^0(\boldsymbol{A})}{\text{maximum}} \left(\underset{x \in \mathbf{K}^0(\boldsymbol{A})}{\text{minimum}}\ \psi(\boldsymbol{A}^*, x, \pi) \right) \quad (4.76)$$

Now, since $\mathbf{\Pi} \subset \mathbf{\Gamma}^0(\boldsymbol{A})$, we have

$$\underset{\pi \in \mathbf{\Gamma}^0(\boldsymbol{A})}{\text{maximum}} \left(\underset{x \in \mathbf{K}^0(\boldsymbol{A})}{\text{minimum}}\ \psi(\boldsymbol{A}^*, x, \pi) \right) \geqq \underset{\pi \in \mathbf{\Pi}}{\text{maximum}} \left(\underset{x \in \mathbf{K}^0(\boldsymbol{A})}{\text{minimum}}\ \psi(\boldsymbol{A}^*, x, \pi) \right) \quad (4.77)$$

and (4.76) and (4.77) together imply (4.72). ∎

THEOREM 4.25 Suppose the regularity conditions (4.47) and (4.48) hold. Let $\boldsymbol{A}^*$ be a perturbation matrix that is selected and fixed. Then $f(\lambda, \boldsymbol{A}^*)$ has the right derivative $f'(0, \boldsymbol{A}^*)$, and it is given by

$$f'(0, \boldsymbol{A}^*) = \underset{\pi \in \mathbf{\Gamma}^0(\boldsymbol{A})}{\text{maximum}} \left(\underset{x \in \mathbf{K}^0(\boldsymbol{A})}{\text{minimum}}\ \psi(\boldsymbol{A}^*, x, \pi) \right) \quad (4.78)$$

Proof: Let α' be a positive number satisfying the result in Theorem 4.20. Let $\alpha_1, \alpha_2, \ldots$ be a monotonic decreasing sequence of positive numbers converging to zero, satisfying $\alpha_1 \leqq \alpha'$. Let $\pi(\alpha_h) \in \mathbf{\Gamma}^0(\boldsymbol{A} + \alpha_h \boldsymbol{A}^*)$, for $h = 1, 2, \ldots$. Let $\bar{x} \in \mathbf{K}^0(\boldsymbol{A})$. From (4.69) we have

$$\frac{f(\alpha_h, \boldsymbol{A}^*) - f(0, \boldsymbol{A}^*)}{\alpha_h} \leqq \psi(\boldsymbol{A}^*, \bar{x}, \pi(\alpha_h)) \quad (4.79)$$

Since (4.79) holds for every $\bar{x} \in \mathbf{K}^0(\boldsymbol{A})$, we have

$$\frac{f(\alpha_h, \boldsymbol{A}^*) - f(0, \boldsymbol{A}^*)}{\alpha_h} \leqq \min_{x \in \mathbf{K}^0(\boldsymbol{A})} \psi(\boldsymbol{A}^*, x, \bar{\pi}(\alpha_h)) \quad (4.80)$$

Choose $\varepsilon > 0$. When h is very large, from the definition of $\mathbf{\Pi}$, we can always find a $\tilde{\pi} \in \mathbf{\Pi}$ such that $\psi(\boldsymbol{A}^*, x, \bar{\pi}(\alpha_h)) \leqq \varepsilon + \psi(\boldsymbol{A}^*, x, \tilde{\pi})$ for all $x \in \mathbf{K}^0(\boldsymbol{A})$. So from (4.80) we have

$$\frac{f(\alpha_h, \boldsymbol{A}^*) - f(0, \boldsymbol{A}^*)}{\alpha_h} \leqq \varepsilon + \min_{x \in \mathbf{K}^0(\boldsymbol{A})} \psi(\boldsymbol{A}^*, x, \tilde{\pi}) \leqq \varepsilon + \underset{\pi \in \mathbf{\Pi}}{\text{maximum}} \left(\underset{x \in \mathbf{K}^0(\boldsymbol{A})}{\text{minimum}}\ \psi(\boldsymbol{A}^*, x, \pi) \right) \quad (4.81)$$

So when h is very large

$$\left(\frac{f(\alpha_h, \boldsymbol{A}^*) - f(0, \boldsymbol{A}^*)}{\alpha_h} \right) - \max_{\pi \in \mathbf{\Gamma}^0(\boldsymbol{A})} \left(\min_{x \in \mathbf{K}^0(\boldsymbol{A})} \psi(\boldsymbol{A}^*, x, \pi) \right)$$
$$\leqq \varepsilon + \underset{\pi \in \mathbf{\Pi}}{\text{maximum}} \left(\underset{x \in \mathbf{K}^0(\boldsymbol{A})}{\text{minimum}}\ \psi(\boldsymbol{A}^*, x, \pi) \right) - \underset{\pi \in \mathbf{\Gamma}^0(\boldsymbol{A})}{\text{maximum}} \left(\underset{x \in \mathbf{K}^0(\boldsymbol{A})}{\text{minimum}}\ \psi(\boldsymbol{A}^*, x, \pi) \right) \leqq \varepsilon$$

by (4.77). Using (4.69) and (4.72) and similar arguments, when h is very large the quantity $((f(\alpha_h, \boldsymbol{A}^*) - f(0, \boldsymbol{A}^*))/\alpha_h - \max_{\pi \in \mathbf{\Gamma}^0(\boldsymbol{A})} (\min_{x \in \mathbf{K}^0(\boldsymbol{A})} \psi(\boldsymbol{A}^*, x, \pi))$ can be shown to be larger than any negative number ε. This implies that the limit of $(f(\alpha_h, \boldsymbol{A}^*) - f(0, \boldsymbol{A}^*))/\alpha_h$ is $f'(0, \boldsymbol{A}^*)$ defined in (4.78). ∎

Clearly the LP (4.4) is stable iff for any perturbation matrix $\boldsymbol{A}^*$ the perturbed LP (4.54) has an optimum feasible solution in an interval of positive length $0 \leqq \lambda \leqq \alpha'$; its optimum objective value is continuous to the right at $\lambda = 0$ and possesses a derivative to its right at $\lambda = 0$. By Theorems 4.18 to 4.25 these properties hold iff the regularity conditions (4.47) and (4.48) hold, or equivalently iff the sets of optimum feasible solutions of the primal-dual pair of LPs (4.4) and (4.5) are both nonempty and bounded.

Hence the regularity conditions (4.47), (4.48) are the necessary and sufficient conditions for the stability of the LP model (4.4) and its dual (4.5).

Comment 4.3: Theorems 4.18 to 4.24 are due to A. C. Williams [4.34]. The result in Theorem 4.25 is due to H. D. Mills [4.20].

In [4.24–4.26], S. M. Robinson has strengthened these results. He has proved that there exists an $\varepsilon_0 > 0$ such that for all perturbation matrices $\boldsymbol{A}^*$ of norm less than ε_0, both $\mathbf{K}^0(\boldsymbol{A} + \boldsymbol{A}^*)$ and $\mathbf{\Gamma}^0(\boldsymbol{A} + \boldsymbol{A}^*)$ are nonempty iff the regularity conditions (4.47) and (4.48) hold, or equivalently, iff both $\mathbf{K}^0$ and $\mathbf{\Gamma}^0$ are nonempty and bounded. He has also proved that under these conditions, there exists a positive numbers $\varepsilon_1 \leqq \varepsilon_0$, γ, such that for any $x^* \in \mathbf{K}^0(\boldsymbol{A} + \boldsymbol{A}^*)$, and $\pi^* \in \boldsymbol{\Gamma}^0(\boldsymbol{A} + \boldsymbol{A}^*)$, the distance of x^* to its nearest point in $\mathbf{K}^0$, or the distance of π^* to its nearest point in $\boldsymbol{\Gamma}^0$, is less than or equal to γ times the norm of the perturbation matrix $\boldsymbol{A}^*$, whenever the norm of the perturbation matrix $\boldsymbol{A}^*$ is less than ε_1.

He has also proved the following. Suppose the primal regularity condition (4.48) is not satisfied, and let ε be any given positive number. Let x' be any given feasible solution for (4.4). Then it is possible to produce a perturbed problem by modifying a single element of b and the entries in the corresponding row of A by amounts less than ε in absolute value, leaving all the rest of the data unchanged, so that the optimum objective value in the perturbed problem is equal to cx'. Analogously, if the dual regularity condition (4.47) is not satisfied, we could produce a perturbed problem with a different optimum objective value by just modifying one element of c and the entries in the corresponding column of A by arbitrarily small amounts.

These results clearly indicate that the LP model (4.4) is well behaved under small perturbations in the data values if the regularity conditions (4.47) and (4.48) hold, and can behave badly otherwise.

Example 4.19

Consider the LP discussed in Example 4.17. There it was shown that this LP is unstable in the neighborhood of $\alpha = 0$. When $\alpha = 0$, it can be verified that the set of optimum solutions of the dual of this problem is unbounded, and hence the primal regularity condition is violated in the neighborhood of $\alpha = 0$.

In [4.26] S. M. Robinson has also given an economic interpretation of the regularity conditions (4.47) and (4.48) as they apply to the activity analysis model that we discuss next.

Economic Interpretation of the Regularity Conditions on an Activity Analysis Model

Consider a firm whose list of possible activities includes n activities, and that produces or consumes m goods. Let:

c_j = cost of carrying activity j at unit level $j = 1$ to n.

a_{ij} = 0 if the ith good is neither produced as an output nor needed as an input when activity j is carried out.
= the positive quantity, net number of units of good i produced when activity j is carried at unit level, if good i is produced as a net output when activity j is carried out.
= the negative quantity whose absolute value is the number of units of good i needed as input when activity j is carried out at unit level, if good i is needed as a net input under activity j.

b_i = minimum number of units of good i required to be produced by this firm.

$A = (a_{ij})$, $b = (b_1, \ldots, b_m)^{\mathrm{T}}$, $c = (c_1, \ldots, c_n)$.

Let x_j be the level at which activity j is carried out, and $x = (x_1, \ldots, x_n)^{\mathrm{T}}$. Then, x is the *production plan*. The *activity analysis model* for this firm, to meet the production requirements at minimum total cost, is the LP (4.4) with these data. This is the problem of selecting nonnegative activity levels $x_1, \ldots, x_n$ to fulfill the production requirements on all the goods at least cost. The condition for (4.4) (the primal problem) to be regular is (4.48). By Tucker's theorem of the alternatives (Theorem 4.11), it can be seen that (4.48) holds iff there exists some nonnegative $x \in \mathbf{R}^n$ such that $Ax > b$. This condition therefore holds iff the firm is able to produce a bill of goods strictly larger than b, or equivalently that the requirement to produce the vector b of goods does not strain the firm's production capacity to its limit.

The condition for the dual of (4.4) to be regular is (4.47); that is, for any semipositive vector y for which $Ay \geqq 0$, we must have $cy > 0$. Economically this means that any production plan that produces a nonnegative bill of goods (i.e., that results is net nonnegative output of each good) is either the zero production plan or is associated with a strictly positive cost.

Thus when applied on the activity analysis model (4.4), the regularity conditions (4.47) and (4.48) both have economically meaningful interpretations, and it seems reasonable to expect that practical economic systems that one might wish to model by linear programming models do satisfy these regularity conditions. The results of this section suggest that natural activity analysis problems give rise to linear programming models that are stable.

Stability in the General Linear Programming Model

Here we consider the general linear programming model in which there may be some equality constraints and some inequality constraints, some unrestricted variables and some variables restricted to be nonnegative. According to the discussion in Chapters 2 and 3, an equality constraints of the form $a_{i1}x_1 + \cdots + a_{in}x_n = b_i$, is equivalent to the pair of inequality constraints

$$\begin{aligned} a_{i1}x_1 + \cdots + a_{in}x_n &\geqq b_i \\ -a_{i1}x_1 - \cdots - a_{in}x_n &\geqq -b_i \end{aligned} \tag{4.82}$$

Also, if x_i is an unrestricted variable, it can be replaced by the difference of two nonnegative variables, $x_i = x_i^+ - x_i^-$, where x_i^+ and x_i^- are both nonnegative variables.

Using these transformations the general LP model can be transformed into an equivalent LP model that is in symmetric form. However, whenever any of these two transformations are carried out, the resulting LP in symmetric form would invariably be unstable, even though the original LP in general form may be stable. As an illustration, suppose the resulting model in symmetric form is (4.4), which contains a pair of constraints of the form (4.82). Let $\tilde{\mu}$ be the multiplier vector corresponding to the constraints in this LP model, in which the multipliers $\tilde{\mu}_i$ associated with the pair of constraints of the form (4.82) are both equal to 1, while the multipliers associated with all the other constraints are zero. Clearly this leads to a semipositive multiplier vector satisfying $\tilde{\mu}A = 0$, $\tilde{\mu}b = 0$, violating the regularity condition (4.48) for the equivalent LP model in symmetric form. In a similar manner, if the transformation of replacing an unrestricted variable by the difference of two nonnegative variables is used, it can be shown that the resulting equivalent LP model in symmetric form does not satisfy the regularity condition (4.47).

Let the LP model we wish to study be

$$\begin{aligned} &\text{Minimize} && c^1x^1 + c^2x^2 \\ &\text{Subject to} && A^{11}x^1 + A^{12}x^2 \geqq b^1 \\ &&& A^{21}x^1 + A^{22}x^2 = b^2 \\ &x^1 \geqq 0 && x^2 \text{ unrestricted} \end{aligned} \tag{4.83}$$

where $x^1 = (x_1^1, \ldots, x_{n_1})^{\mathrm{T}}$, $x^2 = (x_1^2, \ldots, x_{n_2}^2)^{\mathrm{T}}$, A^{11} is a matrix of order $m_1 \times n_1$, A^{21} is a matrix of order $m_2 \times n_1$, etc. The dual of (4.83) is

$$\begin{aligned} &\text{Maximize} && \pi^1b^1 + \pi^2b^2 \\ &\text{Subject to} && \pi^1A^{11} + \pi^2A^{21} \leqq c^1 \\ &&& \pi^1A^{12} + \pi^2A^{22} = c^2 \\ &\pi^1 \geqq 0 && \pi^2 \text{ unrestricted} \end{aligned} \tag{4.84}$$

where $\pi^1 = (\pi_1^1, \ldots, \pi_{m_1}^1)$, $\pi^2 = (\pi_1^2, \ldots, \pi_{m_2}^2)$. The data in this primal dual pair of LPs (4.83) and (4.84) are conveniently represented in matrix form as:

$$\hat{\mathbf{A}} = \left(\begin{array}{c|c|c} A^{11} & A^{12} & b^1 \\ \hline A^{21} & A^{22} & b^2 \\ \hline c^1 & c^2 & 0 \end{array}\right)$$

where in $\hat{\mathbf{A}}$, the bottom right entry is always zero, and except for this entry, all the other entries represent data elements in this LP model. Let

$$x = \begin{pmatrix} x^1 \\ \cdots \\ x^2 \end{pmatrix} \qquad b = \begin{pmatrix} b^1 \\ \cdots \\ b^2 \end{pmatrix} \qquad \pi = (\pi^1 \vdots \pi^2)$$

$$\psi(\hat{\mathbf{A}}, x, \pi) = c^1x^1 + c^2x^2 - \pi\left(\begin{pmatrix} A^{11} & \vdots & A^{12} \\ \cdots & & \cdots \\ A^{21} & \vdots & A^{22} \end{pmatrix} x - b\right)$$

An $\bar{x}$, $\bar{\pi}$ together satisfy the saddle point conditions corresponding to this primal-dual pair of LPs (4.83) and (4.84) if

$$\begin{gathered} \bar{x}^1 \geqq 0 \qquad \bar{\pi}^1 \geqq 0 \\ \psi(\hat{\mathbf{A}}, \bar{x}, \pi) \leqq \psi(\hat{\mathbf{A}}, \bar{x}, \bar{\pi}) \leqq \psi(\hat{\mathbf{A}}, x, \bar{\pi}) \end{gathered} \tag{4.85}$$

for all x, π such that $x^1 \geqq 0$ and $\pi^1 \geqq 0$. Given $\bar{x}$, $\bar{\pi}$ it can be proved that they are optimal to the respective problems in the primal-dual pair (4.83) and (4.84) iff $(\bar{x}, \bar{\pi})$ together satisfy the saddle point conditions (4.85).

The regularity conditions for the primal dual pair of LPs (4.83) and (4.84) are the following.

There exists no $y = \begin{pmatrix} y^1 \\ y^2 \end{pmatrix}$ satisfying

$$\begin{aligned} A^{11}y^1 + A^{12}y^2 &\geqq 0 \\ A^{21}y^1 + A^{22}y^2 &= 0 \\ y^1 \qquad\quad &\geqq 0 \\ c^1y^1 + c^2y^2 &\leqq 0 \\ y &\neq 0 \end{aligned} \tag{4.86}$$

There exists no $\mu = (\mu^1, \mu^2)$ satisfying

$$\begin{aligned} \mu^1A^{11} + \mu^2A^{21} &\leqq 0 \\ \mu^1A^{12} + \mu^2A^{22} &= 0 \\ \mu^1 \qquad\quad &\geqq 0 \\ \mu^1b^1 + \mu^2b^2 &\geqq 0 \\ \mu &\neq 0 \end{aligned} \tag{4.87}$$

A result analogous to Theorem 4.17 can be proved using arguments similar to those in the proof of Theorem 4.17. In particular, the sets of optimum feasible solutions of (4.83) and (4.84) are both nonempty and bounded iff (4.86) and (4.87) hold.

Let $\mathbf{L}^0(\hat{\boldsymbol{A}})$ be the set of optimum feasible solutions of (4.83) and let $\Delta^0(\hat{\boldsymbol{A}})$ be the set of optimum feasible solutions of (4.84).

We study the stability of the LP model (4.83) under small perturbations in the data elements in A^{11}, A^{12}, A^{21}, A^{22}, b^1, b^2, c^1, c^2. We assume that the nonnegativity restrictions on x^1 remain unchanged under this perturbation, and that the variables in x^2 remain unrestricted. Define the perturbation matrix

$$\hat{\boldsymbol{A}}^* = \left(\begin{array}{c|c|c} A^{11*} & A^{12*} & b^{1*} \\ \hline A^{21*} & A^{22*} & b^{2*} \\ \hline c^{1*} & c^{2*} & 0 \end{array}\right)$$

and consider the perturbed problem:

$$\begin{array}{ll} \text{Minimize} & (c^1 + \lambda c^{1*})x^1 + (c^2 + \lambda c^{2*})x^2 \\ \text{Subject to} & (A^{11} + \lambda A^{11*})x^1 + (A^{12} + \lambda A^{12*})x^2 \\ & \quad \geqq b^1 + \lambda b^{1*} \\ & (A^{21} + \lambda A^{21*})x^1 + (A^{22} + \lambda A^{22*})x^2 \\ & \quad = b^2 + \lambda b^{2*} \\ & x^1 \geqq 0 \qquad x^2 \text{ unrestricted} \end{array} \tag{4.88}$$

and its dual. All the results analogous to Theorems 4.18 to 4.25 can be proved for the perturbed problem (4.88) and its dual using arguments similar to those used earlier. In particular, we can prove that there exists a positive number α' such that the perturbed problem (4.88) has an optimum feasible solution for all $0 \leqq \lambda \leqq \alpha'$, and if $f(\lambda, \hat{\boldsymbol{A}}^*)$ is the optimum objective value in (4.88), $f(\lambda, \hat{\boldsymbol{A}}^*)$ is continuous to the right at $\lambda = 0$, and has a right derivative $f'(0, \hat{\boldsymbol{A}}^*)$ where

$$f'(0, \hat{\boldsymbol{A}}^*) = \underset{\pi \in \Delta^0(\hat{A})}{\text{maximum}} \left(\underset{x \in \mathbf{L}^0(\hat{A})}{\text{minimum}} \; \psi(\hat{\boldsymbol{A}}^*, x, \pi) \right)$$

iff the regularity conditions (4.86) and (4.87) hold. Hence the regularity conditions (4.86) and (4.87) provide the necessary and sufficient conditions for the stability of the primal dual pair of LPs (4.83) and (4.84).

Exercises

4.15 Obtain the necessary and sufficient conditions for the LP (4.6) in standard form to be stable.

4.16 Discuss which of the transformations listed in Section 3.18 that take one LP into an equivalent LP preserve the stability of the model, and which do not.

4.17 Consider the LP (4.4). Suppose A and c are fixed, but b is allowed to vary. Let $g(b)$ denote the optimum objective value of (4.4) as a function of b. For a given b, if $\bar{\pi}$ is the unique optimum dual solution, prove rigorously that $(\partial g(b)/\partial b) = \bar{\pi}$. State the corresponding result for the LP (4.83) and prove it rigorously.

Summary In the activity analysis model, we have seen that stability is a natural property in reasonable production systems that give rise to this model. However, stability that might be present can be destroyed by improper formulation (for example, an equality constraint written as the equivalent pair of inequality constraints leads to an unstable model). This points out the importance of exercising great care in formulating LP models of practical problems in order not to destroy any stability that might otherwise be present.

Concluding Remarks

The dual problem was formulated by J. von Neumann. Duality theorems were proved by D. Gale, H. W. Kuhn, and A. W. Tucker. The optimum strategies for the two-person zero sum-matrix game discussed in Section 4.6.6 are known as *minimax strategies* (because they minimize the maximum loss that the player may incur). J. von Neumann first defined these strategies. The complementary slackness theorem is due to G. B. Dantzig and A. Orden [4.5]. D. Gale, H. W. Kuhn, and A. W. Tucker were awarded the 1980 von Neumann Theory Prize of TIMS/ORSA for their research in the mathematical theory of linear inequalities as applied to linear programs, matrix games, and for developing duality. See the last page in *OR/MS Today*, vol. 7, no. 3, May/June 1980.

Exercises

4.18 Let $\alpha > 0$. If $\bar{x}$ is an optimum solution for the following LP when $\lambda = 0$ and α, respectively, prove that $\bar{x}$ remains an optimum solution for all $0 \leqq \lambda \leqq \alpha$.

$$\begin{array}{ll} \text{Minimize} & (c + \lambda c^*)x \\ \text{Subject to} & Ax = b \\ & Dx \geqq d \end{array}$$

4.19 Consider the LP in standard form (4.6) and its dual (4.7). Assume that $(x_1, \ldots, x_m)$ is a basic vector for this LP, associated with the basis B. Let the canonical tableau with respect to this basic vector be the one given in Tableau 2.8 of Section 2.5.1. Denoting the vector of dual variables by $\mu = (\mu_1, \ldots, \mu_m)$ write down the dual of the problem in Tableau 2.8. Obtain the linear transformation between μ and π that transforms a feasible solution of one into the corresponding feasible solution of the other.

4.20 Verify the assertion made in Section 1.6 that all the basic solutions and the dual basic solutions associated with an LP with rational data are rational vectors. Also prove that if an LP with rational data has an optimum solution, then it has an optimum solution that is a rational vector.

4.21 Write the duals of the following problems. Verify that the dual of the dual is the original problem. Also transform each problem into standard form and verify that the dual of the problem in standard form is equivalent to the dual of original problem.

(a) Maximize $4x_1 - 3x_2 + 8x_3$

Subject to

$$-2 \leqq x_1 \leqq 6$$
$$4 \leqq x_2 \leqq 14$$
$$-12 \leqq x_3 \leqq -8$$

(b) Maximize $z'(x) = \sum_{j=1}^{n} c'_j x_j$

Subject to

$$\sum_{j=1}^{n} a_{1j}x_j = b_1$$
$$\sum_{j=1}^{n} a_{2j}x_j \geqq b_2$$
$$\sum_{j=1}^{n} a_{3j}x_j \leqq b_3$$
$$\ell_j \leqq x_j \leqq k_j \quad \text{for all } j$$

(c) Minimize

$$z = 3y_1 + 4y_2$$

Subject to

$$\sum_{j=1}^{n} a_{ij}x_j = b_{i1}y_1 + b_{i2}y_2 \qquad \text{for } i = 1 \text{ to } r$$
$$\sum_{j=1}^{n} a_{ij}x_j \geqq b_{i1}y_1 + b_{i2}y_2 \qquad \text{for } i = r+1 \text{ to } m$$
$$\sum_{j=1}^{n} x_j = 4$$
$$x_j \geqq 0 \qquad \text{for all } j$$
$$y_1, y_2 \text{ unrestricted in sign}$$

(d) The network Exercise 1.19.

(e) The curve-fitting problems discussed in Section 1.2.5. Also prove that the dual problem obtained is feasible, by actually constructing a dual feasible solution.

(f) Maximize $z'(x) = -17x_2 + 83x_4 - 8x_5$

Subject to

$$-x_1 - 13x_2 + 45x_3 + 16x_5 - 7x_6 \geqq 107$$
$$3x_3 - 18x_4 + 30x_7 \leqq 81$$
$$4x_1 - 5x_3 + x_6 = -13$$
$$-10 \leqq x_1 \leqq -2$$
$$-3 \leqq x_2 \leqq 17$$
$$16 \leqq x_3 \qquad x_4 \leqq 0$$
$$x_5 \text{ unrestricted} \qquad x_6, x_7 \geqq 0$$

(g) Minimize $z(x) = 3x_1 - 7x_2 + 6x_4 + 5x_5 - x_6$

Subject to

$$5x_1 + 8x_2 + 3x_3 + 3x_4 + 2x_5 + 11x_6 = 200$$
$$5 \leqq x_j \leqq 20 \quad \text{for all } j$$

$$\begin{array}{lrl}
\text{Minimize} & z(x) = & -2x_1 + 13x_2 + 3x_3 - 2x_4 + 5x_5 + 5x_6 + 10x_7 \\
\text{Subject to} & & x_1 - x_2 + 4x_4 - x_5 + x_6 - 4x_7 = 5 \\
& & x_1 + 7x_4 - 2x_5 + 3x_6 - 3x_7 \geqq -1 \\
& & 5x_2 + x_3 - x_4 + 2x_5 - x_6 - 2x_7 \leqq 5 \\
& & 3x_2 + x_3 + x_4 + x_5 + x_6 - x_7 = 2 \\
& & x_j \geqq 0 \quad \text{for all } j
\end{array}$$

4.22 Write the complementary slackness conditions for optimality for each of the LPs in Exercise 4.21.

4.23 Consider the following LP, where A is an $m \times n$ matrix and $\ell = (\ell_1, \ldots, \ell_m)^{\mathrm{T}}$ and $k = (k_1, \ldots, k_m)^{\mathrm{T}}$ are given vectors. Write the dual of this problem and the complementary slackness conditions for optimality for this primal-dual pair. What conditions on the A matrix would guarantee that the dual problem is feasible?

$$\begin{array}{ll}
\text{Minimize} & z(x) = cx \\
\text{Subject to} & \ell_i \leqq A_{i.}x \leqq k_i \quad \text{for } i = 1 \text{ to } m
\end{array}$$

4.24 Consider the following LP, where A is an $m \times n$ matrix. Write the dual of this problem. Using duality theorem prove that if the problem is feasible, then the minimum value of $z(x)$ in this problem is finite iff c can be expressed as a linear combination of the row vectors of A. Using this prove that either $z(x)$ is a constant or unbounded below on the set of solutions of this problem.

$$\begin{array}{ll}
\text{Minimize} & z(x) = cx \\
\text{Subject to} & Ax = b
\end{array}$$

4.25 Consider the LP in standard form (4.6). Write the corresponding Phase I problem with a full artificial basis. Write the dual of this Phase I problem and the complementary slackness conditions for optimality in this primal-dual pair.

4.26 A, b, c, are given matrices of order $m \times n$, $m \times 1$, and $1 \times n$. Prove that the following LP is either infeasible or has an optimum objective value of zero.

$$\begin{array}{lrr}
\text{Minimize} & cx - b^{\mathrm{T}}y & \\
\text{Subject to} & Ax & \geqq b \\
& -A^{\mathrm{T}}y & \geqq -c^{\mathrm{T}} \\
& x \geqq 0 & y \geqq 0
\end{array}$$

4.27 Consider the following LP, where $b_i > 0$ for all i and $c_j \geqq 0$ for $j = m + 1$ to n. Write the dual of this problem and prove that the dual problem has a unique optimal solution. Find that dual optimal solution.

Minimize

$$\sum_{j=m+1}^{n} c_j x_j$$

Subject to

$$x_i + \sum_{j=m+1}^{n} a_{ij}x_j = b_i \qquad i = 1 \text{ to } m$$

$$x_j \geqq 0 \qquad \text{for } j = 1 \text{ to } n$$

4.28 Consider the LP (4.6), where A is of order $m \times n$ and rank m. Suppose this problem has an optimum nondegenerate feasible basis. Using the results in Exercise 4.27 prove that the dual of this problem has a unique optimal solution.

4.29 For the LP given at the top prove that $x = (6, 0, 1, 0, 1, 0, 0)^{\mathrm{T}}$ is an optimum feasible solution by using the complementary slackness theorem.

4.30 For the LP given at the top of page 223 check whether $x = (7, 0, \frac{5}{2}, 0, 3, 0, \frac{1}{2})^{\mathrm{T}}$ is an optimum feasible solution.

$$\begin{array}{ll} \text{Maximize} & z(x) = x_1 + 2x_2 + x_3 - 3x_4 + x_5 + x_6 - x_7 \\ \text{Subject to} & x_1 + x_2 - x_4 + 2x_6 - 2x_7 \leqq 6 \\ & x_2 - x_4 + x_5 - 2x_6 + 2x_7 \leqq 4 \\ & x_2 + x_3 + x_6 - x_7 \leqq 2 \\ & x_2 - x_4 - x_6 + x_7 \leqq 1 \\ & x_j \geqq 0 \quad \text{for all } j \end{array}$$

4.31 For the transportation problem (4.10) with the following data, find an optimum feasible solution using the complementary slackness conditions, given that $u = (u_i) = (0, 3, -4)$, $v = (v_j) = (7, 2, -5, 7)$ is an optimum dual solution.

$$c = (c_{ij}) = \begin{pmatrix} 7 & 2 & -2 & 8 \\ 19 & 5 & -2 & 12 \\ 5 & 7.2 & -9 & 3 \end{pmatrix}$$

$$a = (a_i) = (3, 3, 4),\ b = (b_j) = (2, 3, 2, 3)$$

4.32 For the LP given at the bottom check whether $x = (3, 4, 0, 0, 0, 0, 0)^{\mathrm{T}}$ is an optimum feasible solution by using the complementary slackness conditions. Using these conditions characterize the set of optimum solutions. What does the fact column vector of $x_5 = -$(column vector of x_1) $-$ 2(column vector of x_2) imply about the shape of the set of optimum solutions of this problem?

4.33 Consider the following LP. Write the dual problem. If $\bar{x}$ is an optimum feasible solution of this problem, prove that there exists a real number $\bar{\pi}$ such that for all j: $\bar{\pi}a_j < c_j$ implies $\bar{x}_j = \ell_j$, $\bar{\pi}a_j > c_j$ implies $\bar{x}_j = k_j$, and $\ell_j < \bar{x}_j < k_j$ implies $\bar{\pi}a_j = c_j$.

$$\begin{array}{ll} \text{Minimize} & z(x) = \sum c_j x_j \\ \text{Subject to} & \sum a_j x_j = b \\ & \ell_j \leqq x_j \leqq k_j \quad \text{for all } j \end{array}$$

4.34 If the following LP has a feasible solution, prove that it has an optimum feasible solution iff c lies in the Pos cone generated by the row vectors of A.

$$\begin{array}{ll} \text{Minimize} & z(x) = cx \\ \text{Subject to} & Ax \geqq b \end{array}$$

4.35 For the LP given below compute the primal and dual solutions associated with the basis corresponding to the basic vector (x_3, x_5, x_1). Is it an optimum basis? Characterize the set of optimum solutions of this problem using the complementary slackness theorem. Do the same for the basic vector (x_5, x_2, x_3, x_8) for the LP given at the top of page 224.

x_1	x_2	x_3	x_4	x_5	x_6	x_7	$-z$	b
1	4	1	-2	2	5	1	0	11
0	-2	0	2	-1	-1	-1	0	-4
1	1	0	-1	1	3	2	0	4
2	7	1	-1	2	13	2	1	0

$x_j \geqq 0$ for all j; minimize z.

4.36 Consider a primal-dual pair of LPs, both of which are feasible. Prove that there exists a pair of primal and dual feasible optimum solutions satisfying the properties: (a) Whenever a constraint in one problem is tight, the associated variable in the other problem is nonzero; (b) whenever a nonnegative variable

$$\begin{array}{ll} \text{Minimize} & z(x) = -x_1 - 2x_2 + 4x_3 + 5x_5 \\ \text{Subject to} & x_1 + x_3 - 2x_4 - x_5 + 2x_6 = 3 \\ & x_1 + x_2 - x_4 - 3x_5 + 3x_6 + x_7 = 7 \\ & -x_1 - 2x_2 - x_3 - x_4 + 5x_5 - 2x_6 \leqq -4 \\ & x_j \geqq 0 \quad \text{for all } j \end{array}$$

$$\begin{array}{ll}\text{Minimize} & 3x_1 - 2x_2 + x_3 - x_4 - 5x_5 + 4x_6 - 2x_7 - 3x_8 \\ \text{Subject to} & x_1 + x_2 + x_3 + x_4 + x_5 + x_6 + x_7 + x_8 = 14 \\ & x_1 + x_2 + x_3 + x_5 + x_6 + x_7 = 9 \\ & x_1 + x_2 + x_5 + x_6 = 5 \\ & x_1 + x_5 = 2 \\ & x_j \geqq 0 \quad \text{for all } j\end{array}$$

in one problem is zero, the associated constraint in the other problem is slack.

4.37 Consider the LP

$$\begin{array}{ll}\text{Minimize} & x_1 + x_2 + x_3 \\ \text{Subject to} & -x_2 + x_3 \geqq -1 \\ & x_1 - x_3 \geqq -1 \\ & -x_1 + x_2 \geqq -1 \\ & x_j \geqq 0 \quad \text{for all } j\end{array}$$

Prove that this problem is equivalent to its dual. Such LPs are known as *self-dual linear programs*. Assuming that A is a square matrix, obtain sufficient conditions on c, A, b, under which the following LP is self-dual:

$$\begin{array}{ll}\text{Minimize} & cx \\ \text{Subject to} & Ax \geqq b \\ & x \geqq 0\end{array}$$

4.38 Let A be a given $m \times n$ matrix. Prove that either (I) has a nonzero solution x or (II) has a solution π, but not both.

$$Ax = 0, \qquad x > 0 \qquad \text{(I)}$$

$$\pi A < 0 \qquad \text{(II)}$$

4.39 After introducing slack variables, the primal-dual pair of LPs (4.4) and (4.5) can be written as below.

Primal

$$\begin{array}{ll}\text{Minimize} & z(x) = cx \\ \text{Subject to} & Ax - v = b \\ x \geqq 0, & v \geqq 0\end{array} \qquad (4.89)$$

Dual

$$\begin{array}{ll}\text{Maximize} & \pi b \\ \text{Subject to} & \pi A + u = c \\ \pi \geqq 0, & u \geqq 0\end{array} \qquad (4.90)$$

Let $\mathbf{K} = \{\binom{x}{v} : \binom{x}{v}$ feasible to (4.89)$\}$, and let $\mathbf{\Gamma} = \{(\pi, u) : (\pi, u)$ feasible to (4.90)$\}$. The complementary pairs of variables in these problems are (x_j, u_j) for $j = 1$ to n and (π_i, v_i) for $i = 1$ to m. Assuming that $\mathbf{K} \neq \emptyset$, $\mathbf{\Gamma} \neq \emptyset$, prove the following results. (1) In the primal-dual pair of LPs (4.89) and (4.90), a variable is zero in all optimum feasible solutions of one of these problems iff its complementary variable is strictly positive in some optimum feasible solution of the other problem. (2) A variable in one of these problems is unbounded above on the set of feasible solutions of that problem iff its complementary variable is bounded above on the set of feasible solutions of the other problem. (3) A variable in one of these problems is unbounded above on the set of optimum feasible solutions of that problem iff its complementary variable is zero in all feasible solutions of the other problem. (4) A variable in one of these problems assumes a strictly positive value in some feasible solution of that problem iff its complementary variable is bounded above on the set of optimum feasible solutions of the other problem. (5) The set of feasible solutions in one of these problems is bounded iff every variable in the other problem is unbounded above on the set of feasible solutions of that problem. (6) For one of these problems there exists a feasible solution in which all the variables are strictly positive iff the set of optimum feasible solutions of the other problem is bounded.

—(*A. C. Williams* [4.34])

4.40 *CLARK'S THEOREM*. Consider a primal-dual pair of LPs, say (4.4) and (4.5). Suppose the set of feasible solutions of at least one of the problems in this pair is nonempty, then prove that the set of feasible solutions of at least one of the problems in this pair is both nonempty and unbounded.

—(*F. E. Clark* [4.3])

4.41 Let $\mathbf{K} \subset \mathbf{R}^n$ be a convex polytope and let $\mathbf{S}$ denote its sets of extreme points. Prove that $x^1, x^2 \in \mathbf{S}$ are adjacent extreme points of $\mathbf{K}$ iff there exists a hyperplane that separates the two sets $\{x^1, x^2\}$ and $\mathbf{S} \backslash \{x^1, x^2\}$. Provide a counterexample to show that this result may not be true if $\mathbf{K}$ is a convex polyhedron but not a convex polytope (i.e., $\mathbf{K}$ is not bounded).

—(*D. Hausmann* [4.10])

4.42 Consider the linear programming formulation of the problem discussed in Exercise 1.36. Using duality theory of linear programming, prove the following results on this problem. $\bar{x}_i = \text{maximum } \{a_j - (\alpha/w_j) : j \leqq i\}$, $\hat{x}_i = \text{minimum } \{a_j + (\alpha/w_j) : j \geqq i\}$. Let $\bar{x} = (\bar{x}_1, \ldots, \bar{x}_n)$, $\hat{x} = (\hat{x}_1, \ldots, \hat{x}_n)$. Prove that both $\bar{x}$, $\hat{x}$ are optimum solutions of this problem. Furthermore, prove that if $x = (x_1, \ldots, x_n)$ satisfies $x_i \leqq x_{i+1}$ for $i = 1, \ldots, n-1$, then x is an optimum solution of this problem iff $\bar{x}_i \leqq x_i \leqq \hat{x}_i$, $i = 1$ to n.

—(*V. A. Ubhaya* [1.32, 1.33])

4.43 Consider any $n+1$ hyperplanes $\mathbf{H}_0, \mathbf{H}_1, \ldots, \mathbf{H}_n$ in $\mathbf{R}^n$ passing through a point x^* and in general position; that is, having the property that any n of these hyperplanes intersect in a single point. Show that (a) any one of the $n+1$ hyperplanes, say $\mathbf{H}_0$, intersects all but two of the 2^n conical regions formed by the remaining n hyperplanes, and (b) devise an efficient procedure for identifying these two regions by an explicit set of inequalities. (*Hint:* Formulate two linear programs for which x^* is shown to be optimal using duality.)

—(*J. May and R.L. Smith* [4.18])

4.44 Consider a problem such as the one discussed in Example 1.1, in which there are lower- and/or upper-bound constraints on the ratio of two affine functions in the variables. Suppose such a problem is transformed into an LP as in Example 1.1. (Sometimes this transformation can be justified if the denominator in a ratio is known to be strictly positive or strictly negative on the set of feasible solutions.) Given the optimum primal and dual solutions associated with this LP, discuss how one can obtain the marginal values corresponding to the bounds on the ratios in the original problem.

—[*W. A. Spivey*]

4.45 *CONSTRAINED TWO-PERSON ZERO-SUM GAME.* Consider the two-person zero-sum game in which $A = (a_{ij})$ of order $m \times n$ is Player I's payoff matrix. So $-A$ is player II's payoff matrix. Let $x = (x_1, \ldots, x_m)^{\mathrm{T}}$ and $y = (y_1, \ldots, y_n)^{\mathrm{T}}$ denote the probability vectors with which Player's I and II, respectively, make their choice. Consider the game in which each player wishes to limit his or her choice of mixed strategies to a specified convex polytope, for some practical reasons. To be specific, suppose Players I and II want to choose their mixed strategies x, y so as to satisfy the constraints given below, where D, F, d, f are all given matrices or vectors. Formulate the problem of determining the optimal mixed strategies for Players I and II that maximize the guaranteed minimum expected payoff to each player, subject to the above constraints, as a primal-dual pair of linear programs.

$Dx \geqq d$	$Fy \geqq f$
$\sum_{i=1}^{m} x_i = 1$	$\sum_{j=1}^{n} y_j = 1$
$x \geqq 0$	$y \geqq 0$

—(*G. Owen* [4.22])

4.46 Consider the following LP

$$\begin{aligned} &\text{Minimize} && z(x) = cx \\ &\text{Subject to} && \sum_{j=1}^{n} a_{ij}x_j \geqq b_i \qquad i = 1 \text{ to } m \\ &&& \sum_{j=1}^{n} d_{pj}x_j \geqq g_p \qquad p = 1 \text{ to } t \\ &&& x_j \geqq 0 \qquad \text{for all } j \end{aligned} \tag{4.91}$$

The constraints in this LP are all inequality constraints, and they are divided into two groups: the first m, called the *a-constraints* are indexed by the subscript i; the last t, called the *d-constraints* are indexed by the subscript p. It is known that $\bar{x} = (\bar{x}_1, \ldots, \bar{x}_n)^{\mathrm{T}}$ is an optimum solution of (4.91) and that $\bar{\pi} = (\bar{\pi}_1, \ldots, \bar{\pi}_m)$, $\bar{\mu} = (\bar{\mu}_1, \ldots, \bar{\mu}_t)$ are an optimum dual solution corresponding to (4.91). Now consider the following LP (4.92). Prove that $\bar{x}$ is an optimum solution for (4.92), and that $\bar{\mu}$ is an optimum dual solution associated with (4.92).

Minimize

$$v(x) = cx - \sum_{i=1}^{m} \bar{\pi}_i \left(\sum_{j=1}^{n} a_{ij} x_j \right)$$

Subject to

$$\sum_{j=1}^{n} d_{pj} x_j \geqq g_p \qquad p = 1 \text{ to } t$$

$$x_j \geqq 0 \qquad \text{for all } j \tag{4.92}$$

4.47 Prove Theorems 4.12, 4.13, 4.14 and 4.15 using the result in Exercise 3.34.

4.48 Consider the LP (P): minimize cx subject to $Ax \geqq c^{\mathrm{T}}$, $x \geqq 0$, where A is an $n \times n$ symmetric matrix. Prove that if $\bar{x}$ satisfies $A\bar{x} = c^{\mathrm{T}}$, $\bar{x} \geqq 0$, then $\bar{x}$ must be optimal to (P).

—(*M. Partovi*)

4.49 Consider the following LPs:

$$\begin{array}{ll} \text{Minimize} & cx \\ \text{Subject to} & Ax \geqq b \\ & x \text{ unrestricted} \end{array} \tag{4.93}$$

$$\begin{array}{ll} \text{Maximize} & cy \\ \text{Subject to} & Ay \leqq b \\ & y \text{ unrestricted} \end{array} \tag{4.94}$$

where A is a matrix of order $m \times n$. If both (4.93) and (4.94) are feasible, prove that if one of these problems has an optimum solution, the other must have an optimum solution, too. Also prove that if (4.93) and (4.94) are both feasible, the objective function in (4.93) is unbounded below iff the objective function is unbounded above in (4.94). Assuming that both problems have optimum solutions, prove that if x and y are feasible to (4.93) (4.94), respectively, then $cx \geqq cy$.

4.50 Derive necessary and sufficient conditions for systems of linear constraints of the forms (4.4) or (4.6) or (4.8) or the primal in Example 4.5, to have a unique feasible solution.

4.51 Let A and b be given matrices of order $m \times n$ and $m \times 1$, respectively, and $b \geqq 0$. Let I be the unit matrix of order n. Suppose there exists a positive number λ such that for each $j = 1$ to n, the system $Ax = b + \lambda I_{.j}$, $x \geqq 0$, has a feasible solution. Prove that this implies that the system $Ax = b$, $x \geqq 0$ has a feasible solution.

4.52 Let A, b, c be matrices of orders $m \times n$, $m \times 1$, and $1 \times n$, respectively, and let d be a real number. Let $\mathbf{K} = \{x: Ax \leqq b\}$. If $\mathbf{K} \neq \emptyset$ and every $x \in \mathbf{K}$ satisfies $cx \leqq d$, prove that there exists a $\pi \geqq 0$ satisfying $\pi A = c$ and $\pi b \leqq d$.

—(*B. C. Eaves and R. M. Freund* [4.6])

4.53 Let $\{x^1, \ldots, x^r\}$, $\{y^1, \ldots, y^{\mathrm{L}}\}$ be two given finite sets of points in $\mathbf{R}^n$. Let $\mathbf{K}_1$ be the convex hull of $\{x^1, \ldots, x^r\}$ and let $\mathbf{K}_2 = \{x: x = \alpha_1 x^1 + \cdots + \alpha_r x^r + \beta_1 y^1 + \cdots + \beta_{\mathrm{L}} y^{\mathrm{L}}$, where $\alpha_1, \ldots, \alpha_r, \beta_1, \ldots, \beta_{\mathrm{L}}$ are all nonnegative real numbers satisfying $\alpha_1 + \cdots + \alpha_r = 1\}$. It is required to find systems of linear inequalities of the form $Ax \geqq b$, whose set of feasible solutions is (a) $\mathbf{K}_1$ and (b) $\mathbf{K}_2$. Develop efficient methods for determining such systems of constraints.

4.54 Let $\mathbf{K}$ denote the set of feasible solutions of (4.6). Assume that $\mathbf{K}$ has two or more extreme points. The point x^1 is a given extreme point of $\mathbf{K}$. Let $x^2, \ldots, x^{\mathrm{L}}$ be all the other extreme points of $\mathbf{K}$ and let $y^1, \ldots, y^p$ be all the extreme homogeneous solutions corresponding to (4.6). Define $\mathbf{K}_1 = \{x: x = \alpha_2 x^2 + \cdots + \alpha_{\mathrm{L}} x^{\mathrm{L}} + \beta_1 y^1 + \cdots + \beta_p y^p$, where $\alpha_2, \ldots, \alpha_{\mathrm{L}}, \beta_2, \ldots, \beta_p$ are all nonnegative real numbers satisfying $\alpha_2 + \cdots + \alpha_{\mathrm{L}} = 1\}$. The convex polyhedron $\mathbf{K}_1$ is said to be the polyhedron obtained by deleting the extreme point x^1 from those of $\mathbf{K}$.

It is required to determine the additional linear constraints that should be included in (4.6) so that the set of feasible solutions of the augmented system is exactly $\mathbf{K}_1$. Discuss an efficient method for generating these additional constraints.

4.55 Read each of the following statements carefully and check whether it is *true* or *false*. Justify your answer by constructing a simple illustrative example, if possible, or by a simple proof. In all these statements, problem (1) and the rest of the notation, is the same as that in Exercise 3.75.

1 In the statement of an LP, if there is a constraint that a variable has to be greater than or equal to a specified lower bound, then in writing the dual problem, that variable is treated as an unrestricted variable.

2 When a problem is formulated as an LP, the dual variables in the associated dual can always be interpreted as the equilibrium prices associated with the *items*, each of which corresponds to a constraint in the original problem.

3 The dual variables are the Lagrange multipliers associated with primal constraints.

4 The dual problem in a primal-dual pair of LPs is always the maximization problem in the pair.

5 Since the dual is a maximization problem and the primal is a minimization problem, the dual objective value is always greater than or equal to the primal objective value.

6 A basis for (1) is primal feasible if all the basic variables are nonnegative in the associated primal basic solution. Likewise, the basis is dual feasible if all the dual variables are nonnegative in the associated dual basic solution.

7 If B is a dual feasible basis for (1), and b is a nonnegative linear combination of the columns in B, B must be an optimum basis for (1).

8 Let $\bar{x}$ be a BFS for (1). If all the bases for which $\bar{x}$ is the associated primal basic solution are dual infeasible, the optimum objective value in (1) must be less than $c\bar{x}$.

9 Let $\bar{\pi}$ be an optimum dual solution corresponding to (1). If $\bar{\pi}_1$ is positive, we can expect that $f(b)$ will increase if b_1 is increased slightly from its present value, while all the other data in the problem remain unchanged.

10 If (1) has a nondegenerate optimum BFS, the dual problem has a unique optimum solution.

11 In solving (1) by the simplex algorithm starting with a feasible basis, a terminal basis is either dual feasible or shows conclusively that the dual constraints are inconsistent.

12 In every primal-dual pair of LPs the primal objective value is always greater than or equal to the dual objective value.

13 If the dual problem in a primal-dual pair is infeasible, the primal objective function must be unbounded in the primal problem.

14 In a primal-dual pair, if the dual objective function is unbounded in the dual problem, the primal problem must be infeasible.

15 If $f(b)$ is $-\infty$, it is possible to change the b vector to a b' such that $f(b')$ is finite.

16 Complementary slackness conditions for optimality exist only for inequality constraints or sign restrictions.

17 Since all the constraints in (1) are equations, there are no complementary slackness conditions for optimality for problem (1).

18 By the complementary slackness theorem, if x_j is positive in an optimum solution of (1), $c_j - \pi A_{.j}$ must be zero in all dual optimum solutions. Conversely, if $c_j - \pi A_{.j}$ is zero in every dual optimum solution, x_j must be positive in every primal optimum solution of (1).

19 If a method for solving systems of linear inequalities can be developed, that method can be used to solve any linear programming problem right away, without doing any optimization.

20 If $\mathbf{S}$ and $\mathbf{T}$ are two disjoint subsets of $\mathbf{R}^n$, there exists a hyperplane separating them.

21 If a system of linear inequalities is inconsistent, there exists a nonnegative linear combination of the inequalities in the system, which is inconsistent by itself.

22 If the dual is the maximization problem and the primal is the minimization problem in a primal-dual pair of LPs, the dual objective value will always be greater than or equal to the primal objective value, at feasible solutions of the respective problems.

23 If the dual of (1) is infeasible, (1) cannot have an optimum solution.

24 The duality theorem states that the primal and dual optimum solutions corresponding to (1) are always the same.

25 The dual of the Phase I problem corresponding to (1) is always feasible.

26 In a primal-dual pair of LP's, each dual variable can be interpreted as the *price* of the item that led to the corresponding primal constraint, so dual variables are always restricted to be nonnegative.

27 If there exists a linear combination of the row vectors of A that is less than or equal to c, the objective function in (1) cannot be unbounded below.

28 If $\bar{\pi}$ is the dual optimum solution corresponding to (1), and $\bar{\pi}_1 < 0$, then the optimum objective value in (1) can be decreased by increasing b_1 from its present value.

29 In a primal-dual pair of LPs, if all the primal variables in the primal problem are restricted to be nonnegative, then all the dual variables in the dual problem will also be restricted to be nonnegative.

30 If there is a cycle of linearly dependent constraints in (1), the dual variable associated with one of the constraints in the cycle can be assumed to be equal to zero in the dual problem.

31 Problem (1) is an LP in n variables subject to m constraints, where $m < n$. The dual of (1) is an LP in $m + n$ variables (m dual variables and n dual slack variables) subject to n constraints. Since the dual problem has more variables, as well as more constraints, solving the dual of (1) is always much harder than solving (1).

32 Let π_i be the dual variable associated with the ith constraint in (1). Eliminating the ith constraint from (1) has the effect of introducing the additional constraint $\pi_i = 0$ on the dual problem.

33 If $\bar{\pi}$ is the unique dual optimum solution associated with (1), and $\pi_i < 0$, then it pays the decision maker to increase the value of b_i marginally, as long as the cost of achieving this increase is less than $|\bar{\pi}_i|$ per unit.

34 If the primal problem in a primal-dual pair of LPs is infeasible, the dual problem must be infeasible too.

35 The zero vector is a feasible solution for the dual of the Phase I problem corresponding to (1).

36 If the dual of (1) is known to be feasible, then the unboundedness criterion will not be satisfied when (1) is solved by the primal simplex algorithm.

37 Let (x_j, u_j) for $j = 1$ to n, (π_i, v_i) for $i = 1$ to m be all the complementary pairs in a primal-dual pair of LPs. If $\bar{x}, \bar{v}, \bar{\pi}, \bar{u}$ are, respectively, primal and dual optimal, then for any j such that $\bar{x}_j = 0$, $\bar{u}_j$ must be >0; and for any i such that $\bar{v}_i = 0$, $\bar{\pi}_i$ must be >0.

38 The termination criteria in the primal simplex algorithm are the dual feasibility and the dual infeasibility criteria.

39 Given an optimum solution of the primal problem in a primal-dual pair of LPs, the complementary slackness conditions for optimality lead to a system of linear equations that the associated optimum dual solution must satisfy, from which the associated optimum dual solution can be uniquely determined.

40 At least one problem in any primal-dual pair of LPs is always feasible.

41 The optimum dual solution corresponding to (1) can be interpreted as the vector of partial derivatives of the optimum objective value in (1) with respect to the right-hand-side constants.

42 If $\bar{x}$ is a primal feasible solution, $\bar{\pi}$ is a dual feasible solution in a primal-dual pair of LPs, and if $\bar{x}$, $\bar{\pi}$ together satisfy the complementary slackness condition, then the primal objective value at $\bar{x}$ is equal to the dual objective value at π.

43 It is impossible for both the problems in a primal-dual pair of LPs to have unbounded objectives.

44 If (1) is infeasible, it can be made feasible by changing the b vector appropriately.

45 The complementary slackness conditions for optimality for (1) state that for each j exactly one quantity in the pair $(x_j, \bar{c}_j)$ is zero and the other is strictly positive.

46 In the optimum dual solution associated with the Phase I problem corresponding to (1), all dual variables will be less than or equal to one.

47 The optimality criterion of the primal simplex algorithm for a given feasible basis for (1) is just the dual feasibility criterion for that basis.

48 The unboundedness criterion in the primal simplex algorithm for solving (1) just verifies that the dual of (1) is infeasible.

49 The relative cost coefficients of the primal simplex algorithm can be interpreted as slack variables in the dual problem.

50 If a point $y \in \mathbf{R}^n$, is not contained in a convex polyhedral cone $\mathbf{K} \subset \mathbf{R}^n$, $\mathbf{K}$ and y can be separated by a hyperplane.

51 If $\bar{\pi}$ is an optimum dual solution corresponding to (1), then $f(b) = \bar{\pi}b$.

52 Let $\bar{\pi}$ be an optimum dual solution corresponding to (1), and $\bar{\pi}$ is a row vector in $\mathbf{R}^m$. If α is a small positive number and (1) remains feasible when b is changed to $b - \alpha\bar{\pi}^{\mathrm{T}}$, $f(b - \alpha\bar{\pi}^{\mathrm{T}}) \leqq f(b)$.

REFERENCES

4.1 M. Akgül, "A Note on Shadow Prices in Linear Programming," Technical Report, Department of Math · Sciences, University of Delaware, Newark, Del. 1982.

4.2 D. C. Aucamp and D. I. Steinberg, "The Computation of Shadow Prices in Linear Programming," *Journal of the Operational Research Society 33* (1982), 557–565.

4.3 F. E. Clark, "Remark on the Constraint Sets in Linear Programming," *American Mathematical Monthly 68* (1961), 351–352.

4.4 R. W. Cottle and G. B. Dantzig, "Complementary Pivot Theory of Mathematical Programming," *Linear Algebra and Its Applications 1* (1968), 103–125.

4.5 G. B. Dantzig and A. Orden, "Notes on Linear Programming: Part II, Duality Theorems," Rand Research Memorandum RM-1265, The Rand Corporation, Santa Monica, Calif., October 30, 1953.

4.6 B. C. Eaves and R. M. Freund, "Optimal Scaling of Balls and Polyhedra," *Mathematical Programming 23* (1982), 138–147.

4.7 D. Gale, H. W. Kuhn, and A. W. Tucker, "Linear Programming and the Theory of Games," in T. C. Koopmans (Ed.), *Activity Analysis of Production and Allocation*, Wiley, New York, 1951.

4.8 A. J. Goldman and A. W. Tucker, "Theory of linear programming", in *Linear Inequalities and Related Systems*, H. W. Kuhn and A. W. Tucker (Eds.), Princeton University Press, Princeton, N.J., 1956.

4.9 N. B. Haaser and J. A. Sullivan, *Real Analysis*, Van Nostrand-Reinhold, New York, 1971.

4.10 D. Hausmann, *Adjacency on Polytopes in Combinatorial Optimization*, Oelgeschlager, Gunn & Hain, Cambridge, Mass., 1980.

4.11 A. J. Hoffman, "On Approximate Solutions of Systems of Linear Inequalities," *Journal Research National Bureau of Standards 49* (1952), 263–265.

4.12 W. Karush, "Minima of Functions of Several Variables with Inequalities as Side Conditions," M. S. dissertation, Department of Mathematics, University of Chicago, December 1939.

4.13 H. W. Kuhn and A. W. Tucker, "Nonlinear Programming," in *Proceedings of Second Berkeley Symposium on Mathematical Statistics and Probability*, J. Neyman (Ed.), University of California Press, Berkeley, (1951), 481–492.

4.14 H. W. Kuhn, "Nonlinear Programming: A Historical View," in *Nonlinear Programming: Proceedings of the Joint SIAM-AMS Symposium on Applied Mathematics*, R. W. Cottle and C. E. Lemke (Eds.), (New York City, March 1975), American Mathematical Society, Providence, R. I., 1976.

4.15 C. E. Lemke, "Bimatrix Equilibrium Points and Mathematical Programming," *Management Science 11* (1965), 681–689.

4.16 C. E. Lemke and J. T. Howson, "Equilibrium Points of Bimatrix Games," *SIAM Journal of Applied Mathematics 12* (1964), 413–423.

4.17 O. L. Mangasarian, "Uniqueness of Solutions in Linear Programming," *Linear Algebra and Its Applications 25* (1979), 151–162.

4.18 J. May and R. L. Smith, "Random Polytopes: Their Definition, Generation, and Aggregate Properties," *Mathematical Programming 24*, 1 (September 1982), 39–54.

4.19 R. R. Meyer, "Continuity Properties of Linear Programs," Computer Science Technical Report 373, University of Wisconsin, Madison, November 1979.

4.20 H. D. Mills, "Marginal Values in Matrix Games and Linear Programs," in *Linear Equalities and Related Systems*, H. W. Kuhn and A. W. Tucker (Eds.), Princeton University Press, Princeton, New Jersey, 1956, pp. 183–194.

4.21 M. Morishima, *The Economic Theory of Modern Society*, Cambridge University Press, Cambridge, England, 1976.

4.22 G. Owen, *Game Theory,* Saunders, Phildelphia, 1968.

4.23 M. S. Ramanujan and E. S. Thomas, *Intermediate Analysis,* Macmillan, New York, 1970.

4.24 S. M. Robinson, "Bounds for Error in the Solution Set of a Perturbed Linear Program," *Linear Algebra and Its Applications 6* (1963), 69–81.

4.25 S. M. Robinson, "Stability Theory for Systems of Inequalities, Part I: Linear Systems," *SIAM Journal Numerical Analysis 12* (1975), 754–769.

4.26 S. M. Robinson, "A Characterization of Stability in Linear Programming," *Operations Research 25*, 3 (May/June 1977), 435–447.

4.27 H. L. Royden, *Real Analysis*, 2nd ed., Macmillan, New York, 1968.

4.28 W. Rudin, *Principles of Mathematical Analysis*, 3rd ed., McGraw-Hill, New York, 1976.

4.29 P. Sengupta and D. Solow, "A Finite Descent Theory for Linear Programming, Piecewise Linear Convex Minimization, and the Linear Complementarity Problem," Technical Report, Department of Operations Research, Case Western Reserve University, Cleveland, Ohio, June 1981.

4.30 R. L. Smith, "An Elementary Proof of the Duality Theory of Linear Programming," *Journal of Optimization Theory and Applications 12*, 2 (August 1973), 129–135.

4.31 J. von Neumann and O. Morgenstern, *Theory of Games and Economic Behavior*, Princeton University Press, Princeton, N.J., 1944.

4.32 J. von Neumann, "Discussion of a maximization problem," manuscript from the Institute for Advanced Study, 1947.

4.33 H. M. Wagner, "Linear Programming Techniques for Regression Analysis," *Journal of the American Statistical Association 54* (1959), 206.

4.34 A. C. Williams, "Marginal Values in Linear Programming," *Journal of SIAM 11*, 1 (March 1963), 82–94.

4.35 A. C. Williams, "Complementarity Theorems for Linear Programming," *SIAM Review 12*, 1 (January 1970), 135–137.

4.36 P. Wolfe, "Errors in the Solution of Linear Programming Problems," in *Error in Digital Computation,* vol. II, L. B. Rau (Ed.), Wiley, New York, 1965, pp. 271–284.

Revised (Primal) Simplex Method

chapter **5**

5.1 INTRODUCTION

In the simplex algorithm using canonical tableaux discussed in Chapter 2, many computations are performed at every pivot step. Every time a pivot is performed, it is carried out on every column of the tableau. This can be very time-consuming. In the revised simplex method, all the necessary computations are carried out using the formulas discussed in Sections 3.11 and 3.12 and restricting the pivot operations to the inverse matrix. The revised simplex method is purely an efficient computational scheme for applying the main ideas of the simplex method. At present, the revised simplex method is still the only approach that has proved to be computationally efficient and robust in practice. To solve an LP by the revised simplex method, first transform it into the *standard form*. Suppose it is

$$\begin{aligned} \text{Minimize} \quad & z(x) = cx \\ \text{Subject to} \quad & Ax = b \\ & x \geqq 0 \end{aligned} \tag{5.1}$$

where $b \geqq 0$ and A is of order $m \times n$. There are two forms of the revised simplex method depending on how the inverse matrix is stored; one uses the explicit inverse and the other uses the product form of the inverse.

5.2 REVISED SIMPLEX ALGORITHM WITH THE EXPLICIT FORM OF INVERSE WHEN PHASE II CAN BEGIN DIRECTLY

If there exists a unit matrix of order m as a submatrix of the matrix A, Phase II can begin directly. Choose an initial basic vector associated with a basis that is the unit matrix of order m. Introduce a unit matrix of order $m + 1$ by the side of the tableau. The *inverse tableau* will be obtained from these column vectors.

Original Tableau (for Starting Phase II Directly)

$x_1 \dots x_n$	$-z$	*Columns for the Inverse Tableau*	b
$a_{11} \dots a_{1n}$	0	$1 \dots 0$	b_1
$\vdots \quad \vdots$	$\vdots$	$\vdots \quad \vdots$	$\vdots$
$a_{m1} \dots a_{mn}$	0		b_m
$c_1 \dots c_n$	1	$0 \dots 1$	0
These column vectors will remain unchanged		Elementary row operations will be performed only on these columns while pivoting	

To obtain the initial inverse tableau perform the row operations required for pricing out the basic column vectors in the cost row. These row operations are performed only on the $m + 1$ column vectors introduced for obtaining the inverse tableau and on the column vector of the right-hand-side constants. The initial inverse tableau will be of the form given below, with the matrix $(\beta_{ij}) = I$, since the initial basis is the unit matrix. If B is the current basis at any stage, in the notation of Section 3.10, the inverse tableau will be

$$T = \begin{pmatrix} B & \vdots & 0 \\ \cdots & \vdots & \cdots \\ c_B & \vdots & 1 \end{pmatrix}^{-1} = \begin{pmatrix} B^{-1} & \vdots & 0 \\ \cdots\cdots & \vdots & \cdots \\ -c_B B^{-1} & \vdots & 1 \end{pmatrix}$$

In a general stage of the simplex algorithm, say, at the end of stage k, let B be the basis under consideration. Let T, the inverse tableau at this stage, be as shown in Tableau 5.1.

Tableau 5.1 Inverse Tableau at the End of Stage k

Basic Variables	*Inverse Tableau* $= T$	*Basic Values*
	$\beta_{11} \dots \beta_{1m} \quad 0$	$\bar{b}_1$
	$\vdots \quad \vdots \quad \vdots$	$\vdots$
	$\beta_{m1} \dots \beta_{mm} \quad 0$	$\bar{b}_m$
$-z$	$-\pi_1 \dots -\pi_m \quad 1$	$-z$

The matrix (β_{ij}) in the inverse tableau is B^{-1}. The $(-\pi)$ in the last row of the inverse tableau is $(-c_B B^{-1})$, and π is the dual solution of (5.1) corresponding to the present basis B. The vector of basic values at this stage is

$$\begin{pmatrix} x_B \\ \cdots \\ -z \end{pmatrix} = \begin{pmatrix} \bar{b} \\ \cdots \\ -\bar{z} \end{pmatrix} = T\begin{pmatrix} b \\ \cdots \\ 0 \end{pmatrix}$$

From the results in Chapters 3 and 4 the relative cost coefficients with respect to the current basis are for $j = 1$ to n

$$\bar{c}_j = c_j - \pi A_{.j} = c_j - \sum_{i=1}^{m} \pi_i a_{ij} = (-\pi \vdots 1)\begin{pmatrix} A_{.j} \\ \cdots \\ c_j \end{pmatrix}$$

The a_{ij}, c_j, b_i, *etc.*, here are the entries in the original tableau. Hence the relative cost coefficient of x_j is the dot product of the last row vector (the one corresponding to the cost row) in the inverse tableau with the column vector of x_j in the original tableau. While the pivot operations are confined to the inverse matrix, the $\bar{c}_j$ have to be computed for all j using this formula in each step.

The Optimality Criterion

The present BFS is an optimum feasible solution if the relative cost coefficients are all nonnegative; that is, $\bar{c}_j \geqq 0$ for each $j = 1$ to n.

Choosing the Pivot Column

If the optimality criterion is not satisfied at this stage, pick as the entering variable a nonbasic variable x_s associated with a negative relative cost coefficient. A convenient pivot column choice rule is to pick x_s with $\bar{c}_s = \text{minimum } \{\bar{c}_j : j = 1 \text{ to } n\}$ as the entering variable, or use any of the other entering variable choice rules discussed in Section 2.5.7. The pivot column is the updated column vector of the entering variable x_s with respect to the current basis. From the results in Chapters 3 and 4, it is

$$\begin{pmatrix} \bar{A}_{.s} \\ \cdots \\ \bar{c}_s \end{pmatrix} = T\begin{pmatrix} A_{.s} \\ \cdots \\ c_s \end{pmatrix}$$

Unboundedness Criterion

If the pivot column is such that $\bar{c}_s < 0$, $\bar{A}_{.s} \leqq 0$, the objective function is unbounded below on the set of feasible solutions of the problem and we terminate the algorithm. The feasible solution $x(\lambda)$ given by

$$\begin{aligned} i\text{th basic variable} &= \bar{b}_i - \lambda \bar{a}_{is} \qquad i = 1 \text{ to } m \\ x_s &= \lambda \\ \text{all other nonbasic variables} &= 0 \\ z &= \bar{z} + \lambda \bar{c}_s \end{aligned}$$

is feasible for all $\lambda \geqq 0$ and as λ tends to $+\infty$, $z(x(\lambda))$ tends to $-\infty$.

Minimum Ratio Test

If the unboundedness criterion is not satisfied, x_s is brought into the basic vector. To determine the pivot row compute $\bar{b}_r/\bar{a}_{rs} = \text{minimum } \{\bar{b}_i/\bar{a}_{is} : i, \text{ such that } \bar{a}_{is} > 0\}$, the pivot row is then the rth row. Ties here are broken arbitrarily, or by any of the dropping variable choice rules discussed in Section 2.5.8. The present rth basic variable is the *leaving variable,* or the *dropping variable.*

Obtaining the New Inverse Tableau by Pivoting

The pivot column is included at the end of the inverse tableau. Performing the pivot transforms the inverse tableau into the inverse tableau with respect to the new basis. Here again the row operations required for the pivot operation are performed only on the column vectors of the inverse tableau and on the updated right-hand-side constant vector; they are not performed on the column vectors of the original tableau. After the pivot, the pivot column is transformed into $I_{.r}$ (where I is the unit matrix of order $m + 1$) and is dropped. The pivot updates the inverse tableau and the column vector of basic values (Tableau 5.2). This pivot operation is also called the operation of *updating the inverse tableau.*

Obtaining the New Inverse Tableau by Using Pivot Matrices

As discussed in Chapter 3, the new inverse tableau can also be obtained by multiplying the present inverse tableau on the left by the *pivot matrix* corresponding to this pivot step. The pivot matrix for this operation is the matrix P^k in equation (3.25), it differs from the unit matrix of order $(m + 1)$ in just the rth column (here r is the number of the pivot row). This special column of the pivot matrix is known as its *eta column* or *eta vector.* Indicating the entries in the new inverse tableau with the superscript*, we get

$$T^* = P^k T \qquad \text{and} \qquad \begin{pmatrix} b^* \\ \cdots \\ -z^* \end{pmatrix} = P^k \begin{pmatrix} \bar{b} \\ \cdots \\ -\bar{z} \end{pmatrix}$$

Tableau 5.2 Old Inverse Tableau (at the End of Stage k)

Basic Variables	Inverse Tableau	Basic Values	Pivot Column (updated column of x_s)	
	$\beta_{11} \dots \beta_{1m}$ 0	$\bar{b}_1$	$\bar{a}_{1s}$	
	$\vdots \quad \vdots$	$\vdots$	$\vdots$	
x_r	$\beta_{r1} \dots \beta_{rm}$ 0	$\bar{b}_r$	$\textcircled{\bar{a}_{rs}}$	Pivot Row
	$\vdots \quad \vdots$	$\vdots$	$\vdots$	
	$\beta_{m1} \dots \beta_{mm}$ 0	$\bar{b}_m$	$\bar{a}_{ms}$	
$-z$	$-\pi_1 \dots -\pi_m$ 1	$-\bar{z}$	$\bar{c}_s$	

Having obtained the new inverse tableau, check whether the new basis is terminal. If not, repeat the iterations until termination occurs.

The Dual Solution

The last row of the inverse tableau provides the negative of the dual solution corresponding to the present basis. If an optimum basis for the problem is obtained at termination, the dual solution obtained from the last row of the terminal inverse tableau is an optimum dual feasible solution corresponding to (5.1).

Starting Phase II from an Arbitrary Primal Feasible Basis

Even if the original input-output coefficient matrix A does not contain a unit matrix of order m as a submatrix, Phase II can begin directly if some primal feasible basis is known. The inverse tableau corresponding to it can be computed using the formulas discussed above. Continue the iterations from there until either the optimality or unboundedness criteria are satisfied.

Example 5.1

Consider the LP given at the bottom. The original tableau is in canonical form, and (x_4, x_6, x_1) provides an initial feasible basic vector. With this, we start Phase II directly.

First Inverse Tableau

Basic Variables	Inverse Tableau				Basic Values	Pivot Volumn $\bar{A}_{.3}$	Ratios
x_4	1	0	0	0	2	2	2/2
x_6	0	1	0	0	1	$\textcircled{1}$	1/1
x_1	0	0	1	0	3	-2	
$-z$	-3	4	-2	1	-8	-2	

The vector of relative cost coefficients is obtained by multiplying the original tableau on the left by $(-3, 4, -2, 1)$. It leads to $\bar{c} = (0, 6, -2, 0, -12, 0, 15)$. The optimality criterion is not satisfied, and either x_3 or x_5 is eligible to be chosen as the entering variable. We choose x_3. The pivot column is the updated column of x_3 and it is obtained by multiplying the column vector of x_3 on the left by the current inverse tableau. It is already entered on the tableau above. By the minimum ratio test, the pivot row can be either row 1 or row 2 and we select row 2. The pivot element is circled. Performing the pivot, we get the next inverse tableau.

x_1	x_2	x_3	x_4	x_5	x_6	x_7	$-z$	b	Columns for the Inverse Tableau				b
0	1	2	1	1	0	-5	0	2	1	0	0	0	2
0	2	1	0	-2	1	0	0	1	0	1	0	0	1
1	1	-2	0	1	0	3	0	3	0	0	1	0	3
2	3	-4	3	1	-4	6	1	0	0	0	0	1	0

$x_j \geqq 0$, for all j; z to be minimized.

Second Inverse Tableau

Basic Variable	Inverse Tableau				Basic Values	Pivot Column $\bar{A}_{.5}$	Ratios
x_4	1	−2	0	0	0	⑤	0/5
x_3	0	1	0	0	1	−2	
x_1	0	2	1	0	5	−3	
$-z$	−3	6	−2	1	−6	−16	

The current vector of relative cost efficients is $\bar{c} = (0, 10, 0, 0, -16, 2, 15)$. We select x_5 as the next entering variable.

Third Inverse Tableau

Basic Variable	Inverse Tableau				Basic Values	Pivot Column $\bar{A}_{.7}$
x_5	$\frac{1}{5}$	$-\frac{2}{5}$	0	0	0	−1
x_3	$\frac{2}{5}$	$\frac{1}{5}$	0	0	1	−2
x_1	$\frac{3}{5}$	$\frac{4}{5}$	1	0	5	0
$-z$	$\frac{1}{5}$	$-\frac{2}{5}$	−2	1	−6	−1

The current vector of relative cost coefficients is $\bar{c} = (0, 2/5, 0, 16/5, 0, -22/5, -1)$. Variables x_6 and x_7 are eligible to enter the basic vector, and we select x_7. The pivot column is the updated column of x_7, and it is entered on the third inverse tableau. There are no positive entries in the pivot column. This implies that the objective function is unbounded below on the set of feasible solutions of this LP. The extreme half-line along which the objective value diverges to $-\infty$ is $(x_1, x_2, x_3, x_4, x_5, x_6, x_7)^{\mathrm{T}} = (5, 0, 1 + 2\lambda, 0, \lambda, 0, \lambda)^{\mathrm{T}}$, $z = 6 - \lambda$, for $\lambda \geqq 0$.

Note 5.1: *THE CHANGES THAT OCCUR IN A PIVOT STEP IN THE PRIMAL SIMPLEX ALGORITHM* Let x_s be the entering variable with its updated column (pivot column) $= (\bar{a}_{1s}, \ldots, \bar{a}_{ms}, \bar{c}_s)^{\mathrm{T}}$, and let row r be the pivot row in a step of the primal simplex algorithm. Let π be the present dual solution, $\boldsymbol{\beta}_{r.}$ be the rth row of the present basis inverse, $\bar{b}$ be the updated right-hand-side constants vector, and $\bar{z}$ be the present objective value. The pivot step is a *degenerate pivot step* if $\bar{b}_r = 0$; a *nondegenerate pivot step* otherwise. If this is a degenerate pivot step, there will be no change in the updated right-hand-side constants vector in this step since $\bar{b}_r = 0$, so the primal solution and the objective value do not change. If it is a nondegenerate pivot step, the objective value changes from $\bar{z}$ into $\bar{z} + \bar{c}_s(\bar{b}_s/\bar{a}_{rs})$ in this pivot step, undergoing a strict decrease. Whether this is a degenerate or nondegenerate pivot step, the relative cost coefficient of the entering variable x_s will change from $\bar{c}_s$ (which is < 0) into 0 in this pivot step. Hence the dual solution changes from π into $\tilde{\pi} = \pi + \boldsymbol{\beta}_{r.}(\bar{c}_s/\bar{a}_{rs})$. This formula is obtained by looking at the present inverse tableau and the inverse tableau obtained after performing this pivot step. Since $\boldsymbol{\beta}_{r.}$ is a row of the basis inverse, $\boldsymbol{\beta}_{r.} \neq 0$. Hence $\tilde{\pi} \neq \pi$, and the dual solution definitely changes, whether this is a degenerate pivot step or not.

5.3 REVISED SIMPLEX METHOD USING PHASE I AND PHASE II

If the original input-output coefficient matrix in (5.1) does not contain a unit matrix of order m as a sub-

Original Tableau for Phase I Problem

x_1		x_n	x_{n+1}		x_{n+m}	$-z$	$-w$	Columns for the Inverse Tableau			
a_{11}	...	a_{1n}	1	...	0	0	0	1	...	0	b_1
⋮		⋮	⋮		⋮	⋮	⋮	⋮		⋮	
a_{m1}	...	a_{mn}	0	...	1	0	0		...		b_m
c_1	...	c_n	0	...	0	1	0	0	...	0	0
0	...	0	1	...	1	0	1	0	...	1	0
These column vectors remain unchanged								This is a unit matrix of order $(m + 2)$. Row operations will be performed only on these column vectors			

matrix, and if a primal feasible basis is not known, begin Phase I with a full artificial basis. Suppose the artificial variables are $x_{n+1}, \ldots, x_{n+m}$ The original Phase I tableau is at the bottom of page 234.

The first inverse tableau during Phase I is obtained by carrying out the row operations required for pricing out of the initial basic columns in the Phase I and II cost rows. These row operations are carried out only on the column vectors introduced for getting the inverse tableau and on the right-hand-side constants vector. During Phase I, inverse tableaux are of order $m + 2$. Let B be a basis and let c_B and d_B be the Phases II and I basic cost vectors, respectively, corresponding to it. The inverse tableau corresponding to the basis B is

$$T = \begin{pmatrix} B & \vdots & 0 \\ \cdots & & \cdots \\ c_B & \vdots & 1 \quad 0 \\ d_B & \vdots & 0 \quad 1 \end{pmatrix}^{-1} = \begin{pmatrix} B^{-1} & \vdots & 0 \\ \cdots & & \cdots \\ -c_B B^{-1} & \vdots & 1 \quad 0 \\ -d_B B^{-1} & \vdots & 0 \quad 1 \end{pmatrix}$$

In a general stage of the algorithm during Phase I, say, at the end of stage k, let B be the basis under consideration. Let the present inverse tableau be

Inverse Tableau during Phase I at the End of Stage k

Current Basic Variables	*Inverse Tableau* $= T$				*Basic Values*
	$\beta_{11} \cdots$	β_{1m}	0	0	$\bar{b}_1$
	$\vdots$	$\vdots$	$\vdots$	$\vdots$	$\vdots$
	$\beta_{m1} \cdots$	β_{mm}	0	0	$\bar{b}_m$
$-z$	$-\pi_1 \quad \ldots$	$-\pi_m$	1	0	$-\bar{z}$
$-w$	$-\sigma_1 \quad \ldots$	$-\sigma_m$	0	1	$-\bar{w}$

The present Phase I objective value is $\bar{w} = \sigma b$. The Phase I relative cost coefficients with respect to the present basis are

$$\bar{d}_j = d_j - \sigma A_{.j} = (-\sigma, 0, 1) \begin{pmatrix} A_{.j} \\ \cdots \\ c_j \\ d_j \end{pmatrix}$$

for $j = 1$ to n. Hence $\bar{d}_j$ is obtained by multiplying the $m + 2$th row vector in the inverse tableau on the right by the column vector of x_j in the original Phase I tableau.

Phase I Termination Criterion

As in Chapter 2, Phase I of the simplex method terminates when $\bar{d}_j \geqq 0$ for all $j = 1$ to n.

Infeasibility Criterion

When the Phase I termination criterion is satisfied, if the Phase I objective value $\bar{w}$ is positive, the original problem has no feasible solution and we terminate. The σ_i's from the terminal Phase I inverse tableau provide the coefficients of a linear combination of the original system of constraints that leads to an impossible constraint. Multiply the ith equation in the original system (5.1) by σ_i and add over $i = 1$ to m. This leads to the constraint

$$\sum_{j=1}^{n} (-\bar{d}_j) x_j = \bar{w} \tag{5.2}$$

Since $\bar{d}_j \geqq 0$ for all $j = 1$ to n and $\bar{w} > 0$, this constraint can never be satisfied by any $x \geqq 0$.

Switching Over to Phase II

When the Phase I termination criterion is satisfied, if $\bar{w} = 0$, the present BFS of the Phase I problem yields a BFS of the original problem when the zero values of the artificial variables are suppressed. In all subsequent iterations, all the original problem variables x_j, where j is such that $\bar{d}_j > 0$, are set equal to zero, and such variables are never considered eligible to enter the basic vector. The $m + 2$th row in the tableau is deleted, as is the $m + 2$th column in the inverse tableau. The inverse tableaux during Phase I are of order $m + 2$, but they will be of order $m + 1$ during Phase II. The computations during Phase II are carried out as in Section 5.2. Any artificial variables in the basic vector at the end of Phase I will be equal to zero in the corresponding basic solution, and they are retained in the basic vector, unless they are displaced from the basic vector during the iterations in Phase II. In all the iterations during Phase II, all the artificial variables in the basic vector will be equal to zero.

Pivot Column and Pivot Row Choice in Phase I

If the Phase I termination criterion is not satisfied, an original problem variable x_s, where s is such that $\bar{d}_s < 0$, is selected as the entering variable. A convenient rule is to pick x_s associated with $\bar{d}_s =$ minimum $\{\bar{d}_j : j = 1 \text{ to } n\}$, as the entering variable, or to select the entering variable by any of the other entering variable choice rules discussed in Chapter 2. The pivot column is the updated column vector of the entering variable x_s, and it is

$$\begin{pmatrix} \bar{A}_{.s} \\ \cdots \\ \bar{c}_s \\ \bar{d}_s \end{pmatrix} = T \begin{pmatrix} A_{.s} \\ \cdots \\ c_s \\ d_s \end{pmatrix}$$

The pivot row is determined by performing the minimum ratio test as in Section 5.2.

Updating the Inverse Tableau During Phase I

The pivot column is entered at the end of the present inverse tableau and the pivot performed. This transforms the present inverse tableau into the inverse tableau with respect to the new basis. The new inverse tableau can also be obtained by multiplying the present inverse tableau on the left by the pivot matrix.

$$rth\ column$$

$$P^k = \begin{bmatrix} 1 \cdots 0 & -\bar{a}_{1s}/\bar{a}_{rs} & 0 \cdots 0 \\ \vdots\quad\vdots & \vdots & \vdots\quad\vdots \\ 0 \cdots 1 & -\bar{a}_{r-1,s}/\bar{a}_{rs} & 0 \cdots 0 \\ 0 \cdots 0 & 1/\bar{a}_{rs} & 0 \cdots 0 \\ 0 \cdots 0 & -\bar{a}_{r+1,s}/\bar{a}_{rs} & 1 \cdots 0 \\ \vdots\quad\vdots & \vdots & \vdots\quad\vdots \\ 0 \cdots 0 & -\bar{a}_{ms}/\bar{a}_{rs} & 0 \cdots 0 \\ 0 \cdots 0 & -\bar{c}_{s}/\bar{a}_{rs} & 0 \cdots 0 \\ 0 \cdots 0 & -\bar{d}_{s}/\bar{a}_{rs} & 0 \cdots 1 \end{bmatrix}$$

Since the inverse matrix is of order $m + 2$, the pivot matrices during Phase I are of order $m + 2$.

Check whether the new basis satisfies the Phase I termination criterion. Otherwise continue the iterations.

Initial Basic Vector in Phase I

If the original input-output coefficient matrix contains some column vectors of the unit matrix of order m, then it is only necessary to introduce as many artificial variables as necessary to complete a unit submatrix of order m in the augmented tableau. The Phase I objective function is always the sum of the artificial variables that are introduced.

Example 5.2

Consider the LP

Minimize $z(x) = 3x_1 - x_2 - 7x_3 + 3x_4 + x_5$

Subject to $5x_1 - 4x_2 + 13x_3 - 2x_4 + x_5 = 20$

$x_1 - x_2 + 5x_3 - x_4 + x_5 = 8$

$x_j \geq 0$ for $j = 1, \ldots, 5$

The artificial variables x_6, x_7 are introduced; the Phase I original tableau, and the first (Phase I) inverse tableau are given at the bottom. The Phase I relative cost coefficient vector is $(\bar{d}_1, \bar{d}_2, \bar{d}_3, \bar{d}_4, \bar{d}_5) = (-6, 5, -18, 3, -2)$. Hence, x_3 is brought into the basic vector. The pivot column is recorded on the tableau. The new inverse tableau is obtained either by pivoting or by multiplying the present inverse tableau on the left by the pivot matrix P^1 given on the next page. In this numerical example, the pivot

Original Phase I Tableau

x_1	x_2	x_3	x_4	x_5	x_6	x_7	$-z$	$-w$	Columns for Inverse Tableau				b
5	−4	13	−2	1	1	0	0	0	1	0	0	0	20
1	−1	5	−1	1	0	1	0	0	0	1	0	0	8
3	−1	−7	3	1	0	0	1	0	0	0	1	0	0
0	0	0	0	0	1	1	0	1	0	0	0	1	0

First Inverse Tableau (Phase I)

Basic Variables	Inverse Tableau				Basic Values	$\bar{A}_{.3}$	Ratios	
x_6	1	0	0	0	20	⑬	$\dfrac{20}{13}$	minimum ratio
x_7	0	1	0	0	8	5	$\dfrac{8}{5}$	
$-z$	0	0	1	0	0	−7		
$-w$	−1	−1	0	1	−28	−18		

$$P^1 = \begin{bmatrix} \frac{1}{13} & 0 & 0 & 0 \\ -\frac{5}{13} & 1 & 0 & 0 \\ \frac{7}{13} & 0 & 1 & 0 \\ \frac{18}{13} & 0 & 0 & 1 \end{bmatrix} \qquad P^2 = \begin{bmatrix} 1 & -\frac{1}{8} & 0 & 0 \\ 0 & \frac{13}{8} & 0 & 0 \\ 0 & -\frac{20}{8} & 1 & 0 \\ 0 & 1 & 0 & 1 \end{bmatrix} \qquad P^3 = \begin{bmatrix} 1 & \frac{3}{7} & 0 \\ 0 & \frac{8}{7} & 0 \\ 0 & \frac{36}{7} & 1 \end{bmatrix}$$

matrices corresponding to all the pivot steps are given for illustration. The new inverse tableau is obtained *either* by performing the pivot operations on the rows of the present inverse tableau *or* by multiplying the present inverse tableau on the left by the pivot matrix.

Second Inverse Tableau (Phase I)

Basic Variables	*Inverse Tableau*				*Basic Values*	$\bar{A}_{.5}$	*Ratios*
x_3	$\frac{1}{13}$	0	0	0	$\frac{20}{13}$	$\frac{1}{13}$	20
x_7	$-\frac{5}{13}$	1	0	0	$\frac{4}{13}$	$\frac{8}{13}$ (circled)	$\frac{4}{8}$
$-z$	$\frac{7}{13}$	0	1	0	$\frac{140}{13}$	$\frac{20}{13}$	
$-w$	$\frac{5}{13}$	-1	0	1	$-\frac{4}{13}$	$-\frac{8}{13}$	

$\bar{d} = (\bar{d}_1, \bar{d}_2, \bar{d}_3, \bar{d}_4, \bar{d}_5) = (12/13, -7/13, 0, 3/13, -8/13)$. Therefore, bring x_5 into the basic vector. The second pivot matrix is P^2.

Third Inverse Tableau (Phase I)

Basic Variables	**Inverse Tableau**				**Basic Values**
x_3	$\mathbf{\frac{1}{8}}$	$\mathbf{-\frac{1}{8}}$	**0**	0	$\mathbf{\frac{3}{2}}$
x_5	$\mathbf{-\frac{5}{8}}$	$\mathbf{\frac{13}{8}}$	**0**	0	$\mathbf{\frac{1}{2}}$
$-z$	$\mathbf{\frac{3}{2}}$	$\mathbf{-\frac{5}{2}}$	**1**	0	**10**
$-w$	0	0	0	1	0

The artificials have all left the basic vector. Also $\bar{d} = 0$ and $\bar{w} = 0$ in this stage. Strike off the last row and the fourth column from the inverse tableau, which leads to the first inverse tableau in Phase II (this consists only of the bold portions of the above inverse tableau).

$\bar{c} = (8, -9/2, 0, 5/2, 0)$ and, hence, x_2 enters the basic vector. The pivot matrix is P^3.

Basic Variables	*Inverse Tableau*			*Basic Values*
x_3	$-\frac{1}{7}$	$\frac{4}{7}$	0	$\frac{12}{7}$
x_2	$-\frac{5}{7}$	$\frac{13}{7}$	0	$\frac{4}{7}$
$-z$	$-\frac{12}{7}$	$\frac{41}{7}$	1	$\frac{88}{7}$

Now $\bar{c} = (2/7, 0, 0, 4/7, 36/7)$ and, hence, this is an optimum basis. The optimum solution is $(x_1, x_2, x_3, x_4, x_5) = (0, 4/7, 12/7, 0, 0)$, with a minimum optimal objective value of $-88/7$.

5.4 INFEASIBILITY ANALYSIS

Suppose Phase I terminates with the conclusion that the system of constraints in (5.1) is infeasible. Let the last row (the one corresponding to the Phase I objective function) in the final Phase I inverse tableau be $(-\hat{\sigma}_1, \ldots, -\hat{\sigma}_m, 0, 1)$. Row $(\hat{\sigma}_1, \ldots, \hat{\sigma}_m)$ is the Phase I dual solution from the terminal Phase I inverse tableau. Multiplying the ith equation in (5.1) by $-\hat{\sigma}_i$ and adding over $i = 1$ to m leads to (5.2), which can never be satisfied by any $x \geqq 0$. There are two cases to consider here.

Case 1 Suppose $\bar{d}_j = 0$ for all $j = 1$ to n. In this case, the left-hand side of (5.2) is zero. This means that the system of equations $Ax = b$ itself does not have a solution. In this case, the set of constraints $\{A_{i.}x = b_i\colon i$, such that $\hat{\sigma}_i \neq 0\}$ together are inconsistent, and unless one of the constraints from this set is eliminated or, at least modified to remove the inconsistency, the problem will remain inconsistent.

Case 2 Suppose $\bar{d}_j > 0$ for at least one j between 1 to n. In this case, (5.1) has a solution but no nonnegative solution. Hence the set of constraints $\{A_{i.}x = b_i\colon i$, such that $\hat{\sigma}_i \neq 0\}$ does not admit a nonnegative solution, and unless one of these constraints is eliminated, or modified to remove the infeasibility, the problem will remain infeasible.

Example 5.3

Consider the system of constraints:

x_1	x_2	x_3	x_4	x_5	x_6	b
0	1	−1	2	1	−2	3
2	1	0	0	1	1	2
1	0	0	−2	1	−1	4
1	1	1	−1	1	0	1

$x_j \geqq 0$ for all j.

Introducing the artificial variables x_7, x_8, x_9, x_{10}, in that order, Phase I was carried out. The terminal Phase I relative cost vector is $\bar{d} = (0, 1, 7/4, 0, 0, 5/4)$, and the final Phase I objective value is $\bar{w} = 15/4$. The final Phase I dual solution is $(1/4, 1/4, 1, -3/2)$. Hence, in the original system of equations, if we multiply the first equation by $-1/4$, the second by $-1/4$, the third by -1, and the fourth by 3/2, and add, we get $x_2 + (7/4)x_3 + (5/4)x_6 = -(15/4)$, clearly this equation can never be satisfied by any $x \geqq 0$. This infeasible equation was obtained by taking a linear combination of the four equations in the original system, with nonzero coefficients for each equation. So the original system of four equations together do not admit a nonnegative solution. In the final Phase I solution, only the artificial variable x_9 (the one introduced in the third equation of the original system) has a nonzero value of 15/4. From the Phase I original tableau, this implies that if the right-hand-side constant in the third equation of the original system is reduced by 15/4 (i.e., changed from 4 to 1/4), the system becomes feasible.

Thus the final Phase I dual solution can be used to determine a subset of the original system of constraints, from which at least one should be eliminated or modified to make the system feasible. The values of the artificial variables in the final Phase I solution can be used to determine how to change the original right-hand-side constant vector, to make the system of constraints feasible. In this manner, the output from Phase I of the revised simplex method can be used to provide guidance for modifying an infeasible problem into a feasible one.

5.5 REVISED SIMPLEX METHOD USING THE PRODUCT FORM OF THE INVERSE

In this method, the inverse tableau is not stored, but the pivot matrices generated at each stage are stored individually. Since each pivot matrix differs from the unit matrix in just its eta column, it can be stored very conveniently by storing that column and its position in the unit matrix. Its storage in this manner occupies very little core space in the computer.

Suppose we start out in Phase I. The initial inverse tableau is a square matrix of order $m + 2$. Suppose it is denoted by P^0. It differs from the unit matrix of order $(m + 2)$ only in the last row. It can be stored easily by storing the last row, and its position in P^0.

Suppose the pivot matrices generated in the course of the algorithm are $P^1, P^2 \cdots$, in that order. Successive inverse tableaux can be obtained by multiplying P^0 on the left by the pivot matrices in the order in which they are generated. Suppose at the kth stage we are still in Phase I, and the column vector of basic values is $(\bar{b}_1, \ldots, \bar{b}_m, -\bar{z}, -\bar{w})^{\mathrm{T}}$. The pivot matrices $P^0, P^1, \ldots, P^k$ generated in the algorithm so far, in the specific order, are known as the *current string of pivot matrices*. The values of the basic variables and the objective value in the next stage are

$$(b_1^*, \ldots, b_m^*, -z^*, -w^*)^{\mathrm{T}} = P^k(\bar{b}_1, \ldots, \bar{b}_m, \bar{z}, -\bar{w})^{\mathrm{T}}$$

Let σ^*, π^* be the Phases I and II dual solutions corresponding to the next stage. Then

$$(-\sigma^*, 0, 1) = (0, \ldots, 0, 1)P^kP^{k-1} \cdots P^1P^0 \qquad (5.3)$$

where $(0, \ldots, 0, 1)$ is the row vector of order $m + 2$ with all zeros excepting the 1 in the $(m + 2)$th place.

Starting from the left, the multiplications in (5.3) can be carried out very quickly. $(0, \ldots, 0, 1)$ is a row vector. When it is multiplied on the right by P^k, we get another row vector. This row vector is then multiplied on the right by P^{k-1}, leading to another row vector, and so on. This procedure is repeated each time the current row vector is multiplied on the right by the next pivot matrix on the right-hand side of (5.3), until the entire product on the right-hand side is completed. This left-to-right string computation of the product on the right-hand side of (5.3) therefore requires $(k+1)$ operations of multiplying a row vector on the right by a pivot matrix. Since the pivot matrix differs from the unit matrix in exactly one column, each of these operations can be carried out very efficiently. Having obtained σ^*, the Phase I relative cost coefficients can be calculated as in Section 5.3.

Since matrix multiplication is not, in general, commutative, to compute σ^* using (5.3), it is necessary to record the pivot matrices in the specific order in which they are generated, and to carry out the product on the right-hand side of (5.3) from left to right using the pivot matrices in the specific order listed in (5.3).

Check whether the Phase I termination criterion is satisfied. If it is not, pick the nonbasic variable x_s with the most negative Phase I relative cost coefficient as the entering variable, or use one of the other entering variable choice rules discussed in Chapter 2. The pivot column is the updated column vector of the entering variable x_s, and it is

$$\begin{pmatrix} \bar{A}_{.s} \\ \cdots \\ \bar{c}_s \\ \bar{d}_s \end{pmatrix} = P^k P^{k-1} \cdots P^1 P^0 \begin{pmatrix} A_{.s} \\ \cdots \\ c_s \\ d_s \end{pmatrix} \tag{5.4}$$

The multiplication in (5.4) is easily carried out by starting from the right-hand side. The column $(a_{1s}, \ldots, a_{ms}, c_s, d_s)^{\mathrm{T}}$ is the rightmost column in (5.4). When this column is multiplied on the left by P^0, we get another column vector. This column vector is then multiplied on the left by P^1, and so on. This procedure is repeated, each time, multiplying the current column vector on the left by the next pivot matrix on the right-hand side of (5.4). This right-to-left string computation of the product on the right-hand side of (5.4) requires $(k+1)$ operations of multiplying a column vector on its left by a pivot matrix. After the pivot column is obtained in this manner, the pivot element in the pivot column is determined by performing the usual minimum ratio test. Let the pivot matrix corresponding to this pivot step be P^{k+1}. The algorithm is now ready for the next stage.

On the other hand, if all the Phase I relative cost coefficients are nonnegative, then Phase I is terminated. If $w^* > 0$, the problem is infeasible and we terminate. If $w^* = 0$, we move to Phase II by deleting the last row and last column in each of the previous pivot matrices. For convenience we continue to denote the pivot matrices [after the $(m+2)$ row and column are struck off] by the same symbols $P^1, \ldots, P^k$ as before.

Suppose we are in the tth stage of Phase II. Let $(\bar{b}_1, \ldots, \bar{b}_m, -\bar{z})^{\mathrm{T}}$ be the column vector of present basic values. Let P^t be the pivot matrix corresponding to the pivot in this step. Suppose the superscript* identifies the quantities after the pivot. Then,

$$\begin{pmatrix} b^* \\ \cdots \\ -z^* \end{pmatrix} = P^t \begin{pmatrix} \bar{b} \\ \cdots \\ -\bar{z} \end{pmatrix}$$

$$(-\pi^*, 1) = (0, \ldots, 0, 1) P^t P^{t-1} \cdots P^1$$

and the updated column vector of any nonbasic variable x_s is

$$\begin{pmatrix} A^*_{.s} \\ \cdots \\ c^*_s \end{pmatrix} = P^t \cdots P^1 \begin{pmatrix} A_{.s} \\ \cdots \\ c_s \end{pmatrix}$$

These multiplications can be carried out conveniently as before. Using these, the Phase II iterations are repeated until termination.

Example 5.4

Minimize

$$z = x_1 + x_2 + x_3 + 3x_4 + 4x_5 + 5x_6 + 2x_7$$

Subject to

$$\begin{aligned} x_1 + x_2 \quad\quad + 2x_4 + x_5 + x_6 + x_7 &= 3/2 \\ x_2 - x_3 + x_4 \quad\quad - x_6 + 5x_7 &= 1 \\ 2x_1 \quad - x_3 + 2x_4 - x_5 + x_6 + x_7 &= 1 \\ x_j \geqq 0 \quad \text{for all } j & \end{aligned}$$

Introducing artificial variables t_1, t_2, t_3, we find that the original tableau for Phase I is as follows.

Original Tableau for Phase I

x_1	x_2	x_3	x_4	x_5	x_6	x_7	t_1	t_2	t_3	$-z$	$-w$	b
1	1	0	2	1	1	1	1	0	0	0	0	$\frac{3}{2}$
0	1	-1	1	0	-1	5	0	1	0	0	0	1
2	0	-1	2	-1	1	1	0	0	1	0	0	1
1	1	1	3	4	5	2	0	0	0	1	0	0
0	0	0	0	0	0	0	1	1	1	0	1	0

Initial Inverse Tableau

Basic Variables	P^0					Basic Values
t_1	1	0	0	0	0	$\frac{3}{2}$
t_2	0	1	0	0	0	1
t_3	0	0	1	0	0	1
$-z$	0	0	0	1	0	0
$-w$	-1	-1	-1	0	1	$-\frac{7}{2}$

At this stage we have $(t_1, t_2, t_3, -z, -w)^{\mathrm{T}} = P^0 b = (3/2, 1, 1, 0, -7/2)^{\mathrm{T}}$. $(-\sigma, 0, 1) = (0, 0, 0, 0, 1)\ P^0 = (-1, -1, -1, 0, 1)$ and $\bar{d} = (-3, -2, 2, -5, 0, 1, -7)$. The variable x_4 has a negative Phase I relative cost coefficient, and it can be the entering variable. The updated column vector of x_4 is P^0 (original column of x_4) $= (2, 1, 2, 3, -5)^{\mathrm{T}}$. Perform the minimum ratio test:

Basic Variables	Present Basic Values	Pivot Column	Ratio
t_1	$\frac{3}{2}$	2	$\frac{3}{4}$
t_2	1	1	1
t_3	1	2	$\frac{1}{2}$ min

The variable x_4 enters the basic vector, replacing t_3. The pivot matrix corresponding to this basis change is P^1 given below. The new basic values vector is $(t_1, t_2, x_4, -z, -w)^{\mathrm{T}} = P^1(3/2, 1, 1, 0, -7/2)^{\mathrm{T}} = (1/2, 1/2, 1/2, -3/2, -1)^{\mathrm{T}}$. For this stage we have $(-\sigma, 0, 1) = (0, 0, 0, 0, 1)P^1P^0 = (0, 0, 5/2, 0, 1)P^0 = (-1, -1, 3/2, 0, 1)$. Therefore, $\bar{d} = (-1, -1, 3/2, 0, 1) \times$ (original tableau) $= (2, -2, -1/2, 0, -5/2, 3/2, -9/2)$. So x_2 is a candidate to enter the basic vector. The pivot column is P^1P^0 (original column of x_2) $= P^1P^0(1, 1, 0, 1, 0)^{\mathrm{T}} = P^1(1, 1, 0, 1, -2)^{\mathrm{T}} = (1, 1, 0, 1, -2)^{\mathrm{T}}$. For the ratio test we have:

Basic Variables	Basic Values	Pivot Column	Ratio
t_1	$\frac{1}{2}$	1	$\frac{1}{2}$
t_2	$\frac{1}{2}$	1	$\frac{1}{2}$
x_4	$\frac{1}{2}$	0	

The variable x_2 can replace either t_1 or t_2 in the basic vector. Suppose we decide to drop t_2 from the basic vector. The new pivot matrix is P^2 given below. The new basic values vector is $(t_1, x_2, x_4, -z, -w)^{\mathrm{T}} = P^2(1/2, 1/2, 1/2, -3/2, -1)^{\mathrm{T}} = (0, 1/2, 1/2, -2, 0)^{\mathrm{T}}$. The new Phase I dual vector is $(-\sigma, 0, 1) = (0, 0, 0, 0, 1)P^2P^1P^0 = (0, 2, 0, 0, 1)P^1P^0 = (0, 2, 3/2, 0, 1)P^0 = (-1, 1, 1/2, 0, 1)$. The new vector of Phase I relative cost coefficients is $\bar{d} = (-1, 1, 1/2, 0, 1) \times$ (original tableau) $= (0, 0, -3/2, 0, -3/2, -3/2, 9/2)$. Since some of the $\bar{d}_j$ are negative, the Phase I optimality criterion is not yet satisfied.

Note 5.2: The present basis associated with the basic vector (t_1, x_2, x_4) is degenerate, since the value of t_1 in the basic solution is 0. Actually the present value of the Phase I objective value is 0. Hence the present BFS of the Phase I problem is an optimum solution to the Phase I problem. However, this is not an optimal basis for the Phase I problem. This situation can only arise under degeneracy. When we continue to apply the algorithm, all subsequent pivots will be degenerate pivots. This will change the basis without any change in the solution, until a basis satisfying the optimality criterion is obtained.

$$P^1 = \begin{bmatrix} 1 & 0 & -\frac{2}{2} & 0 & 0 \\ 0 & 1 & -\frac{1}{2} & 0 & 0 \\ 0 & 0 & \frac{1}{2} & 0 & 0 \\ 0 & 0 & -\frac{3}{2} & 1 & 0 \\ 0 & 0 & \frac{5}{2} & 0 & 1 \end{bmatrix} \quad P^2 = \begin{bmatrix} 1 & -1 & 0 & 0 & 0 \\ 0 & 1 & 0 & 0 & 0 \\ 0 & 0 & 1 & 0 & 0 \\ 0 & -1 & 0 & 1 & 0 \\ 0 & 2 & 0 & 0 & 1 \end{bmatrix} \quad P^3 = \begin{bmatrix} \frac{2}{3} & 0 & 0 & 0 & 0 \\ \frac{1}{3} & 1 & 0 & 0 & 0 \\ \frac{1}{3} & 0 & 1 & 0 & 0 \\ -2 & 0 & 0 & 1 & 0 \\ 1 & 0 & 0 & 0 & 1 \end{bmatrix} \quad P^4 = \begin{bmatrix} 1 & 0 & 0 & 0 \\ 0 & 1 & 1 & 0 \\ 0 & 0 & 1 & 0 \\ 0 & 0 & 1 & 1 \end{bmatrix}$$

Suppose we bring x_3 into the basic vector. The updated column vector of x_3 is $P^2P^1P^0(0, -1, -1, 1, 0)^T = P^2P^1(0, -1, -1, 1, 2)^T = P^2(1, -1/2, -1/2, 5/2, -1/2)^T = (3/2, -1/2, -1/2, 3, -3/2)^T$. The ratio test gives the following values.

Basic Variables	Basic Values	Pivot Column	Ratios
t_1	0	$\frac{3}{2}$	0
x_2	$\frac{1}{2}$	$-\frac{1}{2}$	
x_4	$\frac{1}{2}$	$-\frac{1}{2}$	

The variable x_3 comes into the basic vector replacing t_1, and this involves a degenerate pivot step. The pivot matrix is P^3 given above. Now all the artificial variables have left the basic vector. The present basic vector (x_3, x_2, x_4) is an optimum basic vector of the Phase I problem, and is a feasible basic vector of the original problem. To proceed to Phase II, delete the last row and column from each of the pivot matrices P^0, P^1, P^2, P^3, and for convenience in notation, refer to them by the same name. Drop the Phase I cost row and the w column vector from the original tableau. The present basic value vector is $(x_3, x_2, x_4, -z)^T = P^3(0, 1/2, 1/2, -2)^T = (0, 1/2, 1/2, -2)^T$. The corresponding Phase II dual vector is $(-\pi, 1) = (0, 0, 0, 1)P^3P^2P^1 = (-2, 0, 0, 1)P^2P^1 = (-2, 1, 0, 1)P^1 = (-2, 1, 0, 1)$. Hence, the Phase II relative cost coefficient vector is $\bar{c} = (-\pi, 1) \times$ (original tableau) $= (-1, 0, 0, 0, 2, 2, 5)$. Variable x_1 enters the basic vector. The updated column vector of x_1 is $P^3P^2P^1(1, 0, 2, 1)^T = P^3P^2(-1, -1, 1, -2)^T = (0, -1, 1, -1)^T$.

Basic Variables	Basic Values	Pivot Column	Ratios
x_3	0	0	
x_2	$\frac{1}{2}$	-1	
x_4	$\frac{1}{2}$	1	$\frac{1}{2}$

The variable x_1 comes into the basic vector, replacing x_4. The pivot matrix is P^4.

The new basic value vector is $(x_3, x_2, x_1, -z)^T = P^4(0, 1/2, 1/2, -2)^T = (0, 1, 1/2, -3/2)^T$. The new dual vector is $(-\pi, 1) = (0, 0, 0, 1)P^4P^3P^2P^1 = (0, 0, 1, 1)P^3P^2P^1 = (-5/3, 0, 1, 1)P^2P^1 = (-5/3, 2/3, 1, 1)P^1 = (-5/3, 2/3, 1/3, 1)$. Hence the vector of relative cost coefficients is $\bar{c} = (-5/3, 2/3, 1/3, 1) \times$ (original tableau) $= (0, 0, 0, 1, 2, 3, 4)$. Hence, the present basis is an optimal basis. The optimum solution to the problem is $(x_1, x_2, x_3, x_4, x_5, x_6, x_7) = (1/2, 1, 0, 0, 0, 0, 0)$ with objective value $z = 3/2$.

5.6 ADVANTAGES OF THE PRODUCT FORM IMPLEMENTATION OVER THE EXPLICIT FORM IMPLEMENTATION

For solving small LPs (in which the number of constraints m is small), the product form is cumbersome (when using hand computation, in particular) and the method using the explicit form of the inverse would probably be preferred. However, for solving large practical problems using a digital computer, the product form implementation has several advantages, which are discussed below.

A matrix is said to be a *sparse matrix* if the proportion of nonzero entries in it is small. Likewise, a vector is said to be a *sparse vector* if the proportion of nonzero entries in it is small. The *sparsity* of a matrix or vector is determined by how small is the proportion of nonzero entries in them. In most real-life LPs of the type (5.1), the matrix A is usually sparse. If A is sparse, it usually happens that the eta vectors in the pivot matrices obtained when solving (5.1) by the simplex method, are themselves sparse. But the explicit inverse of any basis obtained while solving (5.1) by the simplex method is usually not sparse (see Exercise 5.6).

When the eta vector is sparse, it can be stored very compactly by storing the nonzero entries in it and their respective positions in the vector. If each eta vector is not sparse, the space required for storing m pivot matrices will be approximately equal to the space required for storing the explicit inverse of any basis for (5.1). However, if A is sparse (as in most LPs arising in practical applications), the tendency of the eta vectors to be sparse leads to a considerable reduction in storage space requirements when the problem is solved using the product form of the inverse.

Also, when the eta vector in a pivot matrix is sparse, the computational effort involved in multiplying that pivot matrix by a vector is small. When all the eta vectors tend to be sparse, the computational effort involved in carrying out the revised simplex algorithm with the product form of the inverse tends to be smaller than that in carrying out the same algorithm using explicit inverses which are usually not sparse. The product form of the inverse is also numerically more stable and has a lower error accumulation.

Example 5.5

Let $u = (2, -1, 3, 4, 7)$, $v = (6, 9, -4, 8, 2)^{\mathrm{T}}$, and $D = vu$. It is a matrix of order 5×5. The matrix D can be stored explicitly as the 5×5 matrix, or in product form as vu, where v is the column vector and u is the row vector given above. To store D explicitly requires 25 locations, but to store D in product form as vu requires only 10 locations. Also, let $b = (1, 2, -1, 3, -2)^{\mathrm{T}}$. Suppose it is required to compute $p = Db$. It can be verified that $p = (-30, -45, 20, -40, -10)^{\mathrm{T}}$. Computing p as Db, using D explicitly requires 25 multiplications and 20 additions. If D is stored in product form as vu, then $p = vub$, and to compute p we first compute $ub = -5$ and then $p = v(ub)$. In this product form, the computation of p requires only ten multiplications and four additions.

This simple example is provided to illustrate that it may be possible to perform matrix computations more efficiently if the matrices generated during the computations are expressed and stored in product form.

In this simple example, the matrix D is stored as the product of two vectors. In the revised simplex method using the product form of the inverse, the basis inverse is stored as the product of a sequence of pivot matrices, where each pivot matrix in the sequence differs from the identity matrix in just its eta vector. Hence, for the purpose of storage in the computer, and for performing the necessary computations, dealing with a pivot matrix is very much like dealing with its eta vector. In solving LPs by the simplex method, the tendency of the explicit basis inverse not to be sparse, and the eta vectors of the pivot matrices to be sparse, in general, results in substantial savings, on an average, in storage space requirements on the computer and in the total number of additions, multiplications, and comparisons to be performed during the algorithm, if the product form implementation is used. Because of these advantages most of the commercially available computer programs for solving LPs are based on the revised simplex method using the product form of the inverse or on one of the newer implementations discussed in Chapter 7.

5.7 REINVERSIONS

In solving (5.1) by the revised simplex method, during Phase II let x_B be the basic vector in some stage of computation; B and c_B the basis and the basic cost vector, respectively, $\bar{x}_B$ the column vector of values of the primal basic variables; and $\bar{\pi}_B$ the dual solution. If all the computations are carried out exactly without any round off, both the quantities $\|b - B\bar{x}_B\|$

and $\|c_B - \tilde{\pi}_B B\|$ will be exactly zero. Due to round-off errors, it is possible that with the present solutions $\|b - B\tilde{x}_B\|$ and $\|c_B - \tilde{\pi}_B B\|$ are not zero. Under finite precision arithmetic, it is unreasonable to expect exact solutions, so some amount of error should be allowed. Hence, a small positive number called the *tolerance* is specified (this may be something like 10^{-6} or lower, depending on the degree of precision required), and as long as both $\|b - B\tilde{x}_B\|$ and $\|c_B - \tilde{\pi}_B B\|$ are less than this tolerance, the present solutions are accepted to be accurate within the limits of precision required, and the execution of the method can be continued. However, if either $\|b - B\tilde{x}_B\|$ or $\|c_B - \tilde{\pi}_B B\|$ exceeds the tolerance, it is an indication that the accumulation of round-off errors has exceeded the error limit specified. Variables may leave the basic vector and enter again several times as pointed out in Note 2.10, and because of this, many pivot steps may have been carried out in the algorithm to reach the present basis. At this time we can expect *reinversion* to reduce the errors in the solution. This operation involves retrieving the actual basis B from the original tableau and computing its inverse afresh by the direct pivotal method discussed in Section 3.3.2, or by any other method. If the revised simplex method with the explicit form of the inverse is being used, the present inverse tableau (which has too many errors due to round-off accumulation) is discarded, and the inverse tableau corresponding to the present basic vector is computed afresh using the freshly computed B^{-1} and the formulas discussed in Section 5.2, and the execution of the algorithm continued with this newly computed inverse tableau.

If the problem is being solved by the revised simplex method with the product form of the inverse, when the inverse of B is computed afresh by the pivotal method discussed in Section 3.3.2, the pivot matrix corresponding to each pivot step carried out in this inversion is stored in the order in which it is generated. The current string of pivot matrices accumulated in the algorithm is then discarded, it is replaced by the string of pivot matrices freshly generated in reinverting B, and the execution of the algorithm is continued with this new string of pivot matrices.

When the problem is solved by the revised simplex method with the product form of the inverse, a new pivot matrix accumulates on the string after each pivot step in the algorithm. So the size of the current string (in terms of the number of pivot matrices in it) can become indefinitely long. This is clearly undesirable, as the amount of space needed to store it may ultimately exceed the space needed to store the explicit inverse if it were computed. Thus whenever the size of the current string is strictly greater than m, it can be shortened to a new string of size at most m by reinverting, thereby reducing both computation time and loss of accuracy in succeeding iterations.

Efficient revised simplex codes have the feature of *sparse storage,* that is, they usually employ a packing scheme to store only the nonzero elements of matrices by column. Such codes employ periodic reinversion to maintain sparsity and numerical accuracy.

5.8 THE REVISED SIMPLEX FORMAT AND THE COLUMN GENERATION PROCEDURE IN MODELING

One of the greatest advantages of the revised version of the simplex algorithm is that it makes it possible for us to solve a linear programming problem by the simplex algorithm, without having all the data in the model explicitly in hand at any time. The canonical tableau version of the simplex algorithm discussed in Chapter 2 requires all the data in the model right from the start and in each pivot step it modifies all the data. On the other hand, in the revised simplex method discussed in this chapter, in any step, we only need the data contained in the present basic columns (this is needed to construct the present inverse tableau) and then we need a *subroutine to check the dual feasibility* of the present dual basic solution. Assuming that the LP model is (5.1) in standard form, this subroutine for checking dual feasibility takes as input the present dual vector, π (read off from the inverse tableau with respect to the present basic vector) and should be able to determine whether "$\pi A_{.j} \leqq c_j$" holds for all j (in which case the present primal BFS is optimal), or produce a nonbasic column $\begin{pmatrix} A_{.j} \\ \cdots \\ c_j \end{pmatrix}$ in (5.1) that violates dual feasibility, that is, it satisfies $\pi A_{.j} > c_j$. This column vector is all that we need to move to the next basic vector (essentially the variable corresponding to this column vector is treated as the entering variable, and the basis change is carried out as discussed in earlier sections) and the whole process is repeated starting with the new basic vector. It is not necessary to know what procedures the subroutine employs to check dual feasibility; in this sense the subroutine for checking dual feasibility can be treated as a *black box* for executing the revised simplex algorithm on (5.1). In some applications, the problem of checking the dual feasibility of a given vector can itself be posed as an *auxiliary problem* and the subroutine for checking dual feasibility may be an algorithm for solving this auxiliary problem. Linear programs arising from combinatorial optimization problems often lead to models with a very large number of variables, each representing some activity or combinatorial pattern. This is because there are usually many possible patterns satisfying the combinatorial restrictions of the problem. In such problems, because of the very large number of variables, it is practically impossible to even generate the entire model, not to mention the task of modifying it in each pivot step. If we have to execute the simplex algorithm using the canonical tableaux, it would be virutally impossible to solve such linear programs. Fortunately, the revised version of the simplex algorithm makes it possible for us to handle such models efficiently using a sub-routine for checking dual feasibility in each step. The algorithm begins with a known feasible basic vector (this itself can be generated by solving the corresponding Phase I problem, initiated with an appropriate artificial basic vector, again using the revised simplex format). In each step it calls the subroutine for checking dual feasibility once. This either confirms the optimality of the present BFS or generates an entering column to continue the revised simplex algorithm. In this procedure, we only maintain the present set of basic columns in the original linear programming model; the entering columns are generated one by one as they are needed, using the subroutine for checking dual feasibility. Hence, this procedure is known as a *column generation procedure*. In writing this section, I benefited greatly from discussions with R. Chandrasekharan.

As an illustration, we present an application of a column generation procedure to solve the one-dimensional cutting stock problem. The problem is the following. A continuous sheet of stock material, in the form of rolls of breadth L (the material may be paper, foil, plastic film, etc.) is to be cut so as to satisfy demands for N_i rolls of strips of breadth ℓ_i, $i = 1$ to m. If $L \geqq \ell_i$, $i = 1$ to m, the problem has a solution. Define a cutting pattern (or a slitting pattern) as a particular way of cutting a roll of breadth L into a_i strips of breadth ℓ_i, $i = 1$ to m, where $a_1, \dots, a_m$ are nonnegative integers satisfying

$$L \geqq \ell_1 a_1 + \cdots + \ell_m a_m \tag{5.5}$$

We will represent this cutting pattern by the column vector $(a_1, a_2, \dots, a_m)^T$. Thus each nonnegative integer feasible solution in $a_1, \dots, a_m$ of (5.5) leads to a cutting pattern and vice versa. For example, if $m = 5$, $L = 200$, $(\ell_1, \ell_2, \ell_3, \ell_4, \ell_5) = (20, 17, 5, 30, 13)$, $a = (4, 2, 0, 2, 2)$ is a cutting pattern that slits the 200 cm width roll into four strips 20 cm wide, two strips 17 cm wide, two strips 30 cm wide, and two strips 13 cm wide. In practical applications encountered in paper, foil, or steel sheet industries, we usually have $m = 40$ or more, $L = 500$, and the ℓ_i's ranging from 50 to 200. The number of cutting patterns easily exceeds 10 or even 100 million.

Let $A_{.j} = (a_{1j}, \dots, a_{mj})^T$ be the column vector representing the jth cutting pattern. That is, a_{ij} is the number of strips of width ℓ_i generated, $i = 1$ to m, when a stock roll is slit by this pattern. Let x_j denote the number of stock rolls slit using cutting pattern j. Then the problem of minimizing the total number stock rolls that must be cut to meet the demand, is to

$$\begin{aligned} \text{Minimize} \quad & \sum_j x_j \\ \text{Subject to} \quad & \sum_j A_{.j} x_j \geqq N \\ & x_j \geqq 0 \quad \text{for all } j \end{aligned} \tag{5.6}$$

$$x_j \text{ integer for all } j \tag{5.7}$$

where $N = (N_1, \dots, N_m)^T$. This is an integer programming problem, since the variables are restricted to take on only integer values. Experience has shown that if we take an optimum solution of the LP (5.6) [ignoring (5.7)] and round it so as to satisfy (5.7), we get a very satisfactory solution of the integer problem.

Hence, we consider the LP (5.6), ignoring the integer requirement (5.7). Note that each variable in (5.6) corresponds to a cutting pattern and vice versa. Even though the number of constraints, m, in (5.6) is reasonably small, (5.6) becomes hard because of the very large number of variables in it. Thus, it is practically impossible to generate explicitly all the columns in the model (5.6). The only practical approach to solving (5.6) is by a column generation procedure. This procedure maintains the $m \times m$ basis inverse, and generates the next column when needed using a subroutine for checking dual feasibility, and never needs more than $(m + 1)$ columns in (5.6) at any one time.

Introduce slack variables y_i, $i = 1$ to m corresponding to the inequality constraints in (5.6). Let x_B denote a feasible basic vector for this problem (some of the basic variables in it may be the slack variables y_i). Let $\bar{\pi}$ denote the corresponding dual basic solution. The dual feasibility (or the optimality) conditions are

$$\pi \geqq 0$$
$$\pi A_{.j} \leqq 1 \quad \text{for each cutting pattern } j \qquad (5.8)$$

So, if any of the $\bar{\pi}_i$ are negative in the present dual solution, choose the corresponding slack variable y_i as the entering variable into the basic vector x_B, and continue.

If the present $\bar{\pi} \geqq 0$, whether it satisfies the dual feasibility condition or not can be determined by solving the following auxiliary problem: Find $(a_1, \ldots, a_m)$ to:

$$\begin{aligned} &\text{Maximize} && \sum_{i=1}^{m} \bar{\pi}_i a_i \\ &\text{Subject to} && \sum_{i=1}^{m} \ell_i a_i \leqq L \qquad (5.9) \\ & && a_i \geqq 0 \text{ and integer for all } i = 1 \text{ to } m \end{aligned}$$

If $(\tilde{a}_1, \ldots, \tilde{a}_m)$ is an optimum feasible solution of (5.9), and $\sum_{i=1}^{m} \bar{\pi}_i \tilde{a}_i \leqq 1$, then $\bar{\pi}$ obviously satisfies (5.8), and the present BFS is optimal. If, on the other hand, $\sum_{i=1}^{m} \bar{\pi}_i \tilde{a}_i > 1$, then $\bar{\pi}$ violates the dual feasibility condition in the column corresponding to the cutting pattern $(\tilde{a}_1, \ldots, \tilde{a}_m)^{\mathrm{T}}$. So associate a variable x_j to this cutting pattern $(\tilde{a}_1, \ldots, \tilde{a}_m)^{\mathrm{T}}$, bring this variable into the present basic vector, x_B, and continue.

The problem (5.9) is itself an integer programming problem, known in the literature as a *knapsack problem* (it is an integer program in which there is only a single constraint on the variables, other than the nonnegative integer requirement on the variables). Many efficient and practical branch and bound algorithms are available for solving it. See books on integer programming, for example, [1.71, 1.92, 1.94, and 1.104].

A feasible basic vector for (5.6) can itself be generated by using the above column generation procedure applied to the Phase I problem initiated with a full artificial basic vector.

Another application of the column generation procedure is discussed in Chapter 12 to solve a master problem generated when an LP with many constraints is transformed into an equivalent master problem with fewer constraints but many more variables. There, the subroutine for checking dual feasibility consists of solving auxiliary problems that are themselves specially generated small linear programs called subproblems. For discussion of column generation procedures in other areas see references [1.84 and 5.8 to 5.17].

EXERCISES

5.1 Solve the following LPs by the revised simplex method.

(a) Maximize

$$5x_1 - x_2 + x_3 - 10x_4 + 7x_5$$

Subject to

$$\begin{aligned} 3x_1 - x_2 - x_3 \qquad\qquad &= 4 \\ x_1 - x_2 + x_3 + x_4 \qquad &= 1 \\ 2x_1 + x_2 + 2x_3 \qquad + x_5 &= 7 \\ x_j \geqq 0 \quad \text{for all } j & \end{aligned}$$

(b) Minimize $\quad -x_1 + 2x_2$

Subject to

$$\begin{aligned} 5x_1 - 2x_2 &\leqq 3 \\ x_1 + x_2 &\geqq 1 \\ -3x_1 + x_2 &\leqq 3 \\ -3x_1 - 3x_2 &\leqq 2 \\ x_j \geqq 0 \quad &\text{for all } j \end{aligned}$$

In this problem, plot the path taken by the algorithm on a diagram.

(c)

x_1	x_2	x_3	x_4	x_5	x_6	$-z$	
1	2	0	1	0	−6	0	11
0	1	1	3	−2	−1	0	6
1	2	1	3	−1	−5	0	13
3	2	−3	−6	10	−5	1	0

$x_j \geqq 0$ for all j; minimize z.

(d)

x_1	x_2	x_3	x_4	x_5	x_6	x_7	$-z$	b
1	−1	1	2	1	−1	0	0	2
−1	1	1	1	2	1	−2	0	4
1	1	−1	1	−1	2	−4	0	6
2	4	−6	4	2	1	−11	1	0

$x_j \geqq 0$ for all j; minimize z.

(e)

x_1	x_2	x_3	x_4	x_5	x_6	x_7	x_8	$-z$	b
2	1	1	−2	−1	−1	0	0	0	5
6	5	−1	−3	−1	0	1	0	0	10
2	3	−3	1	1	0	0	−1	0	3
1	0	−2	3	0	0	0	0	1	0

$x_j \geqq 0$ for all j, minimize z.

(f)

x_1	x_2	x_3	x_4	x_5	x_6	$-z$	b
−1	1	2	1	0	1	0	0
0	1	3	0	1	3	0	5
2	−1	−1	−2	1	1	0	5
−3	3	9	5	0	4	1	0

$x_j \geqq 0$ for all j; minimize z.

5.2 Solve the following LP by the revised simplex algorithm using (x_1, x_2, x_6) as an initial basic vector.

Minimize

$$14x_1 - 19x_2 \qquad + 21x_4 + 52x_5$$

Subject to

$$\begin{aligned} x_1 \qquad\qquad - x_4 + x_5 + x_6 &= 3 \\ x_1 + x_2 \qquad - x_4 + 3x_5 \qquad &= 4 \\ x_1 + x_2 + x_3 - 3x_4 \qquad + x_6 &= 6 \\ x_j \geqq 0 \quad \text{for all } j & \end{aligned}$$

5.3 Solve the following LP using (x_3, x_5) as an initial basic vector by the revised simplex algorithm with the product form of the inverse.

Minimize

$$2x_1 + 5x_2 \qquad + 7x_4 + 15x_5 + 14x_6$$

Subject to

$$\begin{aligned} x_1 + 2x_2 - x_3 + x_4 + 4x_5 + 5x_6 &= 10 \\ x_1 + 3x_2 - 2x_3 + 2x_4 + 5x_5 + 7x_6 &= 12 \\ x_j \geqq 0 \quad \text{for all } j & \end{aligned}$$

5.4 Get an upper bound on the number of additions, subtractions, multiplications, divisions, and comparisons to be performed in a general step of the following, in the course of solving the same given LP, as a function of m and n, where m, n are the number of constraints and the number of variables, respectively, after the problem is expressed in standard form.

(a) The revised simplex algorithm with the explicit form of the inverse.

(b) The simplex algorithm using canonical tableaux.

5.5 Discuss how to implement the algorithm presented in Section 3.16, maintaining the basis inverse in each stage instead of computing the canonical tableaux. State clearly how to update the basis inverse efficiently as you move from one stage to the next in that algorithm, when additional rows get included in the satisfied part of the tableau.

5.6 Let

$$B = \begin{pmatrix} 1 & 0 & 0 & 0 & 1 \\ 1 & 1 & 0 & 0 & 0 \\ 0 & 1 & 1 & 0 & 0 \\ 0 & 0 & 1 & 1 & 0 \\ 0 & 0 & 0 & 1 & 1 \end{pmatrix}$$

For computing B^{-1}, if we choose the pivot elements from columns 1 to 5, pivoting down the main diagonal in that order, show that the various eta columns of the pivot matrices obtained are (in each eta column the pivot position is indicated by the circle):

η_5	η_4	η_3	η_2	η_1
$-\frac{1}{2}$	0	0	0	①
$\frac{1}{2}$	0	0	①	−1
$-\frac{1}{2}$	0	①	−1	0
$\frac{1}{2}$	①	−1	0	0
($\frac{1}{2}$)	−1	0	0	0

The total number of nonzero entries in all these eta columns put together is 13.

Compute B^{-1} explicitly and show that the total number of nonzero entries in it is 25. (This illustrates the fact that when B is sparse, B^{-1} may not be sparse, but the eta vectors in the expression of B^{-1} in product form tend to be sparse.)

—(*W. Orchard-Hays* [1.96])

5.7 *BOTTLENECK LINEAR PROGRAMMING.* Let $c = (c_1, \ldots, c_n) \in \mathbf{R}^n$ be a given vector. Consider the following:

$$\begin{aligned} \text{Minimize} \quad & z(x) = \max\{c_j: j, \text{ such that } x_j > 0\} \\ \text{Subject to} \quad & Ax = b \\ & x \geqq 0 \end{aligned}$$

where A and b are given matrices of orders $m \times n$ and $m \times 1$, respectively. Develop an approach for solving this problem efficiently. Characterize the set of optimum solutions for this problem.

—(*A. M. Frieze* [5.3])

5.8 Consider the one-dimensional cutting stock problem discussed in Section 5.8. Suppose stock material is available in rolls of breadths L_t, $t = 1$ to k. Stock rolls of breadth L_t cost $\$c_t$ each, $t = 1$ to k. As before, the requirement is to meet the demand for N_i rolls of strips of breadth ℓ_i, $i = 1$ to m. Discuss a column generation procedure for solving this problem.

Apply this procedure and solve the numerical problem in which 20 rolls of breadth 2 cm, 10 rolls of breadth 3 cm, and 20 rolls of breadth 4 cm are to be cut from stock rolls of breadth 5, 6, and 9 cm with costs respectively of 6, 7, and 10 dollars per roll. Use the branch and bound method discussed in [1.92, 1.94] to solve the required knapsack problems (P. C. Gilmore and R. E. Gomory [5.12]).

5.9 Consider the one-dimensional cutting stock problem discussed in Section 5.8. Suppose stock material is available in rolls of breadth L cm each. Often customer orders need not be filled exactly but can lie within a certain range. Specificially, suppose we are given positive integers N_i' and N_i'' (satisfying $N_i' < N_i''$ for all i) and are required to deliver N_i rolls of strips of breadth ℓ_i cm, where the N_i is required to satisfy $N_i' \leqq N_i \leqq N_i''$, $i = 1$ to m. With the quantities to be produced confined to ranges rather than exact amounts, the total number of stock rolls used is not a good objective function; it would be better to minimize the fraction of wastage. As discussed in Section 5.8, let $A_{.j} = (a_{1j}, \ldots, a_{mj})^{\mathrm{T}}$ be the column vector defining the jth cutting pattern. The waste per stock roll in this cutting pattern is $w_j = L - \sum_{i=1}^{m} a_{ij}\ell_i$. The fraction of wastage is then $(\sum w_j x_j)/(\sum x_j)$. Discuss a column generation approach for meeting the requirements with minimum fraction wastage, using the approaches discussed in Section 3.20 for handling the ratio objective function (P. C. Gilmore and R. E. Gomory [5.13].)

REFERENCES

5.1 G. B. Dantzig, "Computational Algorithm of the Revised Simplex Method," RM-1266, The RAND Corporation, Santa Monica, Calif., 1953.

5.2 G. E. Forsythe and C. B. Moler, *Computer Solutions of Linear Algebraic Systems*, Prentice-Hall, Englewood Cliffs, N.J., 1967.

5.3 A. M. Frieze, "Bottleneck Linear Programming," *Operational Research Quarterly 26*, 4, ii (1975), 871–874.

5.4 W. Orchard-Hays, "Background, Development and Extensions of the Revised Simplex Method," RM-1433, The RAND Corporation, Santa Monica, Calif. 1954.

5.5 G. M. Roodman, "Post-Infeasibility Analysis in Linear Programming," *Management Science 25*, 9 (September 1979), 916–922.

5.6 D. M. Smith and W. Orchard-Hays, "Computational Efficiency in Product Form LP Codes," in *Recent Advances in Mathematical Programming*, R. L. Graves and P. Wolfe (Ed.), McGraw-Hill, New York, 1963.

5.7 H. M. Wagner, "A Comparison of the Original and Revised Simplex Methods," *Operations Research 5* (1957), 361–369.

REFERENCES ON COLUMN GENERATION PROCEDURES

5.8 E. M. L. Beal P. A. B. Hughes and R. E. Small, "Experiences in Using a Decomposition Program," *Computer Journal*, *8*, 1 (1965) 13–18.

5.9 B. P. Dzielinski and R. E. Gomory, "Optimal Programming of Lot Sizes, inventory and labor allocations," *Management Science*, *11*, 9 (1965), 874–890.

5.10 K. Eisemann "The Trim Problem," *Management Science*, *3* (1957), 279–284.

5.11 L. R. Ford, Jr. and D. R. Fulkerson, "A Suggested Computation for Maximal Multi-Commodity Network Flows," *Management Science*, *5* (1958), 97–101.

5.12 P. C. Gilmore and R. E. Gomory, "A Linear Programming Approach to the Cutting Stock Problem," *Operations Research*, *9*, 6 (1961), 849–859.

5.13 P. C. Gilmore and R. E. Gomory, "A Linear Programming Approach to the Cutting Stock Problem—Part II," *Operations Research*, *11*, 6 (1963), 863–888.

5.14 P. C. Gilmore and R. E. Gomory, "Multistage Cutting Stock Problems of Two and More Dimensions," *Operations Research*, *13*, 1 (1965), 94–120.

5.15 R. E. Gomory and T. C. Hu, "An Application of Generalized Linear Programming to Network Flows," *SIAM J. on Applied Mathematics*, *10*, 2 (1962), 260–283.

5.16 A. S. Manne, "Programming of Economic Lot Sizes," *Management Science*, *4* (1958), 115–135.

5.17 P. Wolfe and G. B. Dantzig, "Linear Programming in a Markov Chain," *Operations Research*, *10* (1962), 702–710.

The Dual Simplex Method

chapter 6

6.1 INTRODUCTION

Consider the LP

$$\begin{aligned} \text{Minimize} \quad & z(x) = cx \\ \text{Subject to} \quad & Ax = b \qquad (6.1) \\ & x \geqq 0 \end{aligned}$$

where A is a given matrix of order $m \times n$ and rank m. Let B be a basis for (6.1). B is a *primal feasible basis* if $B^{-1}b \geqq 0$ and a *dual feasible basis* if $c - c_B B^{-1} A \geqq 0$, and an *optimal basis* for (6.1) if it is both primal and dual feasible. However, in the basic solution of (6.1) with respect to a dual feasible basis, some of the basic variables may be negative (i.e., it may be primal infeasible).

The *primal simplex algorithm* discussed in Chapters 2, 3, and 5, and its variants start with a primal feasible basis and move to a *terminal basis* by walking through a sequence of primal feasible bases along the edges of the set of feasible solutions. All the bases with the possible exception of the terminal basis obtained in the primal algorithm are dual infeasible. At each pivot step, the primal algorithm makes an attempt to reduce dual infeasibility while retaining primal feasibility.

The *dual simplex algorithm* does just the opposite. It starts with a dual feasible, but primal infeasible basis and walks to a terminal basis by moving among adjacent dual feasible bases. At each pivot step, this algorithm tries to reduce primal infeasibility while retaining dual feasibility. If a feasible basis is reached, the dual simplex algorithm terminates by declaring it as an optimum basis. Because of its name, the reader may be under the impression that the dual simplex algorithm works directly on the dual problem. This is not the case. It is an algorithm for solving the original (primal) problem (6.1), dealing with basic vectors for it. In each step, exactly one basic variable is replaced by a nonbasic variable. It may be carried out either using the canonical tableaux (in which case the dual solutions are not computed in any step, but the relative cost coefficients will be nonnegative in all steps), or by using the inverse tableaux.

6.2 DUAL SIMPLEX ALGORITHM WHEN A DUAL FEASIBLE BASIS IS KNOWN INITIALLY

Let B be a known dual feasible basis for (6.1). Let $x_B = (x_1, \dots, x_m)$ be the associated basic vector. If the algorithm is to be carried out using the canonical tableaux, let the canonical tableau with respect to the basis B be Tableau 2.8 of Section 2.5.1. Solving the problem using the inverse tableaux leads to the *revised dual simplex algorithm.* If the problem is to be solved using the revised dual simplex algorithm, let the inverse tableau corresponding to the basis B be Tableau 5.1 of Section 5.2.

In general, if B is any basis with its inverse $B^{-1} = \boldsymbol{\beta} = (\beta_{ij})$, and if $\bar{a}_{rs}$ is the entry in the canonical tableau of (6.1) with respect to the basis B, then for any $r = 1$ to m, and $s = 1$ to n,

$$\bar{a}_{rs} = (\beta_{r1}, \dots, \beta_{rm})A_{.s} \qquad (6.2)$$

Hence the updated rth row is obtained by multiplying the original tableau on the left by the rth row vector in the inverse tableau. All the entries in the rth row of the canonical tableau with respect to the basis B can be conveniently obtained from the original tableau and the inverse tableau corresponding to the basis B by using (6.2).

Since the starting basis B is known to be dual feasible, $\bar{c}_j \geqq 0$ for all $j = 1$ to n.

Optimally Criterion in the Dual Simplex Algorithm

The present basis is optimal if $\bar{b}_i \geqq 0$ for all $i = 1$ to m. If the optimality criterion is satisfied, the current basic solution is primal feasible and optimal; hence we terminate.

Choosing the Pivot Row

If the optimality criterion is not satisfied, pick a row i such that $\bar{b}_i < 0$ and choose the pivot element from it. In the present basic solution, the ith basic variable is equal to $\bar{b}_i < 0$. The purpose of pivoting in this row is to obtain a new basic vector such that the value of the ith basic variable in the basic solution corresponding to that basic vector becomes positive. For this the pivot element $\bar{a}_{ij}$ has to be chosen among the negative entries in the updated pivot row. *Hence, in the dual simplex algorithm, the pivot elements in all pivots will be negative.* This is just contrary to the primal simplex algorithm, where the pivot elements have to be positive in order to retain primal feasibility.

The updated right-hand-side constant in the pivot row becomes positive as a result of the pivot step. The pivot step as it is carried out in the dual simplex algorithm will be called a *dual simplex pivot step.*

At this stage, row i is said to be an *eligible row* if the updated right-hand-side constant $\bar{b}_i$ in it is strictly negative. At each stage, any of the eligible rows can be selected as the pivot row. There are several pivot row choice rules for making this choice (these rules correspond to the pivot column choice rules in the primal simplex algorithm). One commonly used pivot row choice rule is to pick the rth row as the pivot row, where $\bar{b}_r = \text{minimum } \{\bar{b}_i : i = 1 \text{ to } m\}$. Break ties for r arbitrarily.

Primal Infeasibility Criterion

The original problem (6.1) is infeasible if in a canonical tableau, with respect to a basis, there is an i, such that

$$\bar{b}_i < 0 \qquad \bar{a}_{ij} \geqq 0 \qquad \text{for each } j = 1 \text{ to } n$$

Discussion Suppose this criterion is satisfied in the ith row of the canonical tableau. This row corresponds to the constraint $\sum_{j=1}^{n} \bar{a}_{ij}x_j = \bar{b}_i$. Since this comes from a canonical tableau, it is a linear combination of the original constraints in (6.1). Hence, every feasible solution of (6.1) must satisfy it. However, since $\bar{b}_i < 0$ and $\bar{a}_{ij} \geqq 0$ for all j, this cannot be satisfied by any $x \geqq 0$.

To check whether the primal infeasibility criterion is satisfied, one may have to look at virtually all the data in the present canonical tableau. In practice, this is not worthwhile. Usually, if the optimality criterion is not satisfied, one of the eligible rows is chosen as the pivot row, say, row r. If all the entries in the updated rth row are $\geqq 0$, the algorithm is terminated with the conclusion that the LP is infeasible. Otherwise, the algorithm is continued.

If the dual simplex algorithm is carried out using the inverse tableaux, under primal infeasibility it is possible to identify a subset of the primal constraints, which together with the nonnegativity restrictions on the primal variables, are mutually inconsistent. Suppose row r has been selected as the pivot row. Let $(\beta_{r1}, \ldots, \beta_{rm}, 0)$ be the rth row vector of the current inverse tableau. Then the pivot row (updated rth row) corresponds to the constraint $\sum_{j=1}^{n} \bar{a}_{rj}x_j = \bar{b}_r$, where $\bar{b}_r = \sum_{i=1}^{m} \beta_{ri}b_i$. By the choice of the pivot row, $\bar{b}_r < 0$. If $\bar{a}_{rj} = 0$ for all $j = 1$ to n, the above equation is an inconsistent equation and cannot be satisfied by any x. In this case, the set of equality constraints in (6.1) are inconsistent by themselves, and the constraints $\{i: i \text{ between 1 to } m \text{ and such that } \beta_{ri} \neq 0\}$ are mutually inconsistent. Unless one of these constraints is eliminated, or the right-hand-side constraints modified so that $\sum_{i=1}^{m} \beta_{ri}b_i$ becomes equal to zero, the system will remain inconsistent.

If $\bar{a}_{rj} \geqq 0$ for all j, and >0 for at least one j, the constraint represented by the pivot row can never be satisfied by any nonnegative vector x. Hence, the constraints from the original tableau $\{i: i \text{ from 1 to } m, \text{ such that } \beta_{ri} \neq 0\}$ are mutually inconsistent with the nonnegativity restrictions on the primal variables. Unless one of these constraints is either eliminated or modified to change the original right-hand-side constants so that $\sum_{i=1}^{m} \beta_{ri}b_i$ becomes nonnegative, the primal will remain infeasible. For an illustration of primal infeasibility, see Example 6.2.

Terminal Tableau in the Dual Simplex Algorithm

In the dual simplex algorithm, a dual feasible basic vector is said to be a *terminal basic vector* if either the dual simplex optimality criterion or the primal infeasibility criterion is satisfied in the corresponding canonical tableau. The basis corresponding to a terminal basic vector is called a *terminal basis.* The canonical tableau corresponding to a terminal basic vector is called a *terminal canonical tableau.*

Choosing the Pivot Column

If the primal infeasibility criterion is not satisfied in the selected pivot row, we have to choose a pivot column and perform the pivot step. Let the rth row in the tableau be the selected pivot row. The pivot column has to be selected in such a way that this pivot step leads to a new basis, which is also dual feasible. If the pivot element is $\bar{a}_{rs}$, the relative cost coefficients in the next basis will be $\bar{c}_j - \bar{c}_s\bar{a}_{rj}/\bar{a}_{rs}$, $j = 1$ to n. Hence the next basis will remain dual feasible if $\bar{c}_j - \bar{c}_s\bar{a}_{rj}/\bar{a}_{rs} \geqq 0$ for all j. Since the present basis is dual feasible, $\bar{c}_j \geqq 0$ for all $j = 1$ to n. In order to satisfy this, the pivot column is picked as the sth column where s is such that

$$-(\bar{c}_s/a_{rs}) = \text{minimum } \{-\bar{c}_j/\bar{a}_{rj}: j, \text{ such that } \bar{a}_{rj} < 0\} \quad \textbf{(6.3)}$$

Any ties for s in (6.3) are broken arbitrarily. The minimum ratio in (6.3) is known as the *dual simplex minimum ratio*, and the process of computing it and identifying the pivot column is known as the *dual simplex minimum ratio test*. As usual, the pivot element for this pivot step is the element that is in both the pivot row and the pivot column for this step. By the choice of the pivot column, the pivot element is strictly negative.

Pivoting

If the canonical tableaux are being computed, the new canonical tableau is obtained by pivoting on the column vector of the entering variable x_s with $\bar{a}_{rs}$ as the pivot element. By elementary row operations, this column vector is transformed into the rth unit vector and priced out in the cost row. This gives the new canonical tableau. If the inverse tableaux are being used to perform the computations $\bar{A}_{.s}$, the pivot column is the updated column vector of the entering variable x_s and the new inverse tableau is obtained by performing the pivot step as in Tableau 5.2 of Section 5.2.

Change in the Objective Value in a Step of the Dual Simplex Algorithm

Let x_s, $\bar{z}$, $\bar{c}_s$, $\bar{a}_{rs}$, $\bar{b}_r$ be the entering variable, current objective value, relative cost coefficient of the entering variable, pivot element, and updated right-hand-side constant in pivot row, respectively. It can be verified that the objective value after this pivot step is given by

$$\text{New objective value} = \bar{z} + \bar{b}_r\bar{c}_s/\bar{a}_{rs} \quad (6.4)$$

After the pivot step is complete, test whether the new basis satisfies either the optimality or the primal infeasibility criterion. Repeat the iterations until a basis that satisfies either the optimality or the primal infeasibility criterion is obtained.

Note 6.1: *ABOUT UNBOUNDEDNESS* Since the dual problem is known to be feasible, the primal problem cannot be unbounded by the weak duality theorem. The algorithm discussed here terminates with a basis that satisfies either the optimality criterion or the infeasibility criterion.

Note 6.2: Clearly, the dual simplex algorithm can also be carried out using the product form of the inverse, using the ideas discussed in Section 5.4. If $P^0, P^1, \ldots, P^k$ are the pivot matrices generated in that order at some stage of the algorithm, and if the rth row of the inverse tableau is needed for the computations at this stage, then it can be verified that it is $I_{r.}P^kP^{k-1}\cdots P^1P^0$, where $I_{r.}$ is the rth row of the unit matrix of appropriate order.

Example 6.1 Example Using Canonical Tableaux

Solve the following problem by the dual simplex algorithm:

Original Tableau

x_1	x_2	x_3	x_4	x_5	x_6	$-z$	b	
1	0	0	4	−5	7	0	8	
0	1	0	−2	4	−2	0	−2	
0	0	1	1	−3	2	0	2	
0	0	0	1	3	2	1	0	(minimize z)

$x_j \geqq 0$ for all j.

(x_1, x_2, x_3) constitutes a dual feasible basic vector. Here is the canonical tableau with respect to this basic vector.

First Canonical Tableau

Basic Variables	x_1	x_2	x_3	x_4	x_5	x_6	$-z$	*Basic Values*	
x_1	1	0	0	4	−5	7	0	8	
x_2	0	1	0	(−2)	4	−2	0	−2	Pivot Row
x_3	0	0	1	1	−3	2	0	2	
$-z$	0	0	0	1	3	2	1	0	
$-\frac{\bar{c}_j}{\bar{a}_{2j}}$ for $\bar{a}_{2j} < 0$				$\frac{1}{2}$		$\frac{2}{2}$			

Since $\bar{b}_2$ is -2, the second row is picked as the pivot row. The minimum ratio $(-\bar{c}_j/\bar{a}_{2j})$ for j, such that $\bar{a}_{2j} < 0$, occurs in the column of the nonbasic variable x_4. So x_4 is the entering variable. It replaces x_2 from the basic vector. The pivot element has been circled. The new canonical tableau is

Basic Variables	x_1	x_2	x_3	x_4	x_5	x_6	$-z$	Basic Values
x_1	1	2	0	0	3	3	0	4
x_4	0	$-\frac{1}{2}$	0	1	-2	1	0	1
x_3	0	$\frac{1}{2}$	1	0	-1	1	0	1
$-z$	0	$\frac{1}{2}$	0	0	5	1	1	-1

Since all the basic variables are positive, this basis is now optimal. So the optimal feasible solution is $(x_1, x_2, x_3, x_4, x_5, x_6) = (4, 0, 1, 1, 0, 0)$, $z(x) = 1$.

Example 6.2 Example Using Inverse Tableaux

Consider the LP

$$\begin{aligned} \text{Minimize} \quad & z(x) = 8x_1 + 8x_2 + 9x_3 \\ \text{Subject to} \quad & x_1 + x_2 + x_3 \leqq 1 \\ & 2x_1 + 4x_2 + x_3 \geqq 8 \\ & x_1 - x_2 - x_3 \geqq 2 \\ & x_1 \geqq 0 \quad x_2 \geqq 0 \quad x_3 \geqq 0 \end{aligned}$$

Slack variables x_4, x_5, x_6 are introduced. Here is the original tableau (the initial right-hand-side constants are not required to be nonnegative here)

Original Tableau

x_1	x_2	x_3	x_4	x_5	x_6	$-z$	b	
1	1	1	1	0	0	0	1	
-2	-4	-1	0	1	0	0	-8	
-1	1	1	0	0	1	0	-2	
8	8	9	0	0	0	1	0	(minimize z)

$x_j \geqq 0$ for all j.

(x_4, x_5, x_6) constitutes a dual feasible basic vector here. The inverse tableau with respect to this basic vector is

Basic Variables	Inverse Tableau				Basic Values	Pivot Column	
x_4	1	0	0	0	1	1	
x_5	0	1	0	0	-8	-4 (circled)	Pivot Element
x_6	0	0	1	0	-2	1	
$-z$	0	0	0	1	0	8	

Pick the second row as the pivot row. Below are the values of the relative cost coefficients and the updated second row.

		x_1	x_2	x_3	x_4	x_5	x_6
$\bar{a}_{2j}$	=	-2	-4	-1	0	1	0
$\bar{c}_j$	=	8	8	9	0	0	0
$-\frac{\bar{c}_j}{\bar{a}_{2j}}$	for $\bar{a}_{2j} < 0$ =	$\frac{8}{2}$	$\frac{8}{4}$	9			

The variable x_2 enters the basic vector. The pivot column and the pivot element are recorded on the inverse tableau.

Second Inverse Tableau

Basic Variables	Inverse Tableau				Basic Values	Pivot Column x_1	
x_4	1	$\frac{1}{4}$	0	0	-1	$\frac{1}{2}$	
x_2	0	$-\frac{1}{4}$	0	0	2	$\frac{1}{2}$	
x_6	0	$\frac{1}{4}$	1	0	-4	$-\frac{3}{2}$ (circled)	Pivot Element
$-z$	0	2	0	1	-16	4	

	x_1	x_2	x_3	x_4	x_5	x_6
$\bar{a}_{3j}$ =	$-\frac{3}{2}$	0	$\frac{3}{4}$	0	$\frac{1}{4}$	1
$\bar{c}_j$ =	4	0	7	0	2	0
$-\frac{\bar{c}_j}{\bar{a}_{3j}}$ for $\bar{a}_{3j} < 0$ =	$\frac{8}{3}$					

The variable x_1 is the entering variable. The updated column vector of x_1 is computed and recorded on the inverse tableau.

Third Inverse Tableau

Basic Variables	*Inverse Tableau*				*Basic Values*
x_4	1	$\frac{1}{3}$	$\frac{1}{3}$	0	$-\frac{7}{3}$
x_2	0	$-\frac{1}{6}$	$\frac{1}{3}$	0	$\frac{2}{3}$
x_1	0	$-\frac{1}{6}$	$-\frac{2}{3}$	0	$\frac{8}{3}$
$-z$	0	$\frac{8}{3}$	$\frac{8}{3}$	1	$-\frac{80}{3}$

The next pivot row is the first row. The updated first row is $(\bar{a}_{11}, \bar{a}_{12}, \bar{a}_{13}, \bar{a}_{14}, \bar{a}_{15}, \bar{a}_{16}) = (0, 0, 1, 1, 1/3, 1/3)$. Since this contains no negative entries, the infeasibility criterion is satisfied. The row in the final inverse tableau corresponding to the pivot row is (1, 1/3, 1/3, 0). Hence the row obtained by multiplying row 1 of the original tableau by 1 and adding to it row 2 multiplied by 1/3 and row 3 multiplied by 1/3 yields the inconsistent equation $x_3 + x_4 + 1/3x_5 + 1/3x_6 = -7/3$. Since all the variables in the problem are restricted to be nonnegative, the left-hand side of this equation will always be nonnegative, and hence, the above equation can never be satisfied by any $x \geqq 0$. The first row in the inverse tableau at this stage is (1, 1/3, 1/3, 0), and this implies that the first three constraints in the original tableau are mutually inconsistent with the nonnegativity restrictions on the variables.

Note 6.3: It is possible that a basic variable has a positive value in the solution at the beginning of a pivot step in the dual simplex algorithm, and yet the same basic variable may have a negative value in the solution obtained after this pivot step (see Exercise 6.7). In general, it is not possible to execute the dual simplex algorithm to satisfy the property that once the value of a primal variable becomes nonnegative in some iteration of the dual simplex algorithm, then its value stays nonnegative in the primal solutions obtained in all subsequent iterations.

Dual Degeneracy and Nondegeneracy

A basis B for (6.1) is said to be

dual degenerate, if the relative cost coefficient of at least one nonbasic variable is zero in the canonical tableau with respect to the basis B;

dual nondegenerate, if the relative cost coefficient of all the nonbasic variables are nonzero in the canonical tableau with respect to the basis B.

The LP (6.1) is itself said to be *dual nondegenerate* if in every solution $(\pi, \bar{c})$ for $\pi A + \bar{c} = c$ in which the variables are π and $\bar{c}$ at least $n - m$ of the $\bar{c}_j$ are nonzero; otherwise, it is dual degenerate. Thus if (6.1) is dual nondegenerate, the relative cost coefficients of all the nonbasic variables will be nonzero in the canonical tableau with respect to any basis for (6.1).

Sufficient Condition for (6.1) to Have a Unique Optimum Solution By results of Section 3.10 a sufficient condition for (6.1) to have a unique optimum solution is that there exist an optimum basis for (6.1) that is dual nondegenerate.

Nondegenerate or Degenerate Pivot Steps in the Dual Simplex Algorithm

In a pivot step of the dual simplex algorithm for solving (6.1), let x_s, row r, $\bar{a}_{rs}$, x_{j_r}, $\bar{b}_r$, $\bar{z}$, $\boldsymbol{\beta}_{r.} = (\beta_{r1}, \ldots, \beta_{rm})$, and π be the entering variable, pivot row, pivot element, leaving variable, updated right-hand-side constant in the pivot row, objective value, rth row of basis inverse, and dual solution, respectively. Let $\tilde{\pi}$ and $\tilde{z}$ represent corresponding quantities obtained after the pivot step. In the primal basic solution obtained after this pivot step, x_{j_r} will be a nonbasic variable and have value 0, and x_s will be a basic variable with value $(\bar{b}_r/\bar{a}_{rs}) > 0$. Thus, while the present primal basic solution violates the nonnegativity restriction $x_{j_r} \geqq 0$, the primal basic solution obtained after this pivot step satisfies it as an equation. Also, the value of the entering variable x_s will be strictly positive in the primal basic solution obtained after this pivot step.

Since the present basis is dual feasible, we have $\bar{c}_s \geqq 0$. This pivot step is said to be a *degenerate pivot step* in this dual simplex algorithm, if $\bar{c}_s = 0$; a *nondegenerate pivot step* if $\bar{c}_s > 0$. We have $\tilde{\pi} = \pi + \beta_{r.}(\bar{c}_s/\bar{a}_{rs})$, $\tilde{z} = \bar{z} + \bar{b}_r(\bar{c}_s/\bar{a}_{rs})$. These formulas are obtained by looking at the present inverse tableau and the inverse tableau obtained after performing this pivot step. If this is a degenerate pivot step, we have $\bar{c}_s = 0$, so $\tilde{\pi} = \pi$, $\tilde{z} = \bar{z}$. If this is a nondegenerate pivot step, we have $\bar{c}_s > 0$, and hence $\tilde{\pi} \neq \pi$ and $\tilde{z} > \bar{z}$ (since $\bar{b}_r < 0$, and $\bar{a}_{rs} < 0$).

To summarize, in every pivot step of the dual simplex algorithm, the primal basic solution definitely changes. The dual solution and the objective value do not change in a degenerate pivot step; in a nondegenerate pivot step the dual solution changes and the objective value strictly increases.

Exercises

6.1 Suppose (6.1) is solved using the revised dual simplex algorithm. Show that the dual solutions obtained during this algorithm can be interpreted as those that will be obtained when the primal simplex algorithm is applied on the dual problem corresponding to (6.1).

6.2 *FINITE TERMINATION OF THE DUAL SIMPLEX ALGORITHM UNDER DUAL NONDEGENERACY.* Suppose (6.1) is dual nondegenerate. In solving (6.1) by the dual simplex algorithm, starting with a dual feasible basic vector, prove that the objective value strictly increases in each pivot step of the algorithm. Using this, prove that the algorithm terminates in a finite number of pivot steps.

Cycling Under Degeneracy in the Dual Simplex Algorithm

Under dual degeneracy it is possible that the dual simplex algorithm for solving (6.1) may end up going around a cycle of dual degenerate pivot steps without ever terminating. This never seems to occur in practice, but the possibility of its occurrence has been established in specially constructed problems. Methods for resolving this possibility of cycling under dual degeneracy have been developed, and one of them based on a lexicographic ordering is discussed in Section 10.6. With the use of these dual degeneracy resolving methods, the dual simplex method for solving (6.1) is theoretically guaranteed to terminate after a finite number of pivot steps.

6.3 DISADVANTAGES AND ADVANTAGES OF THE DUAL SIMPLEX ALGORITHM

In the dual simplex algorithm all but the final basic solution are primal infeasible. If we run out of computer time in the middle of solving an LP by the dual simplex algorithm, all the effort spent is wasted since we do not even have a feasible solution to the problem until the algorithm is carried out to the end. This is a disadvantage. In the primal simplex algorithm all the basic solutions obtained are primal feasible. Thus, if it becomes necessary to terminate the computation before the optimality criterion is satisfied, we can at least be content with the best among the feasible solutions obtained. Even though it may not be optimal to the problem, it is feasible. That's why, if a primal feasible basis and another dual feasible basis are available for an LP, it is preferable to solve it by the primal simplex algorithm starting with the known primal feasible basis, than to approach it by the dual simplex algorithm starting with the known dual feasible basis.

However, the dual simplex algorithm is very useful in doing sensitivity analysis (i.e., postoptimality analysis). After the optimal basis is obtained, it may become necessary to change the right-hand constant vector b. When the b vector is changed, the previous optimal basis may no longer be primal feasible. However, it is dual feasible, assuming that the cost coefficients remain unchanged. Starting from that basis the dual simplex algorithm can be applied to solve the problem for the new b vector. Similarly, the dual simplex algorithm is used in solving LPs with parametric right-hand-side vectors (Chapter 8). In practical applications of linear programming, it may become necessary to introduce a new constraint into the LP model after it has been solved. The dual simplex algorithm can be used to find an optimum solution of the augmented model starting from an optimum solution of the present model (Chapter 9). The same property makes the dual simplex algorithm very useful in cutting plane approaches for solving integer programs.

6.4 THE PATHS TAKEN BY THE PRIMAL AND DUAL SIMPLEX ALGORITHMS

Discussion of Some Properties of the Dual Simplex Algorithm

First we discuss some properties of the dual simplex algorithm that are needed to understand the illustrative example properly. Let **K** be the set of feasible solutions of the LP (6.1). The primal simplex algorithm starts at an extreme point of **K** and moves along the edges of **K**, with the objective value decreasing in each move until an optimum extreme point of **K** is reached. The dual simplex algorithm begins with an infeasible basic solution for (6.1) in which some of the basic variables are negative. This property is shared by all but the final optimum solution. Let $\bar{x}$ be an intermediate solution obtained during the dual simplex algorithm. Let $\mathbf{J} = \{j: \bar{x}_j < 0\}$. Then the point $\bar{x}$ violates the restriction $x_j \geqq 0$ in (6.1) for each $j \in \mathbf{J}$. The point $\bar{x}$ lies in the set $\mathbf{K_J}$, which is the set of feasible solutions of the problem obtained by relaxing the restrictions $x_j \geqq 0$ for each $j \in \mathbf{J}$ from (6.1), and this set properly contains **K**. Let $\bar{c} = (\bar{c}_1, \ldots, \bar{c}_n)$ be the row vector of relative cost coefficients in this step. We have $\bar{c} \geqq 0$. Since nonbasic variables are always zero in a basic solution, for all $j \in \mathbf{J}$ x_j must be a basic variable in this step, and hence $\bar{c}_j = 0$. If $\bar{z}$ is the objective value of $\bar{x}$, from the objective row in the current canonical tableau, we get

$$z = \bar{z} + \sum_{j \notin \mathbf{J}} \bar{c}_j x_j \tag{6.5}$$

We have $\bar{c}_j \geqq 0$ for all $j \notin \mathbf{J}$, and $x_j \geqq 0$ for each $j \notin \mathbf{J}$ in $\mathbf{K_J}$. Hence the term $\sum_{j \notin \mathbf{J}} \bar{c}_j x_j$ is nonnegative over $\mathbf{K_J}$, and from (6.5) we conclude that the minimum value of $z(x)$ over $\mathbf{K_J}$ is $\bar{z}$. This implies that $\bar{x}$ minimizes the objective function in the set $\mathbf{K_J}$. Thus, in each step of the dual simplex algorithm, the current primal solution is an optimum solution of the problem obtained by relaxing the conditions violated by the current primal solution.

If row r is the pivot row and x_{j_1} the present basic variable in this row, then $\bar{x}$ is on the wrong side of the hyperplane $x_{j_1} \geqq 0$. After this pivot step, x_{j_1} leaves the basic vector and hence this pivot step moves the solution to a point on the hyperplane $x_{j_1} = 0$.

The Illustrative Example

It is convenient to illustrate the paths taken by the primal simplex algorithm and the dual simplex algorithm, using an LP involving only two decision variables x_1 and x_2, in which x_2 has to be minimized subject to several inequality constraints. In this problem, the set of feasible solutions can be plotted easily on the x_1, x_2-Cartesian plane. We chose the x_2-axis as the vertical axis, and hence, minimizing x_2 is equivalent to finding the lowest (i.e., the southernmost) point in the set of feasible solutions.

The set of feasible solutions for our example, **K**, is plotted in Figure 6.1. There are seven inequality constraints, numbered from 1 to 7 on the variables x_1 and x_2 in this problem, and we are required to be

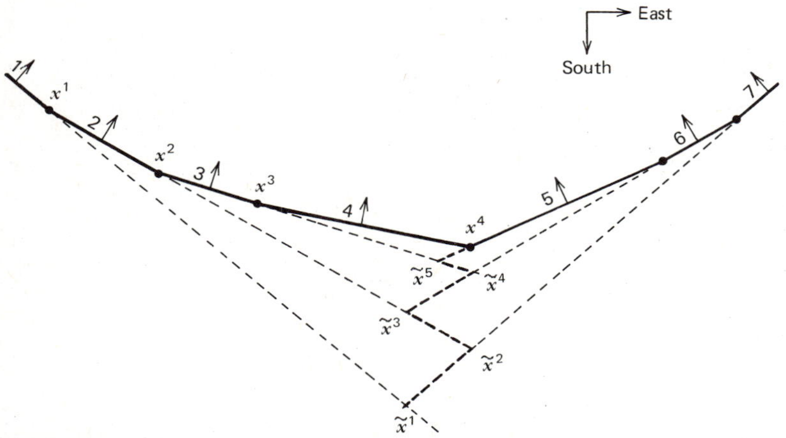

Figure 6.1 The set of feasible solutions is the set of points lying above the continuous lines. The continuous thick path is the path followed by the primal simplex algorithm. The dashed thick path is the path taken by the dual simplex algorithm.

toward the side marked by the arrow on each constraint. Each inequality constraint has an associated slack variable. Let the slack variable corresponding to the ith constraint be denoted by s_i. Points on the side marked by the arrow on constraint i make s_i positive, points lying on the constraint i itself make s_i equal to zero, and points lying on the side opposite to that marked by the arrow on constraint i make s_i negative. The conditions on the variables in the problem are that each s_i should be nonnegative for each $i = 1$ to 7. The set of feasible solutions is the convex polyhedron **K**, bounded by the continuous lines in Figure 6.1.

The primal simplex algorithm begins at an extreme point of **K**, say x^1, and moves along edges of **K**, moving down to the bottom in each step until it stops at the bottommost point in **K**. Hence the primal simplex algorithm moves from x^1 to x^2, then to x^3, and finally to x^4, which is the bottommost point in **K**. The path taken by the primal simplex algorithm is marked by thick continuous lines in **K**.

For solving this LP the dual simplex algorithm might begin at the point $\tilde{x}^1$. At $\tilde{x}^1$, the variables s_2, s_3, s_4, s_5, s_6 all assume strictly negative values and hence $\tilde{x}^1$ is outside of **K**, the set of feasible solutions of this LP. When the nonnegativity restrictions on s_2, s_3, s_4, s_5, s_6 are relaxed, the set of feasible solutions of the relaxed problem is $\mathbf{K}^1$, which is the union of **K** and the triangular region at the bottom bounded by the dashed lines. The point $\tilde{x}^1$ is an extreme point of $\mathbf{K}^1$, and it is the bottommost point in $\mathbf{K}^1$. Hence $\tilde{x}^1$ is the optimum solution in the relaxed problem obtained by relaxing the nonnegativity restrictions on s_2, s_3, s_4, s_5, s_6. If s_2 (which has a negative value in the present solution $\tilde{x}^1$) is chosen to determine the pivot row, then the dual simplex algorithm would move to $\tilde{x}^2$. Notice that although $\tilde{x}^2$ satisfies $s_2 = 0$, it still violates the nonnegativity restrictions on s_3, s_4, s_5, and s_6. Also verify that $\tilde{x}^2$ is the optimum solution of the relaxed problem obtained by relaxing the nonnegativity restrictions on s_3, s_4, s_5, and s_6. The dual simplex algorithm continues in this way, moving from $\tilde{x}^1$ to $\tilde{x}^2$; to $\tilde{x}^3$, $\tilde{x}^4$, $\tilde{x}^5$, and finally to x^4. When it reaches x^4, it terminates because x^4 satisfies all the conditions in the original problem. The point x^4 is an optimum solution of the original problem.

6.5 DUAL SIMPLEX METHOD WHEN A DUAL FEASIBLE BASIS IS NOT KNOWN AT THE START

Suppose a dual feasible basis for (6.1) is not known. By transforming the system of constraints $Ax = b$ into row echelon normal form, obtain an arbitrary basis for (6.1). Let B denote the basis. Let $x_B = (x_1, \ldots, x_m)$ be the basic vector. If this basis turns out to be dual feasible, it can be used as a starting dual feasible basis for the dual simplex algorithm.

However, the basis B may be neither primal nor dual feasible. If B is not dual feasible, let the canonical tableau with respect to the basis B be Tableau 2.8 of Section 2.5.1. Add an artificial variable, x_0 with cost coefficient zero, and an artificial constraint:

$$x_0 + x_{m+1} + \cdots + x_n = \boldsymbol{M} \tag{6.6}$$

where $(x_{m+1}, \ldots, x_n)$ is the nonbasic vector and $\boldsymbol{M}$ is a very large positive number. It is not necessary to give a specific numerical value to $\boldsymbol{M}$; it should always be considered to be strictly larger than any other number with which it is compared. The original problem modified by the addition of the constraint (6.6) will be called the *augmented problem*.

Let $\bar{c}_s = \text{minimum } \{\bar{c}_j : j = 1 \text{ to } n\}$. Since B is not a dual feasible basis of the original problem, $\bar{c}_s < 0$ is the most negative $\bar{c}_j$. Bring x_s into the basic vector, replacing x_0 from it. By the definition of $\bar{c}_s$, this pivot will transform all the entries in the cost row of the

Tableau for the Augmented Problem

Basic Variables	x_0	x_1	...	x_m	x_{m+1}	...	x_s	...	x_n	$-z$	Basic Values
x_0	1	0	...	0	1	...	1	...	1	0	$\boldsymbol{M}$
x_1	0	1	...	0	$\bar{a}_{1,m+1}$	...	$\bar{a}_{1s}$	...	$\bar{a}_{1n}$	0	$\bar{b}_1$
$\vdots$	$\vdots$	$\vdots$		$\vdots$	$\vdots$		$\vdots$		$\vdots$	$\vdots$	$\vdots$
x_m	0	0	...	1	$\bar{a}_{m,m+1}$	...	$\bar{a}_{ms}$	...	$\bar{a}_{mn}$	0	$\bar{b}_m$
$-z$	0	0	...	0	$\bar{c}_{m+1}$	...	$\bar{c}_s$	...	$\bar{c}_n$	1	$-\bar{z}$

tableau into nonnegative numbers. Thus we get a dual feasible basis for the augmented problem. The basic vector $(x_s, x_1, \ldots, x_m)$ is a dual feasible basic vector for the augmented problem. Starting with this dual feasible basic vector, apply the dual simplex algorithm and solve the augmented problem. Termination of this algorithm may occur in any of the three following ways.

Augmented Problem Primal Infeasible In this case, the original problem is also infeasible because if the original problem has a feasible solution $\bar{x} = (\bar{x}_1, \ldots, \bar{x}_n)$, then $(\bar{x}_0, \bar{x}_1, \ldots, \bar{x}_n)$, where $\bar{x}_0 = \boldsymbol{M} - \sum_{j=m+1}^{n} \bar{x}_j$ is a feasible solution of the augmented problem.

Augmented Problem Has an Optimum Basic Vector Containing x_0 Let $(\bar{x}_0, \bar{x}_1, \ldots, \bar{x}_n)$ be an optimal BFS of the augmented problem. In this case, the optimum objective value of the augmented problem is obviously independent of $\boldsymbol{M}$, as long as $\boldsymbol{M}$ is very large. Hence, $(\bar{x}_1, \ldots, \bar{x}_n)$ is an optimum BFS of the original problem (6.1). An optimal basis for (6.1) is obtained by suppressing the row and the column corresponding to x_0 from the optimum basis for the augmented problem.

Augmented Problem Has an Optimum Basic Vector Not Containing x_0 In this case, the values of the basic variables in the final optimum BFS of the augmented problem depend on $\boldsymbol{M}$. If the optimal objective value for the augmented problem depends on $\boldsymbol{M}$, it must tend to $-\infty$ as $\boldsymbol{M}$ tends to $+\infty$. In this case, the original problem is feasible and unbounded. All one needs for this is that π_0, the dual variable associated with the first constraint in the augmented problem, be nonzero in the current dual solution.

If π_0 is zero in the current dual solution, the optimal objective value for the augmented problem is independent of $\boldsymbol{M}$, the original problem is feasible, and has an optimal feasible solution, which can be obtained from the optimal solution of the augmented problem by suppressing x_0 (which is equal to zero in this solution). In this case, an optimal BFS of the original problem may be obtained by decreasing the value of $\boldsymbol{M}$ until one of the basic variables in the final optimum BFS of the augmented problem vanishes.

Example 6.3

Original Tableau

x_1	x_2	x_3	x_4	x_5	$-z$	b
1	0	0	1	−1	0	2
0	1	0	−1	−1	0	−4
0	0	1	−2	2	0	−3
2	1	0	0	0	1	0 (z minimize)

$x_j \geqq 0$ for all j.

Select (x_1, x_2, x_3) as the initial basic vector. Pricing out and introducing the artificial constraint, we get the original tableau for the augmented problem.

Original Tableau for the Augmented Problem

x_0	x_1	x_2	x_3	x_4	x_5	$-z$	b
1	0	0	0	①	1	0	$\boldsymbol{M}$
0	1	0	0	1	−1	0	2
0	0	1	0	−1	−1	0	−4
0	0	0	1	−2	2	0	−3
0	0	0	0	−1	3	1	0

Bringing x_4 into the basic vector leads to a dual feasible tableau. Here is the corresponding inverse tableau.

Basic Variables	*First Dual Feasible Inverse Tableau*					*Basic Values*	*Pivot Column x_0*
x_4	1	0	0	0	0	$\boldsymbol{M}$	1
x_1	−1	1	0	0	0	$2 - \boldsymbol{M}$	(−1)
x_2	1	0	1	0	0	$\boldsymbol{M} - 4$	1
x_3	2	0	0	1	0	$2\boldsymbol{M} - 3$	2
$-z$	1	0	0	0	1	$\boldsymbol{M}$	1

The pivot row is the second row. Below are the updated second and the cost rows.

Updated	x_0	x_1	x_2	x_3	x_4	x_5
Pivot Row	−1	1	0	0	0	−2
Cost Row	1	0	0	0	0	4
Ratios	1 minimum					2

Hence, x_0 enters the basic vector. Its updated column is the pivot column and is entered on the right-hand side of the tableau.

Basic Variables	Inverse Tableau					Basic Values	Pivot Column (x_5)	
x_4	0	1	0	0	0	2	−1	
x_0	1	−1	0	0	0	$\boldsymbol{M} - 2$	2	
x_2	0	1	1	0	0	−2	(−2) circled	Pivot Row
x_3	0	2	0	1	0	1	0	
$-z$	0	1	0	0	1	2	2	

The third row is the pivot row.

Updated	x_0	x_1	x_2	x_3	x_4	x_5
Pivot Row	0	1	1	0	0	−2
Cost Row	0	1	0	0	0	2
Ratios						1

Variable x_5 enters the basic vector replacing x_2. Its updated column vector is the pivot column, and is entered on the right-hand side of the tableau.

Basic Variables	Inverse Tableau					Basic Values
x_4	0	$\frac{1}{2}$	$-\frac{1}{2}$	0	0	3
x_0	1	0	1	0	0	$\boldsymbol{M} - 4$
x_5	0	$-\frac{1}{2}$	$-\frac{1}{2}$	0	0	1
x_3	0	2	0	1	0	1
$-z$	0	2	1	0	1	0

When $\boldsymbol{M}$ is large this tableau is primal feasible and, hence, optimal. $(x_0, x_1, \ldots, x_5)^{\mathrm{T}} = (\boldsymbol{M} - 4, 0, 0, 1, 3, 1)^{\mathrm{T}}$ is optimal to the augmented problem with the optimum objective value being zero. Hence, $(x_1, \ldots, x_5)^{\mathrm{T}} = (0, 0, 1, 3, 1)^{\mathrm{T}}$ is optimal to the original problem with the optimal objective value equal to zero.

Example 6.4

x_1	x_2	x_3	x_4	x_5	x_6	$-z$	b	
1	0	0	1	−3	7	0	−5	
0	1	0	−1	1	−1	0	1	
0	0	1	3	1	−10	0	8	
1	3	0	0	0	−2	1	0	(z minimize)

$x_j \geqq 0$ for all j.

Take (x_1, x_2, x_3) as the initial basic vector. Pricing out and introducing the artificial constraint leads to the following.

Basic Variables	x_0	x_1	x_2	x_3	x_4	x_5	x_6	$-z$	b
x_0	1	0	0	0	1	1	(1) circled	0	$\boldsymbol{M}$
x_1	0	1	0	0	1	−3	7	0	−5
x_2	0	0	1	0	−1	1	−1	0	1
x_3	0	0	0	1	3	1	−10	0	8
$-z$	0	0	0	0	2	0	−6	1	2

Bringing x_6 into the basic vector leads to a dual feasible tableau.

Basic Variables	x_0	x_1	x_2	x_3	x_4	x_5	x_6	$-z$	b	
x_6	1	0	0	0	1	1	1	0	$\boldsymbol{M}$	
x_1	−7	1	0	0	−6	(−10) circled	0	0	$-5 - 7\boldsymbol{M}$	Pivot Row
x_2	1	0	1	0	0	2	0	0	$1 + \boldsymbol{M}$	
x_3	10	0	0	1	13	11	0	0	$8 + 10\boldsymbol{M}$	
$-z$	6	0	0	0	8	6	0	1	$2 + 6\boldsymbol{M}$	
Ratios	$\frac{6}{7}$				$\frac{8}{6}$	$\frac{6}{10}$				

Basic Variables	x_0	x_1	x_2	x_3	x_4	x_5	x_6	$-z$	b
x_6	3/10	1/10	0	0	4/10	0	1	0	$(3\boldsymbol{M} - 5)/10$
x_5	7/10	−1/10	0	0	6/10	1	0	0	$(5 + 7\boldsymbol{M})/10$
x_2	−4/10	2/10	1	0	(−12/10)	0	0	0	$-(4\boldsymbol{M}/10)$
x_3	23/10	11/10	0	1	64/10	0	0	0	$(23\boldsymbol{M} + 25)/10$
$-z$	18/10	6/10	0	0	44/12	0	0	1	$(18\boldsymbol{M} - 10)/10$
Ratios	18/4				44/12				
x_6	5/30	5/30	1/3	0	0	0	1	0	$(\boldsymbol{M} - 3)/6$
x_5	5/10	0	1/2	0	0	1	0	0	$(5 + 5\boldsymbol{M})/10$
x_4	4/12	−2/12	−10/12	0	1	0	0	0	$\boldsymbol{M}/3$
x_3	5/30	13/6	64/12	1	0	0	0	0	$(\boldsymbol{M} + 15)/6$
$-z$	10/30	109/90	44/12	0	0	0	0	1	$(\boldsymbol{M} - 3)/3$

This is an optimum tableau for the augmented problem for large $\boldsymbol{M}$. The optimum objective value is $-(\boldsymbol{M} - 3)/3$, and this diverges to $-\infty$ as $\boldsymbol{M}$ moves toward $+\infty$. Hence, the original problem is feasible and unbounded below. This feasible solution

$$x^{\mathrm{T}} = \left(0, 0, \frac{\boldsymbol{M} + 15}{6}, \frac{\boldsymbol{M}}{3}, \frac{1 + \boldsymbol{M}}{2}, \frac{\boldsymbol{M} - 3}{6}\right)$$

$$\text{Objective value} = -\frac{(\boldsymbol{M} - 3)}{3}$$

is feasible for all $\boldsymbol{M}$ sufficiently large and as $\boldsymbol{M}$ moves toward $+\infty$, the objective value decreases indefinitely.

6.6 COMPARISON OF THE PRIMAL AND THE DUAL SIMPLEX METHODS

Feature of the Primal Simplex Method	Comparative Feature of the Dual Simplex Method
1 The primal simplex algorithm needs a primal feasible basis to start with.	**1** The dual simplex algorithm needs a dual feasible basis to start with.
2 Starting with a primal feasible basis, the primal simplex algorithm tries to attain dual feasibility while maintaining primal feasibility throughout.	**2** Starting with a dual feasible basis, the dual simplex algorithm tries to attain primal feasibility while maintaining dual feasibility throughout.
3 In the primal simplex algorithm the optimality criterion is the dual feasibility criterion.	**3** In the dual simplex algorithm the optimality criterion is the primal feasibility criterion.
4 Starting with a primal feasible basis, the primal simplex algorithm terminates either with an optimum basis or by establishing that the primal objective function is unbounded (this is the dual infeasibility criterion).	**4** Starting with a dual feasible basis, the dual simplex algorithm terminates either with an optimum basis or by establishing primal infeasibility (in this case, the dual objective function is unbounded).
5 Suppose the problem is being solved by the revised primal simplex algorithm, starting with a known primal feasible basis. All the primal solutions obtained during the algorithm are primal feasible. The dual solutions obtained during the algorithm (excepting the final one when the optimality criterion is satisfied) are dual infeasible.	**5** Suppose the problem is being solved by the revised dual simplex algorithm starting with a known dual feasible basis. All the dual solutions obtained during the algorithm are dual feasible. The primal solutions obtained in the algorithm (excepting the final one when the optimality criterion is satisfied) are primal infeasible.

	Feature of the Primal Simplex Method		Comparative Feature of the Dual Simplex Method
6	Pivot elements are positive in all pivot steps, to maintain primal feasibility.	6	Pivot elements are negative in all pivot steps, to move closer to primal feasibility.
7	The pivot column is first selected by the problem solver, among those columns that correspond to a negative relative cost coefficient (i.e., negative dual slack variable). The pivot row is then determined by the algorithm using the primal simplex minimum ratio test, the purpose of which is to gurantee that primal feasibility is maintained in the next basis.	7	The pivot row is first selected by the problem solver, among those rows that have a negative updated right-hand-side constant (i.e., a negative-valued primal basic variable). The pivot column is then determined by the algorithm, using the dual simplex minimum ratio test, the purpose of which is to guarantee that dual feasibility is maintained in the next basis.
8	The primal simplex minimum ratio test uses the updated right-hand-side constants and positive entries in the pivot column.	8	The dual simplex minimum ratio test uses the updated cost coefficients (i.e., the relative cost coefficients) and negative entries in the pivot row.
9	Dual infeasibility is established in the primal simplex algorithm if the updated column vector of the entering variable (i.e., the pivot column) has no positive entries.	9	Primal infeasibility is established in the dual simplex algorithm if the pivot row has no negative entries.
10	The primal solution in an intermediate state of the primal simplex algorithm, is primal feasible, but may not be optimal. Hence the corresponding objective value is an upper bound for the minimum objective value in the problem. Also the objective value steadily decreases as the algorithm progresses.	10	The primal solution in an intermediate stage of the dual simplex algorithm is primal infeasible, because some basic variables have negative values in it, violating the nonnegativity restrictions. It is an optimum solution of the problem obtained by relaxing the nonnegativity restrictions on the primal basic variables. Hence the corresponding objective value is a lower bound for the minimum objective value in the original problem. Also the objective value steadily increases as the algorithm progresses.
11	If a primal feasible basis for the original problem is not known, an augmented problem is created by introducing nonnegative artificial variables, so that the augmented problem has a readily available primal feasible basis.	11	If a dual feasible basis for the original problem is not known, an augmented problem is created by introducing an artificial inequality constraint (which can be written as an equation by including an appropriate nonnegative slack variable), so that the augmented problem has a readily available dual feasible basis.

Remarks 6.1: The dual simplex algorithm has been developed by C. E. Lemke. See reference [6.4]. The device presented in Section 6.5 for initiating the dual simplex method by constructing the augmented problem is due to E. M. L. Beale [6.1].

6.7 ALTERNATE OPTIMUM SOLUTIONS OF THE DUAL PROBLEM

Consider the LP (6.1). Let $x_B = (x_{j_1}, \ldots, x_{j_m})$ be an optimum basic vector for this problem, associated with the optimum BFS $\bar{x}$ and the dual optimum solution $\bar{\pi}$. At this stage the following questions may arise. What are the necessary and sufficient conditions for $\bar{\pi}$ to be the unique optimum dual solution? Characterize the set of all alternate optimum dual solutions if $\bar{\pi}$ is not the unique one. If $\bar{\pi}$ is not the unique optimum dual solution, how can alternate basic optimal dual solutions be obtained, beginning with the inverse tableau, with respect to the basic vector x_B? We provide answers to these questions here.

Let $\mathbf{J} = \{j: \bar{x}_j > 0\}$. Since $\bar{x}$ is the BFS of (6.1) associated with the basic vector x_B, $\mathbf{J} \subset \{j_1, \ldots, j_m\}$. From the complementary slackness theorem (Theorem 4.7), the set of optimum solutions of the dual of (6.1) is the set of feasible solutions of

$$\begin{aligned} \pi A_{.j} &= c_j && \text{for } j \in \mathbf{J} \\ \pi A_{.j} &\leqq c_j && \text{for } j = 1 \text{ to } n, j \notin \mathbf{J} \end{aligned} \tag{6.7}$$

If $\mathbf{J} = \{j_1, \ldots, j_m\}$, then $\bar{x}$ is a (primal) nondegenerate BFS of (6.1), and in this case the equality constraints in (6.7) lead to the unique feasible solution $\bar{\pi}$ for it. These facts imply that a sufficient condition for the dual of (6.1) to have a unique optimum solution is that (6.1) have a nondegenerate optimum BFS.

If $\mathbf{J}$ is a proper subset of $\{j_1, \ldots, j_m\}$, that is, if $\bar{x}$ is degenerate, there are not enough equality constraints in (6.7) to determine a unique solution for (6.7). In this case, the possibility of alternate optimum solutions for the dual of (6.1) exists. Define $\mathbf{\Gamma} = \{j: \bar{\pi}A_{.j} = c_j, j = 1 \text{ to } n, j \notin \mathbf{J}\}$. Now consider the following LP:

$$\begin{aligned} &\text{Minimize} && \sum_{j \in \mathbf{\Gamma}} \pi A_{.j} \\ &\text{Subject to} && (6.7) \end{aligned} \tag{6.8}$$

By the constraints in (6.7), the optimum objective value in (6.8) is $\leqq \sum_{j \in \mathbf{\Gamma}} c_j$. If the optimum objective value in the LP (6.8) is equal to $\sum_{j \in \mathbf{\Gamma}} c_j$, this would imply that the constraints (6.7) together imply that $\pi A_{.j} = c_j$ for all $j \in \mathbf{\Gamma} \cup \mathbf{J}$, and since $\mathbf{\Gamma} \cup \mathbf{J} \supset \{j_1, \ldots, j_m\}$, this implies that $\bar{\pi}$ is the unique solution for (6.7). On the other hand, if the optimum objective value in the LP (6.8) is $< \sum_{j \in \mathbf{\Gamma}} c_j$, this guarantees that (6.7) has solutions besides $\bar{\pi}$ (which makes the objective in (6.8) equal to $\sum_{j \in \mathbf{\Gamma}} c_j$). Thus the dual of (6.1) has a unique optimum solution iff the optimum objective value in (6.8) is $\sum_{j \in \mathbf{\Gamma}} c_j$. If (6.8) has solutions that make $\sum_{j \in \mathbf{\Gamma}} \pi A_{.j} < \sum_{j \in \mathbf{\Gamma}} c_j$, the dual of (6.1) has alternate optimum solutions and vice versa. Thus the question of whether $\bar{\pi}$ is the unique optimum solution for the dual of (6.1) can be settled by solving the LP (6.8), if the optimum BFS $\bar{x}$ of (6.1) is degenerate.

The set of optimum solutions of the dual of (6.1), determined by (6.7), is a convex polyhedral subset of $\mathbf{R}^m$. So convex combinations of optimum dual solutions are also optimal to the dual. If $\mathbf{J}$ is a proper subset of $\{j_1, \ldots, j_m\}$, alternate basic optimum dual solutions associated with (6.1) can be obtained by performing dual simplex pivot steps beginning with the inverse tableau of x_B, in rows corresponding to basic variables x_{j_r}, where $j_r \notin \mathbf{J}$ (i.e., zero-valued basic variables in $\bar{x}$).

6.8 PRIMAL-DUAL METHODS FOR LINEAR PROGRAMS

To explain the main strategy employed by these methods, we consider the following general LP: Find $x \in \mathbf{R}^n$ to

$$\begin{aligned} &\text{Minimize} && z(x) = cx \\ &\text{Subject to} && A_{i.}x = b_i, && i = 1 \text{ to } m \\ &&& \phantom{A_{i.}x} \geqq b_i, && i = m+1 \text{ to } m+p \end{aligned} \tag{6.9}$$

where all the conditions on the variables, including any bound or sign restrictions on the variables are included in the system of constraints in (6.9). Let A be the $(m+p) \times n$ matrix whose rows are $A_{i.}$, $i = 1$ to $m+p$. If $\mu = (\mu_1, \ldots, \mu_{m+p})$ is the row vector of dual variables, the dual problem is

$$\begin{aligned} &\text{Maximize} && \sum_{i=1}^{m+p} b_i \mu_i \\ &\text{Subject to} && \sum_{i=1}^{m+p} \mu_i A_{i.} = c \\ &&& \mu_i \geqq 0, \quad i = m+1 \text{ to } m+p \end{aligned} \tag{6.10}$$

The complementary slackness conditions for this primal-dual pair of LPs are

$$\mu_i(A_{i.}x - b_i) = 0, \qquad i = m+1 \text{ to } m+p \tag{6.11}$$

Primal-dual methods need an initial dual feasible solution, which need not be a BFS of (6.10). Any feasible solution of (6.10), whether basic or not, is adequate to initiate these methods. If μ is any dual feasible solution, let $\mathbf{\Delta}(\mu) = \{i: m+1 \leqq i \leqq m+p$

and $\mu_i > 0\}$. If μ^1 is the initial dual feasible solution, the set of primal feasible solutions that satisfy the complementary slackness conditions (6.11) together with μ^1, is the set of feasible solutions of the following system known as the *restricted primal* corresponding to μ^1.

$$\begin{aligned} A_{i.}x &= b_i, \qquad i = 1 \text{ to } m, \text{ or } i \in \mathbf{\Delta}(\mu^1) \\ &\geqq b_i, \qquad m+1 \leqq i \leqq m+p, \text{ and } i \notin \mathbf{\Delta}(\mu^1) \end{aligned} \tag{6.12}$$

The restricted primal (6.12) is now solved by some method such as Phase I of the primal simplex method. If (6.12) is feasible, and $\bar{x}$ is any feasible solution for it; $\bar{x}$, μ^1 are feasible to (6.9), (6.10), respectively, and satisfy (6.11) and hence by Theorem 4.7, $\bar{x}$ is optimal to (6.9) and μ^1 is optimal to (6.10). If (6.12) is infeasible, let x^1 be a primal vector that minimizes a measure of infeasibility for (6.12), that is, the optimum solution of the corresponding Phase I problem, and let the row vector σ^1 be the corresponding dual optimum for this problem of minimizing the infeasibility measure. By considering the dual of the Phase I problem corresponding to (6.12), we conclude that

$$\begin{aligned} &\sigma^1 A = 0 \\ &\sigma_i^1 \geqq 0 \qquad \text{for } i \in \{m+1, \ldots, m+p\} \backslash \mathbf{\Delta}(\mu^1) \\ &\sigma^1 b > 0 \end{aligned} \tag{6.13}$$

because $\sigma^1 b$ = minimum value of the infeasibility measure, which is >0 since (6.12) is infeasible.

Now the method constructs a new dual feasible solution for (6.10) of the form $\mu^1 + \alpha\sigma^1$ where $\alpha > 0$. Since μ^1 is dual feasible we have $\mu^1 A = c$, and from (6.13) we have $(\mu^1 + \alpha\sigma^1) = c$ for all α. Also, since $\sigma_i^1 \geqq 0$, $\mu_i^1 \geqq 0$ for all $i \in \{m+1, \ldots, m+p\} \backslash \mathbf{\Delta}(\mu^1)$, we have $\mu_i^1 + \alpha\sigma_i^1 \geqq 0$ for all these i, for any $\alpha \geqq 0$. So for the new solution to be dual feasible, it is enough if we choose α to satisfy $\mu_i^1 + \alpha\sigma_i^1 \geqq 0$ for all $i \in \mathbf{\Delta}(\mu^1)$. Since $\mu_i^1 > 0$ for all $i \in \mathbf{\Delta}(\mu^1)$, the maximum value of α that satisfies this condition is given by

$$\begin{aligned} \theta &= +\infty \qquad \text{if } \sigma_i^1 \geqq 0 \quad \text{for all } i \in \mathbf{\Delta}(\mu^1) \\ &= \text{Minimum } \{-\ \mu_i^1/\sigma_i^1 : i \in \mathbf{\Delta}(\mu^1) \text{ and} \\ &\qquad \text{satisfying } \sigma_i^1 < 0\} \text{ otherwise} \end{aligned} \tag{6.14}$$

If $\theta = +\infty$, $\mu^1 + \alpha\sigma^1$ is feasible to (6.10) for all $\alpha \geqq 0$, and $(\mu^1 + \alpha\sigma^1)b$ diverges to $+\infty$ as α tends to $+\infty$, by (6.13). So, in this case, the dual problem (6.10) is feasible, and the dual objective value is unbounded above. By the duality theorem (Theorem 4.5), the primal problem (6.9) is infeasible, and the method terminates.

If θ is finite, the new dual feasible solution (for (6.10)) is $\mu^2 = \mu^1 + \theta\sigma^1$. It can be verified that x^1, μ^2 together satisfy the complementary slackness conditions (6.11) again. Now the restricted primal corresponding to μ^2 is constructed, and the Phase I problem corresponding to this new restricted primal is solved beginning with the present primal vector x^1. Usually, the necessary changes are carried out in the Phase I problem of the current restricted primal so that it becomes the Phase I problem for the new restricted primal, and the application of the method to solve the modified Phase I problem continues beginning with x^1. If the primal simplex algorithm is being employed to solve these Phase I problems, the slack variable corresponding to the ith primal constraint for the i that attains the minimum in (6.14) becomes eligible to enter the basic vector when we begin to solve the new Phase I problem starting with x^1. Thus, after each primal vector change, the infeasibility form for the restricted primal strictly decreases. Also, since $\sigma^1 b > 0$, we have $\mu^2 b > \mu^1 b$. Thus the dual objective value, μb, strictly increases after each dual solution change step. These facts imply that if the Phase I problems corresponding to the restricted primals are solved by the primal simplex algorithm using some method for resolving the problem of cycling under degeneracy, the overall method is guaranteed to terminate after a finite number of steps.

The name *primal-dual method* refers to a variant of the simplex algorithm, which solves an LP by the above strategy. It needs a dual feasible solution initially. It maintains a dual feasible solution, and tries to obtain a primal vector closest to primal feasibility while satisfying the complementary slackness conditions with the present dual feasible solution. If the primal vector is feasible, it is an optimum solution of the primal, and the method terminates.

A Primal-Dual Algorithm for the LP in Standard Form

Now we describe how the primal-dual method is specialized to solve an LP in standard form, using the primal simplex algorithm to solve the Phase I problems. Let the LP to be solved be (6.1) where A is of order $m \times n$. We assume that $b_i \geqq 0$ for all i (multiply both sides of the ith equation in (6.1) by -1 if this condition is not satisfied). Let $\pi = (\pi_1, \ldots, \pi_m)$ be the row vector of dual variables. The dual is

$$\begin{array}{ll} \text{Maximize} & \pi b \\ \text{Subject to} & \pi A \leqq c \end{array} \tag{6.15}$$

Let $\bar{c} = c - \pi A$. $\bar{c} = (\bar{c}_1, \ldots, \bar{c}_n)$ is the row vector of dual slack variables. The complementary slackness conditions are

$$\bar{c}_j x_j = 0 \qquad \text{for all } j = 1 \text{ to } n \tag{6.16}$$

Let π^1 be the initial dual feasible solution and $\bar{c}^1$, the corresponding dual slack vector. So $\bar{c}^1 \geqq 0$. Suppose $\bar{c}_j^1 = 0$ for $j = 1$ to r and $\bar{c}_j^1 > 0$ for $j = r + 1$ to n. Then the restricted primal corresponding to π^1 is

$x_1 \ldots x_r$	
$a_{11} \ldots a_{1r}$	b_1
$\vdots \quad \vdots$	$\vdots$
$a_{m1} \ldots a_{mr}$	b_m

(6.17)

$x_{r+1} = \cdots = x_n = 0$. $x_j \geqq 0$, $j = 1$ to r.

If we use a full artificial basis, denoting the artificial variables by $t_1, \ldots, t_m$, and the Phase I objective function by w, the corresponding Phase I problem is

$x_1 \ldots x_r$	$t_1 \ldots t_m$	$-w$	
$a_{11} \ldots a_{1r}$	$1 \ldots 0$	0	b_1
$\vdots \quad \vdots$	$\vdots \ \ddots \ \vdots$	$\vdots$	$\vdots$
$a_{m1} \ldots a_{mr}$	$0 \ldots 1$	0	b_m
$0 \ldots 0$	$1 \ldots 1$	1	0

Solve this Phase I problem by the revised primal simplex algorithm beginning with the feasible basic vector $(t_1, \ldots, t_m)$. In this process, once an artificial variable t_j drops out of the basic vector, discard it from the problem. Let x^1 be the primal vector obtained at termination in this Phase I problem, and let w^1 be the minimum Phase I objective value. If $w^1 = 0$, x^1, π^1 are optimal to (6.1) and its dual, respectively, terminate. If $w^1 > 0$, let σ^1 be the terminal Phase I dual solution obtained.

To get the new dual feasible solution for (6.15), compute $\bar{d}_j^1 = 0 - \sigma^1 A_{.j}$ for $j = 1$ to n. For solving the Phase I corresponding to the present restricted primal we need only $\bar{d}_1^1, \ldots, \bar{d}_r^1$ (since the variables $x_{r+1}, \ldots, x_n$ do not appear in the present restricted primal) and all these are $\geqq 0$, since the Phase I termination criterion is satisfied. However, to obtain the new dual feasible solution for (6.15) we compute $\bar{d}_{r+1}^1, \ldots, \bar{d}_n^1$ too, and some of these quantities may be negative. The new dual solution for (6.15) will be of the form $\pi^1 + \alpha\sigma^1$. For this to be dual feasible we need $c - (\pi^1 + \alpha\sigma^1)A = \bar{c}^1 + \alpha\bar{d}^1 \geqq 0$. The maximum value for α subject to these conditions is

$$\begin{aligned} \theta &= +\infty \qquad \text{if } \bar{d}_j^1 \geqq 0 \quad \text{for } j = r + 1 \text{ to } n \\ &= \text{Minimum } \{-\bar{c}_j^1/\bar{d}_j^1 : r + 1 \leqq j \leqq n \text{ and} \\ &\qquad \text{satisfying } \bar{d}_j^1 < 0\} \end{aligned} \tag{6.18}$$

If $\theta = +\infty$, we conclude that (6.1) is infeasible and terminate. If θ is finite, $\pi^2 = \pi^1 + \theta\sigma^1$ is the new dual feasible solution for (6.15), and let $\bar{c}^2$ be the corresponding dual slack vector. If j is such that $x_j^1 > 0$ we must have $\bar{d}_j^1 = 0$ (since the Phase I relative cost coefficient of a basic variable is always zero) and $\bar{c}_j^1 = 0$ and hence $\bar{c}_j^2 = 0$ also. Thus x^1, π^2 together satisfy (6.16). The new restricted primal corresponding to π^2 is obtained by deleting all the variables x_j for j satisfying $\bar{c}_j^2 > 0$ from the present restricted primal (these variables have value 0 in the present primal vector x^1); and introducing all variables x_j not in the restricted primal, for j satisfying $\bar{c}_j^2 = 0$. All the variables x_j for j attaining the minimum in (6.18) get introduced into the restricted primal in this process, and since $\bar{d}_j^1 < 0$ for these j, every one of these newly introduced x_j is eligible to enter the basic vector when Phase I process is resumed on the new restricted primal starting with the present basic vector.

If methods for resolving degeneracy are used in this Phase I process, the overall method must terminate after a finite number of steps, since a basic vector can never reappear.

Example 6.5

Consider the following LP

x_1	x_2	x_3	x_4	x_5	x_6	$-z$	b
1	−1	2	−1	1	2	0	1
−1	2	1	−2	−1	1	0	3
2	1	−1	1	−2	1	0	2
3	2	1	2	2	0	1	0

$x_j \geqq 0$ for all j. Minimize z.

Let A denote the coefficient matrix in the constraints in this problem and c the row vector of original cost coefficients. Since $c \geqq 0$, $\pi^1 = (0, 0, 0)$ is a dual feasible solution. We initialize the primal-dual method

Basic Variable	Inverse				$\bar{b}$	Pivot Column x_6	Ratios
t_1	1	0	0	0	1	(2)	1/2 Min.
t_2	0	1	0	0	3	1	3/1
t_3	0	0	1	0	2	1	2/1
$-w$	-1	-1	-1	1	-6	-4	
x_6	1/2	0	0	0	1/2		
t_2	$-1/2$	1	0	0	5/2		
t_3	$-1/2$	0	1	0	3/2		
$-w$	1	-1	-1	1	-4		

with π^1. The corresponding dual slack vector is $\bar{c}^1 = c - \pi^1 A = (3, 2, 1, 2, 2, 0)$. Since $\bar{c}_j^1 = 0$ only for $j = 6$, x_6 is the only possible original problem variable in the initial restricted primal. Using the artificial variables t_1, t_2, t_3, we get the following Phase I problem for this restricted primal.

x_6	t_1	t_2	t_3	$-w$	
2	1	0	0	0	1
1	0	1	0	0	3
1	0	0	1	0	2
0	1	1	1	1	0

All variables $\geqq 0$. Minimize w. x_1 to $x_5 = 0$.

We get the inverse tableaux at the top when we solve the Phase I problem. Pivot elements are circled in each pivot step.

Now the simplex algorithm on the present Phase I problem terminates. The present Phase I dual optimum solution is $\sigma^1 = (-1, 1, 1)$, from the last row of the terminal inverse tableau. The Phase I relative cost vector for all the original problem variables is $\bar{d}^1 = 0 - \sigma^1 A = (0, -4, 2, 0, 4, 0)$. The new dual feasible solution is $\pi^2 = \pi^1 + \theta\sigma^1$ where $\theta = \text{Minimum } \{-\bar{c}_2^1/\bar{d}_2^1 = 2/4\} = 2/4$. So $\pi^2 = (-1/2, 1/2, 1/2)$. The corresponding dual slack vector is $\bar{c}^2 = c - \pi^2 A = (3, 0, 2, 2, 4, 0)$. So the Phase I problem for the new restricted primal is (the artificial t_1 dropped out of the basic vector and is discarded) the one given below.

x_2	x_6	t_2	t_3	$-w$	
-1	2	0	0	0	1
2	1	1	0	0	3
1	1	0	1	0	2
0	0	1	1	1	0

All variables $\geqq 0$. Minimize w. $x_1, x_3, x_4, x_5 = 0$.

Resuming the solution of this Phase I problem beginning with the present basic vector (x_6, t_2, t_3), we find that the Phase I relative cost coefficient of x_2 is -4. Hence x_2 is chosen as the entering variable. Its updated column is $(-1/2, 5/2, 3/2, -4)^T$. Hence it can replace t_3 as the basic variable, leading to the following inverse tableau.

Basic Variable	Inverse				$\bar{b}$
x_6	1/3	0	1/3	0	1
t_2	1/3	1	$-5/3$	0	0
x_2	$-1/3$	0	1	0	1
$-w$	$-1/3$	-1	5/3	1	0

The Phase I objective value has become zero. So the present primal vector $x^2 = (0, 1, 0, 0, 0, 1)^T$ is an optimum solution of the LP and π^2 is a dual optimum solution.

Example 6.6

Consider the following LP.

x_1	x_2	x_3	x_4	x_5	$-z$	b
1	0	1	1	1	0	5
0	1	2	1	-1	0	6
1	1	3	2	0	0	4
0	-1	10	14	3	1	0

$x_j \geqq 0$ for all j. Minimize z.

The initial dual feasible solution $\pi^1 = (-1, -2, 1)$ is given. $\bar{c}^1 = c - \pi^1 A = (0, 0, 12, 15, 2)$. So the Phase I problem for the initial restricted primal (t_1, t_2, t_3 are the artificial variables and w is the Phase I objective function) is:

x_1	x_2	t_1	t_2	t_3	$-w$	
1	0	1	0	0	0	5
0	1	0	1	0	0	6
1	1	0	0	1	0	4
0	1	1	1	1	1	0

All variables $\geqq 0$. Minimize w. $x_3, x_4, x_5 = 0$.

When this Phase I problem is solved by the primal revised simplex method beginning with (t_1, t_2, t_3) as the initial basic vector, we get the following terminal inverse tableau.

Basic Variable	Inverse				$\bar{b}$
t_1	1	0	-1	0	1
t_2	0	1	0	0	6
x_1	0	0	1	0	4
$-w$	-1	-1	1	1	-7

The Phase I dual optimum solution is $\sigma^1 = (1, 1, -1)$. The Phase I relative cost vector $\bar{d}^1 = 0 - \sigma^1 A = (0, 0, 0, 0, 0)$. Since $\bar{d}^1 \geqq 0$, we conclude that the original problem is infeasible.

Usefulness of Primal-Dual Methods

The primal-dual approach can be expected to be efficient if an initial dual feasible solution can be computed readily and the Phase I problems corresponding to the restricted primals can be solved by highly efficient special methods. In single commodity transportation and network flow problems, this turns out to be the case. In these problems, finding an initial dual feasible solution turns out to be a trivial task. And the restricted primal problem can be posed as a maximum value flow problem on a subnetwork, and solved by the very efficient labeling algorithms. Hence the primal-dual approach leads to efficient algorithms for solving many types of network flow problems. In the same manner, using the primal-dual approach, very efficient algorithms have been developed for solving a variety of matching and edge covering problems in networks. See references [6.1, 6.2, 6.3, 6.5, 1.61, 1.68, 1.85, 1.90, 1.92, 1.94].

The Primal-Dual Method When a Dual Feasible Solution is Not Available

If a dual feasible solution is not readily available for (6.1), we can construct an augmented problem as in Section 6.4, for which an initial dual feasible solution is readily available. Then the augmented problem can be solved by the primal-dual method, beginning with the initial dual feasible solution for it. See reference [6.2].

Comment: In [6.3] H. W. Kuhn developed a special primal-dual algorithm called the *Hungarian method* for solving the assignment problem, based on the work of the Hungarian mathematician J. Egervàry. L. R. Ford, Jr. and D. R. Fulkerson generalized this method to a primal-dual algorithm for solving transportation problems [1.68]. J. Edmonds generalized the Hungarian method to a primal-dual algorithm called the *blossom algorithm* for solving weighted matching problems in undirected networks (see references [6.5, 1.85, 1.90, 1.94]). The primal-dual method for general LPs was described by G. B. Dantzig, L. R. Ford, Jr. and D. R. Fulkerson in [6.2].

Exercises

6.3 Solve the following LPs by the dual simplex method.

(a)

x_1	x_2	x_3	x_4	x_5	x_6	x_7	$-z$	b
1	2	0	-3	0	2	-3	0	-4
0	-1	0	2	1	-1	2	0	1
0	2	1	-3	0	1	-2	0	1
0	4	0	5	0	9	9	1	0

$x_j \geqq 0$ for all j; minimize z.

(b)

$$\begin{array}{ll} \text{Minimize} & 3x_1 + 4x_2 + 2x_3 + x_4 + 5x_5 \\ \text{Subject to} & x_1 - 2x_2 - x_3 + x_4 + x_5 \leqq -3 \\ & -x_1 - x_2 - x_3 + x_4 + x_5 \leqq -2 \\ & x_1 + x_2 - 2x_3 + 2x_4 - 3x_5 \leqq 4 \\ & x_j \geqq 0 \quad \text{for all } j \end{array}$$

(c)

$$\begin{array}{ll} \text{Minimize} & -2x_1 - x_2 - 8x_4 + 2x_5 - 3x_7 \\ \text{Subject to} & 3x_1 + x_3 + 16x_4 - 2x_5 + 5x_6 + 4x_7 = 18 \\ & 2x_1 + x_3 + 11x_4 - x_5 + 3x_6 + 3x_7 = 11 \\ & x_1 - x_2 + x_3 + 7x_4 - 2x_5 + 2x_6 + x_7 = 6 \\ & x_j \geqq 0 \quad \text{for all } j \end{array}$$

Use (x_1, x_2, x_3) as the initial basic vector.

(d)

$$\begin{array}{ll} \text{Minimize} & -2x_1 + 4x_2 + x_3 - x_4 + 6x_5 + 8x_6 - 9x_7 - 5x_8 \\ \text{Subject to} & x_1 + x_4 - 2x_5 + x_6 + x_7 - 2x_8 = -3 \\ & x_2 - x_4 + x_5 + x_6 - 3x_7 - x_8 = -14 \\ & x_3 + x_4 - x_5 - 2x_6 - x_7 + x_8 = -5 \\ & x_j \geqq 0 \quad \text{for all } j \end{array}$$

(e)

$$\begin{array}{ll} \text{Minimize} & -2x_1 + 4x_2 + 2x_3 + x_4 - 4x_5 - 10x_6 \\ \text{Subject to} & 5x_2 + 2x_3 + x_4 - 3x_5 - 9x_6 - 4x_7 = -8 \\ & x_1 - 3x_2 - x_3 - x_4 + 2x_5 + 8x_6 = 7 \\ & -2x_1 - x_3 + x_4 - 5x_6 + 6x_7 = -3 \\ & x_j \geqq 0 \quad \text{for all } j \end{array}$$

6.4 Solve the following system using the dual simplex algorithm:

$$\begin{array}{l} 5x_1 + 4x_2 - 7x_3 \leqq 1 \\ -x_1 + 2x_2 - x_3 \leqq -4 \\ -3x_1 - 2x_2 + 4x_3 \leqq 3 \\ x_j \geqq 0 \quad \text{for all } j \end{array}$$

6.5 Show that the method of generating an artificially dual feasible basis by introducing the artificial constraint (6.6) is the dual analog of generating an artificially primal feasible basis by introducing an artificial variable.

6.6 Consider the LP (6.1). Suppose a dual feasible basis for it is known, and suppose the LP is being solved by the dual simplex algorithm. At the kth stage of this algorithm, suppose the primal infeasibility criterion is satisfied. Using the data in the canonical tableau or the inverse tableau at this stage, explain how to construct a half-line of dual feasible solutions, along which the dual objective function diverges to $+\infty$.

6.7 Solve the following LP by the dual simplex algorithm starting with (x_1, x_2, x_3) as the basic vector. Verify that even though x_3 has a positive value in the initial solution, its value becomes negative in the solution obtained after the first pivot step in this algorithm.

x_1	x_2	x_3	x_4	x_5	x_6	$-z$	b
1	0	0	−1	2	2	0	10
0	1	0	−1	−2	3	0	−9
0	0	1	1	1	1	0	3
0	0	0	1	10	15	1	0

$x_j \geqq 0$ for all j, z to be minimized.

6.8 Suppose it is required to find a feasible solution of (6.1). Consider the following procedure. Obtain a canonical tableau with respect to an arbitrary basic vector. If there is a negative updated right-hand-side constant, pivot in its row using a negative element as the pivot element (if no negative element exists, conclude (6.1) is infeasible), and continue in the same way. Can this algorithm cycle? What other difficulties might arise in this algorithm? Discuss how to resolve them.

6.9 Consider the LP (6.1), and assume that $c \geqq 0$. In this case, develop a method for doing the row operations on $Ax = b$ to reduce it to row echelon normal form, so that the basic vector obtained is a dual feasible basic vector for (6.1).

6.10 Consider the LP (6.1). If the set of feasible solutions of (6.1) is nonempty and bounded, prove that the set of feasible solutions of its dual must be nonempty and unbounded.

6.11 Consider the LP (6.1) with $n = m + 2$, and $\text{rank}(A) = m$. In this case, assuming primal nondegeneracy, prove that in solving (6.1) by the primal simplex algorithm beginning with a feasible basic vector, the number of negative relative cost coefficients in the current canonical tableau does not increase as the algorithm progresses. Is this statement also true when $n = m + 3$? Why?

6.12 Let $\mathbf{H}_i = \{x: A_{i.}x = b_i\}$, $i = 1$ to m, be distinct hyperplanes in $\mathbf{R}^n$ with a common point $\bar{x}$, satisfying $\bigcap_{i=1}^{m} \mathbf{H}_i = \{\bar{x}\}$. Let $\mathbf{H} = \{x: A_{m+1.}x = b_{m+1}\}$ be another hyperplane containing $\bar{x}$, where $A_{m+1.}$ is not a linear combination of any set of $n - 1$ rows from $\{A_{1.}, \ldots, A_{m.}\}$. The hyperplanes $\mathbf{H}_1, \ldots, \mathbf{H}_m$ partition $\mathbf{R}^n$ into conical regions, each of which is the set of feasible solutions of $A_{i.}x \gtreqqless b_i$, $i = 1$ to m, for some choice of this inequality for each i. Count the number of these conical regions that have a nonempty interior and lie wholly within the half-space $A_{m+1.}x \geqq b_{m+1}$. Prove that this number is independent of $\mathbf{H}$, but depends only on $\mathbf{H}_1, \ldots, \mathbf{H}_m$.

—(S. Berenguer and R. L. Smith)

REFERENCES

6.1 E. M. L. Beale, "An Alternative Method for Linear Programming," *Proceedings of the Cambridge Philosophical Society, 50*, 4 (1954), 513–523.

6.2 G. B. Dantzig, L. R. Ford, Jr. and D. R. Fulkerson, "A Primal-Dual Algorithm for Linear Programs," pp. 171–181 in reference [1.83].

6.3 H. W. Kuhn, "The Hungarian Method for the Assignment Problem," *Naval Research Logistics Quarterly, 2*, 1 and 2 (1955), 83–97.

6.4 C. E. Lemke, "The Dual Method of Solving the Linear Programming Problem," *Naval Research Logistics Quarterly, 1* (1954), 36–47.

6.5 C. H. Papadimitriou and K. Steiglitz, *Combinatorial Optimization, Algorithms and Complexity*, Prentice-Hall, Englewood Cliffs, New Jersey, 1982.

chapter 7 Numerically Stable Forms of the Simplex Method

7.1 REVIEW

Consider the LP in standard form:

$$\begin{aligned} \text{Minimize} \quad & z(x) = cx \\ \text{Subject to} \quad & Ax = b \qquad (7.1) \\ & x \geqq 0 \end{aligned}$$

where A is a matrix of order $m \times n$. In the process of solving this problem by the simplex algorithm, consider a step in which the basic vector is x_B. Let B be the corresponding basis, c_B the associated basic cost vector, and π_B the corresponding dual solution. The computations to be performed in this step are summarized below.

1 Compute the values of the basic variables in the basic solution of (7.1) corresponding to the basic vector x_B, by solving

$$Bx_B = b \qquad (7.2)$$

In the notation of Chapters 2 and 5, the solution x_B of this system is denoted by $\bar{b}$. Normally (7.2) is not solved afresh in each step. The $\bar{b}$ vector in the present step is normally obtained by updating the $\bar{b}$ vector in the previous step. Hence either (7.2) has to be solved or the $\bar{b}$ vector has to be updated in each step.

2 Compute π_B by solving

$$\pi_B B = c_B \qquad (7.3)$$

3 Compute $\bar{c}_j = c_j - \pi_B A_{.j}$ for each j. If $\bar{c}_j \geqq 0$ for all j, the algorithm terminates in this step. Otherwise find an s such that $\bar{c}_s < 0$ and select x_s as the entering variable.

4 Compute $\bar{A}_{.s}$, the updated column vector of x_s, by solving

$$B\bar{A}_{.s} = A_{.s} \qquad (7.4)$$

5 If $\bar{A}_{.s} \leqq 0$, the algorithm terminates in this step. Otherwise, find an r satisfying $\bar{b}_r/\bar{a}_{rs} = \text{minimum}$ $\{\bar{b}_i/\bar{a}_{is}\colon i, \text{ such that } \bar{a}_{is} > 0\}$. Then drop the rth column vector in the basis B and include $A_{.s}$ in its place. This gives the new basis. Then go to the next step, in which all these computations are repeated with the new basis.

In the revised simplex algorithm all these computations are performed using B^{-1} either explicitly or as a product of a sequence of pivot matrices. The major drawback in this approach is that roundoff errors accumulate as the algorithm moves from step to step. After several steps, the actual basis inverse may be quite different from the explicit inverse in storage now or the product of the sequence of pivot matrices in storage now. To reduce the effect of these roundoff errors, it is often recommended that periodically, after a certain number of steps of the algorithm, the basis should be identified from the original tableau and its inverse actually computed afresh, or the pivot matrices whose product is the basis inverse should be recomputed by pivoting on the basis. Even with such periodic reinversion, round-off error accumulation poses a serious problem, undermining the ability of the simplex algorithm to produce accurate answers.

There are two different methods for carrying out the computations in the simplex algorithm that have been proposed to alleviate this drawback. These are (1) the *LU*-decomposition and (2) the Cholesky factorization. Here we discuss these methods briefly.

We have summarized above the basic computations to be carried out in a pivot step of the *primal simplex algorithm* applied to solve (7.1). Instead, if (7.1) is to be solved by the dual simplex algorithm, we have to carry out essentially the same computations, but use different pivot row and column choice criteria, as discussed in Chapter 6. Thus, even though our discussion in this chapter is centered on the primal simplex algorithm, the same techniques can be used to implement the dual simplex algorithm by making appropriate changes in the pivot row and column choice criteria.

7.2 THE *LU*-DECOMPOSITION

Let B be the basis under consideration. This method generates an upper triangular matrix U, a sequence of lower triangular pivot matrices $\Gamma_1, \Gamma_2, \ldots, \Gamma_k$, and a sequence of permutation matrices $Q_1, \ldots, Q_k$ (some of these permutation matrices may be equal to the unit matrix) satisfying the equation

$$\Gamma_k Q_k \cdots \Gamma_2 Q_2 \Gamma_1 Q_1 B = U \tag{7.5}$$

The name *LU-decomposition* stems from the customary practice of denoting the product $\Gamma_k Q_k \cdots \Gamma_1 Q_1$ in the decomposition by the symbol L^{-1}. If all the permutation matrices $Q_k, \ldots, Q_1$ are equal to the unit matrix, L^{-1} will be a lower triangular matrix, but in general, in the decomposition of a basis obtained in the simplex algorithm after one or more basic variable changes, L^{-1} may not be lower triangular, or even triangular. L^{-1} is never computed explicitly. It is always stored in product form by storing the pivot matrices and the permutation matrices in their proper order. The diagonal entries of all the pivot matrices in the decomposition will always be equal to one. Since each pivot matrix is also lower triangular, it can be stored in a very compact fashion. Clearly, it is only necessary to store the permutation matrices that differ from the unit matrix.

First we discuss how to perform all the computations in the simplex method using such a decomposition, before discussing methods for obtaining and updating the decomposition.

7.2.1 How to Perform the Computations Using the *LU*-Decomposition

Suppose you have to solve a system of equations such as

$$By = d \tag{7.6}$$

for y. Let the basis B have the decomposition as in (7.5). First compute $t = \Gamma_k Q_k \cdots \Gamma_1 Q_1 d$. Since each Q_i is a permutation matrix and each Γ_i is a lower triangular pivot matrix, this multiplication can be carried out very efficiently from the right by including one new matrix in the product at a time, until t is obtained. This is the same string computation from right to left as discussed in Section 5.5. It begins by multiplying the column vector d on the left by Q_1, leading to another column vector. This column vector, in turn, is multiplied on the left by Γ_1, leading to a new column vector, and so on. After the vector t is computed, solve the system $Uy = t$ for y, by back susbtitution. Both systems (7.2) and (7.4) are of the form (7.6); hence, they can be solved as above using the decomposition (7.5).

To solve (7.3), first find the vector h from $hU = c_B$ by back substitution. Then find the vector π_B using the equation $\pi_B = h\Gamma_k Q_k \cdots \Gamma_1 Q_1$, where the multiplication here can be carried out very conveniently by starting from the left and including one matrix in the product at a time on the right-hand side, until π_B is obtained. This is again the string computation from left to right as discussed in Section 5.5. It begins by multiplying the row vector h on its right by Γ_k, leading to another row vector. This row vector, in turn, is multiplied on its right by Q_k, leading to a new row vector, and so on.

7.2.2 How to Obtain the *LU*-Decomposition for a Given Basis

Let the basis be B. When the following steps 1 to $m - 1$ are carried out on B, it gets transformed into U.

In step 1, the present matrix is $B^{(0)} = B$. At the end of the rth step, suppose B is transformed into $B^{(r)}$, $r = 1$ to $m - 1$.

Step *r* At the beginning of this step, the present matrix is $B^{(r-1)} = (b_{ij}^{(r-1)})$. Find i_1 such that $|b_{i_1,r}^{(r-1)}| =$ maximum $\{|b_{i,r}^{(r-1)}|: i = r \text{ to } m\}$ (see Section 3.3.8).

Interchange the rth row in the present matrix with row i_1. This operation is equivalent to multiplying the present matrix $B^{(r-1)}$ on the left by Q_r, which is the identity matrix with row r and row i_1 interchanged. After the interchange, add suitable multiples of row r to row i to transform the entry in row i and column r to zero for each $i \geqq r + 1$. This is the *Gaussian elimination* pivot step as discussed in Section 3.3.8. It is equivalent to multiplying the present matrix on the left by an elementary matrix of the form given in Figure 7.1, where $|g_{i,r}| \leqq 1$ for each $i \geqq r + 1$ (after multiplying $B^{(r-1)}$ on the left by Q_r, let the rth column in the resulting matrix be $(\bar{a}_{1r}, \ldots, \bar{a}_{mr})$. Then $g_{ir} = -\bar{a}_{ir}/\bar{a}_{rr}$, for $i = r + 1$ to m). Since we are only reducing the rth column of $B^{(r-1)}$ to upper triangular form, the matrix Γ_r is not a pivot matrix in the sense of Section 3.11 (pivot matrices discussed in Section 3.11 are Gauss–Jordan pivot matrices. Such a pivot matrix would reduce this column to the rth column vector of the unit matrix). Γ_r is a *Gaussian pivot matrix*. It has the property that it differs from the unit matrix in just one column, like

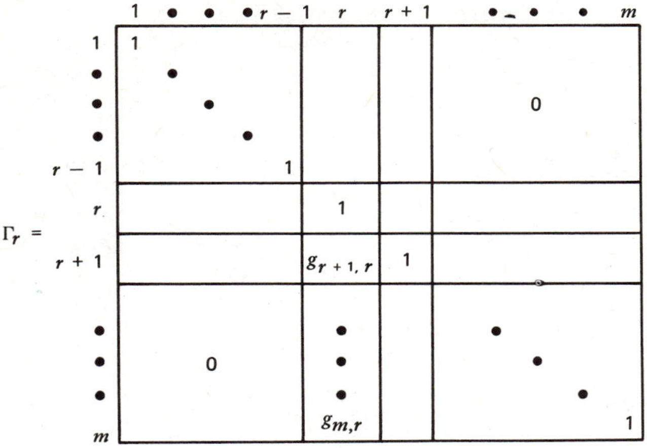

Figure 7.1

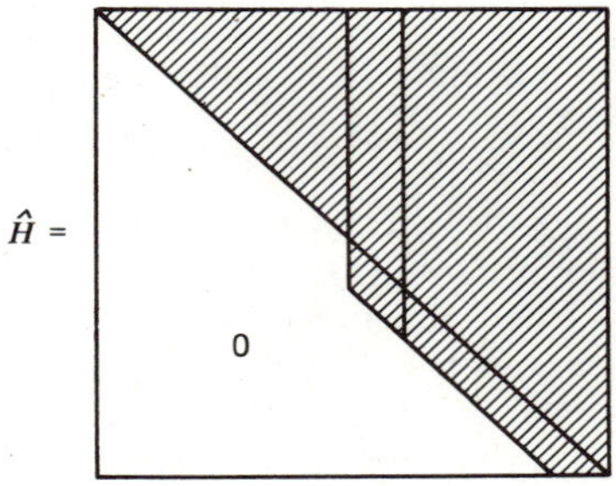

Figure 7.2

the pivot matrices of Section 3.11 and we will also refer to matrices such as Γ_r as pivot matrices.

Then go to step $r + 1$ if $r < m - 1$.

At the end of step $m - 1$, the matrix is transformed into the upper triangular matrix U. Also, from the manner in which these matrices are generated, it is clear that $\Gamma_{m-1}Q_{m-1} \cdots \Gamma_1 Q_1 B = U$, which is the decomposition required for the basis B. Since each of these matrices $\Gamma_{m-1}, \ldots, \Gamma_1, Q_{m-1}, \ldots, Q_1$ is a matrix of special structure, it can be stored in a compact manner. U and the lower triangular pivot matrices $\Gamma_1, \ldots, \Gamma_{m-1}$, can be stored in the space required for storing B. Additional locations may be required for storing the permutation matrices.

7.2.3 Updating the *LU*-Decomposition When One Basic Column Vector Changes

Let the present basis be $B = (A_{.j_1}, \ldots, A_{.j_m})$. Let the present decomposition be $\Gamma_\ell Q_\ell \cdots \Gamma_1 Q_1 B = U$. In this pivot step suppose $A_{.s}$ is the entering column and suppose that $A_{.j_r}$ is the leaving basic column. In updating the decomposition, the entering column is always treated as the rightmost column in the new basis. All the columns in B to the right of the leaving column are recorded in the same order in which they appear in B, but they are shifted by one position to the left, that is, the new basis is recorded as $\hat{B} = (A_{.j_1}, \ldots, A_{.j_{r-1}}, A_{.j_{r+1}}, \ldots, A_{.j_m}, A_{.s})$. Let $f = \Gamma_\ell Q_\ell \cdots \Gamma_1 Q_1 A_{.s}$. Notice that f is obtained as a by-product when (7.4) is solved by the method discussed in Section 7.2.1. Hence the vector f is obtained without doing any additional computation. Then $\Gamma_\ell Q_\ell \cdots \Gamma_1 Q_1 \hat{B} = (U_{.1}, \ldots, U_{.r-1}\ U_{.r+1}, \ldots, U_{.m}, f) = \hat{H}$ where the $U_{.j}$ is the jth column vector of U. $\hat{H}$ is a matrix with zeros below the diagonal in the first $r - 1$ columns and zeros in rows that are two or more rows below the diagonal in columns r to m.

A matrix like this is known as an *upper Hessenberg matrix* (see Figure 7.2). $\hat{H}$ can be transformed into an upper triangular matrix by doing row operations as in Section 7.2.2 to zero out all the elements just below the diagonal in columns r to m. This is achieved by performing the following operations for $k = r$ to $m - 1$ in that order.

1 Let the diagonal entry in column k of the present matrix be h_{kk} and let the entry just below it be $h_{k+1,k}$.

2 Since the previous matrix U was nonsingular, $h_{k+1,k} \neq 0$. Find out which of $|h_{k,k}|$ and $|h_{k+1,k}|$ is larger. If $|h_{k+1,k}| > |h_{k,k}|$, interchange rows k and $k + 1$ of the present matrix. This is equivalent to multiplying the present matrix on the left by $Q_k^{(1)}$, where $Q_k^{(1)}$ is the identity matrix with its rows k and $k + 1$ interchanged. If $|h_{k,k}| \geqq |h_{k+1,k}|$, let $Q_k^{(1)} = I$. Now subtract a suitable multiple of the present kth row from the $k + 1$th row to transform the element in the present $k + 1$th row and column k into zero. This is equivalent to multiplying the present matrix on the left by an elementary matrix of the form given in Figure 7.3, where $g_k^{(1)} = -\bar{h}_{k+1,k}/\bar{h}_{k,k}$, if the kth and $k + 1$th

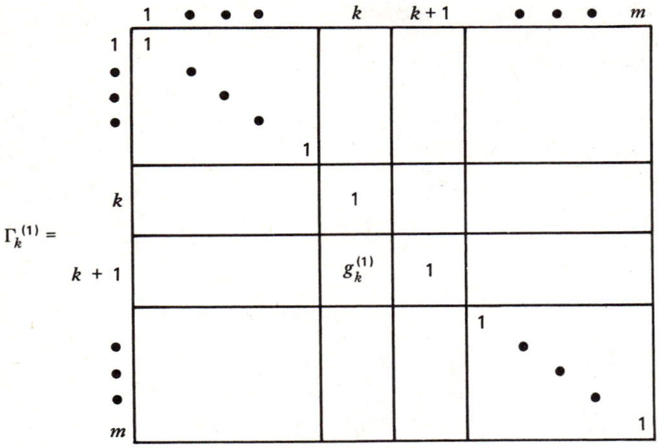

Figure 7.3

entries in column k of the present matrix, after the interchange if necessary, are $\bar{h}_{k,k}$ and $\bar{h}_{k+1,k}$ respectively.

After these operations are performed for $k = r + 1$ to $m - 1$, the matrix $\hat{H}$ is transformed into an upper triangular matrix, $\hat{U}$. The decomposition for the new basis $\hat{B}$ is $\Gamma_{m-1}^{(1)} Q_{m-1}^{(1)} \cdots \Gamma_r^{(1)} Q_r^{(1)} \Gamma_\ell Q_\ell \cdots \Gamma_1 Q_1 \hat{B} = \hat{U}$. The simplex algorithm can now be continued using the decomposition of $\hat{B}$.

If $\tilde{B}$ is the basis under consideration at some stage of the simplex algorithm, the implementation computes the upper triangular matrix $\tilde{U}$, in the decomposition of $\tilde{B}$, explicitly. In each pivot step of the simplex algorithm, several lower triangular pivot matrices such as $\Gamma_j^{(k)}$ and permutation matrices $Q_j^{(k)}$ are generated. All these matrices are stored in the order in which they are generated. All the computations needed in this stage of the simplex algorithm are carried out using $\tilde{U}$, and the lower triangular pivot matrices and the permutation matrices generated so far, as in Section 7.2.1.

The simplex method using the *LU*-decomposition implemented in this manner is called the simplex method using the *Elimination Form of the Inverse* (*EFI*) *implementation*.

Numerical Example

Here we illustrate how to obtain the *LU*-decomposition of a given basis, and how to update the decomposition when a column vector in this basis is changed. Let the initial basis be

$$B = \begin{pmatrix} 1 & -1 & 1 & 2 \\ 2 & 0 & 0 & 1 \\ 1 & 1 & 0 & 1 \\ -1 & 2 & 2 & 1 \end{pmatrix}$$

Going to step 1 of Section 7.2.2, let $B^{(0)} = B$. Rows 1 and 2 have to be interchanged. This leads to

$$Q_1 = \begin{pmatrix} 0 & 1 & 0 & 0 \\ 1 & 0 & 0 & 0 \\ 0 & 0 & 1 & 0 \\ 0 & 0 & 0 & 1 \end{pmatrix} \qquad \Gamma_1 = \begin{bmatrix} 1 & 0 & 0 & 0 \\ -\frac{1}{2} & 1 & 0 & 0 \\ -\frac{1}{2} & 0 & 1 & 0 \\ \frac{1}{2} & 0 & 0 & 1 \end{bmatrix} \qquad B^{(1)} = \Gamma_1 Q_1 B^{(0)} = \begin{bmatrix} 2 & 0 & 0 & 1 \\ 0 & -1 & 1 & \frac{3}{2} \\ 0 & 1 & 0 & \frac{1}{2} \\ 0 & 2 & 2 & \frac{3}{2} \end{bmatrix}$$

$$Q_2 = \begin{pmatrix} 1 & 0 & 0 & 0 \\ 0 & 0 & 0 & 1 \\ 0 & 0 & 1 & 0 \\ 0 & 1 & 0 & 0 \end{pmatrix} \qquad \Gamma_2 = \begin{bmatrix} 1 & 0 & 0 & 0 \\ 0 & 1 & 0 & 0 \\ 0 & -\frac{1}{2} & 1 & 0 \\ 0 & \frac{1}{2} & 0 & 1 \end{bmatrix} \qquad B^{(2)} = \Gamma_2 Q_2 B^{(1)} = \begin{bmatrix} 2 & 0 & 0 & 1 \\ 0 & 2 & 2 & \frac{3}{2} \\ 0 & 0 & -1 & -\frac{1}{4} \\ 0 & 0 & 2 & \frac{9}{4} \end{bmatrix}$$

$$Q_3 = \begin{pmatrix} 1 & 0 & 0 & 0 \\ 0 & 1 & 0 & 0 \\ 0 & 0 & 0 & 1 \\ 0 & 0 & 1 & 0 \end{pmatrix} \qquad \Gamma_3 = \begin{bmatrix} 1 & 0 & 0 & 0 \\ 0 & 1 & 0 & 0 \\ 0 & 0 & 1 & 0 \\ 0 & 0 & \frac{1}{2} & 1 \end{bmatrix} \qquad U = B^{(3)} = \Gamma_3 Q_3 B^{(2)} = \begin{bmatrix} 2 & 0 & 0 & 1 \\ 0 & 2 & 2 & \frac{3}{2} \\ 0 & 0 & 2 & \frac{9}{4} \\ 0 & 0 & 0 & \frac{7}{8} \end{bmatrix}$$

Rows 2 and 4 of $B^{(1)}$ have to be interchanged. The matrices generated in step 2 are Q_2, Γ_2, $B^{(2)}$.
In step 3, we have to interchange rows 3 and 4 in $B^{(2)}$. The matrices generated are Q_3, Γ_3, $B^{(3)}$
The decomposition for B is $\Gamma_3 Q_3 \Gamma_2 Q_2 \Gamma_1 Q_1 B = U$, where the various matrices are given above. Here, for illustration, all the matrices are written out in full. In the implementation, they will be stored in a compact fashion, as discussed earlier.

Now suppose a new basis $\hat{B}$ is obtained by introducing the column vector $A_{.5} = (0, -2, 0, 4)^{\mathrm{T}}$ into B, and dropping the second column vector of B. We compute $f = \Gamma_3 Q_3 \Gamma_2 Q_2 \Gamma_1 Q_1 A_{.5} = (-2, 3, 5/2, 3/4)^{\mathrm{T}}$. So

$$\hat{H} = \begin{bmatrix} 2 & 0 & 1 & -2 \\ 0 & 2 & \frac{3}{2} & 3 \\ 0 & 2 & \frac{9}{4} & \frac{5}{2} \\ 0 & 0 & \frac{7}{8} & \frac{3}{4} \end{bmatrix}$$

We have to reduce columns 2 to 4 of $\hat{H}$ to upper triangular form. We begin by reducing column 2 first, as in Section 7.2.3. No interchange of rows is necessary.

$$Q_2^{(1)} = I, \qquad \Gamma_2^{(1)} = \begin{pmatrix} 1 & 0 & 0 & 0 \\ 0 & 1 & 0 & 0 \\ 0 & -1 & 1 & 0 \\ 0 & 0 & 0 & 1 \end{pmatrix} \qquad \Gamma_2^{(1)} Q_2^{(1)} \hat{H} = \begin{bmatrix} 2 & 0 & 1 & -2 \\ 0 & 2 & \frac{3}{2} & 3 \\ 0 & 0 & \frac{3}{4} & -\frac{1}{2} \\ 0 & 0 & \frac{7}{8} & \frac{3}{4} \end{bmatrix}$$

$$Q_2^{(2)} = \begin{pmatrix} 1 & 0 & 0 & 0 \\ 0 & 1 & 0 & 0 \\ 0 & 0 & 0 & 1 \\ 0 & 0 & 1 & 0 \end{pmatrix} \qquad \Gamma_2^{(2)} = \begin{bmatrix} 1 & 0 & 0 & 0 \\ 0 & 1 & 0 & 0 \\ 0 & 0 & 1 & 0 \\ 0 & 0 & -\frac{6}{7} & 1 \end{bmatrix} \qquad \hat{U} = \Gamma_2^{(2)} Q_2^{(2)} \Gamma_2^{(1)} Q_2^{(1)} \hat{H} = \begin{bmatrix} 2 & 0 & 1 & -2 \\ 0 & 2 & \frac{3}{2} & 3 \\ 0 & 0 & \frac{7}{8} & \frac{3}{4} \\ 0 & 0 & 0 & -\frac{8}{7} \end{bmatrix}$$

Now we have to reduce column 3 of $\Gamma_2^{(1)} Q_2^{(1)} \hat{H}$ to upper triangular form. Rows 3 and 4 of this matrix are interchanged.

The decomposition for the new basis $\hat{B}$ is $\Gamma_2^{(2)} Q_2^{(2)} \Gamma_2^{(1)} Q_2^{(1)} \Gamma_3 Q_3 \Gamma_2 Q_2 \Gamma_1 Q_1 \hat{B} = \hat{U}$, where the various matrices in the decomposition are obtained above.

7.2.4 Advantages

The *LU*-implementation provides much better numerical accuracy in the results obtained in the simplex algorithm compared to the implementation using either the explicit inverse or the product form of the inverse. Also, it helps to preserve the *sparsity* in the basis; that is, the number of nonzero entries in U and the lower triangular pivot matrices generated is not expected to be much larger than the number of nonzero entries in B. The actual inverse of B does not normally preserve sparsity. To keep the discussion brief, we do not go into the reasons for these facts; the interested reader should look up the references listed at the end of the chapter.

The *LU*-decomposition can also be periodically reevaluated using the method in Section 7.2.2 for obtaining even better numerical accuracy.

If the simplex algorithm begins with the unit matrix as the basis, U is I initially, and there are no lower triangular pivot matrices or permutation matrices in the beginning step. If the algorithm begins with a known basis B different from I, its *LU*-decomposition is obtained as in Section 7.2.2 and the algorithm continued from there. If B is sparse, the rows and columns of B can be permuted to make it as close to either upper or lower triangular form as possible (before obtaining the *LU*-decomposition) in order to preserve sparsity in the *LU*-decomposition. See references [7.8, 7.9, 7.10, 7.14, 7.15, 7.23] for generating these row and column permutations to preserve sparsity.

7.3 THE CHOLESKY FACTORIZATION

Let B be a basis. Then BB^{T} is a positive definite matrix. It is a well-known result in linear algebra that there exists a lower triangular matrix L such that $BB^{\mathrm{T}} = LL^{\mathrm{T}}$. L is known as the *Cholesky factor of* BB^{T}. We refer to L as the *Cholesky factor associated with the basis* B.

A square matrix Q is said to be an *orthogonal* matrix if $QQ^{\mathrm{T}} = I$. If B is a nonsingular matrix, it is well known that there exists an orthogonal matrix Q such that $QB^{\mathrm{T}} = R$, where R is a nonsingular upper triangular matrix. $R^{\mathrm{T}} = L$ is the Cholesky factor associated with B. The orthogonal matrix Q is only used in the discussion of the method; it is neither stored nor updated at any stage of the algorithm.

First we discuss how to perform all the computations in the simplex algorithm using the Cholesky factors before discussing methods for obtaining and updating the Cholesky factors.

7.3.1 How to Perform the Computations in the Simplex Algorithm Using Cholesky Factors

Let B be the present basis and let L be the Cholesky factor associated with it. To solve a system of equations of the form (7.6), first find v from $Lv = d$ by back substitution, and then solve for u in $L^{\mathrm{T}}u = v$, again by back substitution. Then $y = B^{\mathrm{T}}u$ is the solution of (7.6).

To solve the system (7.3), first solve for w from $wL^{\mathrm{T}} = c_B B^{\mathrm{T}}$ and then solve for π_B from $\pi_B L = w$. Since L is lower triangular, these are both solvable by back substitution. Hence all the computations in a step of the simplex algorithm can be performed as here, by using the Cholesky factor associated with the basis.

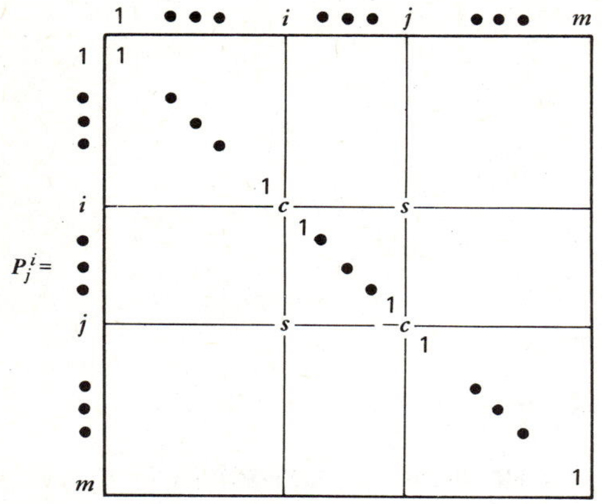

Figure 7.4

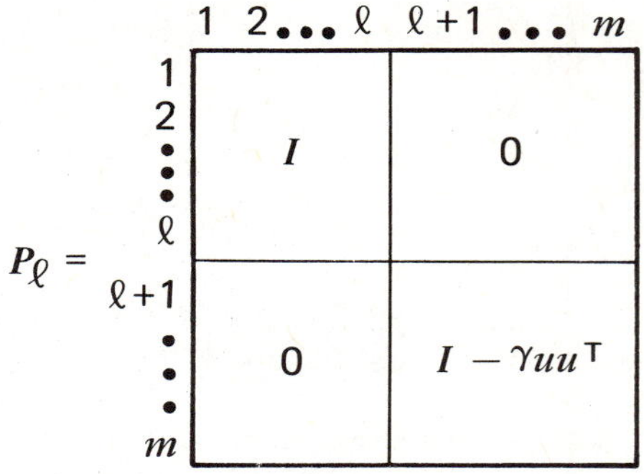

Figure 7.5

7.3.2 To Obtain the Cholesky Factors Associated with a Given Basis

We discuss two commonly used methods for doing this.

Method 1 Using Givens Matrices Let $a = (a_1, \ldots, a_m)^T$ be a given column vector. Consider the element a_i. For some $j \neq i$, if there is an element $a_j \neq 0$, it can be reduced to zero by multiplying the column vector a on the left by an orthogonal matrix P_j^i as in Figure 7.4, where $\gamma = \sqrt{a_i^2 + a_j^2}$, $c = a_i/\gamma$, and $s = a_j/\gamma$. Then $P_j^i a = (a_1, \ldots, a_{i-1}, \gamma, a_{i+1}, \ldots, a_{j-1}, 0, a_{j+1}, \ldots, a_m)^T$. A matrix such as P_j^i is known as a *Givens matrix*. It is a symmetric orthogonal matrix with a very special structure. To find the Cholesky factor associated with a given basis B, start with the matrix B^T. By multiplying the matrix on the left by a sequence of Givens matrices of the form P_j^1, transform all the entries in column 1 below the diagonal into zero. Now by multiplying the resulting matrix on the left by a sequence of Givens matrices of the form P_j^2 transform all the entries in column 2 below the diagonal into zero. Then go to column 3. In a similar manner, repeat the process until the matrix is transformed into an upper triangular matrix. The transpose of the final upper triangular matrix is the Cholesky factor associated with the basis B.

Method 2 Using Householder Matrices Let $a = (a_1, a_2, \ldots, a_m)^T$ be a given column vector. Suppose $(a_{\ell+2}, \ldots, a_m) \neq 0$. Then the bottom $m - \ell - 1$ elements of a can be reduced to zero by multiplying it on the left by a matrix P_ℓ as in Figure 7.5, where,

$$u = (u_1 u_2, \ldots, u_{m-\ell})^T$$
$$u_1 = a_{\ell+1} + s$$
$$u_j = a_{\ell+j}, \qquad \text{for } j = 2, \ldots, m - \ell.$$
$$s = f(a_{\ell+1}) \sqrt{a_{\ell+1}^2 + \cdots + a_m^2}$$
$$\gamma = 1/(u_1 s)$$

$$f(a_{\ell+1}) = \begin{cases} 1 & \text{if } a_{\ell+1} > 0 \\ -1 & \text{if } a_{\ell+1} \leqq 0 \end{cases}$$

A matrix such as P_ℓ is known as a *Householder matrix*. It is a symmetric orthogonal matrix with a special structure. Then $P_\ell a = (a_1, \ldots, a_\ell, -s, 0, \ldots, 0)^T$. To find the Cholesky factor associated with the basis B, start with the matrix $B^{(0)} = B^T$. Go through steps $\ell = 0, 1, \ldots, m - 2$, as discussed below.

Step ℓ If $\ell = 0$, the present matrix is $B^{(0)}$. If $\ell \geqq 1$, suppose this matrix was transformed into $B^{(\ell)}$ at the end of the previous step. Transform the bottom $m - \ell - 1$ entries in the $\ell + 1$th column of $B^{(\ell)}$ into zeros, by multiplying $B^{(\ell)}$ on the left by a Householder matrix of the form P_ℓ. Let $B^{(\ell+1)} = P_\ell B^{(\ell)}$. If $\ell = m - 2$, terminate. Otherwise go to the next step. The matrix $B^{(m-1)}$ at the end of step $m - 2$ is upper triangular. Its transpose is the Cholesky factor associated with the basis B.

Justification: It is well known that a product of orthogonal matrices is an orthogonal matrix. In both these methods, B^T was transformed into an upper triangular matrix R by multiplying B^{T} on the left by a sequence of orthogonal matrices. Let Q be the product (in the proper order) of these orthogonal matrices. So Q is an orthogonal matrix itself. So $QB^{\mathrm{T}} = R$; therefore, $R^{\mathrm{T}}R = BQ^{\mathrm{T}}QB^{\mathrm{T}} = BB^{\mathrm{T}}$. Since R is upper triangular, R^{T} is lower triangular, and this shows that it is the Cholesky factor associated with B.

7.3.3 Updating the Cholesky Factor When One Basic Column Vector Changes

Let B be the present basis and L be the Cholesky factor associated with it. Suppose $A_{.s}$ is the entering column and $A_{.r}$ is the leaving column. Let $\hat{B}$ be the new basis. Then $\hat{B}\hat{B}^{\mathrm{T}} = BB^{\mathrm{T}} - A_{.r}A_{.r}^{\mathrm{T}} + A_{.s}A_{.s}^{\mathrm{T}}$. Each of the matrices $A_{.r}A_{.r}^{\mathrm{T}}$ and $A_{.s}A_{.s}^{\mathrm{T}}$ are matrices of rank one. Hence $\hat{B}\hat{B}^{\mathrm{T}}$ is obtained from BB^{T} by making two changes of rank, one each. The Cholesky factor associated with the new basis is obtained by updating the present Cholesky factor in two stages.

Stage 1 Here the Cholesky factor associated with the matrix obtained by deleting $A_{.r}$ from B is computed. Let $R = L^{\mathrm{T}}$ be the transpose of the Cholesky factor associated with B. Find p by solving $R^{\mathrm{T}}p = A_{.r}$ by back substitution. Let F be the matrix as in Figure 7.6. F is a square matrix of order $m + 1$. By multiplying F on the left by a sequence of Givens matrices (discussed in Section 7.3.2) of the form P_j^1, $j = m + 1, m, \ldots, 2$, in that specific order, transform all the entries in column 1, rows $m + 1$ to 2 of F, into zeros. When the operations are carried out in this specific order, the first row of F gets filled up from right to left. R is modified row by row, but retains its upper triangular structure. Notice that the Givens matrices used here are all of order $m + 1$. At the end, suppose F gets transformed into

$$\begin{pmatrix} \delta & \vdots & v \\ \cdots & \cdots & \cdots \\ 0 & \vdots & \bar{R} \end{pmatrix}$$

The number δ here should be equal to 1. Furthermore, the vector v should be $A_{.r}^{\mathrm{T}}$. Then $\bar{R}^{\mathrm{T}}$ is the Cholesky factor associated with the matrix obtained after the column $A_{.r}$ has been deleted from B. Hence $\bar{R}$ is singular.

Stage 2 Here the Cholesky factor $\bar{R}$, obtained at the end of stage 1, is modified into $\hat{R}$, the Cholesky factor associated with $\hat{B}$. Let E be the $(m + 1) \times m$ matrix

$$E = \begin{pmatrix} \bar{R} \\ \cdots \\ A_{.s}^{\mathrm{T}} \end{pmatrix}$$

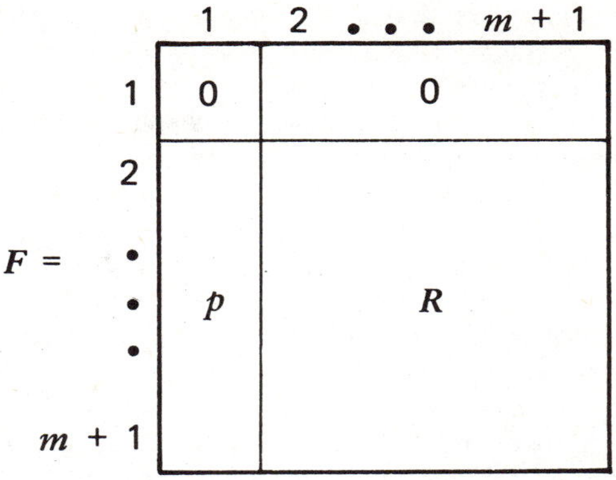

Figure 7.6

By multiplying E on the left by a sequence of Givens matrices (each of order $m + 1$) of the form P_{m+1}^i, in the order $i = 1$ to m, transform all the entries in the last row of E into zero. At the end suppose E is transformed into

$$\begin{pmatrix} \hat{R} \\ \cdots \\ 0 \end{pmatrix}$$

$\hat{R}$ is nonsingular and upper triangular. $\hat{R}$ is the transpose of the Cholesky factor associated with the new basis $\hat{B}$.

Justification for Stage 1: Since R^{T} is the Cholesky factor associated with the basis B, there exists an orthogonal matrix Q such that $QB^{\mathrm{T}} = R$. Since Q is an orthogonal matrix, $Q^{\mathrm{T}} = Q^{-1}$. By the above equation, this implies that $R^{\mathrm{T}}Q = B$. Assume that $A_{.r}$, the dropping column, is the ℓth column of B. Since $R^{\mathrm{T}}p = A_{.r}$ and $R^{\mathrm{T}}Q = B$, we conclude that $p = Q_{.\ell}$, the ℓth column of this orthogonal matrix Q. Since Q is orthogonal, $(Q_{.\ell})^{\mathrm{T}}Q_{.\ell} = 1$, and this implies that $p^{\mathrm{T}}p = 1$. In this stage, F is transformed by multiplying it on the left by an orthogonal matrix, say, $\bar{Q}$. So

$$\bar{Q}\begin{pmatrix} 0 & \vdots & 0 \\ \cdots & \cdots & \cdots \\ p & \vdots & R \end{pmatrix} = \begin{pmatrix} \delta & \vdots & v \\ \cdots & \cdots & \cdots \\ 0 & \vdots & \bar{R} \end{pmatrix}$$

Since $\bar{Q}^{\mathrm{T}}\bar{Q} = I$, we have

$$\begin{pmatrix} 0 & \vdots & p^{\mathrm{T}} \\ \cdots & \cdots & \cdots \\ 0 & \vdots & R^{\mathrm{T}} \end{pmatrix}\begin{pmatrix} 0 & \vdots & 0 \\ \cdots & \cdots & \cdots \\ p & \vdots & R \end{pmatrix} = \begin{pmatrix} \delta & \vdots & 0 \\ \cdots & \cdots & \cdots \\ v^{\mathrm{T}} & \vdots & \bar{R}^{\mathrm{T}} \end{pmatrix}\begin{pmatrix} \delta & \vdots & v \\ \cdots & \cdots & \cdots \\ 0 & \vdots & \bar{R} \end{pmatrix}$$

By writing down the products of these partitioned matrices on both sides, and equating corresponding terms, we conclude that $\delta^2 = p^{\mathrm{T}}p = 1$. From Section

7.3.2, this implies that $\delta = 1$. From the above matrix product, we have $p^{T}R = \delta v = v$, and hence, $v^{T} = A_{.r}$, the dropping column. Also $R^{T}R = v^{T}v + \bar{R}^{T}\bar{R}$, $\bar{R}^{T}\bar{R} = R^{T}R - v^{T}v = BB^{T} - A_{.r}A_{.r}^{T}$. Hence $\bar{R}^{T}$ is the Cholesky factor associated with the matrix obtained after the rank one correction in stage 1.

The justification for stage 2 is based on a similar argument, and it is left to the reader to figure out.

This method of updating is discussed in Saunders [7.21] and it is based on the fact that each Givens matrix used in the updating process is an orthogonal matrix.

There are several other methods for updating the factorization. In some of these methods, the factorization itself is expressed as $BB^{T} = LDL^{T}$, where D is a positive diagonal matrix and L is unit lower triangular, in order to minimize the number of square roots and divisions per iteration. The methods discussed here have been chosen because they are efficient and easy to understand, and because they illustrate the main features of the algorithm using these factorizations, in a simple manner. See the references listed at the end of the chapter for other methods of implementing the factorization.

The simplex algorithm can begin with a unit basis by taking $L = I$ initially. The algorithm can also begin with an arbitrary known basis by computing the Cholesky factor associated with it as in Section 7.3.2. After each step, the Cholesky factor associated with the new basis is obtained by updating in two stages as above.

Cholesky factorization is numerically stable, but it does not preserve sparsity as well as *LU*-decomposition.

7.4 REINVERSIONS

Even under these implementations, the round-off errors accumulate from step to step. At regular intervals (e.g., every 25 or 50 steps) the quantities $\|b - Bx_B\|$ and $\|c_B - \pi_B B\|$ should be computed. If both these numbers are smaller than a specified tolerance (e.g., 10^{-6} or 10^{-7}), the algorithm is continued with the existing *LU*-decomposition or the Cholesky factor. Otherwise the basis corresponding to the present basic vector is read off from the original tableau and its *LU*-decomposition or the Cholesky factor associated with it is computed afresh as in Section 7.2.2 or Section 7.3.2. The present *LU*-decomposition or the Cholesky factor in storage now is replaced by the freshly computed one and the algorithm continued as before. This operation is called *reinversion*. Also in the method using Cholesky factors, reinversion is performed if the quantities $|\delta^2 - 1|$ and $\|A_{.r}^{T} - v\|$ (obtained during stage 1 of updating the Cholesky factor in any step) are not smaller than specified tolerances.

To keep the presentation brief, we have not discussed methods for permuting the rows and columns of the basis at each "reinversion" for preserving sparsity, for stability, and for keeping the computational effort to a minimum. See references [5.2, 7.6, 7.10, 7.14, 7.15, 7.23, 7.25, 7.26] for these methods, and for methods for preprocessing the input-output coefficient matrix A in the problem before applying the simplex method on it, to take advantage of the sparsity of A.

7.5 USE IN COMPUTER CODES

The *LU*-decomposition approach seems to offer the most attractive alternative for implementing the simplex algorithm for solving large-scale linear programming problems. It is numerically stable and helps to preserve sparcity. For this reason, most of the software packages for large-scale linear programming prepared in recent years have adopted this approach.

REFERENCES

7.1 R. H. Bartels and G. H. Golub, "The Simplex Method of Linear Programming Using LU-Decomposition," *Communications of the ACM 12* (1969), 266–268 and 275–278.

7.2 R. H. Bartels, "A Stabilization of the Simplex Method," *Numerische Mathematik 16* (1971), 414–434.

7.3 R. H. Bartels, G. Golub, and M. A. Saunders, "Numerical Techniques in Mathematical Programming," in *Nonlinear Programming*, J. B. Rosen, O. L. Mangasarian, and K. Ritter (Eds.), Academic, New York, 1970, pp. 123–176.

7.4 M. Benichou, J. M. Gauthier, G. Hentges, and G. Ribière, "The Efficient Solution of Large-Scale Linear Programming Problems—Some Algorithmic Techniques and Computational Results," *Mathematical Programming 13* (December 1977), 280–322.

7.5 A. R. Curtis and J. K. Reid, "On the Automatic Scaling of Matrices for Gaussian Elimination," *Journal of the Institute of Mathematics and Its Applications 10* (1972), 118–124.

7.6 J. J. H. Forrest and J. A. Tomlin, "Updating Triangular Factors of the Basis to Maintain Sparsity in the Product Form Simplex Method," *Mathematical Programming 2* (1972), 263–278.

7.7 D. M. Gay, "On Combining the Schemes of Reid and Saunders for Sparse LP Bases," in *Sparse Matrix Proceedings*, I. S. Duff and G. W. Stewart (Eds.), SIAM, Philadelphia, 1978.

7.8 P. E. Gill and W. Murray, "A numerically stable form of the simplex algorithm," *Journal of Linear Algebra and Its Applications 6*, 1973.

7.9 P. E. Gill, G. H. Golub, W. Murray, and M. A. Saunders, "Methods for Modifying Matrix Factorizations," *Mathematics of Computation 28* (1974), 505–535.

7.10 P. E. Gill and W. Murray (Eds.), *Numerical Methods for Constrained Optimization*, Academic, London, 1974.

7.11 P. E. Gill, W. Murray, and M. A. Saunders, "Methods for Computing and Modifying the LDV Factors of a Matrix," *Mathematics of Computation 29* (1975), 1051–1077.

7.12 D. Goldfarb and J. K. Reid, "A Practicable Steepest Edge Simplex Algorithm," *Mathematical Programming 12* (1977), 361–371.

7.13 P. M. J. Harris, "Pivot Selection Methods of the Devex LP Code," *Mathematical Programming Study 4* (1975), 30–57.

7.14 E. Hellerman and D. Rarick, "Reinversion with the Pre-assigned Pivot Procedure," *Mathematical Programming 1* (1971), 195–216.

7.15 E. Hellerman and D. Rarick, "The Partitioned Pre-assigned Pivot Procedure (p4)," in *Sparse Matrices and Their Applications*, D. J. Rose and R. A. Willoughby (Eds.), Plenum, New York, 1972, pp. 67–76.

7.16 H. M. Markowitz, "The Elimination Form of the Inverse and its Application to Linear Programming," *Management Science 3* (1957), 255–269.

7.17 B. A. Murtagh and M. A. Saunders, "MINOS—A Large-Scale Nonlinear Programming System (for Problems with Linear Constraints)—User's Guide," Technical Report SOL 77-9, Department of Operations Research, Stanford University, Stanford, Calif. Feb. 1977.

7.18 IBM Mathematical Programming System Extended/370 (MPSX/370), *Program Reference Manual*, SH 19-1095; Mixed Integer Programming/370 (MIP/370) *Program Reference Manual*, SH19-1099, December 1974.

7.19 A. F. Perold, "A Degeneracy Exploiting *LU*-Factorization for the Simplex Method," *Mathematical Programming 19*, 3 (November 1980) 239–254.

7.20 D. J. Rose and R. A. Willoughby (Eds.), *Sparse Matrices and Their Applications*, Plenum, New York, 1972.

7.21 M. A. Saunders, "Large-Scale Linear Programming Using the Cholesky Factorization," Stan-cs-72-252, Computer Sciences Department, Stanford University, Stanford, Calif. January 1972.

7.22 M. A. Saunders, "Product Form of the Cholesky Factorization for Large Scale Linear Programming," Stan-cs-72-301, Computer Sciences Department, Stanford University, Stanford, Calif., August 1972.

7.23 M. A. Saunders, "A Fast, Stable Implementation of the Simplex Method Using Bartels-Golub Updating," in J. R. Bunch and D. J. Rose (Eds.), *Sparse Matrix Computations*, Academic, New York, 1976, pp. 213–226.

7.24 M. A. Saunders, "MINOS System Manual," Technical Report SOL 77-31, Department of Operation Research, Stanford University, Stanford, Calif. (December 1977).

7.25 J. A. Tomlin, "Pivoting for Size and Sparsity in Linear Programming Inversion Routines," *Journal of the Institute of Mathematics and Its Applications 10* (1972), 289–295.

7.26 J. A. Tomlin, "Modifying Triangular Factors of the Basis in the Simplex Method," in *Sparse Matrices and Their Applications*, D. J. Rose and R. A. Willoughby (Eds.), Plenum, New York, 1972, pp. 77–85.

7.27 J. H. Wilkinson, *The Algebraic Eigenvalue Problem*, Clarendon Press, Oxford, England, 1965.

chapter 8 Parametric Linear Programs

8.1 INTRODUCTION

In practical applications of linear programming, it often happens that the coefficients of the objective function or the right-hand-side constants are not precisely known at the time the problem is being solved. For example, in modeling an automobile manufacturer's problem, the coefficients of the objective function may depend on several parameters, such as the price of steel. The manufacturer may have no control over the values of these parameters, and they may vary from time to time. Each variation changes its objective function, and thus, creates a new problem. In order to plan its strategy well in advance, the manufacturer clearly requires the optimum solution for the model for all possible values of the parameters beforehand.

Example 8.1

Consider the fertilizer product mix problem discussed in Section 1.1. The availability of raw materials 1, 2, 3 at present is $b = (b_1, b_2, b_3)^{\mathrm{T}} = (1500, 1200, 500)^{\mathrm{T}}$. Suppose the company has opened new quarries. When the quarries are in full production, they are expected to provide additional supplies of raw materials 1, 2, 3 in the amounts $b^* = (500, 200, 300)^{\mathrm{T}}$. For several years before they come into full production, the quarries may work only at a fraction of their full production capacity. If the fraction is λ, the availability of raw materials, 1, 2, 3 will be $b + \lambda b^* = (1500 + 500\lambda, 1200 + 200\lambda, 500 + 300\lambda)^{\mathrm{T}}$, $0 \leqq \lambda \leqq 1$. With this the fertilizer product mix problem becomes a parametric right-hand-side problem. Optimum solutions of this problem are required for all values of λ between 0 and 1.

Since only linear models are being considered here, it is assumed that the coefficients of the objective function or the right-hand-side constants vector vary linearly with the parameters. First, consider the case where there is only one parameter, λ. Problems in which the data depend on the values of two or more parameters are much harder to tackle. The problems to be solved are of the following forms:

$$\begin{aligned} \text{Minimize} \quad & z_\lambda(x) = (c + \lambda c^*)x \\ \text{Subject to} \quad & Ax = b \\ & x \geqq 0 \end{aligned} \tag{8.1}$$

and

$$\begin{aligned} \text{Minimize} \quad & z(x) = cx \\ \text{Subject to} \quad & Ax = b + \lambda b^* \\ & x \geqq 0 \end{aligned} \tag{8.2}$$

where A is a matrix of order $m \times n$; b, b^*, c, c^* are vectors of appropriate dimensions; and λ is a parameter that takes real values. The LP (8.1) is known as the *parametric cost problem* and (8.2) is known as the *parametric right-hand-side problem*. LPs in which both cost coefficients and the right-hand-side constants simultaneously depend on the value of a parameter are considered later on.

In the parametric cost problem, the constraints do not involve the parameter, and hence if there are any redundant equality constraints, they can be eliminated. Thus, without any loss of generality, we can assume that the rank of A is m in (8.1).

In the parametric right-hand-side problem, the constraints involve the parameter λ. However, it can be verified from standard results in linear algebra that if either b or b^* or both do not lie in the linear hull of the columns of A, the system $Ax = b + \lambda b^*$ has a solution for at most a single value of the parameter λ. This is a trivial case. Hence we assume that both b and b^* are in the linear hull of the columns of A. Using this, it can be verified that if a constraint in $Ax = b + \lambda b$ is redundant for some value of λ, it is redundant for all values of λ, and hence can be eliminated. Thus even in (8.2), we can, without any loss of generality, assume that the rank of A is m.

In the sequel we therefore assume that the rank of A is m in both (8.1) and (8.2).

Exercises

8.1 Show that every parametric right-hand-side problem is the dual of a parametric cost problem and vice versa.

In (8.1) let $z(x) = cx$, $z^*(x) = c^*x$. Hence, $z_\lambda(x) = z(x) + \lambda z^*(x)$. For any given value of λ, (8.1) is a standard LP and, hence, can be solved by the simplex method or one of its variants. If B is a feasible basis for (8.1), it is an optimum basis for a specified value of λ if all the relative cost coefficients with respect to it are nonnegative for that value of λ. The updating operation can be carried out by treating each cost coefficient as a function of λ, without giving any specific value for it. All the relative cost coefficients will turn out to be functions of λ, and B is an optimum basis for all values of λ that keep all the relative cost coefficients nonnegative. Computing all the relative cost coefficients as functions of λ is made easier by keeping the parametric objective function in two rows of the tableau. One row contains the constant terms of the objective function, namely, the c_j's and this is the row corresponding to $z(x)$. The second row contains all the coefficients of λ in the objective function, namely, the c_j^*'s, and this is the row corresponding to $z^*(x)$. Treating a cost coefficient as a function of λ and pricing it out is equivalent to pricing it out in both the cost rows. If we represent the objective function in this manner, the original tableau for (8.1) is Tableau 8.1. In a similar manner, the original tableau for the parametric right-hand-side problem is Tableau 8.2.

The question of the feasibility of (8.1) does not depend on λ, since λ affects only the objective function in (8.1). Thus, if (8.1) is infeasible for some λ, it is infeasible for all values of λ, and we terminate.

Tableau 8.1 Origin Tableau for the Parametric Cost Problem

x_1	...	x_j	...	x_n	$-z$	$-z^*$	b
a_{11}	...	a_{1j}	...	a_{1n}	0	0	b_1
$\vdots$		$\vdots$		$\vdots$	$\vdots$	$\vdots$	$\vdots$
a_{m1}	...	a_{mj}	...	a_{mn}	0	0	b_m
c_1	...	c_j	...	c_n	1	0	0
c_1^*	...	c_j^*	...	c_n^*	0	1	0

Tableau 8.2 Original Tableau for the Parametric Right-Hand-Side Problem

x_1	...	x_n	$-z$	b	b^*
a_{11}	...	a_{1n}	0	b_1	b_1^*
$\vdots$		$\vdots$	$\vdots$	$\vdots$	$\vdots$
a_{m1}	...	a_{mn}	0	b_m	b_m^*
c_1	...	c_n	1	0	0

If $z(x)$ is unbounded below on the set of feasible solutions of (8.2) for some specified value of λ, it remains unbounded below for every value of λ for which (8.2) is feasible (see Section 4.5.7). Hence, in this case, the only remaining question is to find the set of values of λ for which (8.2) is feasible. This question does not depend on the objective function. Therefore, we can replace the original objective function by the zero objective function $z'(x) = 0$, in which all the cost coefficients are zero. Function $z'(x)$ is always zero and, hence, the modified problem cannot be unbounded. In solving the modified problem, the algorithm will determine the range of values of λ for which (8.2) is feasible.

8.2 THE PARAMETRIC COST SIMPLEX ALGORITHM

This algorithm begins with an optimum basis for (8.1) for some fixed value of the parameter λ, and then solves (8.1) for all values of λ, using only primal simplex pivot steps.

Analysis of the Parametric Cost Problem Given an Optimum Basis for Some λ

Let $x_B = (x_1, \ldots, x_m)$, associated with the basis B, be an optimum basic vector for (8.1) for some specified value of λ, say, λ_0. If the problem is being solved using the canonical tableaux, let the canonical tableau with respect to the basis B be Tableau 8.3. If the problem is being solved using the inverse tableaux, let the inverse tableau corresponding to the basis B be Tableau 8.4.

Here $(\beta_{ij}) = B^{-1}$. If c_B, c_B^* are the cost vectors associated with the basic vector x_B, $\pi = c_B B^{-1}$ and $\pi^* = c_B^* B^{-1}$. The relative cost coefficient of x_j as a function of the parameter λ is $\bar{c}_j + \lambda \bar{c}_j^*$ where $\bar{c}_j = c_j - \pi A_{.j}$ and $\bar{c}_j^* = c_j^* - \pi^* A_{.j}$. Notice that $\bar{c}_j$ and $\bar{c}_j^*$ are the dot products of the row vectors corresponding

Tableau 8.3

Basic Variables	x_1	...	x_m	x_{m+1}	...	x_j	...	x_n	$-z$	$-z^*$	Basic Values
x_1	1	...	0	$\bar{a}_{1,m+1}$	...	$\bar{a}_{1j}$	...	$\bar{a}_{1n}$	0	0	$\bar{b}_1$
$\vdots$	$\vdots$		$\vdots$	$\vdots$		$\vdots$		$\vdots$	$\vdots$	$\vdots$	$\vdots$
x_m	0	...	1	$\bar{a}_{m,m+1}$	...	$\bar{a}_{mj}$	...	$\bar{a}_{mn}$	0	0	$\bar{b}_m$
$-z$	0	...	0	$\bar{c}_{m+1}$	...	$\bar{c}_j$	...	$\bar{c}_n$	1	0	$-\bar{z}$
$-z^*$	0	...	0	$\bar{c}^*_{m+1}$	...	$\bar{c}^*_j$	...	$\bar{c}^*_n$	0	1	$-\bar{z}^*$

Tableau 8.4

Basic Variables	Inverse Tableau					Basic Values
x_1	β_{11}	...	β_{1m}	0	0	$\bar{b}_1$
$\vdots$	$\vdots$		$\vdots$	$\vdots$	$\vdots$	$\vdots$
x_m	β_{m1}	...	β_{mm}	0	0	$\bar{b}_m$
$-z$	$-\pi_1$	...	$-\pi_m$	1	0	$-\bar{z}$
$-z^*$	$-\pi^*_1$	...	$-\pi^*_m$	0	1	$-\bar{z}^*$

to $-z$ and $-z^*$, respectively, in the inverse tableau, with the column vector of x_j in the original tableau. $\bar{z} = c_B B^{-1} b$ and $\bar{z}^* = c^*_B B^{-1} b$.

8.2.1 Optimality Interval or the Characteristic Interval of the Basis *B*

The basis B is an optimum basis for all values of the parameter λ satisfying

$$\bar{c}_j + \lambda \bar{c}^*_j \geqq 0 \qquad \text{for all } j \tag{8.3}$$

Hence a necessary condition for B to be an optimum basis is

$$\bar{c}_j \geqq 0 \qquad \text{for all } j \text{ such that } \bar{c}^*_j = 0 \tag{8.4}$$

If (8.4) is violated, B can never be an optimum basis for (8.1). For B to be an optimum basis for a value of λ, it is not necessary for either $\bar{c}_j$ or $\bar{c}^*_j$ to be nonnegative. Condition (8.3) is the only optimality criterion to be satisfied for B to be optimal. If $\bar{c}^*_j > 0$, (8.3) is equivalent to $\lambda \geqq -\bar{c}_j/\bar{c}^*_j$. If $\bar{c}^*_j < 0$, (8.3) is equivalent to $\lambda \leqq -\bar{c}_j/\bar{c}^*_j$. Let

$$\begin{aligned} \overline{\lambda}_B &= \text{minimum} \left\{ \frac{-\bar{c}_j}{\bar{c}^*_j} : j, \text{ such that } \bar{c}^*_j < 0 \right\} \\ &= +\infty \qquad \text{if} \quad \bar{c}^*_j \geqq 0 \qquad \text{for all } j \end{aligned} \tag{8.5}$$

$$\begin{aligned} \underline{\lambda}_B &= \text{maximum} \left\{ \frac{-\bar{c}_j}{\bar{c}^*_j} : j, \text{ such that } \bar{c}^*_j > 0 \right\} \\ &= -\infty \qquad \text{if} \quad \bar{c}^*_j \leqq 0 \qquad \text{for all } j \end{aligned} \tag{8.6}$$

The value $\underline{\lambda}_B$ is known as the *lower characteristic value* (or the *lower break point*) of the basis B, and $\overline{\lambda}_B$ is known as the *upper characteristic value* (or the *upper breakpoint*) of the basis B. The closed interval $\{\lambda : \underline{\lambda}_B \leqq \lambda \leqq \overline{\lambda}_B\} = [\underline{\lambda}_B, \overline{\lambda}_B]$ is known as the *characteristic interval* or the *optimality interval* of the basis B. For all values of the parameter λ in its characteristic interval, B is an optimum basis for the parametric cost problem (8.1). If $\underline{\lambda}_B > \overline{\lambda}_B$, the optimality interval of the basis B is empty and B can never be an optimal basis. If $\underline{\lambda}_B = \overline{\lambda}_B$, the optimality interval of the basis B contains only a single point. For all values of λ outside the optimality interval of B, B is nonoptimal for (8.1). Hence in exploring for optimal solutions for all λ, once we leave its optimality interval, a basis will never reappear as an optimal basis.

If the optimality interval of the basis B is nonempty, for λ satisfying $\underline{\lambda}_B \leqq \lambda \leqq \overline{\lambda}_B$, an optimal BFS of (8.1) is:

$$\begin{aligned} \text{The basic vector: } x_B &= \bar{b} = B^{-1} b \\ \text{All nonbasic variables} &= 0 \\ \text{Optimum objective value} &= \bar{z} + \lambda \bar{z}^* \\ \text{Optimum dual solution} &= \pi + \lambda \pi^* \end{aligned}$$

Hence in the interval $\underline{\lambda}_B \leqq \lambda \leqq \overline{\lambda}_B$, the optimal feasible solution is the same, but the optimal objective value is a linear function of λ.

For an example illustrating how to compute the optimality interval of a primal feasible basis B, see Example 8.3. If (8.4) is satisfied and $\underline{\lambda}_B < \overline{\lambda}_B$, the optimality interval is $\underline{\lambda}_B \leqq \lambda \leqq \overline{\lambda}_B$, a nonempty interval of positive length; or it may happen that $\underline{\lambda}_B = \overline{\lambda}_B$ (in this case, the optimality interval for the basis B consists of a single point); or $\underline{\lambda}_B > \overline{\lambda}_B$ (in this case, the basis B is not an optimum basis for any value of λ); or $\underline{\lambda}_B = -\infty$ (this case only occurs if $\bar{c}^*_j \leqq 0$ for all j, and the optimality interval for the basis B is the unbounded interval $\lambda \leqq \overline{\lambda}_B$); or $\overline{\lambda}_B = +\infty$ (this case only occurs if $\bar{c}^*_j \geqq 0$ for all j, and the optimality interval for the basis B is the unbounded interval $\lambda \geqq \underline{\lambda}_B$); or (8.4) may not even be satisfied (in this case, B is not an optimum basis for any value of λ).

8.2.2 Solving the Problem for $\lambda > \overline{\lambda}_B$

Suppose B is a primal feasible basis for (8.1) with a nonempty optimality interval $\underline{\lambda}_B \leqq \lambda \leqq \overline{\lambda}_B$ and suppose $\overline{\lambda}_B$ is finite. If it is required to solve (8.1) for $\lambda > \overline{\lambda}_B$, identify a j that attains the minimum in (8.5), breaking ties arbitrarily. Suppose this is $j = s$. Then $\overline{\lambda}_B = -\bar{c}_s/\bar{c}_s^*$. When λ exceeds $\overline{\lambda}_B$, the relative cost coefficient of x_s becomes negative. Hence to solve the problem for $\lambda > \overline{\lambda}_B$, bring x_s into the present basic vector. See Note 8.1. When this is done, two cases might occur.

Case 1 The unboundedness criterion may be satisfied. This happens if the updated column vector of x_s is such that $\bar{a}_{is} \leqq 0$ for all i. In this case, the parametric objective function $z_\lambda(x)$ is unbounded below on the set of feasible solutions of (8.1) whenever $\lambda > \overline{\lambda}_B$.

Case 2 If the unboundedness criterion is not satisfied, bring x_s into the basic vector, as in the simplex algorithm or the revised primal simplex algorithm. Suppose the rth basic variable, x_r, drops out of the basic vector and is replaced by x_s. Remember to price out the pivot column in both the cost rows. Let the new basis be $\tilde{B}$.

Exercises

8.2 Prove that the relative cost coefficients of any variable x_j with respect to the bases B and $\tilde{B}$ are both equal when $\lambda = \overline{\lambda}_B$. Thus show that B, $\tilde{B}$ are two alternate optimal bases for (8.1) when $\lambda = \overline{\lambda}_B$.

8.3 Let $\bar{c}_j$, $\bar{c}_j^*$, $\bar{a}_{ij}$ denote the updated entries with respect to the basis B, and $\tilde{c}_j$, $\tilde{c}_j^*$, $\tilde{a}_{ij}$ denote the updated entries with respect to the basis $\tilde{B}$. Prove that the relative cost coefficient of the dropping variable x_r with respect to the new basis $\tilde{B}$ is $\tilde{c}_r + \lambda\tilde{c}_r^* = -(\bar{c}_s + \lambda\bar{c}_s^*)/\bar{a}_{rs}$. Hence show that $\tilde{B}$ is not an optimum basis for any $\lambda < \overline{\lambda}_B$.

8.4 By using the results in Exercises 8.2 and 8.3, prove that the lower characteristic point of the new basis $\tilde{B}$ is $\overline{\lambda}_B$.

The optimality interval of $\tilde{B}$ is another closed interval that has its left end point coinciding with the right end point ($= \overline{\lambda}_B$) of the optimality interval of the previous basis B. The same procedure can be repeated with the basis $\tilde{B}$, and optimum solutions of (8.1) determined as λ increases further. For an illustrative example, see Example 8.3.

8.2.3 Solving the Problem for $\lambda < \underline{\lambda}_B$

Returning to the original basis B, suppose $\underline{\lambda}_B$ is finite. If it is required to solve (8.1) for $\lambda < \underline{\lambda}_B$, identify a j that attains the maximum in (8.6). Suppose this is $j = t$. Then bring x_t into the basic vector x_B. See Note 8.1. If the unboundedness criterion is satisfied, $z_\lambda(x)$ is unbounded below on the set of feasible solutions of (8.1) for all $\lambda < \underline{\lambda}_B$. Otherwise, bring x_t into the basic vector and suppose the new basis is $\hat{B}$. From results similar to those in Section 8.2.2, it is clear that B, $\hat{B}$ are alternate optimal bases for (8.1) when $\lambda = \underline{\lambda}_B$. The optimality interval of $\hat{B}$ is another closed interval with its right end point equal to the left end point ($= \underline{\lambda}_B$) of the optimality interval of B. Repeating the same procedure, optimum solutions of (8.1) can be obtained as λ decreases further.

Note 8.1 *CYCLING* To obtain optimum solutions of (8.1) for $\lambda > \overline{\lambda}_B$, once the entering variable x_s is determined as in Section 8.2.2., the dropping basic variable that it will replace has to be determined by the primal simplex minimum ratio test in order to retain primal feasibility. When there is a tie, the dropping variable can be chosen arbitrarily among those eligible to drop, but this could result in cycling at this value of $\lambda = \overline{\lambda}_B$. The algorithm could then keep going indefinitely through a cycle of feasible bases, everyone of which is only optimal for $\lambda = \overline{\lambda}_B$, without ever being able to increase the value of λ beyond $\overline{\lambda}_B$, even though optimum solutions of (8.1) may actually exist for $\lambda > \overline{\lambda}_B$. For an example of such cycling, see Section 10.7. This cycling is a rare phenomenon; it does not seem to occur in practical applications, even with an arbitrary choice of the dropping basic variable among those eligible. But the possibility of cycling clearly exists, as indicated by the examples given in Section 10.7. In Section 10.7, we also discuss an efficient method for resolving this problem of cycling. The same comments hold for obtaining optimum solutions of (8.1) for values of $\lambda < \underline{\lambda}_B$. These methods for resolving cycling may

Basic Variables	Inverse Tableau					Basic Values	Pivot Column x_5	Ratios
x_1	−3	2	−1	0	0	6	1	6
x_2	−6	3	−1	0	0	3	−1	
x_3	4	−2	1	0	0	2	(1)	2 minimum
$-z$	−4	1	−1	1	0	−23	3	
$-z^*$	−8	4	−3	0	1	−16	−2	

not be important in practical applications, but they are mathematically important to guarantee that the parametric cost simplex algorithm solves (8.1) for all real values of λ in a finite number of pivot steps.

8.3 TO FIND AN OPTIMUM BASIS FOR THE PARAMETRIC COST PROBLEM

Solve (8.1) for some value of λ, say, $\lambda = \lambda_1(\lambda_1$ could be 0) first, either by the simplex method or a variant of it. If (8.1) has an optimum feasible solution for λ_1, this will terminate with an optimum feasible basis for $\lambda = \lambda_1$. However, it may happen that $z_\lambda(x)$ is unbounded below on the set of feasible solutions of (8.1) when $\lambda = \lambda_1$. In this case, the simplex method terminates with a feasible basis B that satisfies the unboundedness criterion when $\lambda = \lambda_1$. Let $\bar{c}_j$, $\bar{c}_j^*$, $\bar{a}_{ij}$ denote the updated entries with respect to the basis B. Hence, there exists a nonbasic variable x_s, such that $\bar{c}_s + \lambda_1\bar{c}_s^* < 0$ and $\bar{a}_{is} \leqq 0$ for all i. Clearly (8.1) remains unbounded below for all λ satisfying $\bar{c}_s + \lambda\bar{c}_s^* < 0$. Therefore, if $\bar{c}_s^* = 0$, and $\bar{c}_s < 0$, (8.1) is unbounded below for all λ, and we terminate. Otherwise, let $\lambda_2 = -\bar{c}_s/\bar{c}_s^*$. If $\bar{c}_s^* > 0$, (8.1) is unbounded below for all $\lambda < \lambda_2$. If $\bar{c}_s^* < 0$, (8.1) is unbounded below for all $\lambda > \lambda_2$. In either case, change the value of the parameter to λ_2 and try to solve the problem again. This is done by continuing the simplex algorithm starting from the basis B, in which the relative cost coefficients are given by $\bar{c}_j + \lambda_2\bar{c}_j^*$. If the unboundedness criterion is satisfied again, change the value of λ again and continue the iterations of the simplex (or the revised simplex) algorithm with the new value of λ. In a finite number of iterations we will either find a value of λ and an optimal feasible basis corresponding to it, or conclude that the problem is unbounded for all values of λ.

Example 8.2

Consider the following parametric cost LP.

x_1	x_2	x_3	x_4	x_5	x_6	x_7	$-z$	$-z^*$	b
1	0	1	−2	2	−1	1	0	0	8
2	1	3	−5	4	−2	3	0	0	21
0	2	3	−2	1	−1	2	0	0	12
2	1	4	−7	8	−3	9	1	0	0
0	2	5	−1	1	−2	1	0	1	0

$x_j \geqq 0$ for all j, minimize $z + \lambda z^*$.

We had to first find a BFS to this problem by solving a Phase I problem. This led to the feasible basic vector (x_1, x_2, x_3). The inverse tableau corresponding to this basic vector is given at the top.

The updated cost vectors for the problem with respect to this basic vector (computed by the formulas discussed in Section 8.2) are given at the bottom. Since $\underline{\lambda} > \bar{\lambda}$, this basic vector (x_1, x_2, x_3) is not optimal to the problem for any value of λ. So we select a value for λ, say $\lambda = 0$, and try to find an optimum solution when $\lambda = 0$. Since $\bar{c}_4 = -2 < 0$, x_4 is the entering variable for getting an optimum solution

$\bar{c}$		0	0	0	−2	3	0	6	
$\bar{c}^*$		0	0	0	1	−2	1	−1	
$\frac{-\bar{c}_j}{\bar{c}_j^*}$	for $\bar{c}_j^* < 0$					$\frac{3}{2}$		6	minimum $= \bar{\lambda} = \frac{3}{2}$
$\frac{-\bar{c}_j}{\bar{c}_j^*}$	for $\bar{c}_j^* > 0$				2		0		maximum $= \underline{\lambda} = 2$

when $\lambda = 0$. The pivot column is the updated column vector of x_4, and it is $(-2, -1, 0; -2, 1)$. Hence, when $\lambda = 0$, the unboundedness criterion is satisfied, and $z(x)$ is unbounded below on the set of feasible solutions of this problem. Verify that the unboundedness criterion is satisfied in the updated column of x_4 for all values of λ satisfying $-2 + \lambda < 0$, that is, $\lambda < 2$. Hence for all $\lambda < 2$, $z(x) + \lambda z^*(x)$ is unbounded below on the set of feasible solutions of this problem. Now change the value of λ to 2. The vector of relative cost coefficients is $\bar{c} + 2\bar{c}^* = (0, 0, 0, 0, -1, 2, 4)$. So, select x_5 as the entering variable. The pivot column is the updated column of x_5 and it is already on the inverse tableau. The new inverse tableau, obtained after the pivot step, is:

Basic Variables	*Inverse Tableau*					*Basic Values*
x_1	−7	4	−2	0	0	4
x_2	−2	1	0	0	0	5
x_5	4	−2	1	0	0	2
$-z$	−16	7	−4	1	0	−29
$-z^*$	0	0	−1	0	1	−12

It can be verified that condition (8.4) is now satisfied and that (x_1, x_2, x_5) is an optimum feasible basic vector for the problem for all $\underline{\lambda} = 2 \leqq \lambda \leqq 3 = \bar{\lambda}$.

8.4 SUMMARY OF THE RESULTS ON THE PARAMETRIC COST PROBLEM

1 The characteristic intervals of the optimal bases obtained in the course of the algorithm do not overlap (except at end points). The characteristic intervals and the intervals in which the objective function is unbounded form a partition of the real line.

2 All values of λ for which the objective function $z_\lambda(x)$ is unbounded below on the set of feasible solutions of (8.1) are either in open intervals of the form $(-\infty, \lambda^*)$ or $(\lambda^*, +\infty)$, for some λ^*.

3 In exploring for optimum solutions of (8.1) for all λ, once λ leaves the characteristic interval of some optimum basis, that basis never reappears as an optimum basis. Also, as mentioned in Note 8.1, when the methods discussed in Section 10.7 are used, in the algorithmic steps discussed in Sections 8.2.2 and 8.2.3, cycling cannot occur at any fixed value of λ. The total number of bases is finite. These facts together imply that the total number of characteristic intervals obtained in the course of the algorithm will be finite. Thus, (8.1) can be solved for all real values of λ in a finite number of pivot steps, by the parametric cost simplex algorithm discussed previously, with the use of anticycling rules discussed in Section 10.7 for resolving ties in the dropping variable choice in steps discussed in Sections 8.2.2 and 8.2.3.

4 In each characteristic interval, the optimum feasible solution is fixed, but the optimum objective value varies linearly with λ. Also the optimum objective values at the common end point of two consecutive characteristic intervals are equal. Hence, the optimal objective value of a parametric cost problem is a piecewise linear continuous function in the parameter λ.

5 In each characteristic interval, the optimum dual solution obtained is a piecewise linear vector function of the parameter λ.

8.5 THE SLOPE OF THE OPTIMUM OBJECTIVE VALUE FUNCTION IN A PARAMETRIC COST LINEAR PROGRAM

Let $g(\lambda)$ denote the optimum objective value in (8.1) as a function of the parameter λ. When (8.1) is solved, we get the entire curve of $g(\lambda)$ for all λ. In some cases we may be interested in just finding the slope of $g(\lambda)$ at a specified point $\lambda = \lambda_0$. Here we discuss an efficient method for computing just that. As discussed above, $g(\lambda)$ is a piecewise linear function, and hence there may be values of λ where the slope of $g(\lambda)$ changes. If λ_0 is one such point, there may be two slopes for $g(\lambda)$ at λ_0, one for values of λ increasing through λ_0 denoted by $g'(\lambda_{0_+})$, another for values of λ reaching to λ_0 from below denoted by $g'(\lambda_{0_-})$. We want to compute both these slopes efficiently.

First solve (8.1) fixing $\lambda = \lambda_0$. Suppose Tableau 8.3 is an optimum canonical tableau for (8.1) at $\lambda = \lambda_0$, associated with the basic vector $x_B = (x_1, \ldots, x_m)$. Compute the optimality interval of x_B, and suppose it is $[\underline{\lambda}_B, \bar{\lambda}_B]$. If λ_0 is an interior point of this interval, that is, $\underline{\lambda}_B < \lambda_0 < \bar{\lambda}_B$, then $g(\lambda)$ is differentiable at λ_0 and its slope (or derivative) at λ_0 is equal to $\bar{z}^* = c_B^* B^{-1} b$.

Now consider the case where $\lambda_0 = \bar{\lambda}_B$. Let $\mathbf{J} = \{j : \bar{c}_j + \lambda_0 \bar{c}_j^* = 0\}$ and $\bar{\mathbf{J}} = \{j : \bar{c}_j + \lambda_0 \bar{c}_j^* > 0\}$. If there are some present nonbasic variables x_j with $j \in \mathbf{J}$, there may be alternate optimum bases in (8.1) at $\lambda = \lambda_0$, and by the results in Section 3.10, it is enough to restrict our attention to variables among x_j for $j \in \mathbf{J}$ to get such bases. If $x_{\tilde{B}}$ is any basic vector for (8.1) in which all the basic variables are from the set $\{x_j: j \in \mathbf{J}\}$ and $\tilde{c}$, $\tilde{c}^*$ are the updated cost vectors with respect to $x_{\tilde{B}}$, then we have $\tilde{c}_j + \lambda_0 \tilde{c}_j^* = \bar{c}_j + \lambda_0 \bar{c}_j^*$ for all j ($x_{\tilde{B}}$ can be obtained from x_B through a sequence of pivot steps in only columns j for $j \in \mathbf{J}$. Since $\bar{c}_j + \lambda_0 \bar{c}_j^* = 0$ for all $j \in \mathbf{J}$, each of these pivot steps will leave the row vector $\bar{c} + \lambda_0 \bar{c}^*$ unchanged, as the entry in this row in each pivot column is zero).

To compute $g'(\lambda_{0_+})$ in this case, we should obtain a basic vector for (8.1) that is optimal at $\lambda = \lambda_0$ and remains optimal for values of λ slightly greater than λ_0. Let $x_{\tilde{B}}$ be such a basic vector in which all the basic variables are from the set $\{x_j: j \in \mathbf{J}\}$. Let $\tilde{B}$ be the associated basis, and let $\tilde{c}$ and $\tilde{c}^*$ be the updated cost vectors with respect to it. Let $c_{\tilde{B}}$ and $c_{\tilde{B}}^*$ be the row vectors of original cost coefficients of the basic variables in $x_{\tilde{B}}$. By the previous discussion $\tilde{c}_j + \lambda_0 \tilde{c}_j^* = \bar{c}_j + \lambda_0 \bar{c}_j^*$ for all j. So $\tilde{c}_j + \lambda_0 \tilde{c}_j^* > 0$ for all $j \in \bar{\mathbf{J}}$, and this implies that $\tilde{c}_j + \lambda \tilde{c}_j^* \geqq 0$ for all $j \in \bar{\mathbf{J}}$ for λ slightly greater than λ_0. However, $\tilde{c}_j + \lambda_0 \tilde{c}_j^* = \bar{c}_j + \lambda_0 \bar{c}_j^* = 0$ for $j \in \mathbf{J}$, and $\tilde{c}_j + \lambda \tilde{c}_j^* \geqq 0$ for λ slightly larger than λ_0 will only hold if $\tilde{c}_j^* \geqq 0$ for all $j \in \mathbf{J}$. These facts clearly imply that $x_{\tilde{B}}$ is an optimum basic vector for the LP

$$\begin{array}{lll} \text{Minimize} & \sum_{j \in \mathbf{J}} c_j^* x_j & \\ \text{Subject to} & \sum_{j \in \mathbf{J}} A_{.j} x_j = b & \\ x_j \geqq 0 & \text{for all } j \in \mathbf{J} & \end{array}$$

This is equivalent to the LP

$$\begin{array}{ll} \text{Minimize} & c^* x \\ \text{Subject to} & Ax = b \\ x_j = 0 & \text{for all } j \in \bar{\mathbf{J}} \\ x_j \geqq 0 & \text{for all } j \in \mathbf{J} \end{array}$$

These conditions imply that $\tilde{B}$ is an alternate optimal basis for (8.1) at $\lambda = \lambda_0$ and that the upper characteristic point of $\tilde{B}$ is $> \lambda_0$. Hence the slope of $g(\lambda)$ for values of λ increasing through λ_0, $g'(\lambda_{0_+})$ is $c_{\tilde{B}}^* \tilde{B}^{-1} b$, which is the optimum objective value in the above LP. From the results in Section 3.10, it is clear that the set of feasible solutions of the above LP is the set of optimum solutions of (8.1) at $\lambda = \lambda_0$. Thus $g'(\lambda_{0_+})$ is the minimum value of c^*x over the set of optimum solutions of (8.1) at $\lambda = \lambda_0$.

Suppose the objective function in the above LP is unbounded below. This happens if there exists a feasible basic vector, $x_{\hat{B}}$, say, for (8.1), consisting of basic variables from the set $\{x_j: j \in \mathbf{J}\}$ and a nonbasic variable x_s, $s \in \mathbf{J}$, whose updated column with respect to this basic vector $(\hat{a}_{1s}, \ldots, \hat{a}_{ms}, \hat{c}_s, \hat{c}_s^*)^{\mathrm{T}}$ satisfies the property that $\hat{c}_s^* < 0$ and $\hat{a}_{is} \leqq 0$ for all $i = 1$ to m. In this case, continuing the parametric simplex algorithm of Section 8.2 with the feasible basic vector $x_{\hat{B}}$ optimal at $\lambda = \lambda_0$, we conclude that $g(\lambda) = -\infty$ for $\lambda > \lambda_0$, through the column of x_s.

In the same manner, if $\lambda = \underline{\lambda}_B$, to find $g'(\lambda_{0_-})$ we should find a feasible basic vector x_{B^1} for (8.1) in which all the basic variables are from the set $\{x_j: j \in \mathbf{J}\}$, and the updated c_j^* is $\leqq 0$ for all $j \in \mathbf{J}$. This leads to the conclusion that

$$\begin{array}{lll} g'(\lambda_{0_-}) = \text{Maximum value of} & & c^* x \\ \text{Subject to} & & Ax = b \\ x_j = 0 & \text{for all } j \in \bar{\mathbf{J}} & \\ x_j \geqq 0 & \text{for all } j \in \mathbf{J} & \end{array}$$

If the objective value in this LP is unbounded above (which happens when there exists a feasible basic vector, x_{B^2}, say, for (8.1), consisting of basic variables from the set $\{x_j: j \in \mathbf{J}\}$ and a nonbasic variable x_t, $t \in \mathbf{J}$, whose updated column with respect to this basic vector, $(a_{1t}^2, \ldots, a_{mt}^2, c_t^2, c_t^{*2})^{\mathrm{T}}$, satisfies $c_t^{*2} > 0$, $a_{ts}^2 \leqq 0$, for all $i = 1$ to m), we can, as before, conclude that $g(\lambda) = -\infty$ for $\lambda < \lambda_0$. So, we see that $g'(\lambda_{0_-})$ is the maximum value of c^*x over the set of optimum solutions of (8.1) at $\lambda = \lambda_0$.

We summarize these facts in the following theorem.

THEOREM 8.1 Suppose (8.1) is feasible and λ_0 is such that the optimum objective value in (8.1) at $\lambda = \lambda_0$, $g(\lambda_0)$ is finite. Then the following results hold. (1) Let $\alpha =$ minimum value of c^*x over the set of optimum solutions of (8.1) at $\lambda = \lambda_0$. If $\alpha = -\infty$, $g(\lambda) = -\infty$ for $\lambda > \lambda_0$. If α is finite, it is the slope $g'(\alpha_{0+})$ of $g(\lambda)$ as λ increases through λ_0. (2) Let $\gamma =$ maximum value of c^*x over the set of optimum solutions of (8.1) at $\lambda = \lambda_0$. If $\gamma = +\infty$, $g(\lambda) = -\infty$ for $\lambda < \lambda_0$. If γ is finite, it is the slope $g'(\lambda_{0-})$ of $g(\lambda)$ as λ reaches λ_0 from below.

With A, b the same as in (8.1), denote the optimum objective value in the LP: minimize px, subject to $Ax = b$, $x \geqq 0$, as a function of the cost vector p by the symbol G(p). Then, in the notation previously used, we have $g(0) = \text{G}(c + 0c^*) = \text{G}(c)$. The quantity $g'(0_+)$ is the limit of $(\text{G}(c + \varepsilon c^*) - \text{G}(c))/\varepsilon$, as ε tends to zero through positive values. This $g'(0_+)$ is known as the *directional derivative* of the optimum objective value function G(p) in the direction c^* at the point $p = c$.

8.6 PARAMETRIC RIGHT-HAND-SIDE SIMPLEX ALGORITHM

This algorithm begins with an optimum basis for (8.2) for some fixed value of the parameter λ, and then solves (8.2) for all values of λ using only dual simplex pivot steps.

Analysis of the Parametric Right-Hand-Side Problem Given an Optimal Basis for Some λ

Let x_B be an optimum basic vector for (8.2) for some specified value of λ, say λ_0, and let B be the associated basis. Let c_B be the corresponding basic cost vector. Let $\bar{b} = B^{-1}b$, $\bar{b}^* = B^{-1}b^*$, $\bar{z} = c_B B^{-1}b$, and $\bar{z}^* = c_B B^{-1}b^*$. The BFS corresponding to the basis B as a function of λ is tabulated as in Tableau 8.5.

Since B is an optimum basis for (8.2) when $\lambda = \lambda_0$, B must be dual feasible. The parameter λ appears only in the right-hand-side constant vector and, hence, dual feasibility of a basis for (8.2) is a property that is independent of λ. Hence, B is an optimum basis for (8.2) for all values of λ for which it is primal feasible, that is, for all λ satisfying $\bar{b}_i + \lambda\bar{b}_i^* \geqq 0$ for all i. Since B is primal feasible when $\lambda = \lambda_0$, we must have $\bar{b}_i + \lambda_0\bar{b}_i^* \geqq 0$ for all i. This implies that

$$\bar{b}_i \geqq 0 \qquad \text{for all } i, \text{ such that } \bar{b}_i^* = 0 \tag{8.7}$$

If (8.7) is not satisfied, B is not even primal feasible and it can never be optimal. $\bar{b}_i + \lambda\bar{b}_i^* \geqq 0$ for all i implies that $\lambda \geqq -\bar{b}_i/\bar{b}_i^*$ for all i, such that $\bar{b}_i^* > 0$, and that $\lambda \leqq -\bar{b}_i/\bar{b}_i^*$ for all i, such that $\bar{b}_i^* < 0$. Let

$$\begin{aligned} \bar{\lambda}_B &= \text{minimum } \{-\bar{b}_i/\bar{b}_i^* : i, \text{ such that } \bar{b}_i^* < 0\} \\ &= +\infty, \qquad \text{if} \quad \bar{b}_i^* \geqq 0, \text{ for all } i \end{aligned} \tag{8.8}$$

$$\begin{aligned} \underline{\lambda}_B &= \text{maximum } \{-\bar{b}_i/\bar{b}_i^* : i, \text{ such that } \bar{b}_i^* > 0\} \\ &= -\infty, \qquad \text{if} \quad \bar{b}_i^* \leqq 0, \text{ for all } i \end{aligned} \tag{8.9}$$

If $\underline{\lambda}_B > \bar{\lambda}_B$, B is never primal feasible for any λ. If $\underline{\lambda}_B \leqq \bar{\lambda}_B$, then the closed interval $\underline{\lambda}_B \leqq \lambda \leqq \bar{\lambda}_B$; that is, $[\underline{\lambda}_B, \bar{\lambda}_B]$ is the *characteristic interval* or *the optimality interval* of the basis B. For all values of λ in this optimality interval, the BFS listed in the Tableau 8.5 is an optimum feasible solution of (8.2). Hence, in this interval, the optimum objective value and the values of the basic variables in the optimum BFS vary linearly with λ.

For an example illustrating how to compute the optimality interval of a dual feasible basis, see Example 8.4. In general, the optimality interval of a dual feasible basis B_1 may turn out to be a nonempty interval of positive length, or a single point, or empty, or an unbounded interval, depending on the values of $\underline{\lambda}_{B_1}$ and $\bar{\lambda}_{B_1}$ and whether (8.7) is satisfied or not.

To Solve the Problem for $\lambda > \bar{\lambda}_B$

If $\bar{\lambda}_B$ is finite and if it is required to solve (8.2) for values of $\lambda > \bar{\lambda}_B$, identify the value of i that attains the minimum in (8.8). Suppose it is $i = r$. Then $\bar{\lambda}_B = -\bar{b}_r/\bar{b}_r^*$ and for $\lambda > \bar{\lambda}_B$, $\bar{b}_r + \lambda\bar{b}_r^* < 0$. Let $\bar{A}_{r.} = (\bar{a}_{r1}, \ldots, \bar{a}_{rn})$, be the updated rth row vector with respect to the basis B. If $\bar{A}_{r.} \geqq 0$, that is, $\bar{a}_{rj} \geqq 0$ for all j, the problem is clearly infeasible whenever $\lambda > \bar{\lambda}_B$. This is the primal infeasibility criterion. If the primal infeasibility criterion is not satisfied, make a dual simplex pivot step in the rth row (see Note 8.2). Suppose it gives a consecutive optimal basis $\tilde{B}$.

Tableau 8.5 Basic Values as Functions of λ

Basic Variables	*Updated Right-Hand-Side Vectors*		*Basic Values*
	$\bar{b}_1$	$\bar{b}_1^*$	$\bar{b}_1 + \lambda\bar{b}_1^*$
	$\vdots$	$\vdots$	$\vdots$
x_B	$\bar{b}_i$	$\bar{b}_i^*$	$\bar{b}_i + \lambda\bar{b}_i^*$
	$\vdots$	$\vdots$	$\vdots$
	$\bar{b}_m$	$\bar{b}_m^*$	$\bar{b}_m + \lambda\bar{b}_m^*$
$-z$	$-\bar{z}$	$-\bar{z}^*$	$-\bar{z} - \lambda\bar{z}^*$

Exercises

8.5 Prove that both B and $\tilde{B}$ are alternate optimal bases when $\lambda = \bar{\lambda}_B$. Also prove that $\underline{\lambda}_{\tilde{B}} = \bar{\lambda}_{\tilde{B}}$.

The next optimal basis $\tilde{B}$ has its characteristic interval to the right of $\bar{\lambda}_B$. The same procedure is repeated with the new basis $\tilde{B}$ until optimal feasible solutions are obtained for all required $\lambda > \bar{\lambda}_B$. For an illustrative example, see Example 8.4.

To Solve the Problem for $\lambda < \underline{\lambda}_B$

Returning to the original basis B, suppose $\underline{\lambda}_B$ is finite and we have to solve the problem for $\lambda < \underline{\lambda}_B$. Identify the i that attains the maximum in (8.9). Suppose it is $i = t$. If the updated tth row, $\bar{A}_{t.} \geqq 0$, (8.2) is primal infeasible for all $\lambda < \underline{\lambda}_B$. Otherwise a dual simplex pivot step is made in the tth row, and the procedure continued in a similar manner (see Note 8.2).

Note 8.2: *CYCLING* As pointed out in Note 8.1 for the parametric cost simplex algorithm, in trying to get optimum solutions of (8.2) for $\lambda > \bar{\lambda}_B$, if there is a tie in the dual simplex minimum ratio test for the choice of the pivot column, and the choice is made arbitrarily, cycling can occur at that value of $\lambda = \bar{\lambda}_B$ here, too. The same thing can happen when trying to get optimum solutions of (8.2) for $\lambda < \underline{\lambda}_B$. Methods for resolving this problem of cycling are discussed in Section 10.7.

8.7 TO FIND AN INITIAL OPTIMUM BASIS FOR THE PARAMETRIC RIGHT-HAND-SIDE PROBLEM GIVEN A FEASIBLE BASIS FOR SOME λ

Let B be a feasible basis for (8.2) for some specified value of λ, say, λ_1. Fix $\lambda = \lambda_1$ (i.e., treat the right-hand constant vector as equal to $b + \lambda_1 b^*$). Starting with the feasible basis B, minimize $z(x)$ by applying either the simplex (or the revised simplex) algorithm. Two possibilities may occur. (1) An optimal feasible solution exists when $\lambda = \lambda_1$. Then the simplex (or the revised simplex) algorithm will find an optimal basis for $\lambda = \lambda_1$. (2) The simplex algorithm may terminate by satisfying the unboundedness criterion for $\lambda = \lambda_1$. In case 2, $z(x)$ is unbounded below on the set of feasible solutions of (8.2) for all values of the parameter λ for which (8.2) is feasible. To determine the range of values of λ for which (8.2) is feasible, replace all the cost coefficients by zero. From now on, all the relative cost coefficients will be equal to zero. For this zero objective function, the final basis obtained under the simplex algorithm is optimal.

8.8 TO FIND A FEASIBLE BASIS FOR THE PARAMETRIC RIGHT-HAND-SIDE PROBLEM

Pick an arbitrary values for λ, say λ_1 (λ_1 could be 0), and try to find a feasible basis. In this case, (8.2) is an LP with the right-hand-side constant vector $b + \lambda_1 b^*$. Write down all the constraints in (8.2) in such a way that the right-hand-side constants vector is nonnegative when $\lambda = \lambda_1$, that is, if $b_i + \lambda_1 b_i^* < 0$, multiply both sides of the ith constraint in (8.2) by -1, for each i. If the original tableau is canonical, a feasible unit basis is available for the case when $\lambda = \lambda_1$. If the tableau is not canonical, augment the tableau with a full artificial basis and formulate the corresponding Phase I problem as below.

$$\begin{aligned} \text{Minimize} \quad & w(x, y) = \sum_{i=1}^{m} y_i \\ \text{Subject to} \quad & Ax + Iy = b + \lambda b^* \\ & x \geqq 0 \qquad y \geqq 0 \end{aligned} \tag{8.10}$$

Starting with the artificial basic vector, solve the Phase I parametric problem by the procedures in Sections 8.6 and 8.7. There are two possibilities here. (1) It may turn out that the minimum objective value of the parametric Phase I problem is positive for all λ. In this case, the original problem (8.2) is infeasible for all λ. (2) A value of the parameter may be found, say, λ_2, such that when $\lambda = \lambda_2$, the Phase I objective function has a minimum value of 0. In this case, the optimal basis for (8.10) when $\lambda = \lambda_2$ provides a feasible basis for (8.2). For an illustrative example, see Example 8.4.

8.9 SUMMARY OF THE RESULTS ON THE PARAMETRIC RIGHT-HAND-SIDE PROBLEM

1. The characteristic intervals of the optimal bases obtained in the course of the algorithm do not overlap (except at end points).
2. All the values of λ for which (8.2) is infeasible are either in open intervals of the form $(-\infty, \lambda)$ or $(\lambda, +\infty)$, for some λ.
3. The characteristic intervals of the optimal bases obtained in the course of the algorithm and the open intervals in which (8.2) is infeasible, form a partition of the real line.
4. In exploring for optimum solutions of (8.2) for all λ, once λ leaves the characteristic interval of some optimum basis, that basis never reappears as an optimum basis (since it will not even be primal feasible). Also, as mentioned in Note 8.2, when the methods discussed in Section 10.7 are used for choosing the pivot column, cycling cannot

occur at any fixed value of λ. Since the total number of bases is finite, the total number of characteristic intervals obtained in the course of the algorithm will be finite. Thus, (8.2) can be solved for all real values of λ in a finite number of pivot steps, by the parametric right-hand-side simplex algorithm previously discussed, with the use of anticycling rules discussed in Section 10.7.

5 In each interval, the optimal feasible solution and the optimum objective value vary linearly in the parameter λ, but the optimum dual solution is fixed. The optimal objective value is a piecewise linear continuous function in the parameter, λ.

8.10 THE SLOPE OF THE OPTIMUM OBJECTIVE VALUE FUNCTION IN PARAMETRIC RIGHT-HAND-SIDE LINEAR PROGRAMS

Let $f(\lambda)$ denote the optimum objective value in (8.2) as a function of the parameter λ. As previously discussed, $f(\lambda)$ is piecewise linear. So, at any specified value of $\lambda = \lambda_0$, $f(\lambda)$ may have two slopes, one for values of λ increasing through λ_0 denoted by $f'(\lambda_{0_+})$, and another for values of λ reaching to λ_0 from below denoted by $f'(\lambda_{0_-})$. If $f'(\lambda_{0_+}) = f'(\lambda_{0_-})$, this common value is the slope or derivative of $f(\lambda)$ at λ_0, and $f(\lambda)$ is differentiable at λ_0. Here we discuss efficient methods for computing $f'(\lambda_{0_+})$, $f'(\lambda_{0_-})$ for specified λ_0.

First solve (8.2) fixing $\lambda = \lambda_0$. Suppose this leads to the basic vector $x_B = (x_1, \ldots, x_m)$ optimal at λ_0. Let B be the associated basis. Compute the optimality interval $[\underline{\lambda}_B, \overline{\lambda}_B]$ of x_B. If $\underline{\lambda}_B < \lambda_0 < \overline{\lambda}_B$, then $f(\lambda)$ is differentiable at λ_0 and its slope or derivative at λ_0 is $\overline{z}^* = c_B B^{-1} b^*$.

Now consider the case where $\lambda_0 = \overline{\lambda}_B$. Let the canonical tableau of (8.2) with respect to the basis B be

Basic Variables	x_1	...	x_m	x_{m+1}	...	x_n		
x_1	1	...	0	$\bar{a}_{1,m+1}$	...	$\bar{a}_{1n}$	$\bar{b}_1$	$\bar{b}_1^*$
$\vdots$	$\vdots$		$\vdots$	$\vdots$		$\vdots$	$\vdots$	$\vdots$
x_m	0	...	1	$\bar{a}_{m,m+1}$	...	$\bar{a}_{mn}$	$\bar{b}_m$	$\bar{b}_m^*$
$-z$	0	...	0	$\bar{c}_{m+1}$	...	$\bar{c}_n$	$-\bar{z}$	$-\bar{z}^*$

Assume that $\bar{b}_i + \lambda_0 \bar{b}_i^* > 0$ for $i = 1$ to r and $=0$ for $i = r+1$ to m. Using arguments similar to those in Section 8.5, in order to find $f'(\lambda_{0_+})$, we should try to obtain an optimum basic vector $x_{\tilde{B}}$ optimal at λ_0 containing $x_1, \ldots, x_r$ as basic variables, using dual simplex pivot steps in the bottom $m-r$ rows only, beginning with the above canonical tableau, such that the updated $b_i^* \geqq 0$ for all $i = r+1$ to m. This leads to the following LP.

$$\begin{aligned} &\text{Minimize} \quad cx \\ &\text{Subject to} \quad x_i + \sum_{j=m+1}^{n} \bar{a}_{ij} x_j = \bar{b}_i^*, \; i = 1 \text{ to } m \\ &x_1, \ldots, x_r \text{ unrestricted} \qquad x_{r+1}, \ldots, x_n \geqq 0 \end{aligned}$$

This LP is clearly equivalent to the LP

$$\begin{aligned} &\text{Minimize} \quad cx \\ &\text{Subject to} \quad Ax = b^* \\ &x_1, \ldots, x_r \text{ unrestricted} \qquad x_{r+1}, \ldots, x_n \geqq 0 \end{aligned}$$

The dual of this LP is

$$\begin{aligned} &\text{Maximize} \quad \pi b^* \\ &\text{Subject to} \quad \pi A_{.j} = c_j \qquad j = 1 \text{ to } r \\ &\qquad\qquad\qquad \leqq c_j \qquad j = r+1 \text{ to } n \end{aligned}$$

which by the results in Section 6.7, is the problem of maximizing πb^* on $\boldsymbol{\Gamma}$ = the set of dual optimum solutions associated with (8.2) at $\lambda = \lambda_0$.

When such an optimal basic vector at λ_0 is found, again using arguments similar to those in Section 8.5, its upper characteristic point is $> \lambda_0$ and the slope $f'(\lambda_{0_+})$ is the optimum objective in the last LP; that is, the maximum value of πb^* over the set of dual optimum solutions of (8.2) at $\lambda = \lambda_0$. If the system: $Ax = b^*$, $x_1, \ldots, x_r$ unrestricted, $x_{r+1}, \ldots, x_n \geqq 0$, is infeasible (which happens when πb^* is unbounded above on the set $\boldsymbol{\Gamma}$), (8.2) is infeasible when $\lambda > \lambda_0$.

In the same spirit, in order to find $f'(\lambda_{0_-})$, we should find an optimum basic vector for (8.2) at λ_0, containing $x_1, \ldots, x_r$ as basic variables, using dual simplex pivot steps in the bottom $m-r$ rows in the canonical tableau of x_B, in which the updated $b_i^* \leqq 0$ for all $i = r+1$ to m. This leads to the LP

$$\begin{aligned} &\text{Minimize} \quad cx \\ &\text{Subject to} \quad x_i + \sum_{j=m+1}^{n} \bar{a}_{ij} x_j = \bar{b}_i^* \qquad i = 1 \text{ to } m \\ &x_1 \text{ to } x_r \text{ unrestricted} \qquad x_{r+1} \text{ to } x_n \leqq 0 \end{aligned}$$

which is equivalent to

$$\begin{array}{ll} \text{Minimize} & cx \\ \text{Subject to} & Ax = b^* \\ x_1 \text{ to } x_r \text{ unrestricted} & x_{r+1} \text{ to } x_n \leqq 0 \end{array}$$

The dual of this LP is equivalent to the problem of minimizing πb^* on $\boldsymbol{\Gamma}$. So we conclude from arguments similar to those previously made, that $f'(\lambda_{0_-})$ is the minimum value of πb^* over $\boldsymbol{\Gamma}$ if this is finite, and that (8.2) is infeasible for $\lambda < \lambda_0$ if the minimum value of πb^* on $\boldsymbol{\Gamma}$ is $-\infty$.

We summarize these facts in the following theorem.

THEOREM 8.2 Suppose (8.2) is feasible at λ_0 and and has a finite optimum objective value $f(\lambda_0)$. Let δ and γ be the maximum and minimum values, respectively, of πb^* on the set of optimum dual solutions associated with (8.2) at $\lambda = \lambda_0$. If $\delta = +\infty$, (8.2) is infeasible for $\lambda > \lambda_0$. If δ is finite, $f'(\lambda_{0_+}) = \delta$. If $\gamma = -\infty$, (8.2) is infeasible for $\lambda < \lambda_0$. If γ is finite, $f'(\lambda_{0_-}) = \gamma$.

With A and c the same as in (8.2), denote the optimum objective value in the LP: minimize cx, subject to $Ax = d$, $x \geqq 0$, as a function of the right-hand-side constants vector d by the symbol $\text{F}(d)$. Then, in the notation previously used, $f(0) = \text{F}(b + 0b^*) = \text{F}(b)$. The quantity $f'(0_+)$ is the limit of $(\text{F}(b + \varepsilon b^*) - \text{F}(b))/\varepsilon$, as ε tends to zero through positive values. This $f'(0_+)$ is known as the *directional derivative* of the optimum objective value function $\text{F}(d)$ in the direction b^* at the point $d = b$. The same quantity $f'(0_+)$ is also known as the *marginal value* in the LP: minimize cx, subject to $Ax = d$, $x \geqq 0$, at the point $d = b$, *in the direction* b^*.

8.11 PROPERTIES OF THE OPTIMUM OBJECTIVE VALUE FUNCTION IN PARAMETRIC LINEAR PROGRAMS

THEOREM 8.3 The optimum objective value in a parametric right-hand-side LP in which the objective function is being minimized is a piecewise linear convex function of the parameter.

Proof: Let $f(\lambda)$ denote the optimum objective value in (8.2) as a function of λ. If $f(\lambda) = -\infty$ for some value of λ in this interval, by the arguments used earlier, $f(\lambda) = -\infty$, a constant, for all λ in the entire interval $\lambda_* < \lambda < \lambda^*$ of feasibility of this problem. Now consider the case where $f(\lambda)$ is finite. In this case, by the results summarized in Section 8.9, we know that $f(\lambda)$ is a piecewise linear continuous function defined on $[\lambda_*, \lambda^*]$. It now remains to be proved that $f(\lambda)$ is convex. Let B be an optimum basis obtained when solving (8.2) with $[\underline{\lambda}_B, \overline{\lambda}_B]$ as its optimality interval. Let $\bar{b} = B^{-1}b$, $\bar{b}^* = B^{-1}b^*$, $\pi = c_B B^{-1}$, $\bar{z} = c_B B^{-1}b$, $\bar{z}^* = c_B B^{-1}b^*$, and let $(\bar{c}_j)$ be the vector of relative cost coefficients with respect to the basis B. So $\overline{\lambda}_B =$ minimum $\{-\bar{b}_i/\bar{b}_i^*: i$, such that $\bar{b}_i^* < 0\}$. Suppose $\overline{\lambda}_B$ is finite, and let r be such that $\overline{\lambda}_B = -\bar{b}_r/\bar{b}_r^*$. To find optimum solutions for values of $\lambda > \overline{\lambda}_B$ we make a dual simplex pivot step with row r as the pivot row. In the interval $[\underline{\lambda}_B, \overline{\lambda}_B]$, $f(\lambda) = \bar{z} + \lambda\bar{z}^*$, and hence $\bar{z}^*$ is the slope of $f(\lambda)$ in this interval. Suppose the pivot column, the updated column of x_s, is $(\bar{a}_{1s}, \ldots, \bar{a}_{rs}, \ldots, \bar{a}_{ms}, \bar{c}_s)^{\text{T}}$. By dual feasibility, $\bar{c}_s \geqq 0$. The pivot element $\bar{a}_{rs}$ must be <0. Let $\tilde{B}$ be the basis obtained after this pivot step. In this pivot step, the objective value clearly changes to $\tilde{z} + \lambda\tilde{z}^*$, where $\tilde{z}^* = \bar{z}^* + \bar{c}_s(\bar{b}_r^*/\bar{a}_{rs})$. This is a degenerate or nondegenerate dual simplex pivot step, depending on whether $\bar{c}_s = 0$, or >0. The optimality interval for the new basis $\tilde{B}$ is an interval to the right of $\overline{\lambda}_B$ with $\overline{\lambda}_B$ as its lower end point. The slope of the optimum objective value is $\tilde{z}^*$. Since $\bar{c}_s \geqq 0$, $\bar{b}_r^* < 0$, $\bar{a}_{rs} < 0$, we have $\tilde{z}^* \geqq \bar{z}^*$ (actually $\tilde{z}^* = \bar{z}^*$, if $\bar{c}_s = 0$, $\tilde{z}^* > \bar{z}^*$ if $\bar{c}_s > 0$). So if the pivot step just performed is a degenerate dual simplex pivot step, the slopes of $f(\lambda)$ in the optimality intervals corresponding to the basis B and $\tilde{B}$ are the same; if it is a nondegenerate dual simplex pivot step, the slope of $f(\lambda)$ strictly increases as we move from the optimality interval of B to that of $\tilde{B}$. From this it is clear that the slope of $f(\lambda)$ increases with λ. This implies that $f(\lambda)$ is convex. Hence $f(\lambda)$ is a piecewise linear convex function of λ. ■

Alternate Proof of Theorem 8.3: Let $[\lambda_*, \lambda^*]$ be the range of values of λ within which (8.2) is feasible. Assume that $f(\lambda)$ is finite on $[\lambda_*, \lambda^*]$. Let λ_1 and λ_2 be any two points in $[\lambda_*, \lambda^*]$. Let x^1 and x^2 be optimum solutions of (8.2) when $\lambda = \lambda_1, \lambda_2$, respectively. So $Ax^1 = b + \lambda_1 b^*$, $x^1 \geqq 0$; $Ax^2 = b + \lambda_2 b^*$, $x^2 \geqq 0$, and $f(\lambda_1) = cx^1$, $f(\lambda_2) = cx^2$. Let $0 < \alpha < 1$. Let $x(\alpha) = \alpha x^1 + (1 - \alpha)x^2$. From the above we have $Ax(\alpha) = b + (\alpha\lambda_1 + (1 - \alpha)\lambda_2)b^*$, $x(\alpha) \geqq 0$. So $x(\alpha)$ is a feasible solution for (8.2) when $\lambda = \alpha\lambda_1 + (1 - \alpha)\lambda_2$. But $x(\alpha)$ may not be an optimum solution of (8.2) for this value of λ. Since (8.2) is a minimization problem, $f(\alpha\lambda_1 + (1 - \alpha)\lambda_2)$, the optimum objective value in (8.2) will be less than or equal to the objective value at any

		x_1	x_2	x_3	x_4	x_5	x_6	x_7	
$\bar{c}$		0	0	0	4	0	1	3	
$\bar{c}^*$		0	0	0	-1	-2	2	1	
$-\frac{\bar{c}_j}{\bar{c}_j^*}$	for $\bar{c}_j^* < 0$				4	0			Minimum $\bar{\lambda} = 0$
$-\frac{\bar{c}_j}{\bar{c}_j^*}$	for $\bar{c}_j^* > 0$						$-\frac{1}{2}$	-3	Maximum $= -\frac{1}{2} = \underline{\lambda}$

feasible solution for (8.2) for this value of λ. So $f(\alpha\lambda_1 + (1-\alpha)\lambda_2) < cx(\alpha) = \alpha(cx^1) + (1-\alpha)\ (cx^2) = \alpha f(\lambda_1) + (1-\alpha)f(\lambda_2)$. Hence $f(\alpha)$ is a convex function on $[\lambda_*, \lambda^*]$. ■

The following theorems can be proved using identical arguments.

THEOREM 8.4 The optimum objective value in a parametric right-hand-side LP in which the objective function is being maximized, is a piecewise linear concave function of the parameter.

THEOREM 8.5 The optimum objective value in a parametric cost LP is (1) a piecewise linear concave function, if the objective function is being minimized in the problem and (2) a piecewise linear convex function, if the objective function is being maximized in the problem.

Example 8.3 Parametric Cost Problem Using Inverse Tableaux

Original Tableau

x_1	x_2	x_3	x_4	x_5	x_6	x_7	$-z$	$-z^*$	b
1	0	0	1	-2	1	1	0	0	1
0	1	0	1	1	-2	1	0	0	3
0	0	1	3	1	1	-2	0	0	7
0	0	0	4	0	1	3	1	0	0
0	0	0	-1	-2	2	1	0	1	0

$x_j \geqq 0$ for all j; minimize $z + \lambda z^*$ for all λ.

(x_1, x_2, x_3) forms an optimal basic vector when $\lambda = 0$. Below is the inverse tableau corresponding to this basic vector.

Basic Variables	*Inverse Tableau*					*Basic Values*	*Pivot Column* (x_5)
x_1	1	0	0	0	0	1	-2
x_2	0	1	0	0	0	3	①
x_3	0	0	1	0	0	7	1
$-z$	0	0	0	1	0	0	0
$-z^*$	0	0	0	0	1	0	-2

The updated cost rows corresponding to this basis are given in the table at the top.

Hence, the basic vector (x_1, x_2, x_3) is optimal for all $-1/2 \leqq \lambda \leqq 0$. Hence, the optimum feasible solution in this interval is $x = (1, 3, 7, 0, 0, 0, 0)^{\mathrm{T}}$, $z = 0$. To investigate for $\lambda > 0$, bring x_5 into the basic vector. The updated column vector of x_5 is the pivot column, and this is entered on the right-hand side of the first inverse tableau.

Basic Variables	*Inverse Tableau*					*Basic Values*	*Pivot Column* (x_6)
x_1	1	2	0	0	0	7	-3
x_5	0	1	0	0	0	3	-2
x_3	0	-1	1	0	0	4	③
$-z$	0	0	0	1	0	0	1
$-z^*$	0	2	0	0	1	6	-2

Here are the updated cost vectors.

		x_1	x_2	x_3	x_4	x_5	x_6	x_7	
$\bar{c}$		0	0	0	4	0	1	3	
$\bar{c}^*$		0	2	0	1	0	-2	3	
$-\frac{\bar{c}_j}{\bar{c}_j^*}$	for $\bar{c}_j^* < 0$						$\frac{1}{2}$		Minimum $= \bar{\lambda} = \frac{1}{2}$
$-\frac{\bar{c}_j}{\bar{c}_j^*}$	for $\bar{c}_j^* > 0$		0		-4			-1	Maximum $= \underline{\lambda} = 0$

	x_1	x_2	x_3	x_4	x_5	x_6	x_7	
$\bar{c}$	0	$\frac{1}{3}$	$-\frac{1}{3}$	$\frac{10}{3}$	0	0	4	
$\bar{c}^*$	0	$\frac{4}{3}$	$\frac{2}{3}$	$\frac{7}{3}$	0	0	1	
$-\frac{\bar{c}_j}{\bar{c}_j^*}$ for $\bar{c}_j^* < 0$								Minimum $= \bar{\lambda} = +\infty$
$-\frac{\bar{c}_j}{\bar{c}_j^*}$ for $\bar{c}_j^* > 0$		$-\frac{1}{4}$	$\frac{1}{2}$	$-\frac{10}{7}$			-4	Maximum $= \underline{\lambda} = \frac{1}{2}$

The BFS, $x = (7, 0, 4, 0, 3, 0, 0)^T$ and $z = -6\lambda$, is therefore optimal for all $0 \leqq \lambda \leqq 1/2$. The next variable to enter the basic vector is x_6, whose updated column vector is already entered on the right side of the inverse tableau.

Basic Variables	Inverse Tableau					Basic Values
x_1	1	1	1	0	0	11
x_5	0	$\frac{1}{3}$	$\frac{2}{3}$	0	0	$\frac{17}{3}$
x_6	0	$-\frac{1}{3}$	$\frac{1}{3}$	0	0	$\frac{4}{3}$
$-z$	0	$\frac{1}{3}$	$-\frac{1}{3}$	1	0	$-\frac{4}{3}$
$-z^*$	0	$\frac{4}{3}$	$\frac{2}{3}$	0	1	$\frac{26}{3}$

The updated cost rows are given in the table at the top.

The BFS, $x = (11, 0, 0, 0, 17/3, 4/3, 0)^T$ and $z = 4/3 - 26\lambda/3$, is therefore optimal for all $1/2 \leqq \lambda$. To find optimum solutions for $\lambda < -1/2$, return to the tableau corresponding to the basic vectror (x_1, x_2, x_3) and bring x_6 into it. Continuing this way, we get the solutions tabulated in the table at the bottom. The optimum objective value function is drawn in Figure 8.1.

Example 8.4 Parametric Right-Hand-Side Problem

x_1	x_2	x_3	x_4	x_5	x_6	x_7	$-z$	b	b^*
1	0	0	2	1	0	-1	0	-1	3
0	1	0	1	0	1	1	0	-2	1
0	0	1	0	1	1	2	0	-3	2
1	3	2	3	-5	1	3	1	0	0.

$x_j \geqq 0$ for all j; z to be minimized.

To find a feasible basis, try to solve the Phase I problem for some value of λ, say, 0. The original tableau for the Phase I problem is given in page 291. x_8, x_9, x_{10} are the artificial variables and w is the Phase I objective function. This is a feasible tableau for the Phase I problem when $\lambda = 0$. Computing the Phase I relative cost coefficients, we have, $\bar{d} = (1, 1, 1, 3, 2, 2, 2)$. Since $\bar{d} > 0$, this is an optimum basis for the Phase I problem. The value of $w = 6 - 6\lambda = 6$ when $\lambda = 0$. Hence, the original problem is infeasible when $\lambda = 0$. The basic vector (x_8, x_9, x_{10}) is a feasible to the Phase I problem for all $-\infty \leqq \lambda \leqq 1/3$ and in this range, the optimum Phase I objective value is $6 - 6\lambda > 0$. Hence, the original problem is infeasible for all $\lambda \leqq 1/3$. Now try to find optimum solutions for the Phase I problem for $\lambda > 1/3$. For this, make a dual simplex pivot in the first row. The pivot row is generated using the inverse tableau. This leads to the updated rows $\bar{A}_{1.}$, $\bar{d}$ given in page 291.

Basic Vector	Optimality Interval	BFS	Optimum Objective Value	Optimum Dual Solution
(x_6, x_2, x_3)	$-1 \leqq \lambda \leqq -\frac{1}{2}$	(0, 5, 6, 0, 0, 1, 0)	$1 + 2\lambda$	$(1 + 2\lambda, 0, 0)$
(x_6, x_2, x_5)	$-4 \leqq \lambda \leqq -1$	(0, 11, 0, 0, 2, 5, 0)	$5 + 6\lambda$	$\left(\frac{1}{3} + \lambda\left(\frac{4}{3}\right), 0, \frac{2}{3} + \lambda\left(\frac{2}{3}\right)\right)$
	$\lambda < -4$		$-\infty$	

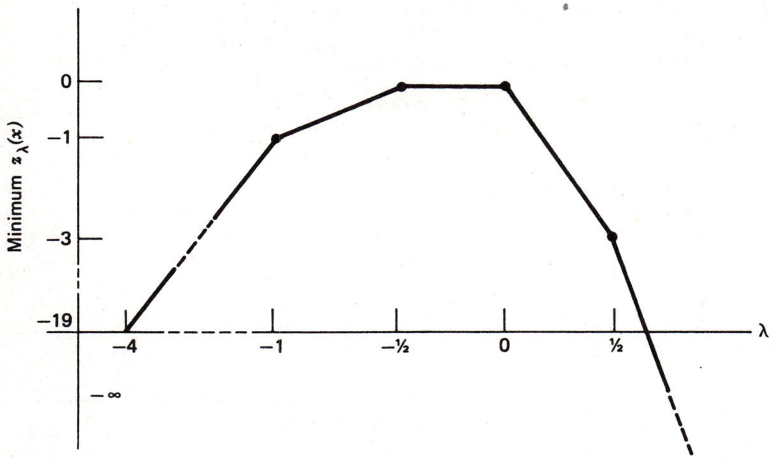

Figure 8.1

Original Tableau for the Phase I Problem

x_1	x_2	x_3	x_4	x_5	x_6	x_7	x_8	x_9	x_{10}	$-z$	$-w$	b	b^*
−1	0	0	−2	−1	0	1	1	0	0	0	0	1	−3
0	−1	0	−1	0	−1	−1	0	1	0	0	0	2	−1
0	0	−1	0	−1	−1	−2	0	0	1	0	0	3	−2
1	3	2	3	−5	1	3	0	0	0	1	0	0	0
0	0	0	0	0	0	0	1	1	1	0	1	0	0

	x_1	x_2	x_3	x_4	x_5	x_6	x_7
$\bar{A}_{1.}$	−1	0	0	−2	−1	0	1
$\bar{d}$	1	1	1	3	2	2	2
$\dfrac{\bar{d}_j}{\bar{a}_{1j}}$ for $\bar{a}_{1j} < 0$	1			$\frac{3}{2}$	2		

Hence, the entering variable is x_1. Its updated column vector is the pivot column and this is entered on the right-hand side of the inverse tableau. The next inverse tableau is in page 292.

The new basic vector (x_1, x_9, x_{10}) is optimal for all $1/3 \leqq \lambda \leqq 3/2$ and this region, minimum $w = 5 - 3\lambda$.

Phase I Initial Inverse Tableau

Basic Variables	*Inverse Tableau*					$\bar{b}$	$\bar{b}_i^*$	*Ratios* $-\bar{b}_i/\bar{b}_i^*$ *for* $\bar{b}_i^* < 0$	*Ratios* $-\bar{b}_i/\bar{b}_i^*$ *for* $b_i^* > 0$	*Pivot Column* (x_1)
x_8	1	0	0	0	0	1	−3	$\frac{1}{3}$		(−1) circled
x_9	0	1	0	0	0	2	−1	2		0
x_{10}	0	0	1	0	0	3	−2	$\frac{3}{2}$		0
$-z$	0	0	0	1	0	0	0			1
$-w$	−1	−1	−1	0	1	−6	6			1
								Minimum $= \bar{\lambda} = \frac{1}{3}$	Maximum $= \underline{\lambda} = -\infty$	

Basic Variables	Phase 1 Inverse Tableau					$\bar{b}$	$\bar{b}^*$	Ratio $-\bar{b}_i/\bar{b}_i^*$, $\bar{b}_i^* < 0$	Ratio $-\bar{b}_i/\bar{b}_i^*$, $\bar{b}_i^* > 0$	Pivot Column x_5
x_1	-1	0	0	0	0	-1	3		$\frac{1}{3}$	1
x_9	0	1	0	0	0	2	-1	2		0
x_{10}	0	0	1	0	0	3	-2	$\frac{3}{2}$		-1 (circled)
$-z$	1	0	0	1	0	1	-3			-6
$-w$	0	-1	-1	0	1	-5	3			1
								Minimum $= \bar{\lambda} = \frac{3}{2}$	Maximum $= \underline{\lambda} = \frac{1}{3}$	

Hence, for all λ in this region, the original problem is still infeasible. To obtain optimum solutions to the Phase I problem when $\lambda > 3/2$, a Phase I dual simplex pivot should be made on the third row. The pivot row and the Phase I relative cost coefficients are as follows:

	x_1	x_2	x_3	x_4	x_5	x_6	x_7
$\bar{A}_{3.}$	0	0	-1	0	-1	-1	-2
$\bar{d}$	0	1	1	1	1	2	3
$-\frac{\bar{d}_j}{\bar{a}_{3j}}$ for $\bar{a}_{3j} < 0$			1		1	2	$\frac{3}{2}$

The entering variable can be either x_3 or x_5, and we pick x_5. Its updated column is the pivot column. The next inverse tableau is given at the top of page 293.

The basic vector (x_1, x_9, x_5) is optimal to the Phase I problem when $3/2 \leqq \lambda \leqq 2$, and in this range the minimum value of w is $2 - \lambda$. Thus, the Phase I objective attains the value zero for the first time when $\lambda = 2$. Hence, when $\lambda = 2$, a feasible solution of the original problem is $x_1 = 2 + 2$, $x_5 = -3 + 4$; all other $x_j = 0$. The Phase II relative cost vector corresponding to this basis is $\bar{c} = (0, 3, 8, 1, 0, 7, 16) \geqq 0$. Hence, the above feasible solution is optimal to the original problem when $\lambda = 2$. To get optimal solutions for $\lambda > 2$, make a Phase II dual simplex pivot in the second row.

	x_1	x_2	x_3	x_4	x_0	x_6	x_7
$\bar{A}_{2.}$	0	-1	0	-1	0	-1	-1
$\bar{c}$	0	3	8	1	0	7	16
$-\frac{\bar{c}_j}{\bar{a}_{2j}}$ for $\bar{a}_{2j} < 0$		3		1		7	16

The pivot column is the updated column of x_4, and x_4 replaces the artificial variable x_9, which is in the present basic vector with zero value when $\lambda = 2$. With this pivot all the artificial variables have left the basis, and we drop the w column and the Phase I relative cost row and go to Phase II. Continuing the same way, we get the solutions tabulated in the second table from the top of page 293. See Figure 8.2 for the optimum objective value function.

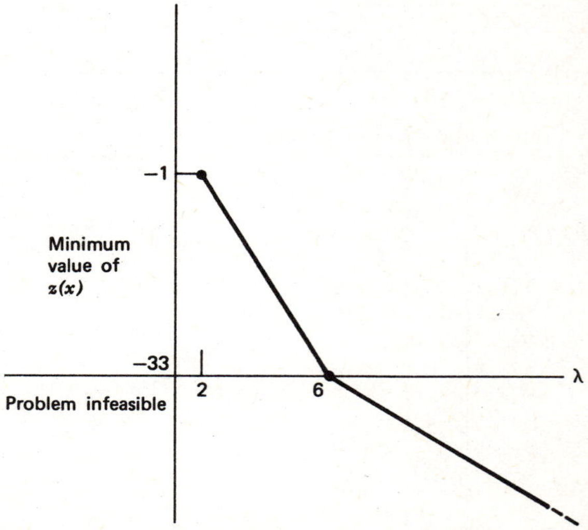

Figure 8.2 Optimum objective value as a function of the parameter

Basic Variables	Phase 1 Inverse Tableau					$\bar{b}$	$\bar{b}^*$	Ratios $-\bar{b}_i/\bar{b}_i^*$: $\bar{b}_i^* < 0$	Ratios $-\bar{b}_i/\bar{b}_i^*$: $\bar{b}_i^* > 0$	Pivot Column x_4
x_1	-1	0	1	0	0	2	1		-2	2
x_9	0	1	0	0	0	2	-1	2		$\boxed{-1}$
x_5	0	0	-1	0	0	-3	2		$\frac{3}{2}$	0
$-z$	1	0	-6	1	0	-17	9			1
$-w$	0	-1	0	0	1	-2	1			1
								Minimum $= \bar{\lambda} = 2$	Maximum $= \underline{\lambda} = \frac{3}{2}$	

Basic Vector	Optimality Interval	BF	Optimum Objective Value	Optimum Dual Solution
(x_1, x_4, x_5)	$2 \leqq \lambda \leqq 6$	$(6-\lambda, 0, 0, -2+\lambda, -3+2\lambda, 0, 0)^T$	$15-8\lambda$	$(-1, -1, 6)$
(x_2, x_4, x_5)	$6 \leqq \lambda$	$(0, -3+\lambda/2, 0, 1+\lambda/2, -3+2\lambda, 0, 0)^T$	$9-7\lambda$	$(0, -3, 5)$

Exercises

8.6 *COMBINED PARAMETRIC COST AND RIGHT-HAND-SIDE PROBLEMS.* In practical applications, we encounter parametric LPs of the following form, where A, b, b^*, c, c^* are all given. Combine the ideas in sections 8.2, 8.3, 8.6, 8.7, and 8.8 and describe an algorithm for solving this problem for all real values of λ. Explain clearly when primal simplex or dual simplex pivots have to be made during the algorithm.

$$\begin{aligned} \text{Minimize} \quad & (c + \lambda c^*)x \\ \text{Subject to} \quad & Ax = b + \lambda b^* \\ & x \geqq 0 \end{aligned}$$

8.7 For the combined parametric problem in Exercise 8.6 show that the optimum objective value may not be piecewise linear in λ and possibly neither convex nor concave. Construct illustrative examples.

8.8 Consider the following multiparameter parametric cost problem where t, the number of parameters is greater than or equal to 2. Let $\lambda = (\lambda_1, \ldots, \lambda_t)$ be the parameter vector. If B is an optimum basis for this problem when the parameter vector is $\bar{\lambda}$, prove that the region of the parameter vector λ for which B remains an optimum basis is a convex polyhedral set in $\mathbf{R}^t$. Find this set

$$\begin{aligned} \text{Minimize} \quad & z(x) = (c + \lambda_1 c^1 + \lambda_2 c^2 + \cdots + \lambda_t c^t)x \\ \text{Subject to} \quad & Ax = b \\ & x \geqq 0 \end{aligned}$$

8.9 Just as in the parametric cost problem, show that the parametric right-hand-side problem becomes very complicated when the right-hand-side constant vector depends on two or more parameters.

This explains the difficulty of tackling parametric problems when there are two or more parameters. With only one parameter, the parameter space is the real line and convex sets in the real line are just intervals. Hence for a one-parameter problem, the parameter space can be covered easily by passing from one interval (the characteristic interval of some optimal basis) to another. On the other hand, convex polyhedral sets in two or higher dimensional spaces can have several facets and, if the parametric cost problem involves two or more parameters, it becomes very complicated to cover the entire parameter space by convex polyhedral sets, which are the optimality regions of the various bases for the problem.

8.12 COMPUTATIONAL COMPLEXITY OF PARAMETRIC LINEAR PROGRAMMING

In this section, the symbols w_r, z_r are used to denote the variables in a parametric linear program, and the reader is cautioned not to confuse these symbols with the Phase I, II objective functions. We consider (8.2), where A is a matrix of order $m \times n$ and rank m. It may be required to solve (8.2) for all real values of λ, or for all values of λ in some specified interval of the real line. Let $f(\lambda)$ denote the optimum objective value in this problem. Let $\phi(A, b, b^*, c)$ denote the total number of intervals on the real line, each of positive length, such that the slope of $f(\lambda)$ is a constant in each interval, and the slopes of $f(\lambda)$ in any two intervals are different. Since each of these intervals must be separately obtained (as the slopes of $f(\lambda)$ in them are different) when (8.2) is solved, $\phi(A, b, b^*, c)$ provides a lower bound on the computational effort required for solving (8.2) for all real values of λ by any algorithm. We construct a class of parametric LPs in which all the data are integer. For positive integer $r \geqq 2$, the rth problem in the class has $m = r$, $n = 2r$, and needs a total of $(7r^2 + 3r + 2)/2$ binary bits for storage of the data in it, and we show that the value of $\phi(A, b, b^*, c)$ for it is 2^r. This conclusively establishes that in the worst case, the computational effort required for solving the parametric linear program (8.2) is not bounded above by any polynomial in the size of the problem. The results in this section are taken from reference [8.4].

The Class of Problems

In the rth problem in our class, the number of constraints is r, and the number of variables is $2r$. For the sake of convenience, we denote the variables in the problem by the symbols $w_1, \ldots, w_r; z_1, \ldots, z_r$. In tableau form, the rth problem in our class is the following:

Tableau 8.6

w	z	$-\psi$	
I_r	$-\tilde{M}_r$	0	$b(r) - e_r\lambda$
0	d^r	1	0

$w = (w_1, \ldots, w_r) \geqq 0$; $z = (z_1, \ldots, z_r) \geqq 0$; minimize ψ.

where I_r is the unit matrix of order r; $-\tilde{M}_r$ is a square lower triangular matrix of order r in which all the diagonal entires are -1, all elements below the diagonal are -2, and all elements above the diagonal are 0; $b(r) = (2^r, 2^{r-1}, \ldots, 2)^T$, $d^r = (4^{r-1}, 4^{r-2}, \ldots, 4, 1)$, and e_r is the vector in $\mathbf{R}^r$, all of whose entries are 1. Let $A_r = (I_r \vdots -\tilde{M}_r)$, $b^*(r) = -e_r$, $c(r) = (0, d^r) \in \mathbf{R}^{2r}$. As discussed above, the computational complexity of this problem is $\phi(A_r, b(r), b^*(r), c(r))$.

We call the pair of variables (w_j, z_j) the jth *complementary pair of variables,* and each variable in this pair is known as the complement of the other. A *complementary vector of variables* for this problem is an ordered vector of variables $(y_1, \ldots, y_r)$, where $y_j \in \{w_j, z_j\}$ for each $j = 1$ to r. Thus, there are 2^r complementary vectors of variables. Since $-\tilde{M}_r$ is lower triangular and because of the special structure of the coefficients in Tableau 8.6, it can be verified that the matrix consisting of the column vectors of any complementary vector of variables in Tableau 8.6 is a lower triangular matrix, with diagonal entries equal to $+1$ or -1. Hence every complementary vector of variables is a basic vector for Tableau 8.6.

The Sequence of Complementary Vectors

In our proofs we encounter the complementary vectors of variables for the rth problem in our class, arranged in a specific order, as a sequence. Here we describe how to generate this sequence. For $r = 2$, the specific order is (w_1, w_2), (w_1, z_2), (z_1, z_2), (z_1, w_2). Suppose the order for the $(r-1)$th problem for our class is known. Let $a(s) = 2^s$, for any $s \geqq 2$. To get the order for the rth problem in the class, let $\boldsymbol{v}_1, \boldsymbol{v}_2, \ldots, \boldsymbol{v}_{a(r-1)}$ be the ordered sequence of complementary basic vectors for the $(r-1)$th problem when the variables in it are treated as $w_2, \ldots, w_r$; $z_2, \ldots, z_r$ (instead of $w_1, \ldots, w_{r-1}$; $z_1, \ldots, z_{r-1}$). For example, $\boldsymbol{v}_1 = (w_2, \ldots, w_r)$, $\boldsymbol{v}_2 = (w_2, \ldots, w_{r-1}, z_r)$, etc. Then the specific order of complementary vectors of variables for the rth problem is: $\boldsymbol{V}_1 = (w_1, \boldsymbol{v}_1), \ldots, (w_1, \boldsymbol{v}_{a(r-1)})$; $(z_1, \boldsymbol{v}_{a(r-1)}), \ldots, (z_1, \boldsymbol{v}_1) = \boldsymbol{V}_{a(r)}$. So $\boldsymbol{V}_1 = (w_1, w_2, \ldots, w_r)$, $\boldsymbol{V}_2 = (w_1, w_2, \ldots, w_{r-1}, z_r)$, $\ldots$, $\boldsymbol{V}_{a(r)} = (z_1, w_2, \ldots, w_r)$.

Example 8.5

Here we illustrate the solution of the problem corresponding to $r = 3$ in our class. The parametric simplex algorithm begins with (w_1, w_2, w_3) as the unique optimum feasible basic vector corresponding to $\lambda = 0$. In each canonical tableau, the pivot element is circled. It indicates the pivot row and the pivot column to make the pivot step that leads to the next optimality interval.

It can be verified that it partitions the parameter space into $2^3 = 8$ optimality intervals, and required $2^3 - 1 = 7$ pivots in all. The relative cost coefficients of the nonbasic variables are strictly positive in each of the canonical tableaux obtained, and this implies that in the interior of each optimality interval obtained,

Basic Variables	w_1	w_2	w_3	z_1	z_2	z_3	$-\psi$	Basic Values	Optimality Interval
w_1	1	0	0	-1	0	0	0	$8-\lambda$	
w_2	0	1	0	-2	-1	0	0	$4-\lambda$	$\lambda \leqq 2$
w_3	0	0	1	-2	-2	-1	0	$2-\lambda$	
$-\psi$	0	0	0	16	4	1	1	0	
w_1	1	0	0	-1	0	0	0	$8-\lambda$	
w_2	0	1	0	-2	-1	0	0	$4-\lambda$	$2 \leqq \lambda \leqq 4$
z_3	0	0	-1	2	2	1	0	$-2+\lambda$	
$-\psi$	0	0	1	14	2	0	1	$2-\lambda$	
w_1	1	0	0	-1	0	0	0	$8-\lambda$	
z_2	0	-1	0	2	1	0	0	$-4+\lambda$	$4 \leqq \lambda \leqq 6$
z_3	0	2	-1	-2	0	1	0	$6-\lambda$	
$-\psi$	0	2	1	10	0	0	1	$10-3\lambda$	
w_1	1	0	0	-1	0	0	0	$8-\lambda$	
z_2	0	-1	0	2	1	0	0	$-4+\lambda$	$6 \leqq \lambda \leqq 8$
w_3	0	-2	1	2	0	-1	0	$-6+\lambda$	
$-\psi$	0	4	0	8	0	1	1	$16-4\lambda$	
z_1	-1	0	0	1	0	0	0	$-8+\lambda$	
z_2	2	-1	0	0	1	0	0	$12-\lambda$	$8 \leqq \lambda \leqq 10$
w_3	2	-2	1	0	0	-1	0	$10-\lambda$	
$-\psi$	8	4	0	0	0	1	1	$80-12\lambda$	
z_1	-1	0	0	1	0	0	0	$-8+\lambda$	$10 \leqq \lambda \leqq 12$
z_2	2	-1	0	0	0	0	0	$12-\lambda$	
z_3	-2	2	-1	0	0	1	0	$-10+\lambda$	
$-\psi$	10	2	1	0	0	0	1	$90-13\lambda$	
z_1	-1	0	0	1	0	0	0	$-8+\lambda$	
w_2	-2	1	0	0	-1	0	0	$-12+\lambda$	$12 \leqq \lambda \leqq 14$
z_3	2	0	-1	0	2	1	0	$14-\lambda$	
$-\psi$	14	0	1	0	2	0	1	$114-15\lambda$	
z_1	-1	0	0	1	0	0	0	$-8+\lambda$	
w_2	-2	1	0	0	-1	0	0	$-12+\lambda$	$14 \leqq \lambda$
w_3	-2	0	1	0	-2	-1	0	$-14+\lambda$	
$-\psi$	16	0	0	0	4	1	1	$128-16\lambda$	

the optimum basis obtained is unique. In the interior of each optimality interval, the optimum BFS is primal nondegenerate. Each pivot step made in the algorithm is a nondegenerate dual simplex pivot step, and hence, after every pivot step, the slope of the optimum objective value strictly increases. So $\phi(A_3, b(3), b^*(3), c(3)) = 8 = 2^3$. Also verify that the sequence of optimum basic vectors obtained is exactly the complementary vectors of variables for this problem, in the specific order discussed above.

THEOREM 8.6 The following results hold when the rth parametric linear program in the class is solved, beginning with $(w_1, \ldots, w_r)$ as the unique optimum basic vector corresponding to $\lambda = 0$, for $r \geqq 2$. (1) The sequence of optimum basic vectors obtained in the parametric algorithm is exactly the complementary vectors of variables for this problem, in the specific order discussed above. (2) The optimality intervals obtained are $[-\infty, 2]$, $[2, 4]$, $[4, 6]$, . . . , $[2^{r+1}-4, 2^{r+1}-2]$, $[2^{r+1}-2, \infty]$. As the algorithm

Tableau 8.7

Basic Variables	w_1	w_2	$w_3 \cdots w_r$	z_1	z_2	$z_3 \quad \cdots \quad z_r$	$-\psi$	
w_1	1	0	$0 \cdots 0$	-1	0	$0 \cdots 0$	0	$2^r - \lambda$
z_2	0	-1	$0 \cdots 0$	2	1	$0 \cdots 0$	0	$-2^{r-1} + \lambda$
w_3	0	-2	$1 \cdots 0$	2	0	$-1 \cdots 0$	0	$2^{r-2} - 2^r + \lambda$
$\vdots$	$\vdots$	$\vdots$	$\vdots \cdots \vdots$	$\vdots$	$\vdots$	$\vdots$	$\vdots$	$\vdots$
w_r	0	-2	$0 \cdots 1$	2	0	$-2 \cdots -1$	0	$2 - 2^r + \lambda$
$-\psi$	0	4^{r-2}	$0 \cdots 0$	$2 \times 4^{r-2}$	0	$4^{r-3} \cdots 1$	1	$4^{r-2}(2^{r-1} - \lambda)$

moves from one interval to the next in this sequence, the slope of the optimum objective value increases strictly. (3) The relative cost coefficients of all the nonbasic variables are strictly positive in every tableau obtained during the algorithm. (4) In the interior of each optimality interval, the optimum basis obtained is primal and dual nondegenerate, and is the unique optimum basis for the problem. (5) The algorithm goes through $2^r - 1$ pivot steps before termination.

Proof: The statement of the theorem can be verified for $r = 2$. From the numerical example given above, we verify it for $r = 3$. Proof is by inducation on r. We now set up an inducation hypothesis.

INDUCTION HYPOTHESIS The statement of the theorem holds for the $(r-1)$th parametric linear program in our class.

Under the induction hypothesis, we now prove that the statement of the theorem must also hold for the rth problem in our class, which is the problem given in Tableau 8.6. When the first row and the columns of w_1, z_1 are eliminated from Tableau 8.6, it can be verified that it becomes the original tableau for the $(r-1)$th problem in our class, with the exception that the variables are called $w_2, \ldots, w_r$; $z_2, \ldots, z_r$ (instead of $w_1, \ldots, w_{r-1}$; $z_1, \ldots, z_{r-1}$). Call this the *principle subproblem* of the rth problem in our class. Let $\boldsymbol{v}_1, \boldsymbol{v}_2, \ldots, \boldsymbol{v}_{a(r-1)}$ be the specific order of complementary vectors of variables for this problem with these names for the variables.

From Tableau 8.6 we see that any pivots performed in the columns of $w_2, \ldots, w_r$; $z_2, \ldots, z_r$ do not change row 1. Also $2^r - \lambda$ remains strictly positive for all $\lambda < 2^r$. It follows that when the rth problem in our class is solved beginning with $(w_1, \ldots, w_r)$ as the initial basic vector, w_1 remains in the basic vector until the value of λ reaches 2^r. Applying the induction hypothesis to the principal subproblem, we see that the entering and leaving variables in the pivot steps that occur in solving the rth problem in our class will be the same as those needed when the principal subproblem is solved, until λ reaches the value $2^r - 2$. By the induction hypothesis this requires $2^{r-1} - 1$ pivot steps, and at the end of these steps we reach the basic vector $(w_1, z_2, w_3, \ldots, w_r)$. The canonical tableau of Tableau 8.6 with respect to this basic vector is given in Tableau 8.7.

By the induction hypothesis applied to the principal subproblem, and the arguments listed above, the sequence of basic vectors obtained before reaching Tableau 8.7 is $\boldsymbol{V}_1 = (w_1, \boldsymbol{v}_1)$, $\boldsymbol{V}_2 = (w_1, \boldsymbol{v}_2), \ldots$, $\boldsymbol{V}_{a(r-1)} = (w_1, \boldsymbol{v}_{a(r-1)})$, and statements (3) and (4) of the theorem hold for each of these basic vectors. Now it can be verfied that the basic feasible solution in Tableau 8.7 is optimal in the interval $2^r - 2 \leqq \lambda \leqq 2^r$, it is primal and dual nondegenerate, and is the unique optimum basis in the interior of this interval. To continue the parametric algorithm for $\lambda > 2^r$, we have to make a dual simplex pivot step in row 1 of Tableau 8.7, and clearly the column vector of z_1 is the pivot column. This leads us to the next canonical tableau, given in Tableau 8.8.

The basic feasible solution in Tableau 8.8 corresponding to the basic vector $\boldsymbol{V}_{a(r-1)+1} = (z_1, \boldsymbol{v}_{a(r-1)})$ is optimal in the interval $2^r \leqq \lambda \leqq 2^r + 2$. The relative cost coefficients associated with all the nonbasic variables are strictly positive in Tableau 8.8, and the basic vector $\boldsymbol{V}_{a(r-1)+1}$ is primal nondegenerate when λ is in the interior of the interval $[2^r, 2^r + 2]$. The updated right-hand-side constant in the first row in Tableau 8.8 is $-2^r + \lambda$, and this remains strictly positive for all $\lambda > 2^r + 2$. So when we continue the solution of the rth problem in our class from Tableau 8.8 (for $\lambda > 2^r + 2$), we will never have to choose row 1 as the pivot row again. Calling $\lambda - 2^r = \nu$, and eliminating row 1 and the columns of w_1, z_1 from

Tableau 8.8

Basic Variables	w_1	w_2	$w_3 \cdots w_r$	z_1	z_2	$z_3 \cdots z_{r-1}$	z_r	$-\psi$	
z_1	-1	0	$0 \cdots 0$	1	0	$0 \cdots 0$	0	0	$-2^r + \lambda$
z_2	2	-1	$0 \cdots 0$	0	1	$0 \cdots 0$	0	0	$2^{r-1} - (\lambda - 2^r)$
w_3	2	-2	$1 \cdots 0$	0	0	$-1 \cdots 0$	0	0	$2^{r-2} - (\lambda - 2^r)$
$\vdots$	$\vdots$	$\vdots$	$\vdots \quad \vdots$	$\vdots$	$\vdots$	$\vdots \quad \vdots$	$\vdots$	$\vdots$	$\vdots$
w_r	2	-2	$0 \cdots 1$	0	0	$-2 \cdots -2$	-1	0	$2 - (\lambda - 2^r)$
$-\psi$	$2 \times 4^{r-2}$	4^{r-2}	$0 \cdots 0$	0	0	$4^{r-3} \cdots 4$	1	1	$4^{r-2}(5 \times 2^{r-1} - 3\lambda)$

Tableau 8.8, we verify that it leads again to the $(r-1)$th problem in the class, with the exception that the variables are now called $z_2, w_3, \ldots, w_r$; $w_2, z_3, \ldots, z_r$ in that order, and the parameter, $\nu(\lambda \geqq 2^r + 2$, corresponds to $\nu \geqq 2)$. Call this the *principal subproblem of Tableau 8.8*. Using the induction hypothesis on this principal subproblem of Tableau 8.8 and the above facts, we conclude that when the solution of the rth problem in our class is continued from Tableau 8.8 for $\lambda > 2^r + 2$, the basic variable changes that occur are exactly the same as those that will occur when this principal subproblem of Tableau 8.8 is solved, and hence it will lead through the following basic vectors: $\boldsymbol{V}_{a(r-1)+2} = (z_1, \boldsymbol{V}_{a(r-1)-1}), \ldots, \boldsymbol{V}_{a(r)} = (z_1, \boldsymbol{V}_1)$. By the induction hypothesis, statements (3) and (4) of the theorem continue to hold. Also, (3) implies that after each pivot step in this algorithm, the slope of the optimum objective value strictly increases.

These facts imply that under the induction hypothesis, the statements of the theorem hold for the rth problem in the class.

The theorem has already been verified for $r = 2, 3$. Hence it is true for all $r \geqq 2$. ■

COROLLARY 8.1 $\phi(A_r, b(r), b^*(r), c(r)) = 2^r$.

Proof: Follows from Theorem 8.6. Thus the complete answer to the rth parametric LP in our class is itself exponentially long. ■

COROLLARY 8.2 In the worst case, the computational effort needed to completely solve the parametric LP (8.2) is not bounded above by any polynomial in the size of this problem.

Proof: As discussed above, $\phi(A, b, b^*, c)$ is a measure of the computational complexity of (8.2). The class of parametric LPs constructed above clearly establishes this corollary. ■

Note 8.3: Since each of the basic vectors obtained when the rth problem in our class is solved is both primal and dual nondegenerate in the interior of its optimality interval, by taking the class of dual problems we conclude that the same results hold for parametric cost linear programs.

Consider parametric LPs that are required to be solved only for values of the parameter λ in some specified finite interval, say the interval $[0, 1]$. Even on these problems, the worst case computational requirements grow exponentially or faster with the size of the problem. A class of parameteric LPs exhibiting this fact can be constructed from the class of parametric LPs discussed above, by proper scaling of the data.

The results obtained are worst case results. However, it has been observed empirically that in almost all the parametric LP models encountered in practical applications, the actual computational effort expended turns out to be a polynomial of small degree in the size. Thus, even though in a worst case sense the algorithm for solving parametric LPs may require an unreasonable amount of computational effort as the problem size grows, on an average it seems to behave very reasonably.

8.13 THE SELF-DUAL PARAMETRIC METHOD (ALSO CALLED CRISS-CROSS METHOD) FOR SOLVING LINEAR PROGRAMS

Consider the following linear program.

$$\begin{aligned} \text{Minimize} \quad & z(x) = cx \\ \text{Subject to} \quad & Ax = b \\ & x \geqq 0 \end{aligned} \tag{8.11}$$

To solve (8.11) by the self-dual parametric method, first transform the system of equations $Ax = b$ into row echelon normal form, as discussed in Sections 3.3.5, and 3.3.6. If any redundant equations are found, eliminate them. Suppose the basic variables selected are $x_1, x_2, \ldots, x_m$, in that order. After pricing out the basic columns, let $(0, \ldots, 0, c'_{m+1}, \ldots, c'_n)$ be the updated cost vector and let $(b'_1, \ldots, b'_m)$ be the updated right-hand-side vector with respect to this basic vector. If either $(b'_1, \ldots, b'_m) \geqq 0$ or $(c'_{m+1}, \ldots, c'_n) \geqq 0$, $(x_1, \ldots, x_m)$ is primal feasible or dual feasible, respectively, and starting with it (8.11) can be solved by the primal simplex or the dual simplex algorithms. If neither of these conditions holds, for each $i = 1$ to m define $b^*_i = 0$ if $b'_i \geqq 0$, 1 otherwise; and for each $j = 1$ to n define $c^*_j = 0$ if $c'_j \geqq 0$, 1 otherwise. Consider the following modified problem.

Minimize

$$z_\theta(x) = z(x) + \theta z^*(x) = \sum_{j=1}^{n} (c'_j + \theta c^*_j)x_j$$

Subject to

$$x_i + \sum_{j=m+1}^{n} a'_{ij}x_j = b'_i + \theta b^*_i \qquad i = 1 \text{ to } m$$

$$x_j \geqq 0 \qquad \text{for all } j \qquad (8.12)$$

It is clear from the definition of b^*_i, c^*_j that the basic vector $(x_1, \ldots, x_m)$ is an optimum basic vector for (8.12) whenever the parameter θ has a large positive value. The smallest value of θ for which the present basic vector is an optimum basic vector for (8.12) is called the *critical value* corresponding to the present basic vector, and it is denoted by the symbol γ. We have $\gamma = 0$ if all $b'_i, c'_j \geqq 0$, or maximum $\{-(b'_i/b^*_i)$ for i such that $b'_i < 0$, $-(c'_j/c^*_j)$ for j such that $c'_j < 0\}$ otherwise. If $\gamma > 0$, the ith row in the present tableau is called a *critical row* if $(-b'_i/b^*_i) = \gamma$, and the jth column in the present tableau is called a *critical column* if $(-c'_j/c^*_j) = \gamma$. Hence, if $\gamma > 0$, the present tableau contains at least a critical row or a critical column.

The self-dual parametric method begins with (8.12) and goes through several stages. In each stage either a dual simplex pivot step is carried out in a critical row, or a primal simplex pivot step is carried out in a critical column. The critical value is nonincreasing during the method, and in a finite number of stages it either decreases to zero, or primal or dual infeasibility will be discovered.

In a general step, suppose we have the canonical tableau of (8.12) with respect to a basic vector. In the canonical tableau the columns of b', b^* and the rows of c', c^* are updated separately, and suppose these updated entities are $\bar{b}$, $\bar{b}^*$, $\bar{c}$, $\bar{c}^*$, respectively. The critical value corresponding to the present basic vector is $\bar{\gamma} = 0$, if $\bar{b} \geqq 0$, $\bar{c} \geqq 0$; or $\bar{\gamma}$ = maximum $\{-(\bar{b}_i/\bar{b}^*_i)$ for i such that $\bar{b}_i < 0$; $-(\bar{c}_j/\bar{c}^*_j)$ for j such that $\bar{c}_j < 0\}$ otherwise. If $\bar{\gamma} = 0$, the present basic vector is an optimum basic vector for the original problem (8.11), and the present solution an optimum BFS for it, terminate. If $\bar{\gamma} > 0$, identify the critical rows and the critical columns in the current tableau. Fix the parameter θ at $\bar{\gamma}$; that is, treat the problem as a regular LP with the updated right-hand-side constants column vector as $\bar{b} + \bar{\gamma}\bar{b}^*$, and the updated cost row as $\bar{c} + \bar{\gamma}\bar{c}^*$. If there is at least one critical row in the current tableau, select one of them, say, row r, and perform a dual simplex pivot step with row r as the pivot row (in selecting the pivot column for this step by the dual simplex minimum ratio test, remember that the current updated cost row is $\bar{c} + \bar{\gamma}\bar{c}^*$). If there are no critical rows in the current tableau, there must be at least one critical column; select one of them, say, column s, and perform a primal simplex pivot step with that as the pivot column (in selecting the pivot row for this step by the primal simplex minimum ratio test, remember that the current updated right-hand-side constants column vector is $\bar{b} + \bar{\gamma}\bar{b}^*$). Go to the next step with the canonical tableau obtained after the pivot step.

Example 8.6

Consider the following LP:

x_1	x_2	x_3	x_4	x_5	$-z$	b
1	0	0	1	−2	0	−3
0	1	0	2	1	0	−7
0	0	1	−1	−3	0	0
0	0	0	−12	10	1	0

$x_j \geqq 0$ for all j; minimize z.

The basic vector (x_1, x_2, x_3), for which the present tableau is the canonical tableau, is neither primal nor dual feasible. Introducing the coefficient vectors b^*, c^* of the parameter θ, the modified problem is given at the top of page 299.

The critical value corresponding to the present basic vector (x_1, x_2, x_3) is 12. There are no critical rows in the present tableau, but the columns of x_4

Basic Variables	x_1	x_2	x_3	x_4	x_5	$-z$	$-z^*$	$\bar{b}$	$\bar{b}^*$	$-\bar{b}_i/\bar{b}_i^*$ for $\bar{b}_i < 0$	$\bar{b}_i + \gamma \bar{b}_i^*$	Ratios for Entering Variable x_4
x_1	1	0	0	(1)	−2	0	0	−10	1	10	2	2 minimum
x_2	0	1	0	2	1	0	0	−7	1	7	5	$\frac{5}{2}$
x_3	0	0	1	−1	−3	0	0	0	0		0	
$-z$	0	0	0	−12	10	1	0	$0 = -\bar{z}$				
$-z^*$	0	0	0	1	0	0	1					
$\frac{-\bar{c}_j}{\bar{c}_j^*}$ for $\bar{c}_j < 0$				12				maximum $= \gamma = 12$				

Basic Variables	x_1	x_2	x_3	x_4	x_5	$-z$	$-z^*$	$\bar{b}$	$\bar{b}^*$	$-\bar{b}_i/\bar{b}_i^*$ for $\bar{b}_i < 0$
x_4	1	0	0	1	(−2)	0	0	−10	1	10
x_2	−2	1	0	0	5	0	0	13	−1	
x_3	1	0	1	0	−5	0	0	−10	1	10
$-z$	12	0	0	0	−14	1	0	$-120 = -\bar{z}$		
$-z^*$	−1	0	0	0	2	0	1			
$\frac{-\bar{c}_j}{\bar{c}_j^*}$ for $\bar{c}_j < 0$					7					maximum $= \gamma = 10$
$\bar{c}_j + \gamma \bar{c}_j^*$	2	0	0	0	6					
Ratios for row 1 as pivot row					$\frac{6}{2}$ minimum					

and x_7 are critical. Performing of primal simplex pivot in the column vector of x_4 leads to the next tableau.

In this tableau there are no critical columns, but rows 1 and 3 are critical. Performing a dual simplex pivot in row 1 leads to the next tableau given at the bottom.

The new critical value is 8. The algorithm can be continued in a similar manner until termination.

Basic Variables	x_1	x_2	x_3	x_4	x_5	$-z$	$-z^*$	$\bar{b}$	$\bar{b}^*$	$-\bar{b}_i/\bar{b}_i^*$ for $\bar{b}_i < 0$
x_5	$-\frac{1}{2}$	0	0	$-\frac{1}{2}$	1	0	0	5	$-\frac{1}{2}$	
x_2	$\frac{1}{2}$	1	0	$\frac{5}{2}$	0	0	0	−12	$\frac{3}{2}$	8
x_3	$-\frac{3}{2}$	0	1	$-\frac{5}{2}$	0	0	0	15	$-\frac{3}{2}$	
$-z$	5	0	0	−7	0	1	0	$-50 = -z$		
$-z^*$	0	0	0	1	0	0	1			
$\frac{-\bar{c}_j}{\bar{c}_j^*}$ for $\bar{c}_j < 0$				7						Maximum $= \gamma = 8$

Exercises

8.10 Complete the solution of the problem in Example 8.6. Also solve the following LP by the self-dual parametric method.

x_1	x_2	x_3	x_4	x_5	x_6	x_7	$-z$	b
1	0	0	−1	1	−1	1	0	1
0	1	0	−2	−1	1	1	0	−2
0	0	1	1	−2	−3	2	0	−4
0	0	0	−4	6	3	−7	1	0

$x_j \geqq 0$, for all j; minimize z.

Note 8.4: The limited amount of computational experience on the self-dual parametric method for solving LPs seems to indicate that it is probably not as efficient as the two-phase simplex method (see reference [8.2] of R. S. Garfinkel and P. L. Yu).

8.14 PROPERTIES OF THE OPTIMUM OBJECTIVE VALUE FUNCTION IN LINEAR PROGRAMS

In this section we consider an LP in which all the data are fixed, except either the right-hand-side constants or the original cost coefficients, which are themselves treated as real valued parameters. We also study the properties of the optimum objective value as a function of these parameters. To study this section, the reader will need some knowledge of the calculus of functions of many variables. We briefly review some of these results first.

Let $f(x)$ be a real-valued function defined over a convex subset Γ of $\mathbf{R}^n$. We assume that the reader is familiar with the definition of continuity of $f(x)$ at a point $x^0 \in \Gamma$, and the definition of the vector of partial derivatives of $f(x)$ at the point x^0, $\nabla f(x^0) = (\partial f(x^0)/\partial x_1, \ldots, \partial f(x^0)/\partial x_n)^{\mathrm{T}}$ when it exists. The function $f(x)$ is said to be *differentiable* at x^0 if $\nabla f(x^0)$ exists, and for any $y \in \mathbf{R}^n$, $(1/\alpha)(f(x^0+\alpha y)-f(x^0)-\alpha(\nabla f(x^0))^{\mathrm{T}}y)$ tends in the limit to zero as α tends to zero. If $f(x)$ is differentiable at x^0, then for any $y \in \mathbf{R}^n$, we can approximate $f(x^0+\alpha y)$ by $f(x^0)+\alpha(\nabla f(x^0))^{\mathrm{T}}y$ for values of α for which $|\alpha|$ is small. This is the *first-order Taylor series expansion* for $f(x^0+\alpha y)$. If $f(x)$ is differentiable at x^0, the partial derivative vector $\nabla f(x^0)$ is known as the *gradient vector* of $f(x)$ at x^0.

The reader should also review Section 1.2 for the definitions of convex, concave functions, and of piecewise linear functions defined over the real line.

Piecewise Linear Functions Defined on $\mathbf{R}^n$

Let $\Gamma \subset \mathbf{R}^n$ be either $\mathbf{R}^n$ or a convex polyhedral subset of $\mathbf{R}^n$. The real-valued function $f(x)$ defined over Γ is said to be a *piecewise linear function* if Γ can be partitioned into convex polyhedral subsets, such that in each of these subsets $f(x)$ is an affine function.

Let $c_0^t + \sum_{j=1}^n c_j^t x_j$, for $t = 1$ to r, be given affine functions defined on $\mathbf{R}^n$. For each $x \in \mathbf{R}^n$, define

$$s(x) = \text{maximum}\left\{c_0^t + \sum_{j=1}^{n} c_j^t x_j : t = 1 \text{ to } r\right\} \quad (8.13)$$

The function $s(x)$ is known as the *pointwise supremum* of the given set of affine functions.

THEOREM 8.7 The function $s(x)$ defined in (8.13) is convex and piecewise linear.

Proof: Let x^1 and x^2 be two points in $\mathbf{R}^n$. $s(x^1)$ is defined by computing the maximum as in (8.13) after substituting x^1 for x. Suppose the maximum is attained by $t = p$. Similarly, to compute $s(x^2)$, let the maximum in (8.13) be attained by $t = q$, when $x = x^2$. So we have:

$$s(x^1) = c_0^p + \sum_{j=1}^{n} c_j^p x_j^1 \geqq c_0^t + \sum_{j=1}^{n} c_j^t x_j^1 \quad \text{for all } t = 1 \text{ to } r \quad (8.14)$$

$$s(x^2) = c_0^q + \sum_{j=1}^{n} c_j^q x_j^2 \geqq c_0^t + \sum_{j=1}^{n} c_j^t x_j^2 \quad \text{for all } t = 1 \text{ to } r \quad (8.15)$$

Let $0 < \alpha < 1$. Suppose the maximum in (8.13) when $x = \alpha x^1 + (1-\alpha)x^2$ is attained by $t = u$. Then $s(\alpha x^1 + (1-\alpha)x^2) = \alpha(c_0^u + \sum_{j=1}^n c_j^u x_j^1) + (1-\alpha)(c_0^u + \sum_{j=1}^n c_j^u x_j^2) \leqq \alpha s(x^1) + (1-\alpha)s(x^2)$ by (8.14) and (8.15). So $s(x)$ is convex. See Figure 8.3, where several affine functions defined on $\mathbf{R}^1$ are illustrated, together with their pointwise supremum.

For each $x \in \mathbf{R}^n$, the maximum in (8.13) is attained by at least one term on the right-hand side. Clearly for all x satisfying $c_0^p + \sum_{j=1}^n c_j^p x_j - (c_0^t + \sum_{j=1}^n c_j^t x_j) \geqq 0$ for each $t = 1$ to r, $t \neq p$, we have $s(x) = c_0^p + \sum_{j=1}^n c_j^p x_j$. This partitions $\mathbf{R}^n$ into at most r convex polyhedral regions, such that in each region $s(x)$ is affine. So $s(x)$ is also piecewise linear. Hence $s(x)$ is piecewise linear and convex. ■

A function defined as in (8.13) is known as a *polyhedral convex function*. The important property of such a function is that the subset of $\mathbf{R}^{n+1}$,

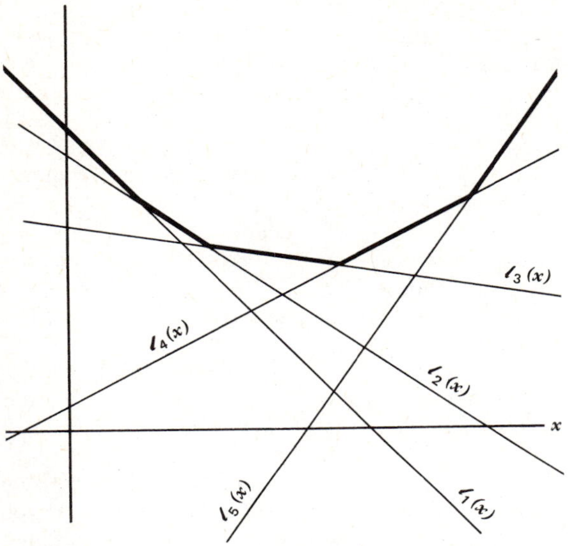

Figure 8.3 Functions $\ell_1(x)$ to $\ell_5(x)$ are five affine functions defined on $\mathbf{R}^1$. Function values are plotted on the vertical axis. Their pointwise maximum is the function marked with thick lines here.

$\{(x_1, \ldots, x_n, \gamma)^{\mathrm{T}}$, where $\gamma \geqq f(x)$, $x = (x_1, \ldots, x_n)^{\mathrm{T}} \in \mathbf{R}^n\}$ is a convex polyhedral subset of $\mathbf{R}^{n+1}$.

Now, consider the following. Let $\mathbf{K} \subset \mathbf{R}^n$ be a closed convex polyhedral set that is partitioned into closed convex polyhedral regions as $\bigcup_{t=1}^{r} \mathbf{K}_t$. So if $u \neq v$, the interiors of $\mathbf{K}_u$ and $\mathbf{K}_v$ have an empty intersection, and $\mathbf{K}_u \cap \mathbf{K}_v$ is itself either empty or is either a face or a subset of a face of each of $\mathbf{K}_u$ and $\mathbf{K}_v$. We assume that each $\mathbf{K}_t$ has a nonempty interior. Suppose the real-valued function $f(x)$ is defined on $\mathbf{K}$ by the following:

$$f(x) = f_t(x) = c_0^t + \sum_{j=1}^{n} c_j^t x_j \qquad \text{if } x \in \mathbf{K}_t \qquad (8.16)$$

where c_0^t, c_j^t are all given constants. Let the row vector $c^t = (c_1^t, c_2^t, \ldots, c_n^t)$. The definition (8.16) assumes that if $\mathbf{K}_u \cap \mathbf{K}_v \neq \emptyset$, then $f_u(x) = f_v(x)$ for all $x \in \mathbf{K}_u \cap \mathbf{K}_v$. The function $f(x)$ defined by (8.16) is a continuous piecewise linear function defined on $\mathbf{K}$.

THEOREM 8.8 The function $f(x)$ defined by (8.16) is convex on $\mathbf{K}$ iff for each $t = 1$ to r

$$f_t(x) = \text{maximum } \{f_1(x), \ldots, f_r(x)\} \quad \text{whenever } x \in \mathbf{K}_t$$

Proof: The "if" part follows from Theorem 8.7. To prove the "only if" part, assume that $f(x)$ is convex. For $u \neq v$, select interior points x^1, x^2 in $\mathbf{K}_u$, $\mathbf{K}_v$, respectively. Let $x(\lambda) = \lambda x^1 + (1-\lambda)x^2$, $0 \leqq \lambda \leqq 1$. Let $g(\lambda) = f(x(\lambda))$. As λ increases from 0 to 1, the point $x(\lambda)$ may cut across several of the polyhedral regions $\mathbf{K}_t$, and whenever it moves into a new region the definition of $f(x(\lambda))$ changes as in (8.16). Since $f(x)$ is convex on $\mathbf{K}$, $g(\lambda)$ must be convex in $0 \leqq \lambda \leqq 1$. As discussed in Section 1.2, $g(\lambda)$ is convex iff its slope is monotone increasing in λ. This theorem follows directly from this condition. ■

Clearly the condition in Theorem 8.6 holds iff for each $t = 1$ to r and $u \neq v$, minimum $\{f_u(x) - f_v(x) : x \in \mathbf{K}_u\} \geqq 0$. Verification of this requires the solution of at most $r(r-1)$ linear programs. This provides a finite, constructive procedure for checking whether the piecewise linear function $f(x)$ is convex.

Note 8.5: If we change the maximum in (8.13) into minimum, we get a function known as the *pointwise infimum* of a given set of affine functions. Using arguments similar to those in Theorem 8.7, it can be shown that the pointwise infimum of a given set of affine functions is piecewise linear and concave.

Subgradients and Subdifferentials of Convex Functions

Let $f(x)$ be a real-valued convex function defined on $\mathbf{R}^n$. Let $x^0 \in \mathbf{R}^n$ be a point where $f(x^0)$ is finite. The vector $d = (d_1, \ldots, d_n)^{\mathrm{T}}$ is said to be a *subgradient* of $f(x)$ at x^0 if

$$f(x) \geqq f(x^0) + d^{\mathrm{T}}(x - x^0) \qquad \text{for all } x \qquad (8.17)$$

Notice that the right-hand side of (8.17) is $\ell(x) = (f(x^0) - d^{\mathrm{T}}x^0) + d^{\mathrm{T}}x$, is an affine function in x, and we have $f(x^0) = \ell(x^0)$. One can verify that $\ell(x)$ is the first-order Taylor expansion for $f(x)$ around x^0, constructed using the vector d in place of the gradient vector of $f(x)$ at x^0. So d is a subgradient of $f(x)$ at x^0 iff this modified Taylor approximation is always an underestimate for $f(x)$ at every x.

Example 8.7

Let $x \in \mathbf{R}^1$, $f(x) = x^2$. The function $f(x)$ is convex. Consider the point $x^0 = 1$, $d = 2$. It can be verified that the inequality (8.17) holds in this case. So $d = (2)$ is a subgradient for $f(x)$ at $x^0 = 1$ in this case. The affine function $\ell(x)$ on the right-hand side of (8.17) in this case is $1 + 2(x-1) = 2x - 1$. See Figure 8.4, where the inequality (8.17) is illustrated.

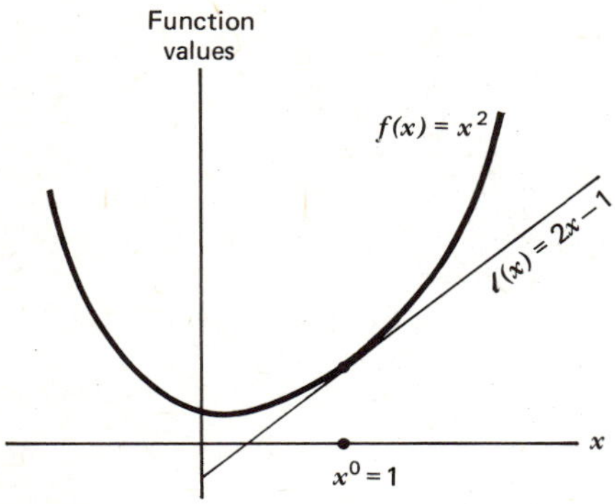

Figure 8.4 A convex function and the affine function below it constructed using a subgradient for it at the point x^0.

The set of all subgradients of $f(x)$ at x^0 is denoted by the symbol $\partial f(x^0)$, and called the *subdifferential set* of $f(x)$ at x^0. It can be proved that if $f(x)$ is differentiable at x^0, then its gradient $\nabla f(x^0)$ is the unique subgradient of $f(x)$ at x^0. Conversely, if $\partial f(x^0)$ contains a single vector, then $f(x)$ is differentiable at x^0 and $\partial f(x^0) = \{\nabla f(x^0)\}$. See the book [8.6] by R. T. Rockafellar for these and other related results.

Subgradients of Concave Functions

Let $g(x)$ be a concave function defined on a convex subset $\Gamma \subset \mathbf{R}^n$. In defining a subgradient vector for $g(x)$ at a point $x^0 \in \Gamma$, the inequality in (8.17) is just reversed; in other words, d is a subgradient for the concave function $g(x)$ at x^0 if $g(x) \leqq g(x^0) + d^T(x - x^0)$ for all x. With this definition, all the results stated above also hold for concave functions.

Optimum Objective Value as a Function of the Right-Hand-Side Constants

Consider the LP

$$\begin{aligned} \text{Minimize} \quad & z(x) = cx \\ \text{Subject to} \quad & Ax = b \\ & x \geqq 0 \end{aligned} \tag{8.18}$$

where A is a matrix of order $m \times n$ and rank m. Assume that A and c are fixed, but that b can vary over $\mathbf{R}^m$. Let $f(b)$ denote the optimum objective value in (8.18) as a function of b. As discussed earlier in Section 3.2, (8.18) has a feasible solution iff $b \in \text{Pos}(A)$. So $f(b)$ is defined only on the cone $\text{Pos}(A)$. If $f(b) = -\infty$ for some $b \in \text{Pos}(A)$, by Corollary 3.3 in Sections 3.7, $f(b) = -\infty$ for every $b \in \text{Pos}(A)$, and in this case there is nothing else to discuss. So we assume that $f(b)$ is finite for some $b \in \text{Pos}(A)$, and in this case it is finite for every $b \in \text{Pos}(A)$.

Let B_1 be a dual feasible basis for (8.18). Let $x_{B_1} = (x_1, \ldots, x_m)^T$, x_{D_1}, c_{B_1} denote the corresponding basic vector, nonbasic vector, and original basic cost vector, respectively. B_1 is therefore an optimum basis for all $b \in \text{Pos}(B_1)$, and inside this simplicial cone $f(b) = c_{B_1}B_1^{-1}b$, which is linear. For $b \in \text{Pos}(B_1)$, let $\bar{x}(b)$ denote the corresponding BFS. Since we assumed $x_{B_1} = (x_1, \ldots, x_m)^T$, $B_1 = (A_{.1}, \ldots, A_{.m})$. Consider the facet $\text{Pos}(A_{.2}, \ldots, A_{.m})$ of $\text{Pos}(B_1)$. Suppose we start at a point b^0 contained in the interior of $\text{Pos}(B_1)$, and move to a point $b^1 \in \text{Pos}(A_{.2}, \ldots, A_{.m})$ and then leave $\text{Pos}(B_1)$ by continuing to move in a straight line (see Figure 8.5). As we move from b^0 to b^1, the value of x_1 decreases from a positive value to zero in the basic solution $\bar{x}(b)$ defined above. As we leave $\text{Pos}(B_1)$ through b^1, the value of x_1 in $\bar{x}(b)$ becomes negative. To keep primal feasibility (and optimality) as we are moving, a dual simplex pivot step must be carried out when we reach b^1, with the pivot row as the row in which x_1 is the basic variable in the canonical tableau of (8.18) with respect to the basic vector x_{B_1}. Two possibilities may occur when this dual simplex pivot step is attempted. (1) The primal infeasibility criterion may be satisfied, which implies that if we change b by leaving $\text{Pos}(B_1)$ through its facet $\text{Pos}(A_{.2}, \ldots, A_{.m})$, (8.18) becomes infeasible.

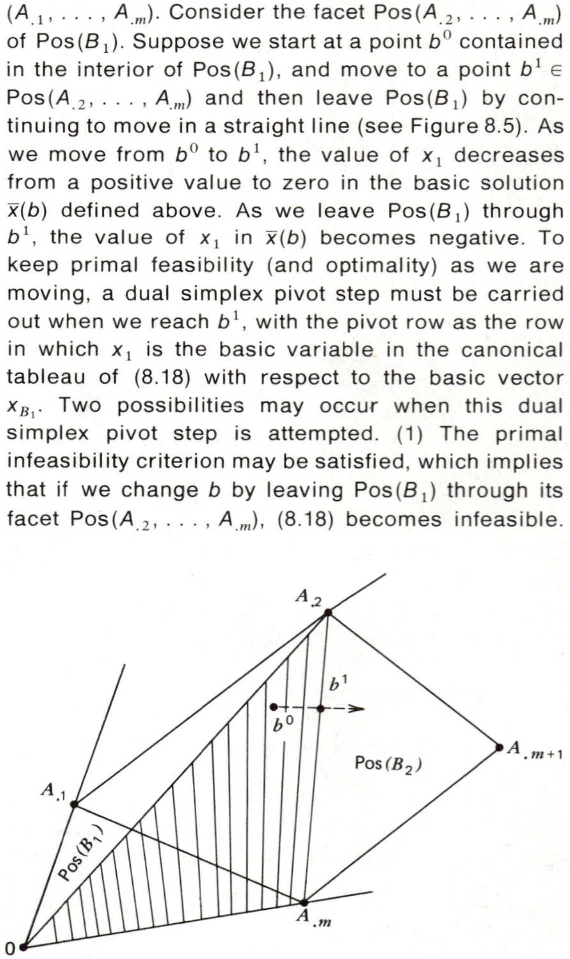

Figure 8.5 The two adjacent simplicial cones $\text{Pos}(B_1)$ and $\text{Pos}(B_2)$. Their common facet $\text{Pos}(A_{.2}, \ldots, A_{.m})$ is marked with dashed lines.

In this case, $\text{Pos}(A_{.2}, \ldots, A_{.m})$ is a boundary face of $\text{Pos}(A)$. (2) A dual simplex pivot step may be possible. Suppose the entering variable is x_{m+1}. This pivot step leads to an adjacent dual feasible basis for (8.18), B_2, say. By the same arguments used earlier, we conclude that $f(b) = c_{B_2}B_2^{-1}b$ whenever $b \in \text{Pos}(B_2)$.

In case (2), $\text{Pos}(A_{.2}, \ldots, A_{.m})$ is a common boundary facet of both $\text{Pos}(B_1)$ and $\text{Pos}(B_2)$, and whenever $b \in \text{Pos}(A_{.2}, \ldots, A_{.m})$, both B_1 and B_2 are alternate optimal bases for (8.18). This is the situation illustrated in Figure 8.5.

Since the total number of dual feasible bases for (8.18) if finite, when this procedure is continued in this manner in different directions repeatedly, we obtain a partition of $\text{Pos}(A)$ as a union of the simplicial cones, say, $\text{Pos}(B_1)$, $\text{Pos}(B_2)$, . . . , $\text{Pos}(B_r)$, where $B_1, \ldots, B_r$ are all dual feasible bases for (8.18), such that inside each of $\text{Pos}(B_t)$, $f(b) = c_{B_t}B_t^{-1}b$ is linear for $t = 1$ to r. Here c_{B_t} is the vector of original basic cost coefficients associated with the basis B_t. Hence $f(b)$ is piecewise linear. In this partition, if two simplicial cones $\text{Pos}(B_p)$ and $\text{Pos}(B_q)$ have a nonempty intersection, they must intersect in a common boundary face, and whenever b is any point in this face, both B_p and B_q are alternate optimum bases for (8.18). This implies that at such points we have $c_{B_p}B_p^{-1}b = c_{B_q}B_q^{-1}b$, hence $f(b)$ is continuous on $\text{Pos}(A)$. We now prove that $f(b)$ is convex on $\text{Pos}(A)$.

THEOREM 8.9 The function $f(b)$ is a piecewise linear convex function defined on $\text{Pos}(A)$.

Proof: We have already shown above that $f(b)$ is piecewise linear. Let b^1 and b^2 be any two points in $\text{Pos}(A)$. Let x^1 and x^2 be optimum solutions for (8.18) when $b = b^1, b^2$, respectively. So $f(b^1) = cx^1$, $f(b^2) = cx^2$, $Ax^1 = b^1$, $x^1 \geqq 0$, $Ax^2 = b^2$, $x^2 \geqq 0$. Let $0 \leqq \alpha \leqq 1$. From these we conclude that $\alpha x^1 + (1 - \alpha)x^2$ is a feasible solution for (8.18) when $b = \alpha b^1 + (1 - \alpha)b^2$. But it may not be optimal. Thus the optimum objective value in (8.18) when $b = \alpha b^1 + (1 - \alpha)b^2$, is then $f(\alpha b^1 + (1 - \alpha)b^2) \leqq c(\alpha x^1 + (1 - \alpha)x^2) = \alpha(cx^1) + (1-\alpha)cx^2 = \alpha f(b^1) + (1-\alpha)f(b^2)$. So $f(\alpha b^1 + (1-\alpha)b^2) \leqq \alpha f(b^1) + (1 - \alpha)f(b^2)$, which proves that $f(b)$ is convex.

We provide here a second method of proof to establish that $f(b)$ is piecewise linear and convex on $\text{Pos}(A)$. The dual problem of (8.18) is

$$\begin{aligned} &\text{Maximize} && \pi b \\ &\text{Subject to} && \pi A \leqq c \end{aligned} \tag{8.19}$$

By our assumptions $f(b)$ is finite on $\text{Pos}(A)$, so by the duality theorem (8.19) is feasible, and $f(b)$ is the optimum objective value in (8.19). Notice that the constraints in (8.19) are independent of b, and they depend only on A and C, which are fixed constants. Let $\pi^1, \pi^2, \ldots, \pi^L$ be all the extreme points of the set of feasible solutions of (8.19). By the results in Chapter 3, one of these extreme points must be optimal to (8.19) for any b. So, for each $b \in \text{Pos}(A)$, $f(b) = \text{maximum } \{\pi^1 b, \pi^2 b, \ldots, \pi^L b\}$. Since $\pi^1, \pi^2, \ldots, \pi^L$ are known vectors independent of b, each of $\pi^1 b, \pi^2 b, \ldots, \pi^L b$ is a linear function in b defined on $\text{Pos}(A)$, and hence $f(b)$ is the pointwise maximum of a set of linear function. By Theorem 8.7 $f(b)$ is convex and piecewise linear. ■

Note 8.6: Theorem 8.9 is only true for an LP of the form (8.18), in which the objective function is required to be minimized. If the objective function is to be maximized, $f(b)$ will turn out to be concave on $\text{Pos}(A)$.

THEOREM 8.10 If $f(b)$ is finite for some $b \in \text{Pos}(A)$, then $f(0) = 0$. Also $f(\lambda b) = \lambda f(b)$, for all $\lambda \geqq 0$.

Proof: The fact that $f(0) = 0$ follows from Corollary 3.5 of Section 3.7. Also, if $\bar{x}$ is an optimum feasible solution for (8.18), $\lambda\bar{x}$ is an optimum feasible solution for the problem obtained when b in (8.18) is replaced by λb, for any $\lambda \geqq 0$. So $f(\lambda b) = \lambda f(b)$ for all $\lambda \geqq 0$. ■

A real-valued function satisfying the property $f(\lambda b) = \lambda f(b)$ for all $\lambda \geqq 0$ and b, is mathematically known as a *positively homogeneous function.* So the optimum objective value in (8.18) considered as a function of b while all the other data remain fixed at present values, is a positively homogeneous function.

THEOREM 8.11 If b^0 is in the interior of $\text{Pos}(B^0)$, where B^0 is a dual feasible basis for (8.18), then $f(b)$ is differentiable at b^0, and its gradient vector at b^0 is $(\pi^0)^T = (c_{B_0}B_0^{-1})^T$, the dual solution of (8.18) corresponding to the basis B_0.

Proof: Here c_{B_0} is the vector of original cost coefficients in (8.18) associated with the basis B_0. Since $f(b) = c_{B_0}B^{-1}b$ for each $b \in \text{Pos}(B^0)$, these results follow directly. ■

THEOREM 8.12 If π^1 is a dual optimum solution corresponding to (8.18) for some $b \in b^1 \in \text{Pos}(A)$,

$$f(b) \geqq \pi^1 b, \qquad \text{for all } b \in \text{Pos}(A)$$

Proof: The dual of (8.18) is (8.19). By the duality theorem, $f(b)$ is the optimum objective value in (8.19) for each $b \in \text{Pos}(A)$. Whatever b may be, π^1 is a feasible, but not necessarily an optimum solution of (8.19). Hence the result follows. ■

THEOREM 8.13 If π^1 is an optimum dual solution associated with (8.2) when $b = b^1$, then $(\pi^1)^T$ is a subgradient of $f(b)$ at b^1.

Proof: By the duality theorem, $f(b^1) = \pi^1 b^1$. So by Theorem 8.12, we have $f(b) \geqq f(b^1) + \pi^1(b - b^1)$, for all $b \in \text{Pos}(A)$. So $(\pi^1)^T$ is a subgradient of $f(b)$ at b^1. ■

So $f(b)$ is a piecewise linear convex function. Its domain Pos(A) is partitioned into the union of simplicial cones $\bigcup_{t=1}^{r} \text{Pos}(B_t)$, where each B_t is a dual feasible basis for (8.18). If b is in the interior of any one of these cones, then $f(b)$ is differentiable at b, and the unique dual optimum solution associated with (8.18) for that b is the gradient vector of $f(b)$ at b. So in a mathematical sense, $f(b)$ is differentiable almost everywhere on Pos(A) (except at points on the boundaries of the simplicial cones Pos(B_t), $t = 1$ to r, which together constitute a set of measure zero). If b is on the boundary of any of these cones Pos(B_t), then $f(b)$ may not be differentiable at that point, but every dual optimum solution associated with (8.18) for that b is a subgradient vector at that point, and hence is contained in the subdifferential set of $f(b)$ at that point b.

THEOREM 8.14 Suppose the column vector $d = (d_1, \ldots, d_m)^T$ is a subgradient vector for $f(b)$ at the point $b^1 \in \text{Pos}(A)$. Then d^T is a dual optimum solution associated with (8.18) when $b = b^1$.

Proof: By the hypothesis we have $f(b) \geqq f(b^1) + d^T(b - b^1)$ for all $b \in \text{Pos}(A)$. Substituting $b = 0 \in \text{Pos}(A)$ in this inequality, we get $db^1 \geqq f(b^1)$, since $f(0) = 0$ by Theorem 8.10. Again, substituting $b = 2b^1$ in the subgradient inequality, we get $f(2b^1) \geqq f(b^1) + db^1$. Since $f(2b^1) = 2f(b^1)$ by Theorem 8.10, this leads to $f(b^1) \geqq db^1$. So $f(b^1) = db^1$. Substituting this in the subgradient inequality, we get $f(b) \geqq d^T b$. Substituting $b = A_{.j}$ in this inequality, we get $f(A_{.j}) \geqq d^T A_{.j}$. But $f(A_{.j}) \leqq c_j$, since $x = I_{.j}$ (where I is the identity matrix of order n) is a feasible but not necessarily optimal solution to (8.18) when $b = A_{.j}$. Combining these we get $c_j \geqq d^T A_{.j}$ for all $j = 1$ to n, that is $d^T A \leqq c$. So d^T is feasible to the dual of (8.18) and $d^T b^1 = f(b^1)$, optimum objective value in (8.18) when $b = b^1$. So by the duality theorem (Theorem 4.5 of Section 4.5.3), d^T is optimal to the dual of (8.18) when $b = b^1$. ■

THEOREM 8.15 For $b \in \text{Pos}(A)$, the subdifferential set of $f(b)$ at b is the set of optimum dual solutions associated with (8.18).

Proof: Follows from Theorems 8.13 and 8.14. ■

Optimum Objective Value as a Function of the Original Cost Coefficients

Consider the LP (8.18) again. Now suppose A and b remain fixed, but the c vector is allowed to vary over $\mathbf{R}^n$. Let $g(c)$ denote the optimum objective value of (8.18) as a function of c. Here we study the properties of $g(c)$. If (8.18) is infeasible, then it remains infeasible whatever c might be and there is nothing else to discuss. So we assume that (8.18) is feasible.

THEOREM 8.16 The function $g(c)$ is piecewise linear and concave.

Proof: By the duality theorem, $g(c)$ is the optimum objective value in (8.19). Using this and arguments similar to those used earlier, this theorem can be proved easily. ■

Let $\bar{x}(c)$ denote an optimum solution for (8.18) as a function of c. Using the results established so far, we conclude that $g(c)$ is differentiable whenever it is finite and c is such that (8.18) is dual nondegenerate, and at such points $\bar{x}(c)$ is the gradient vector of $g(c)$. If $g(c)$ is finite and (8.18) is not dual nondegenerate, then $\bar{x}(c)$ is a subgradient of $g(c)$ at c. The subdifferential set of $g(c)$ at c is the set of optimum solutions of (8.18).

Comment 8.1: For a detailed treatment of the mathematical properties of convex, concave functions, subgradients, etc., see the book [8.6] by R. T. Rockafellar. Theorem 8.8 is taken from [8.5]. Results on the computational complexity of parametric linear programming discussed in Section 8.12 are from [8.4].

8.15 MARGINAL VALUES, OR SHADOW PRICES

Consider the LP (8.18). Let $f(b)$ denote the optimum objective value in it as a function of b, while A and c

remain fixed. We assume that $f(b)$ is finite for $b \in \text{Pos}(A)$. In Chapter 4 we defined the *marginal value* or *shadow price* of the item corresponding to the ith constraint in (8.18), to be the rate of change in the optimum objective value of (8.18) per unit change in the value of b_i, when b is such that the dual of (8.18) has a unique optimum solution. At such points b, $f(b)$ is differentiable and the vector of marginal values is $\partial f(b)/\partial b = \pi$, the unique optimum dual solution, by Theorem 8.11.

Suppose the point b is such that the dual of (8.18) has alternate optimum solutions. At such points b, $f(b)$ is not differentiable. However, since $f(b)$ is piecewise linear, it is possible to define two marginal values corresponding to the ith constraint in (8.18) now, one for measuring the change in the optimum objective value per unit increase from b_i, and the other for measuring the change of the same per unit decrease of b_i. Specifically, the *positive marginal value* (also called *positive shadow price* or the *marginal value* or *shadow price per unit increase*) of constraint i in (8.18) is defined as the limit of $(f(b + \varepsilon I_{.i} - f(b))/\varepsilon$ as ε tends to zero through positive quantities, where $I_{.i}$ is the ith column of the unit matrix of order n. And the *negative marginal value* (also called *negative shadow price*, or the *marginal value* or *shadow price per unit decrease*) of constraint i is defined as the limit of $(f(b - \varepsilon I_{.i} - f(b))/(-\varepsilon)$ as ε tends to zero through positive quantities. So the positive-negative marginal values of constraint i are, respectively, the right-side and left-side partial derivatives of $f(b)$ with respect to b_i at b.

Since $f(b)$ is piecewise linear, both the positive and negative marginal values exists for all $b \in \text{Pos}(A)$ whether $f(b)$ is differentiable at b or not. Let $\mathbf{\Gamma}$ equal the set of optimum solutions of the dual of (8.18). Then, from Section 8.10, we have

$$\text{Positive marginal value of constraint } i = \text{maximum } \{\pi_i : \pi \in \mathbf{\Gamma}\}$$

$$\text{Negative marginal value of constraint } i = \text{minimum } \{\pi_i : \pi \in \mathbf{\Gamma}\}$$

If b is such that $f(b)$ is differentiable at b, $\mathbf{\Gamma}$ contains a single point and both these marginal values are equal. At points b where $f(b)$ is not differentiable, these two marginal values may be different, and they provide information about the rate of change in the optimum objective value in (8.18) as b_i increases or decreases from its present value. $\mathbf{\Gamma}$, which is the set of optimum dual solutions associated with (8.18), can be described through a system of linear constraints as in Section 6.7, and each of these marginal values can be computed by solving one LP on $\mathbf{\Gamma}$. Or these marginal values can also be computed by applying the parametric right-hand-side simplex algorithm, with b in (8.18) replaced by $b + \lambda I_{.i}$ or $b - \lambda I_{.i}$, starting with the known optimum basis for $\lambda = 0$ and applying the algorithm until λ just becomes positive, and then taking the slope obtained as the appropriate marginal value.

Example 8.8

Consider the following LP.

$$\begin{array}{llr} \text{Minimize} & -3x_1 - x_2 & \\ \text{Subject to} & x_1 + x_2 + x_3 & = 1 \\ & 2x_1 + x_2 \qquad + x_4 & = 2 \\ & x_j \geqq 0 \quad \text{for all } j & \end{array}$$

It can be verified that the set of dual optimum solutions corresponding to this problem is $\{\alpha(-3, 0) + (1 - \alpha)(0, -3/2) : 0 \leqq \alpha \leqq 1\}$. So the positive marginal values of both constraints are 0, and the negative marginal values of constraints 1 and 2 are, respectively, -3 and $-3/2$.

Exercises

8.11 Compute the positive-negative marginal values corresponding to each constraint in the following LP, and give a practical interpretation for each:

$$\begin{array}{llr} \text{Minimize} & -x_2 - 2x_3 & \\ \text{Subject to} & x_1 + x_2 + x_3 + x_4 & = 1 \\ & x_3 \qquad + x_5 & = 1 \\ & x_j \geqq 0 \quad \text{for all } j & \end{array}$$

Summary

The positive and negative marginal values of items corresponding to the various constraints in an LP model are well defined and always exist whether the dual optimum solution is unique or not. They provide information about the rate of change in the optimum objective value per unit increase or decrease of the right-hand-side constants, and thus are very useful for planning purposes. The distinction between the positive and negative marginal values is very important and often ignored in technical literature. See the book [2.18] by J. F. Shapiro and the papers [4.2, 4.1] of D. C. Aucamp, D. I. Steinberg, and M. Akgül. Usually commercial software packages for LP provide in the output an optimum primal solution and an optimum dual solution, without bothering to verify the existence of alternate optima. If the optimum primal solution is nondegenerate, the optimum dual solution is unique and it is the marginal value vector. If the optimum primal solution is degenerate, there may be alternate dual optima. In that case, treating the optimum dual solution from the output as the marginal value vector could lead to erroneous conclusions. To avoid such errors when the optimum primal solution is degenerate, the user should determine the positive and negative marginal values for each item as previously described (possibly through a parametric right-hand-side analysis as previously mentioned) and use these to determine profitable resource level changes. Also, in (8.18), if b is required to be changed in a specified direction b^* (i.e., if b in (8.18) is being replaced by $b + \lambda b^*$ for some positive λ), then the directional derivative of $f(b)$ in this direction b^* is the limit of $(f(b + \varepsilon b^*) - f(b))/\varepsilon$ as ε tends to zero through positive values, and it is the maximum value of πb^* over the set of alternate dual optimum solutions associated with (8.18), as shown in Section 8.10. This directional derivative is the *marginal value* in the LP (8.18) *in the direction* b^*, and it can be computed efficiently by solving an LP, using the results in Sections 8.10 and 6.7. For any direction b^*, the marginal value in (8.18) in that direction is well defined and always exists (whether the dual of (8.18) has a unique optimum solution or alternate optimum solutions) and can be computed efficiently.

Exercises

8.12 Solve the following parametric cost problems for all real values of the parameter.

(a)

$$\begin{aligned} \text{Minimize} \quad & (-4 + 4\lambda)x_4 + (3 + 2\lambda)x_5 + (1 - 2\lambda)x_6 \\ \text{Subject to} \quad & x_1 - 2x_4 + x_5 + x_6 = 5 \\ & x_2 - 2x_4 + x_5 + 2x_6 = 4 \\ & x_3 + x_5 - x_6 = 6 \\ & x_j \geqq 0 \quad \text{for all } j \end{aligned}$$

(b)

$$\begin{aligned} \text{Minimize} \quad & \lambda x_1 - x_2 \\ \text{Subject to} \quad & 3x_1 - x_2 \leqq 5 \\ & 2x_1 + x_2 \leqq 3 \\ & x_1, x_2 \text{ unrestricted} \end{aligned}$$

(c)

$$\begin{aligned} \text{Minimize} \quad & 2\lambda x_1 + (1 - \lambda)x_2 - 3x_3 + \lambda x_4 + 2x_5 - 3\lambda x_6 \\ \text{Subject to} \quad & x_1 + 3x_2 - x_3 + 2x_5 = 7 \\ & - 2x_2 + 4x_3 + x_4 = 12 \\ & - 4x_2 + 3x_3 + 8x_5 + x_6 = 10 \\ & x_j \geqq 0 \quad \text{for all } j \end{aligned}$$

(d)

$$\begin{aligned} \text{Minimize} \quad & \lambda x_1 + x_2 + x_3 + (9\lambda - 41)x_4 + (15 - \lambda)x_5 + (20 - 4\lambda)x_6 + (9 - \lambda)x_7 + (\lambda + 2)x_8 \\ \text{Subject to} \quad & x_1 + 2x_2 - 3x_4 + 3x_5 + 3x_6 + 5x_7 + 3x_8 = 22 \\ & x_2 + x_3 - x_4 + 3x_5 + 3x_6 + x_7 + 2x_8 = 12 \\ & 2x_2 + x_3 - 2x_4 + 4x_5 + 5x_6 + 3x_7 + 3x_8 = 21 \\ & x_j \geqq 0 \quad \text{for all } j \end{aligned}$$

(e)

$$\begin{aligned} \text{Minimize} \quad & \lambda x_1 + 2\lambda x_2 + x_3 + (\lambda - 9)x_4 + (5\lambda + 6)x_5 + (9 - 4\lambda)x_6 + (4\lambda - 3)x_7 \\ \text{Subject to} \quad & x_1 + x_3 + x_4 + 2x_5 - 2x_6 + 3x_7 = 5 \\ & x_3 - x_4 + x_5 - x_6 + 2x_7 = 3 \\ & x_1 + x_2 + x_4 + 3x_5 - 2x_6 + 2x_7 = 3 \\ & x_j \geqq 0 \quad \text{for all } j \end{aligned}$$

(f)

$$\begin{aligned} \text{Minimize} \quad & (1 - \lambda)(x_1 + x_2 + x_3) + (6 - 4\lambda)x_4 + (5 - 3\lambda)x_5 + (4 + \lambda)x_6 + (2 + 5\lambda)x_7 \\ \text{Subject to} \quad & x_1 + x_2 + x_3 + 4x_4 + 2x_5 - 3x_7 = 15 \\ & -x_1 - x_2 - 3x_4 - x_5 - 2x_6 + 2x_7 = -12 \\ & x_2 + x_3 + 2x_4 + 3x_5 - x_6 - 2x_7 = 13 \\ & x_j \geqq 0 \quad \text{for all } j \end{aligned}$$

8.13 Consider the parametric cost LP:

$$\begin{aligned} \text{Minimize} \quad & 3x_1 + (4 + 8\lambda)x_2 + 7x_3 - x_4 + (1 - 3\lambda)x_5 \\ \text{Subject to} \quad & 5x_1 - 4x_2 + 14x_3 - 2x_4 + x_5 = 20 \\ & x_1 - x_2 + 5x_3 - x_4 + x_5 = 8 \\ & x_j \geqq 0 \quad \text{for all } j \end{aligned}$$

For what range of values of λ is the basic vector (x_2, x_3) optimal for this problem? Determine optimum feasible solutions for all $\lambda \geqq 1$.

8.14 Consider the parametric cost LP (8.1). The objective function is known to be unbounded below on the set of feasible solutions of this problem when λ is λ_0. Prove that the objective function must be unbounded below on the set of feasible solutions of this problem either in the interval $\lambda \geqq \lambda_0$ or in the interval $\lambda \leqq \lambda_0$ or both.

8.15 Solve for all real values of the parameters λ, μ

$$\begin{aligned} \text{Minimize} \quad & x_1 + \lambda x_2 + \mu x_3 + x_4 + x_5 + x_6 + x_7 + \lambda x_8 + \mu x_9 \\ \text{Subject to} \quad & x_1 - x_4 - 2x_6 + 4x_7 + 2x_8 + x_9 = 4 \\ & x_2 + 2x_4 - 2x_5 + x_6 + 2x_7 + 4x_8 = 2 \\ & x_3 + x_5 + x_6 + x_7 + 2x_9 = 1 \\ & x_j \geqq 0 \quad \text{for all } j \end{aligned}$$

8.16 Solve the following parametric right-hand-side problems for all real values of the parameter:

(a)

$$\begin{aligned} \text{Minimize} \quad & x_1 + 2x_2 - x_3 - 4x_4 - x_5 + 2x_6 + x_7 \\ \text{Subject to} \quad & x_1 + x_2 + x_3 + x_5 = 9 - 2\lambda \\ & x_2 + x_3 + x_4 - x_5 + 2x_6 - x_7 = 4 - \lambda \\ & x_1 + x_2 + 2x_3 + 2x_4 + 2x_5 + x_6 + x_7 = 5 - \lambda \\ & x_j \geqq 0 \quad \text{for all } j \end{aligned}$$

(b)

$$\begin{aligned} \text{Minimize} \quad & 2x_1 + 3x_2 + 4x_3 + 5x_4 \\ \text{Subject to} \quad & 2x_1 - x_2 + x_3 - x_4 \leqq 2 - \lambda \\ & x_1 + 2x_2 + x_3 - x_4 \leqq 10 - 2\lambda \\ & x_1 + x_2 - 2x_3 - x_4 \leqq 3 + \lambda \\ & x_j \geqq 0 \quad \text{for all } j \end{aligned}$$

8.17 Solve the following LP

$$\begin{array}{lrrrrrrrl}
\text{Minimize} & x_1 + & x_2 + 7x_3 + & 3x_4 + & x_5 + & 2x_6 + & x_7 & & \\
\text{Subject to} & x_1 + & 2x_2 - & x_3 - & x_4 + & x_5 + & 2x_6 + & x_7 = & 16 \\
 & & x_2 & & -3x_4 - & x_5 + & 3x_6 - & x_7 = & 2 \\
 & -x_1 & & -3x_3 + & 3x_4 + & x_5 & & -x_7 = & -4 \\
 & & x_j \geqq 0 & \text{for all } j & & & & &
\end{array}$$

In this problem, x_7 is the amount (in tons) of some product manufactured. It is rumored that the federal government will legislate that the company should manufacture some specified amount (which is unknown) of this product in the national interest. To prepare for this possibility, x_7 must be considered as a parameter and the problem solved for all nonnegative values of x_7. Work out the solution.

8.18 Solve the following parametric right-hand-side problem for all real values of the parameter λ and interpret your results geometrically.

$$\begin{array}{lrl}
\text{Minimize} & 3x_1 - 2x_2 & \\
\text{Subject to} & -x_1 + 2x_2 = & 0 - \lambda \\
 & 2x_1 - x_2 = & -1 + \lambda \\
 & x_1, x_2 \geqq 0 &
\end{array}$$

8.19 Consider the LP given at the bottom (λ and μ are two parameters). (1) Assuming that the parameters λ and μ are both equal to zero, compute the primal and dual solutions in the above LP corresponding to the basic vector (x_1, x_2, x_3) and show that it is an optimum basic vector. Check to see whether there are alternate optimum solutions in this case. If so, derive the formula for a general optimum solution in this case. (2) Identify the set of all values of the parameters λ and μ in the λ,μ plane for which the basic vector (x_1, x_2, x_3) remains optimal to this LP. (3) Fix $\lambda = 0$. How much is it worth to the decision maker to increase the parameter μ from zero per unit? For what range of values of μ is this rate valid? (4) Fix $\lambda = 0$. Starting with the basic vector (x_1, x_2, x_3), which is optimal for $\mu = 0$, obtain optimum feasible solutions to this LP for all values of the parameter μ in the range $\mu \geqq 0$.

8.20 Consider the following multiparameter parametric right-hand-side problem.

Minimize

$$z(x) = cx$$

Subject to

$$Ax = b + \lambda_1 b^1 + \lambda_2 b^2 + \cdots + \lambda_t b^t$$
$$x \geqq 0$$

Let $\lambda = (\lambda_1, \ldots, \lambda_t)$ be the parameter vector. Prove that the set of all λ for which the above problem is feasible is a convex subset of $\mathbf{R}^t$, the space of the parameter vector. Let this convex subset be denoted by Γ. Characterize this subset in terms of the data in the problem (namely A, b, $b^1, \ldots, b^t$).

8.21 Consider a multiparameter parametric LP where $\lambda = (\lambda_1, \ldots, \lambda_t)$ is the parameter vector. Prove that the optimum objective value in this problem, as a function of the parameter vector λ, is: (1) A convex function if the problem is a parametric right-hand-side problem and the objective function is to be minimized, or if the problem is a parametric cost problem and the objective function is to be maximized. (2) A concave function if the problem is a parametric right-hand-side problem and the objective function is to be maximized, or if the problem is a parametric cost problem and one objective function is to be minimized.

$$\begin{array}{lrrrrrrl}
\text{Minimize} & 2x_1 + 5x_2 + & (\lambda - 20)x_3 + & (28 - 2\lambda)x_4 + & (57 - \lambda)x_5 + & 13x_6 & & \\
\text{Subject to} & x_1 - x_2 + & x_3 - & 3x_4 - & x_5 - & x_6 = & 2 - 2\mu \\
 & x_1 \quad - & 2x_3 + & x_4 + & 5x_5 + & x_6 = & -7 + 2\mu \\
 & 2x_1 - 2x_2 + & 3x_3 - & 7x_4 - & 4x_5 - & 3x_6 = & 8 - 5\mu \\
 & & x_j \geqq 0 & \text{for all } j & & & &
\end{array}$$

REFERENCES

8.1 T. Gal and J. Nedoma, "Multiparametric Linear Programming," *Management Science 18* (1972), 406–422.

8.2 R. S. Garfinkel and P. L. Yu, "A Primal-Dual Surrogate Simplex Algorithm: Comparison with the Crisscross Method," Working Paper no. 22, College of Business Administration, The University of Tennessee, Knoxville, June 1975.

8.3 S. I. Gass and T. L. Saaty, "The Computational Algorithm for the Parametric Objective Function," *Naval Research Logistics Quarterly* 2, 1 (1955), 39–45.

8.4 K. G. Murty, "Computational Complexity of Parametric Linear Programming," *Mathematical Programming 19* (1980), 213–219.

8.5 K. G. Murty, "Convexity of Piecewise Linear Functions," Technical Report, Department of Industrial and Operations Engineering, The University of Michigan, Ann Arbor, 1980.

8.6 R. T. Rockafellar, *Convex Analysis*, Princeton University Press, Princeton, N.J. 1970.

8.7 H. Valiaho, "A Procedure for One-Parametric Linear Programming," *BIT 19* (Copenhagen, Denmark, 1979), 256–269.

chapter 9 Sensitivity Analysis

9.1 INTRODUCTION

Suppose we take a practical problem and formulate it as the LP

$$\begin{aligned} \text{Minimize} \quad & z(x) = cx \\ \text{Subject to} \quad & Ax = b \\ & x \geqq 0 \end{aligned} \tag{9.1}$$

where A is a matrix of order $m \times n$ and rank m, and obtain an optimum feasible solution for it. In most real-life problems, the coefficients in A, c, and b are estimated from practical considerations. After the final optimal feasible solution has been obtained we may discover that some of the entries in b, c, or A have to be changed or that extra constraints or variables have to be introduced into the model. Solving the modified problem from scratch will be wasteful. *Sensitivity analysis* (also called *post-optimality analysis*) deals with the problem of obtaining an optimum feasible solution of the modified problem starting with the optimum feasible solution of the old problem. We consider the problem of introducing only one change at a time (e.g., how to get the new optimum feasible solution if the value of only one c_j has to be changed or if only one new constraint has to be added, etc.) If several changes have to be made, make them one at a time, or extend the methods discussed here in an obvious manner to take care of several simultaneous changes. In Chapter 8 we discussed some types of postoptimality analysis where the right-hand-side vector or the cost coefficient vector vary linearly as a function of a parameter as it ranges over the real line. Here we discuss various other types of postoptimality analyses. Some of the postoptimality analyses discussed here, for example, changes in cost coefficients or the right-hand-side constants, can be viewed as specializations of the parametric analysis applied to those cases. For an idea of the kind of questions that can be answered using the methods of sensitivity analysis, see Exercise 9.16. In most practical applications using a linear programming model, marginal and sensitivity analysis provide economic information that is very useful in planning.

Let **K** denote the set of feasible solutions of (9.1). Suppose the optimal basis obtained for (9.1) is $\tilde{B}$, associated with the basic vector $x_{\tilde{B}} = (x_1, \ldots, x_m)$, $c_{\tilde{B}} = (c_1, \ldots, c_m)$. Let $\tilde{\pi} = c_{\tilde{B}}\tilde{B}^{-1}$ and let $\tilde{x}$ be the BFS of (9.1) corresponding to the basis $\tilde{B}$. For illustration we will use the following problem, which has $x_B = (x_1, x_2, x_3)$ as an optimum basic vector. Therefore, $\tilde{x} = (3, 4, 2, 0, 0, 0)^{\mathrm{T}}$ is an optimum feasible solution of this problem and the minimum objective value is $z(\tilde{x}) = 11$.

Tableau 9.1 Original Problem

x_1	x_2	x_3	x_4	x_5	x_6	$-z$	b
1	2	0	1	0	−6	0	11
0	1	1	3	−2	−1	0	6
1	2	1	3	−1	−5	0	13
3	2	−3	−6	10	−5	1	0

$x_j \geqq 0$ for all j; z to be minimized.

Tableau 9.2 Optimum Inverse Tableau for the Problem in Tableau 9.1

Basic Variables	*Inverse Tableau*				*Basic Values*
x_1	−1	−2	2	0	3
x_2	1	1	−1	0	4
x_3	−1	0	1	0	2
$-z$	−2	4	−1	1	−11

9.2 INTRODUCING A NEW ACTIVITY

Suppose a new variable called x_{n+1} has to be introduced into the model (9.1). Let $A_{.n+1}$, c_{n+1} be the input-output coefficient vector and the cost coefficient, respectively, of this new variable. Let $X^{\mathrm{T}} = (x^{\mathrm{T}}, x_{n+1})$. The augmented problem is

$$\begin{aligned} \text{Minimize} \quad & Z(X) = cx + c_{n+1}x_{n+1} \\ \text{Subject to} \quad & (A \vdots A_{.n+1})X = b \qquad (9.2) \\ & X \geqq 0 \end{aligned}$$

$\tilde{B}$ remains a feasible basis to the augmented problem, and the BFS of (8.2) corresponding to it is $\tilde{X}^{\mathrm{T}} = (\tilde{x}^{\mathrm{T}}, 0)$. From the optimality criterion, $\tilde{X}$ remains an optimum feasible solution of (9.2) if the relative cost coefficient of the new variable x_{n+1}, with respect to the basis $\tilde{B}$, is nonnegative; that is, if $\bar{c}_{n+1} = c_{n+1} - \tilde{\pi}A_{.n+1} \geqq 0$. On the other hand, if $\bar{c}_{n+1} < 0$, solve (9.2) by using the inverse tableau corresponding to the basis $\tilde{B}$ as an initial tableau. Bring x_{n+1} into the basic vector and complete the solution of (9.2) according to the revised simplex algorithm.

It often happens that one pivot (that of bringing the new variable x_{n+1} into the basic vector $x_{\tilde{B}}$) is all that is necessary to solve (9.2). However, there is no theoretical guarantee that this will be the general case. In some problems, it may be necessary to perform several pivots before the algorithm reaches a terminal basis for (9.2).

Example 9.1

Consider the LP in Tableau 9.1. Suppose a new variable x_7 corresponding to the column vector $A_{.7} = (1, 2, -3)^{\mathrm{T}}$ and cost coefficient $c_7 = -7$ is introduced. The relative cost coefficient of x_7 with respect to the basis in Tableau 9.2 is $\bar{c}_7 = c_7 + (-\tilde{\pi})A_{.7} = -7 + (-2, 4, -1)(1, 2, -3)^{\mathrm{T}} = 2$. Hence, the previous optimum solution with $x_7 = 0$ is still optimal. The optimum feasible solution of the new problem is $\tilde{X} = (3, 4, 2, 0, 0, 0, 0)^{\mathrm{T}}$, with the optimal objective value $= \tilde{Z} = 11$. Thus, if x_7 is the level of some new activity that has become available, we conclude that it is optimal to perform the new activity at zero level.

Example 9.2

Consider the LP in Tableau 9.1 again. Suppose we have to include a new variable x_7 with $A_{.7} = (3, -1, 1)^{\mathrm{T}}$ and $c_7 = 4$. The relative cost coefficient of x_7 with respect to the basis in Tableau 9.2 is $\bar{c}_7 = 4 + (-2, 4, -1)(3, -1, 1)^{\mathrm{T}} = -7 < 0$. We conclude that it is optimal to include the new activity in the basic vector. The updated column vector of x_7 with respect to the basis in Tableau 9.2 is

$$\begin{pmatrix} \bar{A}_{.7} \\ \cdots \\ \bar{c}_7 \end{pmatrix} = \begin{pmatrix} -1 & -2 & 2 & 0 \\ 1 & 1 & -1 & 0 \\ -1 & 0 & 1 & 0 \\ -2 & 4 & -1 & 1 \end{pmatrix} \begin{pmatrix} 3 \\ -1 \\ 1 \\ 4 \end{pmatrix} = \begin{pmatrix} 1 \\ 1 \\ -2 \\ -7 \end{pmatrix}$$

This is the pivot column. The minimum ratio test indicates that x_1 drops out of the basic vector when x_7 enters.

Basic Variables	*Inverse Tableau*				*Basic Values*
x_7	-1	-2	2	0	3
x_2	2	3	-3	0	1
x_3	-3	-4	5	0	8
$-z$	-9	-10	13	1	10

Since the relative cost coefficient of x_6 here is -6, this basis is again not optimal to the augmented problem. Bring x_6 into the basic vector and continue the algorithm until a terminal basis for the augmented problem is obtained.

Exercises

9.1 Complete the solution of this numerical problem.

9.2 Prove that the optimum objective value in (9.2) is less than or equal to the optimum objective value in (9.1). Construct a numerical example to show that the objective function in (9.2) may be unbounded below even though (9.1) has an optimum feasible solution.

9.3 How is the set of feasible solutions of the augmented problem related to **K**?

9.2.1 What Useful Planning Information Can Be Derived from This Type of Sensitivity Analysis?

The sensitivity analysis discussed above is very simple, and yet when a company has a linear programming model of its production operations, it can yield extremely useful planning information. For an example, consider the fertilizer problem (1.1) modeled in Section 1.1.3. From the discussion in Section 4.6.3, we know that the optimum dual solution associated with (1.1) is $\bar{\pi} = (5, 5, 0)$. Suppose a research chemist working for this manufacturer has come up with the formula for a new fertilizer called *Lushlawn*. The manufacture of Lushlawn requires as inputs $A_{.6} = (3, 2, 2)^{\mathrm{T}}$ tons of raw materials 1, 2, and 3, respectively, per ton. Should the manufacturer produce this new fertilizer? How much profit (in dollars per ton) should 1 ton of Lushlawn fetch in the marketplace for the manufacturer to consider it worth producing? To answer these questions (1.1) can be transformed into standard form, and the analysis discussed above is applied. It can be verified that the breakeven profit for Lushlawn is $\bar{\pi}A_{.6} = (5, 5, 0)(3, 2, 2)^{\mathrm{T}} = 25$ \$/ton. That is, Lushlawn is worth producing if it can be sold in the market at a price that leads to a profit $\geqq 25$ \$/ton manufactured. The fertilizer manufacturer can conduct a survey and determine whether the market would accept Lushlawn at a price greater than or equal to this breakeven level. Based on the results of the survey, the decision whether to produce Lushlawn or not can be taken very easily. Also, once the profit level for Lushlawn is set, the new optimum solution for the manufacturer can be determined easily, as discussed above, beginning with an optimum feasible basic vector for the standard form of (1.1).

Many companies use this type of sensitivity analysis to determine whether new products or processes would turn out to be profitable, to set prices on new products, and to estimate how sales volumes of existing products will be effected by the introduction of new products.

9.3 INTRODUCING AN ADDITIONAL INEQUALITY CONSTRAINT

Consider the LP (9.1) and the optimum basis $\tilde{B}$ for it again. Suppose the additional constraint

$$A_{m+1.}x \leqq b_{m+1} \tag{9.3}$$

has to be introduced. Let $\mathbf{K}_1$ denote the set of feasible solutions of the augmented problem. $\mathbf{K}_1 \subset \mathbf{K}$ (see Figure 9.1). If the original problem has an optimum feasible solution, the augmented problem is either infeasible or it has an optimum feasible solution. Also $z(\tilde{x}) = \text{minimum}\ \{z(x) : x \in \mathbf{K}\} \leqq \text{minimum}\ \{z(x) : x \in \mathbf{K}_1\}$. Hence, if $\tilde{x} \in \mathbf{K}_1$, that is, $\tilde{x}$ satisfies (9.3), $\tilde{x}$ remains optimal to the augmented problem. On the other hand, if $\tilde{x}$ does not satisfy (9.3), then

$$\tilde{x}_{n+1} = -A_{m+1.}\tilde{x} + b_{m+1} < 0 \tag{9.4}$$

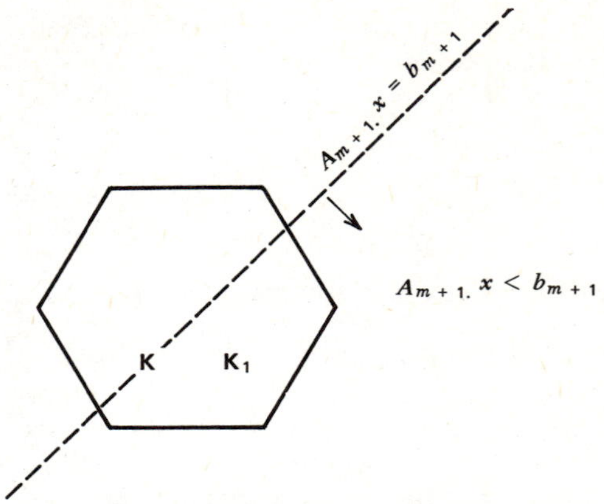

Figure 9.1 Introducing a new inequality constraint.

where x_{n+1} is the slack variable corresponding to (9.3). The augmented problem is

$$\text{Minimize} \quad Z(X) = \sum_{j=1}^{n} c_j x_j + 0x_{n+1}$$

$$\text{Subject to} \quad \begin{pmatrix} A & \vdots & 0 \\ \cdots & \cdots & \cdots \\ A_{n+1} & \vdots & 1 \end{pmatrix} \begin{pmatrix} x \\ \cdots \\ x_{n+1} \end{pmatrix} = \begin{pmatrix} b \\ \cdots \\ b_{n+1} \end{pmatrix} \tag{9.5}$$

$$x_j \geqq 0 \qquad \text{for all } j$$

Let $X^{\mathrm{T}} = (x^{\mathrm{T}}, x_{n+1})$. By including x_{n+1} as an additional basic variable, we can enlarge $\tilde{B}$ into a basis for (9.5), denoted by $\tilde{\mathbf{B}}$. $X_{\tilde{\mathbf{B}}} = (x_1, \ldots, x_m, x_{n+1})^{\mathrm{T}}$. The basic solution of (9.5) corresponding to the basis $\tilde{\mathbf{B}}$ is $\tilde{X}^{\mathrm{T}} = (\tilde{x}^{\mathrm{T}}, \tilde{x}_{n+1})$. By (9.4), $\tilde{\mathbf{B}}$ is an infeasible basis for (9.5). Since $c_{n+1} = 0$, the relative cost coefficient of x_j with respect to the basis $\tilde{\mathbf{B}}$ in (9.5) is equal to the relative cost coefficient of x_j with respect to the basis $\tilde{B}$ in (9.1), which is $\bar{c}_j \geqq 0$ for all $j = 1$ to n. Thus, $\tilde{\mathbf{B}}$ is a dual feasible but primal infeasible basis for (9.5). Using $\tilde{\mathbf{B}}$ as an initial basis, we can solve (9.5) by using the dual simplex algorithm. We now discuss how to obtain the inverse tableau corresponding to the basis $\tilde{\mathbf{B}}$ for (9.5).

$$\tilde{B} = \begin{pmatrix} a_{11} & \cdots & a_{1m} \\ \vdots & & \vdots \\ a_{m1} & \cdots & a_{mm} \end{pmatrix} \qquad \tilde{\mathbf{B}} = \begin{bmatrix} a_{11} & \cdots & a_{1m} & 0 \\ \vdots & & \vdots & \vdots \\ a_{m1} & \cdots & a_{mm} & 0 \\ a_{m+1,1} & \cdots & a_{m+1,m} & 1 \end{bmatrix}$$

Therefore

$$\tilde{\mathbf{B}}^{-1} = \begin{pmatrix} \tilde{B}^{-1} & \vdots & 0 \\ \cdots & \cdots & \cdots \\ -(a_{m+1,1}, \ldots, a_{m+1,m})\tilde{B}^{-1} & \vdots & 1 \end{pmatrix}$$

Basic Variables	*Inverse Tableau*					*Basic Values*	*Pivot Column* (x_4)
x_1	−1	−2	2	0	0	3	−1
x_2	1	1	−1	0	0	4	1
x_3	−1	0	1	0	0	2	2
x_7	5	3	−6	1	0	−12	(−4)
$-z$	−2	4	−1	0	1	−11	1

Thus $\tilde{\boldsymbol{B}}^{-1}$ can easily be obtained from $\tilde{B}^{-1}$. Also obviously $\tilde{\Pi}$ = dual solution of (9.5) corresponding to $\tilde{\boldsymbol{B}}$ is equal to $(\tilde{\pi}, 0)$.

Thus the inverse tableau for (9.5) corresponding to the basis $\tilde{\boldsymbol{B}}$ is easily obtained from the inverse tableau for (9.1) corresponding to the basis $\tilde{B}$. On the other hand, if (9.1) was solved by computing the canonical tableaux after each pivot step, to obtain the canonical tableau of (9.5) with respect to the basis $\tilde{\boldsymbol{B}}$, introduce the $(m + 1)$th constraint row at the bottom of the canonical tableau of (9.1) with respect to the basis $\tilde{B}$, include x_{n+1} as the basic variable in that row, and then price out all the basic column vectors of $\tilde{B}$ in it.

Example 9.3

Returning to the LP in Tableau 9.1, suppose the new constraint, $x_1 - x_2 + 3x_3 \leqq -7$, has to be imposed. The optimum solution $\tilde{x} = (3, 4, 2, 0, 0, 0)^T$ violates this new constraint. Let x_7 be the slack variable. $x_7 = -x_1 + x_2 - 3x_3 - 7$. The inverse tableau for the augmented problem corresponding to the basic vector (x_1, x_2, x_3, x_7) is given at the top. x_7 is the only negative-valued basic variable here. The updated row vector in which x_7 is the basic variable is given at the bottom. The pivot column is the updated column vector of x_4, which is entered as the last column in the inverse tableau at the top. The pivot is performed and the dual simplex algorithm applied in a similar manner until termination.

	x_1	x_2	x_3	x_4	x_5	x_6	x_7	$-z$	b
Updated 4th Row	0	0	0	−4	0	−3	1	0	−12
Updated Cost Row	0	0	0	1	3	8	0	1	−11
$-\dfrac{\bar{c}_j}{\bar{a}_{4j}}$ for $\bar{a}_{4j} < 0$				$\frac{1}{4}$		$\frac{8}{3}$			

Exercises

9.4 Complete the solution of this augmented problem.

9.5 Show that the type of sensitivity analysis discussed here (introducing an additional inequality constraint) is the dual of the one discussed in Section 9.2 (introducing a new sign restricted variable).

9.6 Prove that the set of extreme points of $\mathbf{K}_1$ consists of:

(a) All extreme points of **K** that satisfy (9.3);

(b) Points of intersection of edges of **K** that do not completely lie on the hyperplane **H**, with **H**, where $\mathbf{H} = \{x: A_{m+1.}x = b_{m+1}\}$.

—K. G. Murty [3.35]

9.7 If every optimum feasible solution of (9.1) violates (9.3), and if $\mathbf{K}_1 \neq \emptyset$, prove that every optimum feasible solution of the augmented problem satisfies (9.3) as an equation, that is, lies on the hyperplane **H** defined in Exercise 9.6.

9.4 INTRODUCING AN ADDITIONAL EQUALITY CONSTRAINT

Returning to the LP (9.1), suppose an additional equality constraint has to be introduced: $A_{m+1.}x = b_{m+1}$. The problem (9.1) together with this additional equality constraint is called the *augmented problem*.

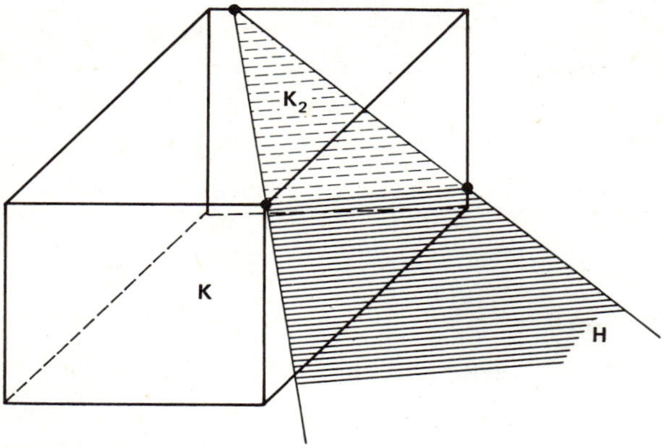

Figure 9.2 Introducing a new equality constraint.

Let $\mathbf{H} = \{x: A_{m+1.}x = b_{m+1}\}$. The set of feasible solutions of the augmented problem is $\mathbf{K}_2 = \mathbf{K} \cap \mathbf{H}$. See Figure 9.2 for an example. If the optimum solution of (9.1), $\tilde{x} \in \mathbf{H}$, obviously $\tilde{x}$ is optimal to the augmented problem. Suppose $\tilde{x} \notin \mathbf{H}$, then $A_{m+1.}\tilde{x} \neq b_{m+1}$. Suppose $A_{m+1.}\tilde{x} > b_{m+1}$, then change the original problem by adding the constraints $A_{m+1.}x - x_{n+1} = b_{m+1}$, $x_{n+1} \geqq 0$. (If, on the other hand, it turned out that $A_{m+1.}\tilde{x} < b_{m+1}$, then the coefficient of x_{n+1} should be $+1$ instead.) Here is the new problem:

Minimize $\quad Z(X) = cx + \boldsymbol{M}x_{n+1}$

$$\text{Subject to} \quad \begin{pmatrix} A & \vdots & 0 \\ \cdots & \cdots & \cdots \\ A_{m+1.} & \vdots & -1 \end{pmatrix} \begin{pmatrix} x \\ \cdots \\ x_{n+1} \end{pmatrix} = \begin{pmatrix} b \\ \cdots \\ b_{m+1} \end{pmatrix}$$

$$X = \begin{pmatrix} x \\ \cdots \\ x_{n+1} \end{pmatrix} \geqq 0 \qquad (9.6)$$

where $\boldsymbol{M}$ is an arbitrarily large positive number. We will refer to (9.6) as the *new problem*. The variable x_{n+1} is an artificial variable in the new problem, and hence it is included in the objective function in (9.6) with a coefficient of $\boldsymbol{M}$, as in the Big-$\boldsymbol{M}$ method. Clearly, $X_{\tilde{\boldsymbol{B}}} = (x_1, \ldots, x_m, x_{n+1})$ is a basic vector for the new problem, and the basic solution of the new problem corresponding to this basic vector can be verified to be $\tilde{X} = (\tilde{x}_1, \ldots, \tilde{x}_n, \tilde{x}_{n+1})^{\mathrm{T}}$, where $\tilde{x} = (\tilde{x}_1, \ldots, \tilde{x}_n)$ is the BFS of the original problem (9.1) corresponding to the optimal basic vector $x_{\tilde{B}}$ for it and $\tilde{x}_{n+1} = A_{m+1.}\tilde{x} - b_{m+1}$. Since $\tilde{x}_{n+1} > 0$ by our assumptions, $\tilde{X} \geqq 0$, and hence $X_{\tilde{\boldsymbol{B}}}$ is a feasible basic vector for (9.6). The associated basis for (9.6), $\tilde{\boldsymbol{B}}$, is

$$\tilde{\boldsymbol{B}} = \begin{bmatrix} a_{11} & \cdots & a_{1m} & 0 \\ \vdots & & \vdots & \vdots \\ a_{m1} & \cdots & a_{mm} & 0 \\ a_{m+1,1} & \cdots & a_{m+1,m} & -1 \end{bmatrix}$$

$$\therefore \quad \tilde{\boldsymbol{B}}^{-1} = \begin{pmatrix} \tilde{B}^{-1} & \vdots & 0 \\ \cdots\cdots & \cdots & \cdots \\ (a_{m+1,1}, \ldots, a_{m+1,m})\tilde{B}^{-1} & \vdots & -1 \end{pmatrix}$$

Define $c_{n+1} = 0$, $c_j^* = 0$, for $j = 1$ to n, and $c_{n+1}^* = 1$. Then the objective function in (9.6) is $Z(X) = \sum_{j=1}^{n+1} (c_j + \boldsymbol{M}c_j^*)x_j$. In the original tableau for (9.6) the cost coefficients can be entered in two separate rows, as under the Big-$\boldsymbol{M}$ method discussed in Section 2.7. One row contains the c_j's, the cost coefficients in $z(x)$ from the original objective function. The second row contains c_j^*'s, the coefficients of $\boldsymbol{M}$ in $Z(X)$. Compute $\Pi_{\tilde{\boldsymbol{B}}} = (\Pi_1, \ldots, \Pi_{m+1}) = c_{\tilde{\boldsymbol{B}}}\tilde{\boldsymbol{B}}^{-1}$, $\Pi_{\tilde{\boldsymbol{B}}}^* = (\Pi_1^*, \ldots, \Pi_{m+1}^*) = \mathbf{c}_{\tilde{\boldsymbol{B}}}^*\tilde{\boldsymbol{B}}^{-1}$. The inverse tableau for (9.6), corresponding to the basic vector $X_{\boldsymbol{B}}$ is:

Basic Variables	*Inverse Tableau*			*Basic Values*
x_1 $\vdots$ x_{n+1}	$\tilde{\boldsymbol{B}}^{-1}$	0		$\bar{b}_1$ $\vdots$ $\bar{b}_{m+1}$
$-Z$	$-\Pi_1 \cdots -\Pi_{m+1}$	1	0	$-\bar{Z}$
	$-\Pi_1^* \cdots -\Pi_{m+1}^*$	0	1	$-\bar{Z}^*$

Letting $\bar{c}_j = c_j - \Pi_{\mathbf{B}}A_{.j}$, $\bar{c}_j^* = c_j^* - \Pi_{\mathbf{B}}^*A_{.j}$, the actual relative cost coefficient of x_j in (9.6) with respect to the current basic vector is $\bar{c}_j + \boldsymbol{M}\bar{c}_j^*$. If $\bar{c}_j + \boldsymbol{M}\bar{c}_j^* \geqq 0$ for each j, the algorithm terminates. If these termination conditions are not satisfied, let $\mathbf{J} = \{j: j$, such that $\bar{c}_j^* < 0$, or $\bar{c}_j^* = 0$ and $\bar{c}_j < 0\}$. The *eligible variables* at this stage are those x_j for $j \in \mathbf{J}$. Select one of the eligible variables as the entering variable, bring it into the basic vector, and continue in this manner until a basic vector satisfying the above termination conditions is obtained. Suppose $\hat{X} = (\hat{x}_1, \ldots, \hat{x}_n, \hat{x}_{n+1})^{\mathrm{T}}$ is the final optimum solution obtained for (9.6). If $\hat{x}_{n+1} > 0$, $\mathbf{K}_2 = \emptyset$, in this case the additional equality constraint has made the original problem infeasible. On the other hand, if $\hat{x}_{n+1} = 0$, $\hat{x} = (\hat{x}_1, \ldots, \hat{x}_n)^{\mathrm{T}}$ is an optimum feasible solution of the augmented problem.

Basic Variables	Inverse Tableau						Basic Values	Pivot Column x_6	Ratios
x_1	−1	−2	2	0	0	0	3	−2	
x_2	1	1	−1	0	0	0	4	−2	
x_3	−1	0	1	0	0	0	2	1	2
x_7	−3	−4	4	−1	0	0	4	②	4/2 minimum
$-Z$	−2	4	−1	0	1	0	−11	8	
	3	4	−4	1	0	1	−4	−2	

Example 9.4

Consider the LP in Tableau 9.1. Suppose the additional constraint $x_1 - 2x_2 + x_4 + 3x_5 = -9$ is to be included. $\tilde{x} = (3, 4, 2, 0, 0, 0)^T$, the optimum solution of the original problem, violates this new constraint. Let $\tilde{x}_7 = \tilde{x}_1 - 2\tilde{x}_2 + \tilde{x}_4 + 3\tilde{x}_5 - (-9) = 4$. The new problem as discussed above, is:

New Problem

	x_1	x_2	x_3	x_4	x_5	x_6	x_7	$-Z$	b
	1	2	0	1	0	−6	0	0	11
	0	1	1	3	−2	−1	0	0	6
	1	2	1	3	−1	−5	0	0	13
	1	−2	0	1	3	0	−1	0	−9
c	3	2	−3	−6	10	−5	0	1	0
c^*	0	0	0	0	0	0	0		0

$x_j \geqq 0$ for all j; minimize Z; x_7 is the artificial variable.

Let $\tilde{\boldsymbol{B}}$ be the basis for the new problem corresponding to the basic vector (x_1, x_2, x_3, x_7). The inverse tableau for the new problem with respect to the basis $\tilde{\boldsymbol{B}}$ is given at the top.

The updated cost rows are $\bar{c} = (0, 0, 0, 1, 3, 8, 0)$ and $\bar{c}^* = (0, 0, 0, 4, -1, -2, 0)$. So x_5 and x_6 are eligible to enter the basic vector. Suppose we choose x_6 as the entering variable. Continuing the application of the Big-$\boldsymbol{M}$ method, it can be verified that the optimum solution of the new problem is $X^T = (0, 8, 7/2, 0, 7/3, 5/6, 0)^T$. Since $x_7 = 0$ in this solution, $x = (0, 8, 7/2, 0, 7/3, 5/6)^T$ is an optimum solution of the augmented problem.

Exercises

9.8 Consider the LP in Tableau 9.1. Suppose the additional constraint $-x_1 - x_2 - x_3 + x_4 = 300$ has to be introduced. Apply the algorithm discussed here to solve the augmented problem. Show that the augmented problem is infeasible.

9.9 Returning to the LP in Tableau 9.1, suppose the additional constraint $x_1 + x_2 + x_3 + x_4 + x_5 + x_6 = 12$ has to be imposed. Obtain an optimum feasible solution of the augmented problem.

9.10 Discuss an approach for solving the augmented problem with the additional equality constraint, using the dual simplex algorithm instead of the Big-$\boldsymbol{M}$ approach described here.

9.11 Prove that the set of extreme points of $\mathbf{K}_2$ consists of (a) all extreme points of $\mathbf{K}$ that are in $\mathbf{H}$, and (b) points of intersection of edges of $\mathbf{K}$ that do not completely lie in $\mathbf{H}$ with $\mathbf{H}$.

9.12 Let $\mathbf{H}$, $\mathbf{K}_2$ be defined as in the beginning of this section. Let $\mathbf{K}_1 = \mathbf{K} \cap \{x: \mathbf{A}_{m+1.}x \leqq b_{m+1}\}$, $\mathbf{K}_3 = \mathbf{K} \cap \{x: \mathbf{A}_{m+1.}x \geqq b_{m+1}\}$. If $\mathbf{K}_2 \neq \emptyset$, prove that $\mathbf{K}_2$ contains an optimum solution of at least one of the following two problems.

(a) Minimize $z(x) = cx$

$$x \in \mathbf{K}_1$$

(b) Minimize $z(x) = cx$

$$x \in \mathbf{K}_3$$

9.5 COST RANGING OF A NONBASIC COST COEFFICIENT

Consider again the LP (9.1). Suppose x_r is a variable that is not in the optimal basic vector $x_{\tilde{B}}$. Assuming that all the other cost coefficients except c_r remain fixed at their specified values, determine the range of values of c_r within which $\tilde{B}$ remains an optimal basis. For this, treat c_r as a parameter. $\tilde{B}$ remains an optimal basis as long as $\bar{c}_r = c_r - \tilde{\pi}A_{.r} \geqq 0$, that is

$c_r \geqq \tilde{\pi}A_{.r}$. The relative cost coefficients of all the other variables are independent of the value of c_r and, hence, they remain nonnegative. If the new value of c_r is not in the closed interval $[\tilde{\pi}A_{.r}, \infty]$, x_r is the only nonbasic variable that has a negative relative cost coefficient with respect to the basis $\tilde{B}$ in the modified problem. Bring x_r into the basic vector and continue the application of the simplex algorithm until a terminal basis for the modified problem is obtained.

Example 9.5

Consider the LP in Tableau 9.1. The range of values of c_5 (whose present value is 10) for which the basic vector $x_B = (x_1, x_2, x_3)$ remains optimal is determined by $c_5 + (-2, 4, -1)(0, -2, -1)^{\mathrm{T}} \geqq 0$, that is, $c_5 \geqq 7$. Suppose the value of c_5 has to be modified to 6. When c_5 is changed from 10 to 6, the updated column vector of x_5 becomes $(2, -1, -1, -1)^{\mathrm{T}}$. With this as the pivot column, bring x_5 into the basic vector. This leads to the following tableau.

Basic Variables		*Inverse Tableau*			*Basic Values*
x_5	$-\frac{1}{2}$	-1	1	0	$\frac{3}{2}$
x_2	$\frac{1}{2}$	0	0	0	$\frac{11}{2}$
x_3	$-\frac{3}{2}$	-1	2	0	$\frac{7}{2}$
$-z$	$-\frac{5}{2}$	3	0	1	$-\frac{19}{2}$

It can be verified that this tableau displays an optimum tableau for the modified problem.

9.6 COST RANGING OF A BASIC COST COEFFICIENT

Consider the LP (9.1). Suppose all the cost coefficients except c_1, which is a basic cost coefficient, remain fixed at their present values. Treating c_1 as a parameter, determine the range of values of c_1 for which $\tilde{B}$ remains an optimal basis. Since c_1 is a basic cost coefficient, any change in c_1 changes the dual solution corresponding to the basis $\tilde{B}$, and all the relative cost coefficients. For simplicity let γ_1 denote the parameter and c_1 its present value. Let $\pi(\gamma_1)$, $\bar{c}_j(\gamma_1)$, etc., denote the dual solution and the relative cost coefficient of x_j corresponding to the basis $\tilde{B}$ as functions of the parameter γ_1. Then $\pi(\gamma_1) = (\gamma_1, c_2, \ldots, c_m)\tilde{B}^{-1}$. Hence, the dual vector is an affine function of γ_1. If we use this dual solution, the relative cost coefficient of x_j is $\bar{c}_j(\gamma_1) = c_j - \pi(\gamma_1)A_{.j}$, for $j = 1$ to n. Hence, each $\bar{c}_j(\gamma_1)$ is itself an affine function of γ_1. The range of values of γ_1 for which $\tilde{B}$ remains an optimum basis is the range of γ_1 within which every $\bar{c}_j(\gamma_1)$ remains nonnegative. This range will turn out to be a nonempty closed interval. If it is necessary to modify the cost coefficient of x_1 to some value c'_1 outside this closed interval, fix $\gamma_1 = c'_1$ and compute the relative cost coefficients $\bar{c}_j(c'_1)$ using the formulas already developed. Bring one of the variables for which $\bar{c}_j(c'_1) < 0$ into the basic vector and continue the applications of the simplex algorithm until a terminal basis for the modified problem is obtained.

Example 9.6

In the LP in Tableau 9.1, let γ_1 represent the cost coefficient of x_1 whose present value is 3. The dual solution corresponding to the basic vector (x_1, x_2, x_3) as a function of γ_1 is $\pi(\gamma_1) = (\gamma_1, 2, -3)\tilde{B}^{-1} = (-\gamma_1 + 5, -2\gamma_1 + 2, 2\gamma_1 - 5)$. Using $\bar{c}_j(\gamma_1) = c_j - \pi(\gamma_1)A_{.j}$, the row of relative cost coefficients is $\bar{c}(\gamma_1) = (0, 0, 0, \gamma_1 - 2, -2\gamma_1 + 9, 2\gamma_1 + 2) \geqq 0$ iff $2 \leqq \gamma_1 \leqq 9/2$. So $\tilde{B}$ is an optimum basis whenever the cost coefficient of x_1 is in the interval [2, 9/2], assuming that all the other cost coefficients remain at their present values.

Suppose γ_1 has to be changed to 5. Changing the cost coefficient of x_1 from the present value of 3 to 5, the modified dual solution is $\pi(\gamma_1 = 5) = (0, -8, 5)$. With this change, the inverse tableau becomes

Basic Variables	*Modified Inverse Tableau for Basis B*				*Basic Values*	*Pivot Column (x_5)*
x_1	-1	-2	2	0	3	② *Pivot Row*
x_2	1	1	-1	0	4	-1
x_3	-1	0	1	0	2	-1
$-z$	0	8	-5	1	-17	-1

The modified relative cost coefficient of x_5 is -1. Hence, bring x_5 into the basic vector. The updated column vector of x_5 is entered as the pivot column in the inverse tableau. Remember that the cost coefficient of x_1 is now 5, and continue the algorithm.

Exercises

9.13 Construct a counter example to the following: "If B is an optimum basis for (9.1), then it remains an optimum basis when some of the entries in c_B are decreased."

9.14 Is the following true? "x_1 is a basic variable in an optimum basic vector for (9.1). If c_1 is decreased and if the modified problem has an optimum feasible solution, then it has an optimum basic vector containing x_1."

9.7 RIGHT-HAND-SIDE RANGING

In the LP (9.1) suppose we wish to determine the range of values of one of the right-hand-side constants, say b_1, for which the basis $\tilde{B}$ remains optimal. Treat this right-hand-side constant as a parameter and denote it by β_1; b_1 is the present value of β_1. We assume that all the other right-hand-side constants stay fixed at their present values. $\tilde{B}$ is an optimum basis when $\beta_1 = b_1$. Hence, $\tilde{B}$ is dual feasible. So $\tilde{B}$ is optimal for all values of β_1 for which it is primal feasible. The values of the basic variables $x_{\tilde{B}}$ are all functions of the parameter β_1, and $\tilde{B}$ is primal feasible as long as all these values are nonnegative. $x_{\tilde{B}}(\beta_1) = \tilde{B}^{-1}(\beta_1, b_2, \ldots, b_m)^{\mathrm{T}} \geqq 0$. Since all these inequalities are linear in β_1, this will determine a closed interval of the form $\underline{\lambda} \leqq \beta_1 \leqq \overline{\lambda}$. For all values of β_1 in this closed interval, $\tilde{B}$ remains an optimal basis. If it is necessary to modify the value of β_1 from its present value b_1 to a value b_1' outside the closed interval $[\underline{\lambda}\ \overline{\lambda}]$, fix β_1 at b_1' and obtain the modified values of the basic variables. $\tilde{B}$ is still dual feasible, but primal infeasible. Starting with $\tilde{B}$, apply the dual simplex routine until a new terminal basis is obtained.

Example 9.7

In the LP in Tableau 9.1 the values of the basic variables in the basic vector (x_1, x_2, x_3) as functions of β_1 are

$$\tilde{B}^{-1}\begin{pmatrix}\beta_1\\6\\13\end{pmatrix} = \begin{pmatrix}-1 & -2 & 2\\1 & 1 & -1\\-1 & 0 & 1\end{pmatrix}\begin{pmatrix}\beta_1\\6\\13\end{pmatrix} = \begin{pmatrix}-\beta_1+14\\\beta_1-7\\-\beta_1+13\end{pmatrix}$$

All these are nonnegative iff $7 \leqq \beta_1 \leqq 13$. Hence, this is the interval within which the basic vector (x_1, x_2, x_3) remains optimal. In this range, an optimum solution of the problem is $x = (14-\beta_1, -7+\beta_1, 13-\beta_1, 0, 0, 0)^{\mathrm{T}}$, with an optimum objective value of $-11 + 2\beta_1$ and an associated optimum dual solution of $\pi = (2, -4, 1)$.

Now, suppose it is required to solve the problem for $\beta_1 = 15$. The inverse tableau for the modified problem corresponding to the present basic vector (x_1, x_2, x_3) is:

Basic Variables	*Inverse Tableau*				*Basic Values*	*Pivot Column*
x_1	-1	-2	2	0	-1	2
x_2	1	1	-1	0	8	-1
x_3	-1	0	1	0	-2	-1 (circled)
$-z$	-2	4	-1	1	-19	3

Applying the dual simplex algorithm starting with this inverse tableau, it can be verified that the modified problem is infeasible.

Exercises

9.15 Obtain the new optimum feasible solution of the LP in Tableau 9.1 after b_1 is changed from its present value of 11 to 6.

9.8 CHANGES IN THE INPUT-OUTPUT COEFFICIENTS IN A NONBASIC COLUMN VECTOR

Let x_j be a variable that is not in the optimum basic vector of $x_{\tilde{B}}$ of (9.1). And a_{ij} is one of the input-output coefficients in the column vector of x_j. If all the other coefficients in the problem except a_{ij} remain fixed at their present values, what is the range of values of a_{ij} within which $\tilde{B}$ remains an optimal basis?

Treat a_{ij} as a parameter. To avoid confusion, call this parameter, α_{ij}, and let a_{ij} be its present value. Since x_j is a nonbasic variable, a change in α_{ij} does not affect the primal feasibility of $\tilde{B}$. A change in α_{ij} can only change the relative cost coefficient of x_j

and this is obtained by $\bar{c}_j(\alpha_{ij}) = c_j + (\sum_{r \neq i}(-\tilde{\pi}_r)a_{rj}) - \tilde{\pi}_i\alpha_{ij}$. As long as $\bar{c}_j(\alpha_{ij}) \geqq 0$, $\tilde{B}$ remains an optimal basis. This determines a closed interval for α_{ij}, and as long as α_{ij} is in this interval $\tilde{B}$ remains an optimal basis. If α_{ij} has to be changed to a value a'_{ij} outside this interval, make the change, bring x_j into the basic vector, and continue with the application of the simplex algorithm until a new terminal basis is obtained.

Example 9.8

In the LP in Tableau 9.1, the present value of α_{25} in the column vector of x_5 is -2. The relative cost coefficient $\bar{c}_5(\alpha_{25}) = 10 + (-2, 4, -1)(0, \alpha_{25} - 1)^{\mathrm{T}} = 11 + 4\alpha_{25} \geqq 0$ iff $\alpha_{25} \geqq -11/4$. Thus $\tilde{B}$ remains an optimal basis for $\alpha_{25} \geqq -11/4$. Suppose we change α_{25} from its present value of -2 to -3. The updated column vector of x_5 changes to $(4, -2, -1, -1)^{\mathrm{T}}$. With this pivot column bring x_5 into the basic vector and continue until termination is reached.

9.9 CHANGES IN A BASIC INPUT-OUTPUT COEFFICIENT

Let x_1 be a basic variable in the optimum basic vector $x_{\tilde{B}}$ for (9.1). Suppose we have to modify one input-output coefficient, say a_{11}, in the column vector of x_1 to a'_{11}. The modified column vector of x_1 will be $A'_{.1} = (a'_{11}, a_{21}, \ldots, a_{m1})^{\mathrm{T}}$. Let x'_1 indicate the level of this activity corresponding to the new column vector $A'_{.1}$. The previous column vector $A_{.1}$ is no longer a part of the problem and it should be eliminated. Physically x'_1 replaces x_1 in the original tableau. We will refer to the altered problem by the name *modified problem.*

Construct a new problem by augmenting the present original tableau with the new variable x'_1 with its column vector $A'_{.1}$ and cost coefficient c_1, and changing the cost coefficient of x_1 to $\boldsymbol{M}$, where $\boldsymbol{M}$ is a very large positive number. This leads to the problem, which we call the *new problem*. In this problem x_1 plays the role of an artificial variable, associated with a cost coefficient of $\boldsymbol{M}$, a very large positive number. The objective function Z, in the new problem is $\sum_{j=2}^{n} c_j x_j + c_1 x'_1 + \boldsymbol{M}x_1$. As in the Big-$\boldsymbol{M}$ method, the cost coefficients are entered in two rows. $\tilde{B}$ is still a feasible basis to this new problem. However, in the inverse tableau corresponding to the basis $\tilde{B}$, the dual vector has to be recomputed, since the cost coefficient of x_1 has been changed to $\boldsymbol{M}$. The new dual solution corresponding to $\tilde{B}$ is $\pi(\boldsymbol{M}) = (\boldsymbol{M}, c_2, \ldots, c_m)\tilde{B}^{-1} = \pi + \boldsymbol{M}\pi^*$, where $\pi = (0, c_2, \ldots, c_m)\tilde{B}^{-1}$, $\pi^* = (1, 0, \ldots, 0)\tilde{B}^{-1}$. Construct the inverse tableau for the new problem, corresponding to the basic vector $x_{\tilde{B}}$ and solve it as discussed in Section 9.4, and the normal termination conclusions of the big-$\boldsymbol{M}$ method apply.

9.10 PRACTICAL APPLICATIONS OF SENSITIVITY ANALYSIS

When studying a system using a linear programming model, techniques of sensitivity analysis can be used to evaluate new products (as in Section 9.2); to determine the breakeven selling price of a new product, at which point it becomes competitive with the existing list of products in terms of profitability (as in Sections 9.2 and 9.5); to evaluate new technologies or processes for making products (as in Sections 9.8 and 9.9); to assess how profitable it is to acquire additional resources and to determine which resources to acquire in what quantities (using the ideas discussed in Sections 4.6.3, 9.7, and Chapter 8); to evaluate the effects of changes in the costs (as in Sections 9.5, 9.6, and Chapter 8); and to determine optimal policies for handling new constraints that might arise. When used in this manner, the linear programming model not only determines an optimum solution to implement but becomes a tool for optimal planning. Examples of some of these uses are discussed in the problems that follow.

Optimization models other than linear programming models (e.g., integer programming models and nonlinear programming models) do not lend themselves that readily to a marginal analysis or sensitivity analysis as the linear programming model. That's why if a practical problem can be approximated reasonably closely by a linear programming model, it is so much easier to study it than otherwise.

Exercises

9.16 Consider the family's diet problem discussed in Chapter 4. The original tableau for it is given at the top of page 319. Here x_1 to x_6 are the kilograms of the primary foods 1 to 6 in the family's diet, and x_7, x_8 are the slack variables representing the excess amounts of the nutrients, vitamins A and C, in the diet over the minimum requirements. The basis B_1,

Original Tableau for the Family's Diet Problem

x_1	x_2	x_3	x_4	x_5	x_6	x_7	x_8	$-z$	b	
1	0	2	2	1	2	-1	0	0	9	
0	1	3	1	3	2	0	-1	0	19	
35	30	60	50	27	22	0	0	1	0	(minimize z)

$x_j \geqq 0$ for all j.

associated with the basic vector (x_5, x_6) is optimal to this problem. The optimum inverse tableau is:

Basic Variables	Inverse Tableau			Basic Values
x_5	$-\frac{1}{2}$	$\frac{1}{2}$	0	5
x_6	$\frac{3}{4}$	$-\frac{1}{4}$	0	2
$-z$	-3	-8	1	-179

Answer each of the following questions with respect to this original problem:

(a) Suppose a new primary food, food 7, is available in the market at 88 cents/kg. One kilogram of this food contains two units of vitamin A and four units of vitamin C. Should the family include this new food in its diet? If not, how much should the cost of this food decrease before the family can consider including it in its diet?

At this breakeven cost show that there is an optimum diet that includes food 7 and another that does not.

What is the optimum diet if food 7 is actually available at 32 cents/kg?

(b) Consider the original family's diet problem again. The family has just read an article in a health magazine that says that another nutrient, vitamin E, is very important for the family's health. The vitamin E content of the six primary foods are 2, 3, 5, 2, 1, and 1 units/kg, respectively. The minimum requirement of vitamin E is 10 units. The family wants to include this additional constraint in the problem. How does this change its optimal diet?

(c) Referring to the original problem, in addition to meeting the minimum requirements on vitamins A and C, suppose the family decides to have the diet consist of a total of 2000 calories exactly. The calorie contents of the six primary foods are 160, 20, 500, 280, 300, and 360 kg, respectively. How does this additional requirement change its optimal diet?

(d) What is the marginal effect of increasing the minimum requirement of vitamin C on the cost of the optimal diet in the original family's diet problem? How much can the minimal requirement of vitamin C increase before the basis B_1 becomes nonoptimal to the problem? What is an optimal diet when this requirement is 39 units?

(e) In the original family's diet problem, suppose that each additional unit of vitamin A in the diet is expected to bring an average savings of 10 cents in medical expenses. How does this alter the optimal diet?

(f) For what range of cost per kilogram of primary foods 4 and 5 does the basis B_1 remain optimal to the original family's diet problem?

(g) What happens to the optimal diet in the original family's diet problem if the vitamin C content of food 4 changes? For what range of values of this quantity does the basis B_1 remain optimal? Suppose a richer version of food 4 containing $1 + \alpha$ units of vitamin C per kilogram is available at a cost of $50 + 4\alpha$ cents/kg for any $\alpha \geqq 0$. What is the minimum value of α at which it becomes attractive for the family to include it in its diet?

9.17 Data on different solvents are given in the table at the top of page 320. Let x_j be the

	Solvent				*Chemical Requirement in Blend per kg*
	1	*2*	*3*	*4*	
Chemical 1 content (units/kg)	180	120	90	60	$\geqq 90$
Chemical 2 content (units/kg)	3	2	6	5	$\leqq 4$
Cost (cent/kg)	16	12	10	11	

proportion of solvent type j in the blend, $j = 1$ to 4. Variables x_5 and x_6 are the slack variables associated with chemicals 1 and 2 requirements, respectively. The basis B_1 associated with the basic vector (x_2, x_3, x_5) is an optimum basis for the problem. Compute B_1^{-1}, and using it, answer the following parts, each of which is independent of the others.

(a) Write down the dual problem and the complementary slackness conditions for optimality. Obtain an optimum feasible solution for the dual problem from the above information.

(b) How much does the optimum objective value change if the minimal chemical 1 requirement is changed to 88 units? Why?

(c) Let β_1 be the minimal chemical 1 requirement. Its present value is 90. For what range of values of β_1 does B_1 remain optimal to the problem? When β_1 is 114, what is an optimum feasible solution to the problem?

(d) How much can the cost per kilogram of solvent 3 change before B_1 becomes nonoptimal to the problem? When the cost per kilogram of solvent 3 becomes 11 cents, what is an optimum solution?

9.18 Consider the LP given at the bottom. Variables x_5, x_6, x_7 are the slack variables corresponding to the various inequalities. The basis B_1 corresponding to the basic vector (x_1, x_3, x_2) is optimal to the problem. Compute B_1^{-1}

(a) If the availability of only one of the raw materials can be marginally increased, which one should be picked? Why?

(b) For what range of values of b_1 (the amount of raw material 1 available) does the basis B_1 remain optimal? What is an optimal solution to the problem if $b_1 = 20$?

(c) If seven more units of raw material 1 can be made available (over the present 8 units), what is the maximum you can afford to pay for it? Why?

(d) The company has an option to produce a new product. Let x_8 be the number of units of this product manufactured. The input-output vector of x_8 will be $(10, 20, 24 - 3\lambda, -25, +\lambda/4)^{\mathrm{T}}$, where λ is a parameter that can be set anywhere from 0 to 6. What is the minimum value of λ at which it becomes profitable to produce the product? What is an optimum solution when $\lambda = 4$?

$$
\begin{aligned}
\text{Minimize} \quad & z(x) = -2x_1 - 4x_2 - x_3 - x_4 \\
\text{Subject to} \quad & x_1 + 3x_2 + x_4 \leqq 8 = \text{available amount of raw material 1} \\
& 2x_1 + x_2 \leqq 6 = \text{available amount of raw material 2} \\
& x_2 + 4x_3 + x_4 \leqq 6 = \text{available amount of raw material 3} \\
& x_j \geqq 0 \quad \text{for all } j
\end{aligned}
$$

9.19 Consider the diet problem with data given at the top of page 321. Let x_1, x_2, and x_3 be the amounts of greens, potatoes, and corn included in the diet, respectively. Let x_4, x_5, and x_6 be the slack variables representing the excess of vitamins A, C, and D, respectively, in the diet. The basis B_1 associated with the basic vector (x_4, x_1, x_6) is optimal to this problem.

(a) Find the optimum primal and dual solutions associated with the basis B_1.

Nutrient	*Nutrient Content in Foods Available (units/kg)* Greens	Potatoes	Corn	*Minimum Daily Requirement (MDR) for Nutrient*
Vitamin A	10	1	9	5
C	10	10	10	50
D	10	11	11	10
Cost (cent/kg)	50	100	51	

(b) A new food (milk) has become available. One liter of milk contains 0, 10, and 20 units, respectively, of vitamins A, C, and D and costs 40 cents. Would you recommend including it in the diet? Why? What is the highest price of milk at which it is still attractive to include it in the diet?

(c) Consider the original problem again. A nutrition specialist claims that the actual MDRs for vitamins A, C, and D should really be 5, $50 + 10\lambda$, and $10 + 15\lambda$, where λ is a nonnegative parameter; and an experiment is proposed to estimate λ. Assuming that the claim is correct, find $\bar{\lambda}$, the maximum value of λ at which B_1 is still an optimum basis to the problem. What is an optimum solution if $\lambda = \bar{\lambda} + 1$?

9.20 Consider the LP (9.1) and its dual. Discuss what effects the following have on the primal and dual feasible solution sets and the respective optimal objective values.

(a) Introducing a new nonnegative primal variable.

(b) Introducing a new inequality constraint in the primal problem.

(c) Introducing a new equality constraint in the primal problem.

9.21 Consider the LP given at the bottom. Construct the inverse tableau for this problem corresponding to the basic vector (x_1, x_2, x_3) and verify that it is an optimum basic vector. Using it, answer the following questions. Each of these questions is independent of the others, and each refers to the original problem given above.

(a) Find the range of values of the parameter c_2 for which the solution obtained above remains optimal to the problem. If the new value of c_2 is slightly greater than the upper bound of the optimality range computed above, which variables become eligible to enter the basic vector? Why?

(b) Find the range of values of b_1 for which the basic vector (x_1, x_2, x_3) remains optimal to the problem. What is the slope of the optimum objective value as a function of b_1 in this optimality range? Why?

(c) Consider the original LP back again. Suppose a new activity has become available. This activity leads to a new nonnegative variable x_7, with the data $A_{.7} = (0, 2, 3)^{\mathrm{T}}$, $c_7 = 18$, in the original tableau for the problem. Is it worthwhile for the decision maker to perform this new activity? Why? If not, determine the value to which c_7 should decrease (assuming that all the other data, including the entries in $A_{.7}$, remain unaltered) before this new activity becomes economically competitive. Find an optimum solution to the augmented problem assuming that $A_{.7}$ is as given above, but that $c_7 = 12$, using the methods of sensitivity analysis.

$$\begin{array}{llrl}
\text{Minimize} & z(x) = -4x_1 & + 9x_3 + 12x_4 - 3x_5 + 7x_6 & \\
\text{Subject to} & x_1 + 2x_2 - x_3 + x_4 + 2x_5 + 2x_6 = 3 & & \\
 & x_2 + x_3 + 2x_4 + x_5 + 2x_6 = 6 & & \\
 & -x_1 + 2x_3 + 2x_4 - 2x_5 + x_6 = 5 & & \\
 & x_j \geqq 0 \quad \text{for all } j = 1 \text{ to } 6. & &
\end{array}$$

$$\begin{array}{lrl} \text{Minimize Cost} = 28x_1 + 67x_2 + 12x_3 + 35x_4 & & \\ \text{Subject to} & \text{Product 1 output} = & x_1 + 2x_2 \qquad + x_4 \geqq 17 \\ & \text{Product 2 output} = & 2x_1 + 5x_2 + x_3 + 2x_4 \geqq 36 \\ & \text{Product 3 output} = & x_1 + x_2 \qquad + 3x_4 \geqq 8 \\ & x_j \geqq 0 & \text{for all } j \end{array}$$

9.22 In the LP given at the top, x_5, x_6, x_7 are the slacks in that order. Vector (x_1, x_2, x_7) is an optimum basic vector. (1) What is the most critical product for the company? Why? (2) Determine the optimality range of c_1 for the present basis. (3) Determine the optimality range of b_3. (4) Find the new optimum when b_3 changes to 16 from 8.

9.23 In (9.1), let $g(b_1)$ denote the optimum objective value as a function of b_1 when all other data remain fixed. Prove that $g(b_1)$ is piecewise linear convex. Discuss how to minimize $g(b_1)$ over b_1. From this, discuss how to find the new optimum when a constraint from (9.1) is eliminated.

9.24 Let $\mathbf{K} = \{x: Ax = b, x \geqq 0\}$. Let $f_1(x) = cx$, $f_2(x) = dx$. If both $f_1(x)$ and $f_2(x)$ are >0 on $\mathbf{K}$, develop an algorithm for minimizing $f_1(x)f_2(x)$ over $x \in \mathbf{K}$, using a parametric right-hand-side LP. Generalize to the case where $f_1(x)$, $f_2(x)$ may not be positive on $\mathbf{K}$.

—[*Y. P. Aneja, V. Aggarwal, and K. P. K. Nair, "On a class of quadratic programs," University of New Brunswick, Fredericton, N. B., Canada*]

9.25 Let (P) be the LP: minimize cx, subject to $A_{i.}x \geqq b_i$, $i = 1$ to m. Let x^1 be optimal to the LP: minimize cx, subject to $A_{i.}x \geqq b_i$, $i = 2$ to m. Let x^2 be optimal to the LP: minimize cx, subject to $A_{1.}x = b_1$, $A_{i.}x \geqq b_i$, $i = 2$ to m. Prove that one of the points x^1 or x^2 must be optimal to (P).

9.26 Let $\mathbf{K}$ be the set of feasible solutions of (9.1), and $\mathbf{K}_1 = \{x: x \in \mathbf{K}, x_n = 0\}$. Assume that (9.1) is nondegenerate and that $\mathbf{K}$, $\mathbf{K}_1$, are both nonempty and bounded. All data are fixed except c_n. Prove that there exists a γ such that for all $c_n > \gamma$, x_n is a nonbasic variable in all optimum bases for (9.1), and for all $c_n < \gamma$, x_n is a basic variable in all optimum bases for (9.1). How can γ be computed?

9.27 Consider the following variant of the simplex algorithm for solving the LP (9.1). It begins with a feasible basic vector x_B for (9.1). Let $\bar{x}$ be the associated BFS, and let $\bar{c} = (\bar{c}_j)$ be the vector of relative cost coefficients with respect to x_B. If $\bar{c} \geqq 0$, $\bar{x}$ is optimal to (9.1), terminate. If $\bar{c} \ngeqq 0$, let $\mathbf{J} = \{j: \bar{c}_j < 0\}$. Let $A_{.n+1} = \sum_{j \in \mathbf{J}} A_{.j}w_j$, $c_{n+1} = \sum_{j \in \mathbf{J}} c_j w_j$, where $w_j > 0$ for each $j \in \mathbf{J}$ are positive weights. Introduce a new variable x_{n+1} into (9.1) with its column vector $A_{.n+1}$ and cost coefficient c_{n+1}. Let $X = (x_1, \ldots, x_n; x_{n+1})$. Bring x_{n+1} into the basic vector x_B for this augmented problem and suppose this leads to the new solution $\tilde{X} = (\tilde{x}_1, \ldots, \tilde{x}_n; \tilde{x}_{n+1})$. Remembering that $x_{n+1} = \sum_{j \in \mathbf{J}} \mathbf{w}_j \mathbf{x}_j$, obtain the feasible solution $\hat{x}$ of the original problem (9.1) corresponding to $\tilde{X}$ for the augmented problem. In general, $\hat{x}$ may not be a BFS for (9.1). Using the method discussed in the proof of Theorem 3.3 of Section 3.5.5, obtain a BFS x^* for (9.1) whose objective values is $\leqq$ the objective value at $\hat{x}$. Repeat the procedure, starting with the new BFS x^*.

Possible choices for the weights in this procedure are either $w_j = 1$ for all $j \in \mathbf{J}$, or $w_j = -\bar{c}_j$ for each $j \in \mathbf{J}$, or some other positive values.

Compare this procedure for solving (9.1), starting with a feasible basic vector for it, with the usual simplex algorithm, in terms of computational efficiency. How does this procedure differ from the simplex algorithm geometrically?

Degeneracy in Linear Programming

chapter 10

10.1 GEOMETRY OF DEGENERACY

Consider the LP

$$\begin{aligned} \text{Minimize} \quad & z(x) = cx \\ \text{Subject to} \quad & Ax = b \qquad (10.1) \\ & x \geqq 0 \end{aligned}$$

where A is a matrix of order $m \times n$ and rank m. A basis B for (10.1), is a *primal nondegenerate basis* iff all the components in the vector $B^{-1}b$ are nonzero, otherwise B is said to be a *primal degenerate basis*. The LP (10.1) is said to be *totally (primal) nondegenerate* iff every basis for (10.1) is nondegenerate.

Exercises

10.1 Prove that (10.1) is totally nondegenerate iff every solution x of $Ax = b$ has at least m nonzero components.

Equivalently, (10.1) is totally nondegenerate iff b cannot be expressed as a linear combination of any set of $m - 1$ or less column vectors of A, that is, iff b does not lie in any subspace that is the linear hull of a set of $m - 1$ or less column vectors of the matrix A.

Example 10.1

Since the objective function does not play any role in determining primal degeneracy or nondegeneracy, we omit it in these examples:

$$A = \begin{pmatrix} 1 & 0 & 0 & 1 \\ 0 & 1 & 0 & 0 \\ 0 & 0 & 1 & -1 \end{pmatrix} \qquad b = \begin{pmatrix} -1 \\ 1 \\ 0 \end{pmatrix}$$

In this case, $m = 3$, and b can be expressed as a linear combination of $A_{.1}$ and $A_{.2}$. Hence, this problem is primal degenerate. Also any basis for this problem that includes $A_{.1}$ and $A_{.2}$ as basic vectors is a degenerate basis. Clearly the problem remains primal degenerate even if b is changed, as long as b lies on any of the coordinate planes, or on the linear hull of $\{A_{.2}, A_{.4}\}$. The problem becomes primal nondegenerate iff b does not lie on any of these four planes (see Figure 10.1). Clearly, if $b = (1, 2, 3)^{\mathrm{T}}$ and A remains the same, the problem is primal nondegenerate.

The b Vector in Primal Degenerate Problems

The LP (10.1) is degenerate iff b is in a subspace of $\mathbf{R}^m$ that is the linear hull of a set of $m - 1$ or less column vectors of A. There are at most $\sum_{r=1}^{m-1} \binom{n}{r}$ subsets of column vectors of A with $m - 1$ or less column vectors in them. The linear hull of each of these subsets is a subspace of $\mathbf{R}^m$ of dimension $m - 1$ or less. The LP (10.1) is degenerate iff b lies in one of

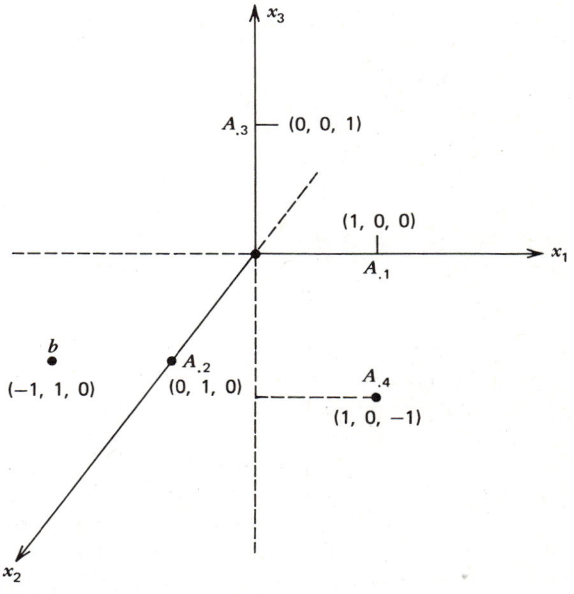

Figure 10.1

these subspaces. Hence, when A is fixed, the set of all $b \in \mathbf{R}^m$ that make (10.1) degenerate is the set of all b that lie in a finite number of subspaces of dimension $m-1$ or less. Thus in a statistical sense *almost all b* will make (10.1) nondegenerate. Also if (10.1) is degenerate, b can be perturbed to a point in its neighborhood that is not in any of the finite number of subspaces discussed earlier, and this would make (10.1) nondegenerate. One such perturbation that makes (10.1) nondegenerate is to replace b by

$$b(\varepsilon) = b + (\varepsilon, \varepsilon^{\mathbf{2}}, \ldots, \varepsilon^{\mathbf{m}})^{\mathrm{T}} \tag{10.2}$$

where ε is an arbitrarily small positive number, and in the symbol $\varepsilon^{\mathbf{r}}$, $\mathbf{r}$ is the exponent, that is, $\varepsilon^{\mathbf{r}} = \varepsilon \times \varepsilon \times \cdots \times \varepsilon$ with $\mathbf{r}$ ε's in this product. Notice the difference in type style between exponents and superscripts.

THEOREM 10.1 Given any $b \in \mathbf{R}^m$, there exists a positive number $\varepsilon_1 > 0$, such that whenever $0 < \varepsilon < \varepsilon_1$, the perturbed problem (10.3) is primal nondegenerate.

$$\begin{aligned} \text{Minimize} \quad & z(x) = cx \\ \text{Subject to} \quad & Ax = b(\varepsilon) \\ & x \geqq 0 \end{aligned} \tag{10.3}$$

Proof: Let $B_1, \ldots, B_{\mathrm{L}}$ be all the bases for (10.3). Consider one of these bases B_r. Let $B_r^{-1} = (\beta_{ij}^r)$. Since B_r^{-1} is nonsingular $(\beta_{i1}^r, \ldots, \beta_{im}^r) \neq 0$ for each $i = 1$ to m. Let $\bar{b}^r = B_r^{-1}b$. Then

$$B_r^{-1}b(\varepsilon) = \begin{pmatrix} \bar{b}_1^r + \beta_{11}^r\varepsilon + \cdots + \beta_{1m}^r\varepsilon^{\mathbf{m}} \\ \vdots \\ \bar{b}_m^r + \beta_{m1}^r\varepsilon + \cdots + \beta_{mm}^r\varepsilon^{\mathbf{m}} \end{pmatrix}$$

For each $i = 1$ to m, and $r = 1$ to L, the polynomial, $\bar{b}_i^r + \beta_{i1}^r\theta + \cdots + \beta_{im}^r\theta^{\mathbf{m}}$ is a nonzero polynomial in θ. Therefore, it has at most m roots. Let the roots be $\theta_{r,i,1}, \ldots, \theta_{r,i,k}$. If ε is not equal to one of these roots, then $\bar{b}_i^r + \beta_{i1}^r\varepsilon + \cdots + \beta_{im}^r\varepsilon^{\mathbf{m}} \neq 0$. Let Γ be the collection of the roots of all these polynomials in θ for $i = 1$ to m and $r = 1$ to L. Γ is a finite set consisting of at most $\mathrm{L}m^2$ real numbers. Hence, it is possible to pick a positive number $\varepsilon_1 > 0$ such that Γ contains no element belonging to the open interval $0 < \varepsilon < \varepsilon_1$. By the construction of the set Γ and the positive number ε_1, it is clear that in any basic solution of (10.3), all the basic variables are nonzero and, hence, that (10.3) is nondegenerate whenever $0 < \varepsilon < \varepsilon_1$. ■

Dual Nondegeneracy

The LP (10.1) is said to be *totally dual nondegenerate* iff for every $\pi \in \mathbf{R}^m$, the row vector $c - \pi A = (c_j - \pi A_{.j})$ has at least $n - m$ nonzero components; that is, if every solution $(\pi, \bar{c})$ of the system of equations $\pi A + \bar{c} = c$ (in which A, c are the data and π, $\bar{c}$ are the variables) has at least $n - m$ of the $\bar{c}_j$ nonzero. In our discussion unless the terms "dual nondegeneracy" or "dual degeneracy" are used, the unqualified terms "degeneracy" or "nondegeneracy" of (10.1) will always mean "primal degeneracy" or "primal nondegeneracy" of (10.1) as defined earlier.

Adjacent Bases

Let $\mathbf{K}$ be the set of feasible solutions of (10.1). Every basis for (10.1) is a nonsingular square matrix of order m. Two feasible bases B_1 and B_2 are said to be *adjacent bases* if the associated basic vectors x_{B_1} and x_{B_2} contain $m - 1$ common basic variables. Thus two consecutive bases obtained in the course of the simplex algorithm are always *adjacent bases*.

Adjacent Extreme Points of K

It was shown in Chapter 3 that if B_1 and B_2 are two feasible adjacent bases for (10.1), and if the BFSs corresponding to these two bases are *distinct* (i.e., not equal), then these two BFSs are adjacent extreme points of $\mathbf{K}$, and the line segment joining them is an *edge* of $\mathbf{K}$. The BFSs corresponding to two feasible adjacent bases B_1 and B_2 are the same iff (1) both B_1 and B_2 are degenerate bases, and (2) B_2 is obtained from B_1 by making a degenerate pivot step and vice versa.

Every extreme point of $\mathbf{K}$ is the BFS corresponding to some feasible basis. Let $\tilde{x}$ and $\hat{x}$ be two distinct extreme points of $\mathbf{K}$. Suppose $\tilde{x}$ is the BFS corresponding to a basis $\tilde{B}$ and $\hat{x}$ is the BFS corresponding to a basis $\hat{B}$. Points $\tilde{x}$ and $\hat{x}$ are adjacent extreme points of $\mathbf{K}$ if either (1) $\tilde{B}$ and $\hat{B}$ are adjacent bases, or (2) there exists a sequence of feasible adjacent bases, $\tilde{B}, B_1, B_2, \ldots, B_r; B_{r+1}, \ldots, B_t, \hat{B}$, such that every pair of consecutive bases in the sequence are adjacent, and the BFSs corresponding to all the bases $B_1, B_2, \ldots, B_r$ are equal to $\tilde{x}$, and the BFSs corresponding to all the bases $B_{r+1}, \ldots, B_t$ are equal to $\hat{x}$.

Adjacent Basis Paths

THEOREM 10.2 Let $\tilde{x}$ be an extreme point of $\mathbf{K}$. Let $\tilde{B}$ and $\bar{B}$ be two distinct bases for (10.1) corresponding to both of which $\tilde{x}$ is the BFS. Then there exists a sequence of bases $\tilde{B}, B_1, \ldots, \bar{B}$, such that consecutive bases in the sequence are adjacent and all the bases in the sequence correspond to $\tilde{x}$.

Proof: This follows from standard results in linear algebra, using the fact that both B and $\bar{B}$ are bases obtained by completing a partial basis. ■

THEOREM 10.3 If $\tilde{B}$ and $\hat{B}$ are a pair of feasible bases for (10.1), then there exists a sequence of bases $\tilde{B}, B_1, B_2, \ldots, B_\ell, \hat{B}$, such that all the bases in the sequence are feasible and every pair of consecutive bases in the sequence is adjacent.

Proof: Suppose $\tilde{x}$ and $\hat{x}$ are the BFSs corresponding to the bases $\tilde{B}$ and $\hat{B}$, respectively. If $\tilde{x} = \hat{x}$ the result follows from the previous theorem. If $\tilde{x} \neq \hat{x}$, let $z^1(x) = \sum d_j x_j$, where $d_j = 1$ if $\hat{x}_j = 0$ and $d_j = 0$, otherwise. Since $\hat{x}$ is a BFS, it is the unique feasible solution of (10.1) that makes $z^1(x) = 0$. Also $z^1(x) \geqq 0$ for all $x \in \mathbf{K}$. Hence, $\hat{x}$ is the unique optimum feasible solution for the problem of minimizing $z^1(x)$ on $\mathbf{K}$. Starting with the feasible basis $\tilde{B}$, apply the simplex algorithm to the problem of minimizing $z^1(x)$ on $\mathbf{K}$. The simplex algorithm walks through a sequence of adjacent bases until it terminates with an optimal basis, the BFS corresponding to which must be $\hat{x}$. Combining this with the previous theorem, the result follows. ■

If (10.1) is nondegenerate, the minimum ratio corresponding to every pivot step is strictly positive. Also, if $\tilde{x}$ is a BFS, there is a unique basis corresponding to it in this case. Hence there are exactly $n - m$ edges of $\mathbf{K}$ that contain the extreme point $\tilde{x}$.

A polyhedron that has the same number (equal to its dimension) of edges going out of each of its extreme points is known as a *simple convex polyhedron*. Hence, if (10.1) is nondegenerate, $\mathbf{K}$ is a simplex convex polyhedron. A bounded convex polyhedron is known as a *convex polytope*. Hence, if (10.1) is nondegenerate and $\mathbf{K}$ is bounded, then $\mathbf{K}$ is a *simple convex polytope*. In this case, each edge of $\mathbf{K}$ is a bounded edge. Hence, every extreme point of $\mathbf{K}$ has exactly $n - m$ adjacent extreme points.

Even if (10.1) is degenerate, if $\tilde{x}$ happens to be a nondegenerate BFS of (10.1), there are exactly $n - m$ edges of $\mathbf{K}$ through $\tilde{x}$. However, if $\tilde{x}$ is a degenerate BFS of (10.1), there are several bases corresponding to $\tilde{x}$. Each of these bases leads to some edges of $\mathbf{K}$ through $\tilde{x}$. In this case, to get all the edges of $\mathbf{K}$ through $\tilde{x}$, examine each basic vector corresponding to $\tilde{x}$ and generate the edges from it obtained by choosing one of the nonbasic variables corresponding to positive minimum ratio as the entering variable. In this case, the number of edges of $\mathbf{K}$ through $\tilde{x}$ and the number of adjacent extreme points of $\tilde{x}$ is likely to be much larger than $n - m$. And these numbers may be different for different extreme points of $\mathbf{K}$.

Exercises

10.2 Show that the following system is degenerate: $\sum_{j=1}^{N} x_{ij} = 1$, for $i = 1$ to N, $\sum_{i=1}^{N} x_{ij} = 1$, for $j = 1$ to N, $x_{ij} \geqq 0$. Prove that $\tilde{x} = I =$ unit matrix of order N, is a BFS of this system. Let $\mathbf{K}$ be the set of feasible solutions of this system. Using the theory of the transportation problem (see Chapter 13), show that the number of adjacent extreme of points of $\tilde{x}$ in $\mathbf{K}$ is $\sum_{r=2}^{N} \binom{N}{r}(r-1)!$. Compare and see that this number is much larger than the dimension of this polytope, $N^2 - (2N - 1)$, the number of variables minus the number of independent equality constraints in this problem.

In general, a degenerate BFS has a much larger number of adjacent BFSs than a nondegenerate BFS.

10.2 RESOLUTION OF CYCLING UNDER DEGENERACY: THE LEXICO PRIMAL SIMPLEX ALGORITHM

10.2.1 Simplex Algorithm Under Nondegeneracy

THEOREM 10.4 Suppose (10.1) is nondegenerate and is being solved by the simplex algorithm starting with a feasible basis:

1 The minimum ratio in every pivot step will be positive.

2 The dropping variable, or equivalently, the pivot row is always uniquely determined in every pivot step during the course of the algorithm; that is, there will be no ties in determining the minimum ratio in every pivot step.

3 The value of the objective function makes a strict decrease after every pivot step.

4 A basis that appeared in the course of the algorithm will never appear later on, that is, there can be no cycling.

5 The algorithm reaches a terminal basis in a finite number of pivot steps.

Proof: By nondegeneracy, the value of every basic variable is nonzero in any basic solution of(10.1). Hence, all the basic values will be positive in every BFS. This implies (1).

To prove (2), let B be the current basis associated with the basic vector $x_B = (x_1, \ldots, x_m)$ and suppose the nonbasic variable x_s is the entering variable. The current vector of basic values is $B^{-1}b = \bar{b}$. The pivot column is $B^{-1}A_{.s} = \bar{A}_{.s}$. The minimum ratio here is minimum $\{\bar{b}_i/\bar{a}_{is}$: i, such that $\bar{a}_{is} > 0\}$. If there is a tie for the i that attains the minimum here, suppose both $i = 1, 2$ are such that $\bar{b}_1/\bar{a}_{1s} = \bar{b}_2/\bar{a}_{2s} =$ minimum ratio here. Hence x_s can enter the basic vector replacing either x_1 or x_2. If x_1 is the dropping variable, the value of the basic variable x_2 is zero in the BFS obtained after the pivot. This contradicts the hypothesis of nondegeneracy of (10.1). Hence, in each pivot step, there can be no ties in the minimum ratio test for determining the pivot row.

If x_s is the entering variable in a pivot step of the algorithm, $\bar{c}_s < 0$. If θ is the minimum ratio in this pivot step, $\theta > 0$ by nondegeneracy. The change in the objective value as a result of this pivot step is $\theta\bar{c}_s < 0$. This implies (3).

Part (4) follows from (3). Part (5) follows from (4) and from the fact that the total number of bases is finite. ■

Hence, if (10.1) is nondegenerate, except for the choice of the pivot column at each stage, the simplex algorithm is fully automatic and the algorithm walks to a terminal basis without ever returning to a basis it has seen already. Under nondegeneracy, there cannot be any problem of cycling. The same claims cannot be made if (10.1) is degenerate (see Chapters 2 and 3). Parts (1) and (4) of the theorem proved here may not hold. Starting from a degenerate feasible basis, the algorithm could go through a sequence of degenerate pivots and return to the basis that started this sequence, and the sequence will be repeated again and again forever. This is the *cycling of the simplex algorithm under degeneracy*. If this happens, part (5) of the theorem cannot be guaranteed. See Example 10.4 for a problem in which cycling is exhibited.

10.2.2 Perturbed Problem

Suppose (10.1) is degenerate. The general strategy for avoiding cycling is as follows. Replace b by $b(\varepsilon)$ from equation (10.2), and get the *perturbed problem* (10.3). From Section 10.1, we know that there exists a positive number ε_1 such that for all $0 < \varepsilon < \varepsilon_1$, (10.3) is nondegenerate. Thus for all ε that are positive but arbitrarily small, (10.3) can be solved by applying the simplex algorithm with no possibility of cycling occurring. It is not necessary to give ε any specific value as long as it is treated as *arbitrarily small*, which will mean that it is a positive number, strictly smaller than any other positive number with which it may be compared during the course of the algorithm.

10.2.3 Feasibility

THEOREM 10.5 If B is a feasible basis for (10.3) when ε is *arbitrarily small*, then B is a feasible basis for (10.1).

Proof: Let $B^{-1} = (\beta_{ij})$ and $\bar{b} = B^{-1}b$. Then $B^{-1}b(\varepsilon) = (\bar{b}_1 + \beta_{11}\varepsilon + \cdots + \beta_{1m}\varepsilon^{\mathbf{m}}, \ldots, \bar{b}_m + \beta_{m1}\varepsilon + \cdots + \beta_{mm}\varepsilon^{\mathbf{m}})^{\mathrm{T}}$. If for some i, $\bar{b}_i < 0$, then $\bar{b}_i + \beta_{i1}\varepsilon + \cdots + \beta_{im}\varepsilon^{\mathbf{m}} < 0$ when ε is arbitrarily small, which is a contradiction. Hence, $\bar{b} \geqq 0$ for all i. Thus B must also be a feasible basis for (10.1). ■

When (10.3) is solved by the simplex algorithm, treating ε as arbitrarily small, a basis that appeared once cannot reappear by nondegeneracy. If (10.3) is feasible, the terminal basis either satisfies the optimality or the unboundedness criteria. These termination criteria are independent of the right-hand-side vector. Hence, the terminal basis obtained for (10.3) is also a terminal basis for (10.1). The final optimal solution of (10.1) is obtained from the corresponding optimum solution of (10.3) by eliminating all the terms involving ε or its positive powers (in effect substituting 0 for ε).

Definition: *LEXICO POSITIVE* Let $\gamma = (\gamma_1, \ldots, \gamma_r)$. It is said to be *lexico positive* if $\gamma \neq 0$ and if the first nonzero component of γ is positive, irrespective of what values the subsequent components may have. We will indicate this by writing $\gamma \succ 0$. The symbol $\succ$ stands for lexicographically greater than. Thus, if $\gamma \succ 0$, exactly one of the following holds:

$$\text{Either } \gamma_1 > 0; \text{ or } \gamma_1 = 0, \gamma_2 > 0; \text{ or } \gamma_1 = 0,$$
$$\gamma_2 = 0, \gamma_3 > 0; \ldots; \text{ or } \gamma_1 = 0,$$
$$\gamma_2 = 0, \ldots, \gamma_{r-1} = 0, \gamma_r > 0$$

Given two vectors $\xi = (\xi_1, \ldots, \xi_r)$ and $\eta = (\eta_1, \ldots, \eta_r)$, we say that $\xi \succ \eta$ iff $\xi - \eta \succ 0$. ξ is said to be *lexico negative* (denoted by $\xi \prec 0$) if $-\xi \succ 0$.

The Lexico Minimum of a Given Set of Vectors from $\mathbf{R}^m$

Let $\gamma^v = (\gamma_1^v, \ldots, \gamma_m^v) \in \mathbf{R}^m$ for $v = 1$ to p. The vector γ^q is said to be the *lexico minimum in the set* $\{\gamma^1, \ldots, \gamma^p\}$ if $\gamma^v - \gamma^q \succeq 0$ for all $v = 1$ to p, $v \neq q$. Similarly, the vector γ^s is said to be the lexico maximum in the set $\{\gamma^1, \ldots, \gamma^p\}$ if $\gamma^s - \gamma^v \succeq 0$ for all $v = 1$ to p, $v \neq s$.

To find the lexico minimum in a set of vectors from $\mathbf{R}^m$, find the minimum among the first components of the vectors and discard all vectors that do not correspond to this minimum. Now find the minimum among the second components of the remaining vectors and discard every vector that does not correspond to this minimum. Continue in the same way with the third component, and so on. At any stage if there is a single vector left, it is the lexico minimum. This procedure terminates after at most m steps. At the end if two or more vectors are left, they are all equal to each other, and each of them is a lexico minimum in the set.

Example 10.2

The vector $(0, 0, 0.1, -1000)$ is lexico positive. The vector $(0, -1, 2000, 10{,}000)$ is lexico negative. In the set of vectors $\{(-2, 0, -1, 0), (-2, 0, -1, 1), (-2, 1, -20, -30), (0, -10, -40, -50)\}$, the vector $(-2, 0, -1, 0)$ is the lexico minimum.

Primal Feasibility for Arbitrarily Small ε

Let B be a basis for (10.3). Let $B^{-1} = (\beta_{ij})$, $\bar{b} = B^{-1}b$.

THEOREM 10.6 B is a feasible basis for (10.3) for *arbitrarily small* ε iff $(\bar{b}_i, \beta_{i1}, \ldots, \beta_{im}) \succ 0$ for all i.

Proof: Since B^{-1} is nonsingular, obviously $(\beta_{i1}, \ldots, \beta_{im}) \neq 0$ for all $i = 1$ to m. So $(\bar{b}_i, \beta_{i1}, \ldots, \beta_{im})$ is nonzero for all i. When ε is positive and arbitrarily small, the sign of $\bar{b}_i + \beta_{i1}\varepsilon + \cdots + \beta_{im}\varepsilon^m$ is obviously the same as the sign of the first nonzero coefficient in this polynomial; that is, the first nonzero component in $(\bar{b}_i, \beta_{i1}, \ldots, \beta_{im})$. So for arbitrarily small ε, $\bar{b}_i + \beta_{i1}\varepsilon + \cdots + \beta_{im}\varepsilon^m$ is positive iff the first nonzero component in the row vector $(\bar{b}_i, \beta_{i1}, \ldots, \beta_{im})$ is positive. Applying this for each i, we have the theorem. ∎

This theorem makes it easy to test the feasibility of a basis for (10.3) for arbitrarily small ε. There is no need to give any specific value to ε. Given the updated right-hand-side constants vector, and the inverse of the basis, testing the basis for feasibility in (10.3) requires the testing of lexico positiveness of m row vectors, which is done without ever mentioning the perturbation variable ε.

Let B be a basis for (10.1), and let $B^{-1} = (\beta_{ij})$, $\bar{b} = B^{-1}b$. The basis B is *feasible* (or *primal feasible*) to (10.1) if $\bar{b} \geqq 0$. The basis B is said to be *lexico feasible* (or *lexico primal feasible*) to (10.1) if $(\bar{b}_i, \beta_{i1}, \ldots, \beta_{im}) \succ 0$ for each $i = 1$ to m. If B is a nondegenerate feasible basis for (10.1), $\bar{b} > 0$, and obviously B is automatically a lexico feasible basis for (10.1). If B is a degenerate feasible basis for (10.1), it is lexico feasible iff the first nonzero entry in $(\beta_{i1}, \ldots, \beta_{im})$ is strictly positive for every i such that $\bar{b}_i = 0$. From the results here we conclude that every feasible basis for (10.3) for arbitrarily small ε is automatically lexico feasible to (10.1) Conversely a basis that is feasible for (10.1) is feasible for (10.3) for arbitrarily small ε iff it is lexico feasible for (10.1).

10.2.4 Pivot Row Choice for Arbitrarily Small ε

Let B be a feasible basis for (10.3) with $B^{-1} = (\beta_{ij})$, and suppose the nonbasic variable x_s has to be brought into the basic vector. $\bar{A}_{.s} = B^{-1}A_{.s}$ is the pivot column. If $(1/\bar{a}_{ps})(\bar{b}_p, \beta_{p1}, \ldots, \beta_{pm}) = (1/\bar{a}_{qs})(\bar{b}_q, \beta_{q1}, \ldots, \beta_{qm})$ for some $p \neq q$, then $(\beta_{p1}, \ldots, \beta_{pm}) = (\bar{a}_{ps}/\bar{a}_{qs}) \times (\beta_{q1}, \ldots, \beta_{qm})$, which is not possible since both $(\beta_{p1}, \ldots, \beta_{pm})$ and $(\beta_{q1}, \ldots, \beta_{qm})$ are row vectors of the nonsingular matrix B^{-1}. Hence, if $\bar{A}_{.s} \nleqq 0$, there is a unique lexico minimum among the nonempty set of vectors $\{(\bar{b}_i, \beta_{i1}, \ldots, \beta_{im})/\bar{a}_{is} : i, \text{ such that } \bar{a}_{is} > 0\}$. The minimum ratio in this pivot step in solving (10.3) is as follows

$$\text{minimum}\ \{(b_i + \beta_{i1}\varepsilon + \cdots + \beta_{im}\varepsilon^m)/\bar{a}_{is} : i, \text{ such that } \bar{a}_{is} > 0\}$$

and clearly when ε is arbitrarily small, this minimum is attained by the same i that corresponds to the unique lexico minimum in the set of vectors $\{(\bar{b}_i, \beta_{i1}, \ldots, \beta_{im})/\bar{a}_{is} : i, \text{ such that } \bar{a}_{is} > 0\}$.

Thus when ε is arbitrarily small, irrespective of its actual value, the pivot row is uniquely determined according to the following steps, and it is independent of the actual value of ε. It turns out that we can describe the pivot row choice rules in terms of the data in the original problem (10.1) itself without making any reference to the perturbation parameter ε.

Lexico Primal Simplex Minimum Ratio Pivot Row Choice Rule

The pivot row is selected according to the following rules.

Step 1 Find out the i that attains the minimum of $(\bar{b}_i/\bar{a}_{is})$ over all i satisfying $\bar{a}_{is} > 0$. If the i attaining this minimum is unique, it determines the pivot row, and we stop. Otherwise go to step 2.

Step 2 If there is a tie for the i attaining the minimum in step 1, find the i that attains the minimum of $(\beta_{i1}/\bar{a}_{is})$ over all i that tied for the minimum in step 1. If the minimum here is attained by a unique i, that determines the pivot row and we stop. Otherwise go to the next step.

The general step is:

Step *t* If the pivot row is not determined unambiguously in step $t-1$, find out the i that attains the minimum of $(\beta_{i,t-1}/\bar{a}_{is})$ over all i that tied for the minimum in step $t-1$. If the minimum here is attained by a unique i, it defines the pivot row and we stop. Otherwise go to the next step.

During or before step m the pivot row will be determined uniquely.

Note 10.1: Basis B is feasible for (10.3) for arbitrarily small positive values of ε iff the first nonzero entry in $(\bar{b}_i, \beta_{i1}, \ldots, \beta_{im})$ is strictly positive for all $i = 1$ to m, as discussed above. If B is such a basis, and the pivot row is selected according to the rules described here, then the next basis obtained after this pivot step will also be a feasible basis for (10.3) for arbitrarily small positive values of ε.

Example 10.3

Consider the inverse tableau given at the bottom, obtained in some stage of solving an LP. The pivot column is entered on the tableau.

The above procedure was carried out through five steps before the pivot row was selected to be row 3 unambiguously. The pivot element is circled. Performing the pivot leads to the following tableau.

Basic Variables	Inverse Tableau							$\bar{b}$
x_1	0	4	4	1	−1	1	0	0
x_2	2	−5	5	2	1	−1	0	0
x_7	$\frac{3}{4}$	$-\frac{2}{4}$	$-\frac{5}{4}$	$-\frac{1}{4}$	0	0	0	$\frac{2}{4}$
x_4	0	0	0	3	1	2	0	0
x_5	1	1	5	3	2	1	0	0
x_6	$-\frac{5}{4}$	7	$-\frac{5}{2}$	$\frac{1}{2}$	−1	−1	0	4
$-z$	−2	−2	−3	−3	1	0	1	−8

The initial basic vector $(x_1, x_2, x_3, x_4, x_5, x_6)$ is lexico feasible to this LP. The pivot row was determined by the lexico minimum ratio pivot row choice rule. The basic vector obtained after this pivot step, $(x_1, x_2, x_7, x_4, x_5, x_6)$, is also lexico feasible to this LP. This example illustrates the fact that the use of the lexico minimum ratio pivot row choice rule determines the

Basic Variables	Inverse Tableau							$\bar{b}$	Pivot Column x_7	Ratios Obtained Using $\bar{b}_s$	Column of Inverse 1st	2nd	3rd	4th
x_1	3	2	−1	0	−1	1	0	2	4	$\frac{2}{4}$	$\frac{3}{4}$	$\frac{2}{4}$		
x_2	5	−7	0	1	1	−1	0	2	4	$\frac{2}{4}$	$\frac{5}{4}$			
x_3	3	−2	−5	−1	0	0	0	2	(4)	$\frac{2}{4}$	$\frac{3}{4}$	$-\frac{2}{4}$	$-\frac{5}{4}$	$-\frac{1}{4}$
x_4	3	−2	−5	2	1	2	0	2	4	$\frac{2}{4}$	$\frac{3}{4}$	$-\frac{2}{4}$	$-\frac{5}{4}$	$\frac{2}{4}$
x_5	4	−1	0	2	2	1	0	2	4	$\frac{2}{4}$	$\frac{4}{4}$			
x_6	−2	8	0	1	−1	−1	0	3	−2					
$-z$	−5	0	2	−2	1	0	1	−10	−4					

pivot row unambiguously and preserves the lexico feasibility of the basis.

Change in Objective Value

Let B be a feasible basis for (10.3) for arbitrarily small ε and let $B^{-1} = (\beta_{ij})$. Let $\bar{b} = B^{-1}b$ and c_B be the row vector of basic cost coefficients. The corresponding dual solution is $\pi = c_B B^{-1}$ and let $\bar{z} = c_B \bar{b}$. The BFS of (10.3) corresponding to the basis B and the objective value $\bar{z}_0$ at this solution are

$$\text{basic vector } x_B = \begin{pmatrix} \bar{b}_1 + \beta_{11}\varepsilon + \cdots + \beta_{1m}\varepsilon^m \\ \vdots \\ \bar{b}_m + \beta_{m1}\varepsilon + \cdots + \beta_{mm}\varepsilon^m \end{pmatrix}$$

all nonbasic variables = 0

$$\bar{z}_0 = \bar{z} + \pi_1\varepsilon + \cdots + \pi_m\varepsilon^m$$

Suppose the nonbasic variable x_s (with relative cost coefficient $\bar{c}_s$) is the entering variable. Suppose the pivot row is the rth row for this pivot step. Let z_1 be the objective value after the pivot. Then

$$\begin{aligned} z_1 &= \bar{z}_0 + \bar{c}_s(\bar{b}_r + \varepsilon\beta_{r1} + \cdots + \varepsilon^m\beta_{rm})/\bar{a}_{rs} \\ &= \bar{z} + \bar{c}_s\frac{\bar{b}_r}{\bar{a}_{rs}} + \sum_{t=1}^{m} \varepsilon^t(\pi_t + \bar{c}_s\beta_{rt}/\bar{a}_{rs}) \end{aligned}$$

For arbitrarily small ε, $z_1 < \bar{z}_0$. This implies that $(\bar{z} + \bar{c}_s\bar{b}_r/\bar{a}_{rs}, \pi_1 + \bar{c}_s\beta_{r1}/\bar{a}_{rs}, \ldots, \pi_m + \bar{c}_s\beta_{rm}/\bar{a}_{rs}) \prec (\bar{z}, \pi_1, \ldots, \pi_m)$.

As long as we do not wish to give a specific value to ε, we can treat the vector $(\bar{b}_i, \beta_{i1}, \ldots, \beta_{im})$ as defining the value of the ith basic variable in the BFS corresponding to the basis B. Given this vector, the actual value of the ith basic variable for any particular value of ε is $\bar{b}_i + \varepsilon\beta_{i1} + \cdots + \varepsilon^m\beta_{im}$. Similarly the vector $(\bar{z}, \pi_1, \ldots, \pi_m)$ may be treated as defining the value of the objective function corresponding to the feasible basis B. Treating the values of the basic variables and the objective value as these vectors, we have the following:

1 A basis is feasible to (10.3) iff the values of all the basic variables are lexico positive. When this happens, we say that the basis is *lexico feasible*.

2 The algorithm starts with a lexico feasible basis and all the bases obtained in the course of the algorithm will be lexico feasible.

3 The termination criteria and the pivot column choice are the same as under the regular simplex method.

4 Once a pivot column has been selected, the pivot row is selected by the rules discussed above. The new pivot row choice rule is known as the *lexico minimum ratio rule.* It determines the pivot row (and, hence, the pivot element) uniquely without any ambiguity.

5 There will be a strict lexico decrease in the vector $(\bar{z}, \pi_1, \ldots, \pi_m)$ after each pivot step. Hence a basis that appeared once can never reappear and cycling cannot occur. The algorithm must terminate after a finite number of pivot steps.

6 When the simplex algorithm using these rules terminates, the terminal tableau is a terminal tableau for (10.3) for all ε arbitrarily small and also for $\varepsilon = 0$.

7 The terminal basis for (10.3) obtained by the simplex algorithm using these rules is also a terminal basis for (10.1).

10.3 SUMMARY OF THE PRIMAL SIMPLEX ALGORITHM USING THE LEXICO MINIMUM RATIO RULE

Consider the LP (10.1). We assume that it is in standard form (i.e., that $b \geqq 0$). The simplex algorithm for solving (10.1) using the lexico minimum ratio pivot row choice rule always begins with a lexico feasible basis.

Case 1 If A contains a unit matrix of order m as a submatrix, pick it as the initial feasible basis. Since b is nonnegative and the basis is the unit matrix, it is obviously lexico feasible. If we start with this basis, the problem is solved using the revised simplex algorithm with the explicit form of the inverse. However, instead of using the regular minimum ratio rule, the lexico minimum ratio rule discussed in Section 10.2.4 is used in each pivot step to determine the pivot row. In the revised simplex algorithm, the inverse tableau is computed and this contains all the data needed to perform the lexico minimum ratio rule instead of the usual minimum ratio rule. It only requires some additional work in identifying the pivot element when a tie occurs in the usual minimum ratio test. This guarantees that there will be strict lexico decrease in the vector $(\bar{z}, \pi_1, \ldots, \pi_m)$ after each pivot step. Hence, a basis that appeared once can never reappear. *There cannot be any cycling.* After a finite number of pivot steps, the algorithm obtains a terminal basis.

Case 2 If A does not contain a unit matrix of order m as a submatrix, provide a full artificial basis and formulate the Phase I problem.

$$\text{Minimize} \qquad w(y) = \sum_{i=1}^{m} y_i$$

$$\text{Subject to} \quad Ax + Iy = b$$

$$x \geqq 0 \qquad y \geqq 0$$

For this Phase I problem, the artificial basic vector is an initial lexico feasible basic vector. Starting with this lexico feasible basic vector, solve this Phase I problem as under Case 1, using the revised simplex algorithm with the explicit form of the inverse using the lexico minimum ratio rule. When Phase I is completed, if the original problem turns out to be feasible, the terminal basis obtained under Phase I will be a lexico feasible basis for (10.1). Go over to Phase II as usual under the simplex algorithm and continue with the revised simplex routine using the lexico minimum ratio pivot row choice rule in each pivot step.

10.4 DO COMPUTER CODES USE THE LEXICO MINIMUM RATIO RULE?

If the LP is solved by the revised simplex method with the explicit form of the inverse, the algorithm in Section 10.3, which guarantees that cycling cannot occur, only requires the replacement of the standard minimum ratio rule by the lexico minimum ratio rule for pivot row choice. It requires very little additional work. However, most computer programs in existence for solving LPs do not include this routine for the following reasons.

1 Cycling does not seem to occur in the course of solving most practical problems even though they are degenerate. Therefore for practical applications it seems unnecessary to include routines for avoiding cycling.

2 Most LP codes use the revised simplex algorithm with the product form of the inverse to conserve core space. Under this routine the actual basis inverse is never actually computed explicitly. It makes it possiblefor them to solve large problems. Consequently, it is very difficult to adopt the lexico minimum ratio rule, as it requires the explicit basis inverse in each pivot step.

3 Even if we have a degenerate problem, computer round-off error will usually change it to a nondegenerate one.

Example 10.4 An Example of Cycling Under Degeneracy

Consider the following by K. T. Marshall and J. W. Suurballe (reference [10.10])

x_1	x_2	x_3	x_4	x_5	x_6	x_7	$-z$	b
1	0	0	1	1	1	1	0	1
0	1	0	$\frac{1}{2}$	$-\frac{11}{2}$	$-\frac{5}{2}$	9	0	0
0	0	1	$\frac{1}{2}$	$-\frac{3}{2}$	$-\frac{1}{2}$	1	0	0
0	0	0	-1	7	1	2	1	0

$x_j \geqq 0$ for all j; z to be minimized.

Use (x_1, x_2, x_3) as the initial basic vector. Verify that the following sequence of pivot steps is such that the incoming variable is eligible to enter the basic vector according to the objective improvement routine and that the outgoing basic variable is determined by the usual minimum ratio test.

Pivot Step Number	*Entering Variable*	*Leaving Variable*
1	x_4	x_2
2	x_5	x_3
3	x_6	x_4
4	x_7	x_5
5	x_2	x_6
6	x_3	x_7

All these are degenerate pivot steps and at the end of pivot step 6, the initial basic vector (x_1, x_2, x_3) is obtained again.

Verify that the lexico minimum ratio rule selects the leaving variable as x_3 in the first pivot step when x_4 is the entering variable into the basic vector (x_1, x_2, x_3). After this pivot we are led to an optimum basis.

How to Resolve Degeneracy when (10.1) is Solved Beginning with a Known Feasible Basis.

Let B_1 be a given feasible basis for (10.1). Suppose we wish to solve (10.1) by the simplex algorithm beginning with the initial feasible basis B_1. Let $B_1^{-1} = (\beta_{ij})$, $\bar{b} = B_1^{-1}b$. If $(\bar{b}_i, \beta_{i1}, \ldots, \beta_{im}) \succ 0$ for each $i = 1$ to m, then B_1 is a lexico feasible basis for (10.1), and the simplex algorithm with the lexico minimum ratio pivot row choice rule discussed in Sections 10.2 and 10.3 can be used to solve (10.1) beginning with the lexico feasible basis B_1.

However, if B_1 is not a lexico feasible basis for (10.1), we cannot use B_1 as the initial basis, since the simplex algorithm with the lexico minimum ratio pivot row choice rule discussed in Sections 10.2 and 10.3 requires the initial basis to be lexico feasible. Let x_{B_1} be the basic vector for (10.1) associated with the basis B_1. Put (10.1) into canonical form with respect to the basis B_1, and eliminate the constant terms in the objective function. This transforms (10.1) into the equivalent problem

$$\begin{aligned} &\text{Minimize} && cx \\ &\text{Subject to} && \bar{A}x = \bar{b} \qquad (10.4) \\ & && x \geq 0 \end{aligned}$$

By the hypothesis $\bar{b} \geqq 0$ and the column vectors associated with the basic vector x_{B_1} in (10.4), form the unit matrix. Hence for the equivalent problem in canonical form, (10.4), x_{B_1} is a lexico feasible basic vector. Thus (10.4) can be solved by the simplex algorithm with the lexico minimum ratio pivot choice rule discussed in Sections 10.2 and 10.3, beginning with x_{B_1} as the initial basic vector. Since (10.4) is equivalent to (10.1), the terminal tableau obtained for (10.4) in this algorithm is also a terminal tableau for (10.1).

Exercises

10.3 Let x_B be any basic vector for (10.1). Let B, $\bar{B}$ denote the bases associated with x_B in (10.1) and (10.4), respectively. Clearly $\bar{B} = B_1^{-1}B$. So $(\bar{B})^{-1} = B^{-1}B_1$. Using this show that solving (10.4), beginning with x_{B_1} as the initial basic vector adopting the lexico minimum ratio dropping variable choice rule of Section 10.2.4 in each pivot step, is equivalent to solving (10.1) directly, beginning with x_{B_1} as the initial basic vector, using the rule that selects the pivot row to be the row obtained by performing the lexico minimum ratio test with basic values defined to be the row vectors in $(B^{-1}b, B^{-1}B_1)$, where B is the basis (from the original data in (10.1) in the pivot step. Thus show that the simplex algorithm with the lexico minimum ratio rule applied to (10.4) beginning with x_{B_1} as the initial basic vector, can be implemented using the original data in (10.1) directly without computing the canonical form (10.4) explicitly.

10.5 RESOLVING DEGENERACY USING ONLY THE UPDATED COST ROW, THE UPDATED RIGHT-HAND-SIDE CONSTANTS VECTOR, AND THE PIVOT COLUMN

In the primal simplex algorithm for solving (10.1) using the lexico minimum ratio rule, any one of the columns in which the current relative cost coefficient is negative may be chosen as the pivot column. Hence selection of the pivot column requires only the data from the updated cost row, and the choice is arbitrary except for the condition that the relative cost coefficient in the pivot column must be negative. However, once the pivot column is chosen, the choice of the pivot row is made unambiguously by the lexico minimum ratio rule itself. One disadvantage of the lexico minimum ratio rule is that it may require the entries in the basis inverse in each step. If the problem is solved using the product form of the inverse, computing the basis inverse is computationally expensive and goes counter to the philosophy of the product form. Because of this disadvantage, it would be nice if a method for resolving degeneracy does not need the data from the current basis inverse for implementing it. One such method for resolving degeneracy has been developed by R. G. Bland [10.3] and we discuss it here. In this method, both the pivot column and the pivot row are selected unambiguously by the rules given in the method itself. The method requires that a specific ordering of the (primal) variables in (10.1) be chosen before the algorithm is initiated (say, the order x_1, $x_2, \ldots, x_n$). This ordering can be arbitrary, but once it is selected, it is fixed and cannot be changed after the algorithm is initiated. The entering and dropping variable choice rules in this method applied to solve (10.1) are the following.

Bland's Entering Variable Choice Rule

In each pivot step, the entering variable is chosen to be the first variable that is eligible to enter the basic vector in that step, when the variables are examined in the specific ordering chosen for the variables.

Bland's Dropping Variable Choice Rule

In each pivot step, the dropping basic variable is chosen to be the first basic variable that is eligible to drop from the basic vector in that step, when the basic variables are examined in the specific ordering chosen for the variables.

We now prove that the simplex alogrithm cannot cycle if both these entering and dropping variable choice rules are adopted in each step of the algorithm.

THEOREM 10.7 Cycling cannot occur when the LP (10.1) is solved by the simplex algorithm using the entering and dropping variable choice rules, given above, in each step.

Proof: The proof is by R. Bland [10.3]. Without any loss of generality we assume that the specific ordering chosen for the variables is the natural order $x_1, x_2, \dots, x_n$. In this case, Bland's entering and dropping variable choice rules are also called the *least subscript* or *least index rules*.

Suppose that cycling occurs. Initiate the algorithm with one of the basic vectors in the cycle. Then the algorithm goes only through basic vectors in the cycle. In each step the rules being used (those described above) identify both the entering and dropping variables uniquely and unambiguously. The following fact is clear: since we are dealing with basic vectors obtained during a cycle, all the pivot steps during the cycle must be degenerate pivot steps, and hence the updated right-hand-side constants vector with respect to every basic vector during the cycle must be identically the same. Let this be $(\bar{b}_1, \dots, \bar{b}_m, \bar{z})^{\mathrm{T}}$. Let $\mathbf{J}$ be the set of subscripts of variables that are the entering variables in pivot steps during the cycle. Hence if $j \notin \mathbf{J}$, x_j must either be a basic variable in all the basic vectors obtained during the cycle, or a nonbasic variable throughout the cycle. Also, since we are dealing with a cycle of basic vectors, every variable that is an entering variable in a pivot step during the cycle must be the dropping variable in some other pivot steps in the cycle. Let $t = \text{maximum } \{j : j \in \mathbf{J}\}$. Let the canonical tableau obtained during the cycle in which x_t is the entering variable be:

Canonical Tableau P

x	$-z$	
A'	0	$\bar{b}$
c'	1	$-\bar{z}$

Since x_t is the entering variable in canonical tableau P, by the entering variable choice rule being used here, $c'_t < 0$ and $c'_j \geqq 0$ for $j \leqq t - 1$. Since x_t is an entering variable at some stage during the cycle, it must be the dropping variable in some other stage of the cycle. Let $(x_j : j \in \{p_1, \dots, p_r = t, p_{r+1}, \dots, p_m\})$ be a basic vector obtained during the cycle, in which x_t is the rth basic variable (i.e., the basic variable in the rth row) into which x_s is the entering variable replacing x_t. Let the canonical tableau with respect to this basic vector be

Canonical Tableau Q

x	$-z$	
A''	0	$\bar{b}$
c''	1	$-\bar{z}$

Define the column vector $y = (y_j) \in \mathbf{R}^{n+1}$ by $y_j = -1$, a''_{is}, c''_s, 0, depending on whether $j = s$, p_i for $i = 1$ to m, $n + 1$, or $\notin \{p_1, \dots, p_m, s, n+1\}$, respectively. Hence $y_t = a''_{rs}$, and since a''_{rs} is the pivot element (since x_s is the entering variable in this stage and x_t, the rth basic variable, is the dropping variable), we have $y_t = a''_{rs} > 0$. Also, from the definition of y and the fact that Q is the canonical tableau with respect to the basic vector $(x_j : j \in \{p_1, \dots, p_m\})$

$$\begin{pmatrix} A'' & \vdots & 0 \\ \cdots & \cdots & \cdots \\ c'' & \vdots & 1 \end{pmatrix} y = 0 \tag{10.5}$$

Now since each canonical tableau during the cycle is obtained by multiplying canonical tableau Q by the appropriate series of nonsingular pivot matrices, (10.5) implies

$$\begin{pmatrix} A' & \vdots & 0 \\ \cdots & \cdots & \cdots \\ c' & \vdots & 1 \end{pmatrix} y = 0 \tag{10.6}$$

From (10.6) we have $\sum_{j=1}^{n} c'_j y_j + c''_s = 0$, because $y_{n+1} = c''_s$. But $c''_s < 0$, since x_s is the entering variable in canonical tableau Q, so $\sum_{j=1}^{n} c'_j y_j > 0$. Hence there must exist a j between 1 and n, say v, such that $c'_v \neq 0$ and $y_v \neq 0$ and $c'_v y_v > 0$. The facts that $y_v \neq 0$ and v is between 1 and n imply from the definition of y that $v \in \{p_1, \dots, p_m, s\}$. Also x_v must be a nonbasic variable in canonical tableau P, since $c'_v \neq 0$. So x_v is either a basic variable or an entering variable in canonical tableau Q, and is a nonbasic variable in canonical tableau P. Hence $v \in \mathbf{J}$, which implies that $v \leqq t$. But $v \neq t$ because $c'_t < 0$, $y_t = a''_{rs} > 0$, and we already established that $c'_v y_v > 0$. Thus $v < t$. So $c'_v \geqq 0$, and since $c'_v \neq 0$, we must have $c'_v > 0$. So $y_v > 0$, too. Since $y_s = -1$ by definition, this implies that $v \neq s$. Hence $v \in \{p_1, \dots, p_m\}$, and hence x_v is a basic variable in canonical tableau Q. Let ℓ be such that $v = p_\ell$. So

$y_v = a''_{\ell s}$ and, since we already established that $y_v > 0$, $a''_{\ell s} > 0$.

We already established that $v \in \mathbf{J}$; hence, x_v is a basic variable in some basic vectors of the cycle, and a nonbasic variable in some canonical tableaux obtained during the cycle. However, since all the pivot steps during the cycle are degenerate pivot steps, this implies that whenever x_v is a basic variable during the cycle, its value in the corresponding basic solution must be zero. Since $v = p_\ell$, x_v is the ℓth basic variable in canonical tableau Q. Hence $\bar{b}_\ell$, the updated ℓth right-hand-side constant in canonical tableau Q, must be zero. Since $\bar{b}_\ell = 0$ and $a''_{\ell s} > 0$, row ℓ is one of the rows in which the minimum ratio occurs for the operation of bringing x_s into the basic vector in canonical tableau Q. The basic variable in row ℓ in canonical tableau Q is x_{p_ℓ} where $p_\ell = v$ has already been established to be strictly less than t. Hence by the dropping variable choice rule being adopted here, $x_t = x_{p_r}$ could not be the dropping variable in canonical tableau Q. The same argument implies that x_t can never be the dropping variable during the cycle, a contradiction. Hence cycling could not have occurred. ∎

The simplex algorithm with Bland's pivot column and row choice rules in each step, is a finite algorithm. However, computationally it seems to be inefficient. In experiments with randomly generated problems it fared poorly in comparison with variants of the simplex algorithm using the other choice rules discussed in Chapter 2 (see [10.1] of D. Avis and V. Chvatal). Even though they may not be of much practical use, Bland's rules are important for theoretical studies on the simplex algorithm.

Note 10.2: The reader should notice the difference in the way the lexicographic minimum ratio version, and the Bland's version of the simplex algorithm operate, to guarantee finite termination. The lexicographic minimum ratio version starts with an initial lexico primal feasible basic vector, leaves the choice of the entering variable in each pivot step to the user by any method he or she may want to use, but requires the dropping variable to be chosen always by the lexico minimum ratio test. On the other hand, the Bland's version can begin with any primal feasible basic vector, but the user has no freedom at all in choosing either the entering or the dropping variables, since in each pivot step these are uniquely and unambiguously made by the rules prescribed in the version.

10.6 RESOLUTION OF CYCLING IN THE DUAL SIMPLEX ALGORITHM: THE LEXICO DUAL SIMPLEX ALGORITHM

10.6.1 Eliminating a Variable from a Positively Dependent Pair of Variables

Consider the LP (10.1). Two variables x_p and x_q in this problem are said to be *positively dependent* if there exists a positive constant α such that the column vector of x_q in the original tableau (including the cost coefficient) is α times the column vector of x_p. In this case, let $y = x_p + \alpha x_q$. Since $\alpha > 0$, and x_p and x_q are nonnegative variables, y defined as above will be nonnegative too. Eliminate both x_p and x_q from the problem and introduce y into the problem in their place with the column vector for y exactly the same as the column vector for x_p. The result is an LP with one variable less than (10.1). If $\bar{y}$ is the value of y in an optimum solution of the reduced problem, then replace $\bar{y}$ in that solution by $\bar{x}_p$ and $\bar{x}_q$, where these numbers are chosen to satisfy $\bar{x}_p + \alpha \bar{x}_q = \bar{y}$, $\bar{x}_p \geqq 0$, and $\bar{x}_q \geqq 0$. The result is clearly an optimum solution of the original problem. In particular, we can take $\bar{x}_q = 0$ and $\bar{x}_p = \bar{y}$. Since the column vector of y in the reduced problem is the same as the column vector of x_p in the original problem, the reduced problem is essentially obtained by deleting the variable x_q from the original problem, which is the same thing as setting $x_q = 0$ in the original problem. Thus if the LP (10.1) contains a positively dependent pair of variables, one of the variables in the pair can be deleted by setting it equal to zero.

If (10.1) is not totally dual nondegenerate, the dual simplex algorithm applied on (10.1), starting with a dual feasible basic vector, can cycle. See Exercise 10.5 for an example of cycling in the dual simplex algorithm.

10.6.2 Lexico Dual Feasible Basis

Let B be a basis for (10.1) and x_B the associated basic vector. Let $\bar{A}$ and $\bar{c}$ be the updated matrix A and the updated cost vector, respectively, with respect to the basis B. The basis B and the corresponding basic vector x_B are *dual feasible* for (10.1) if $\bar{c}_j \geqq 0$ for each $j = 1$ to n. The basis B, and the corresponding basic vector x_B are said to be *lexico dual feasible* for (10.1) iff $(\bar{c}_j, \bar{a}_{1j}, \dots, \bar{a}_{mj}) \succ 0$ for each $j = 1$ to n.

Clearly, if (10.1) is totally dual nondegenerate, every dual feasible basis for (10.1) is lexico dual feasible and vice versa. If (10.1) is not totally dual nondegenerate, a basis B is lexico dual feasible for (10.1) only if all the relative cost coefficients with respect to it are nonnegative, and the topmost nonzero entry in every column of the canonical tableau of (10.1) with respect to the basis B is strictly positive in every column in which the relative cost coefficient is zero.

How to Obtain an Initial Lexico Dual Feasible Basis Suppose B is a known dual feasible basis for (10.1), associated with the basic vector x_B. If B also happens to be lexico dual feasible, we are done. Otherwise a dual feasible basis constructed for an augmented problem obtained by augmenting (10.1) with a constraint of the type discussed in Section 6.5 is lexico dual feasible for that problem.

10.6.3 Pivot Column Choice Rule to Resolve Dual Degeneracy in the Dual Simplex Algorithm

In this method the algorithm is always initiated with a lexico dual feasible basis. All the bases obtained during this algorithm will remain lexico dual feasible.

Let x_B be a lexico dual feasible basic vector, B the corresponding basis for (10.1), π_B the associated dual basic solution, and $\bar{A}, \bar{b}, \bar{c}, \bar{z}$ the updated entities with respect to this basic vector. The pivot row is chosen as discussed in Chapter 6, and suppose it is row r; thus, $\bar{b}_r < 0$. If $(\bar{a}_{r1}, \dots, \bar{a}_{rn}) \geqq 0$, the primal infeasibility criterion is satisfied, and we terminate with the conclusion that (10.1) is infeasible. On the other hand, if there exists at least one j satisfying $\bar{a}_{rj} < 0$, choose the pivot column to be the column of x_s, where s is such that $(\bar{c}_s, \bar{a}_{1s}, \dots, \bar{a}_{ms})/(-\bar{a}_{rs}) =$ lexico minimum $\{(\bar{c}_j, \bar{a}_{1j}, \dots, \bar{a}_{mj})/(-\bar{a}_{rj}): j$, such that $\bar{a}_{rj} < 0\}$. If this lexico minimum is not unique, let $\mathbf{S}$ be the subset of all j that tied for it. If $p, q \in \mathbf{S}$, we must have $(\bar{c}_p/\bar{a}_{rp}) = (\bar{c}_q/\bar{a}_{rq})$ and $(\bar{a}_{tp}/\bar{a}_{rp}) = (\bar{a}_{tq}/\bar{a}_{rq})$ for all $t = 1$ to m. Let $\alpha = (\bar{a}_{rq}/\bar{a}_{rp})$. We must have $\bar{a}_{rp} < 0$, $\bar{a}_{rq} < 0$, and this implies that $\alpha > 0$. These facts imply that $(\bar{a}_{1q}, \dots, \bar{a}_{mq}, \bar{c}_q) = \alpha(\bar{a}_{1p}, \dots, \bar{a}_{mp}, \bar{c}_p)$, which in turn implies $(a_{1q}, \dots, a_{mq}, c_q) = \alpha(a_{1p}, \dots, a_{mp}, c_p)$. Hence x_p, x_q are a positively dependent pair in (10.1). Hence the set of all variables x_j for $j \in \mathbf{S}$ are pairwise positively dependent in this case. So by the discussion above, we select any one of the variables x_s for some $s \in \mathbf{S}$ as the entering variable into the basic vector x_B, in this case, set all other x_j for $j \in S$, $j \neq s$ equal to zero in (10.1), and delete them and the corresponding column vectors from further consideration. The column vector of x_s is the pivot column.

It can be verified that if x_B is a lexico dual feasible basic vector, and the pivot column is chosen by the above procedure, then the basic vector obtained after this pivot step is also lexico dual feasible, and that the vector $(-\bar{z}, \bar{b}_1, \dots, \bar{b}_m)$ undergoes a strict lexico decrease in this pivot step. Using the same lexicographic arguments as before, it can be proved that if (10.1) is solved by the dual simplex algorithm starting with a lexico dual feasible basic vector, determining the pivot column in each pivot step by the procedure discussed here, then cycling cannot occur and the algorithm must terminate after a finite number of pivot steps, by satisfying one of the termination conditions in the dual simplex algorithm.

10.7 RESOLUTION OF CYCLING IN THE PARAMETRIC SIMPLEX ALGORITHMS

Consider the parametric right-hand-side LP

$$\begin{aligned} \text{Minimize} \quad & z(x) = cx \\ \text{Subject to} \quad & Ax = b + \lambda b^* \\ & x \geqq 0 \end{aligned} \tag{10.7}$$

where A is a given matrix of order $m \times n$ and rank m. For solving this problem, the parametric right-hand-side simplex algorithm of Section 8.6 starts with a basis B for (10.7) that is optimal for a fixed value of λ, and determines the optimality interval of B, $\underline{\lambda}_B \leqq \lambda \leqq \bar{\lambda}_B$, as in (8.8) and (8.9). Let $\bar{b} = B^{-1}b$, $\bar{b}^* = B^{-1}b^*$. Then

$$\begin{aligned} \bar{\lambda}_B &= \text{minimum } \{-\bar{b}_i/\bar{b}_i^*: i, \text{ such that } \bar{b}_i^* < 0\} \\ &= +\infty, \text{ if } \bar{b}_i^* \geqq 0 \quad \text{for all } i \end{aligned} \tag{10.8}$$

If $\bar{\lambda}_B \neq +\infty$, let $\mathbf{J} = \{i: i \text{ ties for the minimum in (10.8)}\}$. It selects an $r \in \mathbf{J}$ and performs a dual simplex pivot step in row r in the canonical tableau of (10.7) with respect to the basis B. This yields an adjacent basis $\hat{B}$ that is also optimal for $\lambda = \bar{\lambda}_B$. The lower characteristic value of the basis $\hat{B}$ is $\bar{\lambda}_B$, and hence its optimality interval is to the right of $\bar{\lambda}_B$, and the algorithm continues with $\hat{B}$. However, cycling can occur at

$\lambda = \bar{\lambda}_B$. We illustrate this with a numerical example due to A. Gana [10.8]. The three constraints in this example can be read from the first of the following tableaux. The constant x_j is required to be $\geqq 0$ for all j. The problem is to find feasible solutions for all $\lambda \geqq 0$. As an optimization problem, this is equivalent to minimizing $0x$ subject to the constraints and non-negativity restrictions. Since all the cost coefficients are zero, all the relative cost coefficients are always zero; hence performing a dual simplex pivot step in any row requires the choice of the pivot column to be any column with a negative updated entry in that row. For this reason we do not give the cost row in the following tableaux. The initial basic vector is (x_1, x_2, x_3) feasible for $\lambda = 0$. Pivot elements are circled. The basic vector (x_4, x_5, x_3) has repeated at the exit at $\lambda = 1$, and cycling has occurred. The algorithm can go through this cycle indefinitely without ever being able to increase λ beyond 1, even though feasible solutions exist for all λ.

Similar examples can be constructed to show that cycling can also occur in the parametric cost simplex algorithm at a value of λ. We now discuss how this cycling problem can be resolved.

10.7.1 Resolution of Cycling in the Parametric Right-Hand-Side Simplex Algorithm

Let $\mathbf{B} = \{1, \dots, m\}$ and $x_B = (x_j : j \in \mathbf{B})$ be an optimum basic vector for (10.7) for which the optimality interval is $\underline{\lambda} \leqq \lambda \leqq \bar{\lambda}$. Let $\bar{A}$ and $\bar{c}$ be the updated coefficient matrix and cost vector, respectively, corresponding to x_B. The method that we propose for resolving cycling in the parametric right-hand-side simplex algorithm requires the use of the lexico dual simplex algorithm of Section 10.6. In order to use this algorithm we assume that this initial basic vector x_B is lexico dual

Basic Variables	x_1	x_2	x_3	x_4	x_5	x_6	b	b^*	**J**, *and the Feasibility Interval*
x_1	1	0	0	(−1)	−2	0	1	−1	$\mathbf{J} = \{1, 2, 3\}$
x_2	0	1	0	0	−1	−2	1	−1	$-\infty \leqq \lambda \leqq 1$
x_3	0	0	1	−2	0	−1	1	−1	
x_4	−1	0	0	1	2	0	−1	1	$\mathbf{J} = \{2\}$
x_2	0	1	0	0	(−1)	−2	1	−1	$1 \leqq \lambda \leqq 1$
x_3	−2	0	1	0	4	−1	−1	1	
x_4	(−1)	2	0	1	0	−4	1	−1	$\mathbf{J} = \{1, 3\}$
x_5	0	−1	0	0	1	2	−1	1	$1 \leqq \lambda \leqq 1$
x_3	−2	4	1	0	0	−9	3	−3	
x_1	1	−2	0	−1	0	4	−1	1	$\mathbf{J} = \{3\}$
x_5	0	−1	0	0	1	2	−1	1	$1 \leqq \lambda \leqq 1$
x_3	0	0	1	−2	0	(−1)	1	−1	
x_1	1	−2	4	−9	0	0	3	−3	$\mathbf{J} = \{1, 2\}$
x_5	0	(−1)	2	−4	1	0	1	−1	$1 \leqq \lambda \leqq 1$
x_6	0	0	−1	2	0	1	−1	1	
x_1	1	0	0	(−1)	−2	0	1	−1	$\mathbf{J} = \{1\}$
x_2	0	1	−2	4	−1	0	−1	1	$1 \leqq \lambda \leqq 1$
x_6	0	0	−1	2	0	1	−1	1	
x_4	−1	0	0	1	2	0	−1	1	$\mathbf{J} = \{2, 3\}$
x_2	4	1	−2	0	−9	0	3	−3	$1 \leqq \lambda \leqq 1$
x_6	2	0	(−1)	0	4	1	1	−1	
x_4	−1	0	0	1	2	0	−1	1	$\mathbf{J} = \{2\}$
x_2	0	1	0	0	−1	−2	1	−1	$1 \leqq \lambda \leqq 1$
x_3	−2	0	1	0	4	−1	−1	1	

feasible to (10.7). Let B be the associated basis ($A_{.j}$: $j \in \mathbf{B}$), and $\bar{b} = B^{-1}b$, $\bar{b}^* = B^{-1}b^*$. Suppose $\bar{\lambda} < \infty$, and let $\mathbf{J}$ denote the set of all i that tie for the minimum in (10.8) determining $\bar{\lambda}$. Let $\mathbf{T} = \{i$: i, such that $\bar{b}_i + \bar{\lambda}\bar{b}_i^* = 0\}$. Clearly $\mathbf{J} \subset \mathbf{T}$. As long as λ remains equal to $\bar{\lambda}$, any pivot step performed in rows $i \in \mathbf{T}$, beginning with x_B will not change the primal basic solution of (10.7), because when the basic variable in the pivot row is zero in the basic solution, the pivot step is a primal degenerate pivot step that leaves the primal basic solution unchanged. Let $x_{\hat{B}}$ be any dual feasible basic vector for (10.7) containing all the variables x_j for $j \in \mathbf{B}\backslash\mathbf{T}$ as basic variables. Suppose the updated right-hand-side constant vectors with respect to $x_{\hat{B}}$ are $\hat{b}$ and $\hat{b}^*$. From these arguments, we conclude that the basic solution of (10.7) with respect to $x_{\hat{B}}$ at $\lambda = \bar{\lambda}$ is

$$x_{\hat{B}} = \hat{b} + \lambda\hat{b}^* = \bar{b} + \bar{\lambda}\bar{b}^*$$

all nonbasic variables $= 0$. Thus $\hat{b}_i + \bar{\lambda}\hat{b}_i^* = 0$ for all $i \in \mathbf{T}$ and >0 for all $i \notin \mathbf{T}$. Thus the upper characteristic value associated with $x_{\hat{B}}$ is $>\bar{\lambda}$ iff $\hat{b}_i^* \geqq 0$ for all $i \in \mathbf{T}$. So in order to obtain optimum solutions of (10.7) for values of $\lambda > \bar{\lambda}$, we consider the following problem, which we call the *subproblem* at $\lambda = \bar{\lambda}$.

$$\begin{aligned} \text{Minimize} \quad & \sum_{j \notin \mathbf{B}\backslash\mathbf{T}} \bar{c}_j x_j \\ \text{Subject to} \quad & \sum_{j \notin \mathbf{B}\backslash\mathbf{T}} \bar{a}_{ij} x_j = \bar{b}_i^* \qquad i \in \mathbf{T} \qquad (10.9) \\ & x_j \geqq 0 \qquad \text{for } j \notin \mathbf{B}\backslash\mathbf{T} \end{aligned}$$

(x_j: $j \in \mathbf{B} \cap \mathbf{T}$) is a dual feasible basic vector for (10.9). The subproblem can be solved by the lexico dual simplex algorithm of Section 10.6 in a finite number of steps without the possibility of cycling. If this process terminates by satisfying the primal infeasibility criterion for (10.9) in some row, that same row satisfies the primal infeasibility criterion for the original problem for values of $\lambda > \bar{\lambda}$. On the other hand, if it terminates with an optimum basic vector for (10.9), (x_j: $j \in \mathbf{D}$), then the basic vector (x_j: $\mathbf{D} \cup (\mathbf{B}\backslash\mathbf{T})$) is an optimum basic vector for the original problem (10.7) at $\lambda = \bar{\lambda}$, for which the upper characteristic value is $>\bar{\lambda}$.

It is not necessary to solve the subproblem separately. The pivot steps needed to solve it can be carried out on the original problem (10.7) itself. At each stage, it is necessary to check whether the updated b_i^* are $\geqq 0$ for all $i \in \mathbf{T}$. If so, the subproblem pivots are done and we resume the parametric right-hand-side simplex algorithm. If not, we continue the subproblem pivots by selecting the pivot row as any arbitrary row in which the present updated b_i^* is <0 for $i \in \mathbf{T}$, and the pivot column by the lexico dual simplex minimum ratio test of Section 10.6. After at most a finite number of such subproblem pivots steps at $\lambda = \bar{\lambda}$, this procedure will either determine that the original problem (10.7) is infeasible for $\lambda > \bar{\lambda}$, or find an optimum basis for it at $\lambda = \bar{\lambda}$, for which the upper characteristic value is $>\bar{\lambda}$. And then the parametric right-hand-side simplex algorithm can be continued in the same way.

The method is applied in a similar way to determine optimum solutions for $\lambda < \underline{\lambda}$.

Thus, starting with a basis that is primal feasible for (10.7) for some λ and lexico dual feasible, and choosing the pivot rows arbitrarily according to the rules given in Section 8.6, but using the lexico dual simplex minimum ratio test to choose the pivot column in each pivot step, this method solves (10.7) for all values of λ. The basis stays lexico dual feasible always, cycling cannot occur, and the method terminates after at most a finite number of pivot steps. The reader will notice that the only difference between this version of the parametric right-hand-side simplex algorithm and the one discussed in Section 8.6 is that here the initial optimal basis for (10.7) is required to be lexico dual feasible, and that in each pivot step the pivot column is required to be chosen here by the lexico dual simplex minimum ratio test.

10.7.2 Resolution of Cycling in the Parametric Cost Simplex Algorithm

Consider the parametric cost LP

$$\begin{aligned} \text{Minimize} \quad & (c + \lambda c^*)x \\ \text{Subject to} \quad & Ax = b \qquad (10.10) \\ & x \geqq 0 \end{aligned}$$

Let $\mathbf{B} = \{1, \ldots, m\}$ and $x_B = (x_j$: $j \in \mathbf{B})$ be an optimum basic vector for (10.10) for which the optimality interval is $\underline{\lambda} \leqq \lambda \leqq \bar{\lambda}$. The method that we propose for resolving the problem of cycling in the parametric cost simplex algorithm requires the use of the lexico primal simplex algorithm discussed in Section 10.2. In order to use this algorithm we assume that this initial basic vector x_B is lexico primal feasible for (10.10). Let B be the associated basis ($A_{.j}$: $j \in \mathbf{B}$), c_B, c_{B^*} the associated row vectors of basic coefficients in c and c^*, respectively, and $\bar{c} = c - c_B B^{-1}A$, $\bar{c}^* = c^* - c_B^* B^{-1}A$. Then from Section 8.2 $\bar{\lambda} = +\infty$, if $\bar{c}^* \geqq 0$, or the minimum $\{-\bar{c}_j/\bar{c}_j^*: j$, such that $\bar{c}_j^* < 0\}$, otherwise. Suppose $\bar{\lambda} < \infty$, and let **J** denote the set of j that tie for the minimum here in determining $\bar{\lambda}$. Let $\mathbf{T} = \{j: j, \text{such that } \bar{c}_j + \bar{\lambda}\bar{c}_j^* = 0\}$. Clearly $\mathbf{J} \subset \mathbf{T}$. Let x_B be any primal feasible basic vector for (10.10) in which all the basic variables are x_j for $j \in \mathbf{T}$ only, and let $\hat{c}$ and $\hat{c}^*$ denote the updated cost vectors with respect to it. Clearly $\hat{c}_j + \bar{\lambda}\hat{c}_j^* = 0$ for all $j \in \mathbf{T}$ and $\hat{c}_j - \bar{\lambda}\hat{c}_j^* > 0$ for all $j \notin \mathbf{T}$. So $x_{\hat{B}}$ is another optimum basic vector for (10.10) at $\lambda = \bar{\lambda}$. Clearly the upper characteristic value for $x_{\hat{B}}$ is $> \bar{\lambda}$ iff $\hat{c}_j^* \geqq 0$ for all $j \in \mathbf{T}$. So in order to obtain optimum solutions for (10.10) for $\lambda > \bar{\lambda}$, we consider the following problem, which we call the subproblem at $\lambda = \bar{\lambda}$.

$$\begin{aligned} \text{Minimize} \quad & \sum_{j \in \mathbf{T}} c_j^* x_j \\ \text{Subject to} \quad & \sum_{j \in \mathbf{T}} A_{.j} x_j = b \\ x_j \geqq 0 \quad & \text{for all } j \in \mathbf{T} \end{aligned} \tag{10.11}$$

As $(x_j: j \in \mathbf{B})$ is a lexico primal feasible basic vector for (10.11), this subproblem can be solved by the lexico primal simplex algorithm of Section 10.2 in a finite number of steps without the possibility of cycling. If this process terminates by satisfying the unboundedness criterion for (10.11) in some column, that same column satisfies the unboundedness criterion for the original problem (10.10) for values of $\lambda > \bar{\lambda}$. On the other hand, if it terminates with an optimum basic vector for (10.11), this same basic vector is an optimum basic vector for (10.10) at $\lambda = \bar{\lambda}$, for which the upper characteristic value is $> \bar{\lambda}$.

It is not necessary to solve the subproblem (10.11) separately. The pivot steps needed to carry it out can be carried out on the original problem (10.10) itself. At each stage we check whether the updated c_j^* are $\geqq 0$ for each $j \in \mathbf{T}$, and if so, we resume the parametric cost simplex algorithm. If not, we select a $j \in \mathbf{T}$, such that the present updated c_j^* is < 0, and perform a pivot step in that column using the lexico primal simplex minimum ratio test of Section 10.2 to determine the pivot row. After at most a finite number of such subproblem pivot steps at $\lambda = \bar{\lambda}$, this procedure will either determine that the objective function is unbounded below in the original problem (10.10) for all $\lambda > \bar{\lambda}$, or find an optimum basis for it at $\lambda = \bar{\lambda}$ for which the upper characteristic value is $> \bar{\lambda}$. And then the parametric cost simplex algorithm can be continued in the same way.

The method is applied in a similar way to determine optimum solutions for $\lambda < \underline{\lambda}$.

Thus, starting with a basis for (10.10), which is dual feasible for some λ and lexico primal feasible, and choosing the pivot column arbitrarily according to the rules given in Section 8.2, but using the lexico primal simplex minimum ratio test to choose the pivot row in each step, this method solves (10.10) for all values of λ. The basis stays lexico primal feasible always, cycling does not occur, and termination occurs after at most a finite number of pivot steps. Notice that the only difference between the version here and that in Section 8.2 is that the initial optimum basis is required to be lexico primal feasible, and that in each step the pivot row is required to be chosen by the lexico primal simplex minimum ratio test.

Exercises

10.4 Consider the following LP. In solving this problem by the simplex algorithm starting with the feasible basic vector (x_1, x_2, x_3, x_4), in each step select the entering variable as x_s, where $s = \text{minimum } \{j: j, \text{ such that } \bar{c}_j < 0\}$, and pick the pivot row as the rth row, where $r = \text{minimum } \{i: i$, such that row i is eligible to be the pivot row by the usual minimum ratio test$\}$. Using these rules show that cycling occurs. Verify that the use of the lexico minimum ratio rule avoids the cycling.

x_1	x_2	x_3	x_4	x_5	x_6	x_7	$-z$	b
1	0	0	0	$\frac{3}{5}$	$-\frac{32}{5}$	$\frac{24}{5}$	0	0
0	1	0	0	$\frac{1}{5}$	$-\frac{9}{5}$	$\frac{3}{5}$	0	0
0	0	1	0	$\frac{2}{5}$	$-\frac{8}{5}$	$\frac{1}{5}$	0	0
0	0	0	1	0	1	0	0	1
0	0	0	0	$-\frac{2}{5}$	$-\frac{2}{5}$	$\frac{9}{5}$	1	0

$x_j \geqq 0$ for all j; z to be minimized.

—(*K.T. Marshall and J. W. Suurballe* [10.10])

10.5 Consider the following LP. If the simplex algorithm is applied on this problem with (x_1, x_2, x_3, x_4) as the initial feasible basic vector, using the selection rules described in exercise 10.4, show that cycling occurs, even though the objective function is unbounded below on the set of feasible solutions of this problem.

x_1	x_2	x_3	x_4	x_5	x_6	x_7	$-z$	b
1	0	0	0	$\frac{1}{3}$	$-\frac{32}{9}$	$\frac{20}{9}$	0	0
0	1	0	0	$\frac{1}{6}$	$-\frac{13}{9}$	$\frac{5}{18}$	0	0
0	0	1	0	$\frac{2}{3}$	$-\frac{16}{9}$	$\frac{1}{9}$	0	0
0	0	0	1	0	0	0	0	1
0	0	0	0	-7	-4	15	1	0

$x_j \geqq 0$ for all j; z to be minimized.

—(*J. W. Suurballe*)

10.6 Consider the following LP. Verify that (x_1, x_2, x_3) is a dual feasible (but primal infeasible) basic vector. Starting with this basic vector, apply the dual simplex algorithm using the following choice rules. (a) The pivot row is the row corresponding to the most negative updated right-hand-side constant and also the topmost row among such rows, in case of a tie. (b) The pivot column is the leftmost of the eligible columns as determined by the usual dual simplex minimum ratio test. Verify that cycling occurs.

x_1	x_2	x_3	x_4	x_5	x_6	$-z$	b
1	0	0	$-\frac{1}{3}$	$-\frac{1}{6}$	$-\frac{2}{3}$	0	$-\frac{1}{3}$
0	1	0	$\frac{32}{9}$	$\frac{13}{9}$	$\frac{16}{9}$	0	$\frac{8}{9}$
0	0	1	$-\frac{20}{9}$	$-\frac{5}{18}$	$-\frac{1}{9}$	0	$\frac{4}{9}$
1	1	1	1	1	1	1	0

$x_j \geqq 0$ for all j; z to be minimized.

—(*J. W. Suurballe*)

10.7 Consider the LP (10.1). Let x_{B_1} be a feasible basic vector satisfying the property that exactly one basic variable is zero in the BFS corresponding to it (i.e., if B is the associated basis, then $B^{-1}b \geqq 0$, and exactly one component of the vector $B^{-1}b$ is zero). Prove that x_{B_1} cannot be in a cycle of degenerate basic vectors for this problem.

10.8 Suppose the LP (10.1), where A is a matrix of order $m \times n$ and rank m, is being solved beginning with an initial basis that is the unit matrix. Develop a variant of the lexicographic method to choose the dropping variable in each step, which maintains lexicopositivity of $(\bar{b}_i, \beta_{im}, \beta_{i,m-1}, \ldots, \beta_{i1})$, for each $i = 1$ to m, (where $\boldsymbol{\beta} = (\beta_{ij})$ is the basis inverse and $\bar{b} = (\bar{b}_i)$ is the updated right-hand-side constants vector) and is guaranteed to terminate finitely.

Let $(j_1, \ldots, j_m)$ be a permutation of $(1, 2, \ldots, m)$. Show that in fact a variant of the lexicographic method to choose the dropping variable in each step can be developed by maintaining the lexico positivity of $(\bar{b}_i, \beta_{i,j1}, \ldots, \beta_{i,j_m})$ for each $i = 1$ to m, and that such a method is guaranteed to terminate finitely.

10.9 Develop pivot row selection and entering variable selection rules for the algorithm discussed in Exercise 6.8 to solve the system of linear equations in nonnegative variables (6.1) to guarantee that once a basic vector for (6.1) appears in the algorithm, it cannot reappear in subsequent steps.

10.10 Consider the LP (10.1). Suppose we are solving (10.1) by the self-dual parametric method discussed in Section 8.13. Can cycling occur in this method? Construct an example where cycling occurs in this method, or prove conclusively that cycling cannot occur in this method. If you show that cycling can occur in the self-dual parametric method when applied to (10.1), relate it to primal or dual degeneracy of (10.1). Also, in this case, develop resolution methods that guarantee that cycling will not occur.

10.11 Suppose the LP (10.1) is being solved by the primal simplex algorithm. Consider the following scheme. Continue the application on the simplex algorithm using any one of the available dropping variable choice rules in pivot steps in which the updated right-hand-side constants vector contains at most one zero entry (i.e., pivot steps in which at most one basic variable is zero in the primal feasible solution in that step). The moment a pivot step is reached in which two or more updated right-hand-side constants are zero, go to the following special procedure.

Special Procedure Let r be the number of basic variables that are zero in the present feasible solution. So, by our assumption, $r \geqq 2$. Let b^1 be the present updated right-hand-side constants vector. Construct a matrix $D = (d_{ij})$ of order $m \times r$, in which columns are $I_{.j}$ (here I is the unit matrix of order m) for j such that $b_j^1 = 0$, in some order. Clearly, the vector $(b_i^1, d_{i1}, \ldots, d_{ir})$ is lexico positive for each $i = 1$ to m. Introduce D on the right-hand side of the canonical tableau or the inverse tableau, whichever is being used, and start updating D after each pivot step. As long as degenerate pivot steps continue, choose the dropping variable in each step so that the pivot row corresponds to the row determined by the lexico minimum ratio test performed with the basic values defined to be the row vectors in the updated $(b \vdots D)$. This choice guarantees that the updated rows of $(b_i^1, d_{i1}, \ldots, d_{ir})$, $i = 1$ to m, all stay lexico positive throughout the application of this special procedure. The moment a nondegenerate pivot step occurs, terminate this special procedure, discard the D-matrix, and go back to using any of the available dropping variable choice rules again.

If another pivot step is reached again later on, in which two or more basic variables have again zero value in the solution, start the special procedure again at that time, and continue in the same manner. Prove the following about this scheme.

1. In each degenerate pivot step, the dropping variable is identified uniquely and unambiguously. However, in some nondegenerate pivot steps, the dropping variable choice may not identify the dropping variable uniquely, depending on which available rule is used.
2. Under this scheme, the simplex algorithm will terminate after a finite number of pivot steps (i.e., cycling cannot occur).

Compare and elaborate the main differences between this scheme, and the method based on using the lexico minimum ratio rule for choosing the dropping variable in every pivot step discussed in Section 10.2. Compare also the computational advantages and disadvantages of each.

—(*R. Saigal*, [10.11])

10.12 Consider the LP (10.1). Suppose it is to be solved by the regular simplex method using the canonical tableaux discussed in Chapter 2. In this approach, show that the problem of cycling under degeneracy can be resolved by developing a method of choosing the dropping variable in each pivot step based on the perturbation of the vector b into $(b_i + \sum_{j=1}^{n} a_{ij}\varepsilon^j)$, where ε is treated as a small positive parameter without giving any specific value to it. Express the dropping variable choice rule with this perturbation, in terms of the updated data in the canonical tableau in each step.

—(*A. Charnes* [10.5])

10.13 Let Γ be the set of feasible solutions of the LP (4.8), where A, D, E, F, b, d, c, g, are given matrices of appropriate orders. (1) Develop an appropriate definition for the nondegeneracy or degeneracy of a given extreme point of Γ and justify it. (2) Develop appropriate definitions of primal and dual nondegeneracy or degeneracy of the LP (4.8) and justify them. Discuss what these conditions imply geometrically about the vectors (b, g) and (c, d) in the system.

10.14 Using the results of Section 13.2.4, prove that the problem of checking whether an instance of (10.1) with integer data is degenerate, is NP-complete.

—(*R. Chandrasekharam, S. N. Kabadi, and K. G. Murty,* [10.4])

10.15 Read each of the following statements carefully and check to see whether it is *true* or *false*. Justify your answer by constructing a simple illustrative example, if possible, or by a simple proof. In all these statements, problem (1) and the rest of the notation is the same as that in Exercise 3.75.

1. If the pivot row in a pivot step of the revised simplex method is the rth row,

then the eta column in the pivot matrix corresponding to that pivot step is the rth column.

2 In the dual simplex algorithm for solving (1) starting with a dual feasible basis, there is no criterion for unboundedness of $z(x)$.

3 In solving (1) by the primal simplex algorithm, the primal variables always stay nonnegative. Likewise, in solving (1) by the dual simplex algorithm, the dual variables always remain nonnegative.

4 The objective value is monotone decreasing, as we move from one step to the next during the solution of (1) by the dual simplex algorithm.

5 In solving (1) by the dual simplex algorithm, once a variable becomes nonnegative in one step, it stays nonnegative in all subsequent steps.

6 Suppose (1) is being solved by the dual simplex algorithm. The objective value in any step of this algorithm is a lower bound for the minimum objective value in (1).

7 Cycling under degeneracy occurs only in the primal simplex algorithm. It cannot occur in the dual simplex algorithm.

8 The primal simplex method can be initiated on (1) by introducing a nonnegative artificial variable. Likewise, the dual simplex method can be initiated on (1) by introducing a dummy inequality constraint.

9 The optimum objective value in a parametric right-hand-side minimization problem is a piecewise linear concave function of the parameter.

10 The optimum objective value in a parametric LP is a continuous function of the parameter in the region in which it is finite.

11 In practice, the self-dual parametric method for solving (1), which uses both primal and dual simplex pivot steps, as needed, is much more efficient than either the primal simplex method or the dual simplex method.

12 Suppose in (1), the variables are arranged so that the cost coefficients of the variables are increasing from left to right, that is, $c_1 < c_2 < c_3 < \cdots < c_n$. Since we are required to minimize $\sum_{j=1}^{n} c_j x_j$, if $(x_1, \ldots, x_m)$ is a feasible basic vector of (1), it must be optimal.

13 If $c \geqq 0$ in (1), the dual associated with (1) must be feasible.

14 In a primal dual pair of LPs, the primal problem has alternate optimum solutions iff the same things hold in the dual problem.

15 The pivot matrices in the primal simplex algorithm differ from the unit matrix in just one column. Correspondingly, the pivot matrices associated with pivot steps in the dual simplex algorithm differ from the unit matrix is just one row.

16 The dimensions of the two sets of feasible solutions of the primal and dual problems in a primal dual pair of LPs are always equal.

REFERENCES

10.1 D. Avis and V. Chvatal, "Notes on Bland's Pivoting Rule," in *Mathematical Programming Study 8*, North-Holland, Amsterdam, July 1978, pp. 24–34.

10.2 E. M. L. Beale, "Cycling in the Dual Simplex Algorithm," *Naval Research Logistics Quarterly 2* (1955), 269–276.

10.3 R. G. Bland, "New Finite Pivoting Rules for the Simplex Method," *Mathematics of Operations Research 2* (May, 1977), 103–107.

10.4 R. Chandrasekharan, S. N. Kabadi, and K. G. Murty, "Some NP-Complete Problems in Linear Programming," *Operations Research Letters 1*, 3 (July 1982), 101–104.

10.5 A. Charnes, "Optimality and Degeneracy in Linear Programming," *Econometrica 20*, 2 (April 1952) 160–170.

10.6 W. H. Cunningham, "A Network Simplex Method," *Mathematical Programming 11* (1976), 105–116.

10.7 G. B. Dantzig, A. Orden, and P. Wolfe, "Generalized Simplex Method for Minimizing a Linear Form Under Linear Inequality Constraints," *Pacific Journal of Mathematics 5* (1955), 183–195.

10.8 A. Gana, "Studies in the Complementarity Problem," Ph. D. dissertation, Department of Industrial and Operations Engineering, The University of Michigan, Ann Arbor, 1982.

10.9 A. J. Hoffman, "Cycling in the Simplex Algorithm," National Bureau of Standards Report 2974, Washington, D.C., December 1953.

10.10 K. T. Marshall and J. W. Suurballe, "A Note on Cycling in the Simplex Method," *Naval Research Logistics Quarterly 16,* 1 (March 1969), 121–137.

10.11 R. Saigal, "A Homotopy for Solving Large Sparse and Structured Fixed Point Problems," *Mathematics of Operations Research*, to be published.

10.12 P. Wolfe, "A Technique for Resolving Degeneracy in Linear Programming," *Journal of SIAM 11* (1963), 205–211.

chapter 11

Bounded-Variable Linear Programs

11.1 INTRODUCTION

Here we consider LPs in which some or all of the variables are restricted to lie within specified finite lower and upper bounds, known as *bounded-variable linear programs*. Let p be the number of bound constraints, and let m be the number of other constraints in the problem. This problem can of course be solved by including all the bound restrictions as constraints, but the increased size of the problem is undesirable computationally. Here we develop special techniques that make it possible to solve this problem using bases of order m only.

11.2 BOUNDED-VARIABLE PROBLEMS

11.2.1 Standard Form for Bounded-Variable Linear Programs

We consider LPs of the form

$$\begin{array}{lll} \text{Minimize} & cy & \\ \text{Subject to} & Ay = b' & \\ & \ell_j \leqq y_j \leqq k_j & \text{for } j \in \mathbf{J} = \{1, \ldots, n_1\} \\ & y_j \geqq 0 & \text{for } j \in \bar{\mathbf{J}} = \{n_1 + 1, \ldots, n\} \end{array} \quad (11.1)$$

where A is a matrix of order $m \times n$ and rank m. If $n_1 = n$, then $\bar{\mathbf{J}} = \emptyset$, and all the variables in the problem are bounded. The bounds ℓ_j and k_j are finite and $\ell_j \leqq k_j$ for all $j \in \mathbf{J}$. Substitute $y_j = x_j + \ell_j$ for all $j \in \mathbf{J}$ and $y_j = x_j$ for all $j \in \bar{\mathbf{J}}$ and eliminate the y_j from the problem. Let $b = b' - \sum_{j \in \mathbf{J}} A_{.j}\ell_j$. Transferring all the constant terms to the right-hand side and eliminating the constant terms from the objective function, the LP (11.1) is transformed into the equivalent problem:

$$\begin{array}{lll} \text{Minimize} & z(x) = cx & \\ \text{Subject to} & Ax = b & \\ & x_j \leqq \mathrm{U}_j = k_j - \ell_j & \text{for } j \in \mathbf{J} = \{1, \ldots, n_1\} \\ & x_j \geqq 0 & \text{for all } j \end{array} \quad (11.2)$$

As usual, assume that all b_i are nonnegative. If some b_i is negative, multiply both sides of the ith equation in (11.2) by -1. If $\mathrm{U}_j = 0$ for some $j \in \mathbf{J}$, the corresponding variable x_j should be equal to zero in all feasible solutions of (11.2), and it can be eliminated from further consideration. From now on we will assume that $\mathrm{U}_j > 0$ for all $j \in \mathbf{J}$. Now the problem is in standard form for bounded-variable LPs.

11.2.2 Basic Feasible Solutions

By the results in Section 3.4.5 we know that a feasible solution $\bar{x}$ of (11.2) is a BFS iff the set $\{A_{.j}: j \in \mathbf{J}, j \text{ such that } 0 < \bar{x}_j < \mathrm{U}_j\} \cup \{A_{.j}: j \in \bar{\mathbf{J}}, j \text{ such that } \bar{x}_j > 0\}$ is linearly independent. For the LP (11.2), we define a *working basis* to be a square nonsingular submatrix of A of order m. Given a working basis for (11.2), a variable associated with a column vector of the working basis will be called a *basic variable* when referring to this working basis. All other variables will be called *nonbasic variables.*

From the previous discussion, it is clear that a feasible solution $\bar{x}$ is a BFS (and, hence, an extreme point of the set of feasible solutions) iff there exists a working basis with the property that all nonbasic variables with respect to this working basis are either equal to their lower bound (0 here) or their upper bound (provided this upper bound is finite) in the solution $\bar{x}$. All basic variables must satisfy the bound restrictions on them, but they can have any value within their respective bounds.

Example 11.1

Consider the system of constraints in Tableau 11.1.

Tableau 11.1

x_1	x_2	x_3	x_4	x_5	x_6	x_7	b
1	0	1	−1	1	2	1	6
0	1	0	1	−2	1	−2	4
0	0	1	−1	0	2	1	1

$0 \leqq x_j \leqq 6$ for $j = 1, 2$; $0 \leqq x_4 \leqq 4$, $0 \leqq x_5 \leqq 2$, $0 \leqq x_6 \leqq 10$; and $x_3, x_7 \geqq 0$.

Let $\bar{x}$ be $(3, 4, 5, 4, 2, 0, 0)^{\mathrm{T}}$. Let B_1 be the working basis associated with the basic vector (x_1, x_2, x_3). $\bar{x}$ is a feasible solution. The nonbasic variables x_4, x_5 are at their upper bounds and the nonbasic variables x_6, x_7 are at their lower bounds in the feasible solution $\bar{x}$. The basic variables x_1, x_2, x_3 are strictly within their respective bounds in $\bar{x}$. Hence $\bar{x}$ is a BFS corresponding to the working basis B_1.

Consider the feasible solution $\tilde{x} = (5, 2, 0, 1, 0, 1, 0)^{\mathrm{T}}$ of this system. In this solution, the variables x_1, x_2, x_4, and x_6 are all strictly within their respective bounds. Since the set of four column vectors associated with these four variables is linearly dependent, $\tilde{x}$ is not a BFS.

Note 11.1: For a general LP in standard form, given a basis, the basic solution associated with this basis is uniquely defined as in Section 3.5.1. However, for the bounded variable LP (11.2), given a working basis, we have the freedom to set the nonbasic variables either at their lower bounds or upper bounds. Hence, there may be several basic solutions of (11.2) associated with a given working basis.

Feasible Partitions of Variables

From the above discussion, it is clear that each basic solution for (11.2) corresponds to a partition of the set of variables $\{x_1, \ldots, x_n\}$ into three subsets $(\mathbf{B}, \mathbf{L}, \mathbf{U})$. There will be exactly m variables in $\mathbf{B}$, they are such that $\{A_{.j}: j, \text{ such that } x_j \in \mathbf{B}\}$ is a linearly independent set, and the variables in $\mathbf{B}$ form the associated working basic vector. All the variables $x_j \in \mathbf{L}$ have their value equal to their lower bound, zero, in the basic solution. All the variables $x_j \in \mathbf{U}$ are such that the upper bounds U_j corresponding to them are finite, and these variables are equal to their upper bounds. Throughout this section, whenever we talk of a partition, it refers to a partition $(\mathbf{B}, \mathbf{L}, \mathbf{U})$ of $\{x_1, \ldots, x_n\}$ in this manner. After substituting the values of the variables in $\mathbf{L}$, $\mathbf{U}$, the values of the variables in $\mathbf{B}$ are determined uniquely so as to satisfy the equality constraints in (11.2). Let x_B be the vector of variables in $\mathbf{B}$ arranged in some specific order, and let B be the corresponding working basis. Then the basic solution of (11.2) corresponding to the partition $(\mathbf{B}, \mathbf{L}, \mathbf{U})$ is $\bar{x}$ where

$$\begin{aligned} \bar{x}_j &= 0 && \text{for } j \text{ such that } x_j \in \mathbf{L} \\ &= U_j && \text{for } j \text{ such that } x_j \in \mathbf{U} \qquad (11.3) \\ \bar{x}_B &= B^{-1}(b - \textstyle\sum(A_{.j}U_j: j, \text{ such that } x_j \in \mathbf{U})) \end{aligned}$$

The partition $(\mathbf{B}, \mathbf{L}, \mathbf{U})$ is a *primal feasible partition for (11.2)* if $0 \leqq \bar{x}_j \leqq U_j$ for each $x_j \in \mathbf{B}$. The solution corresponding to a feasible partition is a BFS for (11.2), and conversely every BFS for (11.2) corresponds to a feasible partition.

In the partition $(\mathbf{B}, \mathbf{L}, \mathbf{U})$ associated with the basic solution $\bar{x}$ for (11.2), we denote $\{j: j, \text{ such that } x_j \in \mathbf{L} \cup \mathbf{U}\}$ by the symbol $\mathbf{N}$. The variables $x_j \in \mathbf{B}$ are called the *basic variables*, and the variables x_j for $j \in \mathbf{N}$ are the *nonbasic variables* in this partition. Given $\mathbf{N}$, clearly $\mathbf{L} = \{x_j: j \in \mathbf{N} \text{ and } \bar{x}_j = 0\}$ and $\mathbf{U} = \{x_j: j \in \mathbf{N} \text{ and } \bar{x}_j = U_j\}$.

11.2.3 Optimality Criterion

Assigning the dual variables $\pi_1, \ldots, \pi_m; \mu_1, \ldots, \mu_{n_1}$, to the constraints, in that order, the dual problem of (11.2) is

$$\begin{aligned} \text{Maximize} \quad & \pi b - \mu U \\ \text{Subject to} \quad & \pi A_{.j} - \mu_j \leqq c_j \quad \text{for } j \in \mathbf{J} \\ & \pi A_{.j} \qquad \leqq c_j \quad \text{for } j \in \bar{\mathbf{J}} \\ \pi \text{ unrestricted} \quad & \mu \geqq 0 \end{aligned}$$

The complementary slackness conditions for optimality in this primal-dual pair are

$$\begin{aligned} x_j(c_j - \pi A_{.j} + \mu_j) &= 0 \qquad \text{for } j \in \mathbf{J} \\ x_j(c_j - \pi A_{.j}) &= 0 \qquad \text{for } j \in \bar{\mathbf{J}} \\ \mu_j(U_j - x_j) &= 0 \qquad \text{for } j \in \mathbf{J} \end{aligned}$$

Define $\bar{c}_j = c_j - \pi A_{.j}$. Dual feasibility requires $\bar{c}_j + \mu_j \geqq 0$ for all $j \in \mathbf{J}$ and $\bar{c}_j \geqq 0$ for all $j \in \bar{\mathbf{J}}$. For some $j \in \mathbf{J}$, if $\bar{c}_j < 0$, dual feasibility automatically requires

$\mu_j > 0$ (because $\bar{c}_j + \mu_j \geqq 0$ for $j \in \mathbf{J}$) and the complementary slackness conditions imply that $x_j = U_j$. For $j \in \mathbf{J}$, if $\bar{c}_j > 0$, then $\bar{c}_j + \mu_j > 0$ (because μ_j is a nonnegative dual variable) and, in this case, the complementary slackness conditions imply that $x_j = 0$. Again for $j \in \mathbf{J}$, if x_j is strictly within its bounds, the complementary slackness property will only be satisfied if $\bar{c}_j = 0$ and $\mu_j = 0$.

For $j \in \bar{\mathbf{J}}$, $\bar{c}_j \geqq 0$ for dual feasibility and $\bar{c}_j > 0$ implies that $x_j = 0$ by the complementary slackness property.

Thus, given a feasible solution $\bar{x}$ of (11.2), it is optimal iff there exists a vector π such that when $\bar{c} = c - \pi A$, the variable x_j is at its lower bound 0 whenever the corresponding $\bar{c}_j > 0$, and at its finite upper bound U_j, whenever the corresponding $\bar{c}_j < 0$, in the solution $\bar{x}$.

Let $\bar{x}$ be a BFS of (11.2) associated with a working basis B. If x_j is a basic variable, $\bar{x}_j$ can lie strictly within its respective bounds, and by the complementary slackness property this arrangement requires that $\bar{c}_j$ is 0. Hence, the π vector associated with the working basis B can be obtained by solving $c_j - \pi A_{.j} = 0$ for all j, such that $A_{.j}$ is a column vector in the working basis B. As usual, let c_B represent the basic cost coefficient vector. Then π is obtained by solving $c_B = \pi B$. Therefore, $\pi = c_B B^{-1}$, which is the same as in the general LP model. Having computed π, $\bar{c}$ is obtained from $\bar{c} = c - \pi A$. The optimality criteria are

$$\begin{aligned} &\text{for} \quad j \in \mathbf{J},\ \bar{c}_j < 0 \quad &&\text{implies } \bar{x}_j = U_j \\ &\qquad\qquad\quad \bar{c}_j > 0 \quad &&\text{implies } \bar{x}_j = 0 \\ &\text{and} \quad j \in \bar{\mathbf{J}},\ \bar{c}_j \geqq 0 && \end{aligned}$$

Example 11.2

Let $z(x) = -3x_1 + 4x_2 + 2x_3 - 2x_4 - 14x_5 + 11x_6 - 5x_7$. Consider the problem of minimizing $z(x)$ on the set of feasible solutions of the numerical example in Tableau 11.1. Let B_1 be the working basis associated with the basic vector (x_1, x_2, x_3). It is required to check whether the feasible solution $\bar{x} = (3, 4, 5, 4, 2, 0, 0)^{\mathrm{T}}$ corresponding to the working basis B_1 is optimal to this problem.

$$\pi = c_{B_1} B_1^{-1} = (-3, 4, 2)\begin{pmatrix} 1 & 0 & 1 \\ 0 & 1 & 0 \\ 0 & 0 & 1 \end{pmatrix}^{-1} = (-3, 4, 5)$$

$$\bar{c} = c - \pi A = (0, 0, 0, -4, -3, 3, 1)$$

$\bar{c}_4 = -4$ and $\bar{x}_4$ is 4, the upper bound for x_4. $\bar{c}_5 = -3$ and $\bar{x}_5$ is 2, the upper bound for x_5. $\bar{c}_6, \bar{c}_7$ are both positive and both $\bar{x}_6, \bar{x}_7$ are equal to zero, the lower bound for the variables x_6, x_7. Hence, all the optimality conditions are satisfied and $\bar{x}$ is an optimum solution of this problem.

From this it is clear that all computations required for solving the problem can be performed using the working bases only. This is described in the next section.

Optimal Feasible Partitions

Given the feasible partition $(\mathbf{B}, \mathbf{L}, \mathbf{U})$ associated with the BFS $\bar{x}$ for (11.2), let B be the working basis consisting of the columns $A_{.j}$ for j, such that $x_j \in \mathbf{B}$, and c_B the corresponding row vector of basic cost coefficients. The dual basic solution associated with this partition is $\pi = c_B B^{-1}$. Compute $\bar{c} = c - \pi A$. $(\mathbf{B}, \mathbf{L}, \mathbf{U})$ is an optimal feasible partition for (11.2) if: $x_j \in \mathbf{L}$ implies $\bar{c}_j \geqq 0$, $x_j \in \mathbf{U}$ implies $\bar{c}_j \leqq 0$, and in this case the associated BFS $\bar{x}$ is an optimum solution of (11.2).

11.3 PRIMAL SIMPLEX METHOD USING WORKING BASES

11.3.1 Phase I

Consider the bounded variable LP (11.2), where A is a matrix of order $m \times n$ and b is a nonnegative vector. Phase I of the simplex method finds an initial BFS to this problem. Introduce the artificial variables $x_{n+1}, \ldots, x_{n+m}$, as in Tableau 11.2.

Tableau 11.2 Original Tableau for the Phase I Problem

$x_1 \ldots x_n$	$x_{n+1} \ldots x_{n+m}$	$-z$	$-w$	b
$a_{11} \ldots a_{1n}$	$1 \ldots 0$	0	0	b_1
$\vdots \quad \vdots$	$\vdots \quad \vdots$	$\vdots$	$\vdots$	$\vdots$
$a_{m1} \ldots a_{mn}$	$0 \ldots 1$	0	0	b_m
$c_1 \ldots c_n$	$0 \ldots 0$	1	0	0
$0 \ldots 0$	$1 \ldots 1$	0	1	0

$x_j \leqq U_j$ for $j \in \{1, \ldots, n_1\} = \mathbf{J}$; $x_j \geqq 0$ for all j. Artificial variables are $x_{n+1}, \ldots, x_{n+m}$.

In the BFSs obtained during the algorithm, the nonbasic variable x_j for $j \in \mathbf{J}$ is either equal to zero or equal to its upper bound U_j. However, nonbasic x_j

for $j \in \bar{\mathbf{J}}$ are always equal to zero in the solution. *In each stage of this algorithm, the present value of each nonbasic variable has to be clearly recorded.*

During Phase I, try to minimize w, the sum of all the artificial variables introduced. To discuss the Phase I termination criterion, assume that during Phase I, in a general step, say step k, the inverse tableau is as found in Tableau 11.3

Tableau 11.3

Basic Variables	Inverse Tableau				Basic *Values*
x_{i_1}	$\beta_{11} \cdots$	β_{1m}	0	0	$\bar{b}_1$
$\vdots$	$\vdots$	$\vdots$	$\vdots$	$\vdots$	$\vdots$
x_{i_m}	$\beta_{m1} \cdots$	β_{mm}	0	0	$\bar{b}_m$
$-z$	$-\pi_1 \quad \ldots$	$-\pi_m$	1	0	$-\bar{z}$
$-w$	$-\sigma_1 \quad \ldots$	$-\sigma_m$	0	1	$-\bar{w}$

Values of nonbasic variables $x_j = \bar{x}_j$ for nonbasic x_j.

The working basis B in this step is the submatrix of A consisting of the column vectors corresponding to the present basic variables x_j for $j \in \{i_1, \ldots, i_m\}$. Let c_B, d_B be the row vectors of the Phase II and Phase I cost coefficients of the basic variables in this step, as usual. Let $\mathbf{N} = \{j : j$, such that x_j is a nonbasic variable at present$\}$. Then the entries in the inverse tableau are

$$(\beta_{ij}) = B^{-1}, \pi = c_B B^{-1}, \sigma = d_B B^{-1}$$

$$\bar{b} = \text{vector of basic values} =$$

$$B^{-1}\left(b - \sum_{j \in \mathbf{N} \cap \mathbf{J}} \bar{x}_j A_{.j}\right)$$

$$\bar{z} = c_B \bar{b} + \sum_{j \in \mathbf{N} \cap \mathbf{J}} c_j \bar{x}_j,$$

$$\bar{w} = \text{sum of the values of basic artificials}$$

The Phase I relative cost coefficient $\bar{d}_j = d_j - \sigma A_{.j} = -\sigma A_{.j}$ for each $j = 1$ to n. The updated column vector of x_j with respect to the present working basis is $\bar{A}_{.j} = B^{-1}A_{.j}$, as usual. Phase I termination criteria are obtained by applying the optimality criteria of Section 11.2.3 to the Phase I problem. These are discussed in the next section.

11.3.2 Phase I Termination Criteria

Phase I terminates if in the present BFS the value of

$$\bar{x}_j = U_j \qquad \text{for all } j\text{, such that } j \in \mathbf{J} \text{ and } \bar{d}_j < 0$$
$$= 0 \qquad \text{for all } j\text{, such that } j \in \mathbf{J} \text{ and } \bar{d}_j > 0$$

and

$$\bar{d}_j \geqq 0 \qquad \text{for all } j \in \bar{\mathbf{J}}$$

11.3.3 Choice of Nonbasic Variable for Change of Status During Phase I

Let Tableau 11.3 be the present inverse tableau and let the solution given there be the present solution. If Phase I termination criteria are not satisfied, select a nonbasic variable x_s, which has led to a violation of the termination criteria, for changing its status. This variable will be called the *entering variable*, as in Chapter 2. It belongs to one of the following types.

(1) $s \in \mathbf{J}$, $\bar{d}_s < 0$ and $\bar{x}_s = 0$

or

(2) $s \in \mathbf{J}$, $\bar{d}_s > 0$ and $\bar{x}_s = U_s$

or

(3) $s \in \bar{\mathbf{J}}$, $\bar{d}_s < 0$

Variables x_j for $j \in \mathbf{N}$ satisfying (1) or (2) or (3) are those *eligible to enter the basic vector* in this step. One of these is selected as the entering variable. Based on the discussion in Sections 11.3.4, 11.3.6, criteria for choosing the entering variable similar to those discussed in Sectiion 2.5.7 can be developed.

11.3.4 Changing the Status of x_s Where $s \in \mathbf{J}$ and $\bar{d}_s < 0$

Here the entering variable x_s is such that its present value is zero and $\bar{d}_s < 0$. This implies that increasing the value of this variable from its present value of 0 would reduce the Phase I objective value. As is usual in the simplex method, leave the values of all nonbasic variables other than x_s unchanged, change the value of x_s from 0 to a parameter λ, and reevaluate the values of all the present basic variables as functions of λ.

Clearly the new solution is given by:

$$\begin{aligned} x_j &= \bar{x}_j && \text{for all } j \in \mathbf{N}, j \neq s \\ x_s &= \lambda \\ x_{i_r} &= \bar{b}_r - \bar{a}_{rs}\lambda && r = 1 \text{ to } m \\ z &= \bar{z} + \bar{c}_s\lambda \\ w &= \bar{w} + \bar{d}_s\lambda \end{aligned} \qquad (11.4)$$

For the new solution to remain feasible, the parameter λ has to be chosen so that the nonnegativity restrictions hold on the values of all the variables and the upper-bound restrictions hold on all variables x_j for $j \in \mathbf{J}$. The restriction that all the present basic variables should be nonnegative implies that $\bar{b}_r - \bar{a}_{rs}\lambda \geqq 0$, that is $\lambda \leqq \bar{b}_r/\bar{a}_{rs}$ for all r, such that $\bar{a}_{rs} > 0$. The upper-bound restriction on the present basic variable x_{i_r}, $i_r \in \mathbf{J}$ implies that $\bar{b}_r - \bar{a}_{rs}\lambda \leqq U_{i_r}$, that is, $\lambda \leqq (\bar{b}_r - U_{i_r})/\bar{a}_{rs}$ for all r, such that $i_r \in \mathbf{J}$ and $\bar{a}_{rs} < 0$. Thus the maximum value that λ can have is $\theta = \text{minimum } \{\bar{b}_r/\bar{a}_{rs}$ for all r, such that $\bar{a}_{rs} > 0$, $(\bar{b}_r - U_{i_r})/\bar{a}_{rs}$ for all r, such that $i_r \in \mathbf{J}$ and $\bar{a}_{rs} < 0$, $U_s\}$. This θ is the minimum ratio in this pivot step. It is the value of the entering variable x_s in the next solution.

Case 1 If $\theta = U_s$, x_s remains a nonbasic variable, but its value becomes equal to its upper bound in the next step. Hence, x_s moves from being a nonbasic variable with a value of 0 to one with a value equal to its upper bound. There is no change in the working basis. The only things that change in the inverse tableau are the values of the basic variables that are computed from (11.4) by substituting $\lambda = \theta = U_s$, and the value of the nonbasic variable x_s that is changed from 0 to U_s. Everything else remains the same and the algorithm moves to the next step.

Example 11.3

Consider the case where (x_1, x_2, x_3) is the present basic vector, while solving a bounded variable LP. Let x_4, whose value in the present solution $\bar{x}_4 = 0$, which has $\bar{d}_4 < 0$, be the entering variable. Let U_1, U_2, U_3, U_4—the upper bounds on x_1, x_2, x_3, x_4—be 12, 30, 35, 3, respectively. Let the vector of basic values in the present BFS be $(x_1, x_2, x_3) = \bar{b} = (10, 20, 30)$.

Let $\bar{A}_{.4}$, the updated column vector of x_4, be $(1, -2, 0)^T$. Then the new solution obtained after changing the value of x_4 from 0 to λ is $(x_1, x_2, x_3, x_4) = (10 - \lambda, 20 + 2\lambda, 30, \lambda)$, $x_j = \bar{x}_j$ is its present value, for all other j. The maximum value that λ can have is 3. Hence in this case x_4 moves from being a zero-valued nonbasic variable to a nonbasic variable at upper bound. The solution after the change is obtained by substituting 3 for λ.

Case 2 Suppose $\theta < U_s$ and suppose $\theta = \bar{b}_t/\bar{a}_{ts}$ for some t such that $\bar{a}_{ts} > 0$. In this case, θ is strictly less than the upper bound for x_s, but if λ exceeds θ, the present tth basic variable x_{i_t} becomes negative. When $\lambda = \theta$, the basic variable x_{i_t} attains its lower bound 0. In this case, the basic variable x_{i_t}, is dropped from the basic vector, and x_s becomes the tth basic variable in its place. Ties for t have to be broken using a *dropping basic variable choice rule* similar to those discussed in Section 2.5.8. The updated column vector of x_s is the pivot column and the new inverse tableau is obtained by performing the pivot with $\bar{a}_{ts}$ as the pivot element.

The new inverse tableau could also have been obtained by multiplying the present inverse tableau on the left by the appropriate pivot matrix corresponding to this pivot operation, instead of performing the pivot.

The vector of new basic values is obtained from (11.4) by substituting $\lambda = \theta$. The variable x_s is removed from the list of nonbasic variables and x_{i_t} is listed as a new nonbasic variable whose value in the new solution is zero. This completes this step and the algorithm moves to the next step.

Example 11.4

Consider the problem in which the data are the same as in Example 11.3, with the exception that $\bar{A}_{.4} = (1, -1, 15)^T$. In this case, the new solution obtained after changing the value of x_4 from 0 to λ is $(x_1, x_2, x_3, x_4) = (10 - \lambda, 20 + \lambda, 30 - 15\lambda, \lambda)$, $x_j = \bar{x}_j$ is its present value for all other j. The maximum value that λ can have here is 2. When $\lambda = 2$, x_3 becomes equal to zero. Hence, x_4 replaces x_3 from the basic vector. The new inverse tableau is obtained as in Chapter 5, by performing a pivot with the updated column vector of x_4 as the pivot column and the row in which x_3 is the basic variable as the pivot row. The new solution, however, is obtained by substituting 2 for λ in the above.

Case 3 Suppose $\theta < U_s$ and suppose $\theta = (\bar{b}_t - U_{i_t})/\bar{a}_{ts}$ for some t such that $\bar{a}_{ts} < 0$ and $i_t \in \mathbf{J}$. In this case, when λ exceeds θ, the value of *the present* tth basic variable x_{i_t} exceeds its specified upper bound U_{i_t} in the new solution, so the variable x_{i_t} is made into a nonbasic variable whose value is equal to its upper bound in the new solution. The variable x_s becomes the tth basic variable in its place. Ties for t have to be broken using a *dropping basic variable choice rule* similar to those discussed in Section 2.5.8. The new inverse tableau is computed as in

Case 2 above. The values of the basic variables in the new solution are obtained from (11.4) by substituting $\lambda = \theta$. The algorithm then moves to the next step.

Example 11.5

Consider the problem in which the data are the same as in Example 11.3, with the exception that $\bar{A}_{.4} = (-1, 1, -1)^T$. The new solution obtained after changing the value of x_4 from 0 to λ is $(x_1, x_2, x_3, x_4) = (10 + \lambda, 20 - \lambda, 30 + \lambda, \lambda)$, $x_j = \bar{x}_j$ is its present value for all other j. In this case, the maximum value that λ can have is 2. When $\lambda = 2$, x_1 reaches its upper bound of 12. Hence x_4 replaces x_1 from the basic vector. Variable x_1 becomes a nonbasic variable at its upper bound. The new solution is obtained by substituting $\lambda = 2$.

11.3.5 Changing the Status of x_s Where $s \in \mathbf{J}$ and $\bar{d}_s > 0$

Here the entering variable x_s is such that its present value is equal to its upper bound and $\bar{d}_s > 0$. This implies that decreasing x_s from its present value U_s, leads to a reduction in the Phase I objective function. Leave the values of all the nonbasic variables other than x_s unchanged, change the value of x_s from U_s to a parameter λ, and obtain the values of the present basic variables as functions of λ. The new solution is

$$
\begin{aligned}
x_j &= \bar{x}_j && \text{its present value for all } j \neq s,\ j \in \mathbf{N} \\
x_s &= \lambda \\
x_{i_r} &= \bar{b}_r + \bar{a}_{rs}U_s - \bar{a}_{rs}\lambda && r = 1 \text{ to } m \qquad (11.5) \\
z &= \bar{z} - \bar{c}_s U_s + \bar{c}_s\lambda \\
w &= \bar{w} - \bar{d}_s U_s + \bar{d}_s\lambda
\end{aligned}
$$

Since $\bar{d}_s > 0$, the maximum decrease in the value of w can be achieved by making λ as low as possible consistent with the requirements for the feasibility of the new solution in (11.5). The restriction that all the present basic variables should be nonnegative implies that $\bar{b}_r + \bar{a}_{rs}U_s - \bar{a}_{rs}\lambda \geqq 0$, that is, $\lambda \geqq (\bar{b}_r + \bar{a}_{rs}U_s)/\bar{a}_{rs}$ for all r, such that $\bar{a}_{rs} < 0$. The restriction that all the present basic variables x_{i_r}, $i_r \in \mathbf{J}$ should be less than or equal to their corresponding upper bounds U_{i_r} implies that $\bar{b}_r + \bar{a}_{rs}U_s - \bar{a}_{rs}\lambda \leqq U_{i_r}$, that is, $\lambda \geqq (\bar{b}_r + \bar{a}_{rs}U_s - U_{i_r})/\bar{a}_{rs}$ for all r, such that $i_r \in \mathbf{J}$ and $\bar{a}_{rs} > 0$. Thus, the minimum value that λ can have is $\theta = \max\{(\bar{b}_r + \bar{a}_{rs}U_s)/\bar{a}_{rs}$ for all r such that $\bar{a}_{rs} < 0$, $(\bar{b}_r + \bar{a}_{rs}U_s - U_{i_r})/\bar{a}_{rs}$ for all r such that $i_r \in \mathbf{J}$ and $\bar{a}_{rs} > 0$, $0\}$. This θ is the value of the entering variable x_s in the new solution. There are several cases to consider here.

Case 1 Suppose $\theta = 0$. Then x_s moves from a nonbasic variable, whose value is currently equal to its upper bound, into a nonbasic variable, whose value is equal to 0 in the next solution. There is no change in the working basis. The only changes in the inverse tableau occur in the values of the basic variables that are computed from (11.5) by substituting $\lambda = 0$. The algorithm then moves to the next step.

Example 11.6

Consider a problem in which the data are the same as in Example 11.3, with the exception that the entering variable x_4 is currently a nonbasic variable at its upper bound 3, with $\bar{d}_4 > 0$, and its updated column vector, $\bar{A}_{.4} = (-1,\ 2,\ -3)^T$. Hence, x_4 should be decreased from its present value of 3. If the value of x_4 is changed to λ, the new solution is

$$
\begin{aligned}
x_1 &= 10 + 3(-1) - \lambda(-1) \\
x_2 &= 20 + 3(2) - \lambda(2) \\
x_3 &= 30 + 3(-3) - \lambda(-3) \\
x_4 &= \lambda \\
x_j &= \bar{x}_j && \text{its present value for all other } j
\end{aligned}
$$

The minimum value that λ can have is zero. Hence x_4 is made into a zero-valued nonbasic variable in the next step. The new solution is obtained by substituting 0 for λ.

Case 2 Suppose $\theta > 0$, and is equal to $(\bar{b}_t + \bar{a}_{ts}U_s)/\bar{a}_{ts}$ for some t, such that $\bar{a}_{ts} < 0$. When $\lambda = \theta$ in (11.5), the present tth basic variable x_{i_t} assumes a value of 0 and it becomes a nonbasic variable whose value is 0 in the new solution. x_s becomes the tth basic variable in its place. Ties for t have to be broken using a *dropping basic variable choice rule* similar to those discussed in Section 2.5.8. The new inverse tableau is obtained as under Case 2 of Section 11.3.4. The basic values in the new solution are obtained from (11.5) by substituting $\lambda = \theta$.

Example 11.7

Consider the problem in which the data are the same as in the Example 11.6, with the exception that the updated column vector of x_4 is $\bar{A}_{.4} = (-5, -6, -2)^T$.

If the value of x_4 is changed to λ, the new solution here is

$$x_1 = 10 + 3(-5) - \lambda(-5)$$
$$x_2 = 20 + 3(-6) - \lambda(-6)$$
$$x_3 = 30 + 3(-2) - \lambda(-2)$$
$$x_4 = \lambda$$
$$x_j = \bar{x}_j \quad \text{its present value for all other } j$$

The minimum value that λ can have is 1. When $\lambda = 1$, x_1 becomes equal to zero. So x_4 replaces x_1 in the basic vector. The variable x_1 becomes a zero-valued nonbasic variable. The new solution is obtained by substituting $\lambda = 1$.

Case 3 Suppose $\theta > 0$ and is equal to $(\bar{b}_{ts} + \bar{a}_{ts}U_s - U_{i_t})/\bar{a}_{ts}$ for some t, such that $i_t \in \mathbf{J}$ and $\bar{a}_{ts} > 0$. In this case, when $\lambda = \theta$ in (11.5), the basic variable x_{i_t} is equal to its upper bound and it becomes a nonbasic variable whose value is equal to its upper bound in the next solution. The variable x_s becomes the tth basic variable in its place. Ties for t have to be broken using a *dropping basic variable choice rule* similar to those discussed in Section 2.5.8. The new inverse tableau is obtained as in Case 2 of Section 11.3.4 and the values of the basic variables in the new solution are computed by substituting $\lambda = \theta$ in (11.5). Go now to the next step.

Example 11.8

Consider the problem in which the data are the same as in Example 11.6, with the exception that the updated column vector of the entering variable x_4 is $\bar{A}_{.4} = (-1, 5, 0)^{\mathrm{T}}$. If the value of x_4 is changed to λ, the new solution in this case is

$$x_1 = 10 + 3(-1) - \lambda(-1)$$
$$x_2 = 20 + 3(5) - \lambda(5)$$
$$x_3 = 30$$
$$x_4 = \lambda$$
$$x_j = \bar{x}_j \quad \text{its present value for all other } j$$

The minimum value that λ can have is 1. When $\lambda = 1$, x_2 becomes equal to 30, its upper bound. So x_4 replaces x_2 in the basic vector. The variable x_2 becomes a nonbasic variable at its upper bound. The new solution is obtained by substituting $\lambda = 1$.

11.3.6 Changing the Status of x_s Where $s \in \bar{\mathbf{J}}$

Here the entering variable x_s is such that $s \in \bar{\mathbf{J}}$. Clearly $\bar{d}_s$ must be strictly negative. Here, there is no upper bound restriction on the entering variable x_s and increasing its value from its present value of 0 leads to a decrease in the Phase I objective function. Keep the values of all the nonbasic variables other than x_s unchanged. Making $x_s = \lambda$, compute the values of the basic variables as functions of λ, so that the new solution satisfies the equality constraints in the problem. The new solution as a function of λ is the same as in (11.4).

Since the Phase I objective function decreases as the value of x_s is increased, λ should be made as large as possible consistent with feasibility. Using arguments as in Section 11.3.4, it can be seen that the maximum value that λ can have is $\theta = \min\{\bar{b}_r/\bar{a}_{rs}$ for all r such that $\bar{a}_{rs} > 0$, $(\bar{b}_r - U_{i_r})/\bar{a}_{rs}$ for all r such that $i_r \in \mathbf{J}$ and $\bar{a}_{rs} < 0$, $\infty\}$. There are several cases to consider here.

Case 1 If $\theta = \bar{b}_t/\bar{a}_{ts}$ for some t such that $\bar{a}_{ts} > 0$, then x_{i_t} drops from the basic vector and becomes a nonbasic variable with value 0 in the next step, and x_s becomes the tth basic variable in its place. The new inverse tableau and the new feasible solution are all obtained as in Case 2 of Section 11.3.4.

Case 2 If $\theta = (\bar{b}_t - U_{i_t})/\bar{a}_{ts}$ for some t such that $i_t \in \mathbf{J}$ and $\bar{a}_{ts} < 0$, then x_{i_t} drops from the basic vector and becomes a nonbasic variable whose value is equal to its upper bound in the next step. The variable x_s becomes the tth basic variable in its place. The new solution and the new inverse tableau are obtained as in Case 3 of Section 11.3.4.

The Phase I objective function is bounded below by zero. As the value of λ increases from 0, the Phase I objective value decreases linearly. Hence, during Phase I, the value of θ will never be equal to ∞, and one of the cases discussed above should occur in each step before Phase I termination.

Note 11.2: Notice that the pivot elements in the pivot steps to be performed in Case 3 of Section 11.3.4, Case 2 of Section 11.3.5, and Case 2 of this section are all negative.

11.3.7 Infeasibility

The algorithm is continued until the Phase I termination criteria are satisfied, which will occur in a finite number of steps. When the Phase I termination criteria are satisfied, if the value of w is strictly positive, the original problem must be infeasible and the algorithm terminates.

11.3.8 Switching over to Phase II

When the Phase I termination criteria are satisfied, if $\bar{w} = 0$, the present solution is a feasible solution of the original problem and we switch over to Phase II.

In the terminal Phase I step, if $\bar{d}_j < 0$ for some j, then $j \in \mathbf{J}$ and the value of x_j must be U_j in the present solution. Decreasing the value of x_j from its present U_j will increase the Phase I objective function from its present value of 0 and, hence, leads to an infeasible solution of the original problem. Hence, in all feasible solutions of the original problem, the variable x_j must be equal to U_j. Therefore, this variable x_j is permanently set equal to U_j, and all such variables are prevented from being chosen for a change in status during Phase II iterations.

Similarly, if $\bar{d}_j > 0$ for some j in the Phase I terminal step, the associated variable x_j must be equal to zero in all feasible solutions of the original problem. Hence, all such variables are permanently set equal to zero, and these variables are never again considered for change of status during Phase II iterations.

Hence all the remaining variables that are eligible for change in status during Phase II iterations must have $\bar{d}_j$ equal to 0 in the terminal Phase I step. Any changes made in the values of these variables will not alter the Phase I objective value from 0; hence, all subsequent solutions obtained in the algorithm will be feasible solutions of the original problem. If any artificial variable is in the basic vector at this stage, its value in the present solution must be equal to zero and it can be left as a basic variable until it is removed from the basic vector during some Phase II iteration. The Phase I objective row and the last unit column vector from the inverse tableau are deleted. (If the problem is being solved by the revised simplex method using the product form of the inverse, this is equivalent to deleting the last row and the last column from each pivot matrix obtained so far.) The Phase I objective row, the column vector of $-w$, and the column vectors of all nonbasic artificial variables are also deleted from the original tableau. Then the algorithm moves to Phase II.

Phase II Termination Criteria

As in the usual simplex algorithm, define $\bar{c}_j = c_j - \pi A_{.j}$ for each j such that x_j is eligible for change of status during Phase II. The optimality criteria for the Phase II problem are discussed in Section 11.2.3.

11.3.9 Choice of the Entering Variable During Phase II

If Phase II optimality criteria are not satisfied, select a nonbasic variable x_s which has led to a violation of the optimality criteria, and make it the entering variable. It can belong to one of the following types.

(i) $s \in \mathbf{J}$ $\bar{c}_s < 0$ and the present value of x_s is 0

In this case, increasing the value of x_s from its present value of 0 leads to a reduction in the value of $z(x)$. The necessary computations for obtaining the next solution are carried out as in Section 11.3.4.

(ii) $s \in \mathbf{J}$ $\bar{c}_s > 0$ and the present value of x_s is U_s

In this case, the value of $z(x)$ will decrease if the value of x_s is decreased from its present value of U_s. The computation for obtaining the next solution are performed as in Section 11.3.5.

(iii) $s \in \bar{\mathbf{J}}$ and $\bar{c}_s < 0$

In this case, the value of $z(x)$ decreases as the value of x_s is increased from its present value of 0. Making x_s equal to λ, the new solution as a function of λ is the same as in (11.4). As in Section 11.3.6 we conclude that the maximum value that λ can have is $\theta = \min\{\bar{b}_r/\bar{a}_{rs}$ for all r such that $\bar{a}_{rs} > 0$, $(\bar{b}_r - U_{i_r})/\bar{a}_{rs}$ for all r, such that $i_r \in \mathbf{J}$ and $\bar{a}_{rs} < 0$, $\infty\}$. There are several cases to be considered here.

Case 1 If $\theta = \infty$, $z(x)$ is unbounded below in this problem. As λ increases from 0, we obtain a feasible solution, the value of $z(x)$ corresponding to which diverges to $-\infty$. Hence if this happens, the algorithm terminates.

Case 2 If $\theta = \bar{b}_t/\bar{a}_{ts}$ for some t, such that $\bar{a}_{ts} > 0$, the computations for obtaining the next solution are performed as in Case 1 of Section 11.3.6.

Case 3 If $\theta = (\bar{b}_t - U_{i,t})/\bar{a}_{ts}$ for some t, the computations for obtaining the next solution are performed as under Case 2 of Section 11.3.6.

Terminal Feasible Partitions

A feasible partition (**B**, **L**, **U**) for (11.2) is said to be a *terminal feasible partition* for it, if, either the optimality criterion is satisfied by it; or if there exists a nonbasic variable x_j in it, in the column of which the unboundedness criterion is satisfied as in Case 1 discussed above.

Original Phase I Tableau

x_1	x_2	x_3	x_4	x_5	x_6	x_7	x_8	x_9	x_{10}	x_{11}	$-z$	$-w$	b
1	0	0	−1	1	2	2	$-\frac{1}{5}$	1	0	0	0	0	$\frac{16}{5}$
0	1	0	1	−2	1	−2	$\frac{2}{5}$	0	1	0	0	0	$\frac{28}{5}$
0	0	1	−1	0	2	1	$\frac{1}{5}$	0	0	1	0	0	$\frac{39}{5}$
−1	2	−2	4	−4	−5	1	$\frac{18}{5}$	0	0	0	1	0	0
0	0	0	0	0	0	0	0	1	1	1	0	1	0

$x_j \geqq 0$ for all j; and $x_j \leqq 4$ for all $j \in \{1, 2, 3, 4, 5, 6, 8\}$.

Example 11.9

Consider the following bounded variable LP.

Original Tableau

x_1	x_2	x_3	x_4	x_5	x_6	x_7	x_8	$-z$	b
1	0	0	−1	1	2	2	$-\frac{1}{5}$	0	$\frac{16}{5}$
0	1	0	1	−2	1	−2	$\frac{2}{5}$	0	$\frac{28}{5}$
0	0	1	−1	0	2	1	$\frac{1}{5}$	0	$\frac{39}{5}$
−1	2	−2	4	−4	−5	1	$\frac{18}{5}$	1	0

z to be minimized; $x_j \geqq 0$ for all j; and $x_1, x_2, x_3, x_4, x_5, x_6$, and x_8 all $\leqq 4$.

In this problem **J** is $\{1, 2, 3, 4, 5, 6, 8\}$ and U_j is 4 for all $j \in \mathbf{J}$. Because of the upper-bound restrictions, the usual solution $x_1 = 16/5$, $x_2 = 28/5$, $x_3 = 39/5$, and all other $x_j = 0$ is not feasible to this problem. (It violates the upper bound restrictions on x_2, x_3.) The original Phase I tableau is given at the top.

Variables x_9, x_{10}, x_{11} are the artificial variables. In Phase I, $w = x_9 + x_{10} + x_{11}$ is to be minimized. The initial inverse tableau is given at the bottom. d is $(-1, -1, -1, 1, 1, -5, -1, -2/5)$. The variable x_8 is entering. Its updated column is $(-1/5, 2/5, 1/5, 18/5, -2/5)^T$. It becomes a nonbasic at the upper bound. The new vector of basic values is entered by the side of the inverse tableau. Next, x_7 is entering. Its updated column is $(2, -2, 1, 1, -1)^T$. Variable x_7 becomes a basic variable at value 2, replacing x_9 from the basic vector. The pivot column is entered by

Initial Inverse Tableau (Phase I)

Basic Variables	*Inverse Tableau*					*Basic Values*	*Basic Values After* $x_8 = 4$	*Pivot Column* x_7
x_9	1	0	0	0	0	$\frac{16}{5}$	4	(2)
x_{10}	0	1	0	0	0	$\frac{28}{5}$	4	−2
x_{11}	0	0	1	0	0	$\frac{39}{5}$	7	1
$-z$	0	0	0	1	0	0	$-\frac{72}{5}$	1
$-w$	−1	−1	−1	0	1	$-\frac{83}{5}$	−15	−1
All nonbasics = 0.								

Basic Variables	Inverse Tableau					Basic Values	Basic Values After $x_3 = 4$	Pivot Column x_6
x_7	$\frac{1}{2}$	0	0	0	0	2	2	1
x_{10}	1	1	0	0	0	8	8	3
x_{11}	$-\frac{1}{2}$	0	1	0	0	5	1	⓵
$-z$	$-\frac{1}{2}$	0	0	1	0	$-\frac{82}{5}$	$-\frac{42}{5}$	-6
$-w$	$-\frac{1}{2}$	-1	-1	0	1	-13	-9	-4
	$x_3 = 4$. All other nonbasics $= 0$.							

the side of the inverse tableau and the pivot element is circled. The next inverse tableau is given at the top. The new $\bar{d}$ is $(-1/2, -1, -1, 1/2, 3/2, -4, 0, -1/2)$. The variable x_3 is entering. Its updated column is $(0, 0, 1, -2, -1)^T$. It becomes a nonbasic variable at upper bound. The new vector of basic values is entered by the side of the inverse tableau. Next x_6 is entering. It enters the basic vector at value 1, replacing x_{11} from the basic vector. The pivot column is already entered by the side of the inverse tableau and the pivot element is circled. The next inverse tableau is given at the bottom. The new $\bar{d}$ is $(-5/2, -1, 3, -3/2, -1/2, 0, 0, 7/10)$. The variable x_3 has to be decreased from its present value of 4. Its updated column vector is $(-1, -3, 1, 4, 3)^T$. It becomes a basic variable at value 3, replacing x_7 from the basic vector. The variable x_7 becomes a zero-valued nonbasic variable. The pivot column is entered by the side of the inverse tableau and the pivot element is circled. The new vector of basic values is obtained as in Case 2, Section 11.3.5. The new inverse tableau is given at the top of page 352. The new $\bar{d}$ is $(1/2, -1, 0, -3/2, 5/2, 0, 3, -1/2)$. The variable x_4 is entering. It becomes a basic variable replacing x_{10} from the basic vector. The pivot column is entered on the inverse tableau and the pivot element is circled. The next inverse tableau is the second tableau from the top of page 352. Now all the artificial variables have become nonbasic variables with values equal to zero. $\bar{w} = 0$ and $\bar{d} = 0$. So Phase I terminates. $\bar{c}$ is $(0, 1, 0, 0, -1, 0, 7, 17/5)$. x_5 is entering. Its updated column is $(-1, -5/3, -1/3, -1)^T$. It becomes a basic variable at value 1, replacing x_3 from the basic vector. The variable x_3 becomes a nonbasic variable at its upper bound. The pivot column is entered by the side of the inverse tableau and the pivot element is circled. The new vector of basic values is obtained as in Case 2 of

Basic Variables	Inverse Tableau					Basic Values	Pivot Column x_3	
x_7	1	0	-1	0	0	1	⊖1 (circled -1)	Pivot Row
x_{10}	$\frac{5}{2}$	1	-3	0	0	5	-3	
x_6	$-\frac{1}{2}$	0	1	0	0	1	1	
$-z$	$-\frac{7}{2}$	0	6	1	0	$-\frac{12}{5}$	4	
$-w$	$-\frac{5}{2}$	-1	3	0	1	-5	3	

$x_3 = x_8 = 4$. All other nonbasics $= 0$.

Basic Variables	Inverse Tableau					Basic Values	Pivot Column x_4	
x_3	-1	0	1	0	0	3	0	
x_{10}	$-\frac{1}{2}$	1	0	0	0	2	$\frac{3}{2}$ (circled)	Pivot Row
x_6	$\frac{1}{2}$	0	0	0	0	2	$-\frac{1}{2}$	
$-z$	$\frac{1}{2}$	0	2	1	0	$\frac{8}{5}$	$\frac{3}{2}$	
$-w$	$\frac{1}{2}$	-1	0	0	1	-2	$-\frac{3}{2}$	

$x_8 = 4$. All other nonbasics $= 0$.

Basic Variables	Inverse Tableau					Basic Values	Pivot Column x_5
x_3	-1	0	1	0	0	3	-1 (circled)
x_4	$-\frac{1}{3}$	$\frac{2}{3}$	0	0	0	$\frac{4}{3}$	$-\frac{5}{3}$
x_6	$\frac{1}{3}$	$\frac{1}{3}$	0	0	0	$\frac{8}{3}$	$-\frac{1}{3}$
$-z$	1	-1	2	1	0	$-\frac{2}{5}$	-1
$-w$	0	0	0	0	1	0	

$x_8 = 4$. All other nonbasics $= 0$.

Section 11.3.4. The new inverse tableau is:

Basic Variables	Inverse Tableau				Basic Values	Pivot Column x_8
x_5	1	0	-1	0	1	$-\frac{2}{5}$ (circled)
x_4	$\frac{4}{3}$	$\frac{2}{3}$	$-\frac{5}{3}$	0	3	$-\frac{1}{3}$
x_6	$\frac{2}{3}$	$\frac{1}{3}$	$-\frac{1}{3}$	0	3	$-\frac{1}{15}$
$-z$	2	-1	1	1	$\frac{3}{5}$	3

$x_3 = x_8 = 4$. All other nonbasic $= 0$.

Since Phase I has terminated, the last column of the inverse tableau and the row corresponding to w have been removed from the inverse tableau. The new $\bar{c}$ is $(1, 1, -1, 0, 0, 0, 8, 3)$. Since $\bar{c}_8 = 3 > 0$ and x_8 is a nonbasic at its upper bound in the current feasible solution, x_8 is entering. The updated column of x_8 is $(-2/5, -1/3, -1/15, 3)^{\mathrm{T}}$. Variable x_8 becomes a basic variable at value 3/2, replacing x_5 from the basic vector. Variable x_5 becomes a nonbasic variable at zero value. The pivot column is entered by the side of the inverse tableau and the pivot element is circled. The new vector of basic values is obtained as in Case 2 of Section 11.3.5. The new inverse tableau is:

Basic Variables	Inverse Tableau				Basic Values
x_8	$-\frac{5}{2}$	0	$\frac{5}{2}$	0	$\frac{3}{2}$
x_4	$\frac{1}{2}$	$\frac{2}{3}$	$-\frac{5}{6}$	0	$\frac{13}{6}$
x_6	$\frac{1}{2}$	$\frac{1}{3}$	$-\frac{1}{6}$	0	$\frac{17}{6}$
$-z$	$\frac{19}{2}$	-1	$-\frac{13}{2}$	1	$\frac{81}{10}$

$x_3 = 4$. All other nonbasics $= 0$.

The new $\bar{c}$ is $(17/2, 1, -17/2, 0, 15/2, 0, 31/2, 0)^T$. It is clear that the present feasible solution is optimal.

11.4 RESOLUTION OF CYCLING UNDER DEGENERACY IN THE BOUNDED-VARIABLE PRIMAL SIMPLEX ALGORITHM

Consider the bounded-variable LP (11.2), where A is a matrix of order $m \times n$ and rank m. For $j \notin \mathbf{J}$, define $U_j = +\infty$. In the sequel, we assume that $U_j > 0$ for all j.

Degenerate and Nondegenerate Partitions

The feasible partition (**B**, **L**, **U**) associated with the BFS $\bar{x}$ for (11.2) is *primal nondegenerate* if $0 < \bar{x}_j < U_j$ for each j, such that $x_j \in \mathbf{B}$, *primal degenerate* otherwise. The BFS $\bar{x}$ is said to be (*primal*) *degenerate* or *nondegenerate* according to whether the associated partition (**B**, **L**, **U**) is (primal) degenerate or nondegenerate, respectively. If (**B**, **L**, **U**) is a primal nondegenerate partition associated with the BFS $\bar{x}$, then clearly $\mathbf{B} = \{x_j: j$, such that $0 < \bar{x}_j < U_j\}$, $\mathbf{L} = \{x_j: j$, such that $\bar{x}_j = 0\}$ and $\mathbf{U} = \{x_j: j$, such that $\bar{x}_j = U_j\}$. Hence every nondegenerate BFS of (11.2) is associated with a unique feasible partition. If (**B**, **L**, **U**) is a feasible partition associated with a degenerate BFS $\bar{x}$ of (11.2), it may be possible to obtain other feasible partitions associated with the same BFS $\bar{x}$ by replacing some $x_j \in \mathbf{B}$, satisfying $\bar{x}_j = 0$ or U_j with variables from **L** or **U**. Thus a degenerate BFS of (11.2) may be associated with many feasible partitions for (11.2).

Interior and Boundary Basic Variables

If (**B**, **L**, **U**) is a feasible partition for (11.2) associated with a BFS $\bar{x}$, a variable $x_j \in \mathbf{B}$ is said to be an *interior basic variable* in this partition if $0 < \bar{x}_j < \bar{U}_j$, or a *boundary basic variable* if $\bar{x}_j = 0$ or U_j. If x_j is a boundary basic variable in this partition, then it is a *lower boundary basic variable* if $\bar{x}_j = 0$, or an *upper boundary basic variable* if $\bar{x}_j = U_j$.

Nondegenerate and Degenerate Pivot Steps in the Bounded Variable Primal Simplex Algorithm

A pivot step is said to be a *nondegenerate pivot step* if the value of the entering variable undergoes a change as a result of this step, a *degenerate pivot step* otherwise. During each nondegenerate pivot step of this algorithm, the solution changes and the objective value undergoes a strict decrease. During a degenerate pivot step, there is no change in the objective value or even in the primal solution, but only the working basic vector and the current feasible partition change. The bounded variable primal simplex algorithm can also cycle under degeneracy just as does the primal simplex algorithm for general LPs. Here we discuss a lexicographic method for resolving the problem of cycling under degeneracy in the bounded variable primal simplex algorithm, using only the working basis inverse, taken from [11.9].

Strongly Feasible Partitions

Let (**B**, **L**, **U**) be a feasible partition for (11.2) associated with the BFS $\bar{x}$. Let $x_B = (x_{i_1}, \ldots, x_{i_m})$ be the associated working basic vector and B the associated working basis for (11.2). Let $\boldsymbol{\beta} = (\beta_{ij}) = B^{-1}$ and let $\boldsymbol{\beta}_{i.}$ denote the ith row vector of $\boldsymbol{\beta}$, for $i = 1$ to m. The feasible partition (**B**, **L**, **U**) is said to be a *strongly feasible partition* for (11.2) if, for each $r = 1$ to m,

$$\bar{x}_{i_r} = 0 \text{ implies } \boldsymbol{\beta}_{r.} \text{ is lexico positive}$$

$$\bar{x}_{i_r} = U_{i_r} \text{ implies } \boldsymbol{\beta}_{r.} \text{ is lexico negative}$$

Dropping Variable Selection Rules to Preserve Strong Feasibility

Let (**B**, **L**, **U**) be the current strongly feasible partition for (11.2) associated with the BFS $\bar{x}$, dual solution π, and relative cost vector $\bar{c}$. Let $x_B = (x_j: j \in \{i_1, \ldots, i_m\})$, B, $\boldsymbol{\beta} = (\beta_{ij}) = B^{-1}$, x_s, $\bar{A}_{.s} = B^{-1}A_{.s}$ be the present working basic vector, working basis, working basis inverse, entering variable, and its updated column, respectively, in this step.

Case 1 $x_s \in \mathbf{L}$: In this case $\bar{x}_s = 0$, $\bar{c}_s < 0$, and hence the value of x_s should be increased. The minimum ratio $\theta = \text{minimum}\ (\{\bar{x}_{i_r}/\bar{a}_{rs}: r$, such that $\bar{a}_{rs} > 0\} \cup \{(\bar{x}_{i_r} - U_{i_r})/\bar{a}_{rs}: r$, such that U_{i_r} is finite and $\bar{a}_{rs} < 0\} \cup \{U_s\})$. If $\theta = +\infty$ (this only happens if θ is the minimum in the empty set), the objective value is unbounded below, terminate. If θ is finite, let $\mathbf{F}_1 = \{x_{i_r}: r$, such that either (1) $\bar{a}_{rs} > 0$ and $\theta = \bar{x}_{i_r}/\bar{a}_{rs}$, or (2) $\bar{a}_{rs} < 0$ and $\theta = (\bar{x}_{i_r} - U_{i_r})/\bar{a}_{rs}\}$, and $\mathbf{F} = \mathbf{F}_1$ if $U_s > \theta$ or $\mathbf{F} = \mathbf{F}_1 \cup \{x_s\}$ if $U_s = \theta$. The variables $x_j \in \mathbf{F}$ are those that are eligible to drop from $\mathbf{B} \cup \{x_s\}$ in this step. If **F** is a singleton set, the dropping variable is identified uniquely and unambigously. On the other hand, if $|\mathbf{F}| \geqq 2$, choose x_s as the dropping variable if $x_s \in \mathbf{F}$, and $\boldsymbol{\beta}_{r.} \succ 0$ for all r, such that $x_{i_r} \in \mathbf{F}_1$ and $\bar{a}_{rs} > 0$, and $\boldsymbol{\beta}_{r.} \prec 0$ for all r such that

$x_{i_r} \in \mathbf{F}_1$ and $\bar{a}_{rs} < 0$. If at least one of these conditions is violated, choose the dropping variable to be $x_{i_t} \in \mathbf{F}_1$, where t is determined to satisfy $(\boldsymbol{\beta}_{t.}/\bar{a}_{ts}) =$ lexico minimum $\{\boldsymbol{\beta}_{r.}/\bar{a}_{rs}$: r, such that $x_{i_r} \in \mathbf{F}_1\}$. Since $\boldsymbol{\beta}_{r.}$ are the rows of B^{-1}, this equation determines t uniquely.

Case 2 $x_s \in \mathbf{U}$. In this case $\bar{x}_s = \mathrm{U}_s$, $\bar{c}_s > 0$, and the value of x_s should be decreased. The minimum ratio is $\theta =$ maximum $(\{(\bar{x}_{i_r} + \bar{a}_{rs}\mathrm{U}_s)/\bar{a}_{rs}$: r such that $\bar{a}_{rs} < 0\} \cup \{(\bar{x}_{i_r} + \bar{a}_{rs}\mathrm{U}_s - \mathrm{U}_{i_r})/\bar{a}_{rs}$: r such that $\bar{a}_{rs} > 0\} \cup \{0\})$. Let $\mathbf{F}_1 = \{x_{i_r}$: r, such that either (1) $\bar{a}_{rs} < 0$ and $\theta = (\bar{x}_{i_r} + \bar{a}_{rs}\mathrm{U}_s)/\bar{a}_{rs}$, or (2) $\bar{a}_{rs} > 0$ and $\theta = (\bar{x}_{i_r} + \bar{a}_{rs}\mathrm{U}_s - \mathrm{U}_{i_r})/\bar{a}_{rs}\}$, and $\mathbf{F} = \mathbf{F}_1$ if $\theta > 0$, or $\mathbf{F}_1 \cup \{x_s\}$ if $\theta = 0$. The variables $x_j \in \mathbf{F}$ are those eligible to drop from the set $\mathbf{B} \cup \{x_s\}$ in this step. If $\mathbf{F}$ is a singleton, the dropping variable is identified uniquely and unambiguously. If $|\mathbf{F}| \geqq 2$, choose the dropping variable to be x_s if $x_s \in \mathbf{F}$, and $\boldsymbol{\beta}_{r.} \succ 0$ for all r such that $x_{i_r} \in \mathbf{F}_1$ and $\bar{a}_{rs} < 0$, and $\boldsymbol{\beta}_{r.} \prec 0$ for all r such that $x_{i_r} \in \mathbf{F}_1$ and $\bar{a}_{rs} > 0$. If at least one of these conditions is violated, choose the dropping variable to be $x_{i_t} \in \mathbf{F}_1$, where t is determined to satisfy $(\boldsymbol{\beta}_{t.}/\bar{a}_{ts}) =$ lexico maximum $\{\boldsymbol{\beta}_{r.}/\bar{a}_{rs}$: r, such that $x_{i_r} \in \mathbf{F}_1\}$.

Note 11.3: The rules discussed above identify the dropping variable unambiguously in each pivot step. Also, if these rules are used, even though x_s is a boundary variable in the new partition, there may be a change in the working basic vector (compare this with the usual primal algorithm discussed in Section 11.3, where x_s is introduced into the working basic vector only when it enters as an interior basic variable).

THEOREM 11.1 If the partition at the beginning of a pivot step is strongly feasible and the dropping variable is determined as above, the new partition obtained at the end of this pivot step will also be strongly feasible.

Proof: These are several cases to consider, depending on whether the entering variable $x_s \in \mathbf{L}$ or $\mathbf{U}$, whether x_s is the dropping variable or not, and whether $x_s \in \mathbf{F}$ or not. In each case, the theorem can be directly verified to be true. ■

THEOREM 11.2 Suppose the partition $(\mathbf{B}, \mathbf{L}, \mathbf{U})$ at the beginning of a pivot step is strongly feasible, and the dropping variable choice rule discussed above is used in this pivot step, leading to the partition $(\mathbf{B}', \mathbf{L}', \mathbf{U}')$. Let π, π' be the dual basic solutions associated with the partitions $(\mathbf{B}, \mathbf{L}, \mathbf{U})$, $(\mathbf{B}', \mathbf{L}', \mathbf{U}')$, respectively. If this is a degenerate pivot step, $\pi' \prec \pi$.

Proof: Here again there are several cases to consider, depending on whether the entering variable $x_s \in \mathbf{L}$ or $\mathbf{U}$, and whether the t in the dropping variable choice satisfies $\bar{a}_{ts} > 0$ or < 0. In each ease, the theorem can be directly verified to be true. ■

Note 11.4: In Theorem 11.2 we have only proved that the dual solution undergoes a strict lexico decrease in a *degenerate pivot step* if strong feasibility is being preserved. The dual solution may not undergo a lexico decrease (it may, in fact, undergo a lexico increase) in a nondegenerate pivot step.

THEOREM 11.3 Suppose (11.2) is solved beginning with a strongly feasible partition for it, and using the dropping variable choice rules for preserving strong feasibility throughout the algorithm. Let $\bar{z}$ denote the objective value, and $\pi = (\pi_1, \ldots, \pi_m)$ the dual solution in any step of this algorithm. Then the vector $(\bar{z}, \pi_1, \ldots, \pi_m)$ undergoes a strict lexico decrease in every pivot step of this algorithm, and the algorithm itself terminates after a finite number of pivot steps.

Proof: In every nondegenerate pivot step, $\bar{z}$ decreases strictly. In every degenerate pivot step the value of $\bar{z}$ remains unaltered, but the vector π undergoes a strict lexico decrease. These facts imply that the vector $(\bar{z}, \pi)$ undergoes a strict lexico decrease in every pivot step of this algorithm. Hence a partition that is obtained in a step of this algorithm can never reappear in subsequent steps. Since there are only a finite number of possible partitions, the algorithm must terminate by satisfying one of the termination conditions after at most a finite number of pivot steps. ■

The Method of Strongly Feasible Partitions

If a strongly feasible partition is available for (11.2) initially, begin with it and apply the bounded variable primal simplex algorithm using the dropping variable choice rules discussed above to preserve strong feasibility throughout the algorithm.

If a strongly feasible partition for (11.2) is not available, set up the Phase I problem by introducing the nonnegative artificial variables $x_{n+1}, \ldots, x_{n+m}$ as in Tableau 11.2 of Section 11.3.1. For each artificial variable the upper bound is $+\infty$. Let $\mathbf{B}_1 = \{x_{n+1}, \ldots, x_{n+m}\}$, $\mathbf{L}_1 = \{x_1, \ldots, x_n\}$, $\mathbf{U}_1 = \varnothing$. Then $(\mathbf{B}_1, \mathbf{L}_1, \mathbf{U}_1)$ is clearly a strongly feasible partition for the Phase I problem in Tableau 11.2 of Section 11.3.1. Starting with this strongly feasible partition, solve the Phase I problem by the bounded variable primal simplex algorithm using the dropping variable choice rules discussed above in every pivot step, to preserve strong feasibility throughout the algorithm. When Phase I terminates either we conclude that the original problem (11.2) is primal infeasible, or we obtain a strongly feasible partition for (11.2). Now switch over to Phase II as in Section 11.3.8, and solve the original problem (11.2) using the dropping variable choice rules discussed above to preserve strong feasibility in each pivot step.

Suppose a feasible partition $(\mathbf{B}_1, \mathbf{L}_1, \mathbf{U}_1)$ for (11.2) is available, but it is not strongly feasible. Let $\bar{x}$ be the BFS of (11.2) associated with $(\mathbf{B}_1, \mathbf{L}_1, \mathbf{U}_1)$. Let B_1 be the working basis consisting of the columns $A_{.j}$ for j such that $x_j \in \mathbf{B}_1$. Obtain the canonical tableau for (11.2) with respect to the basis B_1. For each $i = 1$ to m, multiply row i in the canonical tableau by -1 on both sides if the ith basic variable in $\mathbf{B}_1$ is an upper boundary basis variable in the BFS $\bar{x}$. Let Q be the matrix obtained by multiplying the ith row of the unit matrix of order m by -1 for each i such that the ith basic variable in $\mathbf{B}_1$ is an upper boundary basic variable in $\bar{x}$. As a result of these operations, (11.2) gets transformed into the following problem, which we denote by (P).

$$
\begin{array}{ll}
\text{minimize} & z(x) = cx \\
\text{Subject to} & QB_1^{-1}Ax = QB_1^{-1}b \\
x_j \leqq U_j & \text{for } j \in \mathbf{J} \\
x_j \geqq 0 & \text{for all } j
\end{array}
$$

Now verify that $(\mathbf{B}_1, \mathbf{L}_1, \mathbf{U}_1)$ is a strongly feasible partition for this problem (P). Solve (P) starting with the initial strongly feasible partition $(\mathbf{B}_1, \mathbf{L}_1, \mathbf{U}_1)$ for it, using the dropping variable choice rules discussed above to preserve strong feasibility throughout the algorithm. The terminal partition obtained at the end, is also a terminal partition for the original problem (11.2). Let $(\mathbf{B}, \mathbf{L}, \mathbf{U})$ be any strongly feasible partition for (P) obtained during the execution of this method. Let B, $\bar{B}$ be the working bases associated with $\mathbf{B}$ in (11.2) and (P), respectively. Then $\bar{B} = QB_1^{-1}B$, and thus $(\bar{B})^{-1} = B^{-1}B_1Q$. This method of strongly feasible partitions for (P) can be executed without obtaining all the data in (P) explicitly, using the original data in (11.2) itself (see Exercise 10.3 in Section 10.4).

11.5 THE DUAL SIMPLEX METHOD FOR BOUNDED-VARIABLE LINEAR PROGRAMS

Any LP in which the constraints are a system of linear equations, and each variable is restricted to lie within specified lower and upper bounds both of which are finite, can be transformed as in Section 11.2.1 into a problem of the following form:

$$
\begin{array}{lll}
\text{Minimize} & z(x) = cx & \\
\text{Subject to} & Ax = b & \\
& 0 \leqq x_j \leqq U_j & \text{for all } j = 1 \text{ to } n
\end{array}
\tag{11.6}
$$

where A is a matrix of order $m + n$ and rank m, and U_j are finite nonnegative numbers for each $j = 1$ to n. For solving such bounded variable problems, the dual simplex method seems to offer a particularly attractive and efficient algorithm, and we present it here.

By transforming the system of constraints $Ax = b$ into row echelon normal form, as in Sections 3.3.5, and 3.3.6, obtain an arbitrary working basis B for (11.6). Let $x_B = (x_j\colon\ j \in \{p_1, \ldots, p_m\})$ be the basic vector and c_B be the original basic cost vector. Let $\pi = c_B B^{-1}$, $\bar{c} = c - \pi A$. Let $\mathbf{N} = \{1, \ldots, n\} \setminus \{p_1, \ldots, p_m\}$ be the set of subscripts of current nonbasic variables. Define a solution, $\bar{x} = (\bar{x}_j)$ for (11.6) by

$$
\begin{aligned}
x_j = \bar{x}_j &= U_j && \text{if } \bar{c}_j < 0 \\
&= 0 && \text{if } j \in \mathbf{N} \text{ and } \bar{c}_j \geqq 0 \\
x_B = \bar{x}_B &= B^{-1}\Big(b - \sum_{j \in \mathbf{N}} \bar{x}_j A_{.j}\Big) && \\
z = \bar{z} &= c_B \bar{x}_B + \sum_{j \in \mathbf{N}} c_j \bar{x}_j &&
\end{aligned}
\tag{11.7}
$$

From the manner in which the values for the nonbasic variables are defined in (11.7), it is clear that the current solution satisfies the optimality criterion (i.e., the dual feasibility criterion), but it may be infeasible to (11.6) because in $\bar{x}$ the values of some of the basic variables may violate the nonnegativity or the upper-bound restrictions on those variables. Similar to $\bar{x}$, all the solutions obtained during this algorithm will satisfy the optimality criterion. Hence the moment a solution obtained during this algorithm satisfies the primal feasibility criterion (namely, that the values of the basic variables in that solution satisfy the bound restrictions on those variables), that solution is an optimum feasible solution for (11.6), and the method terminates. If the current

	Nonbasic Variables x_j, Such That $\bar{x}_j = 0$			*Nonbasic Variables* x_j, Such That $\bar{x}_j = U_j$			*Basic Variables*			*Values in $\bar{x}$ of the Basic Variable in the Pivot Row* ($\bar{x}_7$)
	x_1	x_2	x_3	x_4	x_5	x_6	x_7	x_8	x_9	
Pivot Row	3	0	5	−1	−2	0	1	0	0	−2

solution does not satisfy the primal feasibility criterion, go to the pivot row choice rule.

In describing each of these procedures, we assume that $x_B = (x_j\colon j \in \{p_1, \ldots, p_m\})$ is the present basic vector associated with the present basis B, $\mathbf{N} = \{1, \ldots, n\} \setminus \{p_1, \ldots, p_m\}$, and $\bar{x}$ is the present BFS. $\bar{A} = B^{-1}A$, $\bar{b} = B^{-1}b$, $\bar{z}$ is the objective value $c\bar{x}$, and $(\bar{c}_1, \ldots, \bar{c}_n)$ is the present updated cost row.

Pivot Row Choice Rule

Row i is *eligible to be chosen as the pivot row* at this stage if the ith basic variable in the current basic vector x_B has a value in the current solution $\bar{x}$, which is either strictly negative or strictly greater than the upper bound for this variable. Select any of the eligible rows as the pivot row and go to the step of checking for primal infeasibility.

Checking for Primal Infeasibility

Let row r be the selected pivot row in this step. $\bar{A}_{r.} = (\bar{a}_{r1}, \ldots, \bar{a}_{rn})$ is the pivot row. Check whether the following (i) or (ii) hold.

i $\bar{x}_{p_r} < 0$, $\quad \bar{a}_{rj} \geqq 0$ for all $j \in \mathbf{N}$ satisfying $\bar{x}_j = 0$, and $\bar{a}_{rj} \leqq 0$ for all $j \in \mathbf{N}$ satisfying $\bar{x}_j = U_j$.

ii $\bar{x}_{p_r} > U_{p_r}$, $\quad \bar{a}_{rj} \leqq 0$ for all $j \in \mathbf{N}$ satisfying $\bar{x}_j = 0$, and $\bar{a}_{rj} \geqq 0$ for all $j \in \mathbf{N}$ satisfying $\bar{x}_j = U_j$.

If either (i) or (ii) holds, (11.6) is infeasible, terminate. Otherwise, go to the entering variable selection procedure.

Example 11.10

To illustrate the primal infeasibility criterion, only the pivot row is given at the top.

The primal infeasibility criterion (i) is satisfied in this pivot row. This implies that the LP being solved is infeasible.

Example 11.11

In this example also, only the data in the pivot row are given at the bottom, for illustrating primal infeasibility.

The current basic variable in the pivot row is x_7, and suppose $U_7 = 8$. The value of x_7 in the current solution is $10 > U_7$. The primal infeasibility criterion (ii) is satisfied in this pivot row. This implies that the LP being solved is infeasible.

Justification

Since x_{p_t} is the tth basic variable, we clearly have $\bar{a}_{rj} = 0$ for $j \in \{p_1, \ldots, p_{r-1}, p_{r+1}, \ldots, p_m\}$ and $\bar{a}_{r,p_r} = 1$. So the constraint corresponding to the present pivot row is $x_{p_r} + \sum(\bar{a}_{rj}x_j\colon \text{over } j \in \mathbf{N}) = \bar{b}_r$. Hence every solution x of (11.6) satisfies

$$x_{p_r} = \bar{b}_r - \sum_{j \in \mathbf{N}} \bar{a}_{rj}x_j \tag{11.8}$$

We have $\bar{x}_{p_r} = \bar{b}_r - \sum(\bar{a}_{rj}\bar{x}_j\colon \text{over } j \in \mathbf{N})$. From (11.8) we conclude that the maximum value of x_{p_r} in solutions x satisfying $Ax = b$, and $0 \leqq x_j \leqq U_j$ for all $j \in \mathbf{N}$ is $\bar{b}_r - \sum(\bar{a}_{rj}U_j\colon \text{over } j \in \mathbf{N} \text{ satisfying } \bar{a}_{rj} < 0)$. If (i) holds, this is just equal to $\bar{x}_{pr} < 0$. So, if (i) holds, the constraint $x_{p_r} \geqq 0$ can never be satisfied in (11.6), and hence (11.6) is infeasible.

	Nonbasic Variable x_j, *Such That* $\bar{x}_j = 0$			*Nonbasic Variable* x_j, *Such That* $\bar{x}_j = U_j$			*Basic Variables*			*Value in $\bar{x}$ of the Basic Variable in the Pivot Row* ($\bar{x}_7$)
	x_1	x_2	x_3	x_4	x_5	x_6	x_7	x_8	x_9	
Pivot Row	−2	−3	0	0	1	2	1	0	0	10

Similarly, from (11.8) we verify that the minimum value of x_{p_r} in solutions x satisfying $Ax=b$, $0 \leqq x_j \leqq U_j$ for all $j \in \mathbf{N}$ is $\bar{b}_r - \sum(\bar{a}_{rj}U_j$: over $j \in \mathbf{N}$ satisfying $\bar{a}_{rj} > 0)$. If (ii) holds, this is just equal to $\bar{x}_{p_r} > U_{p_r}$. So, if (ii) holds, the constraint $x_{p_r} \leqq U_{p_r}$ can never be satisfied in (11.6), and hence (11.6) is infeasible.

Entering Variable Selection Procedure

Let row r be the pivot row. In this procedure the nonbasic variable that will replace x_{p_r} from the basic vector, called the *entering variable in this pivot step*, is determined to satisfy the following properties.

1. The new solution obtained after the pivot step also satisfies the optimality criterion.
2. In the new solution, both the dropping variable x_{p_r} and the entering variable have values that are within their respective bounds (i.e., in effect, in the new solution all the variables x_j for $j \in \mathbf{N} \cup \{p_r\}$ have values satisfying their respective bound restrictions).

Note 11.5: It is possible that some of the other basic variables, x_j for $j \in \{p_1, \ldots, p_{r-1}, p_{r+1}, \ldots, p_m\}$, that have values within their respective bounds in the current solution, may attain values that violate these bound restrictions in the new solution obtained after this pivot step. This possibility also exists in the unbounded dual simplex method.

In order to satisfy properties (1) and (2), the entering variable is selected according to a special procedure. It may be necessary to perform a series of pivots in immediate succession using the same row r as the pivot row until condition (2) is satisfied. The effect of a series of pivots in the same row is the same as the effect of the last pivot in the series. The special procedure allows us to perform the intermediate pivot steps updating the data in the pivot row and the cost rows only (since these are the only data needed to perform these intermediate pivot steps), and to update the inverse tableau only when the last pivot step in this series is reached.

Thus the procedure may go through several steps before selecting the final entering variable that should replace x_{p_r} from x_B. In each step a tentative entering variable is selected according to some rules, and if this leads to a new solution that satisfies (1) and (2), then the tentative entering variable in this step is declared as the final entering variable to replace x_{p_r} from x_B and the algorithm moves to the pivot step. Otherwise we move to the next step in the entering variable selection procedure itself. In all the steps of this special procedure the pivot row remains the same (row r). Each step of the special procedure replaces the present basic variable in the pivot row (row r) by a nonbasic variable, and this continues until we reach a step in which either the updated pivot row establishes primal infeasibility or the present solution satisfies (1) and (2). So, the variables x_j for $j \in \{p_1, \ldots, p_{r-1}, p_{r+1}, \ldots, p_m\}$ remain as basic variables throughout this special procedure; their values in the solution are updated in each step of the procedure, but not used in executing the procedure.

Since $\bar{A}_{r.} = (\bar{a}_{r1}, \ldots, \bar{a}_{rn})^{\mathrm{T}}$ is the updated row r with respect to the basic vector x_B, we have $\bar{a}_{rj} = 0$ for each $j \in \{p_1, \ldots, p_{r-1}, p_{r+1}, \ldots, p_m\}$. These $m-1$ entries in the pivot row will remain equal to zero throughout this special procedure.

General Step in the Special Procedure

Let $\hat{x} = (\hat{x}_j), (\hat{a}_{r1}, \ldots, \hat{a}_{rm}), (\hat{c}_1, \ldots, \hat{c}_n), \hat{b}_r, \hat{z}, \hat{x}_v$, be the present solution, updated pivot row (row r) updated cost row, updated rth right-hand-side constant, objective value, and basic variable in the pivot row, respectively. So the present nonbasic variables are x_j for $j \in \hat{\mathbf{N}} = (\mathbf{N} \cup \{p_r\} \backslash \{v\})$. We have $\hat{a}_{rv} = 1$, since x_v is the present basic variable in the pivot row, and $\hat{a}_{rj} = 0$ for $j \in \{p_1, \ldots, p_{r-1}, p_{r+1}, \ldots, p_m\}$, since these are subscripts of present basic variables. So the constraint corresponding to the present pivot row is $x_v + \sum_{j \in \mathbf{N}} \hat{a}_{rj}x_j = \hat{b}_r$. From this we see that if $\hat{x}_v < 0$, the value of x_v can be increased by either decreasing the value of a nonbasic variable x_j for j satisfying $\hat{a}_{rj} > 0$ and $\hat{x}_j = U_j$, or by increasing the value of a nonbasic variable x_j for j satisfying $\hat{a}_{rj} < 0$ and $\hat{x}_j = 0$. So, if $\hat{x}_v < 0$, choose x_t as the tentative entering variable to replace x_v, where t is one of the j's that achieves the maximum in maximum $\{\hat{c}_j/\hat{a}_{rj}$: $j \in \hat{\mathbf{N}}$ and either $\hat{x}_j = U_j$ and $\hat{a}_{rj} > 0$, or $\hat{x}_j = 0$ and $\hat{a}_{rj} < 0\}$. In the same manner, if $\hat{x}_v > U_v$, the tentative entering variable to replace x_v should be x_t, where t is one of the j's that ties for the minimum in: minimum $\{\hat{c}_j/\hat{a}_{rj}$: $j \in \hat{\mathbf{N}}$ and either $\hat{x}_j = 0$ and $\hat{a}_{rj} > 0$, or $\hat{x}_j = U_j$ and $\hat{a}_{rj} < 0\}$. This selection guarantees that the new basic vector obtained satisfies the dual feasibility criterion. Let $\tilde{a}_{rj} = \hat{a}_{rj}/\hat{a}_{rt}$, $j = 1$ to n. And $\tilde{b}_r = \hat{b}_r/\hat{a}_{rt}$, $\tilde{c}_j = \hat{c}_j - (\hat{a}_{rj}\hat{c}_t/\hat{a}_{rt})$, $j = 1$ to n. These $\tilde{a}_{rj}$, $\tilde{b}_r$, $\tilde{c}_j$ are the updated entries in the pivot and cost rows after x_t replaces x_v as the basic variable in the pivot row.

Present Solution $\bar{x}_j$	10			0	0	2	2	$\bar{b}$
Variable x_j	x_1	x_2	x_3	x_4	x_5	x_6	x_7	
Updated pivot row $(\bar{a}_{1j})$	1	0	0	-2	3	1	-1	10
Updated cost row $(\bar{c}_j)$	0	0	0	4	10	-2	-3	

Present basic variable in pivot row is x_1.

$\dfrac{\bar{c}_j}{\bar{a}_{1j}}$	For $j \in \hat{N}$ satisfying $\bar{a}_{1j} > 0$ and $\bar{x}_j = 0$			$\dfrac{10}{3}$		Min. $=3$
	For $j \in \hat{N}$ satisfying $\bar{a}_{1j} < 0$ and $\bar{x}_j = 2$				3	

The new solution will be $\tilde{x} = (\tilde{x}_j)$, where

$$\tilde{x}_j = \hat{x}_j \quad \text{for } j \in \hat{N}/\{t\}$$
$$= 0 \text{ or } U_v$$
depending on whether $\hat{x}_v < 0$
or $> U_v$; $\quad$ for $j = v$
$$= (\hat{x}_v + \hat{a}_{rt}\hat{x}_t)/\hat{a}_{rt} \quad \text{or} \quad \hat{x}_t + ((\hat{x}_v - U_v)/\hat{a}_{rt})$$
depending on whether $\hat{x}_v < 0$
or $> U_v$; $\quad$ for $j = t$
$$= \hat{x}_{p_\sigma} + \bar{a}_{\sigma t}(\hat{x}_t - \tilde{x}_t)$$
for $j = p_\sigma$, $\sigma = 1, \ldots, r - 1, r + 1, \ldots, m$
$$\tilde{z} = \hat{z} + \bar{c}_t(\tilde{x}_t - \hat{x}_t)$$

If $0 \leq \tilde{x}_t \leq U_t$, the new solution obtained as a result of choosing x_t as the entering variable in the pivot row (row r) satisfies both properties (1) and (2). Hence x_t is chosen as the final entering variable to replace x_{p_r} from x_B, the entering variable selection procedure terminates, and we move to the pivot step.

If either $\tilde{x}_t < 0$ or $> U_t$, x_t is only a tentative variable replacing x_v as the basic variable in the pivot row.

Check whether the primal infeasibility criteria (i) or (ii) are satisfied by the present data in the pivot row (row r), and if so terminate. Otherwise continue by going to the next step in the special entering variable selection procedure.

Example 11.12

Consider a bounded-variable LP in which $0 \leq x_j \leq 2$ for all $j = 1$ to 7. So, $U_j = 2$ for all j. Suppose (x_1, x_2, x_3) is the current basic vector, and the pivot row is the row in which x_1 is the basic variable. To keep the example simple, the values of the other basic variables x_2, x_3 in the solution and the objective value are not shown, and their updating is not included in the illustration. Data in the pivot row, the updated cost row, and the solution are given at the top.

Step 1 Since $\bar{x}_1 > U_1 = 2$, x_7 is the tentative entering variable. The updated data are given below. Since $x_7 = -6$ in the new solution, and the primal infeasibility criteria are not satisfied, we move to Step 2.

Present Solution $\bar{x}_j$	2			0	0	2	-6	$\bar{b}$
Variable x_j	x_1	x_2	x_3	x_4	x_5	x_6	x_7	
Updated pivot row $(\bar{a}_{1j})$	-1	0	0	2	-3	-1	1	-10
Updated cost row $(\bar{c}_j)$	-3	0	0	10	1	-5	0	

Present basic variable in pivot row is x_7.

$\dfrac{\bar{c}_j}{\bar{a}_{1j}}$	For $j \in \hat{N}$ satisfying $\bar{a}_{1j} > 0$ and $\bar{x}_j = 2$					Max; $= -\dfrac{1}{3}$
	For $j \in \hat{N}$ satisfying $\bar{a}_{1j} < 0$ and $\bar{x}_j = 0$			$-\dfrac{1}{3}$		

Step 2 Since $\bar{x}_7 = -6 < 0$, the tentative entering variable is x_5. It can be verified that the new solution after x_5 replaces x_7 as the basic variable in the pivot row is $(x_1, x_4, x_5, x_6, x_7) = (2, 0, 2, 2, 0)$. Since the value of x_5 in this new solution satisfies the bound constraints on this variable, x_5 is the final entering variable replacing x_1 from the basic vector (x_1, x_2, x_3). The entering variable selection procedure now terminates, and the algorithm should move to the pivot step.

Pivot Step Let x_s be the entering variable (as finally determined by the entering variable selection procedure). The pivot column is then $\bar{A}_{.s} = B^{-1}A_{.s}$ where $A_{.s}$ is the original column of x_s in A of (11.6). Enter the pivot column on the right-hand side of the inverse tableau with respect to the basic vector x_B, and by performing a pivot on it with $\bar{a}_{rs}$ as the pivot element, the new inverse tableau is obtained. The new solution, however, is the one obtained in the final step of the entering variable selection procedure. Check whether the new solution satisfies the primal feasibility criterion; otherwise continue the algorithm in the same manner.

Advantages

The dual simplex method constructs a dual feasible basic solution for (11.6) directly after putting the system of equality constraints in (11.6) in row echelon normal form (which is computationally much simpler than solving a Phase I problem), and from there moves toward attaining primal feasibility while maintaining dual feasibility throughout. Thus, on the whole, the dual simplex method seems to be computationally more efficient than the primal simplex method for solving bounded variable LPs in which all the variables are restricted to lie within specified finite lower and upper bounds. See the papers [11.12 and 11.13] of H. M. Wagner and C. Witzgall. In practical applications of linear programming, each decision variable in the LP model can be assumed to be bounded above by some "practical limit" or some "arbitrarily large number." By introducing such practical bounds on all the variables, any LP model encountered in practical applications can be solved by the dual simplex method discussed in this section.

11.6 SENSITIVITY ANALYSIS IN THE BOUNDED-VARIABLE LINEAR PROGRAMMING PROBLEM

Suppose (11.2) is solved and an optimum feasible partition (**B**, **L**, **U**) has been obtained for it.

If a new bounded variable, say x_{n+1}, has to be introduced into the model, compute its relative cost coefficient $\bar{c}_{n+1}$ with respect to the present optimum working basis. If $\bar{c}_{n+1} \geqq 0$, define $\mathbf{L}_1 = \mathbf{L} \cup \{x_{n+1}\}$, $\mathbf{U}_1 = \mathbf{U}$. On the other hand, if $\bar{c}_{n+1} < 0$, define $\mathbf{L}_1 = \mathbf{L}$, $\mathbf{U}_1 = \mathbf{U} \cup \{x_{n+1}\}$. Then ($\mathbf{B}$, $\mathbf{L}_1$, $\mathbf{U}_1$) can be verified to be a dual feasible partition for the augmented problem with the new variable x_{n+1} included. Beginning with ($\mathbf{B}$, $\mathbf{L}_1$, $\mathbf{U}_1$), the augmented problem can be solved by the dual simplex algorithm of Section 11.5.

If changes occur in the bounds on the variables, or in the b vector, or in the column vector associated with a nonbasic variable in $\mathbf{L} \cup \mathbf{U}$, it can be verified that (**B**, **L**, **U**) remains dual feasible to the modified problem. So, beginning with (**B**, **L**, **U**), the modified problem can be solved by the dual simplex algorithm of Section 11.5.

If changes occur in the cost vector c in (11.2), the modified problem can be solved by the primal simplex algorithm of Section 11.3.8, beginning with the primal feasible partition (**B**, **L**, **U**).

11.7 GENERALIZED UPPER BOUNDING (GUB)

So far we have discussed methods for handling constraints of the form $x_j \leqq U_j$, without having to include rows corresponding to them in the working basis that has to be inverted. This approach has been generalized to handle a set of constraints satisfying the following properties (1) and (2) efficiently in the same manner.

1 Each constraint in the set is an upper-bound constraint on the sum of a subset of variables (i.e., a constraint of the form $\sum(x_j\colon j \in \{j_1, j_2, \ldots, j_k\}) \leqq U_0$, which gets transformed into the equation $\sum(x_j\colon j \in \{j_1, \ldots, j_k\}) + s_1 = U_0$, after introducing the slack variable s_1 corresponding to it.

2 Every variable in the model appears in at most one constraint in this set of constraints.

A set of constraints satisfying properties (1) and (2) are called *generalized upper-bound constraints*, or GUB constraints for short; the approach for solving the LP without having to include the rows corresponding to the GUB constraints in the working basis to be inverted is called the *generalized upper-bounding technique*, or GUB or GUBT for short. GUBT has been developed by G. B. Dantzig and R. M. Vanslyke [11.7].

If GUBT can handle GUB constraints satisfying properties (1) and (2), by scaling of the variables, it can obviously be made to handle a set of constraints satisfying property (2) only. We discuss this generalized version of GUBT here.

Consider an LP in standard form, in which the constraints fall into two sets. The first set of m constraints is of an arbitrary nature. The last set of p constraints satisfies property (2) listed above, and is known as the *generalized upper-bound*, or GUB *constraints*. GUBT is a specialization of the revised simplex method for solving such an LP by maintaining the inverse of a *working basis* of order m only (where m is the number of non-GUB constraints), instead of maintaining the inverse of the basis of order $m+p$ as required by the usual revised simplex method. All the quantities needed for carrying out the simplex method on this LP are derived using the inverse of the working basis and the original data in the problem, and after each pivot step only the inverse of the working basis is updated. When p (the number of GUB constraints)is large, GUBT results in substantial savings in the memory space requirements and the total computational effort over the implementation of the conventional revised simplex method applied on the same LP.

In order to take advantage of GUBT, one must be able to identify the subset of GUB constraints in the LP to be solved. If the LP under consideration is a general linear programming model about whose structure the user has no prior knowledge, he may have to write a special procedure for identifying and isolating as large a subset of constraints as possible in the model satisfying property (2) discussed above, which will be the set of GUB constraints. GUBT cannot be applied unless the GUB constraints are specified beforehand.

By rearranging the rows and columns if necessary, an LP in standard form that can be solved by GUBT can be represented as in Figure 11.1. In Figure 11.1 all the coefficients outside of the blocks are zero.

Each of the thin blocks in the bottom of Figure 11.1 represents one GUB constraint. Let n_0 be the number of decision variables that do not appear in any GUB constraint at all. Let n_i be the number of variables that appear (i.e., appear with a nonzero coefficient) in the ith GUB constraint, $i = 1$ to p. Let $\mathbf{S}_0 = \{1, \ldots, n_0\}$ and $\mathbf{S}_t = \{n_0 + \cdots + n_{t-1} + 1, \ldots, n_0 + \cdots + n_{t-1} + n_t\}$, $t = 1$ to p. The variables x_j for $j \in \mathbf{S}_0$ are those that do not appear in any GUB constraint, and the variables x_j for $j \in \mathbf{S}_t$ are those that appear in the tth GUB constraint $t = 1$ to p. Let x^t be the vector of variables x_j for $j \in \mathbf{S}_t$, $t = 0$ to p.

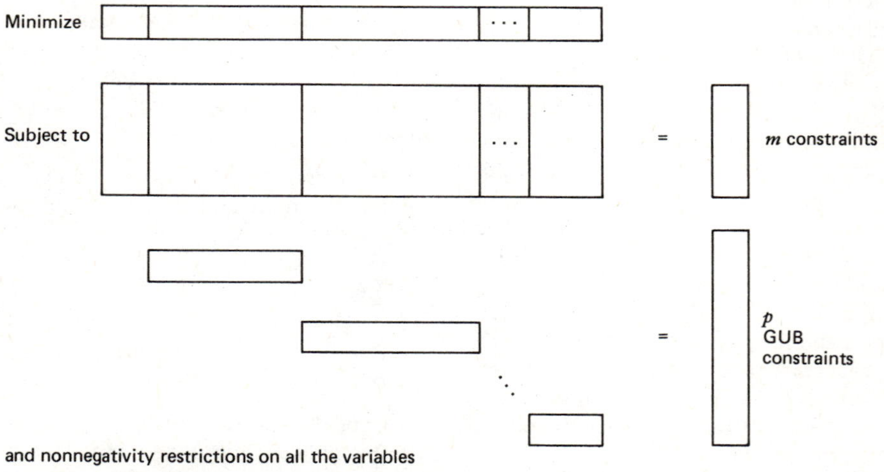

Figure 11.1 Structure of an LP model that can be solved by GUBT.

Then the LP under consideration here can be written in the following manner.

Minimize

$z(x) = c^0x^0 + c^1x^1 + c^2x^2 + \cdots + c^px^p$

Subject to

$$\begin{aligned} A^0x^0 + A^1x^1 + A^2x^2 + \ldots + A^px^p &= b \\ d^1x^1 \qquad\qquad\qquad &= d_1 \\ d^2x^2 \qquad\qquad &= d_2 \\ &\vdots \\ d^px^p &= d_p \\ x^t \geqq 0 \qquad \text{for all } t = 0 \text{ to } p & \end{aligned} \tag{11.9}$$

Here A^j is a matrix consisting of m rows and the appropriate number of columns, and d^j is a row vector of appropriate order for each j. Without any loss of generality, we assume that the systems of constraints in this model is linearly independent. Hence a basic vector for (11.9) consists of $m + p$ basic variables, and every basis for (11.9) is of order $m + p$.

Most textbooks that cover GUBT discuss it only for the special case when all the coefficients in all the GUB constraints are equal to one. Here we present the general version of GUBT, where each GUB constraint may be a general linear equation, and hence our discussion of GUB is slightly different from (and is more general than) that in other textbooks.

Note 11.6: Before applying GUBT, the constraints have to be rearranged so that the bottom p constraints are the GUB constraints, as in (11.9).

Notation Used in This Section

We use the phrase *the variable x_j is from the set* $\mathbf{S}_t$ to imply that $j \in \mathbf{S}_t$. We denote by $\boldsymbol{A}_{.j}$ the column vector of order $m + p$ of the coefficients of the variable x_j in the system of constraints in (11.9). The top m entries in $\boldsymbol{A}_{.j}$ come from the non-GUB constraints, and the bottom p entries in $\boldsymbol{A}_{.j}$ come from the GUB constraints in (11.9). Hence among the bottom p entries in each $\boldsymbol{A}_{.j}$, at most one of them is nonzero. We use the phrase $\boldsymbol{A}_{.j}$ *is a column from the set* $\mathbf{S}_t$ to imply that $j \in \mathbf{S}_t$. Also, if all the bottom p entries in $\boldsymbol{A}_{.j}$ are zero, $\boldsymbol{A}_{.j}$ must be a column from $\mathbf{S}_0$. If the $m + t$th entry in $\boldsymbol{A}_{.j}$ is nonzero, $\boldsymbol{A}_{.j}$ must be a column from $\mathbf{S}_t$.

For any real positive integer n, we denote the unit matrix of order n by I_n in this section.

Phase I and Phase II

To find a BFS for (11.9) we set up the associated Phase I problem with a full artificial basis and solve it first. The Phase I problem has the same GUB structure as the original problem and has an initial artificial feasible basis (feasible for the Phase I problem) to start with. At Phase I termination, if the original problem is not infeasible, a feasible basic vector for the original problem will be obtained, and starting with it, the original problem can be solved in Phase II. So we only discuss how to solve (11.9) starting with a known feasible basic vector for it. If Phase I has to be used, it can be solved by adopting the same type of procedure for it.

Computations to Be Performed in a Step of the Simplex Algorithm

Given a feasible basic vector for (11.9), the computations to be performed in the simplex algorithm using it are the following.

1 Compute the values of the basic variables in the BFS corresponding to it. This may be done by setting all the nonbasic variables equal to zero and then solving the remaining system for the values of the basic variables. In the usual implementation of the simplex algorithm, however, after the first iteration, the values of the basic variables in the current BFS are normally found out by updating the values of the basic variables in the BFS from the previous iteration.

2 Compute the dual basic solution corresponding to the present basis.

3 Compute the vector of relative cost coefficients using the current dual solution obtained in (2) and the original tableau that is stored.

4 Check the relative cost vector and determine whether all the relative cost coefficients are nonnegative (in which case the present BFS is optimal and the algorithm terminates), or select a variable associated with a strictly negative relative cost coefficient as the entering variable.

5 Compute the updated column vector of the entering variable, which is the representation of the original column vector of the entering variable in (11.9) as a linear combination of the basic column vectors.

6 If the updated column vector of the entering variable is nonpositive, the objective function is unbounded below in the problem and the algorithm terminates. Otherwise the minimum ratio test is performed and the dropping basic variable is determined.

7 Update the basic vector, the basis inverse, etc., and go to the next iteration, where the same operations are repeated with the new basic vector.

In solving (11.9) by GUBT starting with an initial feasible basic vector, a working basis of order m is first constructed from the initial basis. All the computations to be performed in this step are performed using the inverse of the working basis. After each change in the basic vector, the inverse of the working basis is updated.

Useful Results on Basic Vectors for (11.9)

THEOREM 11.4 Every basic vector for (11.9) must contain at least one basic variable x_j from the set $\mathbf{S}_t$ for each $t = 1$ to p.

Proof: Let $x_{\boldsymbol{B}}$ be a basic vector and $\boldsymbol{B}$ the corresponding basis for (11.9). Basis $\boldsymbol{B}$ is of order $m + p$ and nonsingular. Hence the $m + t$th row of $\boldsymbol{B}$ must contain at least one nonzero entry for each $t = 1$ to p. The coefficients in the $m + t$th row of (11.9) for all the variables x_j for $j \notin \mathbf{S}_t$ are zero. Hence $\boldsymbol{B}$ contains a nonzero entry in its $m + t$th row only if $x_{\boldsymbol{B}}$ contains at least one variable x_j for some $j \in \mathbf{S}_t$. Hence $x_{\boldsymbol{B}}$ must contain at least one variable from the set $\mathbf{S}_t$ for each $t = 1$ to p. ■

Essential and Inessential Sets

Let $x_{\boldsymbol{B}}$ be a basic vector for (11.9). For $1 \leqq t \leqq p$, the set $\mathbf{S}_t$ is said to be an *essential set with respect to the basic vector* $x_{\boldsymbol{B}}$, if $x_{\boldsymbol{B}}$ contains two or more basic variables from the set $\mathbf{S}_t$. For $1 \leqq t \leqq p$, the set $\mathbf{S}_t$ is said to be an *inessential set with respect to the basic vector* $x_{\boldsymbol{B}}$, if $x_{\boldsymbol{B}}$ contains exactly one basic variable from the set $\mathbf{S}_t$.

A set $\mathbf{S}_t$ that is an essential set with respect to one basic vector for (11.9) might become an inessential set with respect to another basic vector for (11.9). Also, notice that only the sets $\mathbf{S}_t$ for $1 \leqq t \leqq p$ are classified into essential and inessential sets. This classification does not apply to the set $\mathbf{S}_0$.

THEOREM 11.5 Let $x_{\boldsymbol{B}}$ be a basic vector for (11.9). The number of essential sets $\mathbf{S}_t$ with respect to $x_{\boldsymbol{B}}$ is at most m.

Proof: By Theorem 11.4, $x_{\boldsymbol{B}}$ must contain at least one basic variable from the set $\mathbf{S}_t$, for each $t = 1$ to p. Also there are exactly $m + p$ basic variables in $x_{\boldsymbol{B}}$. Hence the number of sets $\mathbf{S}_t$ from which $x_{\boldsymbol{B}}$ contains two or more variables can be at most m. ■

Key and Nonkey Variables

Let $x_{\boldsymbol{B}}$ be a basic vector for (11.9) and $\boldsymbol{B}$ the corresponding basis. By Theorem 11.4, $x_{\boldsymbol{B}}$ contains at least one variable from the set $\mathbf{S}_t$ for every $1 \leqq t \leqq p$. For each $1 \leqq t \leqq p$, select one basic variable in $x_{\boldsymbol{B}}$ from the set $\mathbf{S}_t$ and call it the *key variable from the set* $\mathbf{S}_t$ in the basic vector $x_{\boldsymbol{B}}$. If the set $\mathbf{S}_t$ is essential, then the set $\{x_j : j \in \mathbf{S}_t\}$ contains two or more basic variables in the basic vector $x_{\boldsymbol{B}}$, and the choice of the key variable among them can be arbitrary. Notice that if the set $\mathbf{S}_t$ is inessential, then the key variable in it is the only basic variable in that set. The working basis derived from $\boldsymbol{B}$ depends on the choice of the key variable from essential sets. The original column vector in (11.9) corresponding to a key variable is called a *key column in the basis.* The key column associated with the key variable from the set $\mathbf{S}_t$ is known as the tth key column in the basis, for $t = 1$ to p. Notice that key variables are only selected in sets $\mathbf{S}_t$ for $1 \leqq t \leqq p$, and there is no key variable from $\mathbf{S}_0$. Basic variables that are not key variables are called *nonkey variables*, and their associated columns are called *nonkey columns*. The total number of nonkey variables is m.

Derivation of the Working Basis

Let $x_{\boldsymbol{B}}$ be the initial feasible basic vector and $\boldsymbol{B}$ the corresponding basis for (11.9). Suppose the key variables in $x_{\boldsymbol{B}}$ from each set $\mathbf{S}_t$, $t = 1$ to p, have been selected. For the initial basis, the working basis is derived by the method discussed here.

Rearrange the variables in $x_{\boldsymbol{B}}$ (and correspondingly the columns in the basis $\boldsymbol{B}$), so that the key variables in the sets $\mathbf{S}_t$ appear first in the order $t = 1$ to p. After the key variables, arrange all the nonkey variables in the sets $\mathbf{S}_t$, $t = 0$ to p in some order. When arranged in this manner, the associated basis $\boldsymbol{B}$ can be partitioned as in equation (11.10):

$$\boldsymbol{B} = \left(\begin{array}{c|c} \overset{\text{Key columns in order } t = 1 \text{ to } p}{P} & \overset{\text{Nonkey columns}}{R} \\ \hline D & Q \end{array}\right) \begin{array}{l} \text{First } m \text{ rows} \\ \text{Last } p \text{ rows} \end{array} \tag{11.10}$$

where D is a nonsingular diagonal matrix or order $p \times p$, and Q is a matrix that contains at most one nonzero entry in each of its columns. Let

$$T = \begin{pmatrix} D^{-1} & -D^{-1}Q \\ 0 & I_m \end{pmatrix} \tag{11.11}$$

So T is a square nonsingular matrix of order $m + p$, constructed using D^{-1} and the submatrix Q in the basis $\boldsymbol{B}$, and it is upper triangular. Then

$$\boldsymbol{B}T = \begin{pmatrix} PD^{-1} & B \\ I_p & 0 \end{pmatrix} \tag{11.12}$$

where the matrix $B = -PD^{-1}Q + R$ is a square matrix of order m. If B is singular, then the set of last m columns of $\boldsymbol{B}T$ will be a linearly dependent set, and hence $\boldsymbol{B}T$ will be a singular matrix, which is a contradiction since $\boldsymbol{B}$ and T are both nonsingular matrices. Hence B is a nonsingular square matrix of order m. The matrix B is defined to be the *working basis*. Given the basis $\boldsymbol{B}$ and the choice of the key variables, $\boldsymbol{B}$ is partitioned by rearranging its columns as in (11.10), and the working basis B can be computed easily from $B = -PD^{-1}Q + R$. The working basis B can also be computed by the procedure discussed in the following section.

Procedure for Computing the Working Basis

As mentioned earlier, the constraints in the LP have to be arranged so that the bottommost p constraints are the GUB constraints. Also write down the basis in such a way that the first p columns in it are the key columns in the order $t = 1$ to p. After these rearrangements, let $\boldsymbol{B}$ be the basis. The last m columns in $\boldsymbol{B}$ are the nonkey columns. Starting with the first nonkey column in $\boldsymbol{B}$, do the following computation on each nonkey column. Suppose this nonkey column belongs to the set $\mathbf{S}_{t_1}$. If $t_1 = 0$, there is nothing to be done in this column, so go to the next nonkey column. If $1 \leqq t_1 \leqq p$, subtract a suitable multiple of the key column of this set (i.e., the t_1th key column) from this nonkey column, so that the entry in this nonkey column in the $m + t_1$th row is transformed into zero. Since each nonkey column contains at most one nonzero entry among the bottom p rows, after this subtraction, all the entries in this column among the bottom p rows will be equal to zero.

After completing the above operation in each nonkey column, for $t = 1$ to p, divide all the entries in the tth key column by the entry in it in the $(m + t)$th row.

It is clear that all the above operations on the basis have transformed it into the matrix $\boldsymbol{B}T$ as in equation (11.12). Hence the working basis is the submatrix contained among the first m rows and the last m columns of this resulting matrix.

Example 11.13

Here we only illustrate how to derive the working basis from a given basis. Hence in this illustration, only the basis is given, and not the whole LP. In this illustration $m = 4$, $p = 5$. Let $x_{\boldsymbol{B}} = (x_1, x_2, x_3, x_4, x_5, x_6, x_7, x_8, x_9)$ be the basic vector, where x_t for $t = 1$ to 5 is the tth key variable in this basic vector. Variables x_6, x_7, x_8, x_9 are the nonkey variables in this basic vector. Let the associated basis be

$$\boldsymbol{B} = \left(\begin{array}{ccccc|cccc} \multicolumn{5}{c|}{\text{Key columns}} & \multicolumn{4}{c}{\text{Nonkey columns}} \\ 1 & 2 & 2 & 4 & -1 & 2 & -\frac{7}{3} & 0 & 1 \\ 2 & 3 & 1 & 3 & -2 & 2 & 0 & 1 & 2 \\ 0 & 1 & 1 & 2 & -3 & 0 & 2 & 1 & 1 \\ 0 & 2 & 1 & -1 & 2 & 1 & 0 & 1 & 0 \\ \hline 3 & 0 & 0 & 0 & 0 & 1 & -2 & 0 & 0 \\ 0 & 2 & 0 & 0 & 0 & 0 & 0 & 4 & 0 \\ 0 & 0 & -1 & 0 & 0 & 0 & 0 & 0 & -3 \\ 0 & 0 & 0 & 4 & 0 & 0 & 0 & 0 & 0 \\ 0 & 0 & 0 & 0 & -5 & 0 & 0 & 0 & 0 \end{array}\right) \begin{array}{l} \\ \text{entries from non-GUB rows} \\ \\ \\ \\ \\ \text{Entries from GUB rows} \\ \\ \\ \end{array} \tag{11.13}$$

The sixth and seventh columns in the basis (i.e., the columns corresponding to x_6, x_7) contain nonzero entries in the first GUB row (i.e., the fifth row) in the basis, and hence x_6, x_7 must come from the set $\mathbf{S}_1$ in this problem. Among the basic variables x_1, x_6, x_7 from this set, x_1 has been chosen as the key variable. In a similar manner, it is clear that the nonkey variables x_8, x_9 associated with the eighth and nineth columns in the basis, respectively, are from the sets $\mathbf{S}_2$ and $\mathbf{S}_3$, respectively, in this problem. Also x_4, x_5 are the only basic variables from the sets $\mathbf{S}_4$, $\mathbf{S}_5$, respectively, in this problem. To get the working basis we have to do the following operations on B: subtract (1/3) times column 1 from column 6; subtract (−2/3) times column 1 from column 7; subtract (4/2) times column 2 from column 8; subtract (−3/4) times column 3 from column 9; divide columns 1, 2, 3, 4, 5 by 3, 2, −1, 4, −5, respectively. After these operations, $\boldsymbol{B}$ is transformed into the following matrix.

Use of the Inverse of the Wor ing Basis to Compute the Basic Feasible Solution, and the Updated Column Vectors

We now discuss how to solve a system of equations of the form

$$\boldsymbol{B}y = q \tag{11.14}$$

where $q = (q_1, \ldots, q_m, q_{m+1}, \ldots, q_{m+p})^{\mathrm{T}}$ is a given vector, using the inverse of the working basis B. Let $\xi = (\xi_1, \ldots, \xi_p, \xi_{p+1}, \ldots, \xi_{m+p})^{\mathrm{T}} = T^{-1}y$, where T is the transformation matrix discussed earlier in equation (11.11). Then the system of equations (11.14) is equivalent to $\boldsymbol{B}T(T^{-1}y) = q$; that is,

$$\left(\begin{array}{c|c} PD^{-1} & B \\ \hline I_p & 0 \end{array}\right) \xi = q \tag{11.15}$$

that is, $(\xi_1, \ldots, \xi_p)^{\mathrm{T}} = (q_{m+1}, \ldots, q_{m+p})^{\mathrm{T}}$, and $(\xi_{p+1}, \ldots, \xi_{m+p})^{\mathrm{T}} = B^{-1}((q_1, \ldots, q_m)^{\mathrm{T}} - PD^{-1}(q_{m+1}, \ldots, q_{m+p})^{\mathrm{T}})$. Hence using the information in the basis $\boldsymbol{B}$,

Key | Nonkey; Non-GUB rows (first four), GUB rows (last five)

$$\boldsymbol{B}T = \left(\begin{array}{ccccc|cccc} \frac{1}{3} & 1 & -2 & 1 & \frac{1}{5} & \frac{5}{3} & -\frac{5}{3} & -4 & -5 \\ \frac{2}{3} & \frac{3}{2} & -1 & \frac{3}{4} & \frac{2}{5} & \frac{4}{3} & \frac{4}{3} & -5 & -1 \\ 0 & \frac{1}{2} & -1 & \frac{2}{4} & \frac{3}{5} & 0 & 2 & -1 & -2 \\ 0 & 1 & -1 & -\frac{1}{4} & -\frac{2}{5} & 1 & 0 & -3 & -3 \\ \hline 1 & 0 & 0 & 0 & 0 & 0 & 0 & 0 & 0 \\ 0 & 1 & 0 & 0 & 0 & 0 & 0 & 0 & 0 \\ 0 & 0 & 1 & 0 & 0 & 0 & 0 & 0 & 0 \\ 0 & 0 & 0 & 1 & 0 & 0 & 0 & 0 & 0 \\ 0 & 0 & 0 & 0 & 1 & 0 & 0 & 0 & 0 \end{array}\right) \begin{array}{l} \\ \\ \text{Non-GUB} \\ \\ \\ \\ \text{GUB} \\ \\ \\ \end{array}$$

Hence the working basis here is B given as follows:

$$B = \begin{pmatrix} \frac{5}{3} & -\frac{5}{3} & -4 & -5 \\ \frac{4}{3} & \frac{4}{3} & -5 & -1 \\ 0 & 2 & -1 & -2 \\ 1 & 0 & -3 & -3 \end{pmatrix} \qquad B^{-1} = \begin{pmatrix} 84 & 33 & 48 & -183 \\ 15 & 6 & 9 & -33 \\ 26 & 10 & 15 & -\frac{170}{3} \\ 2 & 1 & 1 & -\frac{14}{3} \end{pmatrix}$$

the vector q, and B^{-1}, the inverse of the working basis, the vector $(\xi_1, \ldots, \xi_{m+p})^{\mathrm{T}}$ can be computed as in (11.15). Having computed $(\xi_1, \ldots, \xi_{m+p})^{\mathrm{T}}$, the solution y of (11.14) is obtained from $y = T(\xi_1, \ldots, \xi_{m+p})^{\mathrm{T}}$.

Example 11.14

Let $m = 4$, $p = 5$, and $x_{\boldsymbol{B}}$, $\boldsymbol{B}$ be the basic vector and the associated basis, respectively, given in Example 11.13, with the key columns and nonkey columns as selected there. Then PD^{-1} is the $m \times p$ matrix in the upper left corner of $\boldsymbol{B}T$ (i.e., the submatrix contained among the non-GUB rows and key columns in $\boldsymbol{B}T$). The transformation matrix T is

$$T = \left(\begin{array}{ccccc|cccc}
\frac{1}{3} & 0 & 0 & 0 & 0 & -\frac{1}{3} & \frac{2}{3} & 0 & 0 \\
0 & \frac{1}{2} & 0 & 0 & 0 & 0 & 0 & -2 & 0 \\
0 & 0 & -1 & 0 & 0 & 0 & 0 & 0 & -3 \\
0 & 0 & 0 & \frac{1}{4} & 0 & 0 & 0 & 0 & 0 \\
0 & 0 & 0 & 0 & -\frac{1}{5} & 0 & 0 & 0 & 0 \\
\hline
0 & 0 & 0 & 0 & 0 & 1 & 0 & 0 & 0 \\
0 & 0 & 0 & 0 & 0 & 0 & 1 & 0 & 0 \\
0 & 0 & 0 & 0 & 0 & 0 & 0 & 1 & 0 \\
0 & 0 & 0 & 0 & 0 & 0 & 0 & 0 & 1
\end{array}\right)$$

Suppose we wish to solve (11.14) with $q = (20, 18, 8, 3, 4, 4, 2, 12, 0)^{\mathrm{T}}$. Let $y = (y_j\colon j = 1 \text{ to } 9)^{\mathrm{T}}$ and $\xi = (\xi_j\colon j = 1 \text{ to } 9)^{\mathrm{T}} = T^{-1}y$. From the above method we get $(\xi_1, \xi_2, \xi_3, \xi_4, \xi_5)^{\mathrm{T}} = (4, 4, 2, 12, 0)^{\mathrm{T}}$ and $(\xi_6, \xi_7, \xi_8, \xi_9)^{\mathrm{T}} = B^{-1}((20, 18, 8, 3)^{\mathrm{T}} - PD^{-1}(4, 4, 2, 12, 0)^{\mathrm{T}}) = B^{-1}(20/3, 7/3, 2, 4)^{\mathrm{T}} = (1, 0, 0, -1)^{\mathrm{T}}$. Hence $\xi = (4, 4, 2, 12, 0, 1, 0, 0, -1)^{\mathrm{T}}$ and $y = T\xi = (1, 2, 1, 3, 0, 1, 0, 0, -1)^{\mathrm{T}}$ is the solution of (11.14) in this case.

The system of equations to compute the BFS corresponding to a given feasible basis for (11.9) and the system of equations to compute the updated column vector of the entering variable in a step of the simplex algorithm are both of the type (11.14), and hence they can both be solved by the method discussed here using the inverse of the working basis.

Use of the Inverse of the Working Basis to Compute the Dual Solution

Let $x_{\boldsymbol{B}}$ be a given basic vector for (11.9), $\boldsymbol{B}$, the associated basis, and $c_{\boldsymbol{B}}$, the row vector of the original cost coefficients corresponding to the basic vector $x_{\boldsymbol{B}}$ in (11.9). Let $\pi = (\pi_1, \ldots, \pi_m, \pi_{m+1}, \ldots, \pi_{m+p})$ be the dual basic solution of (11.9) corresponding to the basis $\boldsymbol{B}$. Then π is obtained by solving $\pi\boldsymbol{B} = c_{\boldsymbol{B}}$, or equivalently $\pi\boldsymbol{B}T = c_{\boldsymbol{B}}T$, that is

$$\pi\left(\begin{array}{c|c} PD^{-1} & B \\ \hline I_p & 0 \end{array}\right) = v = c_{\boldsymbol{B}}T \qquad (11.16)$$

So $v = (v_1, \ldots, v_p, v_{p+1}, \ldots, v_{m+p}) = c_{\boldsymbol{B}}T$ is easily computed. Then from (11.16) we see that

$$\begin{aligned}
(\pi_1, \ldots, \pi_m) &= (v_{p+1}, \ldots, v_{m+p})B^{-1} \\
(\pi_{m+1}, \ldots, \pi_{m+p}) &= (v_1, \ldots, v_p) \\
&\quad -(\pi_1, \ldots, \pi_m)PD^{-1}
\end{aligned} \qquad (11.17)$$

Hence, by using the data in $\boldsymbol{B}$, $c_{\boldsymbol{B}}$, and B^{-1}, π, the dual basic solution corresponding to the basis $\boldsymbol{B}$, is easily computed as in (11.17).

Example 11.15

Let $m = 4$, $p = 5$, and $\boldsymbol{B}$ be the basis given in Example 11.13, with the key and nonkey columns as selected there. Let $c_{\boldsymbol{B}} = (-1, -1, 2, -5, 5, 2, -7/3, -7, -4)$. $\pi = (\pi_i\colon i = 1 \text{ to } 9)$. We compute $v = c_{\boldsymbol{B}}T = (-1/3, -1/2, -2, -5/4, -1, 7/3, -3, -5, -10)$. Hence by the above procedure $(\pi_1, \pi_2, \pi_3, \pi_4) = (7/3, -3, -5, -10)B^{-1} = (1, -1, 0, 2)$; $(\pi_5, \pi_6, \pi_7, \pi_8, \pi_9) = (-1/3, -1/2, -2, -5/4, -1) - (1, -1, 0, 2)PD^{-1} = (0, 2, 1, -1, 0)$. Hence the dual basic solution π here is $(1, -1, 0, 2, 0, -2, 1, -1, 0)$.

How to Do the Other Computations Using the Inverse of the Working Basis

Now compute $\bar{c}_j$, the relative cost coefficient of the variable x_j with respect to the basis $\boldsymbol{B}$, from the usual formula $\bar{c}_j = c_j - \pi\boldsymbol{A}_{.j}$. If all $\bar{c}_j$ are nonnegative, $\boldsymbol{B}$ is an optimum basis and the present BFS is an optimum solution of (11.9). Terminate. Otherwise select a nonbasic variable x_s, where s is such that $\bar{c}_s < 0$, as the entering variable into the basic vector $x_{\boldsymbol{B}}$. $\boldsymbol{A}_{.s}$ is the original column vector corresponding to x_s in (11.9). The updated column vector of x_s is $\bar{A}_{.s}$, which is obtained by solving the system of equations $\boldsymbol{B}\bar{\boldsymbol{A}}_{.s} = \boldsymbol{A}_{.s}$. This system of equations for obtaining $\bar{\boldsymbol{A}}_{.s}$ is similar to the system (11.14), and hence can be solved using B^{-1}, the inverse of the working basis, as discussed earlier.

Having obtained the updated column vector of the entering variable $\bar{\mathbf{A}}_{.s}$, if $\bar{\mathbf{A}}_{.s} \leqq 0$, then z is unbounded below on the set of feasible solutions of (11.9). Terminate. Otherwise determine the dropping variable in $x_{\mathbf{B}}$ by the usual minimum ratio test. Denote the vector of basic values in the BFS of (11.9) corresponding to the present basic vector by $\bar{q}$. The minimum ratio in this step is: $\theta = \text{minimum } \{\bar{q}_i/\bar{a}_{is}$: $i = 1$ to $m + p$, and i, such that $\bar{a}_{is} > 0\}$. If the minimum ratio is attained by $i = r$, the rth basic variable in $x_{\mathbf{B}}$ is the dropping variable, and x_s replaces it to yield the new basic vector. Ties for the $i = r$, which attains the minimum ratio can be broken arbitrarily.

Updating the Basic Values
When x_s replaces the rth basic variable in the basic vector $x_{\mathbf{B}}$, we get the next basic vector. The vector of values of the basic variables in the BFS corresponding to the new basic vector can be updated from: the value of x_s in the new BFS $= \theta$, and the value of $x_{\mathbf{B}}$ in the new BFS $= \bar{q} - \theta\bar{\mathbf{A}}_{.s}$.

Updating the Inverse of the Working Basis
Let $x_{\mathbf{B}}$ be the present feasible basic vector and $\mathbf{B}$ be the associated basis for (11.9), and B the present working basis. Let x_s be the entering variable into the basic vector $x_{\mathbf{B}}$ and x_r be the basic variable dropping out of $x_{\mathbf{B}}$. Let σ be such that $\mathbf{S}_\sigma$ contains s, and let ρ be such that $r \in \mathbf{S}_\rho$. Hence the entering variable in this pivot step is from the set $\mathbf{S}_\sigma$ and the dropping variable is from the set $\mathbf{S}_\rho$. There are several cases to be considered here.

Case 1 The Dropping Variable x_r Is not a Key Variable In this case, the column leaving the basis $\mathbf{B}$ is one of the last m columns of $\mathbf{B}$. Suppose x_r is the $p + g$th variable in $x_{\mathbf{B}}$ in the order in which the basic variables are recorded in $x_{\mathbf{B}}$. So the column of x_r is the $(p + g)$th column in $\mathbf{B}$. This column is to be replaced by the column of x_s in (11.9). Let the current basis in partitioned form be the one given in (11.10).

$$\text{Leaving column} = g\text{th column of} \begin{pmatrix} R \\ \cdots \\ Q \end{pmatrix},$$

$$\text{Entering column} = \begin{pmatrix} A_{.s} \\ \cdots \\ U_{.s} \end{pmatrix}$$

where the entering column is the original column vector of the entering variable x_s, and $U_{.s}$ contains at most one nonzero entry (if $\sigma \neq 0$, since $s \in \mathbf{S}_\sigma$, the σth entry in $U_{.s}$ is nonzero. If $\sigma = 0$, $U_{.s} = 0$). The new working basis is obtained by replacing the gth column in B by $A_{.s} - PD^{-1}U_{.s}$. Let $h = (h_1, \ldots, h_m)^{\mathrm{T}} = B^{-1}(A_{.s} - PD^{-1}U_{.s})$. Hence, in this case, the inverse of the working basis can be updated by entering the column vector h on the right-hand side of the current B^{-1}, and then performing a pivot with h as the pivot column and the gth row as the pivot row.

Note 11.7 Notice that in this case, the vector h, which is the pivot column for updating the inverse of the working basis, is computed using the original column vector of the entering variable x_s in (11.9), and not the updated column vector of x_s.

Example 11.16

Let $m = 4$, $p = 5$, and $x_{\mathbf{B}}$, $\mathbf{B}$ be the same as those in Example 11.13. Suppose the variable x_{10}, associated with its original column vector $\mathbf{A}_{.10} = (-1/3, 5/3, 6, 2;$ $0, 2, 0, 0, 0)$, is the entering variable into the basic vector $x_{\mathbf{B}}$. Clearly, x_{10} is from the set $\mathbf{S}_2$ in this problem. Suppose the dropping variable from $x_{\mathbf{B}}$ is x_9. The new basic vector will be $x_{\mathbf{B}_1} = (x_1, x_2, x_3, x_4, x_5,$ $x_6, x_7, x_8, x_{10})$. And $h = (h_1, h_2, h_3, h_4) = B^{-1}(-1/3,$ $5/3, 6, 2)^{\mathrm{T}} - PD^{-1}(0, 2, 0, 0)^{\mathrm{T}} = B^{-1}(-7/3, -4/3, 5,$ $0)^{\mathrm{T}} = (0, 2, 1, -1)^{\mathrm{T}}$. Since the dropping variable is the fourth nonkey variable in $x_{\mathbf{B}}$, the pivot row is the fourth row.

	B^{-1}			Pivot Column h
84	33	48	-183	0
15	6	9	-33	2
26	10	15	$-\frac{170}{3}$	1
2	1	1	$-\frac{14}{3}$	$\boxed{-1}$

Performing the pivot we get

	B_1^{-1}			Pivot Column h Becomes
84	33	48	-183	0
19	8	11	$-\frac{127}{3}$	0
28	11	16	$-\frac{184}{3}$	0
-2	-1	-1	$\frac{14}{3}$	1
				and is discarded after the pivot

Hence B_1^{-1}, the inverse of the new working basis, is the matrix on the left-hand side of the above tableau.

Case 2 The Dropping Variable Is the Key Variable in an Essential Set In this case, the set $\mathbf{S}_\rho$ that contains the dropping variable x_r is an essential set with respect to the present basic vector $x_{\mathbf{B}}$. The entering variable x_s is from the set $\mathbf{S}_\sigma$, and ρ may or may not be the same as σ.

In this case, by the hypothesis, the set $\mathbf{S}_\rho$ contains some nonkey variables in the present basic vector $x_{\boldsymbol{B}}$. The updating of the inverse of the working basis is done in two stages in this case. In stage 1 we exchange the current key variable x_r in the set $\mathbf{S}_\rho$ with some nonkey variable in the same set in the present basic vector $x_{\boldsymbol{B}}$. After this rearrangement, the basic vector $x_{\boldsymbol{B}}$ becomes $x_{\boldsymbol{B}_0}$, say. Notice that the set of basic variables in both $x_{\boldsymbol{B}}$ and $x_{\boldsymbol{B}_0}$ are the same; they just appear in different orders in the two vectors. Let $\boldsymbol{B}_0$ be the basis associated with the basic vector $x_{\boldsymbol{B}_0}$. Notice that $\boldsymbol{B}_0$ is obtained from $\boldsymbol{B}$ by interchanging the ρth key column vector in $\boldsymbol{B}$ with the column vector of the nonkey basic variable from the set $\mathbf{S}_\rho$, which is made the new key variable in this set. Since the key columns in $\mathbf{B}$ and $\mathbf{B}_0$ are different, the working basis B_0, computed from $\boldsymbol{B}_0$, may be different from the present working basis B. In stage 1 we compute B_0^{-1}, the inverse of this working basis B_0 associated with the basis $\boldsymbol{B}_0$. In stage 2 we replace x_r, which is now a nonkey variable in the rearranged basic vector $x_{\boldsymbol{B}_0}$, by x_s.

Stage 1 Since x_r is the current key column from the $\mathbf{S}_\rho$, the ρth column of $\boldsymbol{B}$ is the column of x_r in (11.9). Let the columns $p + j_1, \ldots, p + j_f$ of $\boldsymbol{B}$ be all the nonkey columns from the set $\mathbf{S}_\rho$ in $\boldsymbol{B}$. Let the entries in the $m + \rho$th row and columns ρ, $p + j_1$, $p + j_2, \ldots, p + j_f$ of $\boldsymbol{B}$ be, respectively, d_0^ρ, d_1^ρ, $d_2^\rho, \ldots, d_f^\rho$. Clearly, these are the only nonzero entries in the $m + \rho$th row of $\boldsymbol{B}$.

Select one of the nonkey columns in $\boldsymbol{B}$ from the set $\mathbf{S}_\rho$, to be exchanged with the current key column from $\mathbf{S}_\rho$. Suppose it is the column $p + j_1$. In this stage we want to exchange the columns ρ and $p + j_1$ in $\boldsymbol{B}$. Define the following elementary matrix T_2 of order $m \times m$

$$T_2 = \begin{pmatrix} & & \text{Column} & & \text{Column} & & \text{Column} & \\ & & J_2 & & J_1 & & J_3 & \\ 1 & 0\cdots & 0 & \cdots 0 & 0 & 0\cdots & 0 & \cdots 0 \\ 0 & 1\cdots & 0 & \cdots 0 & 0 & 0\cdots & 0 & \cdots 0 \\ \vdots & \vdots & \ddots & \vdots & \vdots & \vdots & \vdots & \vdots \\ 0 & 0\cdots & 0 & \cdots 1 & 0 & 0\cdots & 0 & \cdots 0 \\ 0 & 0\cdots & -\dfrac{d_2^\rho}{d_1^\rho} & \cdots 0 & -\dfrac{d_0^\rho}{d_1^\rho} & 0\cdots & -\dfrac{d_3^\rho}{d_1^\rho} & \cdots 0 \\ 0 & 0\cdots & 0 & \cdots 0 & 0 & 1\cdots & 0 & \cdots 0 \\ \vdots & \vdots & \vdots & \vdots & \vdots & \vdots & & \vdots \\ 0 & 0\cdots & 0 & \cdots 0 & 0 & 0\cdots & & \cdots 1 \end{pmatrix}$$

Matrix T_2 is the same as the identity matrix of order m, except for its row j_1 (j_1 is defined by the nonkey column $p + j_1$, in the current basis $\boldsymbol{B}$, which is being exchanged with the key column ρ). The diagonal entry in row j_1 of T_2 is $(-d_0^\rho/d_1^\rho)$. The entries in column j_u, row j_1 of T_2 are $(-d_u^\rho/d_1^\rho)$, for $u = 2$ to f. All other entries in T_2 in row j_1 are zero.

By going over the procedure for generating the working basis carefully, it can be verified that exchanging the columns ρ and $p + j_1$ in $\boldsymbol{B}$ has the effect of replacing the working basis B by $B_0 = BT_2$. Hence the effect of this exchange of key and nonkey columns from the set $\mathbf{S}_\rho$ is to change the inverse of the working basis from B^{-1} to $B_0^{-1} = T_2^{-1}B^{-1}$.

Since T_2 is an elementary matrix it can be verified that T_2^{-1} is obtained from T_2 by dividing all the elements in its row j_1 other than the diagonal element by the negative of the diagonal element in row j_1 and then replacing the diagonal element in row j_1 by its own inverse.

Example 11.17

Let $m = 4$, $p = 5$, and consider the basic vector $x_{\boldsymbol{B}}$ and the associated basis $\boldsymbol{B}$ discussed in Example 11.14. Suppose the variable x_{10} has to be brought into the basic vector $x_{\boldsymbol{B}}$ to replace x_1. Variable x_1 is the key variable in set $\mathbf{S}_1$, and x_6, x_7 are the nonkey variables in $x_{\boldsymbol{B}}$ from this same set $\mathbf{S}_1$. In this case, we first make x_1 a nonkey variable and choose x_6 or x_7 as the new key variable in the set $\mathbf{S}_1$. Suppose we decide on x_7 as the key variable from the set $\mathbf{S}_1$ to replace x_1. After this rearrangement the basic vector $x_{\boldsymbol{B}}$ becomes $x_{\boldsymbol{B}_0} = (x_7, x_2, x_3, x_4, x_5, x_6, x_1, x_8, x_9)$ with the associated basis $\boldsymbol{B}_0$, which is obtained by interchanging columns 7 and 1 in $\boldsymbol{B}$. The matrix B^{-1}, the inverse of the working basis with the original choice of key variables in $x_{\boldsymbol{B}}$, was obtained in Example 11.14. The elementary matrix T_2, as defined above, for exchanging x_1 and x_7 is given below, and also its inverse.

$$T_2 = \begin{pmatrix} 1 & 0 & 0 & 0 \\ \frac{1}{2} & \frac{3}{2} & 0 & 0 \\ 0 & 0 & 1 & 0 \\ 0 & 0 & 0 & 1 \end{pmatrix} \qquad T_2^{-1} = \begin{pmatrix} 1 & 0 & 0 & 0 \\ -\frac{1}{3} & \frac{2}{3} & 0 & 0 \\ 0 & 0 & 1 & 0 \\ 0 & 0 & 0 & 1 \end{pmatrix}$$

So B_0^{-1}, the inverse of the new working basis, is

$$B_0^{-1} = T_2^{-1}B^{-1} = \begin{pmatrix} 84 & 33 & 48 & -183 \\ -\frac{54}{3} & -7 & -10 & 39 \\ 26 & 10 & 15 & -\frac{170}{3} \\ 2 & 1 & 1 & -\frac{14}{3} \end{pmatrix}$$

This completes stage 1 in this pivot step.

Stage 2 Now x_r is a nonkey variable in the basic vector x_{B_0}. Now replace x_r by x_s in x_{B_0}, as under Case 1.

Case 3 The Dropping Variable Is the Key Variable in an Inessential Set By Theorem 11.4, every basic vector from (11.9) must contain at least one variable from each of the sets $\mathbf{S}_t$ for $1 \leqq t \leqq p$. Since the dropping variable x_r is the only basic variable from the set $\mathbf{S}_\rho$ in the present basic vector x_B, the entering variable x_s must also be from the same set $\mathbf{S}_\rho$, as otherwise the vector obtained by replacing x_r by x_s in x_B will not be a basic vector for (11.9). Hence $\rho = \sigma$ in this case.

From the procedure for computing the working basis, it can be verified that no changes occur in the working basis as a result of this pivot step. Hence B^{-1}, the inverse of the working basis, remains unchanged in this case.

Continuing the Algorithm

After the basic vector, basic solution, and the inverse of the working basis are updated, we perform the computations all over again with the new basic vector. This is repeated until termination occurs.

Advantage of GUB Over the Usual Revised Simplex Method

The major computational effort in a step of the simplex method is that of updating the inverse of the basis. To solve (11.9) by the usual revised simplex method requires the updating of the inverse of order $(m + p)$ in each step. GUBT enables us to solve (11.9) by updating the inverse of a working basis of order m in each step. Also, to solve (11.9) by the usual revised simplex method, we have to store the inverse of the basis that is a square matrix of order $m + p$. Solving the same problem using GUBT requires the storage of the inverse of the working basis, which is a square matrix of order m only. Hence, to solve LPs with the structure of (11.9), GUBT is computationally much more efficient than the usual revised simplex method.

Methods for solving LPs that exploit the special structure of the coefficient matrix of the LP, and thus enable us to carry out all the computations in the simplex algorithm by maintaining the inverse of a working basis that is of much smaller order than the order of a basis of this LP, are known as *compact inverse methods*. GUBT is an example of a compact inverse method. The area dealing with methods for solving LPs by taking advantage of the special structure of the coefficient matrix is known as *structured linear programming*. LPs involving a large number of constraints and variables are known as *large-scale linear programs*. Most large-scale linear programs encountered in practical applications tend to have coefficient matrices in which the proportion of nonzero coefficients is very very small, and these nonzero coefficients appear in the coefficient matrix according to some special structure. *Large-scale linear programming* is the area dealing with efficient methods for solving such large-scale linear programs.

Computer codes for solving LPs using GUBT are currently available commercially. An LP in which the constraints and variables are arranged as in Figure 11.1, or as in (11.9), is said to have the GUB *structure*. For solving LPs with the GUB structure, a computer code based on GUBT is likely to be much more efficient than one based on the usual revised simplex method.

An LP in which the constraints and the variables are arranged as below is known to have the *block angular structure*.

$$\begin{array}{lrcl}
\text{Minimize} & z = c^0x^0 + c^1x^1 + c^2x^2 + \cdots + c^px^p & & \\
\text{Subject to} & A^0x^0 + A^1x^1 + A^2x^2 + \cdots + A^px^p & = & b \\
& D^1x^1 \qquad\qquad\qquad & = & d^1 \\
& D^2x^2 \qquad\qquad & = & d^2 \\
& \ddots \qquad & & \vdots \\
& D^px^p & = & d^p \\
& x^t \geqq 0 \quad \text{for all } t = 1 \text{ to } p & &
\end{array}$$

where D^t is a matrix of order $m_t \times n_t$ and rank m_t for $t = 1$ to p and A^t is a matrix of order $m_0 \times n_t$ for $t = 0$ to p, and the rest of the vectors have appropriate dimensions. An extension of GUBT to solve such problems by maintaining and updating a working basis of order m_0, and basis matrices in each D^t for $t = 1$ to p, has been developed in the papers [11.11 and 11.8] of M. Sakarovitch and R. Saigal, and R. N. Kaul.

Exercises

11.1 By using upper-bounding techniques, find a feasible solution of the following:

x_1	x_2	x_3	x_4	x_5	b
1	2	0	−1	2	2
2	−1	2	−5	2	0

$0 \leqq x_j \leqq 1$ for all j.

11.2 Solve

$$\begin{array}{ll} \text{Minimize} & -12x_1 - 12x_2 - 9x_3 - 15x_4 - 90x_5 - 26x_6 - 112x_7 \\ \text{Subject to} & 3x_1 + 4x_2 + 3x_3 + 3x_4 + 15x_5 + 13x_6 + 16x_7 \leqq 35 \\ & 0 \leqq x_j \leqq 1 \quad \text{for all } j \end{array}$$

11.3 Solve the following bounded variable LPs:

(a)

x_1	x_2	x_3	x_4	x_5	x_6	x_7	$-z$	b
1	0	0	1	2	−1	1	0	13
0	1	0	−1	1	1	2	0	9
0	0	1	2	2	2	−1	0	5
3	4	−2	−5	3	2	−1	1	0

$0 \leqq x_j \leqq 5$ for all j; z to be minimized.

(b)

x_1	x_2	x_3	x_4	x_5	x_6	x_7	$-z$	b
1	0	0	2	−2	1	−8	0	0
0	1	0	1	1	−1	1	0	11
0	0	1	3	−1	−2	2	0	6
1	2	1	−1	2	1	−1	1	0

$0 \leqq x_j \leqq 4$ for all j; z to be minimized.

11.4 Discuss how to perform marginal analysis for the right-hand-side constants b_i and the bounds l_j, and k_j in the LP (11.1) when it is solved using working bases as in this chapter.

11.5 Consider the bounded variable LP (11.1). Let $\bar{x}$ and $\hat{x}$ be two BFSs for (11.1) associated with the partitions ($\bar{\mathbf{B}}$, $\bar{\mathbf{L}}$, $\bar{\mathbf{U}}$) and ($\hat{\mathbf{B}}$, $\hat{\mathbf{L}}$, $\hat{\mathbf{U}}$), respectively. Develop necessary and sufficient conditions for $\bar{x}$ and $\hat{x}$ to be adjacent extreme points on the set of feasible solutions of (11.1) in terms of the partitions associated with them.

11.6 Develop an implementation of GUBT to solve (11.9) with additional upper-bound constraints on all the variables of the form $x_j \leqq U_j$ for each j (some of the U_j's may be $+\infty$).

11.7 Consider the following LP, where A is a given matrix of order $m \times n$, and b', b'' are given vectors of appropriate dimension satisfying $b' \leqq b''$. Because of the special nature of the constraints, LPs of this type are known as *interval linear programming problems.* (1) Show that it can be transformed into a bounded variable LP involving m equality constraints in at most $m + n$ variables, and bound restrictions on all the variables. (2) Develop an efficient algorithm for solving it that makes use of its special structure. (See references [11.4, 11.6, 11.10]).

$$\begin{array}{ll} \text{Minimize} & z(x) = cx \\ \text{Subject to} & b' \leqq Ax \leqq b'' \\ & 0 \leqq \ \ x \leqq U \end{array}$$

11.8 Let $n, m \geqq 3$. Consider the following system of constraints in $(mn + 1)$ variables, where $a_{ij} > 0$ for all i, j. Let $\mathbf{X}_j = \{x_{1j}, x_{2j}, \ldots, x_{mj}\}$. Introduce slacks s_j, $j = 1$ to n to convert the inequality constraints into equations. In any feasible basic vector for this problem, show that at most two variables in $\mathbf{X}_j$ can be basic, and that this can occur for at most one value of j. Also show that in any feasible basic vector, if two variables from some $\mathbf{X}_j$ are basic, then $s_j = 0$ in the corresponding BFS, and the y must also be a basic variable in that basic vector.

$$\sum_{i=1}^{m} x_{ij} = 1 \qquad \text{for } j = 1 \text{ to } n$$

$$\sum_{i=1}^{m} a_{ij}x_{ij} - y \geqq 0 \qquad \text{for } j = 1 \text{ to } n$$

$$x_{ij} \geqq 0 \qquad \text{for all } i, j;\ y \geqq 0$$

11.9 Consider the following LP, in which b, a_j, and c_j are all > 0 for all j. Besides the bound restrictions on these variables, there is only one additional constraint. (1) If $\tilde{x}$ is an optimum solution for this problem, prove that there exists a $\tilde{\pi}$ such that $\tilde{\pi}a_j < c_j$ implies $\tilde{x}_j = \ell_j$, $\tilde{\pi}a_j > c_j$ implies $\tilde{x}_j = k_j$, and $\ell_j < \tilde{x}_j < k_j$ implies $\tilde{\pi}a_j = c_j$. (2) Suppose the variables are listed in the problem in such a way that $(c_1/a_1) < (c_2/a_2) < \cdots < (c_n/a_n)$. Prove that if a feasible solution exists, then there is an optimum solution $\bar{x}$ satisfying the property that for some s: $\bar{x}_j = k_j$, $j = 1$ to $s - 1$, $\ell_s \leqq \bar{x}_s \leqq k_s$, and $\bar{x}_j = \ell_j$ for $j = s + 1$ to n.

$$\begin{array}{ll} \text{Minimize} & \sum_{j=1}^{n} c_j x_j \\ \text{Subject to} & \sum_{j=1}^{n} a_j x_j = b \\ & \ell_j \leqq x_j \leqq k_j \qquad j = 1 \text{ to } n \end{array}$$

REFERENCES

11.1 A. V. Aho, J. E. Hopcroft, and J. D. Ullman, *The Design and Analysis of Computer Algorithms,* 3rd printing, Addison-Wesley, Reading, Mass., 1976.

11.2 R. D. Armstrong and J. W. Hultz, "A Computational Study on Solving a Facility Location Problem with Least Absolute Value Criterion," Abstracts for the *International Symposium on Extremal Methods and Systems Analysis,* Austin, 1977.

11.3 I. Barrodale and F. D. K. Roberts, "An Improved Algorithm for Discrete ℓ_1 Linear Approximation," *SIAM Journal of Numerical Analysis 10* (1973), 839–848.

11.4 A. Ben-Israel and A. Charnes, "An Explicit Solution of a Special Class of Linear Programming Problems," *Operations Research 16* (1968), 1166–1175.

11.5 G. G. Brown and D. S. Thomen, "Automatic Identification of Generalized Upper Bounds in Large-Scale Optimization Models," *Management Science 26*, 11 (November 1980), 1166–1184.

11.6 A. Charnes, F. Granot, and F. Phillips, "An Algorithm for Solving Interval Linear Programming Problems," *Operations Research 25*, 4 (July/August 1977), 688–695.

11.7 G. B. Dantzig and R. M. Vanslyke, "Generalized Upper Bounding Techniques," *Journal of Computer and System Sciences 1*, 3 (October 1967), 213–226.

11.8 R. N. Kaul, "An Extension of Generalized Upper Bounding Techniques for Linear Programming," Technical Report ORC 65–27, Operations Research Center, University of California, Berkeley, Calif., 1965.

11.9 K. G. Murty, "Resolution of Degeneracy in the Bounded Variable Primal Simplex Algorithm," Technical Report 78-1, Department of Industrial and Operations Engineering, The University of Michigan, Ann Arbor, 1978.

11.10 P. D. Robers and A. Ben-Israel, "Interval Programming: New Approach to Linear Programming with Applications to Chemical Engineering Problems," *I & EC Process Design Development 8* (1969), 496–501.

11.11 M. Sakarovitch and R. Saigal, "An Extension of Generalized Upper Bounding Techniques for Structured Linear Programs," *SIAM Journal of Applied Mathematics 15*, 4 (July 1967), 906–914.

11.12 L. Schrage, "Implicit Representation of Variable Upper Bounds in Linear Programming," *Mathematical Programming Study, 4* (1975), 118–132.

11.13 L. Schrage, "Implicit Representation of Generalized Variable Upper Bounds in Linear Programming," *Mathematical Programming, 14*, 1 (1978), 11–20.

11.14 M. J. Todd, "An Implementation of the Simplex Method for Linear Programming Problems with Variable Upper Bounds," *Mathematical Programming, 23*, 1 (1982), 34–49.

11.15 H. M. Wagner, "The Dual Simplex Algorithm for Bounded Variables," *Naval Research Logistics Quarterly 5* (1958), 257–261.

11.16 C. Witzgall, "One Row Linear Programs," in *Extremal Methods and Systems Analysis, Proceedings of the International Symposium on the Occasion of Professor Abraham Charnes 60th Birthday* (Austin, 1977), *Lecture Notes in Economics and Mathematical Systems*, No. 174, Springer, New York, 1980, pp. 384–414.

The Decomposition Principle of Linear Programming

chapter 12

12.1 PROBLEMS WITH A BLOCK ANGULAR STRUCTURE

This decomposition approach can be used to solve LPs with a large number of constraints, by solving a finite sequence of smaller size problems. It is based on the resolution theorem for convex polyhedra discussed in Chapter 3. Consider the LP

$$\begin{array}{lr} \text{Minimize} & z(x) = cx \\ \text{Subject to} & Ax = b \\ & x \geqq 0 \end{array} \qquad (12.1)$$

where A is a matrix of order $m \times n$. To solve this problem by the simplex method, we have to deal with bases of order m. If m is very large, the problem becomes unwieldy and it cannot be handled even by large computers. The decomposition principle can be used to break up such a large problem into smaller problems, each of which contains fewer numbers of constraints.

While the decomposition principle can be applied on a general LP, its application can be expected to be fruitful only when the coefficient matrix is a matrix with a special structure. It has been observed that it gives most satisfactory results when the matrix A can be blocked as in Figure 12.1, such that all the entries in it outside of those blocks are zero. This *block angular structure* arises in the optimization of a system consisting of coupled subsystems. For example, consider the problem of modeling the operations of a corporation with many plants that are mostly autonomous. In constructing the product mix problem for such a corporation, the smaller blocks (in the bottom) represent the constraints of the plants that are independent of the activities of the other plants. The top block consists of the linking constraints at the corporate level that tie the plants together. The same structure also appears in modeling a problem dealing with different geographical regions, each with its own production operations, which are coupled by common resource constraints. Similarly, the subsystems may be national economies coupled by international trade. In such problems, when each subsystem is already large and complex, the overall problem becomes very difficult to solve. The decomposition principle allows us to solve LPs with such block angular structure by solving a finite coordinated sequence of independent subproblems corresponding to the subsystems. We discuss the decomposition principle only as it relates to problems of this

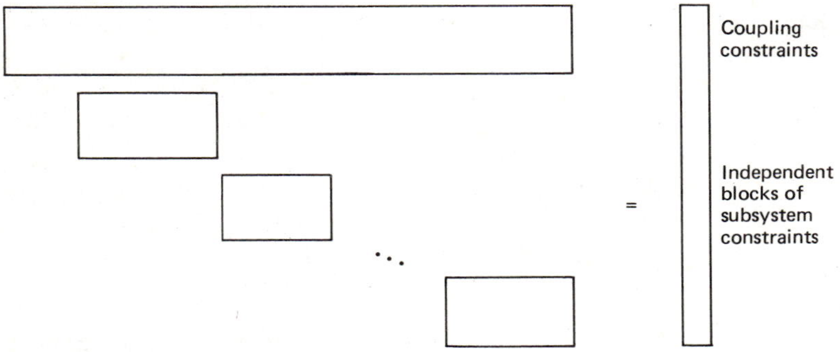

Figure 12.1 The block angular structure.

structure. In particular we consider the problem

$$
\begin{array}{ll}
\text{Minimize} & c^0x^0 + c^1x^1 + c^2x^2 + \cdots + c^kx^k \\
\text{Subject to} & A^{00}x^0 + A^{01}x^1 + A^{02}x^2 + \cdots + A^{0k}x^k = b^0 \\
\hline
& A^1x^1 \qquad\qquad\qquad\qquad = b^1 \\
& \qquad A^2x^2 \qquad\qquad\quad = b^2 \\
& \qquad\qquad \ddots \qquad\qquad\quad \vdots \\
& \qquad\qquad\qquad A^kx^k = b^k \\
& x^t \geqq 0 \qquad \text{for all } t
\end{array}
\tag{12.2}
$$

where A^{0t} is of order $m_0 \times n_t$ and A^t is of order $m_t \times n_t$ for each t. The number of variables in the vector x^t corresponding to the tth independent block is n_t (these can be interpreted as the levels of the activities at the tth plant of the corporation) and the number of constraints in this block is m_t, for $t = 1$ to k, where k is the number of independent blocks (plants). The number of linking constraints (which can be interpreted as the constraints at the corporate level that express the interconnections among the plants) is m_0.

For applying the decomposition principle the constraints in the overall model should be divided into two groups. One group constitutes the constraints of a problem called the *subprogram*. The other group *leads to* a problem called the *master program.* The division of the constraints into the two groups has to be done carefully, taking the structure of the matrix into account, so that the number of constraints in both the subprogram and the master program can be handled by the available computer. For (12.2), it is most advantageous to have the linking constraints on the top generate the master program and all the remaining constraints constitute the subprogram. An advantage of decomposing the problem this way is that the subprogram breaks up into k independent smaller size problems (one corresponding to each plant), called the *subproblems.* When the decomposition principle is applied in this manner, the problem is said to have been decomposed along the dashed line drawn in (12.2). Then the constraints in the subproblem are

$$
\begin{array}{l}
A^1x^1 \qquad\qquad\quad = b^1 \\
\qquad A^2x^2 \qquad\quad = b^2 \\
\qquad\qquad \ddots \qquad\ \vdots \\
\qquad\qquad\quad A^kx^k = b^k \\
x^t \geqq 0 \qquad \text{for } t = 1 \text{ to } k
\end{array}
$$

This breaks up into k independent subproblems, the constraints in the tth subproblem being

$$
\begin{aligned}
A^tx^t &= b^t \\
x^t &\geqq 0
\end{aligned}
\tag{12.3}
$$

for $t = 1$ to k. Since the number of constraints in it is m_t (compared to the original problem (12.2), which had $\sum_{t=0}^{k} m_t$ constraints), it should be much easier to handle.

While the constraints and the variables in the subprogram are all taken from those in the original problem, both the constraints and the variables in the master program are derived from an analysis outlined below. When decomposed as above, the constraints of the original problem that generate the master program are

$$
\begin{aligned}
A^{00}x^0 + A^{01}x^1 + \cdots + A^{0k}x^k &= b^0 \\
x^0 &\geqq 0
\end{aligned}
\tag{12.4}
$$

There are m_0 constraints in (12.4). The master program will be a problem with $m_0 + k$ constraints (k here is the number of independent subproblems into which the subprogram breaks up).

Derivation of the Master Program

If any of the subproblems (12.3) is infeasible, the original problem is infeasible, and we terminate. So we assume that each subproblem is feasible. Let $x^{t,1}, \ldots, x^{t,p_t}$ be all the BFSs of (12.3) and $y^{t,1}, \ldots, y^{t,q_t}$ be all the extreme homogeneous solutions corresponding to (12.3). By Resolution Theorem II (see Section 3.7), every feasible solution x^t of this subproblem can be expressed as

$$
\begin{gathered}
x^t = \alpha_{t,1}x^{t,1} + \cdots + \alpha_{t,p_t}x^{t,p_t} + \beta_{t,1}y^{t,1} + \cdots + \beta_{t,q_t}y^{t,q_t} \\
\text{where } \alpha_{t,1} + \cdots + \alpha_{t,p_t} = 1 \\
\text{and all } \alpha_{t,1}, \ldots, \alpha_{t,p_t}, \beta_{t,1}, \ldots, \beta_{t,q_t} \geqq 0
\end{gathered}
\tag{12.5}
$$

The variables in the master program are those in x^0 (these are the original variables that do not appear in any subproblem) and the multipliers $\alpha_{t,j}$ and $\beta_{t,j}$ corresponding to the extreme points and the extreme homogeneous solutions of the subproblems.

Every feasible solution of the original problem must satisfy (12.5) and (12.4). Substituting (12.5) into

(12.4) and the original objective function leads to

Minimize

$$z=c^0x^0+\sum_{j=1}^{p_1}(c^1x^{1,j})\alpha_{1,j}+\cdots+\sum_{j=1}^{p_k}(c^kx^{k,j})\alpha_{k,j}$$
$$+\sum_{j=1}^{q_1}(c^1y^{1,j})\beta_{1,j}+\cdots+\sum_{j=1}^{q_k}(c^ky^{k,j})\beta_{k,j}$$

Subject to

$$A^{00}x^0+\sum_{j=1}^{p_1}(A^{01}x^{1,j})\alpha_{1,j}+\cdots+\sum_{j=1}^{p_k}(A^{0k}x^{k,j})\alpha_{k,j}$$
$$+\sum_{j=1}^{q_1}(A^{01}y^{1,j})\beta_{1,j}+\cdots+\sum_{j=1}^{q_k}(A^{0k}y^{k,j})\beta_{k,j}=b^0$$
$$\sum_{j=1}^{p_t}\alpha_{t,j}=1 \qquad \text{for } t=1 \text{ to } k$$
$$x^0 \geqq 0 \text{ and } \alpha_{t,j} \text{ and } \beta_{t,j} \geqq 0 \qquad \text{for all } t, j \tag{12.6}$$

Problem (12.6) is the master program. Every feasible solution (x^0, α, β) of (12.6) corresponds to a unique feasible solution of the original problem (12.2) obtained by using (12.5). Also every feasible solution of (12.2) corresponds to at least one feasible solution of (12.6) (possibly many, because there may be many different ways in which a feasible solution x^t of (12.3) can be expressed as in (12.5)). Hence the master program is equivalent to the original problem.

The decomposition principle solves the original problem by solving the master program. The number of constraints in (12.6) is m_0+k. So (12.6) can be solved by the revised simplex method, using bases of order m_0+k each. Since m_0+k is much smaller than the number of constraints in the original problem (which is $\sum_{t=0}^{k} m_t$), it should be much easier to handle the master program than the original problem. However, the number of variables in the master program is $n_0+\sum_{t=1}^{k} p_t+\sum_{t=1}^{k} q_t$, and this is likely to be a very large number. If we have to deal with all the variables in the master program explicitly, this approach would be useless. But, to solve (12.6) by the revised simplex method, we need only the m_0+k basic columns at each stage, and the other columns can be generated as they are needed. Hence this kind of approach is also known as a *column generation approach*. See Section 5.8.

12.2 THE ALGORITHM

12.2.1 Phase I

If $x^{t,j}$ is a BFS of (12.3), we denote $A^{0t}x^{t,j}$ by $S^{t,j}$, and $c^tx^{t,j}$ by $u^{t,j}$. If $y^{t,j}$ is an extreme homogeneous solution corresponding to (12.3), we denote $A^{0t}y^{t,j}$ by $T^{t,j}$ and $c^ty^{t,j}$ by $v^{t,j}$.

Find a BFS for each subproblem (by a Phase I approach, if necessary). If any of the subproblems is infeasible, the original problem is infeasible. Terminate. Otherwise let $x^{t,1}$, $t=1$ to k be the BFSs obtained for the subproblems. As before $S^{t,1}=A^{0,t}x^{t,1}$ and $u^{t,1}=c^tx^{t,1}$. Let $\xi_1,\ldots,\xi_{m_0}$ be artificial variables. Let w be the Phase I objective. Here is the part of the master program corresponding to the initial basis for the Phase I problem

Tableau 12.1 Restricted System Consisting of Only the Initial Basic Variables for Phase I of the Master Program

$\alpha_{1,1}$	$\alpha_{2,1}\ \ldots\ \alpha_{k,1}$	$\xi_1\ \ldots\ \xi_{m_0}$	$-z$	$-w$	
		$\pm1\ \ldots\ 0$	0	0	
$S^{1,1}$	$S^{2,1}\ \ldots\ S^{k,1}$	0			b^0
		$\vdots\ \ddots\ \vdots$	$\vdots$	$\vdots$	
		$0\ \ldots\ \pm1$	0	0	
1	$\ldots\ 0$	$0\ \ldots\ 0$	0	0	1
$\vdots$	$\vdots$	$\vdots\ \ \ \vdots$	$\vdots$	$\vdots$	$\vdots$
0	$\ldots\ 1$	$0\ \ldots\ 0$	0	0	1
$u^{1,1}$	$\ldots\ u^{k,1}$	$0\ \ldots\ 0$	1	0	0
0	$\ldots\ 0$	$1\ \ldots\ 1$	0	1	0

In Tableau 12.1, transform the entries in the columns of $\alpha_{1,1},\ldots,\alpha_{k,1}$ in the first block of m_0 constraints (the original entries here are $S^{1,1}$, $S^{2,1},\ldots,S^{k,1}$, as recorded above) into zero by pivoting in each of these columns with the 1 entry in it in the second block of k rows, as the pivot element. Suppose this transforms the right-hand-side column b^0 into $\bar{b}^0$. Variable ξ_i is the artificial variable in the ith constraint. Choose its coefficient there to be $+1$ or -1, depending on whether $\bar{b}_i^0$ is $\geqq 0$ or <0, respectively (i.e., why these coefficients are recorded as ±1 in the above tableau), for $i=1$ to m_0. With this choice all the artificial variables are nonnegative in the initial basic solution. Obtain the inverse tableau corresponding to this basis as in Chapter 5.

In this decomposition approach, for each α or β variable in the basic vector, always, remember to store the associated BFS or the extreme homogeneous solution of the subproblem carefully. These are needed to get the solution of the original problem corresponding to the present solution of the master program (using equation (12.5)).

In a general stage during Phase I suppose the inverse tableau is

Basic Variables	Inverse Tableau				Basic Values
	B^{-1}		0		$\bar{b}_1$ $\vdots$ $\bar{b}_{mo+k}$
$-z$	$-\pi_1$	$-\pi_{mo+k}$	1	0	$-\bar{z}$
$-w$	$-\sigma_1$	$-\sigma_{mo+k}$	0	1	$-\bar{w}$

1 The Phase I relative cost coefficient of x_j^0 with respect to the present basis in (12.6) is $\bar{d}_j^0 = (-\sigma_1, \ldots, -\sigma_{mo})A_{.j}^{00}$. If $\bar{d}_j^0 < 0$, x_j^0 can be chosen as the entering variable into the basic vector for the master program. The pivot column is its updated column and it is obtained by multiplying the present inverse tableau on the right by the column vector of x_j^0 in (12.6). If $\bar{d}_j^0 \geqq 0$, x_j^0 is not a candidate to enter the basic vector. If $\bar{d}_j^0 \geqq 0$ for all $j = 1$ to n_0, go to 2.

2 Check whether any of the α or β variables corresponding to one of the subproblems can be the entering variable. The Phase I relative cost coefficient of $\beta_{t,j}$ is $(-\sigma_1, \ldots, -\sigma_{mo})T^{t,j} = (-\sigma_1, \ldots, -\sigma_{mo})A^{0t}y^{t,j}$. If this is negative, $\beta_{t,j}$ can be taken as the entering variable. In this case, by the results in Section 3.7, the objective function in the LP (12.7) must be unbounded below.

$$\begin{aligned} \text{Minimize} \quad & (-\sigma_1, \ldots, -\sigma_{mo})A^{0t}x^t \\ \text{Subject to} \quad & A^t x^t = b^t \\ & x^t \geqq 0 \end{aligned} \qquad (12.7)$$

Conversely, when the subproblem (12.7) is solved, if the objective function is unbounded below, find $y^{t,j}$, the extreme homogeneous solution corresponding to the extreme half-line along which the objective function diverges to $-\infty$ in (12.7), associate $y^{t,j}$ with a multiplier $\beta_{t,j}$ and choose $\beta_{t,j}$ as the entering variable into the basic vector for the master program. The column vector corresponding to this $\beta_{t,j}$ in the Phase I problem is given in (12.8), where the **0** in this vector is the zero vector in $\mathbf{R}^k$ (since the β's do not appear in the constraints that the α-multipliers in each subproblem should sum to 1 in (12.6)), $v^{t,j}$ is the Phase II original cost coefficient of $\beta_{t,j}$ in (12.6). The bottom entry in this column is the Phase I original cost coefficient of $\beta_{t,j}$ in the Phase I problem, and since $\beta_{t,j}$ is a variable in (12.6) this is zero.

The updated column vector is obtained by multiplying this column vector on the left by the inverse tableau, and this is the pivot column.

If (12.7) has a finite optimum solution, then none of the β variables corresponding to this subproblem are eligible to be the entering variable. Let the optimum solution of (12.7) be $x^{t,j}$. Associate the multiplier $\alpha_{t,j}$ with $x^{t,j}$. The Phase I relative cost coefficient of $\alpha_{t,j}$ is $-\sigma_{mo+t} + (-\sigma_1, \ldots, -\sigma_{mo})A^{0t}x^{t,j}$. If this is negative, choose $\alpha_{t,j}$ as the entering variable into the basic vector for the master program. The column vector of $\alpha_{t,j}$ in the Phase I problem is given in (12.9), where $I_{.t}$ is the tth column vector of the unit matrix of order k (since $\alpha_{t,j}$ appears with a coefficient of one in the constraint that all the α-multipliers associated with the tth subproblem should sum to one in (12.6)). $u^{t,j}$ is the Phase II original cost coefficient of $\alpha_{t,j}$ in (12.6), and the bottom entry in this column is zero, the Phase I original cost coefficient of $\alpha_{t,j}$ in the Phase I problem.

$$\begin{bmatrix} A^{0t}y^{t,j} \\ \hdashline \mathbf{0} \\ \hdashline v^{t,j} \\ \hdashline 0 \end{bmatrix} \quad (12.8), \qquad \begin{bmatrix} A^{0t}x^{t,j} \\ \hdashline I_{.t} \\ \hdashline u^{t,j} \\ \hdashline 0 \end{bmatrix} \quad (12.9)$$

The updated column vector is obtained by multiplying this column vector on the left by the inverse tableau, and this is the pivot column.

If $-\sigma_{mo+t} + (-\sigma_1, \ldots, -\sigma_{mo})A^{0t}x^{t,j} \geqq 0$, then obviously none of the α or β variables corresponding to this subproblem are eligible to be the entering variables in this stage.

Begin solving the tth subproblem (12.7) from $t = 1$ to k. If an entering variable into the basic vector for the master program is identified, stop solving the subproblems. Get the new inverse tableau corresponding to the new basis for the master program and go to the next stage.

At any stage of the algorithm, the part of the master program remaining after dropping the columns of all the variables except those of the present basic variables and the entering variable is known as the *restricted master program*.

Notice that the objective functions for the subproblems depend on the Phase I dual vector for the master program during the stage and hence may change from stage to stage.

Phase I: Termination Criteria

Phase I for the master program terminates when the following conditions are satisfied.

1 $\bar{d}_j^0 \geqq 0$ for all $j = 1$ to n_0.

2 All the subproblems (12.7) for $t = 1$ to k have finite optimum solutions.

3 For the tth subproblem (12.7), $-\sigma_{m_0+t}$ + the minimum value of $(-\sigma_1, \ldots, -\sigma_{m_0})A^{0t}x^t$ in (12.7) is nonnegative for all $t = 1$ to k.

When Phase I terminates, if the value of w is positive, the original problem is infeasible, and we terminate. If the value of w is zero, and if all the artificial variables have left the basic vector, the Phase I objective row and the w column are deleted from the inverse tableau and we go to Phase II.

If some of the artificial variables are still in the basic vector at Phase I termination, they are left in the basic vector, but care must be taken to see that all the entering variables during Phase II have zero Phase I relative cost coefficients. Do the following before entering Phase II for this.

1 All x_j^0 that have $\bar{d}_j^0 > 0$ in the final Phase I iteration should be set equal to zero and eliminated from consideration during Phase II.

2 Let $(-\bar{\sigma}_1, \ldots, -\bar{\sigma}_{m_0+k}, 0, 1)$ be the last row in the final Phase I inverse tableau. Add the constraint

$$(-\bar{\sigma}_1, \ldots, -\bar{\sigma}_{m_0})A^{0t}x^t + (-\bar{\sigma}_{m_0+t}) = 0$$

to the system of constraints (12.3) in the tth subproblem for each t. This will make all the α and β variables that have positive Phase I relative cost coefficients ineligible from becoming entering variables during Phase II. However, for notational convenience we continue to refer to the system of constraints in the tth subproblem by (12.3) itself. Now delete the Phase I objective row and the w column from the final Phase I inverse tableau and go over to Phase II.

12.2.2 Computations During Phase II

The computations during Phase II are similar to those in Phase I with the only change that Phase II relative cost coefficients should replace Phase I relative cost coefficients. Let the Phase II objective row in the inverse tableau during some stage in Phase II be $(-\pi_1, \ldots, -\pi_{m_0+k}, 1)$. In this stage.

1 The Phase II relative cost coefficient of x_j^0 is $\bar{c}_j^0 = c_j^0 + (-\pi_1, \ldots, -\pi_{m_0})A_j^0$

2 If x_j^0 have nonnegative Phase II relative cost coefficients for all $j = 1$ to n_0, then solve the subproblem:

$$\begin{aligned} \text{Minimize} \quad & (c^t + (-\pi_1, \ldots, -\pi_{m_0})A^{0t})x^t \\ \text{Subject to} \quad & A^t x^t = b^t \\ & x^t \geqq 0 \end{aligned} \tag{12.10}$$

If the objective function is unbounded below in (12.10), the entering column is the column in (12.6) corresponding to the extreme homogeneous solution associated with the extreme half-line along which the objective function in (12.10) diverges to $-\infty$. If (12.10) has a finite optimum solution, let $x^{t,j}$ be an optimal BFS of it and let $\alpha_{t,j}$ be the multiplier associated with it. The Phase II relative cost coefficient of $\alpha_{t,j}$ is

$$(-\pi_{m_0+t}) + (c^t + (-\pi_1, \ldots, -\pi_{m_0})A^{0t})x^{t,j}$$

If this is negative, $\alpha_{t,j}$ is chosen as the entering variable. If this is nonnegative, all the α and β variables corresponding to this subproblem have nonnegative Phase II relative cost coefficients, and are not eligible to be entering variables.

Solve the subproblems from $t = 1$ to k until an entering variable with a negative Phase II relative cost coefficient is identified.

Phase II: Termination Criteria

If at some stage of Phase II, the updated column vector of the entering variable is nonpositive, the objective function is unbounded below in the original problem. If unboundedness did not occur so far, Phase II terminates in this stage if:

1 $\bar{c}_j^0 \geqq 0$ for all $j = 1$ to n_0.

2 All the subproblems (12.10) for $t = 1$ to k have finite optimum solutions.

3 $c^t x^{t,j} + (-\pi_{mo+t}) + (-\pi_1, \ldots, -\pi_{mo}) A^{0t} x^{t,j} \geqq 0$ in subproblem (12.10) for all $t = 1$ to k. (Here $x^{t,j}$ is the optimum solution of subproblem (12.10) in this stage).

Notice that the objective function in the subproblem (12.10) depends on the Phase II dual solution in that stage. Hence as we move from stage to stage, the objective functions in the subproblems change, but the subproblem constraints remain unaltered. Hence after the first stage, the subproblems can be solved efficiently by parametric cost methods. Each subproblem may have to be solved several times, each time with a different objective function.

From the optimal Phase II inverse tableau an optimum solution of the original problem is obtained using (12.5). This is why the BFS or the extreme homogeneous solution associated with each α or β variable in the basic vector should be stored in each stage.

Example 12.1

Consider the following problem.

x_1^1	x_2^1	x_3^1	x_4^1	x_1^2	x_2^2	x_3^2	x_4^2	x_5^2	
3	−1	−3	2	1	2	0	−1	1	1
1	0	−1	1						3
0	1	−1	−1						4
				1	1	1	1	0	1
				2	1	−2	0	1	2
1	−1	−3	3	20	30	7	1	1	minimize

All variables nonnegative. Blank entries in tableau are zero.

We decompose along the dashed line. The subprogram breaks up into two independent subproblems. We use the following BFSs of the subproblems initially.

Associated Multiplier	Solution
$\alpha_{1,1}$	$x^{1,1} = (3, 4, 0, 0)^T$
$\alpha_{1,2}$	$x^{1,2} = (0, 7, 0, 3)^T$
$\alpha_{2,1}$	$x^{2,1} = (0, 0, 0, 1, 2)^T$

We can verify that $(\alpha_{1,1}, \alpha_{1,2}, \alpha_{2,1})$ constitutes a feasible basic vector for the master program; therefore there is no need to do a Phase I solution. The portion of the master program corresponding to this basic vector is

$\alpha_{1,1}$	$\alpha_{1,2}$	$\alpha_{2,1}$	$-z$	
5	−1	1	0	1
1	1	0	0	1
0	0	1	0	1
−1	2	3	1	0

The inverse tableau corresponding to this basic vector is

First Inverse Tableau for the Master Program

Basic Variables	*Inverse Tableau*				*Basic Values*	*Pivot Column* $\beta_{1,1}$
$\alpha_{1,1}$	$\frac{1}{6}$	$\frac{1}{6}$	$-\frac{1}{6}$	0	$\frac{1}{6}$	$-\frac{1}{18}$
$\alpha_{1,2}$	$-\frac{1}{6}$	$\frac{5}{6}$	$\frac{1}{6}$	0	$\frac{5}{6}$	$\frac{1}{18}$ (circled)
$\alpha_{2,1}$	0	0	1	0	1	0
$-z$	$\frac{1}{2}$	$-\frac{3}{2}$	$-\frac{7}{2}$	1	$-\frac{9}{2}$	$-\frac{21}{18}$

The objective function in subproblem 1 in this stage is: $(1/2)(3x_1^1 - x_2^1 - 3x_3^1 + 2x_4^1) + (x_1^1 - x_2^1 - 3x_3^1 + 3x_4^1) = (5/2)x_1^1 - (3/2)x_2^1 - (9/2)x_3^1 + 4x_4^1$. With this objective function, the canonical tableau for this subproblem with (x_1^1, x_2^1) as the basic vector is

x_1^1	x_2^1	x_3^1	x_4^1		
1	0	−1	1	3	
0	1	−1	−1	4	
0	0	$-\frac{7}{2}$	0	$-\frac{3}{2}$	*Objective row*

From the column vector of x_3^1 we see that this problem is unbounded below. The corresponding extreme homogeneous solution is

Associated Multiplier	*Extreme Homogeneous Solution*
$\beta_{1,1}$	$y^{1,1} = \left(\frac{1}{3}, \frac{1}{3}, \frac{1}{3}, 0\right)$

$\beta_{1,1}$ is the entering variable. Its column vector in the master tableau is $(-1/3, 0, 0, -1)^T$. The updated column vector is the pivot column and this is entered on the first inverse tableau.

Second Inverse Tableau for the Master Program

Basic Variables	Inverse Tableau				Basic Values	Pivot Column $\alpha_{2,2}$
$\alpha_{1,1}$	0	1	0	0	1	0
$\beta_{1,1}$	−3	15	3	0	15	−9
$\alpha_{2,1}$	0	0	1	0	1	①
$-z$	−3	16	0	1	13	−1

The objective function in subproblem 1 in this stage is $-3(3x_1^1 - x_2^1 - 3x_3^1 + 2x_4^1) + (x_1^1 - x_2^1 - 3x_3^1 + 3x_4^1) = -8x_1^1 + 2x_2^1 + 6x_3^1 - 3x_4^1$. It can be verified that the optimum solution for subproblem 1 with this objective function is $x^{1,1}$ with an objective value of -16. The relative cost coefficient in its column is $-16 + 16 = 0$. This implies that none of the α or β variables from this subproblem are eligible to enter the basic vector at this stage. Therefore move to subproblem 2.

The objective function in subproblem 2 in this stage is $-3(x_1^2 + 2x_2^2 - x_4^2 + x_5^2) + (20x_1^2 + 30x_2^2 + 7x_3^2 + x_4^2 + x_5^2) = 17x_1^2 + 24x_2^2 + 7x_3^2 + 4x_4^2 - 2x_5^2$. Hence subproblem 2 at this stage is

x_1^2	x_2^2	x_3^2	x_4^2	x_5^2	
1	1	1	1	0	1
2	1	−2	0	1	2
17	24	7	4	−2	minimize

Here is the optimum solution of this subproblem with an objective value of -1.

Associated Multiplier	Optimum Solution
$\alpha_{2,2}$	$x^{2,2} = (0, 0, 1, 0, 4)^T$

So the relative cost coefficient of $\alpha_{2,2}$ is $-1 + 0 = -1$. Hence $\alpha_{2,2}$ is the entering variable. Its column vector in the master program is $(4, 0, 1, 11)^T$ and the updated column vector is the pivot column entered on the second inverse tableau.

Third Inverse Tableau for the Master Program

Basic Variables	Inverse Tableau				Basic Values
$\alpha_{1,1}$	0	1	0	0	1
$\beta_{1,1}$	−3	15	12	0	24
$\alpha_{2,2}$	0	0	1	0	1
$-z$	−3	16	1	1	14

It can be verified that the optimality criterion is now satisfied. Hence the solution in the third inverse tableau is the optimum solution for the master program. The optimum solution of the original problem is:

$$x^1 = (3, 4, 0, 0)^T + 24\left(\frac{1}{3}, \frac{1}{3}, \frac{1}{3}, 0\right)^T = (11, 12, 8, 0)^T$$

$$x^2 = (0, 0, 1, 0, 4)^T$$

$$z = -14$$

12.3 INTERPRETATION IN SUPPORT OF DECENTRALIZED PLANNING

Consider a corporation with k plants that are mostly autonomous and let problem (12.2) be the model that optimizes the operations of this corporation. *Centralized planning* requires solving this overall model and then handing over the optimum solution to the plants to be implemented.

However, the manner in which this model is solved by applying the decomposition principle can be interpreted as suggesting a decentralized planning system that also leads to an optimum solution for the overall model. In this setup the corporate headquarters (HQ) would handle the master program. Each plant would handle the subproblem corresponding to it. In each stage of the algorithm the HQ generates an objective function for each subproblem (corresponding to that stage of the master program) and sends a directive to each plant that it should come up with an optimum solution of its subproblem that minimizes that objective function. The HQ does not need any knowledge as to how the plant solves the subproblem (it does not even need any knowledge of the constraints in the subproblem, that is, the constraints under which the plant is operating). When the optimum solutions from all the plants are received, the HQ combines this information, checks for optimality to the master program, and moves to the next stage in solving the master program, if optimality is not reached yet. In this setup the HQ is acting purely as a coordinating agency giving guidance. The plants do the optimization for their subproblems all by themselves according to the guidance obtained from the HQ. (This guidance comes in the form of the objective function in that stage, which the plant is asked to minimize. No quotas or any such additional constraints are ever imposed.) They then send their optimum plans to the HQ. When this process is repeated, after a finite number of iterations the system moves to a solution that is optimal for the whole corporation.

Comment 12.1: The decomposition principle has been developed by G. B. Dantzig and P. Wolfe, inspired by the work of L. R. Ford, D. R. Fulkerson, and W. S. Jewell on the multicommodity flow problem.

12.4 APPLICABILITY

Initial attempts at testing the decomposition approach produced inconclusive and often less than encouraging results in terms of empirical computational efficiency. When solving a problem using the decomposition approach, substantial improvements in objective value seem to occur at the beginning, followed a lengthy creep of iterations in which the objective value improves very slowly. Subsequently, because of advances in sparsity techniques for variants of the revised simplex method, it has been recognized that decomposition is, in general, unlikely to be significantly more efficient than the revised simplex approach whenever the latter can produce a solution at a reasonable cost. So decomposition cannot compete with the modern variants of the revised simplex approach in its domain of application, but can extend their capabilities for truly large problems. It is only very recently that general-purpose, easily accessible, easy to use, and computationally robust implementations of decomposition compatible with state-of-the-art LP software, in the form of a simple to use option in commercial LP codes, have become available. See the paper [12.4] of J. K. Ho and E. Loute. With the availability of such implementations that build upon existing LP software, it is expected that decomposition will come into increasing use for large-scale practical problem solving.

Exercises

12.1 *CYCLING UNDER DEGENERACY IN THE DECOMPOSITION PRINCIPLE.* Suppose (12.2) is being solved by decomposing it along the dashed line. The BFS of the master program at some stage is degenerate if some of the basic variables in the current basic vector for the master program have zero values in that BFS. As in any other LP, cycling may occur if the master program encounters a degenerate BFS during the algorithm.

Let $B^{-1} = (\beta_{ij})$ be the inverse of a feasible basis for the master program, and let $\bar{b} = (\bar{b}_1, \ldots, \bar{b}_{m_0+k})^{\mathrm{T}}$, be the vector of basic values in the BFS of the master program associated with this basis. The basis (and the associated basic vector) is said to be *lexico feasible to the master program* if each of the row vectors $(\bar{b}_i, \beta_{i1}, \ldots, \beta_{im_0+k})$ is lexico positive for all $i = 1$ to $m_0 = k$ (see Chapter 10 for these definitions).

1 Starting with a lexico feasible basic vector for the master program, suppose we go through one pivot step in the master program. If the dropping variable is determined by the lexico minimum ratio rule as in Chapter 10, prove that the basic vector obtained after this pivot step will also be lexico feasible for the master program.

2 Starting with the Phase I using a full artificial basis for the master program, suppose (12.2) is solved by the decomposition principle using the lexico minimum ratio rule of Chapter 10 to determine the dropping basic variable in the basic vector for the master program in each pivot step. Also suppose some method for resolving degeneracy is used in solving each of the subproblems in every step. Under these conditions prove that this algorithm must terminate after at most a finite number of steps.

REFERENCES

12.1 I. Adler and A. Ülkücü, "On the Number of Iterations in Dantzig-Wolfe Decomposition," in D. M. Himmelblau (Ed.), *Decomposition of*

Large Scale Problems, North-Holland, Amsterdam, The Netherlands, 1973, pp. 181–187.

12.2 E. Beale, P. Huges, and R. Small, "Experiences in Using a Decomposition Program," *Computer Journal 8* (1965), 13–18.

12.3 G. B. Dantzig and P. Wolfe, "The Decomposition Principle for Linear Programs," *Operations Research 8* (1960), 101–111.

12.4 J. K. Ho and E. Loute, "An Advanced Implementation of the Dantzig-Wolfe Decomposition Algorithm for Linear Programming," *Mathematical Programming 20* (May 1981), 303–326.

12.5 D. M. Himmelblau (Ed.), "Decomposition of Large-Scale Problems," in *Proceedings of a NATO Advanced Study Institute*, North-Holland, Amsterdam, The Netherland, 1973.

12.6 M. R. Rao and S. Zionts, "Allocation of Transportation Units to Alternative Trips—A Column Generation Scheme with Out-of-Kilter Subproblems," *Operations Research 16* (1968), 52–63.

chapter 13 The Transportation Problem

13.1 THE BALANCED TRANSPORTATION PROBLEM: THEORY

We consider the transportation problem in which all the constraints are equality constraints.

$$\text{Minimize} \quad z(x) = \sum_{i=1}^{m} \sum_{j=1}^{n} c_{ij} x_{ij}$$

$$\text{Subject to} \quad \sum_{j=1}^{n} x_{ij} = a_i \qquad i = 1 \text{ to m} \qquad (13.1)$$

$$\sum_{i=1}^{m} x_{ij} = b_j \qquad j = 1 \text{ to n}$$

$$x_{ij} \geqq 0 \qquad \text{for all } i, j$$

where $a_i > 0$, $b_j > 0$ for all i, j, and $\sum a_i = \sum b_j$, and c_{ij} may be arbitrary real numbers. Here a_i, b_j are, respectively, the amount (in units) of some commodity available at source i, and required at demand center j. Also c_{ij}, x_{ij} are, respectively, the cost per unit of shipping, and amount shipped from source i to demand center j. The problem is known as the *uncapacitated balanced transportation problem* because there are no specified upper bounds on x_{ij} (although the equality constraints and the nonnegativity restrictions automatically imply an upper bound on x_{ij}, namely minimum $\{a_i, b_j\}$).

Until now we have used the symbols m, n, to denote the number of constraints and the number of variables, respectively, in an LP model. In this chapter we use the symbols m, n, to denote the number of sources and demand centers, respectively, in a transportation model. Notice that the number of constraints in this transportation model is m + n and the number of variables is mn.

The nice structure of this problem makes it possible to compute an initial BFS by very simple techniques without the need to solve any complicated Phase I problem; to compute the dual solution and the relative cost coefficients corresponding to any basis very easily without the need to compute the basis inverse; and to compute the new BFS when a nonbasic variable is brought into the basic vector by making simple adjustments along a loop without doing any complicated pivoting. This simplified version of the primal revised simplex algorithm for the balanced transportation problem is discussed here. First, we review the main results on this problem.

13.1.1 Necessary Conditions for Feasibility

Let $\hat{x}$ be feasible to (13.1). Summing the set of first m constraints and the set of last n constraints in (13.1) separately, we see that $\sum_i a_i = \sum_i \sum_j x_{ij} = \sum b_j$. So

$$\sum_{i=1}^{m} a_i = \sum_{j=1}^{n} b_j \qquad (13.2)$$

is a necessary condition for feasibility of (13.1). Later we show that this condition is also sufficient for the feasibility of (13.1). *We assume that data in (13.1) satisfy this condition.*

Redundancy in the Constraints

The constraints in (13.1) can be written down in the equivalent form

$$\sum_{j=1}^{n} -x_{ij} = -a_i \qquad i = 1 \text{ to m} \qquad (13.3)$$

$$\sum_{i=1}^{m} x_{ij} = b_j \qquad j = 1 \text{ to n} \qquad (13.4)$$

THEOREM 13.1 There is exactly one redundant equality constraint in (13.3) and (13.4). When any one of the constraints in (13.3) or (13.4) is dropped, the remaining is a linearly independent system of constraints.

Proof: Let $x = (x_{11}, \ldots, x_{1n}, x_{21}, \ldots, x_{mn})^T$ be the column vector of all the mn variables in this problem. Let $\boldsymbol{E}$ denote the (m + n) × mn coefficient matrix corresponding to the constraints in (13.3) and (13.4). The column vector in $\boldsymbol{E}$ associated with the variable x_{ij} contains only two nonzero entries: −1 in row i

and a $+1$ in row $m+j$. Hence if all the equations in (13.3) and (13.4) are summed, we get the equation $0=0$ by (13.2), which implies that there is a redundant constraint in (13.3) and (13.4) and that *any one of the constraints in them may be treated as a redundant constraint.*

Let $\bar{\mathbf{E}}$ be the matrix obtained by striking off some row, say, the last row, from $\mathbf{E}$. $\bar{\mathbf{E}}$ is an $(m+n-1)\times mn$ matrix. It is the coefficient matrix of the system of constraints:

$$\sum_{j=1}^{n} -x_{ij} = -a_i \qquad i = 1 \text{ to } m \qquad (13.5)$$

$$\sum_{i=1}^{m} x_{ij} = b_j \qquad j = 1 \text{ to } n-1 \qquad (13.6)$$

Suppose the rank of $\bar{\mathbf{E}}$ is strictly less than $m+n-1$. Then there must exist a nonzero linear combination of the equations in (13.5) and (13.6) in which all the coefficients of the variables on the left-hand side are zero. Suppose this is obtained by multiplying the equations in (13.5) by $\alpha_1, \ldots, \alpha_m$ and the equations in (13.6) by $\alpha_{m+1}, \ldots, \alpha_{m+n-1}$, respectively, and summing. Since this is a nonzero linear combination $(\alpha_1, \ldots, \alpha_m, \ldots, \alpha_{m+n-1}) \neq 0$. The variable x_{in} for any i from 1 to m occurs in exactly one equation among (13.5) and (13.6) with a nonzero coefficient, namely, the ith equation in (13.5). Since the coefficient of each x_{in} is zero in the linear combination discussed above, $\alpha_1 = \cdots = \alpha_m = 0$. However, the variable x_{ij} for any j from 1 to $n-1$ occurs with a nonzero coefficient in exactly one equation in (13.6), namely, the jth equation. Thus, if $\alpha_1 = \cdots = \alpha_m = 0$ and if the linear combination discussed above were to produce an equation in which the coefficients of all the variables on the left-hand side are zero, we must have $\alpha_{m+1} = \cdots = \alpha_{m+n-1} = 0$. This contradicts the assumption that $(\alpha_1, \ldots, \alpha_{m+n-1}) \neq 0$. Hence, the rank of $\bar{\mathbf{E}}$ must be $m+n-1$.

Hence, when any one of the constraints in (13.3) and (13.4) is dropped, the remaining constraint is a linearly independent system of constraints. ■

COROLLARY 13.1 A basic vector for the balanced transportation problem (13.1) consists of $(m+n-1)$ basic variables.

13.1.2 Triangular Matrix

Let $D = (d_{ij})$ be a nonsingular square matrix of order N. If $d_{ij} = 0$ for $j \geqq i+1$, D is said to be a *lower triangular matrix*. See Section 3.3.7. A square matrix D is said to be an *upper triangular matrix,* if D^{T} is a lower triangular matrix.

If D is a nonsingular square matrix, which becomes a lower triangular matrix after a permutation of its columns and rows, D is said to be a *triangular matrix*. A triangular matrix satisfies the following properties:

1 The matrix has a row that contains only a single nonzero entry.

2 The submatrix obtained from the matrix by striking off the row containing a single nonzero entry and the column in which this nonzero entry lies also satisfies property (1). The same process can be repeated until all the rows and columns in the matrix are struck off.

Conversely, any matrix satisfying these properties is a triangular matrix. For example, the following matrix A can be verified to be triangular, while the matrix B is not.

$$A = \begin{pmatrix} 1 & 1 & 0 & 0 & 0 & 0 \\ 0 & 0 & 1 & 1 & 1 & 0 \\ 0 & 0 & 0 & 0 & 0 & 1 \\ 0 & 0 & 1 & 0 & 0 & 1 \\ 1 & 0 & 0 & 0 & 0 & 0 \\ 0 & 0 & 0 & 1 & 0 & 0 \end{pmatrix}, \quad B = \begin{pmatrix} 1 & 1 & 0 & 0 & 0 & 1 \\ 0 & 0 & 1 & 1 & 0 & 0 \\ 0 & 0 & 0 & 0 & 1 & 1 \\ 1 & 0 & 0 & 0 & 1 & 0 \\ 0 & 0 & 1 & 0 & 0 & 0 \\ 0 & 1 & 0 & 1 & 0 & 0 \end{pmatrix}$$

If D is a lower triangular matrix, the systems of equations $Dy = d$ or $\pi D = h$ are solved efficiently by back substitution, as discussed in Section 3.3.7. If D is any triangular matrix, the system $Dy = d$ can still be solved by back substitution. Identify the equation containing a single nonzero entry on the left-hand side; solve that equation for the value of the variable associated with the nonzero coefficient in that equation; substitute the value of this variable in all the remaining equations and continue in the same manner with the remaining equations. The same method can be applied to solve the system $\pi D = h$ when D is triangular.

13.1.3 Triangularity of the Basis in the Transportation Problem

THEOREM 13.2 Every basis for the balanced transportation problem is triangular.

Proof: The constraints in (13.1) can be written as (13.5) and (13.6). The input-output coefficient matrix in these constraints is $\bar{\mathbf{E}}$. Let B be a nonsingular square submatrix of $\bar{E}$ of order $(m + n - 1)$. Every column vector in $\bar{E}$ contains at most two nonzero entries: a -1 and a $+1$, and hence the total number of nonzero entries in B can never exceed $2(m+n-1)$. If the total number of nonzero entries in B is equal to $2(m + n - 1)$, then every column in B must contain two nonzero entries, and since they are a -1 and a $+1$, the sum of all the rows in B must be zero, which is a contradiction to the nonsingularity of B. Thus, the total number of nonzero entries in B must be $< 2(m + n - 1)$. Since B is of order $(m + n - 1)$, there must exist a row of B with a single nonzero entry. By a similar argument, the submatrix of B obtained by striking off this row and the column containing this single nonzero entry must again have the same property, that is, it contains a row with a single nonzero entry and is nonsingular. The same argument can be repeated again and again. Hence, B is triangular. ■

Exercises

13.1 If B is any basis for (13.5) and (13.6), prove that the determinant of B is either $+1$ or -1.

13.2 Let F be any square submatrix of $\bar{E}$. By using an argument similar to that in the proof of Theorem 13.2, prove that the determinant of F is either 0, $+1$, or -1.

Given a basis for (13.5) and (13.6), since it is triangular, the basic solution of (13.5) and (13.6) corresponding to it can be computed by back substitution. Also the corresponding dual solution can be computed by back substitution. *This makes it unnecessary to keep the basis inverse while solving the transportation problem*. Also, since the nonzero entries in $\bar{E}$ are all equal to -1 or $+1$, finding the primal and dual solutions corresponding to a given basis by back substitution involves only additions and subtractions. No multiplications or divisions have to be performed. This leads to the following corollary (a result stronger than this Corollary 13.2 is proved later on in Theorem 13.10).

COROLLARY 13.2 In any basic solution of (13.5) and (13.6) the values of the basic variables are of the form $\sum_{i=1}^{m} \gamma_i a_i + \sum_{j=1}^{n-1} \psi_j b_j$, where all the γ_i, ψ_j are equal to 0 or $+1$ or -1.

INTEGER PROPERTY If all the a_i and b_j are positive integers, then every basic solution of (13.1) is an integer vector. Hence, if all a_i, b_j are positive integers, and if the transportation problem is feasible, it has an optimum solution (x_{ij}) that is an integer vector.

13.1.4 Transportation Array

Clearly, the variables (x_{ij}) in the transportation problem can be arranged in an $m \times n$ array (known as the $m \times n$ *transportation array*) such that the constraints specify the sums of the variables in each row and column of the array. The variable x_{ij} corresponds to the cell (i, j) in such an array and vice versa. With the value of each variable x_{ij} entered in the appropriate cell, the array is denoted by x. See Array 13.1 of Section 13.2.2. The transportation array is not the same as the *simplex tableau* (the tableau of input-output coefficients used in the standard simplex algorithm, which will consist of $m + n - 1$ rows and mn columns for the transportation problem). While the simplex tableau displays the coefficients of the constraints, the transportation array is just a convenient method for displaying a feasible solution of the transportation problem. All the computations required for solving the transportation problem can be organized using the transportation array only without ever requiring the simplex tableau.

13.1.5 Linear Independence of a Subset of Cells of the Array

Definition 13.2: A subset of $m + n - 1$ cells of portation array is said to be *linearly independent* if the set of column vectors in (13.1) corresponding to variables associated with cells in the subset is linearly independent.

Definition 13.2: A subset of $m + n - 1$ cells of the $m \times n$ transportation array is a *basic set* if it is linearly independent.

Test for a Basic Set of Cells

Suppose a subset $\mathscr{B}$ of $m + n - 1$ cells of the $m \times n$ transportation array is given. Let B be the square matrix of order $m + n - 1$ whose column vectors are the column vectors in (13.5) and (13.6), corresponding to variables associated with cells in the subset $\mathscr{B}$. The subset $\mathscr{B}$ is a basic set iff B is nonsingular. From Theorem 13.2, if B is nonsingular, it is a triangular matrix. Hence, there exists a row vector of B with a single nonzero element. Each row in (13.5) and (13.6) represents a constraint that the sum of all the variables in a row or column of the transportation array is a specified number. Hence, there exists a row vector of B with a single nonzero element, iff there exists a row or column of the array that contains exactly one cell from $\mathscr{B}$. Also, B is a triangular matrix iff the same property again holds for the submatrix of B obtained by striking off the row of B containing a single nonzero element and the column of B containing that nonzero element. This leads to the following simple test to check whether $\mathscr{B}$ forms a basic set.

Initially, none of the cells of the array are struck off. In each stage of the test all the cells in a selected row, or column of the array will be struck off. At any stage of the test, a row or column of the array that has some cells not yet struck off is known as a *remaining row or column*. Subset $\mathscr{B}$ forms a basic set iff the following properties hold.

1. At each stage, there exists a remaining row or column of the array that contains exactly one cell of $\mathscr{B}$ that has not yet been struck off.
2. Identify the remaining row or column of the array that contains exactly one cell of $\mathscr{B}$ that has not yet been struck off. Strike off all the cells in that row or column. The remaining part of the array must again satisfy property (1). Repeat until all the cells in the array are struck off.

Exercises

13.3 Do the subsets of marked cells in the following arrays constitute basic sets?

	y				
	y	y			
		y	y		
			y	y	y

(a)

					y
y			y	y	
y	y				
	y	y	y		

(b)

How to Compute the Basic Solution Corresponding to a Basic Set

Let $\mathscr{B}$ be a basic set. When considering $\mathscr{B}$, cells in $\mathscr{B}$ are called *basic cells*, variables associated with them are called *basic variables*, all other cells are *nonbasic cells*, and these are associated with *nonbasic variables*. To get the basic solution of $\mathscr{B}$, set all the nonbasic variables to zero and solve for basic values from the remaining system. This can be done by back substitution. In this method the basic variables are evaluated one by one. In each step the basic variable evaluated is the unique basic variable among the set of unevaluated basic variables at that stage, either in its row or in its column in the array; its value is computed from the constraint corresponding to that row or column, using the values of basic variables computed so far. The basic set $\mathscr{B}$ is primal feasible if all basic variables are $\geqq 0$; primal infeasible otherwise.

Example 13.1 To Compute the Basic Solution

Consider the transportation problem with a_i, b_j as given below:

	$j = 1$	2	3	4	5	6	a_i
$i = 1$			□			□	7
2	□				□		17
3		□	□				5
4	□			□		□	24
b_j	15	10	9	3	8	8	

Compute the basic solution corresponding to the basic set marked by an "□" in the array. Every cell that is not marked by an □ in the array is a nonbasic cell and, hence, the value of x_{ij} corresponding to all such cells is zero in the basic solution.

There must exist either a row or a column of the transportation array that contains exactly one basic cell. In this example, it is column 4 that contains only the basic cell (4, 4). The sum of all the variables corresponding to cells in this column must equal $b_4 = 3$. This implies that $x_{44} = 3$ in the basic solution. Likewise, $x_{25} = 8$. A similar argument shows that $x_{32} = 10$. This and the fact that (3, 3) is the only other basic cell in the third row imply that $x_{33} = 5 - x_{32} = -5$. Consequently, $x_{13} = 9 - x_{33} = 14$. Hence, $x_{16} = 7 - x_{13} = -7$. Thus $x_{46} = 8 - x_{16} = 15$. And $x_{41} = 24 - x_{44} - x_{46} = 6$, implying that $x_{21} = 15 - x_{41} = 9$. Thus, the basic solution corresponding to this basic set is given below. The values of all the nonbasic variables in the basic solution are 0, and these are not entered. This is an infeasible basis to the problem, since some of the basic values violate the nonnegativity restrictions.

	$j = 1$	2	3	4	5	6
$i = 1$			14			−7
2	9				8	
3		10	−5			
4	6			3		15

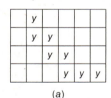

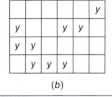

13.1.6 θ-Loops

A θ-loop is a subset of cells of the transportation array with the following properties: (1) it is nonempty; (2) it contains either 0 or 2 cells from each row and column of the array; and (3) no proper subset of it satisfies both properties (1) and (2). If a subset of cells form a θ-loop, entries of $+\theta$ and $-\theta$ can be put alternately in the cells of the subset, such that if a row or column of the array contains a cell from the subset with a $+\theta$ entry, then it also contains one other cell from the subset, and it has a $-\theta$ entry. When this is done, every cell in the subset will end up having either a $+\theta$ entry or a $-\theta$ entry.

Example 13.2

In array 1 given below we have a θ-loop marked with $+\theta$ and $-\theta$ entries. The subset of cells marked with y in array 2 satisfies properties (1) and (2), but not (3); hence it is not a θ-loop.

Array 1

θ				$-\theta$
	θ	$-\theta$		
$-\theta$		θ		
	$-\theta$			θ

Array 2

		y	y		
		y	y		
				y	y
				y	y

THEOREM 13.3 LINEAR DEPENDENCE OF A θ-LOOP A subset of cells of the transportation array that forms a θ-loop is linearly dependent.

Proof: Put entries of $+\theta$ and $-\theta$ alternately among the cells in the θ-loop, and entries of 0 in all the cells of the array, not in the θ-loop. When $\theta = 1$, the sum of all the entries in each row and column of the array is zero. This implies that if we take the column vectors in (13.1) corresponding to the cells in the θ-loop and multiply them by $+1$ if the cell had a $+\theta$ entry and by -1 if the cell had a $-\theta$ entry, and sum, we get the 0 vector. Hence, the set of these column vectors is linearly dependent. ∎

THEOREM 13.4 TRIANGULARITY OF A COLLECTION OF CELLS WITHOUT A θ-LOOP A nonempty collection of cells Δ of the transportation array that contains no θ-loop satisfies the following properties.

1. There exists a row or column of the array that contains exactly one cell from Δ.
2. Every nonempty subset of such a collection of cells Δ satisfies property (1).

Proof: Suppose Δ does not satisfy property (1). Go through the following procedure, which involves several steps. Pick any cell from Δ, say (i_1, j_1) and put a $+\theta$ entry in it. This completes Step 1.

Now consider the general step $r + 1$ in this procedure, for $r \geqq 1$. Let (i_r, j_r) be the last cell from Δ allocated in step r (that cell has a $+\theta$ entry). By our assumptions, row i_r of the array must contain at least one cell from Δ other than the cell (i_r, j_r). Two cases can occur. If all the cells in row i_r from Δ other than (i_r, j_r) appear in one of the columns dropped already in the procedure, select one of them, say (i_r, j_t), put a $-\theta$ entry in it, and terminate the procedure. On the other hand, if there exists a cell in row i_r from Δ other than (i_r, j_r) that does not lie in any of the columns dropped so far, select one of them, say (i_r, j_{r+1}), and put a $-\theta$ entry in it. This is the *first cell* from Δ allocated in this step $r + 1$. Drop row i_r from further consideration. By our assumptions, column j_{r+1} of the array must contain some cells from Δ other than the cell (i_r, j_{r+1}). Again two cases can occur. If all the cells in column j_{r+1} from Δ other than (i_r, j_{r+1}) appear in one of the rows that has been already dropped in the procedure, select one of them, say (i_t, j_{r+1}), put a $+\theta$ entry in it, and terminate the procedure. On the other hand, if there exists a cell in column j_{r+1} from Δ other than (i_r, j_{r+1}) that does not lie in any of the rows dropped so far, select one of them, say (i_{r+1}, j_{r+1}), and put a $+\theta$ entry in it. This is the *second cell* from Δ allocated in this step $r + 1$. Drop column j_{r+1} from further consideration and continue by going to the next step.

In Step g, for $g \geqq 2$, if the procedure did not terminate, we would have dropped a new row and a new column of the array from further consideration. Since there are only m rows and n columns in the array, the procedure must terminate after at most 2 + minimum $\{m, n\}$ steps.

If the last entry made in the procedure is a $-\theta$ entry, and if it is made in column j_1 of the array, then all the cells from Δ that have either a $+\theta$ or a $-\theta$ entry in them, form a θ-loop. If the last entry made in the procedure is a $-\theta$ entry made in column j_t, where $t \neq 1$, all the cells beginning with the second cell allocated in Step t and those allocated subsequently in the procedure, form a θ-loop. If the last entry made in the procedure is a $+\theta$ entry in row i_t, all the cells beginning with the first cell allocated in Step t and those allocated subsequently in the procedure form a θ-loop.

Thus in any case, if Δ does not satisfy property (1), Δ contains a θ-loop as a subset, and the above procedure finds it, contradicting the hypothesis. Hence property (1) must hold. Since Δ contains no θ-loop, every nonempty subset of Δ cannot contain a θ-loop itself, and hence, it must satisfy property (1) too. ■

THEOREM 13.5 Let Δ be a subset of cells of the transportation array. If Δ does not contain a θ-loop as a subset, Δ is linearly independent.

Proof: If Δ is empty, it is obviously linearly independent. Suppose Δ is nonempty. If it is linearly dependent, there must exist a nonzero linear combination of the column vectors in (13.1) corresponding to the cells in Δ, which is equal to 0.

Enter the multipliers of the column vectors in this linear combination, in the corresponding cells in the m × n transportation array and enter 0 in all the cells of the array, not in Δ. Since the linear combination of the column vectors is the zero vector, the sum of all these entries in each row and column of the transportation array must be 0. The subset of all the cells with nonzero entries is a nonempty subset of Δ, and since the sum of these entries in each row and column of the transportation array is 0, this subset of cells cannot satisfy property (1) of Theorem 13.2. This is a contradiction to the hypothesis that Δ contains no θ-loop. ■

Test for Linear Independence

From these results it is clear that a subset of cells Δ of the transportation array is linearly independent iff there is no subset of Δ that is a θ-loop. Equivalently, if Δ is nonempty, it is linearly independent iff Δ and every nonempty subset of Δ satisfies the property that there exists a row or column of the array that contains exactly one cell of the subset. This leads to the following simple test to check whether a subset of cells of the transportation array, Δ, is linearly independent. Strike off all the cells in rows and columns of the transportation array that contain no cells of Δ. The subset of cells Δ is linearly independent iff properties (1) and (2) of Section 13.1.5 hold when $\mathscr{B}$ is replaced by Δ.

A θ-Loop is a Minimal Linearly Dependent Set

A collection of cells of the transportation array is said to be a *minimal linearly dependent set* iff (1) the collection is linearly dependent, and (2) no proper subset of the collection is linearly dependent. From the definition of a θ-loop and from the results discussed here, it is clear that a collection of cells from the transportation array is a minimal linearly dependent set iff the collection is a θ-loop.

13.1.7 θ-Loop in $\mathscr{B} \cup \{(p, q)\}$

THEOREM 13.6 Suppose $\mathscr{B}$ is a basic set of m + n − 1 cells from the m × n transportation array and (p, q) is a nonbasic cell. Then the collection of cells $\mathscr{B} \cup \{(p, q)\}$ contains exactly one θ-loop and this θ-loop contains the nonbasic cell (p, q).

Proof: Since $\mathscr{B}$ forms a basic set, obviously $\mathscr{B}$ cannot contain any θ-loop. Thus, if there is a θ-loop in $\mathscr{B} \cup \{(p, q)\}$, that θ-loop must contain the cell (p, q). Since the rank of $\bar{E}$ is m + n − 1, no subset of m + n cells from the m × n transportation array can be linearly independent. So $\mathscr{B} \cup \{(p, q)\}$ must be linearly dependent, and by Section 13.1.6 it must contain at least one θ-loop. It is a standard result in linear algebra that a set consisting of a basis and exactly one nonbasic column vector contains a unique minimal linearly dependent set. Applying this result to the set of column vectors in (13.5) and (13.6), corresponding to the cells in $\mathscr{B} \cup \{(p, q)\}$ we conclude that $\mathscr{B} \cup \{(p, q)\}$ contains exactly one minimal linearly dependent set. Thus, $\mathscr{B} \cup \{(p, q)\}$ contains exactly one θ-loop. ■

How to Find the θ-Loop in $\mathscr{B} \cup \{(p, q)\}$

Place an entry of $+\theta$ in the nonbasic cell (p, q). Make alternately entries of $-\theta$ and $+\theta$ among the basic cells, such that in the end each row and column of the array contains either a $-\theta$ and $+\theta$ entry, or none at all. The set of all the cells marked with $-\theta$ and $+\theta$ entries at the end is the unique θ-loop in $\mathscr{B} \cup \{(p, q)\}$. When the $+\theta$ and $-\theta$ entries are entered in the cells of the θ-loop in $\mathscr{B} \cup \{(p, q)\}$ in this manner, beginning with a $+\theta$ entry in the nonbasic cell (p, q), the cells with the $+\theta$ entry are known as the *recipient cells*, and the cells with the $-\theta$ are known as the *donor cells* in the θ-loop in $\mathscr{B} \cup \{(p, q)\}$. Notice that (p, q) is a recipient cell in this θ-loop. The variable x_{ij} is called a *donor* or *recipient basic variable* if the corresponding cell is a *donor* or *recipient basic cell*, respectively.

Example 13.3

Consider the 4 × 6 transportation array, and the basic set $\mathscr{B}$ consisting of cells in the array with a □ in them. Suppose it is required to find the θ-loop in $\mathscr{B} \cup \{(2, 2)\}$. Begin by entering a $+\theta$ entry in the cell (2, 2). Now, in row 2, we have to make exactly one $-\theta$ entry among the basic cells in row 2. Suppose the basic cell (2, 5) is selected for this $-\theta$ entry. Then in column 5, we have to introduce exactly one $+\theta$ entry among some other basic cell in that column, and since (2, 5) is the only basic cell in column 5, this is impossible. So the $-\theta$ entry in row 2 cannot be entered in the basic cell (2, 5); hence it has to be entered in the other basic cell (2, 1). Continuing in this manner, the θ-loop in $\mathscr{B} \cup \{(2, 2)\}$ is obtained. So the recipient cells in the θ-loop in $\mathscr{B} \cup \{(2, 2)\}$ in this transportation array are (2, 2), (4, 1), (1, 6), and (3, 3). The donor cells are (2, 1), (4, 6), (1, 3), and (3, 2).

$j =$	1	2	3	4	5	6
$i = 1$			$\square - \theta$			$\square + \theta$
2	$\square - \theta$	$+\theta$			$\square$	
3		$\square - \theta$	$\square + \theta$			
4	$\square + \theta$			$\square$		$\square - \theta$

There is another method for finding the θ-loop in $\mathscr{B} \cup \{(p, q)\}$. At the outset all the m + n cells in $\mathscr{B} \cup \{(p, q)\}$ are uncrossed. Look for a row or a column of the transportation array with a single uncrossed cell from the set $\mathscr{B} \cup \{(p, q)\}$. If such a row or column exists, cross out the single cell that it contains from the set $\mathscr{B} \cup \{(p, q)\}$. Repeat this process again and again until there is no row or column in the transportation array that contains exactly one uncrossed cell from the set $\mathscr{B} \cup \{(p, q)\}$. When this happens, all the remaining uncrossed cells from the set $\mathscr{B} \cup \{(p, q)\}$ form the θ-loop. Starting with a $+\theta$ entry in the cell (p, q), put alternately $-\theta$ and $+\theta$ entries in the remaining uncrossed cells from the set $\mathscr{B} \cup \{(p, q)\}$, until each $+\theta$ entry in any row or column is cancelled by a $-\theta$ entry.

There are other efficient labeling methods for finding the θ-loop in $\mathscr{B} \cup \{(p, q)\}$. They are discussed in Section 13.4.

Exercises

13.4 Consider transportation problem (13.1) in which all the a_i and b_j are given positive integers. Let $\bar{x}$ be a given solution for (13.1). Let $\mathbf{\Delta} = \{(i, j)$: i, j, such that $\bar{x}_{ij}$ is not an integer$\}$. If $\mathbf{\Delta} \neq \emptyset$, prove from first principles that $\mathbf{\Delta}$ must contain at least one θ-loop as a subset.

13.1.8 Useful Results About the Canonical Tableaux for the Balanced Transportation Problem

The equality constraints in (13.1) can be written in the following form

$$\sum_{j=1}^{n} x_{ij} = a_i \qquad \text{for } i = 1 \text{ to m}$$
$$\sum_{i=1}^{m} x_{ij} = b_j \qquad \text{for } j = 1 \text{ to n} - 1 \qquad (13.7)$$

Let T denote the tableau obtained by entering (13.7) in detached coefficient form. Tableau T has m + n − 1 rows. As an illustration, we provide Tableau T when m = 3, n = 4, at the end of this Chapter, in page 432. Let $\hat{x}$ be an m × n matrix with the property that the sum of all the entries in the ith row of $\hat{x}$ is d_i for $i = 1$ to m, and the sum of all the entries in the jth column of $\hat{x}$ is d_{m+j} for $j = 1$ to n. And $\hat{x}$ can be viewed as the matrix obtained by entering the entry $\hat{x}_{ij}$ in the (i, j)th cell of the m × n transportation array for $i = 1$ to m, $j = 1$ to n. Then, with these entries, the row sums in the array are $d_1 \ldots, d_m$ and the column sums are $d_{m+1}, \ldots, d_{m+n}$, respectively. These facts imply that if the column vector corresponding to the variable x_{ij} in tableau T is multiplied by $\hat{x}_{ij}$ and the resulting vectors summed over $i = 1$ to m, $j = 1$ to n, the result will be the column vector $(d_1, \ldots, d_m, d_{m+1}, \ldots, d_{m+n-1})^T$.

Let $\mathscr{B}$ be a basic set of cells for the m × n transportation array, with the cells in $\mathscr{B}$ arranged in some specific order. We refer to $\mathscr{B}$ as a *basic vector of cells* for the transportation problem (13.7). If $x_{\mathscr{B}}$ is the vector of variables in (13.7) corresponding to $\mathscr{B}$, then $x_{\mathscr{B}}$ is a *basic vector* for (13.7), as discussed in Chapters 2 and 3.

THEOREM 13.7 Let $\mathscr{B} = ((i_r, j_r)$: $r = 1$ to m + n − 1) be a basic vector of cells and $x_{\mathscr{B}} = (x_{ij}$: $(i, j) \in \mathscr{B})$ the corresponding basic vector for (13.7). Let (p, q) be a nonbasic cell. Define $(\alpha_1, \ldots, \alpha_{m+n-1})^T$ by $\alpha_r = -1, +1$, or 0, depending on whether (i_r, j_r) is a recipient cell, donor cell, or not contained in the θ-loop in $\mathscr{B} \cup \{(p, q)\}$, respectively. Then the column vector of x_{pq} in tableau T $= \sum_{r=1}^{m+n-1} \alpha_r$ (the column vector of x_{ij} for $(i, j) = (i_r, j_r)$ in tableau T).

Proof: Define $\hat{x} = (\hat{x}_{ij})$, where $\hat{x}_{ij} = 1$ if $(i, j) = (p, q)$, $-\alpha_r$ if $(i, j) = (i_r, j_r)$ for $r = 1$ to m + n − 1, 0 otherwise. From the definition of α_r, it is clear that the sum of all the entries in each row and column of the m × n matrix $\hat{x}$ is zero. Hence by the observation made earlier, if the column vector of x_{ij} in tableau T is multiplied by $\hat{x}_{ij}$ and summed over all i, j, we get 0. From the definition of $\hat{x}$, this implies our result. ■

THEOREM 13.8 Using the notation of Theorem 13.7, let $\bar{T}$ denote the canonical tableau for tableau T with respect to the basic vector $x_{\mathscr{B}}$. Then the updated column vector of the nonbasic variable x_{pq} in $\bar{T}$ is $(\alpha_1, \ldots, \alpha_{m+n-1})^T$, where the α_r are as defined in Theorem 13.7.

Proof: From Section 3.12, we know that the updated column vector of the nonbasic variable x_{pq} in the canonical tableau $\bar{T}$ is the *representation* of the column vector of x_{pq} in tableau T, as a linear combination of the column vectors of the basic variables in tableau T. Hence the result follows from Theorem 13.7. ■

Example 13.4

Consider Array 13.2 and the basic set of cells $\mathscr{B}$, for it marked by a □ in Example 13.7. The θ-loop in $\mathscr{B} \cup \{(4, 2)\}$ is also entered. The donor basic cells in this θ-loop are (4, 4) (1, 6), (2, 1), (3, 2) and the recipient basic cells are (1, 4), (2, 6), (3, 1). It can be verified that the column vector of x_{42} in tableau T corresponding to this problem is equal to the sum of column vectors of the basic variables x_{44}, x_{16}, x_{21} and x_{32} minus the sum of the column vectors of basic variables x_{14}, x_{26}, and x_{31}. This verifies the result in Theorem 13.7 for this case. Also, if we obtain the canonical tableau of tableau T of this problem with respect to the basic vector $x_{\mathscr{B}} = (x_{14}, x_{16}, x_{21}, x_{25}, x_{26}, x_{31}, x_{32}, x_{43}, x_{44})$, it can be verified that the updated column vector of x_{42} in this canonical tableau is $\alpha = (-1, 1, 1, 0, -1, -1, 1, 0, 1)^T$. This is the result of Theorem 13.8 in this case.

Dantzig Property

From Theorem 13.8, we know that the *representation* of any nonbasic column vector in the system of transportation constraints (13.7), as a linear combination of a basic set of column vectors for (13.7), has only entries of 0, +1, or −1 in it. Thus, in the simplex canonical tableau for (13.7) with respect to any basis, the entries in the updated column vectors among the constraint rows will be either 0, +1, or −1. This property is known as the *Dantzig property*.

THEOREM 13.9 Let $\bar{x} = (\bar{x}_{ij})$ be the BFS of (13.7) associated with the basic vector of cells $\mathscr{B}$ and the basic vector $x_{\mathscr{B}}$. Let (p, q) be a nonbasic cell. Make $x_{pq} = \lambda$, all other nonbasic variables = 0, and reevaluate the values of basic variables. This leads to the solution $x(\lambda) = (x_{ij}(\lambda))$, where $x_{ij}(\lambda) = \bar{x}_{ij} + \lambda$, or $\bar{x}_{ij} - \lambda$, or λ, or $\bar{x}_{ij}$, depending on whether (i, j) is a basic recipient cell, basic donor cell, the cell (p, q), or not contained in the θ-loop in $\mathscr{B} \cup \{(p, q)\}$, respectively. If x_{pq} is chosen as the entering variable into $x_{\mathscr{B}}$, the new BFS obtained is $x(\theta_1)$, where θ_1 = minimum $\{\bar{x}_{ij}$: (i, j) a donor basic cell in the θ-loop in $\mathscr{B} \cup \{(p, q)\}\}$.

Proof: This result is analogous to the result in equation (2.6), Section 2.5.3 for the general LP model, and it follows by applying the arguments similar to those used in Sections 2.5.3 and 2.5.4 on (13.7) with the result in Theorem 13.8. From $x(\lambda)$, the maximum value of λ that keeps all $x_{ij}(\lambda) \geqq 0$ is seen to be θ_1. θ_1 is the *primal simplex minimum ratio* for entering x_{pq} into $x_{\mathscr{B}}$, and so $x(\theta_1)$ is the new BFS obtained when this is done. The dropping variable from $x_{\mathscr{B}}$ when x_{pq} enters is any donor basic variable x_{rs}, such that $\bar{x}_{rs} = \theta_1$. For an example, see Section 13.2.3. ■

13.2 REVISED PRIMAL SIMPLEX ALGORITHM FOR THE BALANCED TRANSPORTATION PROBLEM

13.2.1 To Obtain an Initial Basic Feasible Solution

Here we discuss some very efficient special procedures for obtaining an initial BFS for (13.1) using operations on the transportation array only. We assume that the data in the problem satisfy the necessary condition for feasibility (13.2). In our first method, initially, all cells of the transportation array are declared *admissible*. As this method progresses, all the cells in a selected row or a column of the transportation array will be made *inadmissible* in each stage. Also the row and column totals will be modified after each step. The modified row and column totals will be denoted by a'_i and b'_j, respectively. Initially a'_i is the original a_i and b'_j is the original b_j for all i, j. There are several steps in each stage of this procedure.

Step 1 Pick any admissible cell at this stage. It may be worthwhile to pick a cell (r, s) of the form c_{rs} = minimum $\{c_{i,j}$: over all the cells (i, j) that are admissible at this stage$\}$. This may help in producing a BFS that is nearly optimal, even though there is no theoretical guarantee that this will occur.

Step 2 Make the cell (r, s) picked in Step 1 a *basic cell*. Make $x_{rs} = \text{minimum } \{a'_r, b'_s\}$, where a'_r, b'_s are the current values of the modified row and column totals. It is possible that this value is zero. Then, x_{rs} is still treated as a *basic variable* whose value in the BFS is zero.

Step 3 In Step 2, one of the following cases may occur.

Case 1 $a'_r < b'_s$. Make all the cells in the rth row inadmissible for all subsequent stages. Modify b'_s to $(b'_s - a'_r)$. Keep all the other row and column totals unchanged. This completes the stage. Return to Step 1.

Case 2 $a'_r > b'_s$. In this case, make all the cells in the sth column inadmissible for all subsequent stages. Modify a'_r to $(a'_r - b'_s)$ and return to Step 1.

Case 3 $a'_r = b'_s$ and the set of admissible cells at this stage lie in two or more rows. Then make all the cells in the rth row inadmissible for all subsequent stages and modify b'_s to zero. Return to Step 1.

Case 4 $a'_r = b'_s$ and the set of admissible cells at this stage lie in a single row (namely, the rth row), but in two or more columns. Make all the cells in the sth column inadmissible for all subsequent stages. Modify a'_r to zero and return to Step 1.

Case 5 $a'_r = b'_s$ and at this stage (r, s) is the only admissible cell left. In this case, terminate.

Discussion The procedure in this section can only terminate when Case 5 occurs. In each stage before termination, all the cells in either a row or a column of the array are made inadmissible. In the final stage during which termination occurs, only one admissible row and one admissible column are left. In each stage exactly one basic cell is picked. From these it is clear that the number of basic cells picked is exactly $m + n - 1$. From the manner in which the basic cells are picked and the admissible cells are defined for the next stage, it is clear that the set of $(m + n - 1)$ cells picked cannot contain a θ-loop. Hence by the results in Section 13.1, the set of $(m + n - 1)$ cells picked constitute a basic set of cells for the transportation array. From the way the row and column totals are modified in each stage, it is clear that a'_i and b'_j always remain nonnegative for all i, j. Thus, the basic solution obtained is a nonnegative solution and, hence, is a BFS. Some of the variables associated with basic cells may have value equal to zero in the BFS, but they are clearly recorded as basic cells in order to distinguish them from nonbasic cells.

Example 13.5

Find an initial BFS of the transportation problem in which the row and column totals are $a = (a_1, a_2, a_3, a_4) = (6, 25, 20, 13)$, and $b = (b_1, b_2, b_3, b_4, b_5, b_6) = (4, 16, 10, 9, 7, 18)$. In this numerical example, we only wish to illustrate how to get an initial BFS for the transportation problem. Hence, to keep the illustration simple, the objective function is not provided here. Thus, whenever Step 1 has to be executed here, an admissible cell is picked arbitrarily without any regard to the objective function. Suppose (4, 3) is picked as the initial basic cell, the value of x_{43} should be set equal to minimum $\{13, 10\} = 10$. Then modify a'_4 to 3 and make all cells in column 3 inadmissible for the next stage. In the arrays that follow, we indicate the basic cells by putting a small square in the cell with the value of basic variable entered in the square. All inadmissible cells have a dotted line drawn through them. Continuing, we have the following arrays.

	$j = 1$	2	3	4	5	6	a'_i
$i = 1$							6
2							25
3							20
4			$\boxed{10}$				3
b'_j	4	16		9	7	18	

Next cell selected is (2, 5)

	$j = 1$	2	3	4	5	6	a'_i
$i = 1$							6
2					$\boxed{7}$		18
3							20
4			$\boxed{10}$				3
b'_j	4	16		9		18	

Select (4, 4) next

	$j = 1$	2	3	4	5	6	a'_i
$i = 1$							6
2					$\boxed{7}$		18
3							20
4			$\boxed{10}$	$\boxed{3}$			
b'_j	4	16		6		18	

Select (3, 2) next

	$j = 1$	2	3	4	5	6	a_i'
$i = 1$							6
2					[7]		18
3		[16]					4
4			[10]	[3]			
b_j'	4			6		18	

Select (3, 1) next

	$j = 1$	2	3	4	5	6	a_i'
$i = 1$							6
2					[7]		18
3	[4]	[16]					
4			[10]	[3]			
b_j'	0			6		18	

Select (2, 1) next

	$j = 1$	2	3	4	5	6	a_i'
$i = 1$							6
2	[0]				[7]		18
3	[4]	[16]					
4			[10]	[3]			
b_j'				6		18	

Select (2, 6) next

	$j = 1$	2	3	4	5	6	a_i'
$i = 1$							6
2	[0]				[7]	[18]	
3	[4]	[16]					
4			[10]	[3]			
b_j'				6		0	

Select (1, 4) next

	$j = 1$	2	3	4	5	6	a_i'
$i = 1$				[6]			0
2	[0]				[7]	[18]	
3	[4]	[16]					
4			[10]	[3]			
b_j'							0

Select (1, 6) next

	$j = 1$	2	3	4	5	6
$i = 1$				[6]		[0]
2	[0]				[7]	[18]
3	[4]	[16]				
4			[10]	[3]		

All the cells without an entry are nonbasic cells and the values of the corresponding variables in the BFS are all zero.

Vogel's Method

This is a commonly used method for finding an initial BFS for the uncapacitated balanced transportation problem. Except for a special rule for choosing the basic cell among the admissible cells in each step, this method is the same in all other respects as the method discussed above. In any step, let *line* refer to a row or column of the array that contains at least one admissible cell at that stage. Let α and β be the minimum and the second minimum among cost coefficients of admissible cells in this line (after the minimum is excluded, the minimum among the remaining is the second minimum). Then $\beta - \alpha$ is known as the *cost difference* in this line at this stage. Identify the line that has the maximum cost difference at this stage. Select the least cost admissible cell in that line as the basic cell to be selected in this step. The rationale for this selection is the following. If that cell is not selected, any remaining supply or demand in this line has to be shipped using a second minimum cost cell or a higher cost cell in that line, and hence results in the highest increase in unit cost at this stage.

In this method, computation of cost differences and finding the maximum among them imposes additional computational effort in each step. This additional effort seems to be very worthwhile, as Vogel's method seems to produce near optimal BFSs quite often. This has been empirically verified. When used to produce a near optimum BFS, Vogel's method is called *Vogel's approximation method* (or VAM for short).

For an example of Vogel's method, consider the transportation problem with the following data. The original cost coefficient c_{ij}, is entered in the lower right hand corner of cell (i, j). a_i is the units of supply available at source i, and b_j is the units of demand required at demand center j. α, β always denote the minimum and the second minimum cost coefficients among admissible cells at that stage in that row or column. Basic cells are indicated by putting a small square in the center with the value of the basic variable entered inside it. a_i', b_j' are the current modified row and column totals. Admissible cells at any stage are those without a square inside them or a dotted line drawn through them.

$j =$	1	2	3	4	5	6	a_i	α	β	Cost Difference $\beta - \alpha$
$i = 1$	10	20	10	30	15	15	30	10	10	0
2	0	[40] 5	12	13	8	20	60	0	5	5
3	5	40	31	10	22	16	40	5	10	5
4	8	16	21	2	17	4	50	2	4	2
b_j	30	40	13	12	45	40				
α	0	7	10	2	8	3				
β	5	16	12	10	15	4				
Cost Difference $\beta - \alpha$	5	11	2	8	7	1				

$j =$	1	2	3	4	5	6	a_i'	α	β	Cost Difference $\beta - \alpha$
$i = 1$	10	⋮ 20	10	30	15	13	30	10	10	0
2	[20] 0	[40] 5	12	13	8	20	20	0	8	8
3	5	⋮ 40	31	10	22	16	40	5	10	5
4	8	⋮ 16	21	2	17	4	50	2	4	2
b_j'	30		13	12	45	40				
α	0		10	2	8	3				
β	5		12	10	15	4				
Cost Difference $\beta - \alpha$	5		2	8	7	1				

$j =$	1	2	3	4	5	6	a'_i	α	β	Cost Difference $\beta - \alpha$
$i = 1$	10	20	[13] 10	30	15	13	30	10	10	0
2	[20] 0	[40] 5	12	13	8	20				
3	5	40	31	10	22	16	40	5	10	5
4	8	16	21	2	17	4	50	2	4	2
b'_j	10		13	12	45	40				
α	5		10	2	15	4				
β	8		21	10	17	13				
Cost Difference $\beta - \alpha$	3		11	8	2	9				

Initially the maximum cost difference occurs in Column 2, and hence a minimum cost admissible cell in it; cell (2, 2), has been selected as the first basic cell. This leads to the array appearing at bottom of page 390. Row 2 and Column 4 both have the maximum difference of 8. Selecting Row 2, we choose the next basic cell as the minimum cost admissible cell in it, (2, 1). This leads to the array at top of this page. The maximum cost difference at this stage occurs in Column 3. So we select the minimum cost admissible cell in it, (1, 3), as the next basic cell.

Continuing in the same manner, we obtain the following BFS for this problem:

$j =$	1	2	3	4	5	6
$i = 1$	10	20	[13] 10	30	[17] 15	13
2	[20] 0	[40] 5	12	13	8	20
3	[10] 5	40	31	[2] 10	[28] 22	16
4	8	16	21	[10] 2	17	[40] 4

Exercises

13.5 Consider the balanced transportation problem with the following data. (1) Obtain an initial BFS for this problem using the first method discussed above, choosing the basic cell in each step to be the one with minimum c_{ij} over all cells (i, j) admissible at that stage. (2) Obtain an initial BFS for this problem using Vogel's method. (3) Compare the costs of the two BFS's obtained.

	c_{ij}						a_i
$j =$	1	2	3	4	5	6	
$i = 1$	6	2	3	4	5	5	38
2	10	1	15	15	14	16	75
3	8	10	5	7	11	9	40
4	13	1	4	16	12	25	90
b_j	27	13	18	60	25	100	

13.2.2 Finding the Dual Solution

In this chapter we use the symbols u_i, v_j to denote the dual variables. By the results in Chapter 4, the dual of the transportation problem (13.1) is:

$$\begin{aligned} &\text{Maximize} \quad w(u, v) = \sum a_i u_i + \sum b_j v_j \\ &\text{Subject to} \quad u_i + v_j \leqq c_{ij} \qquad \text{for all } i, j \qquad (13.8) \\ &u_i, v_j \qquad \text{unrestricted in sign for all } i, j \end{aligned}$$

The dual constraint $u_i + v_j \leqq c_{ij}$ is associated with the primal variable x_{ij}. So the nonnegative dual slack variable associated with the nonnegative primal variable x_{ij} is $c_{ij} - u_i - v_j$, and the pair $(x_{ij}, c_{ij} - u_i - v_j)$ form a complementary pair in this primal dual pair of problems (see Chapter 4). The dual variable u_i is associated with the primal constraint that requires that the sum of the variables in the ith row of the transportation array should equal a_i. Likewise, the dual variable v_j is associated with the jth column in the transportation array for $j = 1$ to n. Hence the primal constraints, together with the associated dual variables, can be recorded on the transportation array as below (we illustrate the array for the case m = 2, n = 3).

Array 13.1

$j =$	1	2	3	*Availability*	*Associated Dual Variable*
$i = 1$	x_{11}	x_{12}	x_{13}	a_1	u_1
2	x_{21}	x_{22}	x_{23}	a_2	u_2
Requirement	b_1	b_2	b_3		
Associated dual variable	v_1	v_2	v_3		

If $\mathscr{B}$ is a basic set of cells for the m × n transportation array, the dual solution corresponding to this basis is obtained by solving

$$u_r + v_s = c_{rs} \qquad \text{for all basic cells } (r, s) \quad (13.9)$$

The *relative cost coefficient* of any variable x_{ij} in the problem is the associated dual slack variable, namely, $c_{ij} - (u_i + v_j)$.

The transportation problem (13.1) contains exactly one redundant equation. Any one of the constraints in (13.1) can be deleted, and the remaining is an equivalent linearly independent system of constraints. The dual of this equivalent problem is obtained by setting the dual variable corresponding to the deleted constraint equal to zero in (13.8).

To simplify the discussion, assume that the constraint corresponding to $j = \text{n}$ in (13.1) is deleted. This is equivalent to setting $v_\text{n} = 0$ in (13.8). Given a basis consisting of (m + n − 1) basic cells, the corresponding dual solution satisfies (13.9), which is a system of m + n − 1 equality constraints in m + n variables, and it, together with the constraint, $v_\text{n} = 0$, uniquely determines the dual solution. From the results in Section 13.1, it is clear that the dual basic solution can be found by back substitution. There will be at least one basic cell in column n of the transportation array. Suppose (t, n) is a basic cell. Then $v_\text{n} = 0$ and $u_t + v_\text{n} = c_{t,\text{n}}$ together provide the value of the dual variable u_t in the dual basic solution. Substitute the value of u_t in each equation in (13.9) in which it occurs. By the triangularity of the basis this will help in determining the value of another dual variable, and so on.

After all the u_i and v_j are computed, the relative cost coefficients in nonbasic cells (i, j) are calculated from $\bar{c}_{ij} = c_{ij} - u_i - v_j$.

If (u, v) is the dual basic solution corresponding to the basic set $\mathscr{B}$, it is a dual feasible solution iff $u_i + v_j \leqq c_{ij}$, that is, $\bar{c}_{ij} = c_{ij} - u_i - v_j \geqq 0$ for all i and j. The basic set of cells $\mathscr{B}$, for the m × n transportation array is said to be a *dual feasible basic set of cells for (13.1)* if the dual basic solution corresponding to it is dual feasible (i.e., iff all the relative cost coefficients are $\geqq 0$), *dual infeasible* otherwise.

Example 13.6

Consider the transportation problem with the cost matrix given at the top of page 393. *The c_{ij} are entered in the lower right corner in the cell.* It is required to find the dual solution corresponding to the basic set consisting of all the cells with a square marked in the cell. In determining the dual solution, only the cost coefficients of the basic cells are used. We set $v_\text{n} = 0$ and find the rest of the dual solution by back substitution. Setting $v_6 = 0$ and looking at the equations (13.9) corresponding to the basic cells (2, 6) and (1, 6) leads to $u_2 + v_6 = -1$, so $u_2 = -1$; and $u_1 + v_6 = 8$, so $u_1 = 8$. Now using the values of u_1, u_2 in the equations (13.9) corresponding to basic cells (2, 1), (1, 4), and (2, 5), we get $v_1 = -2$, $v_4 = -7$, and $v_5 = 4$. In a similar way, we proceed and obtain the dual solution indicated on the following array. *The relative cost coefficients $\bar{c}_{ij}$ are entered in the upper left corner of the cell.* In each basic cell, the relative cost coefficient is zero, and this is not entered in the array.

	$j = 1$	2	3	4	5	6	u_i
$i = 1$	$4 = \bar{c}_{11}$ $c_{11} = 10$	-4 4	1 12	□ 1	0 12	□ 8	8
2	□ -3	3 2	4 6	10 2	□ 3	□ -1	-1
3	□ 2	□ 4	-5 2	13 10	-6 2	-3 1	4
4	10 10	-4 -2	□ 5	□ -5	1 7	3 5	2
v_j	-2	0	3	-7	4	0	

Optimality criterion

The present BFS is optimal if the relative cost coefficients $\bar{c}_{ij}$ are nonnegative for all i, j.

Hence a basic set of cells $\mathscr{B}$ for the m × n transportation array is an *optimal basic set of cells* if $\mathscr{B}$ is both a primal feasible and dual feasible basic set for (13.1). The primal simplex algorithm for the balanced transportation problem starts with a primal feasible basic set of cells (obtained by the methods discussed in Section 13.2.1, or by some other method) and moves to an optimal basic set of cells, changing one basic cell in each pivot step, while retaining primal feasibility throughout the algorithm.

13.2.3 Improving a Nonoptimal Basic Feasible Solution

If the optimality criterion is not satisfied, pick some nonbasic cell whose relative cost coefficient is negative, and try to bring it into the basic set. Suppose the nonbasic cell to be brought into the basic set is (p, q), $\bar{c}_{pq} < 0$.

Find the θ-loop in the set of cells consisting of the cell (p, q) and the basic cells. Place an entry of $+\theta$ in the cell (p, q) and entries of $-\theta$ and $+\theta$ alternately among the cells in the θ-loop. Let $\tilde{x}$ be the present BFS, and $x(\theta) = (x_{ij}(\theta))$, where $x_{ij}(\theta) = \tilde{x}_{ij} + \theta$, or $\tilde{x}_{ij} - \theta$, or $\tilde{x}_{ij}$, depending on whether (i, j) is a recipient cell, donor cell, or not in the θ-loop, respectively. $x(\theta)$ is the solution obtained if we make $x_{pq} = \theta$, all other nonbasic variables = 0, and reevaluate the basic variables (see Theorem 13.9). The maximum value of θ that keeps $x(\theta) \geqq 0$ is the *minimum ratio*; $\theta_1 = \text{minimum } \{x_{rs}: (r, s) \text{ a donor basic cell}\}$. Any donor basic cell that ties for this minimum in the definition of θ_1 is eligible to be the *dropping cell* when (p, q) enters the basic set. (p, q) can replace any one of these basic cells eligible to drop, leading to the new basic set. The new BFS is $x(\theta_1)$ with an objective value of $z(\tilde{x}) + \bar{c}_{pq}\theta_1$. This process of obtaining the new BFS $x(\theta_1)$, and the new basic set associated with it is known as a *pivot step* in this algorithm.

Iterative Process

The procedure can be repeated with the new basic set, until a basic set satisfying the optimality criterion is obtained.

Exercises

13.6 Let $\mathscr{B}$ be a feasible basic set for (13.1). Let (p, q) be a nonbasic cell. Obtain the θ-loop in $\mathscr{B} \cup \{(p, q)\}$. Define the net unit cost of this θ-loop to be $\sum(c_{ij}: (i, j) \text{ a recipient cell}) - \sum(c_{ij}: (i, j) \text{ a donor cell})$. It is the cost of making a unit adjustment along this θ-loop, increasing shipments in recipient cells by one unit, and reducing shipments in donor cells by one unit. Prove that this net unit cost in this θ-loop is $\bar{c}_{pq}$, the relative cost coefficient of (p, q) with respect to $\mathscr{B}$. Illustrate with a numerical example.

Example 13.7

Solve the transportation problem with the cost matrix given in Example 13.6 and the a, b vectors given in Example 13.5. We pick the BFS to this problem obtained in Example 13.5. The relevant data are:

Array 13.2

	$j = 1$	2	3	4	5	6
$i = 1$	$4 = \bar{c}_{11}$; $c_{11} = 10$	-4 ; 4	-1 ; 12	$x_{14} = \boxed{6} + \theta$; 1	0 ; 12	$\boxed{0} - \theta$; 8
2	$\boxed{0} - \theta$; -3	3 ; 2	4 ; 6	10 ; 2	$\boxed{7}$; 3	$\boxed{18} + \theta$; -1
3	$\boxed{4} + \theta$; 2	$\boxed{16} - \theta$; 4	-5 ; 2	13 ; 10	-6 ; 2	-3 ; 1
4	10 ; 10	-4 θ ; -2	$\boxed{10}$; 5	$\boxed{3} - \theta$; -5	1 ; 7	3 ; 5

The optimality criterion is not satisfied, since in particular $\bar{c}_{42} = -4 < 0$. Therefore, bring (4, 2) into the basic set. This leads to the θ-loop entered in the array. The recipient basic cells are (1, 4), (2, 6), and (3, 1). The donor basic cells are (1, 6), (2, 1), (3, 2), and (4, 4). The new solution is therefore $(x_{14}, x_{16}, x_{21}, x_{25}, x_{26}, x_{31}, x_{32}, x_{42}, x_{43}, x_{44}) = (6 + \theta, 0 - \theta, 0 - \theta, 7, 18 + \theta, 4 + \theta, 16 - \theta, \theta, 10, 3 - \theta)$ and all other $x_{ij} = 0$. The maximum value that θ can have is $\theta_1 = \text{minimum } \{16, 0, 0, 3\} = 0$. So x_{42} enters the basic vector with a value $\theta_1 = 0$ and it can replace either x_{21} or x_{16} from the basic vector. Suppose we drop x_{21} from the basic vector. This is a *degenerate pivot step* and the new basis corresponds to the same BFS.

Compute the dual solution and the relative cost coefficients corresponding to the new basis as before. This leads to the next array appearing at the bottom of this page.

Bring the cell (3, 3) with relative cost coefficient $\bar{c}_{33} = -9$ into the basic set. The θ-loop generated is indicated in the array. The maximum possible value for θ is minimum $\{16, 10\} = 10$. When $\theta = 10$, the basic variable x_{43} becomes equal to zero and is dropped from the basic vector. The new BFS is:

	$j = 1$	2	3	4	5	6
$i = 1$				$\boxed{6}$		$\boxed{0}$
2					$\boxed{7}$	$\boxed{18}$
3	$\boxed{4}$	$\boxed{6}$	$\boxed{10}$			
4		$\boxed{10}$		$\boxed{3}$		

Compute the dual solution corresponding to the new basis and the new relative cost coefficients, and test whether the new basis satisfies the optimality criterion. Otherwise continue the algorithm until the optimality criterion is satisfied.

Entering Cell Choice Rules

Let $\bar{c} = (\bar{c}_{ij})$ be the matrix of relative cost coefficients with respect to the feasible basic set $\mathscr{B}$. If $\bar{c} \ngeq 0$, every nonbasic cell (i, j) for which $\bar{c}_{ij} < 0$ is an *eligible cell* for being chosen as the entering cell into

Array 13.3

	$j = 1$	2	3	4	5	6	u_i
$i = 1$	$8 = \bar{c}_{11}$; $c_{11} = 10$	0 ; 4	1 ; 12	$\boxed{6}$; 1	0 ; 12	$\boxed{0}$; 8	8
2	4 ; -3	7 ; 2	4 ; 6	10 ; 2	$\boxed{7}$; 3	$\boxed{18}$; -1	-1
3	$\boxed{4}$; 2	$\boxed{16} - \theta$; 4	-9 θ ; 2	9 ; 10	-10 ; 2	-7 ; 1	8
4	14 ; 10	$\boxed{0} + \theta$; -2	$\boxed{10} - \theta$; 5	$\boxed{3}$; -5	1 ; 7	3 ; 5	2
v_j	-6	-4	3	-7	4	0	

the present basic set. If there are many eligible cells, the following rules may be used to select the entering cell from them.

Rule 1 Most Negative Relative Cost Coefficient Rule Here the entering cell is selected as a nonbasic cell (p, q) satisfying $\bar{c}_{pq}$ = minimum $\{\bar{c}_{ij}$: overall $i, j\}$. As discussed in Section 2.5.7, this selection does not guarantee that the maximum decrease in the objective value is achieved in each pivot step. This selection rule seems to work well in practice, but it does require several comparisons to identify the most negative relative cost coefficient in the current array.

Example 13.8

In array 13.3 of Example 13.7 the entering cell selected by this rule is (3, 5), with a relative cost coefficient of -10. It can be verified that this actually leads to a degenerate pivot step with no change in the objective value at all.

Rule 2 Entering Cell Choice to Achieve the Greatest Decrease in Objective Value in Each Pivot Step In this rule the entering cell is chosen as that eligible cell that leads to the maximum decrease in the objective value. To adopt this rule, one has to obtain the θ-loop corresponding to each eligible cell in the array, which is computationally burdensome. Hence this rule is rarely used in practical applications.

Rule 3 The LRC (Least Recently Considered) Rule To apply this rule, it is necessary to arrange all the cells in the transportation array in a specific order and fix this order before the algorithm is initiated. Suppose this order is $((i_1, j_1), \ldots, (i_{mn}, j_{mn}))$. In any pivot step the entering cell is selected as follows: Let (i_r, j_r) be the entering cell in the previous step. Examine the cells in the specific order $((i_{r+1}, j_{r+1}), \ldots, (i_{mn}, j_{mn}), (i_1, j_1), \ldots, (i_r, j_r))$. The first cell in this order that is eligible to enter the basic set (i.e., the first cell (i_s, j_s) in this order satisfying $c_{is,js} < 0$) is the entering cell in this step. So this rule circles through the list of cells looking for the entering cell.

Rule 4 The LRB (Least Recently Basic) Rule This rule chooses as the entering cell among those eligible to enter the basic set, the one that was least recently in the basic set.

Rule 5 The IMN (Inward Most Negative) Rule To apply this rule, it is necessary to arrange the columns of the array in a specific order and fix this order before the algorithm is initiated. Suppose this order is $(j_1, \ldots, j_n)$. In any pivot step, the entering cell is selected by this rule as follows: Let (p, j_t) be the entering cell in the previous step. Examine the columns in the specific order $(j_{t+1}, \ldots, j_n, j_1, \ldots, j_t)$. Let j_s be the first column of the array in this specific order that contains a cell eligible to enter the basic set in this step. Choose r such that $\bar{c}_{r,j_s}$ = minimum $\{\bar{c}_{i,j_s}; i = 1 \text{ to } m\}$. Then (r, j_s) is the entering cell in this pivot step.

Rule 6 The AIMN (Altered Inward Most Negative) Rule This rule is the same as IMN, except that the specific order in which the columns are examined is $(j_t, j_{t+1}, \ldots, j_n, j_1, \ldots, j_{t-1})$.

See [13.5, 13.11, 13.12, 13.13, 13.24] for a discussion of various entering cell, dropping cell choice rules, and for comparisons of their computational efficiencies.

Degenerate and Nondegenerate Pivot Steps

Let $\mathscr{B}$ be a feasible basic set of cells of the $m \times n$ transportation array for (13.1), associated with the BFS $\tilde{x} = (\tilde{x}_{ij})$. $\tilde{x}$ is said to be a *nondegenerate* BFS of (13.1) if $\tilde{x}_{ij} > 0$ for all basic cells (i, j). If $\tilde{x}_{ij} = 0$ for at least one basic cell (i, j), $\tilde{x}$ is said to be a *degenerate* BFS of (13.1). Notice that whenever a degenerate BFS is obtained during the algorithm, it is necessary to distinguish between basic cells with zero value and nonbasic cells very clearly, as otherwise the dual solution cannot be computed.

Suppose $\tilde{x}$ is a degenerate BFS, and suppose (p, q) is selected as the entering cell in this step. Let θ_1 = minimum $\{\tilde{x}_{ij}$: (i, j) a donor basic cell in the θ-loop in $\mathscr{B} \cup \{(p, q)\}\}$. The operation of bringing (p, q) into the basic set $\mathscr{B}$ is a *degenerate pivot step* if $\theta_1 = 0$, *nondegenerate pivot step* otherwise. In a degenerate pivot step, the entering cell replaces a zero-valued basic cell from the basic set, and there is no change in the objective value or the current BFS. In every nondegenerate pivot step a new BFS is obtained, accompanied by a strict decrease in the objective value.

In the process of solving a transportation problem, if a degenerate pivot step appears, it is carried out as usual and the algorithm continued.

	$j = 1$	2	3	4	5	6	a_i
$i = 1$	$c_{11} = 2$	7	(4) 6	(5) 7	13	15	9
2	(3) 0	10	(6) 5	15	15	(7) 8	16
3	5	(11) 2	9	(3) 5	(10) 5	(8) 7	32
4	0	(1) 1	(2) 3	8	5	(9) 6	12
b_j	3	12	12	8	10	24	

Exercises

13.7 Complete the solution of the transportation problem in array 13.3. Also solve the following transportation problem:

								a_i
	3	18	5	3	9	6	−2	51
$c =$	4	0	4	6	6	8	18	16
	2	8	3	17	−4	18	16	17
	−6	7	19	9	3	0	12	11
b_j	9	12	25	12	10	20	7	

Solve the following transportation problem. Starting with $\{(1, 2), (1, 5), (2, 1), (2, 4), (3, 2), (3, 3), (3, 6), (4, 4), (4, 5)\}$ as the initial basic set.

							a_i
	2	0	3	5	6	3	10
$c =$	1	3	4	0	9	7	24
	3	7	9	1	8	1	9
	4	8	7	3	6	8	36
b_j	24	4	1	19	23	8	

13.8 In the balanced transportation problem (13.1), prove that the set of optimum feasible solutions is unaffected if the cost matrix is modified by adding or subtracting a constant to all elements in a row or column of (c_{ij}). What happens to the set of optimum feasible solutions if all the elements in a row or column of the cost matrix are multiplied by a positive constant?

13.9 Consider the transportation problem discussed in Exercise 13.5. Are any of the BFSs obtained there optimal? If not, solve this problem beginning with the best of the two BFSs obtained there.

13.10 Consider the following transportation problem. If $\tilde{u} = (-7, 8, 3, 0)$ and $\tilde{v} = (-4, -3, 2, 1, 4, 0)$, is $(\tilde{u}, \tilde{v})$ an optimum feasible solution of the dual problem? Test by using the complementary slackness theorem. At the same time, obtain all the optimal BFSs of this transportation problem.

							a_i
	11	−10	1	−6	1	−7	17
$c =$	11	11	10	12	12	10	7
	2	0	7	8	7	6	9
	−4	−2	2	3	4	2	12
b_j	10	12	7	3	8	5	

13.11 Consider the balanced transportation problem and the solution $(\tilde{x}_{ij})$ tabulated within small circles inside the cells given at the top. (a) Is $(\tilde{x}_{ij})$ a BFS for this transportation problem? If not, using θ-loops, obtain a BFS from it. (b) Is $(\tilde{x}_{ij})$ an optimum feasible solution?

13.2.4 Degeneracy in the Transportation Problem

Some degenerate pivot steps might be encountered in the process of solving (13.1) by the primal algorithm. As in solving the general LP model, there may be a possibility of *cycling under degeneracy* (see Chapter 10). An example of a balanced transportation problem in which cycling has been shown to occur was constructed by L. Johnson and is reported in the paper by B. J. Gassner [13.9]. This example is a balanced transportation problem with the following data:

	c_{ij}				
$j =$	1	2	3	4	a_i
$i = 1$	3	5	5	11	1
2	9	7	9	15	1
3	7	7	11	13	1
4	13	13	13	17	1
b_j	1	1	1	1	

Clearly this is an assignment problem of order 4. The cycle of degenerate pivot steps in this example goes through the following basic sets. All pivot steps are degenerate pivots steps, and every one of these basic sets corresponds to the BFS $\hat{x} = (\hat{x}_{ij}) = I_4$, the unit matrix of order 4.

It can be verified that the relative cost coefficient in the entering cell in each of the pivot steps here is -2. The basic set of cells at the end of Step 12 is the same as the initial basic set of cells at the beginning of Step 1.

This clearly establishes that under degeneracy, cycling can occur in the primal algorithm for the balanced transportation problem. When cycling occurs, the algorithm keeps going around the basic sets of cells in the cycle, and never terminates. We now describe a method using perturbations of the a_i, b_j, with which it can be mathematically guaranteed that an optimum solution of the problem will be obtained after a finite number of pivot steps of the primal algorithm. Since cycling does not seem to be a problem in practical applications, this perturbation method is mostly of theoretical interest.

THEOREM 13.10 Let $\bar{x} = (\bar{x}_{ij})$ be the basic solution of (13.1) associated with the basic set $\mathscr{B}$. For each $(i, j) \in \mathscr{B}$, there exists proper subsets $\mathbf{I} \subset \{1, \ldots, m\}$, $\mathbf{J} \subset \{1, \ldots, n\}$, such that $\bar{x}_{ij} = (\pm 1)(\sum_{i \in \mathbf{I}} a_i - \sum_{j \in \mathbf{J}} b_j)$.

Proof: In $\bar{x}$, all nonbasic variables are zero. The basic values are computed by the back-substitution method discussed in Section 13.1.5. In this procedure, a basic cell, say (r, s), that is the only basic cell in its row or column is identified. If (r, s) is the only basic cell in row r, then $x_{r,s} = a_{rj}$ modify b_s into $b'_s = b_s - a_r$, and eliminate row r from further consideration. On the other hand, if (r, s) is the only basic cell in column s, $x_{rs} = b_{sj}$ modify a_r into $a' = a_r - b_s$ and eliminate column s from further consideration. Repeat in the same way with the remaining basic cells, rows, columns, and the modified row and column totals in them.

At some state in this procedure suppose row i, column j remain, and a'_i, b'_j are the modified totals in them. Suppose the values of the basic variables in basic cells $(i_1, j_1), \ldots, (i_t, j_t)$ have been evaluated so far. Then it can be shown that there exist subsets $\mathbf{M}_1, \mathbf{M}_2 \subset \{i_1, \ldots, i_t\}$, and $\mathbf{N}_1, \mathbf{N}_2 \subset \{j_1, \ldots, j_t\}$, such that $a'_i = a_i + \sum (a_g\colon\ g \in \mathbf{M}_1) - \sum (b_h\colon\ h \in \mathbf{N}_1)$ and

The Basic Sets of Cells in the Various Steps of a Cycle of Degenerate Pivot Steps

Step	Basic Set of Cells	*Entering Cell*	*Dropping Cell*
1	{(1, 1), (2, 2), (3, 3), (4, 4); (1, 2), (2, 3), (3, 4)}	(1, 3)	(2, 3)
2	{(1, 1), (2, 2), (3, 3), (4, 4); (1, 2), (1, 3), (3, 4)}	(4, 2)	(1, 2)
3	{(1, 1), (2, 2), (3, 3), (4, 4); (4, 2), (1, 3), (3, 4)}	(3, 2)	(3, 4)
4	{(1, 1), (2, 2), (3, 3), (4, 4); (4, 2), (1, 3), (3, 2)}	(4, 1)	(4, 2)
5	{(1, 1), (2, 2), (3, 3), (4, 4); (4, 1), (1, 3), (3, 2)}	(4, 3)	(1, 3)
6	{(1, 1), (2, 2), (3, 3), (4, 4); (4, 1), (4, 3), (3, 2)}	(2, 1)	(4, 1)
7	{(1, 1), (2, 2), (3, 3), (4, 4); (2, 1), (4, 3), (3, 2)}	(3, 1)	(3, 2)
8	{(1, 1), (2, 2), (3, 3), (4. 4); (2, 1), (4, 3), (3, 1)}	(2, 4)	(2, 1)
9	{(1, 1), (2, 2), (3, 3), (4, 4); (2, 4), (4, 3), (3, 1)}	(2, 3)	(4, 3)
10	{(1, 1), (2, 2), (3, 3), (4, 4); (2, 4), (2, 3), (3, 1)}	(1, 4)	(2, 4)
11	{(1, 1), (2, 2), (3, 3), (4, 4); (1, 4), (2, 3), (3, 1)}	(3, 4)	(3, 1)
12	{(1, 1), (2, 2), (3, 3), (4, 4); (1, 4), (2, 3), (3, 4)}	(1, 2)	(1, 4)
13	{(1, 1), (2, 2), (3, 3), (4, 4); (1, 2), (2, 3), (3, 4)}		

$b'_j = b_j + \sum(b_h: h \in \mathbf{N}_2) - \sum(a_g: g \in \mathbf{M}_2)$. To prove this, verify that it holds after the first step in which the first basic variable is evaluated. Suppose it holds in some step. Let $x_{k\ell}$ be the basic variable evaluated in this step, with a'_k, b'_ℓ being the current modified row and column totals. Then in this step either a'_k is modified to $a'_k - b'_\ell$ or b'_ℓ is modified to $b'_\ell - a'_k$, so it continues to hold after this step. So, by induction, it holds in all steps.

By the nature of the procedure, the value of each basic variable is equal to either the modified row or column total at the stage when it is evaluated. Hence, our theorem follows from the result in the above paragraph. ■

THEOREM 13.11 The balanced transportation problem (13.1) has a primal degenerate basic solution iff there exist proper subsets $\mathbf{M} \subset \{1, \ldots, m\}$, $\mathbf{N} \subset \{1, \ldots, n\}$, such that $\sum_{i \in \mathbf{M}} a_i = \sum_{j \in \mathbf{N}} b_j$.

Proof: A basic solution $\bar{x}$ of (13.1) is degenerate iff a basic value $\bar{x}_{ij} = 0$. The "only if" part of this theorem follows from this and Theorem 13.10. To prove the "if" part, suppose we are given the proper subsets **M** and **N**, satisfying the above property. Apply the algorithm discussed in Section 13.2.1, restricting the choice of basic cells to those with $i \in \mathbf{M}$ and $\mathbf{j} \in \mathbf{N}$ initially until $|\mathbf{M}| + |\mathbf{N}| - 1$ basic cells have been selected. Verify that the resulting BFS is degenerate. ■

The Perturbation Method for Resolving Degeneracy

Consider the transportation problem (13.1), in which a_i and b_j are all strictly positive and $\sum a_i = \sum b_j$. Perturb a_i to $f_i = a_i + \varepsilon$ for $i = 1$ to m. Leave $b_1, \ldots, b_{n-1}$ unchanged, but perturb b_n to $g_n = b_n + m\varepsilon$. Let $g_j = b_j$, $j = 1$ to $n - 1$. Here ε is a small positive number. If a_i, b_j are all positive integers, ε can be taken to be any positive number less than 1/(2m). If a_i and b_j are all positive rational numbers that have been rounded off at the level 10^{-r}, the ε can be taken to be any positive number less than $(10^{-r})/(2m)$. However, it is not necessary to give a specific value to ε, if can be left as a parameter that is treated as a positive number smaller than any positive number with which it is compared. From the definition of ε, f_i, g_j, it is clear that there exist no proper subsets $\mathbf{M} \subset \{1, \ldots, m\}$, $\mathbf{N} = \{1, \ldots, n\}$ satisfying $\sum_{i \in \mathbf{M}} f_i = \sum_{j \in \mathbf{N}} g_j$. This implies by Theorem 13.11 that the perturbed problem is a nondegenerate transportation problem. From the definition of ε, it can also be verified that any basic set of cells for the m × n transportation array that is feasible to the perturbed problem is also feasible to (13.1). If a basic set of cells for the m × n transportation array is optimal to the perturbed problem, it can also be verified to be optimal to (13.1).

If ε is retained as a parameter, the basic solution $\bar{x}$ of the perturbed problem corresponding to any basic set of cells for the m × n transportation array is of the form $(\bar{x}_{ij} + \varepsilon\bar{x}^*_{ij})$, with an objective value of the form $\bar{z} + \varepsilon\bar{z}^*$, where $(\bar{x}_{ij})$, $\bar{z}$ correspond to the basic solution and objective value for the original right-hand-side constants a_i, b_j, and $(\bar{x}^*_{ij})$, $\bar{z}^*$ correspond to the basic solution and objective value for the right-hand-side constants equal to the coefficients of ε in the perturbed problem. In this case, the solution of the perturbed problem can be carried out by maintaining and updating $\bar{x}_{ij}$, $\bar{x}^*_{ij}$'s separately and always remembering that the value of x_{ij} in the solution is $\bar{x}_{ij} + \varepsilon\bar{x}^*_{ij}$, where ε is a small positive parameter.

The perturbed problem can now be solved by the primal algorithm discussed earlier. Since it is totally (primal) nondegenerate, the primal algorithm will terminate in a finite number of pivot steps, with an optimum BFS of the perturbed problem. The BFS of (13.1) corresponding to the same basic set of cells is an optimum solution of (13.1).

Comment 13.1: This perturbation method is due to A. Orden. See [13.23 pp. 242–243] and [13.18].

13.3 UNIMODULAR AND TOTALLY UNIMODULAR MATRICES

Let A be a real matrix of order $m \times n$ and rank r. The matrix A is said to be a *unimodular matrix* (or, to have the *unimodularity property*) if the determinant of every square submatrix of A of order r is either 0, +1, or −1. Notice that here there is no condition imposed on the determinants of square submatrices of A of order $r - 1$ or less (here remember that the number r is the rank of A).

The matrix A is said to be a *totally unimodular matrix* (or to have the *total unimodularity property*) if the determinant of *every* square submatrix of A is either 0, +1, or −1. This automatically implies that all the entries in A itself are either 0, $+1$, or -1 if it is totally unimodular. Notice the difference in the conditions characterizing unimodularity and total unimodularity. Every totally unimodular matrix is unimodular, but the converse may not be true.

Example 13.9

Consider the matrix $B_n = (\tilde{M}(n) \vdots -\tilde{M}(n))$ of order $n \times 2n$ and rank n, where $\tilde{M}(n)$ is defined in Section 8.12. For $n = 3$, this matrix is

$$B_3 = \begin{pmatrix} 1 & 0 & 0 & -1 & 0 & 0 \\ 2 & 1 & 0 & -2 & -1 & 0 \\ 2 & 2 & 1 & -2 & -2 & -1 \end{pmatrix}$$

It can be verified that every nonsingular square submatrix of B_n of order n is a lower triangular matrix with its diagonal entries equal to $+1$ or -1, and hence its determinant is either $+1$ or -1. Hence this matrix B_n is unimodular. Since some of the entries in B_n are equal to -2, B_n is not totally unimodular.

Consider the LP: minimize cx, subject to $Ax = b$, $x \geqq 0$. From Cramer's rule (Section 3.3.9), if A is totally unimodular and b is an integer vector, every basic solution of this LP will be an integer vector. Hence if A is totally unimodular and we are required to solve the integer program: minimize cx, subject to $Ax = b$, $x \geqq 0$ and x an integer vector; we can neglect the integer constraints and solve the remaining as an ordinary LP by the simplex method. The optimum BFS obtained will actually be an optimum solution of the integer program.

It was proved in Section 13.1 that the input-output coefficient matrix of the transportation problem is totally unimodular. Hence, as long as a_i and b_j are integers for all i and j, the optimum solution of (13.1) obtained by the simplex algorithm will be an integer solution. This is what makes integer programs of the transportation problem type so trivial.

Exercises

13.12 Consider the following assignment problem. Show that this can be solved by the algorithm discussed in Section 13.2. Prove that this algorithm leads to an optimum solution to the assignment problem in which all the x_{ij} are either 0 or 1.

Minimize

$$z(x) = \sum_{i=1}^{n} \sum_{j=1}^{n} c_{ij} x_{ij}$$

Subject to

$$\sum_{i=1}^{n} x_{ij} = 1 \qquad \text{for all } j = 1 \text{ to n}$$

$$\sum_{j=1}^{n} x_{ij} = 1 \qquad \text{for all } i = 1 \text{ to n}$$

$$x_{ij} \geqq 0$$

13.13 Prove the following result by A. J. Hoffman and D. Gale by an inductive argument. Let A be an $m \times$ n matrix whose rows can be partitioned into two disjoint sets **S** and $\bar{\mathbf{S}}$ with the following properties: (1) every column vector of A contains at most two nonzero entries; (2) every entry in A is 0, $+1$, or -1; (3) if two nonzero entries in a column vector of A have the same sign, then one of these nonzero entries is in **S** and the other is in $\bar{\mathbf{S}}$; (4) if two nonzero entries in a column vector of A have opposite signs, then both the rows in which these entries lie are in either **S** or $\bar{\mathbf{S}}$. Then A is a unimodular matrix.

Comment 13.2: Recently, tremendous theoretical advances have been made in characterizing total unimodularity of matrices. Based on the earlier work of J. Edmonds and W. H. Cunningham, Paul D. Seymour developed a polynomially bounded algorithm for testing whether a given matrix is totally unimodular. See reference [13.22]. As shown by K. Truemper (see Exercise 13.14 and reference [13.28]), these results can also be used to check whether a matrix is unimodular if it turns out to be not totally unimodular.

Exercises

13.14 (1) Let A be a given matrix of order $m \times n$. If A contains an identity matrix of order m as a submatrix, prove that it is unimodular iff it is totally unimodular. (2) Let A be a given matrix of order $m \times n$ and rank m. Let B be any basis for A. If the determinant of B is $\neq \pm 1$, A is not unimodular. If the determinant of A is ± 1, prove that A is unimodular iff the matrix $B^{-1}A$ is totally unimodular. Use this result to develop an efficient algorithm to check whether a given matrix is unimodular, based on a subroutine for checking whether a given matrix is totally unimodular.

13.4 LABELING METHODS IN THE PRIMAL TRANSPORTATION ALGORITHM

In Sections 13.1 and 13.2, we discussed only a trial and error method for finding the θ-loop in $\mathscr{B} \cup \{(p, q)\}$. Unless the problem being solved is very small, this trial and error method is likely to be time consuming. Besides, it is very difficult to program it on a computer. The labeling method discussed in this section provides an efficient and systematic method for finding the θ-loop in $\mathscr{B} \cup \{(p, q)\}$. Most of the commercially available computer programs for solving transportation problems use this labeling method. The development of this labeling method has resulted in tremendous improvements in the efficiency with which large transportation problems can be solved using computers.

13.4.1 Review of Graph Theory Concepts

A *graph* is a pair of sets. One of them is a finite set $\mathscr{N}$ of points (also called *nodes* or *vertices*). The other set is a set of lines $\mathscr{A}$ called *edges*, each edge joining a pair of distinct points in $\mathscr{N}$. If $i \in \mathscr{N}$, $j \in \mathscr{N}$, $i \neq j$, the edge joining i and j will be denoted by $(i; j)$ or equivalently $(j; i)$, and this edge is said to be *incident* at points i and j. The graph itself is $G = (\mathscr{N}, \mathscr{A})$. There is at most one edge between any pair of points, and every edge contains exactly two points of $\mathscr{N}$.

Given two nodes i_0 and i_* in $\mathscr{N}$, a path from i_0 to i_* in the graph G is a connected sequence of nodes and edges of the form: i_0, $(i_0; i_1)$, i_1, $(i_1; i_2)$, $i_2, \ldots, i_{k-1}$, $(i_{k-1}; i_*)$, i_*, such that each edge in the sequence is in $\mathscr{A}$. For brevity, we represent the above path by the ordered set of edges $[(i_0; i_1), \ldots, (i_{k-1}; i_*)]$. The length of a path is defined to be equal to the number of edges in the sequence. Thus the length of the path from i_0 to i_* discussed above is k, even though some edges may appear more than once in the sequence.

A path is called a *simple path* if every node along the path appears in the sequence only once. A *cycle* is a path from a node to itself that contains at least two edges. A cycle is a *simple cycle* if it is a simple path from a node to itself.

A graph is a *connected graph* if there exists a path in the graph between every pair of points in the graph. Otherwise it is a *disconnected graph*. A connected graph that contains no cycles is called a *tree*. Given a set of points $\mathscr{N}$, the graph $G = (\mathscr{N}, \mathscr{A})$ is said to be a *spanning tree* if G is a tree and if it contains an edge incident at every point in $\mathscr{N}$. If $G = (\mathscr{N}, \mathscr{A})$ is a tree, a point in $\mathscr{N}$ is said to be a *terminal node* (also called *end node* or *pendant node*) if there is exactly one edge in $\mathscr{A}$ incident at it (see Figure 13.1).

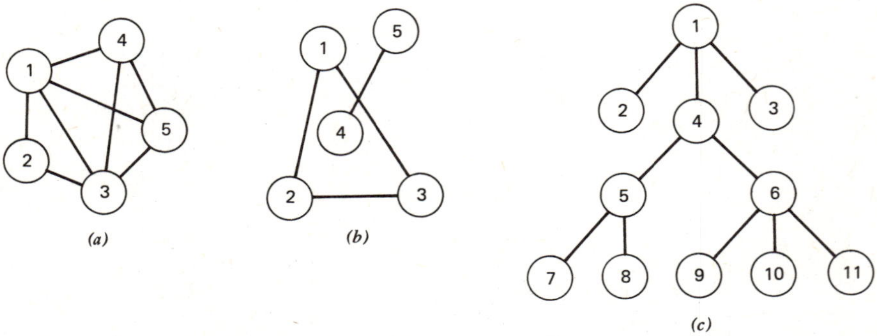

Figure 13.1 (*a*) A connected graph with 5 points and 8 edges. (*b*) A disconnected graph. (*c*) A tree with 11 points. Points 2, 3, 7, 8, 9, 10, 11 are terminal nodes.

THEOREM 13.12 Let i_0, i_* be a pair of points on a tree G. Then there exists a unique simple path in G from i_0 to i_*.

Proof: Since G is a tree, it is connected and, hence, there exists at least one path from i_0 to i_*. If there exist two distinct simple paths from i_0 to i_* in G, these paths together constitute a cycle from i_0 to i_0, contradicting the hypothesis that G is a tree. ■

Exercises

13.15 Prove that every tree has at least one terminal node.

13.16 If $\mathscr{N} = \{1, \ldots, r\}$ and $G = (\mathscr{N}, \mathscr{A})$ is a spanning tree, prove that the number of edges in $\mathscr{A}$ is $r - 1$.

13.17 If $\mathscr{N} = \{1, \ldots, r\}$ and $G = (\mathscr{N}, \mathscr{A})$ is a graph consisting of $(r - 1)$ edges (i.e., $|\mathscr{A}| = r - 1$) containing no cycles, prove that G is a connected graph and hence is a spanning tree spanning the nodes in $\mathscr{N}$.

13.4.2 Applications to the Transportation Problem

Consider a balanced transportation problem with m sources and n markets. The *i*th row of the transportation array is associated with the *i*th source, $i = 1$ to m, and we represent it by a point whose *serial number* is *i*. The *j*th column of the transportation array is associated with the *j*th market, $j = 1$ to n, and we will represent it by a point whose serial number is $m + j$. Let $\mathscr{N} = \{1, \ldots, m + 1, \ldots, m + n\}$. Let Δ be a subset of cells of the transportation array. Determine the corresponding set of edges $\mathscr{A}_\Delta = \{(i; m + j)$: the cell (i, j) is in $\Delta\}$. Edges in the graph are denoted by unordered pairs $(i; j)$ with a semicolon in the middle, and here i, j are serial numbers of points in the graph. Cells in the transportation array are denoted by ordered pairs (i, j) with a comma in the middle, this being the cell in row *i* and column *j* of the array. Then $G_\Delta = (\mathscr{N}, \mathscr{A}_\Delta)$ is the graph associated with the subset Δ.

THEOREM 13.13 Simple cycles and θ-Loops Every θ-loop in Δ corresponds to a simple cycle in G_Δ and vice versa.

Proof: This follows from the definitions of θ-loops and simple cycles. ■

Hence $\boldsymbol{\Delta}$ contains a θ-loop iff there is a simple cycle in $G_{\boldsymbol{\Delta}}$.

$G_{\boldsymbol{\Delta}} = (\mathscr{N}, \mathscr{A}_{\boldsymbol{\Delta}})$ is an undirected network, where the set of points $\mathscr{N}$ contains $m + n$ points. From the results in Exercises 13.16 and 13.17 it follows that if $\mathscr{A}_{\boldsymbol{\Delta}}$ consists of $m + n - 1$ edges and $G_{\boldsymbol{\Delta}}$ contains no simple cycles, then $G_{\boldsymbol{\Delta}}$ is connected and is also a spanning tree. From these and from the results in Section 13.1, it is clear that $\boldsymbol{\Delta}$ forms a basic set of cells for the $m \times n$ transportation array iff $G_{\boldsymbol{\Delta}} = (\mathscr{N}, \mathscr{A}_{\boldsymbol{\Delta}})$ is a spanning tree.

To Draw $G_{\mathscr{B}}$ As a Rooted Tree

Suppose $\mathscr{B}$ is a basic set of cells for the $m \times n$ transportation array. Draw the graph $G_{\mathscr{B}} = (\mathscr{N}, \mathscr{A}_{\mathscr{B}})$ as follows: Pick $m + n$ as the initial node. When drawn in the manner discussed below, the tree $G_{\mathscr{B}}$ is called a *rooted tree* with $m + n$ as the *root*. Any other point in $\mathscr{N} = \{1, \ldots, m, m + 1, \ldots, m + n\}$ could have been picked as the root, but to keep the discussion here compatible with Section 13.2, we pick $m + n$ as the root. The *predecessor index* $P(m + n)$ of the root node $m + n$ is always $= \emptyset$, the empty set. This completes *stage* 0. We will now discuss what happens in a general stage, stage *k*, where $k \geqq 1$.

Stage k In this stage include in the tree all points $i \in \mathscr{N}$ that have not yet been included, satisfying $(i; j) \in \mathscr{A}_{\mathscr{B}}$ for some *j* with the property that *j* is a point included in the graph in stage $k - 1$. The predecessor index $P(i)$ of such a node *i* is defined to be *j*. Draw the edge $(i; j)$. In this case, the point *j* is said to be the *immediate predecessor* or the *parent* of point *i*; and point *i* is said to be an *immediate successor* or a *son* or a *child* of point *j*. The edge $(i; j)$ is said to be an *in-tree edge*. If the tree now contains all the points in $\mathscr{N}$, terminate. Otherwise go to stage $k + 1$.

Example 13.10

Consider the basic set, $\mathscr{B}$, of cells of the transportation array marked by a *T* in the array at the top of page 402.

Column *j* of this transportation array corresponds to the point $4 + j$ in the graph. Row *i* corresponds to point *i*. The rooted tree corresponding to $\mathscr{B}$ is the graph in Figure 13.2, excluding the dashed line (1; 9).

	$j = 1$	2	3	4	5	6	Associated point serial number	Predecessor index	Successor index	Younger brother index	Elder brother index
$i = 1$			T	T			1	8	7	$\varnothing$	$\varnothing$
2	T					T	2	10	5	4	$\varnothing$
3		T			T		3	6	9	$\varnothing$	$\varnothing$
4		T		T		T	4	10	6	$\varnothing$	2
Associated point serial number	5	6	7	8	9	10					
Predecessor index	2	4	1	4	3	$\varnothing$					
Successor index	$\varnothing$	3	$\varnothing$	1	$\varnothing$	2					
Younger brother index	$\varnothing$	8	$\varnothing$	$\varnothing$	$\varnothing$	$\varnothing$					
Elder brother index	$\varnothing$	$\varnothing$	$\varnothing$	6	$\varnothing$	$\varnothing$					

The predecessor indices are entered on the transportation array itself. The tree is drawn in such a way that if points i and j are connected by an edge and j is below i, then $P(j) = i$. If $\Delta = \mathscr{B} \cup \{(1, 5)\}$, the graph G_Δ is the graph in Figure 13.2, including the dashed line (1; 9), and it contains the simple cycle corresponding to the θ-loop in $\mathscr{B} \cup \{(1, 5)\}$.

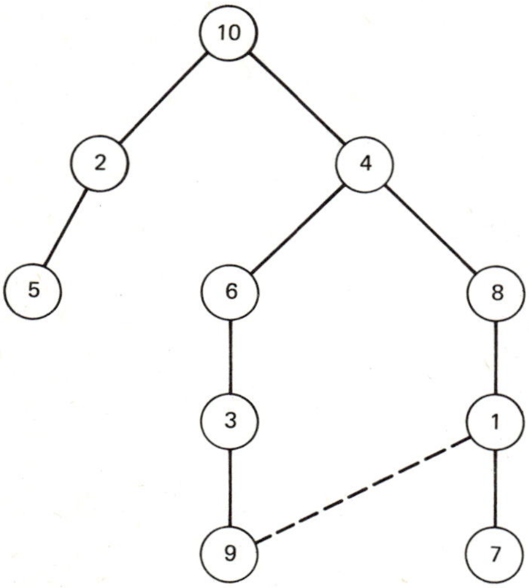

Figure 13.2

Properties of the Rooted Tree

The rooted tree corresponding to a basic set of cells of the transportation array has the following properties. (1) All points included in the tree during a stage of even number in the procedure discussed above are associated with columns of the transportation array and all points included in the tree during a stage of odd number in the procedure are associated with rows of the transportation array. (2) A cell of the transportation array always corresponds to an edge in the graph between two points that are included in the tree at different stages in the procedure. (3) All points have a predecessor index, except the root node. The predecessor index of a point associated with a row of the array is always the serial number of a point associated with a column of the array and vice versa.

13.4.3 The Successor and Brother Indices

Let i be any point in the rooted tree $G_{\mathscr{B}}$. The set of all points in this graph of the form $\{j\colon P(j) = i\}$ is known as the set of *immediate successors* or the set of *sons* or *children* of the point i. Clearly the edge $(i; j)$ is the last edge in the unique simple path from the root node to j for all points j in this set.

If i is a point in G, the set of points $\{j\colon j \neq i$ and j is such that i is a point on the unique simple path from

Point	1	2	3	4	5	6	7	8	9	10	11	12	13	14	15	16	17
Predecessor	13	7	17	15	17	11	5	4	2	3	2	4	6	1	3	3	∅
Successor	14	9	10	8	7	13	2	∅	∅	∅	6	∅	1	∅	4	∅	3
Younger brother	∅	∅	5	∅	∅	∅	∅	12	11	15	∅	∅	∅	∅	16	∅	∅
Elder brother	∅	∅	∅	∅	3	∅	∅	∅	∅	∅	9	8	∅	∅	10	15	∅

the root node to $j\}$ is known as the set of all *descendants* of i. Clearly the set of all descendants of i is the union of the set of immediate successors of i and the sets of all descendants of j as j ranges over the set of immediate successors of i.

If i is not the root node and is a terminal node of $G_{\mathscr{B}}$, its set of immediate successors and its set of all descendants are empty.

In the course of the algorithm, it is necessary to compute the set of immediate successors and the set of descendants. For doing this computation, the successor and brother indices of points are defined. For all terminal nodes not equal to the root node, the *successor index* is defined to be the empty set, since these points have no successors. If the point i has a unique immediate successor j, the successor index of i, S(i), is defined to be j. If a point i has two or more immediate successors, these successors are all *brothers* of each other. Thus, j and k are brothers iff they have the same predecessor index. These successors are arranged in some order as a sequence from left to right. This order can be chosen arbitrarily. If j and k are two points in this sequence and j appears to the left of k, then k is known as a *younger brother* of j and j is an *elder brother* of k. The leftmost of all these brothers is the *eldest son* of i. The *successor index* S(i) is the serial number of the eldest son of i.

The *younger brother index* of a point j, denoted by YB(j), is the serial number of the eldest among the set of younger brothers of j (i.e., the leftmost of the younger brothers of j in the ordering chosen for the brothers of j), if the set of younger brothers of j is nonempty. YB(j) is the empty set if j has no younger brothers.

The *elder brother index* of a point j, EB(j), is the serial number of the youngest (i.e., the rightmost of the elder brothers of j in the sequence of brothers) of the set of elder brothers of j, if this set is nonempty. EB(j) is the empty set if j has no elder brothers.

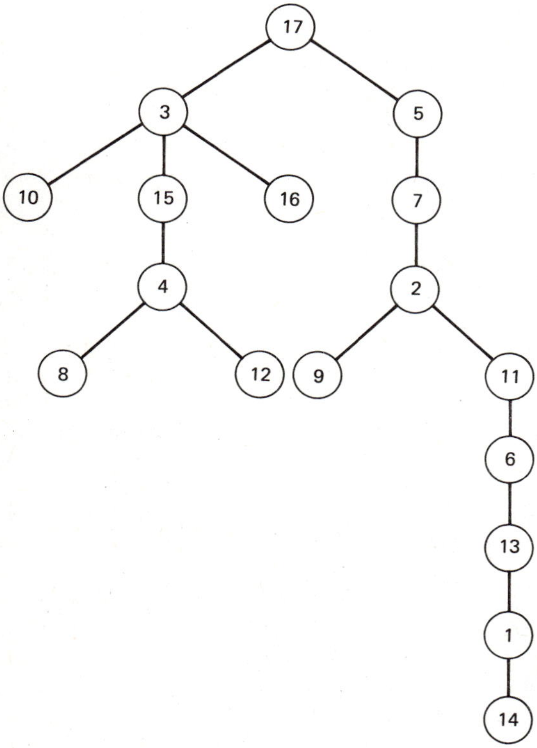

Figure 13.3

Example 13.11

For a transportation problem with $m = 6$, $n = 11$, the rooted tree corresponding to a basis is given in Figure 13.3. The root node is 17. Assuming that the brothers are ordered from left to right as they are recorded in the diagram above, the various indices are tabulated at the top.

Example 13.12

For Example 13.10 in Section 13.4.2, the successor and brother indices are recorded on the transportation array.

To Obtain the Sets of Immediate Successors, Younger Brothers, and Descendants

When the various indices are defined as in the previous sections, the set of younger brothers of a point j is empty if $\mathrm{YB}(j) = \varnothing$. If $\mathrm{YB}(j) \neq \varnothing$, the set of younger brothers of j is the union of $\{\mathrm{YB}(j)\}$ and the set of younger brothers of $\mathrm{YB}(j)$. Using this recursively, the set of younger brothers of any point in the graph can be found by using only the younger brother indices.

The set of immediate successors of a point i is empty if $\mathrm{S}(i) = \varnothing$. If $\mathrm{S}(i) \neq \varnothing$, the set of immediate successors of i is the union of $\{\mathrm{S}(i)\}$ and the set of younger brothers of $\mathrm{S}(i)$.

The set of descendants of a point i is empty if the successor index of i is empty. Otherwise it is the union of the set of immediate successors of i and the sets of descendants of each of the immediate successors of i.

13.4.4 To Obtain the Simple Path Between a Point and the Root Node

Suppose i is a point in the rooted tree $\mathrm{G}_{\mathscr{B}}$, which is not the root node. The following procedure finds the unique simple path between it and the root node.

Step 1 The edge $(i; \mathrm{P}(i))$ is the first edge in the path. If $\mathrm{P}(i)$ is the root node, terminate. Otherwise pick $\mathrm{P}(i)$ as the *current point* and go to Step 2.

Step 2 Suppose the current point is j. $(j, \mathrm{P}(j))$ is the next edge in the path. Go to Step 3.

Step 3 If $\mathrm{P}(j)$ is the root node, terminate. Otherwise, change the current point to $\mathrm{P}(j)$ and go back to Step 2.

The path from any node i in the rooted tree $\mathrm{G}_{\mathscr{B}}$, to the root node is known as the *predecessor path of node i* in $\mathrm{G}_{\mathscr{B}}$. It can be obtained by starting at node i and going up recursively, using only the predecessor indices of the nodes. When the path obtained here is written in the reverse order, it becomes the simple path from the root node to i in the rooted tree $\mathrm{G}_{\mathscr{B}}$.

Example 13.13

Consider the basis for the 4×6 transportation problem discussed in Example 13.10. The root node in the rooted tree corresponding to this basis is 10. Suppose it is required to find the simple path from 9 to the root node. Point 9 corresponds to column 5 in the transportation array and its predecessor index is 3. So the edge (9; 3) is in the path. The predecessor index of row 3 is 6. So the edge (3; 6) is in the path. The predecessor index of point 6 (column 2) is 4, which puts the edge (6; 4) in the path. The predecessor index of point 4 (row 4) is 10, which puts (4; 10) in the path. Since 10 is the root node, the procedure terminates here. Hence the path is ((9; 3), (3; 6), (6; 4), (4; 10)). Verify this from Figure 13.2.

13.4.5 To Determine the θ-Loop in $\mathscr{B} \cup \{(p, q)\}$

Suppose $\mathscr{B}$ is a basic set of cells for the $\mathrm{m} \times \mathrm{n}$ transportation array and (p, q) is a nonbasic cell. The unique θ-loop in $\mathscr{B} \cup \{(p, q)\}$ can be determined easily using the predecessor indices.

The column q of the $\mathrm{m} \times \mathrm{n}$ transportation array corresponds to the point $\mathrm{m} + q$ in the rooted tree $\mathrm{G}_{\mathscr{B}}$. Hence the cell (p, q) corresponds to the edge, $(p; \mathrm{m} + q)$ in $\mathrm{G}_{\mathscr{B}}$. The θ-loop in $\mathscr{B} \cup \{(p, q)\}$ corresponds to the unique simple cycle in the graph $(\mathscr{N}, \mathscr{A}_{\mathscr{B}} \cup \{(p; \mathrm{m} + q)\})$.

Find the unique simple path from $\mathrm{m} + q$ to the root node. Also find the unique simple path from p to the root node. Eliminate all common edges in these two paths. Suppose the remaining portions of these paths are $((p; i_1), (i_1; i_2), \ldots, (i_k; i_*))$ and $((\mathrm{m} + q; j_1), (j_1; j_2), \ldots, (j_t; i_*))$. The common point i_* in these two is known as the *apex* of the simple cycle. The simple cycle itself is $((p; \mathrm{m} + q), (\mathrm{m} + q; j_1), \ldots, (j_t; i_*), (i_*; i_k), \ldots, (i_2; i_1), (i_1; p))$.

An edge $(i; j)$ of the rooted tree where i is a point corresponding to a row of the transportation array and j is a point corresponding to a column of the array corresponds to the cell $(i, j - \mathrm{m})$ in the array. The cells in the θ-loop are the cells of the transportation array corresponding to the edges in the simple cycle. When all the cells in the θ-loop are determined, entries of $+\theta$ and $-\theta$ can be put alternately in them, starting with a $+\theta$ entry in the cell (p, q).

In the primal algorithm for solving the transportation problem, the present basic cell that drops off from the basic set when (p, q) is the entering cell is determined as in Section 13.2.3, once the θ-loop corresponding to (p, q) is determined. Suppose the cell (p, q) replaces the cell (r, s) from the basic set. Let the new basic set be $\mathscr{B}'$.

The graph $\mathrm{G}_{\mathscr{B}'}$ corresponding to the new basic set is obtained from the graph $\mathrm{G}_{\mathscr{B}}$ by deleting the edge $(r; \mathrm{m} + s)$ and adding the edge $(p; m + q)$.

In the pivot step of bringing the cell (p, q) into the basic set $\mathscr{B}$, the edge corresponding to (p, q) is known as the *entering edge*, and the edge corresponding to the cell dropping off from $\mathscr{B}$ is known as the *dropping edge*.

Example 13.14

Consider the basis for the 6×11 transportation problem discussed in Example 13.11. Suppose the θ-loop corresponding to the nonbasic cell (4, 5) has to be found. The edge corresponding to cell (4, 5) is (4; 11). The simple path from the point 4 to the root node is ((4; 15), (15; 3), (3; 17)), and the simple path from the point 11 to the root node is ((11; 2), (2; 7), (7; 5), (5; 17)). These two paths have no common edges and 17 is the common point on these two paths. Hence 17 is the apex of this simple cycle and the simple cycle is ((4; 11), (11; 2), (2; 7), (7; 5), (5; 17), (17; 3), (3; 15), (15; 4)).

So the θ-loop consists of the cells (4, 5), (2, 5), (2, 1), (5, 1), (5, 11), (3, 11), (3, 9), (4, 9). The cells with $+\theta$ entries are (4, 5), (2, 1), (5, 11), (3, 9), and the cells with $-\theta$ entries are (2, 5), (5, 1), (3, 11), (4, 9).

Example 13.15

For the same 6×11 transportation problem and the basic set $\mathscr{B}$ discussed in Example 13.11, find the θ-loop corresponding to the nonbasic cell (1, 3).

The cell (1, 3) corresponds to the edge (1; 9) in the rooted tree. The simple path from the point 1 to the root node is ((1; 13), (13; 6), (6; 11), (11; 2), (2; 7), (7; 5), (5; 17)), and the path from the point 9 to the root node is ((9; 2), (2; 7), (7; 5), (5; 17)). Eliminating the common edges between these two paths, it is seen that the simple cycle in this problem is ((9; 1), (1; 13), (13; 6), (6; 11), (11; 2), (2; 9)) and the apex is 2. This corresponds to the θ-loop consisting of the cells ((1, 3), (1, 7), (6, 7), (6, 5), (2, 5), (2, 3)). The cells with $+\theta$ entries are (1, 3), (6, 7), (2, 5) and the cells with $-\theta$ entries are (1, 7), (6, 5), (2, 3).

13.4.6 To Determine the Simple Path Between Two Points in the Rooted Tree

Suppose i_1 and i_* are two points in the rooted tree $G_{\mathscr{B}}$. To find the simple path in $G_{\mathscr{B}}$ between i_1 and i_*, find the simple path from i_1 to the root node and the simple path from i_* to the root node in $G_{\mathscr{B}}$ by the method discussed in Section 13.4.4. Eliminate the common edges in these two paths. The set of remaining edges on these two paths consists of the edges on the simple path from i_1 to i_* in $G_{\mathscr{B}}$.

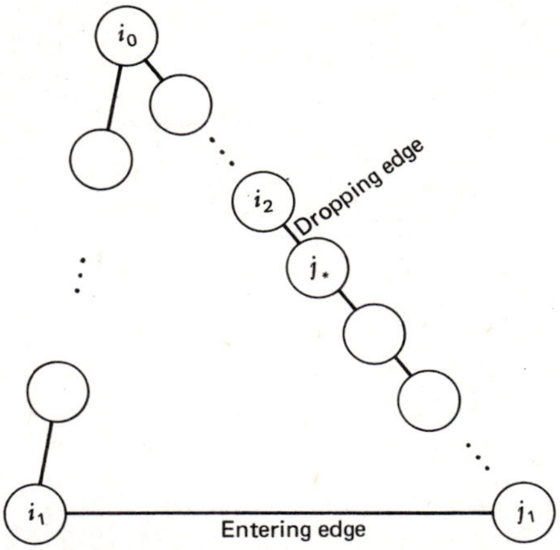

Figure 13.4 The simple cycle, where i_0 is the Apex, $(i_1; j_1)$ and $(i_2; j_*)$ are the entering and dropping edges, respectively.

13.4.7 Updating the Rooted Tree and the Various Indices

Suppose $\mathscr{B}$ is a basic set of cells for the $m \times n$ transportation array and let $G_{\mathscr{B}}$ be the corresponding rooted tree. Suppose the rooted tree $G_{\mathscr{B}'}$ of an adjacent basic set $\mathscr{B}'$ is obtained by dropping the edge $(i_2; j_*)$ from $G_{\mathscr{B}}$ and adding the edge $(i_1; j_1)$. The names of the nodes on these edges are given by the specific rules given below. The dropping edge consists of a parent node and a son node, call them i_2, j_*, respectively. So $P(j_*) = i_2$.

Suppose i_0 is the apex of the simple cycle in $(\mathscr{N}, \mathscr{A}_{\mathscr{B}} \cup \{(i_1; j_1)\})$. The simple cycle consists of the edge $(i_1; j_1)$, a path from i_0 to j_1 in reverse order, and a path from i_0 to i_1. The edge $(i_2: j_*)$ being deleted will be contained on exactly one of these two paths. Denote by j_1 the node on the entering edge whose predecessor path contains the dropping edge (i_2, j_*), and call the other node on the entering edge i_1 (see Figure 13.4).

Note 13.1: We explain the notation used in Figure 13.4 here. Let (p, q) be the entering cell in the array in this pivot step. So the entering edge is the out-of-tree edge $(p; m + q)$. Let point i_0 be the apex. The leaving edge when $(p; m + q)$ is brought in is

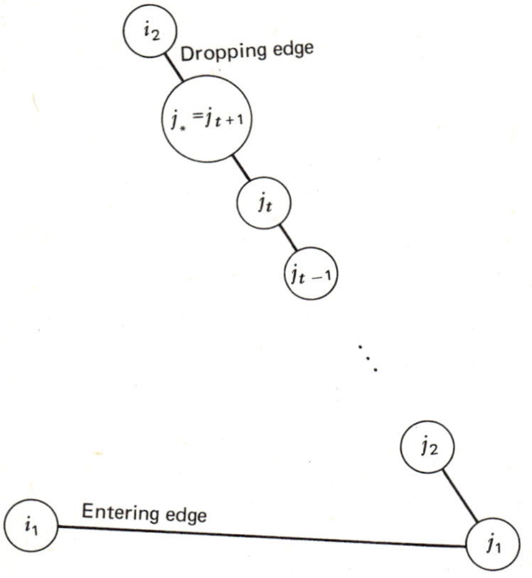

Figure 13.5 Portion of the simple cycle of Figure 13.4, between points i_1 and i_2.

an in-tree edge, and hence it consists of a parent node and a son node; let them be i_2, j_*, respectively. So i_2 is the predecessor index of j_*, $(i_2; j_*)$ is the leaving edge, and it must lie either on the path from p to i_0 or on the path from $m + q$ to i_0 in the present tree. Let $j_1 = p$, $i_1 = m + q$ if the leaving edge lies on the path from p to i_0 in the present tree; otherwise, let $j_1 = m + q$, $i_1 = p$.

The path from j_1 to i_0 on this simple cycle contains the path from j_1 to j_*. Suppose this path is $((j_1; j_2), (j_2; j_3), \ldots, (j_t; j_*))$ where the edges in this path are written in such a way that $P(j_u) = j_{u+1}$ for $u = 1$ to $t - 1$ and $P(j_t) = j_*$. We refer to j_* as j_{t+1} (see Figure 13.5).

Compute the set of all descendants of j_* in the present rooted tree $G_{\mathscr{B}}$ using the procedure in Section 13.4.3. Let **H** be the set containing j_* and the set of all descendants of j_*. We use the symbols $P(i)$, $S(i)$, $YB(i)$, $EB(i)$ to indicate present indices, and $P'(i)$, $S'(i)$, $YB'(i)$, $EB'(i)$ to indicate the updated indices (i.e., the new indices after the basis change). It is convenient to do the updating in the order indicated below.

These formulas for updating can be viewed as those arising in a gravity model of the rooted tree. Think of the rooted tree $G_{\mathscr{B}}$ standing with the root at the top and successor nodes going down. $G_{\mathscr{B}'}$ is obtained by joining the points i_1, j_1 by an edge in $G_{\mathscr{B}}$, and then snipping off the current edge-connecting points i_2 and j_*. When this is done, the points along the chain connecting $i_1, j_1, \ldots, j_*$ in Figure 13.4 fall down by gravity, revolving around each edge as the chain falls down, giving us the new rooted tree $G_{\mathscr{B}'}$. In this process, if some points acquire a new immediate successor, we assume that this immediate successor joins at the left of the existing immediate successors of that point (i.e., as an elder brother of all the existing immediate successors of that point).

Updating the Predecessor Indices

The predecessor indices of all the points except $j_1, \ldots, j_t, j_{t+1}$ remain unchanged. Set $P'(j_1) = i_1$, $P'(j_u) = j_{u-1}$, for $u = 2$ to $t + 1$.

Updating the Successor Indices

The successor indices for all points other than $i_1, j_1, \ldots, j_t, j_{t+1}, i_2$ remain unchanged. Set $S'(i_1) = j_1$, $S'(j_u) = j_{u+1}$, for $u = 1$ to t. If $S(i_2) = j_*$, set $S'(i_2) = YB(j_*)$. If $S(i_2) \neq j_*$, make $S'(i_2) = S(i_2)$. If $S(j_*) = j_t$, set $S'(j_*) = YB(j_t)$. If $S(j_*) \neq j_t$, make $S'(j_*) = S(j_*)$.

Updating the Brother Indices

To update the brother relationships, we will make the assumption that any new immediate successor of a point joins the previous immediate successors of this point as their eldest brother (i.e., any new immediate successor always joins at the left end of the sequence of brothers). Also, if a point is to be removed from the set of immediate successors, we assume that the elder-younger brotherly relationship among the remaining points in this set remains unchanged. This is compatible with what has been done in updating the successor indices.

The brother indices may change only among the set **H** (the set containing j_* and the descendants of j_* in $G_{\mathscr{B}}$), the point $S(i^1)$, and $YB(j_*)$, $EB(j_*)$, if these are not empty. For all other points, the brother indices remain unchanged.

Set $YB'(j_1) = S(i_1)$. For each $u = 1$ to $t + 1$, if $EB(j_u) \neq \emptyset$, set $YB'(EB(j_u)) = YB(j_u)$; and if $YB(j_u) \neq \emptyset$, set $EB'(YB(j_u)) = EB(j_u)$. If $S(i_1) \neq \emptyset$, set $EB'(S(i_1)) = j_1$. For each $u = 1$ to $t + 1$, set $EB'(j_u) = \emptyset$ because these points join as the eldest among their new set of brothers. If $S(j_1) \neq \emptyset$, set $EB'(S(j_1)) = j_2$. For each $u = 2$ to t, if $S(j_u) \neq j_{u-1}$, set $EB'(S(j_u)) = j_{u+1}$. $YB'(j_2) = S(j_1)$. For each $u = 3$ to $t + 1$,

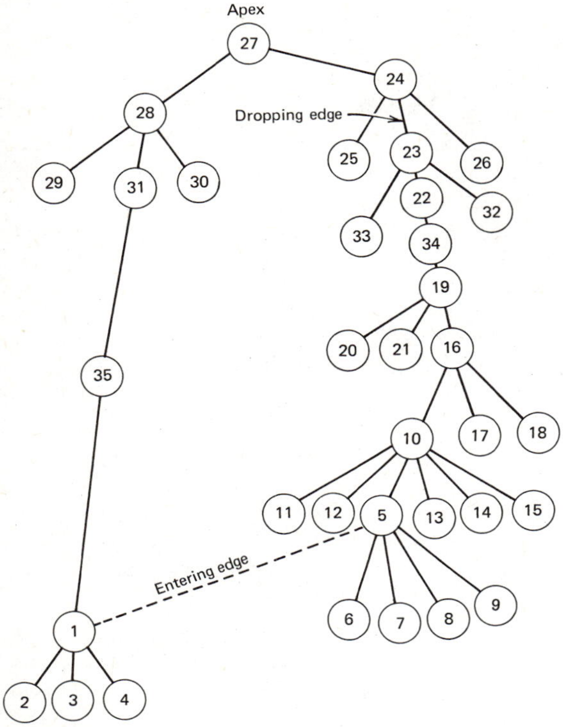

Figure 13.6 Simple cycle of the entering edge (1, 5).

if $S(j_{u-1}) \neq j_{u-2}$, set $YB'(j_u) = S(j_{u-1})$, and if $S(j_{u-1}) = j_{u-2}$, set $YB'(j_u) = YB(j_{u-2})$. If $YB(j_*) \neq \emptyset$, $EB'(YB(j_*)) = EB(j_*)$. If $EB(j_*) \neq \emptyset$, $YB'(EB(j_*)) = YB(j_*)$. Leave all other brother indices unchanged.

Illustration of Updating the Indices

In Figure 13.6, (1; 5) (the dashed edge), (23; 24) are the entering and dropping edges, respectively. Only points on the simple cycle and their immediate successors are shown, since index changes occur only for them. Each in-tree edge (continuous lines in Figure 13.6) is drawn so that the parent node on it is above the son node. Immediate successors of each point are arranged from left to right in decreasing order of age. In Figure 13.6, in the notation of Figure 13.4, we have $j_1 = 5$, $i_1 = 1$, $j_* = 23$, $i_2 = 24$, $i_0 = 27$. The position of the points on this simple cycle in the rooted tree obtained after this pivot step is indicated in Figure 13.7, using the same conventions.

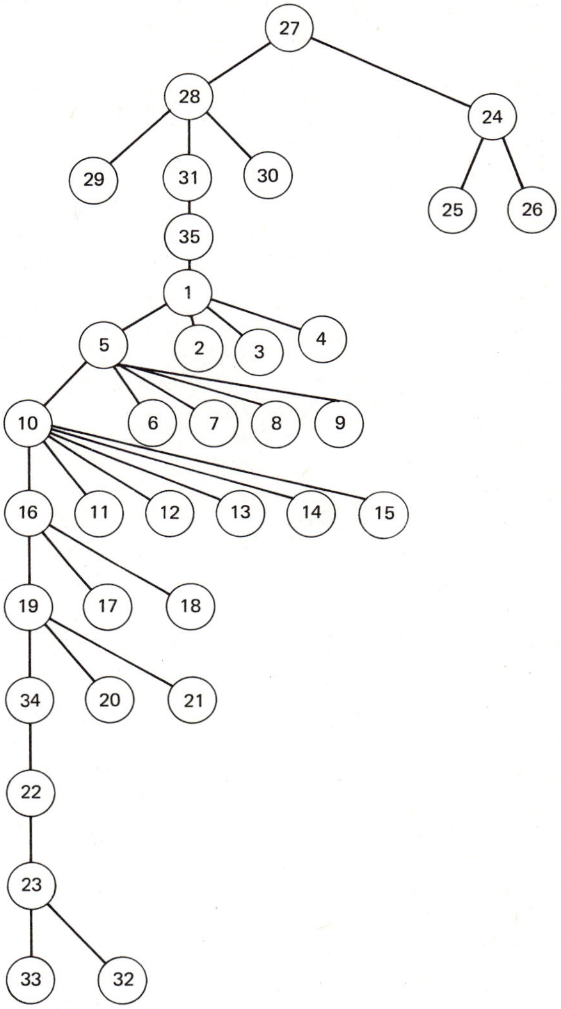

Figure 13.7 Position of the points in Figure 13.6 in the rooted tree obtained after the pivot step.

Example 13.16

Consider the basis for the 6 × 11 transportation problem discussed in Example 13.11. Suppose the edge (3; 17) is being deleted and the edge (4; 11) is being added. The rooted tree corresponding to the new basis is shown in Figure 13.8. The new indices are obtained from the previous indices tabulated in Example 13.11, using the updating procedure. They are tabulated at the top of page 408.

Point	1	2	3	4	5	6	7	8	9	10	11	12	13	14	15	16	17
Predecessor	13	7	15	11	17	11	5	4	2	3	2	4	6	1	4	3	∅
Successor	14	9	10	15	7	13	2	∅	∅	∅	4	∅	1	∅	3	∅	5
Younger brother	∅	∅	∅	6	∅	∅	∅	12	11	16	∅	∅	∅	∅	8	∅	∅
Elder brother	∅	∅	∅	∅	∅	4	∅	15	∅	∅	9	8	∅	∅	∅	10	∅

13.4.8 Updating the Dual Vector and the Relative Cost Coefficients

Suppose the basic set $\mathscr{B}'$ for the m × n transportation problem is obtained by bringing the cell (p, q) into the basic set $\mathscr{B}$. Suppose $(u, v, \bar{c})$ refer to the dual solution and the relative cost coefficients with respect to the basic set $\mathscr{B}$. Let $(u', v', \bar{c}')$ refer to the dual solution and the relative cost coefficients with respect to the new basic set $\mathscr{B}'$.

Suppose that the cell being dropped from the basic set when (p, q) is inserted corresponds to the edge $(i_2; j_*)$ in $G_{\mathscr{B}}$, where $P(j_*) = i_2$. As before, let **H** be the set containing j_* and all the descendants of j_* in the rooted tree $G_{\mathscr{B}}$. Let the entering cell (p, q) correspond to the edge $(i_1; j_1)$ in the graph, where i_1 and j_1 are such that the dropping edge $(i_2; j_*)$ is on the path from the apex to j_1. Let the path from j_1 to the apex be as in Section 13.4.7. Here is the new dual solution.

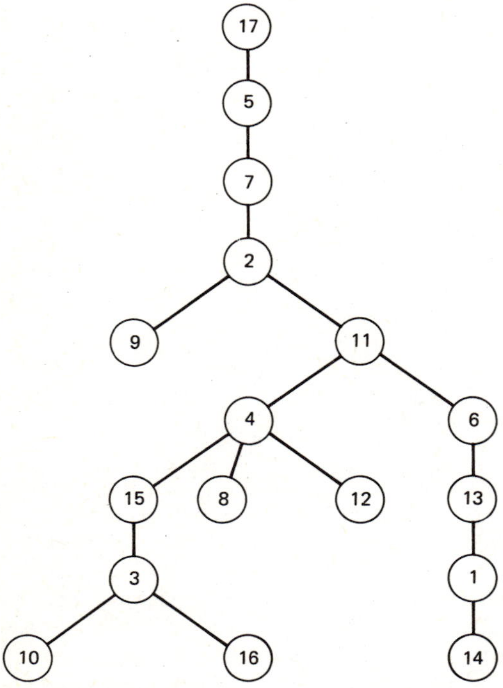

Figure 13.8

$$\begin{aligned} u_i' &= u_i - \alpha\bar{c}_{pq} && \text{for } i \in \mathbf{H} \\ &= u_i && \text{for } i \notin \mathbf{H} \\ v_j' &= v_j + \alpha\bar{c}_{pq} && \text{for m} + j \in \mathbf{H} \\ &= v_j && \text{for m} + j \notin \mathbf{H} \end{aligned}$$

$$\begin{aligned} \bar{c}_{ij}' &= \bar{c}_{ij} + \alpha\bar{c}_{pq} && \text{for } i \in \mathbf{H} \quad \text{and} \quad \text{m} + j \notin \mathbf{H} \\ &= \bar{c}_{ij} - \alpha\bar{c}_{pq} && \text{for } i \notin \mathbf{H} \quad \text{and} \quad \text{m} + j \in \mathbf{H} \\ &= \bar{c}_{ij} && \text{for } i \text{ and } m + j \text{ are both in } \mathbf{H} \text{ or not in } \mathbf{H} \end{aligned}$$

where $\alpha = 1$ if i_1 is the node corresponding to a row of the transportation array, and $= -1$, otherwise.

This follows from the fact that the new dual solution satisfies $u_i' + v_j' = c_{ij}$ for all basic cells (i, j) in the basic set $\mathscr{B}'$ and $\bar{c}_{ij}' = c_{ij} - (u_i' + v_j')$ for all i, j.

Organizing the Computations

Clearly all the computations can be performed using the various indices. There is no need to have a picture of the rooted tree for doing the computations. The indices can be recorded on the transportation array itself and all the computations performed using them.

Example 13.17

Consider the following balanced transportation problem. In this problem m = 5 and n = 8. So rows 1 to 5 correspond to points 1 to 5, respectively, in the graph and columns 1 to 8 correspond to points 6 to 13, respectively, in the graph. All the indices are serial numbers of the points. An initial BFS for the problem is given in array 13.4. Every cell with a small square in the middle is a basic cell and the value of that basic variable is entered in the middle of that square. The dual solution corresponding to this basis, u_i and v_j for $i = 1$ to m, $j = 1$ to n, is tabulated. The original cost coefficient c_{ij} is recorded in the lower right corner of the cell (i, j) and the relative cost coefficient

Array 13.4

	$j = 1$	2	3	4	5	6	7	8	a_i	u_i	P(i)	S(i)	YB(i)	EB(i)
$i = 1$	−6 / 8	5 / 10	1 / 7	−5 / 2	[4] / 8	[5] / 11	0 / 2	−1 / 8	9	9	10	11	∅	∅
2	1 / 10	[3] / 0	0 / 1	6 / 8	7 / 10	−1 / 5	4 / 1	0 / 4	3	4	7	∅	∅	∅
3	[1] / 20	3 / 14	3 / 15	[12] / 13	6 / 20	−7 / 10	[9] / 8	−10 / 5	22	15	12	6	4	∅
4	4 / 25	18 / 30	2 / 15	1 / 15	0 / 15	2 / 20	[17] / 9	4 / 20	17	16	12	∅	∅	3
5	9 / 25	[8] / 7	[19] / 8	11 / 20	[2] / 10	2 / 15	[8] / 4	[6] / 11	43	11	13	7	∅	∅
b_j	1	11	19	12	6	5	34	6						
v_j	5	−4	−3	−2	−1	2	−7	0						
P(m + j)	3	5	5	3	5	1	5	∅						
S(m + j)	∅	2	∅	∅	1	∅	3	5						
YB(m + j)	9	8	10	∅	12	∅	∅	∅						
EB(m + j)	∅	∅	7	6	8	∅	10	∅						

of a nonbasic cell is entered in the upper left corner. The predecessor indices P(i), the successor indices S(i), the younger brother indices YB(i), and the elder brother indices EB(i) of the various points corresponding to the rows and columns of the transportation array are entered. In determining the initial brother indices, it is assumed that all the brothers are arranged from left to right in increasing order of their serial number (i.e., initially we assume that if points i and j are brothers, i is elder brother of j if $i < j$). A diagram of the rooted tree corresponding to this basis is given in Figure 13.9 to aid in understanding how the various indices are obtained. But all the indices could be directly read out from the array itself, starting with the root node, which is the point corresponding to column 8. Since some of the $\bar{c}_{ij} < 0$, the present basis is not optimal. Suppose we decide to bring the nonbasic cell (3, 6) into the basic set. This entering cell (3, 6) corresponds to the edge (3, 11).

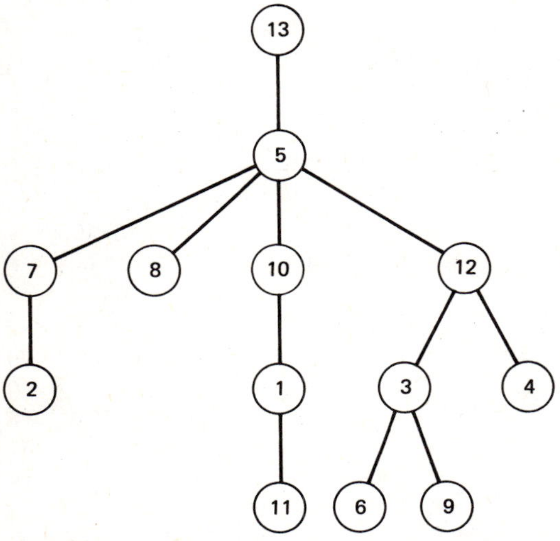

Figure 13.9 Rooted tree corresponding to the present basis.

From the predecessor indices, the path from point 3 to the root node is ((3; 12), (12; 5), (5; 13)) and the path from point 11 to the root node is ((11; 1), (1; 10), (10; 5), (5; 13)). Eliminating the common edges of these two paths, notice that the apex is the point 5 and the simple cycle is ((3; 11), (11; 1), (1; 10), (10; 5), (5; 12), (12; 3)). The simple cycle corresponds to the θ-loop consisting of the cells ((3, 6), (1, 6), (1, 5), (5, 5), (5, 7), (3, 7)). The cells (3, 6), (1, 5), and (5, 7) have a $+\theta$ entry in the θ-loop and the cells (1, 6), (5, 5), (3, 7) have a $-\theta$ entry. The minimum

Array 13.5

	$j=1$	2	3	4	5	6	7	8	u_i	$P(i)$	$S(i)$	$YB(i)$	$EB(i)$
$i=1$	−13 / 8	−2 / 10	−6 / 7	−12 / 2	[6] / 8	[3] / 11	−7 / 2	−8 / 8	16	11	10	∅	∅
2	1 / 10	[3] / 0	0 / 1	6 / 8	14 / 10	6 / 5	4 / 1	0 / 4	4	7	∅	∅	∅
3	[1] / 20	3 / 14	3 / 15	[12] / 13	13 / 20	[2] / 10	[7] / 8	−10 / 5	15	12	11	4	∅
4	4 / 25	18 / 30	2 / 15	1 / 15	7 / 15	9 / 20	[17] / 9	4 / 20	16	12	∅	∅	3
5	9 / 25	[8] / 7	[19] / 8	11 / 20	7 / 10	9 / 15	[10] / 4	[6] / 11	11	13	7	∅	∅
v_j	5	−4	−3	−2	−8	−5	−7	0					
$P(m+j)$	3	5	5	3	1	3	5	∅					
$S(m+j)$	∅	2	∅	∅	∅	1	3	5					
$YB(m+j)$	9	8	12	∅	∅	6	∅	∅					
$EB(m+j)$	11	∅	7	6	∅	∅	8	∅					

basic value (in the present BFS) in the cells with a $-\theta$ entry is minimum $\{5, 2, 9\} = 2$, and it occurs in cell (5, 5). So the cell (5, 5) leaves the basic set.

The leaving cell (5, 5) corresponds to the edge (10; 5). This edge appears on the path from point 11 to the apex. So the path corresponding to $(j_1; j_2), \ldots, (j_t; j_*)$ of Section 13.4.7 in this example is (11; 1), (1; 10). The set **H**, containing 10 and all the descendants of 10, is $\{10, 1, 11\}$. Since the entering cell is (3, 6), the $\bar{c}_{pq}$, corresponding to the discussion above is $\bar{c}_{36} = -7$.

The BFS, the various indices, the dual solution, and the relative cost coefficients are updated using the procedures discussed earlier. The new transportation array is array 13.5. The new basis is again nonoptimal, since some of the relative cost coefficients are still negative. The algorithm can be continued in a similar manner until optimality is attained.

Comment 13.3: Methods for labeling trees were first discussed by H. I. Scoins and E. Johnson (see reference [13.16]). F. Glover, D. Karney, and D. Klingman (reference [13.10]) developed the application of tree labels to the transportation problem and called it the *augmented predecessor index method* (the method discussed in this section is a minor variant of it). See references [13.11, 13.12, 13.13, 13.24] for other variants of this method.

13.5 SENSITIVITY ANALYSIS IN THE TRANSPORTATION PROBLEM

Consider the transportation problem (13.1). Let $\mathscr{B}$ be an optimum basic set of cells for it. Let $\tilde{x}$ be the corresponding optimum solution and let $\tilde{u}$, $\tilde{v}$, $\tilde{\bar{c}}$ be the corresponding dual solution and the relative cost vector for it.

13.5.1 Alternate Optimum Solutions in the Transportation Problem

As in the general LP model (see Section 3.10), $\tilde{x}$ is the unique optimum solution if $\tilde{\bar{c}}_{ij} > 0$ for all nonbasic cells (i, j) not in $\mathscr{B}$.

If $\tilde{\bar{c}}_{ij} = 0$ for some nonbasic (i, j), alternate optimum basic feasible solutions can be obtained by bringing into the basic set $\mathscr{B}$ any nonbasic cell (i, j) for which $\tilde{\bar{c}}_{ij} = 0$. Also if $\tilde{x}$ and $\hat{x}$ are two optimum feasible solutions, then all convex combinations of them, $\alpha\tilde{x} + (1-\alpha)\hat{x}$, $0 \leqq \alpha \leqq 1$, are also optimal to the problem.

The set of optimum solutions of the transportation problem is the set of all feasible solutions x in which $x_{ij} = 0$, whenever (i, j) is such that $\tilde{\bar{c}}_{ij} > 0$.

Array 13.6

$j =$	1	2	3	4	5	6	a_i	$\tilde{u}_i$
$i = 1$	$14 = \bar{c}_{11}$; $c_{11} = 10$	5; 20	[13]; 11	29; 30	0; 13	[17]; 3	30	3
2	12; 0	[25]; 7	9; 12	20; 13	[35]; 5	25; 20	60	−5
3	[30]; 5	16; 40	11; 31	0; 10	[10]; 22	4; 16	40	12
4	11; 8	[15]; 16	9; 21	[12]; 2	3; 17	[23]; 4	50	4
b_j	30	40	13	12	45	40		
$\tilde{v}_j$	−7	12	8	−2	10	0		

As an example, consider the optimum BFS for a transportation problem, given at the top. Basic cells are marked with a □, with the basic value in the corresponding BFS entered inside it. The original cost coefficients c_{ij} and the relative cost coefficients $\bar{c}_{ij}$, are entered in the lower-right and upper-left corners of each cell, respectively. Let $\tilde{\mathscr{B}}$ refer to be the basic set of cells, $\tilde{x}$ to the corresponding BFS, and $(\tilde{u}, \tilde{v})$ to corresponding dual basic solution in Array 13.6 Since all the $\bar{c}_{ij} \geqq 0$, $\tilde{x}$ is an optimum BFS. Since $\bar{c}_{15} = 0$, an alternate optimum BFS can be obtained by bringing the nonbasic cell (1, 5) into the basic set $\tilde{\mathscr{B}}$. This leads to the BFS given in the following array, which we denote by $\hat{x}$.

$j =$	1	2	3	4	5	6
$i = 1$			[13]		[15]	[2]
2		[40]			[20]	
3	[30]				[10]	
4				[12]		[38]

The convex combination $\alpha\tilde{x} + (1 - \alpha)\hat{x}$ is given in the array at the bottom; it is also optimal to this problem for all $0 \leqq \alpha \leqq 1$. Another alternate optimum BFS for this problem can be obtained by bringing the nonbasic cell (3, 4) into the basic set $\tilde{\mathscr{B}}$ in Array 13.6. And this can be continued.

From Array 13.6, we conclude that the set of all optimum solutions of this transportation problem is the set of all feasible solutions of this problem with $x_{11} = x_{12} = x_{14} = x_{21} = x_{23} = x_{24} = x_{26} = x_{32} = x_{33} = x_{36} = x_{41} = x_{43} = x_{45} = 0$ (these are all the nonbasic variables in Array 13.6, with $\bar{c}_{ij} > 0$).

Change in a Nonbasic Cost Coefficient

Let (i, j) be a cell that is not in $\mathscr{B}$. Suppose the cost coefficient c_{ij} is subject to change, while all the remaining data in the problem remain unchanged. Since (i, j) is a nonbasic cell, a change in c_{ij} changes only the relative cost coefficient of the cell (i, j). Hence $\mathscr{B}$ remains an optimum basic set for all c_{ij} satisfying $c_{ij} - (\tilde{u}_i + \tilde{v}_j) \geqq 0$, that is, $c_{ij} \geqq \tilde{u}_i + \tilde{v}_j$. If the new value of c_{ij} violates this constraint, bring (i, j) into the basic set and continue the application of the algorithm until optimality is reached again.

$j =$	1	2	3	4	5	6
$i = 1$			$x_{13} = 13$		$15(1 - \alpha)$	$17\alpha + 2(1 - \alpha)$
2		$25\alpha + 40(1 - \alpha)$			$35\alpha + 20(1 - \alpha)$	
3	30				10	
4		15α		12		$23\alpha + 38(1 - \alpha)$

13.5.2 Change in a Basic Cost Coefficient

Let (r, s) be a cell in $\mathscr{B}$ and suppose c_{rs} is subject to change. Denote the cost coefficient in the cell (r, s) by γ_{rs} and let c_{rs} be its present value. Since (r, s) is a basic cell, a change in γ_{rs} changes the dual solution. Treating γ_{rs} as a parameter, recompute the dual solution corresponding to the basic set $\mathscr{B}$, as a function of γ_{rs}. Let $u(\gamma_{rs})$, $v(\gamma_{rs})$ be the new dual solution. The new relative cost coefficients are obtained from $\bar{c}_{ij}(\gamma_{rs}) = c_{ij} - u_i(\gamma_{rs}) - v_j(\gamma_{rs})$ for all i, j. All $\bar{c}_{ij}(\gamma_{rs})$ are affine functions of γ_{rs}. And $\mathscr{B}$ remains an optimum basic set for all values of γ_{rs} for which $\bar{c}_{ij}(\gamma_{rs}) \geqq 0$ for all i, j. This defines the *optimality interval* of the basic set $\mathscr{B}$. The solution of (13.1) corresponding to the basic set $\mathscr{B}$ does not change as a result of a change in γ_{rs} and, hence, $\tilde{x}$ remains an optimum solution of (13.1) for all values of γ_{rs} in the optimality interval. If the new value of γ_{rs} lies outside the optimality interval of $\mathscr{B}$, replace c_{rs} by the new value of γ_{rs} and compute all the relative cost coefficients $\bar{c}_{ij}(\gamma_{rs})$. Bring a nonbasic cell with a negative $\bar{c}_{ij}(\gamma_{rs})$ into the basic set and continue the application of the simplex algorithm until optimality is reached again.

Example 13.18

Consider the following transportation problem. An optimum BFS is entered in the array. Basic cells are marked with a □ in the middle and the values of the basic variables are entered inside it. The original cost coefficients c_{ij} are entered in the lower right-hand corner of each cell, and the relative cost coefficients of nonbasic cells are entered in the upper left corner of the cell. The u_i and v_j obtained as in Section 13.2.2 are also entered.

$j =$	1	2	3	4	u_i
$i = 1$	$11 = \bar{c}_{11}$; $c_{11} = 15$	8; 11	[3]; 0	[4]; 1	1
2	9; 14	[9]; 4	[5]; 1	6; 8	2
3	[7]; 6	[6]; 5	5; 7	7; 10	3
v_j	3	2	-1	0	

Suppose c_{14} is subject to change. If we denote it by γ_{14}, the new dual solution and the relative cost coefficients are as follows.

$j =$	1	2	3	4	u_i
$i = 1$	11; 15	8; 11	[3]; 0	[4]; γ_{14}	γ_{14}
2	9; 14	[9]; 4	[5]; 1	$7 - \gamma_{14}$; 8	$1 + \gamma_{14}$
3	[7]; 6	[6]; 5	5; 7	$8 - \gamma_{14}$; 10	$2 + \gamma_{14}$
v_j	$4 - \gamma_{14}$	$3 - \gamma_{14}$	$-\gamma_{14}$	0	

All the relative cost coefficients are nonnegative iff $\gamma_{14} \leqq 7$. Hence for all values of $\gamma_{14} \leqq 7$, the present solution is optimal.

13.5.3 Parametric Cost Transportation Problem

Consider the transportation problem of the form (13.1) with the objective function $z(x) = \sum_i \sum_j (c_{ij} + \lambda c_{ij}^*) x_{ij}$, where λ is a parameter; suppose you have to solve the problem for all real values of the parameter λ.

Fix λ at 0 and solve the problem. Let $\mathscr{B}$ be the optimum basic set obtained. Let (u_i), (v_j) be the dual solution corresponding to $\mathscr{B}$ when (c_{ij}) is the cost matrix. Similarly let (u_i^*), (v_j^*) be the dual solution corresponding to $\mathscr{B}$ when (c_{ij}^*) is the cost matrix. Enter u_i, u_i^* in separate columns on the array. Similarly, enter v_j, v_j^* in separate rows on the array. The dual solution corresponding to $\mathscr{B}$ as a function of λ is $(u_i + \lambda u_i^*)$, $(v_j + \lambda v_j^*)$. The relative cost coefficient of the cell (i, j) as a function of λ is $(c_{ij} - u_i - v_j) + \lambda(c_{ij}^* - u_i^* - v_j^*) = \bar{c}_{ij} + \lambda \bar{c}_{ij}^*$. Enter $\bar{c}_{ij}$ and $\bar{c}_{ij}^*$ separately in the cell. The basic set $\mathscr{B}$ is optimum for all values of λ satisfying $\bar{c}_{ij} + \lambda \bar{c}_{ij}^* \geqq 0$. Thus, if

$$\underline{\lambda} = \text{maximum} \left\{ -\frac{\bar{c}_{ij}}{\bar{c}_{ij}^*} : \quad (i, j), \text{ such that } \bar{c}_{ij}^* > 0 \right\}$$

$$= -\infty \qquad \text{if all } \bar{c}_{ij}^* \leqq 0 \tag{13.10}$$

and

$$\bar{\lambda} = \text{minimum} \left\{ -\frac{\bar{c}_{ij}}{\bar{c}_{ij}^*} : \quad (i, j), \text{ such that } \bar{c}_{ij}^* < 0 \right\}$$

$$= +\infty \qquad \text{if all } \bar{c}_{ij}^* \geqq 0 \tag{13.11}$$

then $\mathscr{B}$ is an optimum basic set for all λ in the interval $\underline{\lambda} \leqq \lambda \leqq \bar{\lambda}$.

If we have to find an optimum solution of the problem for $\lambda > \bar{\lambda}$, identify the cell (i, j), which attains

	Cost Matrix						a_i
	$3-\lambda$	$2+\lambda$	$-1+2\lambda$	3	7λ	-8	14
	4	6	5	-2λ	4	-2	10
	0	$3+8\lambda$	-3	$-7-2\lambda$	1	5	5
	$2+\lambda$	λ	0	$8+\lambda$	4	3	5
b_j	6	9	4	3	7	5	

the minimum in (13.11), bring it into the basic set, then repeat the analysis with the new basis.

Exercises

13.18 Solve the parametric cost transportation problem with data given at the top for all real values of λ.

13.5.4 Changes in Material Supplies and Requirements

Marginal Analysis in the Balanced Transportation Problem

Marginal values in a linear programming model are the rates of change in the optimum objective value when a small change from its current value occurs in one of the right-hand-side constants.

If (13.1) is feasible, we must have $\sum_{i=1}^{m} a_i = \sum_{j=1}^{n} b_j$, and we assume that this condition holds. If a change occurs in the value of exactly one a_i or one b_j, while the values of all the remaining right-hand-side constants remain unaltered, the condition for feasibility will be violated and the problem becomes infeasible. To retain feasibility of the problem, whenever a change is made in one of the a_i's, a corresponding change has to be made in one of the b_j's. Thus marginal analysis for the balanced transportation problem (13.1) consists of evaluating the rate of change in the optimum objective value when one of the a_i's *and* one of the b_j's are both changed simultaneously by the same small amount while all the remaining data in the problem remain unchanged.

The dual of (13.1) is (13.8). Let $\hat{u}$, $\hat{v}$ be the unique optimum dual solution for the problem with the present data. Suppose a_p and b_q change to $a_p + \delta$ and $b_q + \delta$, respectively, while all the remaining data in the problem remain unchanged. If $\hat{u}$, $\hat{v}$ remains an optimum dual solution of the modified problem, by the duality theorem, the optimum objective value of the modified problem is $\sum_{i=1}^{m} a_i\hat{u}_i + \sum_{j=1}^{n} b_j\hat{v}_j + \delta(\hat{u}_p + \hat{v}_q)$, and hence the change in the optimum objective value as a result of this change in the data is $\delta(\hat{u}_p + \hat{v}_q)$. Thus the marginal value corresponding to a simultaneous change of the same amount in the values of a_p and b_q is $\hat{u}_p + \hat{v}_q$, and this marginal value is valid as long as $(\hat{u}, \hat{v})$ remains the unique optimum dual solution for the problem with the modified data.

From the result in Exercise 13.37, it can be verified that for each i, j, $u_i + v_j$ is uniquely determined by the basic set $\mathscr{B}$ and does not depend on which constraint in (13.1) is treated as a redundant constraint and deleted.

As an example, consider the transportation problem in Array 13.6 in Section 13.5.1. The optimum BFS, $\tilde{x}$, for this problem displayed in Array 13.6, is primal nondegenerate, since $\tilde{x}_{ij} > 0$ in all basic cells (i, j), which implies by the results in Section 6.7 that the optimum dual solution $(\tilde{u}, \tilde{v})$ given in Array 13.6 is unique. (That is, unique with the value of $\tilde{v}_n$ fixed at 0. As explained in Exercise 13.37, a different dual vector will be obtained if the value of v_n is different from zero, but this does not change the value of $u_i + v_j$, or the value of any relative cost coefficient, for any i, j.) So if the supplies and demands are changed to a $(\delta) = (30, 60 + \delta, 40, 50)$ and $b(\delta) = (30, 40, 13, 12 + \delta, 45, 40)$, the marginal value per unit change in δ is $\tilde{u}_2 + \tilde{v}_4 = -5 - 2 = -7$ (it is the rate of change in the optimum objective value per unit change in δ, at $\delta = 0$).

Exercises

13.19 Let $\mathscr{B}$ be an optimum basic set of cells and x the associated optimum BFS for (13.1). Suppose $\tilde{u}$, $\tilde{v}$ is the associated optimum dual solution obtained as in Section 13.2.2. Let r, p be two integers between 1 to m. Suppose a_r and a_p are likely to change to $a_r + \delta$ and $a_p - \delta$, respectively, where δ is a parameter, while all the other data in the problem remain unchanged. This type of change occurs when some supply from one of the sources is moved to another, keeping the overall amount of supply of the material constant. Find the marginal value corresponding to this change. Discuss how to determine the range of values of δ for which $\mathscr{B}$ remains optimal, and an optimum solution of the modified problem as a function of δ for values of δ in this range. Apply these methods on the transportation problem in Example 13.7 when a_1, a_3 change to $6 + \delta$ and $20 - \delta$, respectively, while all the other data in the problem remain unaltered.

Discuss how to carry out a similar analysis when the change that occurs is in the values of two of the b_j's. Apply these methods on the transportation problem discussed in Example 13.7, when b_2 and b_3 change to $16 + \delta$ and $10 - \delta$, respectively, while all the other data in the problem remain unaltered.

Right-hand-Side Ranging in the Balanced Transportation Problem

Let $\mathscr{B}$ be an optimum basic set of cells for (13.1), and let $\bar{x}$ be the associated optimum BFS. Suppose a_p, b_q change to $a_p + \delta$, $b_q + \delta$, respectively, while all the remaining data in the problem remain unchanged. The pth row corresponds to the point p and the qth column corresponds to the point $m + q$ in the rooted tree $G_{\mathscr{B}}$. Find the unique simple path from p to $m + q$ in $G_{\mathscr{B}}$ as in Section 13.4.6. Suppose the edges in the path correspond to the cells (p, j_1), (i_1, j_1), (i_1, j_2), . . . , (i_t, q) in that order. Add amounts of $+\delta$ and $-\delta$ alternately, starting with $+\delta$ in the cell (p, j_1), to the present basic values in the cells on this path. A cell on this path with a $-\delta$, $+\delta$ alteration is called a *donor cell*, *recipient cell*, respectively. The values of the variables in the cells not on this path remain unchanged. Let $\bar{x}(\delta)$ be the new solution obtained as a result of these alterations. Then $\bar{x}(\delta)$ is the basic solution of the problem with the modified data, corresponding to the basic set of cells $\mathscr{B}$. Let $\delta_2 = \text{minimum } \{\bar{x}_{ij}: (i, j) \text{ a recipient cell}\}$, and $\delta_1 = \text{maximum } \{-\bar{x}_{ij}: (i, j) \text{ a donor cell}\}$. Clearly $\bar{x}(\delta) \geqq 0$ as long as $\delta_1 \leqq \delta \leqq \delta_2$. Hence $\mathscr{B}$ remains an optimum basic set of cells for the problem with the modified data, for all δ satisfying $\delta_1 \leqq \delta \leqq \delta_2$, and in this range $\bar{x}(\delta)$ is an optimum solution of the problem as a function of δ.

Example 13.19

Consider the following optimum BFS for a transportation problem.

$j =$	1	2	3	4	5	6
$i = 1$	3		9			
2		6		7	4	
3				5		3
4	8				13	

In this problem suppose a_3 has to be changed to $8 + \delta$ and b_1 has to be changed to $11 + \delta$. The new BFS as a function of δ is as follows. This solution is optimal for all δ satisfying $-4 \leqq \delta \leqq 7$.

$j =$	1	2	3	4	5	6
$i = 1$	3		9			
2		6		$7 - \delta$	$4 + \delta$	
3				$5 + \delta$		3
4	$8 + \delta$				$13 - \delta$	

The Dual Simplex Algorithm for the Balanced Transportation Problem Starting with a Given Dual Feasible Basic Set

Consider the balanced transportation problem (13.1). Let $\mathscr{B}$ be a given basic set for the $m \times n$ transportation array, with respect to which the relative cost coefficients in (13.1) are all nonnegative. So $\mathscr{B}$ is a *dual feasible basic set* of cells for (13.1). However, suppose $\mathscr{B}$ is primal infeasible for (13.1). Then it is possible to solve (13.1) starting with $\mathscr{B}$ and using the dual simplex algorithm (see Chapter 6). Here we describe how to carry out the dual simplex algorithm using the transportation arrays and not the simplex canonical tableaux.

To Obtain the Set of All Nonbasic Cells for Which a Specified Basic Cell Is a Recipient Cell Let $(r; s) \in \mathscr{B}$, where $\mathscr{B}$ is a basic set of cells for (13.1) associated with the rooted tree $G_{\mathscr{B}}$. The edge $(r; m + s)$ in $G_{\mathscr{B}}$ corresponds to the basic cell (r, s). If $(r; m + s)$ is deleted from $G_{\mathscr{B}}$, it becomes disconnected into two disjoint partial trees. Among the points on the in-tree edge $(r; m + s)$, one of them is a parent node, call it t_1, and the other its son, call it t_2. So $P(t_2) = t_1$ and $\{t_1, t_2\} = \{r, m + s\}$. Let $\bar{\mathbf{X}} = \{t_2\} \cup \{i: i \text{ is a descendent of } t_2\}$, $\mathbf{X} = \mathscr{N} \backslash \bar{\mathbf{X}}$. If $i_1 \in \mathbf{X}$, $i_2 \in \bar{\mathbf{X}}$, the simple path between i_1 and i_2 in $G_{\mathscr{B}}$ must include the edge $(r; \mathrm{m}+s)$. Hence the set $\{(i, j)$: either $i \in \mathbf{X}$, $\mathrm{m}+j \in \overline{\mathbf{X}}$ or $i \in \overline{\mathbf{X}}$, $\mathrm{m}+j \in \mathbf{X}$ and $(i, j) \neq (r, s)\}$ is the set of all nonbasic cells (p, q), such that (r, s) is on the θ-loop in $\mathscr{B} \cup \{(p, q)\}$. Let $\mathbf{L}_1 = \{(i, j): i \in \mathbf{X}, \mathrm{m}+j \in \overline{\mathbf{X}}, (i, j) \neq (r, s)\}$; $\mathbf{L}_2 = \{(i, j): i \in \overline{\mathbf{X}}, \mathrm{m}+j \in \mathbf{X}, (i, j) \neq (r, s)\}$. $\mathbf{L}_1$, $\mathbf{L}_2$ is the set of all nonbasic cells for which (r, s) is a donor basic cell, depending on whether $t_2 = \mathrm{m}+s$ or r, respectively. Similarly $\mathbf{L}_2$, $\mathbf{L}_1$ is the set of all nonbasic cells for which (r, s) is a recipient cell, depending on whether $t_2 = r$ or $\mathrm{m}+s$, respectively.

Example 13.20

Consider the initial basis for the 5×8 transportation problem in Example 3.17 in array 13.4. The rooted tree corresponding to this basic set of cells is in Figure 13.9. Consider the basic cell (5, 7). This corresponds to the edge (5; 12) in the rooted tree. In this case, 5 is the predecessor of 12. Hence $\overline{\mathbf{X}} = \{12, 3, 4, 6, 9\}$, $\mathbf{X} = \{13, 5, 7, 8, 10, 2, 1, 11\}$. Hence the set of all nonbasic cells for which (5, 7) is a donor cell is $\{(1, 7), (1, 1), (1, 4), (2, 7), (2, 1), (2, 4), (5, 1), (5, 4)\}$. The set of all nonbasic cells for which (5, 7) is a recipient cell is $\{(3, 8), (3, 2), (3, 3), (3, 5), (3, 6), (4, 8), (4, 2), (4, 3), (4, 5), (4, 6)\}$.

The Dual Simplex Algorithm Let $\mathscr{B}$ be a dual feasible basic set of cells for (13.1) associated with the basic solution $\overline{x}$. If $\overline{x} \geqq 0$, $\mathscr{B}$ is also primal feasible, and hence optimal to (13.1); terminate. If $\overline{x} \ngeqq 0$, select a basic cell $(r, s) \in \mathscr{B}$, such that $\overline{x}_{rs} < 0$. Find $\mathbf{N}_1$, the set of all nonbasic cells for which (r, s) is a recipient cell. Let (p, q) be a nonbasic cell in $\mathbf{N}_1$, which has the least relative cost coefficient among all the cells in $\mathbf{N}_1$. Select (p, q) as the entering cell to replace (r, s) from the basic set $\mathscr{B}$. Find the θ-loop in $\mathscr{B} \cup \{(p, q)\}$, as in Section 13.1.7 or 13.4.5. Let $\hat{x} = (\hat{x}_{ij})$, where $\hat{x}_{ij} = \overline{x}_{ij} + \overline{x}_{rs}$, or $\overline{x}_{ij} - \overline{x}_{rs}$, or $\overline{x}_{ij}$, depending on whether (i, j) has a $-\theta$ entry, a $+\theta$ entry, or is not contained at all in the θ-loop in $\mathscr{B} \cup \{(p, q)\}$. Then $\hat{x}$ is the basic solution of (13.1) corresponding to the new basic set of cells $\{(p, q)\} \cup (\mathscr{B} \backslash \{(r, s)\})$. Also by the choice of (p, q), the new basic set of cells $\{(p, q)\} \cup (\mathscr{B} \backslash \{(r, s)\})$ will also be dual feasible to (13.1).

We leave it to the reader to verify that this is the implementation of the dual simplex pivot step in the balanced transportation problem (for performing a dual simplex pivot step in the row in which x_{rs} is the basic variable in the canonical tableau for (13.1) with respect to $x_{\mathscr{B}}$ as the basic vector) using the results in Chapter 6 and Section 13.1.8. Also, the relative cost coefficient of the entering cell (p, q) can be verified to be equal to the dual simplex minimum ratio in this pivot step.

Continue until primal feasibility is achieved. When primal feasibility is achieved, we have a basic set of cells that is both dual feasible and primal feasible, and hence optimal, for (13.1). Therefore, terminate the algorithm when primal feasibility is achieved.

Example 13.21

Consider the transportation array with the following data. Cells marked with □ are basic cells and the value of the corresponding basic variable in the basic solution is entered in the □. Cells without a □ are nonbasic cells, and the value of the corresponding variables in the basic solution is zero. Original cost coefficients are entered in the lower right corner of each cell; relative cost coefficients are entered in the upper left corner of each cell. Variables u_i and v_j are the dual variables.

$j =$	1	2	3	4	5	6	u_i
$i = 1$	9 / 22	[25] / 2	[17] / 2	13 / 22	17 / 25	14 / 17	3
2	[−16] $+\theta$ / 7	12 / 8	11 / 7	[26] $-\theta$ / 3	21 / 23	13 / 10	−3
3	[24] $-\theta$ / 15	[−12] / 4	15 / 19	7 / 18	6 / θ / 16	[13] / 5	5
4	9 / 17	8 / 5	12 / 9	[−10] $+\theta$ / 4	[14] $-\theta$ / 3	5 / 3	−2
v_j	10	−1	−1	6	5	0	

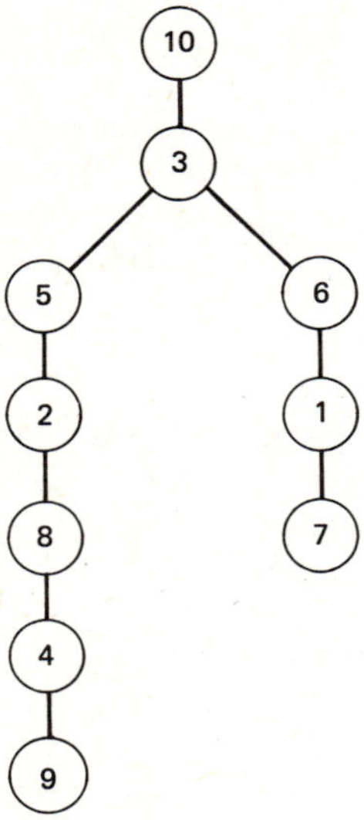

Figure 13.10

The rooted tree corresponding to this basic set is in Figure 13.10. Select the basic cell (2, 1), corresponding to the edge (2; 5) in the rooted tree for the pivot. The set of nonbasic cells for which (2, 1) is a recipient cell is $\{(3, 4), (3, 5), (1, 4), (1, 5)\}$. Among these, (3, 5) has the least relative cost coefficient of 6. Hence (3, 5) is the entering cell. The θ-loop is already marked in the above array. (2, 1) is the dropping cell. The new solution is marked in the array appearing at the bottom of the page. The current solution is still not primal feasible. The algorithm can be continued in a similar manner until primal feasibility is attained, at which point we have an optimum solution.

Changes in Availabilities at the Sources and Requirements at Demand Centers in the Balanced Transportation Problem

Suppose the balanced transportation problem (13.1) is solved and an optimum basic set of cells $\mathscr{B}$ for it is obtained. Afterward, suppose it becomes necessary to change the amount available at source i from the present a_i to a'_i, for $i = 1$ to m, and to change the amount required at demand center j from the present b_j to b'_j, for $j = 1$ to n, where a'_i, b'_j are given positive numbers satisfying $\sum_{i=1}^{m} a'_i = \sum_{j=1}^{n} b'_j$, while all the cost data in the problem remain unchanged. Compute the primal basic solution corresponding to the basic set $\mathscr{B}$, with a'_i, $i = 1$ to m, and b'_j, $j = 1$ to n, as the availabilities and requirements, respectively, using the method discussed in Section 13.1.5. If this solution is nonnegative, it is clearly an optimum solution of the modified problem, and we terminate. Otherwise $\mathscr{B}$ is a dual feasible, but primal infeasible basic set of cells for the modified problem. Starting with $\mathscr{B}$, apply the dual simplex algorithm until the new optimum solution is obtained.

13.6 THE DUAL SIMPLEX METHOD FOR THE BALANCED TRANSPORTATION PROBLEM

Several methods to obtain an initial dual feasible basic set of cells for (13.1) are discussed in the paper [13.17] of K. G. Murty and C. Witzgall. One of them is the following: select any row, say row p. Define $\tilde{v}_j = c_{pj}$ for $j = 1$ to n. For each $r \neq p$, define $\tilde{u}_r = \text{minimum}\ \{c_{rj} - \tilde{v}_j\colon j = 1 \text{ to n}\}$ and let j_r be any one of the j's that attains the minimum here. Define

$j =$	1	2	3	4	5	6	u_i
$i = 1$	9 / 22	[25] / 2	[17] / 2	7 / 22	11 / 25	14 / 17	11
2	6 / 7	18 / 8	17 / 7	[10] / 3	21 / 23	19 / 10	−1
3	[8] / 15	[−12] / 4	15 / 19	1 / 18	[16] / 16	[13] / 5	13
4	15 / 17	14 / 5	18 / 9	[6] / 4	[−2] / 3	11 / 3	0
v_j	2	−9	−9	4	3	−8	

$\tilde{u}_p = 0$. Let $\mathscr{B} = \{(r, j_r): r = 1$ to m, $r \neq p$; $(p, j): j = 1$ to n$\}$. It can be verified that $(\tilde{u}, \tilde{v})$ defined here is the dual basic solution of (13.1) corresponding to the basic set $\mathscr{B}$, and that $\mathscr{B}$ is dual feasible. Similarly, one can modify this procedure to generate a dual feasible basic set of cells by selecting a column initially.

Another method that goes through $(m + n - 1)$ steps is the following. This method selects one basic cell per step. It maintains vectors u, v and updates them in each step. We denote these vectors in the rth step by the symbols u^r, v^r. Let $c_{ij}^r = c_{ij} - u_i^r - v_j^r$ and $c^r = (c_{ij}^r)$. Then c^r will be $\geqq 0$ for all r and once a cell (i, j) is selected as a basic cell, we will have $c_{ij}^r = 0$ for all subsequent r. At termination, the set of cells selected will be a dual feasible basic set associated with the terminal u, v vectors as the dual feasible basic solution. Let $\mathbf{I}_r$, $\mathbf{J}_r$ denote the set of rows and columns, respectively, in which basic cells have been selected by the end of step r. $\overline{\mathbf{I}}_r$, $\overline{\mathbf{J}}_r$, denote the complements of $\mathbf{I}_r$, $\mathbf{J}_r$, respectively. Initially define u^1, v^1 by $u_i^1 = \text{minimum } \{c_{ij}: j = 1 \text{ to n}\}$, $v_j^1 = \text{minimum } \{c_{ij} - u_i^1: i = 1 \text{ to m}\}$. Select any cell (i_1, j_1) satisfying $c_{i_1,j_1}^1 = 0$ as the first basic cell. In step $(r + 1)$, if $\overline{\mathbf{J}}_r \neq \emptyset$, define (s_1, t_1) to be a cell satisfying $c_{s_1,t_1}^r = \text{minimum } \{c_{ij}^r: i \in \mathbf{I}_r,\ j \in \overline{\mathbf{J}}_r\}$, and if $\overline{\mathbf{I}}_r \neq \emptyset$, define (s_2, t_2) to be any cell satisfying $c_{s_2,t_2}^r = \text{minimum } \{c_{ij}^r: i \in \overline{\mathbf{I}}_r, j \in \mathbf{J}_r\}$. Select (s_1, t_1) or (s_2, t_2) as the basic cell (i_{r+1}, j_{r+1}) in this step. Define $\delta = c_{i_{r+1},j_{r+1}}^r$; $u_i^{r+1} = u_i^r + \alpha\delta$ for $i \in \mathbf{I}_r$, u_i^r for $i \in \overline{\mathbf{I}}_r$; $v_j^{r+1} = v_j^r - \alpha\delta$ for $j \in \mathbf{J}_r$, v_j^r for $j \in \overline{\mathbf{J}}_r$; where $\alpha = +1$ or -1, depending on whether $(i_{r+1}, j_{r+1}) = (s_1, t_1)$ or (s_2, t_2). Continue until $(m + n - 1)$ basic cells are selected.

The above method can be simplified by changing the definitions of (s_1, t_1), (s_2, t_2) as follows. If $\overline{\mathbf{J}}_r \neq \emptyset$, select any $t_1 \in \overline{\mathbf{J}}_r$ arbitrarily and let $s_1 \in \mathbf{I}_r$ be a row satisfying $c_{s_1,t_1}^r = \text{minimum } \{c_{i,t_1}^r: i \in \mathbf{I}_r\}$. If $\overline{\mathbf{I}}_r \neq \emptyset$, select any $s_2 \in \overline{\mathbf{I}}_r$ arbitrarily and let $t_2 \in \mathbf{J}_r$ be a column satisfying $c_{s_2,t_2}^r = \text{minimum } \{c_{s_2,j}^r: j \in \mathbf{J}_r\}$. Among these, either (s_1, t_1) or (s_2, t_2) can be selected as the basic cell chosen in step $r + 1$. However, if (s_2, t_2) is selected as the basic cell, define $u_i^{r+1} = u_i^r - c_{s_2,t_2}^r$ for all $i \neq s_2$, $u_{s_2}^r$ for $i = s_2$; $v_j^{r+1} = v_j^r + c_{s_2,t_2}^r$ for $j \in \mathbf{J}_r$, v_j^r for $j \in \overline{\mathbf{J}}_r$. On the other hand, if (s_1, t_1) is selected as the basic cell, define $u_i^{r+1} = u_i^r + c_{s_1,t_1}^r$ for $i \in \mathbf{I}_r$, u_i^r for $i \in \overline{\mathbf{I}}_r$; $v_j^{r+1} = v_j^r - c_{s_1,t_1}^r$ for $j \neq t_1$, $v_{t_1}^r$ for $j = t_1$.

Having obtained a dual feasible basic set for (13.1), say $\mathscr{B}_1$, by any of these methods, check whether it is also primal feasible. If not, beginning with $\mathscr{B}_1$, apply the dual simplex algorithm discussed in Section 13.5.4. to solve (13.1).

Exercises

13.20 Discuss whether it is possible to combine the features of the algorithms discussed here for obtaining a dual feasible basic set of cells for (13.1), and the algorithm for obtaining a primal feasible basic set of cells for (13.1) discussed in section 13.2.1, to obtain directly an optimum basic set of cells for (13.1), without the need to change the basic set once it is obtained.

13.7 TRANSPORTATION PROBLEMS WITH INEQUALITY CONSTRAINTS

13.7.1 Transportation Model with Demand Constraints as Equations

$$\text{Minimize} \quad z(x) = \sum_{i=1}^{m} \sum_{j=1}^{n} c_{ij}x_{ij}$$

$$\text{Subject to} \quad \sum_{i=1}^{n} x_{ij} \leqq a_i \qquad i = 1 \text{ to m}$$

$$\sum_{j=1}^{m} x_{ij} = b_j \qquad j = 1 \text{ to n}$$

$$x_{ij} \geqq 0 \qquad \text{for all } i, j$$

This transportation problem is feasible iff $\sum a_i \geqq \sum b_j$. If we introduce slack variables, the problem may be rewritten as

$$\text{Minimize} \quad z(x) = \sum_{i=1}^{m} \sum_{j=1}^{n+1} c_{ij}x_{ij}$$

$$\text{Subject to} \quad \sum_{j=1}^{n+1} x_{ij} = a_i \qquad i = 1 \text{ to m}$$

$$\sum_{i=1}^{m} x_{ij} = b_j \qquad j = 1 \text{ to n} + 1$$

$$x_{ij} \geqq 0 \qquad \text{for all } i, j$$

Here $b_{n+1} = \sum a_i - \sum b_j$, $c_{i,n+1} = 0$ for all i, and the slack variable $x_{i,n+1}$ represents the amount of material left unutilized at the ith source. The problem with the slack variables is a balanced transportation problem and, hence, can be solved by the algorithm discussed earlier.

13.7.2 The Unbalanced Transportation Problem

Consider this transportation problem:

Minimize

$$z()x = \sum_{i=1}^{m} \sum_{j=1}^{n} c_{ij}x_{ij}$$

Subject to

$$\sum_{j=1}^{n} x_{ij} \leqq a_i \qquad i = 1 \text{ to m} \qquad (13.12)$$

$$\sum_{j=1}^{m} x_{ij} \geqq b_j \qquad j = 1 \text{ to n} \qquad (13.13)$$

$$x_{ij} \geqq 0 \qquad \text{for all } i, \text{j}$$

We refer to this problem as Transportation Problem I. It is feasible iff $\sum a_i \geqq \sum b_j$.

Case 1 If $\sum a_i = \sum b_j$, every feasible solution of Transportation Problem I satisfies all the inequality constraints as equations. Hence, in this case, replace all the inequalities in (13.12) and (13.13), by equations and solve it by the algorithm discussed earlier.

Case 2 Suppose $\sum a_i > \sum b_j$. In this case, Transportation Problem I is known as an *unbalanced transportation problem*. Also, suppose $c_{ij} \geqq 0$ for all i, j.

Exercises

13.21 In this case, prove that there exists an optimum feasible solution $x = (x_{ij})$ of Transportation Problem I that satisfies all the constraints in (13.13) as equations. Also, if $c_{ij} > 0$ for all i, j, prove that every optimum solution of this problem must satisfy (13.13) as equations.

Hence, in this case, an optimum feasible solution of Transportation Problem I can be obtained by replacing all the inequalities in (13.13) by equations and solving the remaining problem as in Section 13.7.1.

Case 3 Suppose $\sum a_i > \sum b_j$ and some c_{ij} are negative. In this case, it may be profitable to oversupply (beyond the minimal demand b_j) the markets j for which some c_{ij} are negative. For each i = 1 to m, let $c_{i,\mathrm{n}+1}$ = minimum $\{0, c_{i1}, \ldots, c_{in}\}$. For each i such that $c_{i,\mathrm{n}+1} < 0$, let $j = r_i$ satisfy $c_{i,\mathrm{n}+1} = c_{i,r_i}$ = minimum $\{c_{i1}, \ldots, c_{in}\}$. If there is a tie for r_i, break it arbitrarily. Also, let $b_{\mathrm{n}+1} = \sum a_i - \sum b_j$. Consider the balanced transportation problem:

$$\text{Minimize} \quad \sum_{i=1}^{m} \sum_{j=1}^{n+1} c_{ij}x_{ij}$$

$$\text{Subject to} \quad \sum_{j=1}^{n+1} x_{ij} = a_i \qquad i = 1 \text{ to m} \qquad (13.14)$$

$$\sum_{i=1}^{m} x_{ij} = b_j \qquad j = 1 \text{ to n+1}$$

$$x_{ij} \geqq 0 \qquad \text{for all } i, j$$

Suppose $\tilde{x} = (\tilde{x}_{ij})$ is an optimum feasible solution of (13.14). For j = 1 to n, let

$$\begin{aligned} \hat{x}_{ij} &= \tilde{x}_{ij} && \text{for all } i \text{ such that } c_{i,\mathrm{n}+1} \geqq 0, \\ &= \tilde{x}_{ij} && \text{for all } i \text{ such that } c_{i,\mathrm{n}+1} < 0, j \neq r_i \\ &= \tilde{x}_{ij} + \tilde{x}_{i,\mathrm{n}+1} && \text{for all } i \text{ such that } c_{i,\mathrm{n}+1} < 0, j = r_i \end{aligned}$$

Then $\hat{x} = (\hat{x}_{ij})$ is an optimum feasible solution of Transportation Problem I in this case.

Exercises

13.22 Prove that $(\hat{x}_{ij})$ is an optimum feasible solution of Transportation Problem I in this case.

—(*O. Merrill and R. Tobin*)

13.23 If all $c_{ij} \leqq 0$, prove that there is an optimum feasible solution for Transportation Problem I that satisfies all the inequalities (13.12) as equations. And if $c_{ij} < 0$ for all i, j, every optimum solution must satisfy (13.12) as equations.

Consider Transportation Problem II:

$$\text{Minimize} \quad \sum_{i=1}^{m} \sum_{j=1}^{n} c_{ij}x_{ij}$$

$$\text{Subject to} \quad \sum_{j=1}^{n} x_{ij} = a_i \qquad i = 1 \text{ to m} \qquad (13.15)$$

$$\sum_{i=1}^{m} x_{ij} \geqq b_j \qquad j = 1 \text{ to n} \qquad (13.16)$$

$$x_{ij} \geqq 0 \qquad \text{for all } i, j$$

It is feasible iff $\sum a_i \geqq \sum b_j$. If $\sum b_j = \sum a_i$, every feasible solution of the problem must satisfy all the inequalities in (13.16) as equations. In this case, replace all inequality signs in (13.16) by equation signs and solve the problem as a balanced transportation problem. If $\sum a_i > \sum b_j$, let r_i be defined by c_{i,r_i} = minimum $\{c_{i1}, \ldots, c_{in}\}$ for each $i = 1$ to m. Break ties for r_i in the above equation arbitrarily. Let $c_{i,n+1} = c_{i,r_i}$ for all $i = 1$ to m, and $b_{n+1} = \sum a_i - \sum b_j$. Consider the balanced transportation problem:

$$\text{Minimize} \quad \sum_{i=1}^{m} \sum_{j=1}^{n+1} c_{ij} x_{ij}$$

$$\text{Subject to} \quad \sum_{j=1}^{n+1} x_{ij} = a_i \qquad i = 1 \text{ to } m$$

$$\sum_{i=1}^{m} x_{ij} = b_j \qquad j = 1 \text{ to } n+1$$

$$x_{ij} \geqq 0 \qquad \text{for all } i, j$$

Suppose $\tilde{x} = (\tilde{x}_{ij})$ is an optimum solution of this balanced problem. Let

$$\hat{x}_{ij} = \tilde{x}_{ij} \qquad \text{for all } i = 1 \text{ to m}, \; j = 1 \text{ to n}, j \neq r_i$$

$$= \tilde{x}_{ij} + \tilde{x}_{i,n+1} \qquad \text{if } j = r_i, i = 1 \text{ to m}$$

Then $\hat{x} = (\hat{x}_{ij})$ is an optimum solution of Transportation Problem II.

Exercises

13.24 Solve the following transportation problems.

(a)

							Maximum amount available (a_i)
$c =$	6	−4	10	3	11	12	8
	1	10	3	1	2	−7	11
	−3	−11	8	5	0	8	7
	8	2	7	19	3	4	23
Exact requirement (b_j)	4	7	3	9	2	10	

(b)

							Maximum amount available (a_i)
$c =$	3	2	1	1	4	3	19
	6	−3	−7	0	11	2	23
	3	0	4	0	1	3	17
	−4	−8	−8	−3	−2	0	29
Minimal requirement (b_j)	8	15	11	3	13	9	

(c)

							Exact supply to be drawn (a_i)
$c =$	13	9	7	4	4	4	39
	4	3	0	0	6	7	14
	−1	−3	−3	−3	2	7	17
	8	0	−9	0	−9	7	21
Minimal requirement (b_j)	18	8	5	9	13	19	

13.8 BOUNDED VARIABLE TRANSPORTATION PROBLEMS

Consider this transportation problem:

$$\text{Minimize} \quad z(x) = \sum\sum c_{ij} x_{ij}$$

$$\text{Subject to} \quad \sum_{j=1}^{n} x_{ij} = a_i \qquad i = 1 \text{ to m} \qquad (13.17)$$

$$\sum_{i=1}^{m} x_{ij} = b_j \qquad j = 1 \text{ to n}$$

$$x_{ij} \geqq 0 \qquad \text{for all } i, j$$

$$x_{ij} \leqq U_{ij} \qquad \text{for } (i, j) \in \mathbf{J}$$

Bounded variable transportation problems are also known as *capacitated transportation problems.* Here **J** is the set of all cells (i, j) in the transportation array for which there is an upper-bound restriction (or capacity restriction) on the amount that can be shipped from supply point i to demand point j. $\bar{\mathbf{J}}$ is the set of all the remaining cells in the transportation array. And $a_i > 0$, $b_j > 0$ for all i, j. A necessary condition for feasibility is that $\sum a_i = \sum b_j$ let this quantity be denoted by s.

A *working basis* for this problem is a basis for the m × n transportation problem as discussed in Section 13.1. A feasible solution of (13.17) is a BFS if it is associated with a working basis with the property that all nonbasic variables x_{ij} for $(i, j) \in \mathbf{J}$ are either equal to 0 or U_{ij} in the solution.

Applying the results of Section 13.2.3 to this problem, the optimality criteria can be derived. Let $\bar{x}$ be a BFS of (13.17) associated with a working basic set $\mathscr{B}$. Compute $u = (u_1, \ldots, u_m)$, $v = (v_1, \ldots, v_n)$ (as in Section 13.2) from

$$u_i + v_j = c_{ij} \qquad \text{for all } (i, j) \text{ in the basic set } \mathscr{B}$$
$$v_n = 0$$

Compute $\bar{c}_{ij} = c_{ij} - u_i - v_j$ for all i, j. $\bar{x}$ is an optimum solution of (13.17) if

$$\bar{c}_{ij} > 0 \qquad \text{implies } \bar{x}_{ij} = 0$$
$$\bar{c}_{ij} < 0 \qquad \text{implies } (i, j) \in \mathbf{J} \qquad \text{and } \bar{x}_{ij} = U_{ij}$$

13.8.1 To Obtain an Initial Basic Feasible Solution

For the uncapacitated transportation problem, it is very easy to obtain an initial BFS directly by the methods discussed in Section 13.2. However, for the capacitated transportation problem, there is no such direct method available, and it may be necessary to solve a Phase I problem. One such method is discussed here. Consider the Phase I problem:

Minimize

$$w(x) = \sum_{i=1}^{m+1} \sum_{j=1}^{n+1} d_{ij} x_{ij}$$

Subject to

$$\sum_{j=1}^{n+1} x_{ij} = a_i \qquad \text{for } i = 1 \text{ to } m+1$$

$$(13.18)$$

$$\sum_{i=1}^{m+1} x_{ij} = b_j \qquad \text{for } j = 1 \text{ to } n+1$$

$$x_{ij} \geqq 0 \qquad \text{for all } i, j$$
$$x_{ij} \leqq U_{ij} \qquad \text{for all } (i, j) \in \mathbf{J}$$

where $a_{m+1} = b_{n+1} = s$, d_{ij} is 0 for all $1 \leqq i \leqq m$ and $1 \leqq j \leqq n$; $d_{m+1,j} = d_{i,n+1} = 1$ for all $1 \leqq i \leqq m$ and $1 \leqq j \leqq n$ and $d_{m+1,n+1} = 0$. Variables $x_{m+1,j}$ and $x_{i,n+1}$ are the artificial variables in this problem.
An initial feasible basis for (13.18) is marked in the transportation array at the top of the next column. Since (13.18) is a transportation problem of order $(m+1) \times (n+1)$, a working basic set for it consists of $m+n+1$ basic cells. Let $\bar{x}$ be the present BFS of (13.18) associated with the present working basic set $\mathscr{B}$. Compute the dual vector $\sigma = (\sigma_1, \ldots, \sigma_{n+1})$, $\mu = (\mu_1, \ldots, \mu_{m+1})$ from

Initial Basic Feasible Solution of (13.18)

	$j = 1$	2	...	n	n + 1
$i = 1$					$\boxed{a_1}$
2 ⋮	Nonbasic cells All equal to zero				$\boxed{a_2}$ ⋮
m					$\boxed{a_m}$
m + 1	$\boxed{b_1}$	$\boxed{b_2}$	...	$\boxed{b_n}$	$\boxed{0}$

$$\mu_i + \sigma_j = d_{ij} \qquad \text{for all cells } (i, j) \text{ in } \mathscr{B}$$
$$\mu_{n+1} = 0$$

and compute $\bar{d}_{ij} = d_{ij} - \mu_i - \sigma_j$ for all i, j. So $\bar{x}$ is an optimum solution of (13.18) if

$$\bar{d}_{ij} > 0 \text{ implies } \bar{x}_{ij} = 0$$
$$\bar{d}_{ij} < 0 \text{ implies } (i, j) \in \mathbf{J} \qquad \text{and} \qquad \bar{x}_{ij} = U_{ij}$$

Choice of Nonbasic Cell for Change of Status during Phase I

Let $\bar{x}$ be the present BFS of (13.18). If Phase I termination criteria are not satisfied, pick a nonbasic cell (p, q) that has led to a violation of the termination criteria, for change of status. It may belong to one of the following types.

1 $(p, q) \in \mathbf{J}$, $\bar{d}_{pq} < 0$ and $\bar{x}_{pq} = 0$
2 $(p, q) \in \mathbf{J}$, $\bar{d}_{pq} > 0$ and $\bar{x}_{pq} = U_{pq}$
3 $(p, q) \in \bar{\mathbf{J}}$ and $\bar{d}_{pq} < 0$

Find out the θ-loop associated with (p, q) with respect to the present working basic set $\mathscr{B}$ by starting with a $+\theta$ entry in the cell (p, q).

13.8.2 Changing the Status of (p, q) Such That $\bar{d}_{pq} < 0$

In this case the value of the Phase I objective function can be decreased by increasing the value of x_{pq} from its present value of 0. The maximum value that can be given to x_{pq} is min$\{\bar{x}_{ij}$ for all basic cells (i, j), such that they are in the θ-loop with a $-\theta$ entry; $U_{ij} - \bar{x}_{ij}$ for all basic cells (i, j), such that $(i, j) \in \mathbf{J}$ and (i, j) is in the θ-loop with a $+\theta$ entry; U_{pq}, if $(p, q) \in \mathbf{J}\}$. There are several cases to consider here.

Case 1 If $(p, q) \in \mathbf{J}$ and the maximum value that x_{pq} can assume turns out to be U_{pq}, make x_{pq} into a nonbasic variable whose value is equal to its upper bound in the next step. Putting θ equal to U_{pq}, revise the values of all the basic variables in the θ-loop and then erase the θ-loop. Keep the same working basis. Move to the next step.

Case 2 If the maximum value that x_{pq} can assume is $< U_{pq}$ and turns out to be some $\bar{x}_{i_1,j_1}$ where (i_1, j_1) is a basic cell with a $-\theta$ entry in the θ-loop, change the working basic set by dropping the cell (i_1, j_1) from it and making the cell (p, q) a basic cell. Make the value of $x_{pq} = \bar{x}_{i_1,j_1}$ and revise the values of all the other basic variables in the θ-loop by substituting $\bar{x}_{i_1,j_1}$ for θ. Make (i_1, j_1) a zero-valued nonbasic cell.

Case 3 If the maximum value that x_{pq} can take is $< U_{pq}$ and turns out to be $U_{i_1,j_1} - \bar{x}_{i_1,j_1}$ for a basic cell (i_1, j_1), such that $(i_1, j_1) \in \mathbf{J}$ and (i_1, j_1) is in the θ-loop with a $+\theta$ entry, drop (i_1, j_1) from the working basic set and make it into a nonbasic cell at its upper bound in the next solution. Make x_{pq} equal to $U_{i_1,j_1} - \bar{x}_{t_1,j_1}$, and revise the values of all the other basic variables in the θ-loop by substituting the same value for θ. Make (p, q) a basic cell and go to the next step.

13.8.3 Changing the Status of (p, q) Such That $\bar{d}_{pq} > 0$

In this case, (p, q) must be in $\mathbf{J}$ and the present value of x_{pq} must be U_{pq}. The Phase I objective value can be decreased by decreasing the value of x_{pq} from its present value of U_{pq}. Since the new value of x_{pq} will be $U_{pq} + \theta$, θ should be negative and as small as possible. The smallest value that θ can have is $\max\{-U_{pq};\ -\bar{x}_{ij}$ for all basic cells (i, j) with a $+\theta$ entry in the θ-loop; $\bar{x}_{ij} - U_{ij}$ for all basic cells (i, j), such that $(i, j) \in \mathbf{J}$ and (i, j) appears with a $-\theta$ entry in the θ-loop$\}$. There are several cases to consider here.

Case 1 If the smallest value that θ can take is $-U_{pq}$, then x_{pq} becomes a nonbasic variable whose value is 0 in the next step. There is no change in the working basis. Revise the values of all the basic variables in the θ-loop by substituting $-U_{pq}$ for θ and go to the next step.

Case 2 If the smallest value that θ can take happens to be is $> -U_{pq}$ and is $-\bar{x}_{i_1,j_1}$ for some basic (i_1, j_1) in the θ-loop, then (i_1, j_1) is dropped from the working basic set and made into a nonbasic cell with zero value. The variable x_{pq} becomes a basic variable, whose value is $U_{pq} - \bar{x}_{i_1,j_1}$ in the next solution. Revise the values of all the other basic variables in the θ-loop by substituting the value $-\bar{x}_{i_1,j_1}$ for θ and go to the next step.

Case 3 If the smallest value that θ can take happens to be is $> -U_{pq}$ and is $\bar{x}_{i_1,j_1} - U_{i_1,j_1}$ for some basic cell (i_1, j_1) in the θ-loop with $(i_1, j_1) \in \mathbf{J}$, then x_{i_1,j_1} is made a nonbasic variable whose value is equal to its upper bound in the next solution. The variable x_{pq} becomes a basic variable in its place with its value equal to $U_{pq} + \theta$ in the next solution. The values of all the other basic variables in the θ-loop are revised by substituting $\bar{x}_{i_1,j_1} - U_{i_1,j_1}$ for θ, and the algorithm then goes to the next step.

13.8.4 Phase I Termination

After a finite number of steps, the Phase I termination criteria will be satisfied. At Phase I termination if any of $x_{m+1,j}$ for $1 \leqq j \leqq n$ or $x_{i,n+1}$ for $1 \leqq i \leqq m$ are strictly positive, the original problem (13.17) must be infeasible and the algorithm terminates.

If $x_{m+1,j}$ for $1 \leqq j \leqq n$ and $x_{i,n+1}$ for $1 \leqq i \leqq m$ are all zero, the original problem is feasible. In this case, if the present working basic set has $m + n - 1$ basic cells (i, j) among those in the range $1 \leqq i \leqq m$ and $1 \leqq j \leqq n$, these cells form a working basic set for the original problem (13.17). Therefore, erase the $(m + n)$th row and the $(n + 1)$th column, which are all cells corresponding to the artificial variables. Go over to Phase II.

On the other hand, suppose the working basic set at Phase I termination has less than $m + n - 1$, say $(m + n - 1) - r$, basic cells (i, j) among those in the range $1 \leqq i \leqq m$ and $1 \leqq j \leqq n$. In this case, pick r nonbasic cells in this range that together with the present $(m + n - 1) - r$ basic cells in this range provide a working basic set for (13.7) and make these r cells into basic cells. Go over to Phase II.

13.8.5 Phase II

Given a BFS $\bar{x}$ associated with a working basis for (13.17), compute the corresponding u, v, $\bar{c}$, and check whether the optimality criterion is satisfied. If not, for change of status pick a nonbasic variable x_{pq} that has led to a violation of the optimality criterion. It can belong to the following types.

1 $\bar{c}_{pq} < 0$ and the present value of x_{pq} is 0. In this case, the objective value can be decreased by increasing the value of x_{pq} from its present value of 0. These computations are carried out as in Section 13.8.2.

2 $(p, q) \in \mathbf{J}$, $\bar{c}_{pq} > 0$ and the present value of x_{pq} is U_{pq}. In this case, the objective value can be

decreased by decreasing the value of x_{pq} from its present value of U_{pq}. These computations are carried out as in Section 13.8.3.

The algorithm will terminate in a finite number of steps with an optimum solution of the problem.

Example 13.22

Consider the following transportation problem.

					a_i
	1	1	2	15	21
$c =$	9	8	17	5	11
	4	2	12	2	28
b_j	8	16	25	11	

The amounts shipped from any supply point to any demand point should be less than or equal to 15. In the following transportation arrays basic cells are always marked with a small square in the middle, with the value of the basic variable entered inside the small square. Nonbasic cells at upper bound are marked with a small circle in the middle, with the value of that variable entered in the middle of the circle. Below is the initial Phase I array.

$j =$	1	2	3	4	5	μ_i
$i = 1$	$-2 = \bar{d}_{11}$; $d_{11} = 0$	−2; 0	−2; 0	−2; 0	[21]; 1	1
2	−2; 0	−2; 0	−2; 0	−2; 0	[11]; 1	1
3	−2; 0	−2; 0	−2; 0	−2; 0	[28]; 1	1
4	[8]; 1	[16]; 1	[25]; 1	[11]; 1	[0]; 0	0
σ_j	1	1	1	1	0	

First, x_{33} is entering. It becomes a nonbasic at the upper bound. Next, x_{24} is entering. It replaces x_{25} from the basic vector. At the top of the next column is the transportation array after these changes.

$j =$	1	2	3	4	5	μ_i
$i = 1$	−2; 0	−2; 0	−2; 0	−2; 0	[21]; 1	1
2	0; 0	0; 0	0; 0	[11]; 0	1	−1
3	−2; 0	−2; 0	−2; (15); 0	−2; 0	[13]; 1	1
4	[8]; 1	[16]; 1	[10]; 1	[0]; 1	[26]; 0	0
σ_j	1	1	1	1	0	

Next the following changes are made in the order listed at the bottom of the page. Notice that the dual solution and the relative cost coefficients have to be recomputed after every basis change.

The transportation array at the end of these changes is:

$j =$	1	2	3	4	5
$i = 1$		[11]	[10]		[0]
2				[11]	
3	[8]	[5]	(15)	[0]	
4					[60]

Clearly Phase I terminates in this stage and the basic cells among the cells of the original transportation array constitute a working basic set for it. Thus the initial Phase II array is:

$j =$	1	2	3	4	u_i
$i = 1$	$-2 = \bar{c}_{11}$; $c_{11} = 1$	[11] $+ \theta$; 1	[10] $- \theta$; 2	14; 15	1
2	2; 9	3; 8	11; 17	[11]; 5	5
3	[8]; 4	[5] $- \theta$; 2	9; (15) $+ \theta$; 12	[0]; 2	2
v_j	2	0	1	0	

Entering Variable	*Leaving Variable*	*Status of Entering Variable*	*Status of Leaving Variable*
x_{31}	x_{41}	Basic at value 8	Zero-valued nonbasic
x_{32}	x_{35}	Basic at value 5	Zero-valued nonbasic
x_{12}	x_{42}	Basic at value 11	Zero-valued nonbasic
x_{13}	x_{43}	Basic at value 10	Zero-valued nonbasic
x_{34}	x_{44}	Basic at value 0	Zero-valued nonbasic

Since $\bar{c}_{33} = 9$, x_{33} should be reduced from its present value of 15. The θ-loop with an entry of $+\theta$ in the cell (3, 3) is entered. When θ is -5, x_{13} reaches its upper bound and leaves the basic vector to become a nonbasic at upper bound. Thus the new array is:

$j =$	1	2	3	4	μ_i
$i = 1$	-2; θ; 1	[6] $- \theta$; 1	-9; (15); 2	14; 15	1
2	2; 9	3; 8	2; 17	[11]; 5	5
3	[8] $- \theta$; 4	[10] $+ \theta$; 2	[10]; 12	[0]; 2	2
v_j	2	0	10	0	

Now, x_{11} is entering. The θ loop is entered. When θ is 5, x_{32} reaches its upper bound. Thus the new array is:

$j =$	1	2	3	4	μ_i
$i = 1$	[5]; 1	[1]; 1	-7; (15); 2	16; 15	-1
2	2; 9	1; 8	2; 17	[11]; 5	5
3	[3]; 4	-2; (15); 2	[10]; 12	[0]; 2	2
v_j	2	2	10	0	

The optimality criterion is now satisfied and, hence, the solution in this transportation array is optimal.

Comment 13.4: The algorithms discussed in this chapter for solving transportation problems have been generalized to solve network flow problems on general networks. See references [13.5, 13.12, 1.68, 1.80, 1.81, 1.85, 1.90] for a discussion of these network flow methods.

13.9 INTERVAL TRANSPORTATION PROBLEM

Let a_i', a_i'', b_j', b_j'' be given positive numbers satisfying $a_i' \leqq a_i''$, $b_j' \leqq b_j''$ for all $i = 1$ to m and $j = 1$ to n. Consider the following transportation model.

$$\text{Minimize} \quad z(x) = \sum_{i=1}^{m} \sum_{j=1}^{n} c_{ij} x_{ij}$$

$$\text{Subject to} \quad a_i' \leqq \sum_{j=1}^{n} x_{ij} \leqq a_i'', \qquad i = 1 \text{ to m}$$

$$b_j' \leqq \sum_{i=1}^{m} x_{ij} \leqq b_j'', \qquad j = 1 \text{ to n} \qquad (13.19)$$

$$x_{ij} \geqq 0 \quad \text{for all } i, j$$

In this model x_{ij} is the amount of material shipped from source i to demand center j. And $a_i = \sum_{j=1}^{n} x_{ij}$ is the total amount of material shipped out of source i, for $i = 1$ to m; $b_j = \sum_{i=1}^{m} x_{ij}$ is the total amount of material shipped to demand center j, for $j = 1$ to n. In model (13.19), a_i can vary within the interval a_i' to a_i'', and b_j can vary within the interval b_j' to b_j''. Hence this model is known as the *interval transportation problem*. In most practical transportation problems the a_i's and b_j's can be varied within reasonable bounds, as in (13.19), and by optimizing under a_i's, b_j's varying within these allowable bounds we can obtain the best possible solution in these problems. Model (13.19) is feasible if $\sum_{i=1}^{m} a_i'' \geqq \sum_{j=1}^{n} b_j'$ and $\sum_{i=1}^{m} a_i' \leqq \sum_{j=1}^{n} b_j''$. For $i = 1$ to m, let $x_{i,n+1}$ be the slack variable corresponding to the constraint $\sum_{j=1}^{n} x_{ij} \leqq a_i''$. For $j = 1$ to n, let $x_{m+1,j}$ be the slack variable corresponding to the constraint $\sum_{i=1}^{m} x_{ij} \leqq b_j''$. Consider the following bounded variable balanced transportation problem.

$$\text{Minimize} \quad \sum_{i=1}^{m} \sum_{j=1}^{n} c_{ij} x_{ij}$$

$$\text{Subject to} \quad \sum_{j=1}^{n+1} x_{ij} = a_i'' \qquad \text{for } i = 1 \text{ to m}$$

$$= \sum_{j=1}^{n} b_j'' \qquad \text{for } i = m + 1$$

$$\sum_{i=1}^{m+1} x_{ij} = b_j'' \qquad \text{for } j = 1 \text{ to n}$$

$$= \sum_{i=1}^{m} a_i'' \qquad \text{for } j = n + 1 \qquad \textbf{(13.20)}$$

$$x_{ij} \geqq 0 \qquad \text{for all } i = 1 \text{ to m} + 1 \text{ and } j = 1 \text{ to n} + 1$$

$$x_{i,n+1} \leqq a_i'' - a_i' \qquad \text{for } i = 1 \text{ to m}$$

$$x_{m+1,j} \leqq b_j'' - b_j' \qquad \text{for } j = 1 \text{ to n}$$

Market	1	2	3	4	Available Gas (units of 10^6 gal.)
Refinery 1	4	7	9	10	8
2	6	4	3	6	10
3	9	6	4	8	6
Requirement (in units of 10^6 gal.)	5	3	8	4	

Clearly (13.20) is equivalent to (13.19). If

$$\bar{X} = \begin{pmatrix} \bar{x}_{11} & \cdots & \bar{x}_{1n} & \bar{x}_{1,n+1} \\ \vdots & & \vdots & \vdots \\ \bar{x}_{m1} & \cdots & \bar{x}_{mn} & \bar{x}_{m,n+1} \\ \bar{x}_{m+1,1} & \cdots & \bar{x}_{m+1,n} & \bar{x}_{m+1,n+1} \end{pmatrix}$$

is an optimum solution of (13.20), then its submatrix $\bar{x}$ obtained by striking off the last row and last column from $\bar{X}$ is an optimum solution of (13.19). Since (13.20) is a bounded variable balanced transportation problem, it can be solved by the algorithm discussed in Section 13.8.

Exercises

In all these exercises, the entries inside the tableau are the unit transportation cost coefficients, c_{ij}.

13.25 There are three refineries that have gas available and four markets that require gas. The data are given at the top. Determine an optimum transportation policy.

13.26 In the following transportation model, the total requirements at the market exceeds the amounts of material available at the plants. The deficits have to be made up by imports. All the markets have the same priority standing. Determine how much of the demand in each market should be fulfilled by each plant to utilize the existing available material at minimal transportation cost.

Market	1	2	3	4	5	Available at Plant (tons)
Plant 1	3	9	5	6	7	10
2	2	1	8	10	13	25
3	3	12	6	5	2	13
4	1	9	14	3	2	33
Required at markets (tons)	30	22	17	19	12	

13.27 For the following transportation problem, is $\{(1, 1)\ (1, 4), (2, 2), (2, 5), (3, 3), (3, 4), (3,5)\}$ a basic set?

$j=$	1	2	3	4	5	a_i
$i=1$	9	16	4	11	19	8
2	8	6	8	12	8	10
3	1	12	3	23	6	30
b_j	5	4	9	8	22	

If so, find the primal and dual solutions corresponding to it. Starting with this solution find all the alternate optimum solutions of the problem.

If a_3 changes to $30 + \delta$ and b_4 changes to $8 + \delta$ find the range of values of δ for which the first optimal basic set obtained remains optimal.

Returning to the original problem (with $\delta = 0$), find the range of values of c_{12} for which the first optimal BFS obtained remains optimal. What is an optimum solution when c_{12} changes to 2?

Returning to the original problem (with $\delta = 0$ and $c_{12} = 16$) find the range of values of c_{35} for which the first optimal BFS obtained remains optimal. What is an optimum solution when c_{35} changes to 20?

13.28 Consider the following transportation problem. Let $f(\delta)$ be the minimum objective value in this problem as a function of δ. Find all the alternate optimum solutions of the problem when $\delta = 0$. Draw $f(\delta)$ in the range $\delta \geqq 0$ as a curve. At what values of δ does $f(\delta)$ change slope? Explain the slopes and the changes in it in terms of the optimum dual solution as a function of δ.

$j =$	1	2	3	a_i
$i = 1$	1	3	2	10
2	2	1	3	15
3	3	2	1	$5 + \delta$
b_j	5	$10 + \delta$	15	

13.29 Solve the following transportation problem for $\lambda = 0$ and get two distinct optimum solutions. Solve for all $\lambda \geqq 0$.

$j =$	1	2	3	a_i
$i = 1$	5	$14 - 2\lambda$	7	15
2	6	7	8	20
3	$14 + \lambda$	$24 - 3\lambda$	9	4
b_j	7	11	21	

13.30 Use $\{(1, 4), (2, 3), (2, 4), (3, 2), (3, 5), (4, 1), (4, 2), (4, 3)\}$ as an initial basic set for $\delta = 0$, and solve the problem. Find the optimality range of the basic set obtained.

$j =$	1	2	3	4	5	a_i
$i = 1$	6	11	14	11	4	$3 + \delta$
2	7	17	9	10	5	15
3	9	5	7	18	0	14
4	9	13	13	15	7	10
b_j	3	$9 + \delta$	14	7	9	

13.31 Show that the basic set $\{(1, 2), (1, 4), (2, 1), (2, 3), (3, 2), (3, 5), (4, 1), (4, 4)\}$ is an optimum basic set for the following problem when $\delta = 0$. Using it, solve the problem for all $0 \leqq \delta \leqq 8$.

13.32 *BOTTLENECK TRANSPORTATION PROBLEM* There are m plants manufacturing a product and the ith plant has a_i units available. There are n markets where there is a demand for this, and the demand in the jth market is b_j units. $\sum a_i = \sum b_j$. The time needed to make a delivery from the ith plant to the jth market is t_{ij} days. $i = 1$ to m, $j = 1$ to n. The activity of shipping from i to j is done independently for each i, j. If x_{ij} is the number of items shipped from i to j, and $x = (x_{ij})$ is a feasible solution of this problem, show that the time needed to implement the solution x is $\theta(x) =$ maximum $\{t_{ij}$: (i, j), such that $x_{ij} > 0\}$. Develop a method for minimizing $\theta(x)$. Solve the problem with the following data:

t_{ij}

$j =$	1	2	3	4	5	6	a_i
$i = 1$	13	18	7	23	47	5	23
2	39	55	3	45	12	29	9
3	23	9	21	32	23	9	17
4	72	11	16	14	35	19	11
b_j	7	8	19	5	13	8	

13.33 Solve the following bounded variable balanced transportation problems:

							a_i
	31	23	20	22	17	14	48
$c =$	17	18	19	5	16	7	43
	16	15	17	29	9	14	41
	22	9	3	22	25	2	46
b_j	19	37	26	30	30	36	

$0 \leqq x_{1j} \leqq 10$; $0 \leqq x_{ij} \leqq 20$ for $i = 2, 3$ and for all j; and $x_{4j} \geqq 0$ for all j.

$j =$	1	2	3	4	5	a_i
$i = 1$	4	4	8	6	10	$14 - \delta$
2	3	8	2	16	14	$12 + 2\delta$
3	9	7	9	18	12	7
4	8	18	11	13	18	18
b_j	$17 + 4\delta$	$11 - \delta$	4	$17 - 2\delta$	2	

(b)

							a_i
	3	4	11	9	8	13	32
$c =$	2	7	6	14	5	2	102
	1	3	7	1	13	17	22
	5	9	4	2	3	18	20
b_j	36	3	17	36	34	50	

$0 \leqq x_{ij} \leqq 20$ for all i, j.

13.34 $a_1, a_2, \ldots, a_m; b_1, b_2, \ldots, b_n$ are given positive integers. $\alpha = \sum a_i$, $\beta = \sum b_j$, $v =$ minimum $\{\alpha, \beta\}$. δ is a positive integer between 1 and v. Let **K** be the set of feasible solutions of the following problem (a) Prove that all the extreme points of **K** are integer points. (b) Show that the above problem can be transformed into a transportation problem. Develop an algorithm for solving it, as δ varies from 1 to v.

$$\text{Minimize} \quad \sum\sum c_{ij}x_{ij}$$

$$\text{Subject to} \quad \sum_j x_{ij} \leqq a_i \qquad i = 1 \text{ to m}$$

$$\sum_i x_{ij} \leqq b_j \qquad j = 1 \text{ to n}$$

$$\sum_i \sum_j x_{ij} = \delta$$

$$x_{ij} \geqq 0 \qquad \text{for all } i, j$$

13.35 Consider the following transportation model, where a_i, b_j are real numbers (some of them may be negative) satisfying $\sum_{i=1}^{m} a_i = \sum_{j=1}^{n} b_j$, and **J** is a specified subset of cells of the m × n transportation array. Develop a primal algorithm for solving it using the ideas discussed in Section 13.8

$$\text{Minimize} \quad z(x) = \sum_{i=1}^{m} \sum_{j=1}^{n} c_{ij}x_{ij}$$

$$\text{Subject to} \quad \sum_{j=1}^{n} x_{ij} = a_i \qquad \text{for } i = 1 \text{ to m}$$

$$\sum_{i=1}^{m} x_{ij} = b_j \qquad \text{for } j = 1 \text{ to n}$$

$$x_{ij} \geqq 0 \quad \text{for } (i, j) \in \mathbf{J} \qquad x_{ij} \leqq 0 \quad \text{for } (i, j) \notin \mathbf{J}$$

13.36 *THE TRANSPORTATION PARADOX.* Let $a = (a_1, \ldots, a_m)$, $b = (b_1, \ldots, b_n)$ and let $f(a, b)$ be the minimum value of $z(x)$ in the balanced transportation problem (13.1). Suppose $\tilde{a}$, $\tilde{b}$, $\hat{a}$, $\hat{b}$ are all positive vectors satisfying

$$\sum_{i=1}^{m} \tilde{a}_i = \sum_{j=1}^{n} \tilde{b}_j \qquad \text{and} \qquad \sum_{i=1}^{m} \hat{a}_i = \sum_{j=1}^{n} \hat{b}_j$$

and $\tilde{a} \geqq \hat{a}$, $\tilde{b} \geqq \hat{b}$. $f(\tilde{a}, \tilde{b})$ is the minimum transportation cost when the supplies at sources are $\tilde{a}_i$ and the demands at the markets are $\tilde{b}_j$. Likewise, $f(\hat{a}, \hat{b})$ is the minimum transportation cost when the supplies at the sources are $\hat{a}_i$ and the demands at the markets are $\hat{b}_j$. Suppose $c_{ij} > 0$ for all i, j. It seems reasonable to expect that the total cost of transportation will increase if the supplies available at the sources and the demands at the markets both go up. Thus in this case, since $\tilde{a}_i \geqq \hat{a}_i$ for all i and $\tilde{b}_j \geqq \hat{b}_j$ for all j, one intuitively expects that $f(\tilde{a}, \tilde{b}) \geqq f(\hat{a}, \hat{b})$. However, this may not be true, as the following example illustrates. This is known as the *transportation paradox.* (1) Consider the balanced transportation problem with the following data. Compute $g(\delta)$, the minimum objective value in this problem, as a function of δ, in the range $0 \leqq \delta \leqq 22$. Show that $g(\delta)$ strictly decreases as δ increases in this range.

	Unit Transportation Costs ($ ton)						*Available at Source i (tons)*
$j =$	1	2	3	4	5	6	a_i
$i = 1$	30	11	5	35	8	29	30
2	2	5	2	5	1	9	$10 + \delta$
3	35	20	6	40	8	33	45
4	19	2	4	30	10	25	30
Required at market j (tons) b_j	25	$20 + \delta$	6	7	22	35	

(2) Give an explanation of the transportation paradox using the results in Section 13.5.4. (3) The transportation paradox indicates that in a practical transportation model, one may be able to decrease the total transportation cost by increasing the supplies at some sources and making corresponding increases in the demands at some markets, from present levels. To take advantage of this phenomenon, we should examine the transportation model of the following form, where a_i, b_j are given positive numbers. Develop an efficient algorithm for solving this transportation model. (Use the algorithm developed in Exercise 13.35) Discuss its uses in practical applications.

$$\text{Minimize} \quad \sum_{i=1}^{m} \sum_{j=1}^{n} c_{ij}x_{ij}$$

$$\text{Subject to} \quad \sum_{j=1}^{n} x_{ij} \geqq a_i \qquad i = 1 \text{ to m}$$

$$\sum_{i=1}^{m} x_{ij} \geqq b_j \qquad j = 1 \text{ to n}$$

$$x_{ij} \geqq 0 \qquad \text{for all } i, j$$

—(*W. Szwarc*[13.26])

13.37 Let $\mathscr{B}$ be a basic set of cells for (13.1). In Section 13.2.2, we computed the associated dual basic solution $(\tilde{u}, \tilde{v})$ by solving: $u_i + v_j = c_{ij}$ for all $(i, j) \in \mathscr{B}$, $v_n = 0$. Let $(u(\alpha), v(\alpha))$ denote the solution of $u_i + v_j = c_{ij}$ for all $(i, j) \in \mathscr{B}$, $v_n = \alpha$. Prove that $v_j(\alpha) = \tilde{v}_j + \alpha$ for all j and $u_i(\alpha) = \tilde{u}_i - \alpha$ for all i. So $u_i(\alpha) + v_j(\alpha) = \tilde{u}_i + \tilde{v}_j$ for all i, j and α, and hence $\bar{c}_{ij} = c_{ij} - \tilde{u}_i - \tilde{v}_j = c_{ij} - u_i(\alpha) - v_j(\alpha)$. Therefore, the values of the relative cost coefficients are unaffected by the value given for v_n in computing the dual solution. Also prove that if $\mathscr{B}$ is an optimum basic set for (13.1), the conclusions drawn from marginal analysis remain unaffected by the value given for v_n in computing the dual solution corresponding to $\mathscr{B}$.

From these arguments, show that the dual solution corresponding to any basic set $\mathscr{B}$ can be computed by giving an arbitrary real value to any one dual variable and then computing the values of the others from the system $u_i + v_j = c_{ij}$ for $(i, j) \in \mathscr{B}$.

13.38 Let $\mathscr{B}$ be a basic set of cells for the m × n transportation array. If $(i, j) \in \mathscr{B}$, prove that it is impossible for (i, j) to be the only basic cell in both the ith row and the jth column of the array (i.e., prove that either row i of the array, or column j of the array, or both of them contain at least one cell from $\mathscr{B}$ other than (i, j)).

13.39 *APPLICATION IN ACCOUNTING.* This application arose in a practical accounting problem in the electrical power industry of Norway. SKN is the association that coordinates and maintains the transmission grid for the various subscribers in that country. The *subscribers* are the various local companies that use the grid. A point that feeds electricity into the grid is called a *source*, and a point that takes electricity from the grid is called a *sink*. Each subscriber may have several sources and sinks on the grid. In each period, the net flow of electrical power to the grid from each source, and into each sink, is metered and recorded. Using these data, SKN is required to charge each subscriber a transmission fee for the use of transmission lines during that period, at the rate of α dollars per unit power per unit distance transmitted, where α is a given number. Let $i = 1$ to m be the sources, $j = 1$ to n be the sinks. Let c_{ij} be the length of the shortest route between source i and sink j along the power line network, for $i = 1$ to m, $j = 1$ to n, and assume that the matrix $c = (c_{ij})$ is given. For $i = 1$ to m, let a_i be the units of electricity that were fed into the system at source i during this period, and for $j = 1$ to n, let b_j be the units of electricity that were drawn from the system by sink j. Usually there are power losses along transmission lines, so $\sum a_i$ will be strictly greater than $\sum b_j$, but SKN compensates for this by multiplying each source amount by the same number less than 1, so that the sum of the modified inputs equals $\sum b_j$. The only data available are $c = (c_{ij})$, α and $a = (a_i)$, $b = (b_j)$ for the period.

The transmission charge for each subscriber should actually depend on the unit distance used by each subscriber, but this information is not available. Let x_{ij} denote the units of power reaching sink j from source

i, for $i = 1$ to m, $j = 1$ to n. If all the x_{ij} are known, we can easily compute the transmission charge for each subscriber. Electric power chooses the shortest path along power lines from source to skink, regardless of the ownership of the lines. So for determining the (x_{ij}), it is assumed that the power transmission occurs in such a way that $\sum\sum x_{ij}c_{ij}$ is minimized. Under this assumption, formulate the problem of determining the transmission charge to be levied for each subscriber as a transportation problem.

For the simple problem in which there are two subscribers, sinks 1, 2, 3 belong to subscriber 1, sinks 4, 5, 6 belong to subscriber 2, $\alpha = 0.1$, and the remaining data are as in the table below, determine the transmission charges for the two subscribers using our formulation.

$j =$	1	2	3	4	5	6	a_i
$i = 1$	3	9	7	19	18	12	33
2	4	12	15	11	4	18	29
3	5	18	13	10	3	9	24
4	7	13	12	8	4	18	24
b_j	30	17	18	10	16	9	

—(*O. Aarvik and P. Randolph* [13.1])

13.40 Consider the balanced transportation problem with the following data. (1) Compute the basic solution corresponding to the basic set of cells $\mathscr{B}_1 = \{(1, 1), (1, 4), (2, 2), (2, 5), (3,1), (3, 3), (4, 3), (4, 5), (4, 6)\}$. Is $\mathscr{B}_1$ primal feasible? Is it dual feasible? If $\mathscr{B}_1$ is optimal, is the primal solution corresponding to it the unique optimum solution of this problem? (2) The demand for material at demand center 4 is going up by an amount of δ tons over the present amount, where δ is an unknown nonnegative parameter. To keep the balance between supply and demand, the company has to make arrangements to increase the supply at some source by δ tons. Use marginal analysis to determine the source (1, 2, 3 or 4) that is the best location for augmenting the supply in this case. Call this best source p. Change a_p to $a_p + \delta$, and b_8 to $16 + \delta$, and obtain the optimum solution for the modified problem as a function of δ for all $\delta \geqq 0$. (3) Consider the original problem again. Treating c_{45} as a parameter (whose present value is 18), while all the data in the problem stay fixed, find out optimum solutions of the problem for all real values of c_{45}.

Demand Center j	c_{ij} = \$/ton, Unit Cost of Shipping from i to J: 1	2	3	4	5	6	a_i = Tons Available at i
Source $i = 1$	8	10	8	2	10	6	34
2	12	3	5	14	4	9	17
3	9	12	5	7	13	1	35
4	22	20	15	16	18	8	60
b_j = tons required at j	38	7	45	16	15	25	

13.41 A company has five plants at sites 7 to 11, respectively. Workers are given free living quarters, and these are located at sites 1 to 6. The company provides free transportation to the workers from their living sites to the plant sites everyday. The work at all the plants is of the same type, and all the workers can do it, and hence a worker accommodated at any living site can be taken to work at any plant site. Data are given at the top of page 429. (1) Obtain primal and dual solutions for this transportation problem, corresponding to the basic set of cells $\mathscr{B}_1 = \{(1, 7), (1, 10), (1, 11), (2, 7), (3, 7), (3, 9), (4, 11), (5, 9), (6, 8), (6, 11)\}$. Is $\mathscr{B}_1$ primal feasible? Dual feasible? Draw the rooted tree corresponding to $\mathscr{B}_1$, and obtain the tree labels corresponding to it. Starting with $\mathscr{B}_1$, solve this problem. Let $\mathscr{B}_2$ denote the final optimum basic of cells obtained. (2) The company has just realized that the worker requirements at plant site 8 are likely to go up

$j =$	c_{ij} = *Company's Cost (cents/ Workers/Day) to Transport from Living Site to Plant Site* 7	8	9	10	11	a_i = *Number of Workers Living Site i Can Accommodate*
$i = 1$	16	23	28	10	7	60
2	15	22	21	11	10	21
3	22	30	23	20	15	25
4	20	21	25	15	8	13
5	15	25	15	4	8	25
6	15	16	20	12	4	17
b_j = Number of workers needed at j daily	50	7	40	24	40	

from the present requirement of 7 workers. Using marginal analysis arguments and the information available from the current optimum array above, discuss which living site is the best location to build living quarters for these new workers (cost of building new quarters is the same at all living sites and, hence, you can ignore that). (3) The new worker requirements at plant site 8 are $7 + \alpha$. where α is a nonnegative integer. Using your answer for (2), determine at which living site the additional α workers should be accommodated (this implies that you should increase the number of workers that can be accomodated at that living site by α from present level), and obtain the basic solution of the modified problem, corresponding to the basic set of cells $\mathscr{B}_2$, as a function of α. Determine the range of values of α for which this solution remains optimal to the modified problem. As α increases from zero, discuss how the optimum site to house the additional workers may change. Obtain optimum solutions for the problem, as functions of α, for all nonnegative integer values of α.

13.42 Consider (13.1) with the additional restrictions $x_{ij} \leqq 1$ for all i, j. (1) Derive simple necessary and sufficient conditions for this problem to be feasible. (2) Develop a simple method for obtaining a BFS to this problem when these feasibility conditions are satisfied. (3) Discuss an efficient method for solving this problem. (4) Apply this method on the problem with the following data.

								a_i
$c =$	3	12	11	2	15	16	21	6
	9	13	19	11	10	6	17	4
	13	15	3	16	9	12	18	3
	5	6	17	13	18	11	22	2
	8	12	14	4	13	7	23	2
b_j	3	2	4	2	1	2	3	

$0 \leqq x_{ij} \leqq 1$ for all i, j.

13.43 Consider the following transportation problem. Verify that $\{(1, 4), (1, 6), (2, 3), (2, 5), (3, 1), (3, 3), (3, 6), (4, 1), (4, 2)\}$ is a dual feasible but primal infeasible basic set of cells for this problem. Starting with this basic set, solve this problem using the dual simplex algorithm.

							a_i
$c =$	13	18	18	1	17	3	6
	19	24	10	19	19	19	12
	14	21	6	10	15	10	62
	2	5	7	12	14	8	9
b_j	10	19	6	22	18	14	

13.44 There are n locations available for storage in a warehouse. There are m different chemicals to be stored. In each location we can store at most one chemical (they cannot be mixed). The amount of chemical i available is expected to require a_i locations to store for $i = 1$ to m. The a_i are given positive integers for $i = 1$ to m, and $\sum_{i=1}^{m} a_i \leqq$ n. c_{ij} is the cost incurred if the jth location is allocated for storing chemical i; and all the c_{ij}'s are given.

Formulate the problem of allocating the available locations to store the chemicals optimally and discuss how it can be solved.

13.45 Consider the following problem. There are m machines and n jobs. For $i = 1$ to m, $j = 1$ to n, t_{ij} is the time it takes to process the jth job on the ith machine. Each machine can process any job, and a machine can process any number of jobs. Beginning at clock-time point 0, we find the time that elapses before each job is processed and turned over. Let p_j be the clock time at which the processing of job j is completed. It is required to assign jobs to machines so that $\sum_{j=1}^{n} p_j$ is minimized ($(\sum_{j=1}^{n} p_j)/n$ is called the *mean flow time* in scheduling literature). Let r_i be the number of jobs assigned to be processed on the ith machine. Number the jobs processed on each machine in reverse order, that is, for machine i, the last job processed on it is numbered 1, the last but one is numbered 2, etc. Let $x_{ijk} = 1$ if job j is the kth on the ith machine, 0 otherwise. Show that $\sum_{j=1}^{n} p_j = \sum_{i=1}^{m} \sum_{j=1}^{n} \sum_{k=1}^{r_i} kx_{ijk}$. Formulate the problem of assigning jobs to machines to minimize $\sum_{j=1}^{n} p_j$ as a transportation problem (it will be of order $(m + 1)n \times mn^2$).

—(*W. A. Horn* [13.15])

13.46 The following table specifies a set of persons, their salaries, and the jobs they can do.

Person	*Salary ($)*	*Jobs Person Can Do*
1	4,000	5
2	6,000	1, 4
3	7,000	2
4	10,000	4
5	12,000	2, 3
6	14,000	1, 4
7	15,000	3
8	17,000	5, 6
9	20,000	5, 6

It is required to assign distinct persons to the six jobs so as to minimize the total salary paid. Formulate this problem as an optimization problem, using the simplest formulation possible.

13.47 Solve the 3×3 interval transportation problem with the constraints $10 \leqq a_1 \leqq 25$, $15 \leqq a_2 \leqq 20$, $20 \leqq a_3 \leqq 35$; $17 \leqq b_1 \leqq 30$, $22 \leqq b_2 \leqq 25$, $11 \leqq b_3 \leqq 20$, and the cost matrix

$$c = \begin{pmatrix} 9 & 19 & 16 \\ 18 & 17 & 11 \\ 13 & 12 & 10 \end{pmatrix}$$

13.48 Consider the transportation problem with the following data, where λ, δ are nonnegative parameters. (1) For $\lambda = \delta = 0$, find an optimum BFS and check for alternate optima. If so find a second optimum BFS. Let $\mathscr{B}_1$ refer to the optimum basic set for this case in which (2, 4) is not a basic cell. (2) Keeping $\delta = 0$, find optimum solutions for all $\lambda \geqq 0$. (3) Keeping $\lambda = 0$, find optimality range for $\mathscr{B}_1$ in δ, say $0 \leqq \delta \leqq \delta_1$. Find an optimum solution when δ just exceeds δ_1.

$$\begin{array}{lccccccc} & & & & & & & a \\ c = & \left(\begin{array}{c} 5 \\ 3 \\ 5 \end{array}\right. & \begin{array}{c} 23 + 2\lambda \\ 35 + \lambda \\ 26 + 2\lambda \end{array} & \begin{array}{c} 13 + \lambda \\ 5 \\ 15 + 2\lambda \end{array} & \begin{array}{c} 2 \\ 3 \\ 9 + \lambda \end{array} & \left.\begin{array}{c} 13 + 2\lambda \\ 12 + \lambda \\ 9 + 2\lambda \end{array}\right) & & \begin{array}{c} 14 + \delta \\ 8 \\ 11 \end{array} \\ b & 5 + \delta & 12 & 3 & 6 & 7 & & \end{array}$$

13.49 Develop an efficient method for computing the inverse of a given basis for (13.7).

13.50 Let $\bar{x}$ be an optimum BFS of (13.1) and suppose it is degenerate. Adapt the results of Sections 6.7, 8.10, and 8.15 to problem (13.1) to develop efficient methods for computing the marginal values in (13.1) in the specified directions discussed at the beginning of Section 13.5.4 and Exercise 13.19.

REFERENCES

13.1 O. Aarvik and P. Randolph, "The Application of Linear Programming to the Determination of Transmission Fee in an Electrical Power Network," *Interfaces* 6, 1 (November 1975), 47–49.

13.2 P. Camion, "Characterization of Totally Unimodular Matrices," *Proceeding of the*

American Mathematical Society 16 (1965), 1068–1073.

13.3 R. Chandrasekharan, "Total Unimodularity of Matrices," *SIAM Journal of Applied Mathematics 17* (1969) 1032–1034.

13.4 R. Chandrasekharan and S. Subba Rao, "A Special Case of the Transportation Problem," *Operations Research 25*, 3 (May/June 1977), 525–528.

13.5 W. H. Cunningham, "Theoretical Properties of the Network Simplex Method," *Mathematics of Operations Research 4*, 2 (May 1979), 196–208.

13.6 J. B. Dennis, "A High-Speed Computer Technique for the Transportation Problem," *Journal of the ACM 8* (1958), 257–276.

13.7 G. Finke and J. H. Ahrens, "A Variant of the Primal Transportation Algorithm," *INFOR 16*, 1 (February 1978), 35–46.

13.8 M. M. Flood, "A Transportation Algorithm and Code," *Naval Research Logistics Quarterly 8*, (1971), 257–276.

13.9 B. J. Gassner, "Cycling in the Transportation Problem," *Naval Research Logistics Quarterly 11*, 1 (March 1964), 43–58.

13.10 F. Glover, D. Karney, and D. Klingman, "The Augmented Predecessor Index Method for Locating Stepping Stone Paths and Assigning Dual Prices in Distribution Problems," *Transportation Science 6*, 1 (1972), 171–180.

13.11 F. Glover, D. Karney, and D. Klingman, "A Note on Computational Studies for Solving Transportation Problems," *Proceedings of the ACM*, (1973), 180–187.

13.12 F. Glover, D. Karney, and D. Klingman, "Implementation and Computational Comparisons of Primal, Dual, and Primal-Dual Computer Codes for Minimum Cost Flow Network Problems," *Networks 4* (1974), 191–212.

13.13 F. Glover, D. Karney, D. Klingman, and A. Napier, "A Computation study on Start Procedures, Basis Change Criteria and Solution Algorithms for Transportation Problems," *Management Science 20*, 5 (January 1974), 793–813.

13.14 I. Heller and C. B. Tompkins, "An Extension of a Theorem of Dantzig," in *Linear Inequalities and Related Systems*, H. W. Kuhn and A. W. Tucker (Ed.), Princeton University Press, Princeton N.J., 1956, pp. 247–254.

13.15 W. A. Horn, "Minimizing Average Flow Time with Parallel Machines," *Operations Research 21* (1973), 846–847.

13.16 E. Johnson, "Networks and Basic Solutions," *Operations Research 14*, 4 (1966), 89–95.

13.17 K. G. Murty and C. Witzgall, "Duall Simplex Method for the Uncapacitated Transportation Problem Using Tree Labellings," Technical Report 77–9, Department of Industrial and Operations Engineering, University of Michigan, Ann Arbor, 1977

13.18 A. Orden, "The Transshipment Problem," *Management Science* 2, 3 (1956), 276–285.

13.19 M. W. Padberg, "Characterizations of Totally Unimodular, Balanced and Perfect Matrices," Working Paper, New York University, Graduate School of Business Administration, 1974.

13.20 K. R. Rebman, "Total Unimodularity and the Transportation Problem: A Generalization," *Linear Algebra and Its Applications 8* (1974), 11–24.

13.21 N. V. Reinfield and W. R. Vogel, *Mathematical Programming,* Prentice-Hall, Englewood Cliffs, N.J., 1958.

13.22 P. D. Seymour, "Decomposition of Regular Matroids," *Journal of Combinatorial Theory B*, *28* (1980), 305–359.

13.23 M. Simonnard, *Linear Programming*, translated from French by W. S. Jewell, Prentice-Hall, Englewood Cliffs, N.J., 1966.

13.24 V. Srinivasan and G. L. Thompson, "Benefit Cost Analysis of Coding Techniques for the Primal Transportation Algorithm," *Journal of the ACM 20* (1973), 194–213.

13.25 V. Srinivasan, "Network Models for Estimating Brand-Specific Effects in Multi-attribute Marketing Models," *Management Science 25*, 1 (January 1979), 11–21.

13.26 W. Szwarc, "The Transportation Paradox," *Naval Research Logistics Quarterly 18*, 2 (June 1971), 185–202.

13.27 A. Tamir, "On Totally Unimodular Matrices," *Networks 6* (1976), 373–382.

13.28 K. Truemper, "Algebraic Characterizations of Unimodular Matrices," *SIAM Journal of Applied Mathematics 35* (1978), 328–332.

13.29 A. F. Veinott and G. B. Dantzing, "Integral Extreme Points," *SIAM Review 10* (1968), 371–372.

13.30 N. Zadeh, "A Bad Network Problem for the Simplex Method and Other Minimum Cost Flow Algorithms," *Mathematical Programming 5* (1973), 255–266.

Tableau T discussed in Section 13.1.8 when m = 3, n = 4

x_{11}	x_{12}	x_{13}	x_{14}	x_{21}	x_{22}	x_{23}	x_{24}	x_{31}	x_{32}	x_{33}	x_{34}	
1	1	1	1	0	0	0	0	0	0	0	0	a_1
0	0	0	0	1	1	1	1	0	0	0	0	a_2
0	0	0	0	0	0	0	0	1	1	1	1	a_3
1	0	0	0	1	0	0	0	1	0	0	0	b_1
0	1	0	0	0	1	0	0	0	1	0	0	b_2
0	0	1	0	0	0	1	0	0	0	1	0	b_3

$x_{ij} \geqq 0$ for all i, j.

Computational Complexity of the Simplex Algorithm

chapter **14**

14.1 HOW EFFICIENT IS THE SIMPLEX ALGORITHM?

Since the development of the simplex algorithm, one of the main questions studied in the theory of linear programming is the following. Given a general LP

$$\begin{aligned} \text{Minimize} \quad & z(x) = cx \\ \text{Subject to} \quad & Ax = b \\ & x \geqq 0 \end{aligned} \tag{14.1}$$

where A is an $m \times n$ matrix, how much computational effort is involved in solving it by the simplex method?

The Size of a Linear Program

As discussed in Section 1.6, we assume that all the data in (14.1) in A, c, and b are rational. In this case, we can multiply both sides of each equation in (14.1) by a suitable integer and scale each variable appropriately to transform (14.1) into an equivalent problem in which all the data are integer.

The LP obviously becomes larger as m and n increase. Even when m and n are fixed, when the data are integer, the problem becomes larger as the number of digits in each data element increases. The *size* of (14.1), defined when all the data in it are integer, should take into account all these factors; for this reason, it is defined to be the total number of digits in all the data in (14.1). The number of digits in the binary encoding of an integer α is $\lceil(1 + \log_2(1 + |\alpha|)\rceil$ (when y is a real number, $\lceil y \rceil$, called the *ceiling of y*, is the smallest integer that is $\geqq y$). So, when all the data in (14.1) are integer, its size is defined to be

$$\begin{aligned} 1 &+ \lceil(1 + \log_2(1 + m))\rceil + \lceil(1 + \log_2(1 + n))\rceil \\ &+ \sum_i \sum_j \lceil(1 + \log_2(1 + |a_{ij}|))\rceil \\ &+ \sum_i \lceil(1 + \log_2(1 + |b_i|))\rceil \\ &+ \sum_j \lceil(1 + \log_2(1 + |c_j|))\rceil \end{aligned}$$

Computational Complexity

The main computational step in the simplex algorithm is the pivot step. Hence the effort involved in solving (14.1) by the simplex method can be measured by the number of pivots made before the algorithm terminates. If a feasible basis for (14.1) is not known, the simplex method applies the simplex algorithm twice, first on a Phase I problem that decides whether (14.1) is feasible, and provides a feasible basis for it if it is feasible, and then on (14.1) itself, starting with the feasible basis provided by the Phase I problem. So from a practical point of view the computational effort required for solving an LP for which a feasible basis is not known, is approximately twice the effort required for solving a comparable problem in which a feasible basis is available for starting the simplex algorithm. Hence in subsequent discussion we assume that an initial feasible basis for (14.1) is given. The question that we study here is: Starting with a known feasible basis for (14.1), how many pivot steps are required before the simplex algorithm terminates?

The answer depends on m and n, the actual numbers in the matrices A, b, c and the initial feasible basis used in the algorithm. The number of pivot steps required before termination is expected to go up as m and n go up. Even when m and n are fixed, the number of pivot steps before termination depends on the actual data in A, b, c. A practitioner who uses the simplex algorithm in his daily work would probably pose the question in the following manner: As a function of m and n, and the size of the LP, what is the average number of pivot steps required before termination, when the simplex algorithm is applied on the class of LPs encountered in practical applications? This function is called the *average computational complexity* of the simplex algorithm.

Determining the average computational complexity is a statistical question that is hard to answer precisely, because the kind of LPs that practical applications do generate is hard to specify. However, from the computational experience on a large number of LPs solved over several years, an empirical answer to this question has emerged. It indicates that the average number of pivot steps required before the termination of the simplex algorithm is a linear function of m that seems to be less than $3m$. See Dantzig's book [1.61, p. 160]. Experience gained after this book was written supports this observation.

A more challenging question is: Given the size, what is a mathematical upper bound on the number of pivot steps required for solving any LP of type (14.1) of that size by the simplex algorithm, starting with a known feasible basis? Probably this question is not of much practical significance, but it is theoretically more interesting. This upper bound determines what is known as the *worst case computational complexity* of the simplex algorithm (as opposed to the average computational complexity discussed above). The worst case computational complexity provides a guaranteed upper limit on the computational effort needed to solve any LP by the simplex algorithm, as a function of its size.

14.2 WORST CASE COMPUTATIONAL COMPLEXITY

An algorithm for solving a problem is said to be a *polynomially bounded algorithm* or a *good algorithm* if the worst case computational effort required for solving the problem using it, is bounded above by a polynomial in the size of the problem. To be specific, we study the following problem: Is the simplex algorithm a *good algorithm* in this sense?

We briefly review how the simplex algorithm solves (14.1). Let **K** be the set of feasible solutions of (14.1). Starting with a BFS of (14.1), the algorithm moves to an adjacent BFS in each step, such that the objective value decreases monotonically as the path progresses. A path such as this on **K** is known as an *isotonic path* with respect to the objective function $z(x)$. It is a sequence of points in **K**; $x^0, x^1, \ldots, x^\ell$ satisfying:

1 Each point in the sequence is a BFS.
2 Consecutive points in the sequence are adjacent BFSs.
3 $z(x^0) > z(x^1) > z(x^2) > \cdots > z(x^\ell)$.

The length of this path is defined to be ℓ, the number of points in it excluding the initial point. Clearly, if an isotonic path is found on **K**, the simplex algorithm can be made to follow that path by making an appropriate entering and dropping variable selection in each pivot step. So an answer to our question can be obtained by studying the following question: "What is the maximum length of an isotonic path in a problem of type (14.1) as a function of the size of this LP?

Klee and Minty [14.10] have provided an answer to this question. First, we discuss a sequence of examples constructed by them. Let $0 < \varepsilon < 1/2$.

Consider the following problem, the set of feasible solutions of which is the slightly perturbed unit square in $\mathbf{R}^2$.

$$\begin{aligned} \text{Minimize} \quad & -x_2 \\ \text{Subject to} \quad & x_1 \geqq 0 \\ & x_1 \leqq 1 \\ & x_2 \geqq \varepsilon x_1 \\ & x_2 \leqq 1 - \varepsilon x_1 \end{aligned} \qquad (14.2)$$

The set of feasible solutions of this problem is plotted in Figure 14.1. The BFSs of this problem are:

$$x^0 = (0, 0) \qquad x^2 = (1, 1-\varepsilon)$$
$$x^1 = (1, \varepsilon) \qquad x^3 = (0, 1)$$

Clearly x^0, x^1, x^2, x^3 is an isotonic path for this problem. The length of this path is $3 = 2^2 - 1$. Consider the next problem:

$$\begin{aligned} \text{Minimize} \quad & -x_3 \\ \text{Subject to} \quad & x_1 \geqq 0 \\ & x_1 \leqq 1 \\ & x_2 \geqq \varepsilon x_1 \\ & x_2 \leqq 1 - \varepsilon x_1 \\ & x_3 \geqq \varepsilon x_2 \\ & x_3 \leqq 1 - \varepsilon x_2 \end{aligned} \qquad (14.3)$$

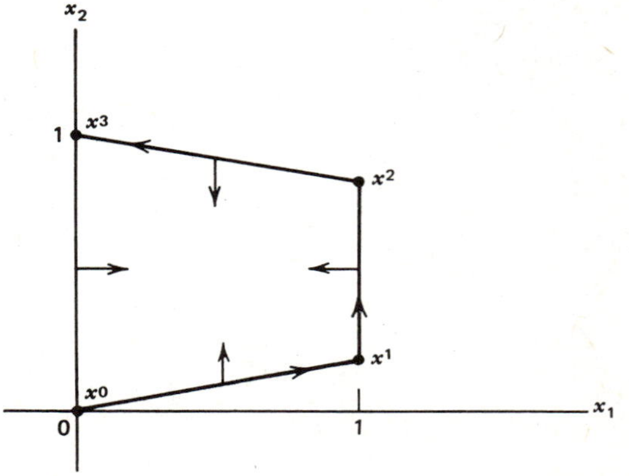

Figure 14.1

The set of feasible solutions of this problem is the slightly perturbed cube in $\mathbf{R}^3$ (Figure 14.2). The BFSs of this problem are:

$x^0 = (0, 0, 0)$	$x^4 = (0, 1, 1 - \varepsilon)$
$x^1 = (1, \varepsilon, \varepsilon^2)$	$x^5 = (1, 1 - \varepsilon, 1 - \varepsilon(1 - \varepsilon))$
$x^2 = (1, 1 - \varepsilon, \varepsilon(1 - \varepsilon))$	$x^6 = (1, \varepsilon, 1 - \varepsilon^2)$
$x^3 = (0, 1, \varepsilon)$	$x^7 = (0, 0, 1)$

Clearly $x^0, x^1, x^2, x^3, x^4, x^5, x^6, x^7$ is an isotonic path for this problem. It includes all the BFSs of this problem. Its length is $7 = 2^3 - 1$.

In general, consider the dth problem in the sequence

$$\begin{aligned} \text{Minimize} \quad & -x_d \\ \text{Subject to} \quad & x_1 \geqq 0 \\ & x_1 \leqq 1 \\ & x_2 \geqq \varepsilon x_1 \\ & x_2 \leqq 1 - \varepsilon x_1 \\ & x_3 \geqq \varepsilon x_2 \\ & x_3 \leqq 1 - \varepsilon x_2 \\ & \vdots \\ & x_d \geqq \varepsilon x_{d-1} \\ & x_d \leqq 1 - \varepsilon x_{d-1} \end{aligned} \tag{14.4}$$

The set of feasible solutions of (14.4) is the slightly perturbed unit cube in $\mathbf{R}^d$. Its BFSs can be arranged in an isotonic path of the form $x^0, x^1, \ldots, x^{2^d - 1}$, where the first $d - 1$ coordinates in $x^0, x^1, \ldots, x^{2^{d-1}-1}$ are the same as those in the points in the isotonic path for the $(d - 1)$th problem in the sequence. The dth coordinate for the points $x^0, x^1, \ldots, x^{2^{d-1}-1}$ is ε times the $(d - 1)$th coordinate. The first $d - 1$ coordinates in the remaining points written in reverse order, that is, in $x^{2^d-1}, \ldots, x^{2^{d-1}+1}, x^{2^{d-1}}$ are again the same as those in the points in the isotonic path for the $(d - 1)$th problem in the sequence. The dth coordinate in each of these points is obtained by subtracting ε times its $(d - 1)$th coordinate from 1. This isotonic path contains all the BFSs of (14.4) and its length is $2^d - 1$.

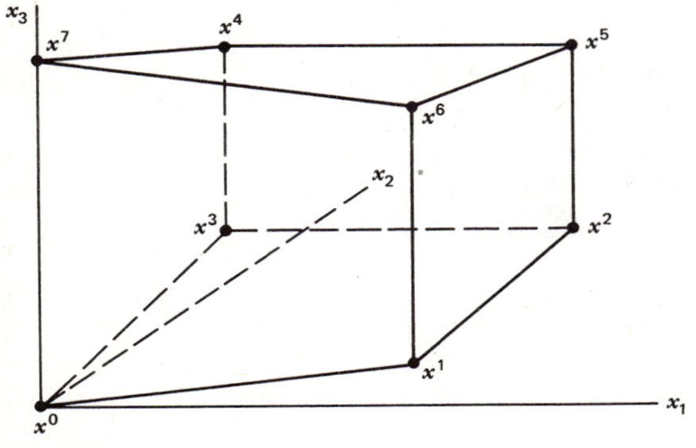

Figure 14.2

When posed in standard form by introducing slack variables and eliminating unrestricted variables from the system, (14.4) becomes a problem of the form (14.1) with $m = d, n = 2d$. If $\varepsilon = 1/4$, for example, the data in the problem are all rational, and when it is scaled so that it gets transformed into a problem of the type (14.1), in which all the data are integer, its size can be verified to be $\leqq 4(d + 1)d$. The isotonic path constructed above for this problem has length $2^d - 1$, and this number is not bounded above by any polynomial in the size of this problem. Thus, when the simplex algorithm is applied on (14.4), and the entering variable selection rule is adopted so that the algorithm follows the isotonic path constructed above, the algorithm passes through all the BFSs and the number of pivot steps required before termination is not bounded above by any polynomial in the size of this problem.

Effect of the Entering Variable Choice Rule on the Computational Complexity of the Simplex Algorithm

If the LP (14.1) is degenerate, the b vector in it can be perturbed slightly to make it nondegenerate. So for the purpose of studying the computational complexity of the simplex algorithm, we can, without any loss of generality, assume that (14.1) is nondegenerate. We make this nondegeneracy assumption in the following. Hence the usual minimum ratio test identifies the dropping basic variable uniquely and unambiguously in every pivot step. In this case, the only freedom we have in executing the simplex algorithm is the entering variable choice rule to be used in each pivot step. It turns out that the entering variable choice rule has a tremendous effect on the average computational complexity of the simplex algorithm. For example, the report of a computational study [10.1] finds that the average number of pivot steps when Bland's entering variable choice rule is used is several times the average when some other rule, such as the steepest descent rule, is used. There are also reasons to believe that the entering variable choice rule may even have an effect on the worst case computational complexity of the simplex algorithm.

$j =$	1	2	3	4	5	6	7	8	9
Pivot step 1	0	0	0	−1	−2	1	2	0	−1
2	2	0	0	0	−2	−1	−1	−1	−2
3	−3	1	0	0	0	−4	−5	−6	−8
4	−4	−2	2	0	0	0	3	−1	−1
5	0	−4	−1	1	0	0	−2	−1	−1
6	1	0	−2	−2	0	0	−3	−2	0

Consider the sequence of pivot steps encountered in the simplex algorithm while solving (14.1), and suppose these are numbered 1, 2, . . . , ℓ in the order in which they occurred. Let $\bar{c}^r = (\bar{c}_1^r, \ldots, \bar{c}_n^r)$ be the row vector of relative cost coefficients in pivot step r, $r = 1$ to ℓ. Define the numbers $t(k)$ by the following: $t(0) = 0$. For any $k \geqq 1$, when $t(k-1)$ is defined, define $t(k)$ to be the least integer greater than $t(k-1)$, if one exists, satisfying the property that maximum $\{\bar{c}_j^r: t(k-1) + 1 \leqq r \leqq t(k)\} \geqq 0$ for all $j = 1$ to n. If $t(k)$ exists, define *stage* k in the simplex algorithm to be the subsequence of pivot steps $t(k-1) + 1, \ldots, t(k)$, $k = 1, \ldots$. Thus a stage in the simplex algorithm applied on (14.1) is a smallest segment of consecutive pivot steps satisfying the property that after it begins, the relative cost coefficient of every variable in (14.1) becomes nonnegative at least once during the stage. In other words, the essential attribute of a *stage* is that no variable remains a nonbasic variable eligible to enter the basic vector in all pivot steps of the stage. The definition of stages in the simplex algorithm as given here is due to W. H. Cunningham [13.5].

Example 14.1

We give at the top the relative cost vectors obtained in various pivot steps of the simplex algorithm when applied to an LP such as (14.1). $t(0) = 0$. It can be verified that $t(1) = 6$. ($\bar{c}_9 < 0$ in pivot steps 1 to 5; it became nonnegative only in pivot step 6.) So pivot steps 1 to 6 (inclusive) constitute a stage here.

Stages in the Klee–Minty Examples Let us examine what the stages are in the pivot steps discussed above for the dth problem in the Klee–Minty examples. Let the nonnegative slack variables associated with the constraints $x_d \geqq \varepsilon x_{d-1}$ and $x_d \leqq 1 - \varepsilon x_{d-1}$ in this problem be s_{d1}, s_{d2}, respectively. It can be verified that the first 2^{d-1} BFSs obtained in the isotonic path developed above are on the hyperplane $x_d = \varepsilon x_{d-1}$ (and hence $s_{d1} = 0$ and $s_{d2} > 0$ in all these BFSs) and all the remaining BFSs in this path are on the hyperplane $x_d = 1 - \varepsilon x_{d-1}$ (and hence $s_{d1} > 0$ and $s_{d2} = 0$ in all these BFSs). So in the first 2^{d-1} pivot steps, s_{d2} is a basic variable, and s_{d-1} is a nonbasic variable. Choosing s_{d1} as the variable to enter the basic vector in any of these pivot steps would lead to s_{d1} replacing s_{d2} as the basic variable, moving from the current extreme point on the hyperplane $x_d = \varepsilon x_{d-1}$ vertically in the direction of the x_d axis, to the corresponding extreme point on the hyperplane $x_d = 1 - \varepsilon x_{d-1}$, and this would decrease the objective value $-x_d$ strictly. Thus in all of the first 2^{d-1} pivot steps in this sequence, the variable s_{d1} is a nonbasic variable eligible to enter the basic vector, and it is not chosen as the entering variable until the $(2^{d-1} + 1)$th pivot step in this path. Thus stage 1 in the sequence of pivot steps on the dth problem in the Klee–Minty example consists of 2^{d-1} pivot steps, which is extremely long (its length is an exponential function in the number of variables in the problem). Continuing in a similar way, it can be verified that there are d stages in the isotonic path developed for the dth Klee–Minty problem, whose lengths (in terms of the number of pivot steps in them) are $2^{d-1}, 2^{d-2}, \ldots, 2, 1$, respectively. Thus, even though the number of stages in this path is small (less than the number of variables in the problem when it is expressed in standard form), some of the stages are extremely long, so that the path is very long.

This property holds in *all* the examples constructed so far to establish that the paths generated by the simplex algorithm may have exponentially growing lengths. Each of them consists of only a small number of stages, but the lengths of some of the stages are very large.

Upper bound on Stage Lengths While solving the LP (14.1) by the simplex algorithm, it is easy to guarantee that the number of pivot steps in any stage remains small. This can be done by choosing an appropriate entering variable choice rule. For example, if the LRC entering variable choice rule discussed in Section 2.5.7 is used, since it circles through the list of variables looking for an entering variable, the length of any stage cannot exceed n. So, long isotonic paths of the type constructed by Klee–Minty (in which some stages are very long in themselves) cannot occur if the simplex algorithm is executed using the LRC entering variable choice rule. At the moment, the worst case computational complexity of the simplex algorithm applied to non-degenerate LPs, and executed with the LRC entering variable choice rule and others similar to it that guarantee that stage lengths remain small, is still an open question.

Conclusion On specially constructed classes of LPs, it has been established that the simplex algorithm takes a very large number of pivot steps before termination if it is executed with arbitrary entering and dropping variable choice rules, or with some of the rules discussed in Chapter 2. See the papers [14.6 and 14.7] of D. Goldfarb and W. Y. Sit, and R. Jeroslow. Exponential growth of the computational requirements in the worst case, of variants of the simplex algorithm for solving transportation and network flow problems has been shown by N. Zadeh [13.30]. The worst case computational complexity of the simplex algorithm executed with some of the other entering variable choice rules such as the LRC rule discussed in Chapter 2, that guarantee that stage lengths remain small, is still unresolved. However, on LPs encountered in practical applications, the simplex algorithm seems to perform very well when implemented with entering and dropping variable choice rules that turned out to be superior in computational tests. With these implementations, the simplex algorithm is still the most practically efficient algorithm known for solving LPs.

Computational Complexity When Both the Primal and Dual Simplex Methods Are Used

So far, we have only discussed the computational complexity of the approach of solving (14.1) by the primal simplex method. We can, of course, begin applying both the primal and dual simplex methods on (14.1) simultaneously, but separately. If this approach is used, we will stop the computation whenever termination occurs in one of the two methods first. This is equivalent to applying the primal simplex method on both the LP and its dual separately but simultaneously. The procedure stops whenever termination occurs in one of these two problems. What is the worst case computational complexity of this joint approach on the LP and its dual?

Let us examine the Klee–Minty class of problems again. We have already seen that for solving the dth problem in this class the primal simplex method may take a very large number of pivot steps. The dual of the dth Klee–Minty problem turns out to be an LP with a very special structure. Its set of feasible solutions is the translate of a cone, and has a unique extreme point. This extreme point can be quickly and easily identified by simply examining the dual constraints. Even if this is not done, the unique extreme point solution for the dual problem is obtained in no more than $2d$ pivot steps on the standard Phase I problem corresponding to the dual problem. Once the unique dual extreme point solution is obtained, it is optimal to the dual problem, and the primal optimum solution is obtained directly using the complementary slackness conditions. So, on the Klee–Minty class of problems, the joint approach of applying the simplex method on both the LP and its dual turns out to be polynomially bounded (see the paper [14.4] of A. Charnes, W. W. Cooper, S. Duffuaa, and M. Kress). However, examples have been constructed to establish that this joint approach is not polynomially bounded under arbitrary entering and dropping variable choice rules (see Exercise 14.1).

Exercises

14.1 Consider the following class of LPs constructed by J. Clausen [14.5]. Let $\alpha = \frac{4}{5}$, $\beta = \frac{5}{4}$, $c^d = (\alpha, \alpha^2, \dots, \alpha^d)$, and $b^d = (1, 5, 5^2, \dots, 5^{d-1})^T$. Let A^d be the $d \times d$ lower triangular matrix whose jth column is $(0, \dots, 0, 1, 2\beta, 2\beta^2, \dots, 2\beta^{d-j})^T$, $j = 1$ to d. The dth LP in the class is: minimize $c^d x$, subject to $A^d x \leqq b^d$, $x \geqq 0$. Study the computational complexity of the joint approach on the primal and dual problems discussed above, on this class of problems.

14.2 Consider the following LP

$$\text{Minimize} \quad y_1 + y_2 + y_3 + y_4$$

$$\text{Subject to} \quad \begin{pmatrix} 7 & 4 & 1 & 0 \\ 4 & 7 & 0 & 1 \\ 43 & 53 & 2 & 5 \\ 53 & 43 & 5 & 2 \end{pmatrix} \begin{pmatrix} y_1 \\ y_2 \\ y_3 \\ y_4 \end{pmatrix} \leqq \begin{pmatrix} 1 \\ 1 \\ 8 \\ 8 \end{pmatrix}$$

$$y \geqq 0$$

Obviously the set of feasible solution **K** is bounded. $y^1 = 0$ is the unique optimum solution of this LP. Let $y^2 = \frac{1}{19}(1, 1, 8, 8)^T$; y^2 is an extreme point of **K**. Show that every edge path from y^2 to y^1 along which the objective value is nonincreasing (i.e., isotonic path) consists of at least five edges. This provides a counterexample to the isotonic path Hirsch conjecture. The dimension of **K** is 4. The number of facets of **K** is 8. Yet every isotonic path from y^2 to y^1 has $5(>8-4)$ edges or more.

—(*M. J. Todd* [14.20])

14.3 *COMPUTATIONAL COMPLEXITY OF THE SIMPLEX ALGORITHM UNDER BLAND'S ENTERING AND DROPPING VARIABLE CHOICE RULES.* The *Fibonacci numbers* are a sequence of positive integers $\{F_n: n = 1, 2, \ldots\}$ constructed by the following method: $F_1 = F_2 = 1$, $F_n = F_{n-1} + F_{n-2}$, for $n \geqq 3$. Now consider another sequence of numbers $\{s_n: n = 1, 2, \ldots\}$ constructed by the following method: $s_1 = 1$, $s_2 = 3$, $s_n = s_{n-1} + s_{n-2} - 1$, for $n \geqq 3$. Prove that $s_n \geqq F_n$ for all n. Actually, it can be shown that $s_n = (2/\sqrt{5})(((1+\sqrt{5})/2)^{n+1} - ((1-\sqrt{5})/2)^{n+1}) - 1$.

Let ε be a real number such that $0 < \varepsilon < 1/2$. Consider the following LP:

$$\text{Maximize} \quad \sum_{j=1}^{n} \varepsilon^{n-j} x_j$$

$$\text{Subject to} \quad 2\left(\sum_{j=1}^{i-1} \varepsilon^{i-j} x_j\right) + x_i + x_{n+i} = 1$$

$$i = 1, \ldots, n$$

$$x_j \geqq 0 \qquad j = 1 \text{ to } 2n$$

Prove that the simplex algorithm, using Bland's entering and dropping variable choice rules, beginning with the solution $\bar{x} = (\bar{x}_j)$, where $\bar{x}_j = 0$ or 1, depending on whether $1 \leqq j \leqq n$ or $n + 1 \leqq j \leqq 2n$, respectively, and using the natural ordering of the variables, goes through s_n pivot steps before reaching the optimum solution of this problem.

—(*D. Avis and V. Chavatal* [10.1])

14.4 *STALLING IN THE SIMPLEX ALGORITHM UNDER BLAND'S RULES.* Consider the following LP, where ε is a real number satisfying $0 < \varepsilon < 1/2$.

$$\text{Maximize} \quad \sum_{j=1}^{n} \varepsilon^{n-j} x_j$$

Subject to

$$2\left(\sum_{j=1}^{i-1} \varepsilon^{i-j} x_j\right) + x_i + x_{n+i} = 0 \qquad i = 1 \text{ to } n$$

$$x_{2n+1} = 1$$

$$x_j \geqq 0 \qquad \text{for all } j = 1 \text{ to } 2n+1$$

Verify that $\bar{x} = (\bar{x}_j)$, where

$$\bar{x}_j = 0 \qquad j = 1 \text{ to } 2n$$
$$= 1 \qquad j = 2n+1$$

is the unique solution for this LP. Suppose we apply the simplex algorithm for solving this LP, using Bland's entering and dropping variable choice rules, initiated with $(x_{n+1}, \ldots, x_{2n}, x_{2n+1})$ as the initial feasible basic vector, and the natural ordering of the variables.

Prove that the algorithm goes through the same sequence of pivot steps (i.e., with the same entering and dropping variables in each step) as encountered while solving the LP discussed in Exercise 14.3. All these pivot steps are degenerate pivot steps. So the algorithm makes s_n (defined as in Exercise 14.3) degenerate pivot steps before finding a basic vector satisfying the optimality criterion.

This phenomenon is called *stalling*. Stalling is said to have occurred in the simplex algorithm whenever the total number of consecutive degenerate pivot steps made between two nondegenerate pivot steps, or between a nondegenerate pivot step or the initial step and termination, is very large (i.e., an exponential function in the number of variables in the LP or some large number such as that).

This class of LPs for $n = 2, 3, \ldots$ clearly illustrates that stalling can occur in the simplex algorithm using Bland's entering and dropping variable choice rules (see the paper [10.1] of D. Avis and V. Chvatal). On specially structured LPs such as minimum cost pure network flow problems, it has been shown that stalling and cycling can both be avoided by using specially designed entering and dropping variable choice rules in each pivot step

[13.5]. However, it is not yet known whether there exist any entering and dropping variable choice rules that guarantee that stalling cannot occur in the simplex algorithm applied on a general LP.

14.5 Prove that a procedure for avoiding both cycling and stalling under degeneracy, while solving (14.1) by the simplex algorithm, exists iff there exists a pivotal algorithm for solving linear inequalities, for which the computational effort required is bounded above by a polynomial in the number of variables and the number of inequalities and the size of the problem.

14.6 Given an instance of (14.1) and a specified objective value, prove that the problem of checking whether there exists a BFS of (14.1) with exactly that objective value, is NP-complete.

—(*R. Chandrasekharan, S. N. Kabadi, and K. G. Murty* [10.4])

14.7 *ANOTHER BAD EXAMPLE FOR BLAND'S ENTERING AND DROPPING VARIABLE CHOICE RULES.* Let $\tilde{M}_n$ be the square lower triangular matrix of order n in which all the diagonal entries are 1, all elements below the diagonal are 2, and all elements above the diagonal are zero, defined in Section 8.12. Let I_n denote the unit matrix of order n, and e_n the column vector in $\mathbf{R}^n$, all of whose entries are 1. Let $\hat{q}_n = (-2^n, -2^n - 2^{n-1}, -2^n - 2^{n-1} - 2^{n-2}, \ldots, -2^n - 2^{n-1} - \cdots - 2^2 - 2)^{\mathrm{T}} \in \mathbf{R}^n$. Consider the following class of LPs, containing one problem for each $n \geqq 1$. The nth problem has $2n + 1$ variables denoted by the symbols $w = (w_1, \ldots, w_n)$, $z = (z_1, \ldots, z_n)$, z_0, for the sake of convenience. The nth problem is

$$\begin{aligned} &\text{Minimize} && z_0 \\ &\text{Subject to} && I_n w - \tilde{M}_n z - e_n z_0 = \hat{q}_n \\ & && w, z, z_0 \geqq 0 \end{aligned}$$

For example, the third problem in the class is

w_1	w_2	w_3	z_1	z_2	z_3	z_0	q
1	0	0	−1	0	0	−1	− 8
0	1	0	−2	−1	0	−1	−12
0	0	1	−2	−2	−1	−1	−14

$z_0, w_j, z_j \geqq 0$ for all j; minimize z_0.

As in Section 8.12, for each j, the pair of variables (w_j, z_j) is known as the jth complementary pair of variables in these problems; that is, each variable in this pair is known as the complement of the other. A complementary vector of variables for the nth problem is a vector $(y_1, \ldots, y_n)$, where $y_j \in \{w_j, z_j\}$ for each $j = 1$ to n. An almost complementary basic vector (abbreviated as ACBV) for the nth problem is a basic vector satisfying the following properties: (1) z_0 is a basic variable in it, (2) it contains exactly one variable from each of $n - 1$ complementary pairs as a basic variable, and (3) there is exactly one complementary pair, both of whose variables are nonbasic. For example $(w_1, \ldots, w_{n-1}, z_0)$ is an ACBV and $(z_1, w_2, \ldots, w_n)$ is a complementary basic vector for the nth problem.

(a) Verify that the ACBV $(w_1, \ldots, w_{n-1}, z_0)$ is a feasible basic vector for the nth problem. Starting with it, perform a sequence of pivot steps on the nth problem by the following rules: The entering variable in step 1 is z_n. Terminate the sequence whenever z_0 drops out of the basic vector. From Step 2 onward, if termination does not occur in it, choose as the entering variable the complement of the variable that was the dropping variable in the previous step. Verify that all the basic vectors obtained in this sequence are ACBVs, except the terminal one, which is the complementary basic vector $(z_1, w_2, \ldots, w_n)$. Also verify that the sequence terminates after $2^n - 1$ pivot steps, and that in the corresponding sequence of BFSs, z_0 starts with a value of $2^{n+1} - 2$, decreases by 2 in each step, and becomes zero at the end. Thus the sequence of BFSs generated here is an isotonic path of length $2^n - 1$ for the nth problem.

(b) If the nth problem is solved by the simplex algorithm beginning with the feasible basic vector $(w_1, \ldots, w_{n-1}, z_0)$, using Bland's entering and dropping variable choice rules, with the specific ordering of the variables in the problem to be $z_n, w_n, z_{n-1}, w_{n-1}, \ldots, z_2, w_2, z_1, w_1, z_0$, prove that it goes through exactly the same sequence of BFSs as that generated in (a), thereby requiring $2^n - 1$ pivot steps before termination.

This problem arose in the study of the computational complexity of Lemke's complementary pivot method for the linear complementarity problem (see [14.15]). The sequence of ACBVs obtained in (a) is exactly the sequence generated when the linear complementarity problem $(\hat{q}_n, \tilde{M}_n)$ is solved by Lemke's complementary pivot method.

Summary

The exponential growth of the computational requirements of the simplex method, in the worst case, has been established, when the method is implemented with certain entering variable choice rules. The worst case computational complexity of the simplex method implemented with other entering variable choice rules is still unknown. See the list of outstanding theoretical problems given below. It has been conjectured that the simplex method may be polynomially bounded when operated with some entering variable choice rules such as the LRC, and dropping variable choice rules such as the lexico minimum ratio rule to resolve the problem of cycling under degeneracy.

14.3 AVERAGE COMPUTATIONAL COMPLEXITY

The simplex algorithm is extensively used in practical applications and it has always been extremely efficient. Even though it has been shown to run in exponential time in the worst case, it has always behaved like a highly efficient polynomial time algorithm in practice. For a long time, the question of why it performed so well on real-world problems remained a mystery. A partial theoretical explanation for this efficiency in terms of average data was provided recently by S. Smale [14.19] and K. Borgwardt [14.2, 14.3]. Using the spherical uniform Lebesgue measure on sets of rays in Cartesian spaces to average the data, S. Smale [14.19] has studied the average number of steps $\rho(m, n)$ to solve the LP of the form: minimize cx, subject to $Ax \geqq b$, $x \geqq 0$, where A is of order $m \times n$, by the self-dual parametric algorithm (see Section 8.13). He has proved that for any positive integer p, there is a positive constant $c(m, p)$, such that $\rho(m, n) \leqq c(m, p)n^{1/p}$, for all n. Thus for a fixed number of constraints, the average number of steps taken to solve the LP by this variant of the simplex method grows slower than any prescribed root of the number of variables. Taking $p = 1$ in this result, we conclude that for fixed m (the number of constraints), the average number of iterations grows in proportion to n (the number of variables).

In [14.2, 14.3] K. H. Borgwardt studied the LP: maximize vx subject to $A_{1.}x \leqq 1, \ldots, A_{m.}x \leqq 1$, where $v, A_{1.}, \ldots, A_{m.} \in \mathbf{R}^n$ are assumed to be distributed independently, identically, and symmetrically under rotations on $\mathbf{R}^n \backslash \{0\}$. He has shown that the expected number of pivot steps by a variant of the simplex method to process this LP is no more than $0(n^4 m)$.

Outstanding Theoretical Problems

1. Consider (14.1) with integer data and assume that it is nondegenerate. In solving this problem by the simplex algorithm, is the number of stages bounded above by a polynomial in m, n, or the size of the problem? Either prove that it is polynomially bounded or construct a class of problems providing a counterexample.

2. Consider again (14.1) with integer data and assume that it is nondegenerate. For solving this problem, is the simplex algorithm with the LRC entering variable choice rule polynomially bounded? Either prove it is or construct a class of problems providing a counterexample.

3. Assume that (14.1) is nondegenerate. In this case, for solving (14.1), is the simplex algorithm not polynomially bounded, irrespective of which entering variable choice rule is used in each pivot step?

4. Let A be of rank m in (14.1) and let $\mathbf{K}$ be the set of feasible solutions of (14.1). Assume that $\mathbf{K}$ is bounded. Let B_0 and B_* be two feasible bases for (14.1). In 1957, W. M. Hirsch conjectured that by using pivot steps of the type used in the simplex algorithm, one can get from the feasible basis B_0 to the feasible basis B_* in at most m pivot steps. See Dantzig's book [1.61, p. 160, 168]. This conjecture is known as the *Hirsch conjecture* or the *m-step conjecture*.

Let V_0 and V_* be any two extreme points of **K**. What does the Hirsch conjecture say about the lengths of edge paths connecting V_0 and V_*?

Let Γ be a convex polytype in $\mathbf{R}^d$ with n facets. Prove that the Hirsch conjecture is equivalent to stating that any two extreme points of Γ are connected by an edge path consisting of at most $n - d$ edges of Γ.

Klee and Walkup [14.11] have constructed examples to show that the Hirsch conjecture is false if **K** is not bounded. They also prove the conjecture for all polytopes for which $n - m \leqq 5$.

Prove this conjecture or construct a counterexample for it.

5 *POLYNOMIAL HIRSCH CONJECTURE*. Consider the LP (14.1) again. Using pivot steps of the type used in the simplex algorithm, is it possible to move from any feasible basis to any other feasible basis along a path of adjacent feasible bases, such that the total number of pivot steps required is bounded above by a polynomial in m, n? Treat the cases where the set of feasible solutions of (14.1) is bounded, or unbounded separately.

Prove that if this polynomial Hirsch conjecture is true for the case where the set of feasible solutions of (14.1) is bounded, then it must also be true for the case where this set is unbounded.

If the above polynomial Hirsch conjecture is false, would it become true if we assumed that all the data in (14.1) are integer and allowed paths in which the number of pivot steps is bounded above by a polynomial in m, n and the size of this LP?

6 Consider the LP (14.1). Suppose it is known that $z(x)$ is unbounded below (i.e., it diverges to $-\infty$) on the set of feasible solutions of (14.1). However, (14.1) has only a finite number of BFSs. Suppose it is required to find a BFS of (14.1) at which $z(x)$ assumes its minimum value among the set of all BFSs of (14.1). Develop an efficient algorithm for doing it. (The famous traveling salesman problem, which is an NP-hard combinatorial optimization problem, is a special case of this problem.)

7 Consider the LP (14.1), where A is a matrix of order $m \times n$ and rank m. Let **K** be the set of feasible solutions of this LP. Let $\tilde{x}$ be a BFS of (14.1).

If $\tilde{x}$ is a nondegenerate BFS of (14.1), the number of adjacent extreme points of $\tilde{x}$ on **K** is at most $n - m$, and they can be found efficiently by the algorithm discussed in Section 3.7.1. If $\tilde{x}$ is a degenerate BFS of (14.1), the number of adjacent extreme points of $\tilde{x}$ on **K** may be much larger than $n - m$. In this case, to find all the adjacent extreme points of $\tilde{x}$ on **K**, we can use the algorithm discussed in Section 3.7.1, but it may be a computationally difficult job.

Suppose $\tilde{x}$ is a degenerate BFS of (14.1). It is required to find an adjacent extreme point of $\tilde{x}$ on **K**, at which $z(x)$ assumes its minimum value among all the adjacent extreme points of $\tilde{x}$ on **K**. Develop an efficient algorithm for doing it. (The traveling salesman problem is a special case of this problem too [14.14]).

8 Let **K** be the set of feasible solutions of

$$Ax \geqq b \tag{14.5}$$

where A is a matrix of order $m \times n$. Assume that $\mathbf{K} \neq \varnothing$ and is unbounded. Let $\mathbf{K}^\Delta$ be the convex hull of all the extreme points of **K**. $\mathbf{K}^\Delta$ is the set of feasible solutions of a system obtained by including some more inequality constraints in (14.5), where each of these additional inequalities corresponds to a facet of $\mathbf{K}^\Delta$. Develop an efficient method for generating these additional inequalities one by one, using the data in A, b.

9 Consider the system of equations

$$Ax = b \tag{14.6}$$

where A is a matrix of order $m \times n$ and rank m. Define a solution $\bar{x}$ of (14.6) to be a *basic solution* if $\{A_{.j}$: j, such that $\bar{x}_j \neq 0\}$ is linearly independent. Obviously, there are only a finite number of basic solutions. Let Γ denote the convex hull of all basic solutions of (14.6). Γ is clearly the set of feasible solutions of a system of linear constraints obtained by adding additional linear inequality or equality constraints to (14.6). Develop an efficient method for generating these additional constraints for characterizing Γ, one by one, using the data in (14.6) (one can easily see that this problem is a special case of problem 8 discussed above).

—(*R. Chandrasekharan*)

10 Let e_r be the column vector in $\mathbf{R}^r$ all of whose entries are 1. Consider the problem:

$$\begin{aligned} \text{Minimize} \quad & e_n^{\mathrm{T}} x \\ \text{Subject to} \quad & Ax \geqq e_m \\ & x \geqq 0 \end{aligned}$$

where A is an $m \times n$, 0-1 matrix. Develop simple necessary and sufficient conditions on the matrix A under which this problem has an optimum solution that is an integer vector.

11 *THE CONE CONJECTURE*. Consider the system of constraints $A'x = b' + \lambda b^*$, $x \geqq 0$, where A' is a matrix of order $m' \times n'$ and rank m', and λ is a nonnegative parameter. And B_1 is a given basis for this problem, which is feasible when $\lambda = 0$. Beginning with B_1, the value of λ is steadily increased from 0, making dual simplex pivot steps whenever we cross the feasibility interval of the current basis, as in the parametric right-hand-side simplex algorithm, until either λ reaches the value 1 or a value $\lambda_1 < 1$, and it can clearly be established that the above system is infeasible whenever $\lambda > \lambda_1$. Does there always exist a choice of pivot rows and pivot columns among those eligible in each stage of this method, such that the total number of pivot steps required before termination is bounded above by some fixed polynomial in m', n' and the size of this system?

Note 14.1: Problems 11 and 5 are related. Consider the system $A'x - x_{n'+1}b^* = b'$, $x_{n'+1} + x_{n'+2} = 1$, $x \geqq 0$, $x_{n'+1} \geqq 0$, $x_{n'+2} \geqq 0$. Let A be the $(m'+1) \times (n'+2)$ matrix of constraint coefficients in this system. It has rank $m'+1$. Clearly, $(x_{B_1}, x_{n'+2})$ is a feasible basic vector for this system, with $x_{n'+1}$ having a value of 0 in the corresponding BFS, since B_1 is a feasible basis for the system given in problem 11, when $\lambda = 0$. Let x_B be an optimum basic vector for the problem of maximizing $x_{n'+1}$, subject to the constraints in this system. The system in problem 11 has a feasible basic vector when $\lambda = 1$ iff the maximum value of $x_{n'+1}$ is equal to 1. Also the answer to problem 11 is in the affirmative iff it is possible to move from the feasible basic vector $(x_{B_1}, x_{n'+2})$ to the feasible basic vector x_B along a path of adjacent feasible basic vectors in this system, such that the total number of pivot steps for this traversal is bounded above by a polynomial in m', n' and the size of this system.

Conversely consider (14.1). Let x_B and $x_{\hat{B}}$ be two feasible basic vectors for (14.1). Let $\hat{x}$ be the BFS of (14.1) corresponding to the feasible basic vector $x_{\hat{B}}$. Let $\mathbf{J} = \{j : x_j \text{ is not in the basic vector } x_{\hat{B}}\}$. Consider the system

$$\begin{aligned} Ax &= b \\ \sum_{j \in \mathbf{J}} x_j &= 0 \\ x &\geqq 0 \end{aligned} \tag{14.7}$$

Let A'' be the $(m+1) \times n$ coefficient matrix in (14.7). The unique feasible solution of (14.7) is $\hat{x}$. Let $j_1 \in \mathbf{J}$ be such that $x_{B_2} = (x_B, x_{j_1})$ is a basic vector for (14.7). Let B_2 be the basis for (14.7) corresponding to the basic vector x_{B_2}. Let $b'' \in \operatorname{Pos}(B_2)$, $b^* = \begin{pmatrix} b \\ \cdot \\ 0 \end{pmatrix} - b''$.

Consider the parametric system

$$\begin{aligned} A''x &= b'' + \lambda b^* \\ x &\geqq 0 \end{aligned} \tag{14.8}$$

x_{B_2} is a feasible basic vector for (14.8) when $\lambda = 0$. The problem of moving between the feasible basic vectors x_B, $x_{\hat{B}}$ in (14.1) along an adjacent feasible basic vector path of (14.1), is clearly the same as the problem of moving from the basic vector x_{B_2} in (14.8) (which is feasible when $\lambda = 0$) to a basic vector for (14.8) that is feasible for (14.8) when $\lambda = 1$, along a path of adjacent basic vectors using the method discussed in Question 11 above. This show that the answer to Question 5 is the affirmative, iff the answer to Question 11 is.

REFERENCES

14.1 D. G. Barman, "Paths of Polytopes," in *Proceedings of the London Mathematical Society* *20* (1970), 161–178.

14.2 K. H. Borgwardt, "Some Distribution Independent Results About the Asymptotic Order of the Average Number of Pivot Steps in the Simplex Method," *Mathematics of Operations Research* *7*, 3 (August 1982), 441–462.

14.3 K. H. Borgwardt, "The Average Number of Pivot Steps Required by the Simplex Method is Polynomial,'' *Journal of Operations Research*, to be published.

14.4 A. Charnes, W. W. Cooper, S. Duffuaa, and M. Kress, ''Complexity and Computability of Solutions to Linear Programming Systems,'' *International Journal of Computer and Information Sciences 9*, 6 (December 1980), 483–506.

14.5 J. Clausen, ''A Tutorial Note on the Complexity of the Simplex-Algorithm,'' Technical Report NR79/16, Institute of Datalogy, University of Copenhagen, Copenhagen, Denmark, 1979.

14.6 D. Goldfarb and W. Y. Sit, ''Worst Case Behavior of Steepest Edge Simplex Method,'' *Discrete Applied Mathematics 1* (1979), 277–285.

14.7 R. G. Jeroslow, ''The Simplex Algorithm with the Pivot Rule of Maximizing Criterion Improvement,'' *Discrete Mathematics 4* (1973), 367–378.

14.8 R. M. Karp, ''Reducibility Among Combinatorial Problems,'' in R. E. Miller and J. W. Thatcher (Eds.), *Complexity of Computer Computations*, Plenum, New York, 1972.

14.9 V. L. Klee ''Paths on Polytopes: A Survey,'' in B. Grunbaum *Convex Polytopes*, Wiley, New York, 1966.

14.10 V. Klee and G. J. Minty, ''How Good Is the Simplex Algorithm?'' in O. Shisha (Ed.), *Inequalities III*, Academic, New York, 1972, 159–175.

14.11 V. Klee and D. W. Walkup ''The *d*-Step Conjecture for Polyhedra of Dimension $d < 6$,'' *Acta Mathematica 117* (1967), 53–78.

14.12 T. M. Liebling, ''On the Number of Iterations of the Simplex Method,'' *Operations Research Verfahren 17* (1972), 248–264.

14.13 P. Mani and D. W. Walkup, ''A 3-Sphere Counterexample to the W_v-Path Conjecture,'' *Mathematics of Operations Research 5*, 4 (November 1980), 595–598.

14.14 K. G. Murty, ''A Fundamental Problem in Linear Inequalities with Applications to the Traveling Salesman Problem,'' *Mathematical Programming 2*, 3 (June 1972), 296–308.

14.15 K. G. Murty, ''Computational Complexity of Complementary Pivot Methods,'' *Mathematical Programming Studies 7*, (1978), 61–73.

14.16 A. Orden, ''A Step Toward Probabilistic Analysis of Simplex Method Convergence,'' *Mathematical Programming 19*, 1 (July 1980), 3–13.

14.17 R. Saigal, ''On the Computational Complexity of a Piece-Wise Linear Homotopy Algorithm,'' *Mathematical Programming*, to be published.

14.18 S. Smale, ''The Fundamental Theorem of Algebra and Complexity Theory,'' *Bulletin (New Series) of the American Mathematical Society 4*, 1 (1981), 1–36.

14.19 S. Smale, ''On the Average Speed of the Simplex Method of Linear Programming,'' Technical Report, Department of Mathematics, University of California, Berkeley, 1982.

14.20 M. J. Todd, ''The Monotonic Bounded Hirsch Conjecture is False for Dimension at Least 4,'' *Mathematics of Operations Research 5*, 4 (November 1980), 599–601.

14.21 J. Traub and H. Wozniakowski, ''Complexity of Linear Programming,'' *Operations Research Letters 1* (1982), 59–62.

14.22 D. W. Walkup, ''The Hirsch Conjecture Fails for Triangulated 27-Spheres,'' *Mathematics of Operations Research 3* (1978), 224–230.

chapter 15 The Ellipsoid Method

15.1 INTRODUCTION

In this chapter we consider LPs and systems of linear constraints with integer data. As in Chapter 14, we define the *size* of such a problem to be the total number of digits in all the data elements. An algorithm for solving such a problem is said to be *polynomially bounded* if the (worst case) computational effort required to solve an instance of such a problem can be proved to be bounded above by a polynomial in the size of that instance. Ever since the time it was established that the simplex algorithm, with many of the commonly used entering variable choice rules, is not polynomially bounded, the question "Is there a polynomially bounded algorithm for solving linear programs?" remained a challenge for researchers in algorithmic mathematics. This challenge was settled in the affirmative by the young Russian mathematician L. G. Khachiyan in the fundamental papers [15.12, 15.19]. There he showed how a method for convex optimization developed earlier by the Russian mathematician N. Z. Shor and others [15.30–15.37, 15.15, 15.16] can be adapted to devise a polynomially bounded algorithm for LP. This method constructs a sequence of ellipsoids in $\mathbf{R}^n$ of decreasing n-dimensional volume, and hence it has come to be known as the *ellipsoid algorithm* or the *ellipsoid method*. When news of its development became known in 1979, it appeared in the headlines of newspapers all over the world. Khachiyan suddenly became famous. The resolution of this outstanding theoretical question concerning linear programming resulted in tremendous excitement among researchers. Journalists inferred that the theoretical criterion for efficiency (polynomial boundedness in the worst case) of the ellipsoid method would immediately translate into a major practical advance, resulting in a profound breakthrough in the solution of real-world problems. However, an algorithm that is superior by the worst case criterion may not necessarily be good in practice on an average. Computational experience with the ellipsoid method has led to the general concensus that at present it is not a practical alternative to the simplex method. Even though it is a tremendous theoretical breakthrough, "it is unlikely to replace the simplex method as the computational workhorse of linear programming," unless there are further breakthroughs in implementation (see the paper [15.5] of R. G. Bland, D. Goldfarb, and M. J. Todd).

An *open* or *strict linear inequality* is of the form $ax < \gamma$, where $a \neq 0$. The ellipsoid algorithm is a method for solving a system of open linear inequalities. In Section 15.2 we present this algorithm and proofs of its convergence and polynomial boundedness. In linear programming problems we generally have *closed linear inequalities* (those of the form $ax \leqq \alpha$) and are required to optimize a linear objective function subject to these constraints. In Section 15.3 we discuss two schemes for solving systems of closed linear inequalities using the ellipsoid method. Then in Section 15.4 we discuss how linear programming problems can be solved using the ellipsoid method.

Definition: An *ellipsoid* in $\mathbf{R}^n$ is uniquely specified by its *center* $\bar{x}$, and a positive definite matrix (a square matrix D of order n is said to be positive definite if $x^{\mathrm{T}}Dx > 0$ for all $x \neq 0$. See [1.92, 1.93]) D of order n. Given x, D, the ellipsoid specified by them is: $\mathbf{E}(\bar{x}, D) = \{x\colon (x - \bar{x})^{\mathrm{T}}D^{-1}(x - \bar{x}) \leqq 1\}$. If $D = I$, the ellipsoid $\mathbf{E}(\bar{x}, I)$ is the unit spherical ball with its center at $\bar{x}$ and radius 1 in $\mathbf{R}^n$. When D is positive definite, for $x, y \in \mathbf{R}^n$, the function

$$f(x, y) = (x - y)^{\mathrm{T}}D^{-1}(x - y)$$

is called the *distance* between x and y with the matrix D^{-1} as the *metric matrix* (when $D = I$, this becomes the usual *Euclidean distance*). The ellipsoid method obtains a new ellipsoid in each step by changing the metric matrix in each step. Hence it is known as a *variable metric method*. Also, the formula for updating the metric matrix from step to step is of the form $D_{r+1} = (\text{constant})(D_r + B_r)$, where B_r is a square matrix of order n and rank 1 obtained by multiplying a column vector in $\mathbf{R}^n$ by its transpose. Methods that update the metric matrix by such a formula are called *rank-one methods* in nonlinear programming. Rank-one methods, and variable metric methods, are used extensively in nonlinear programming for finding the unconstrained minimum of a convex function defined over $\mathbf{R}^n$ (see references [2.7, 2.8, 2.10, 2.12]). The ellipsoid method belongs to these classes of methods.

Why Can't The Ellipsoid Method Solve Systems of Closed Linear Inequalities Directly?

Consider the system of open linear inequalities $Ax < b$, where A is a matrix of order $m \times n$. Let $\mathbf{K}$ be the set of its feasible solution. If $\mathbf{K} \neq \emptyset$, it is a nonempty open subset of $\mathbf{R}^n$ and hence has positive n-dimensional volume. The ellipsoid algorithm begins with an initial ellipsoid $\mathbf{E}_0$ (which is really a large n-dimensional spherical ball) satisfying the property that if $\mathbf{K} \neq \emptyset$, the n-dimensional volume of $\mathbf{E}_0 \cap \mathbf{K}$ is greater than or equal to a known positive number α. It obtains the sequence of ellipsoids $\mathbf{E}_0$, $\mathbf{E}_1, \ldots,$. For all r it is guaranteed that $\mathbf{E}_r \supset \mathbf{E}_0 \cap \mathbf{K}$ and that volume of $\mathbf{E}_{r+1} \leqq \beta$ (volume of $\mathbf{E}_r$), where β is a predetermined positive number strictly less than 1. Let x^r denote the center of $\mathbf{E}_r$. Let N be the smallest positive integer satisfying β^{N} (volume of $\mathbf{E}_0$) $< \alpha$. Then $\mathrm{N} = \lceil (\log \alpha - \log(\text{volume of } \mathbf{E}_0))/\log \beta \rceil$. If $x^r \in \mathbf{K}$, the algorithm terminates in step r with it. If the algorithm goes through N steps without termination, and $x^{\mathrm{N}} \notin \mathbf{K}$, then volume of $\mathbf{E}_{\mathrm{N}} < \alpha \leqq$ volume of $(\mathbf{E}_0 \cap \mathbf{K})$. This implies that $\mathbf{K} = \emptyset$ in this case.

So, either the center x^r becomes a feasible solution in step r for some $r \leqq \mathrm{N}$, or in step N we terminate with the conclusion that there exists no feasible solution. For this conclusion to be valid we need the fact that the n-dimensional volume of $\mathbf{K}$ is strictly positive if $\mathbf{K} \neq \emptyset$. A system of closed linear inequalities, say, $Fx \leqq f$, may be feasible, and yet the n-dimensional volume of its set of feasible solutions may be zero (if one of the constraints holds as an equation at all feasible solutions, the set of feasible solutions lies on the hyperplane in $\mathbf{R}^n$ corresponding to that equation, and hence has zero n-dimensional volume). Hence if the ellipsoid algorithm is applied to solve this closed system directly, we cannot construct any finite number h satisfying the property that if a feasible solution is not obtained in h steps of the algorithm, there exists no feasible solution. However, it is possible to process closed inequality systems by using the ellipsoid method with some special techniques. These techniques will be discussed in Section 15.3.

15.2 THE ELLIPSOID METHOD FOR SOLVING A SYSTEM OF OPEN LINEAR INEQUALITIES

Let the system to be solved be

$$Ax < b \tag{15.1}$$

where $A = (a_{ij})$ and $b = (b_i)$ are the given integer matrices of orders $m \times n$, and $m \times 1$, respectively. We assume that $A_{i.} \neq 0$ for all i. If $n = 1$, or $m = 1$, the problem is trivially solved, so we assume that $m, n \geqq 2$. The size of this system is

$$\mathrm{L} = \Big\lceil (1 + \log_2 n + \log_2 m + \sum_i \sum_j (1 + \log_2(1 + |a_{ij}|)) + \sum_i (1 + \log_2|1 + |b_i|)) \Big\rceil \tag{15.2}$$

Let $\mathbf{K}$ denote the set of feasible solutions of (15.1). The algorithm constructs a finite sequence of ellipsoids, the rth one denoted by $\mathbf{E}_r = \mathbf{E}(x^r, D_r)$, with center x^r and the positive definite matrix D_r constructed at the end of step r. Initially we take $x^0 = 0$, $D_0 = 2^{2\mathrm{L}} I$, where I is the unit matrix of order n.

General Step $r+1$ for $r \geqq 0$ Let x^r, D_r be the center and the positive definite matrix, and $\mathbf{E}_r = \mathbf{E}(x^r, D_r)$, be the ellipsoid, obtained at the end of step r. If x^r is feasible to (15.1), terminate. If x^r violates at least one of the constraints in (15.1), and $r = 6(n+1)^2\mathrm{L}$, terminate, (15.1) is infeasible. If x^r violates at least one of the constraints in (15.1), and $r < 6(n+1)^2\mathrm{L}$, identify a violated constraint. It is advantageous to take this to be the constraint that is most violated by x^r, that is, the ith, where i satisfies $A_{i.}x^r - b_i =$ maximum $\{A_{p.}x^r - b_p : p, \text{ such that } A_{p.}x^r - b_p \geqq 0\}$. Define $v_r = (b_i - A_{i.}x^r)/(+\sqrt{A_{i.}D_rA_{i.}^{\mathrm{T}}})$. If $v_r \leqq -1$, terminate, (15.1) is infeasible. If $v_r > -1$, define

$$x^{r+1} = x^r - \left(\frac{1 - v_r n}{1 + n}\right)\left(\frac{D_rA_{i.}^{\mathrm{T}}}{+\sqrt{A_{i.}D_rA_{i.}^{\mathrm{T}}}}\right) \tag{15.3}$$

$$D_{r+1} = \left(\frac{(1 - v_r^2)n^2}{n^2 - 1}\right)\left(D_r - \frac{2}{n+1}\left(\frac{1 - nv_r}{1 - v_r}\right) \times \left(\frac{(D_rA_{i.}^{\mathrm{T}})(D_rA_{i.}^{\mathrm{T}})^{\mathrm{T}}}{A_{i.}D_rA_{i.}^{\mathrm{T}}}\right)\right) \tag{15.4}$$

With x^{r+1}, D_{r+1}, and $\mathbf{E}_{r+1} = \mathbf{E}(x^{r+1}, D_{r+1})$, go to the next step.

Geometric Interpretation and Proof of Validity

THEOREM 15.1 Let B be a square submatrix of (A, I, b). Then $|B|$, the absolute value of the determinant of B, is $< 2^{\mathbf{L}}/n$.

Proof: This proof is due to S. J. Chung. If B contains some column vectors of the unit matrix, expand the determinant of B using those columns, and reduce it to the determinant of a square submatrix of (A, b), which we denote by B itself for convenience in notation. Let $B = (\alpha_{ij})$ of order $t \times t$. Let $\mathbf{P}$ denote the set of all $t!$ permutations of $(1, \ldots, t)$ and let $p = (p_1, \ldots, p_t) \in \mathbf{P}$. Then $|B| = |\sum_{p \in \mathbf{P}} (\pm 1)\alpha_{1,p_1} \cdots \alpha_{t,p_t}| \leqq \sum_{p \in \mathbf{P}} |\alpha_{1,p_1}| \cdots |\alpha_{t,p_t}|$. Each of these terms appears in the expansion of the product $(|\alpha_{11}| + \cdots + |\alpha_{1t}|) \cdots (|\alpha_{t1}| + \cdots + |\alpha_{tt}|)$. So we have $|B| \leqq (|\alpha_{11}| + \cdots + |\alpha_{1t}|) \cdots (|\alpha_{t1}| + \cdots + |\alpha_{tt}|) \leqq (|\alpha_{11}| + 1) \quad (|\alpha_{12}| + 1) \cdots (|\alpha_{tt}| + 1) \leqq \Pi_{j=1 \text{ to } n}^{i=1 \text{ to } m} (|a_{ij}| + 1) \ \Pi_{i=1 \text{ to } m} (|b_i| + 1)$, where "$\Pi$" indicates the product sign, since the entries in B are only a subset of the entries in (A, b). Taking the antilogs with base 2 on both sides of (15.2), we get $|B| < 2^{\mathbf{L}}/nm < 2^{\mathbf{L}}/n$ from the above. ■

THEOREM 15.2 Every extreme point $\bar{x}$ of $Ax \leqq b$, $x \geqq 0$ satisfies $\|\bar{x}\| < 2^{\mathbf{L}}$. When each nonzero component of $\bar{x}$ is expressed on a ratio of two relatively prime integers, the denominator is at most $2^{\mathbf{L}}$.

Proof: From the definition of a BFS for the system: $Ax + Is = b$, $x \geqq 0$, $s \geqq 0$, and Cramer's rule (see Section 3.3.9), each nonzero component $\bar{x}_j$ of $\bar{x}$ is of the form (determinant of B_1)/(determinant of B), where B is nonsingular and both B_1, B are submatrices of (A, I, b). By Theorem 15.1, the denominator and numerator here have absolute value $< 2^{\mathbf{L}}/n < 2^{\mathbf{L}}$ and both are integers, since all the entries in A, I, b are integers. By nonsingularity, $|B| \geqq 1$, and so $|\bar{x}_j| \leqq |B_1| < 2^{\mathbf{L}}/n$. Thus $\|\bar{x}\| < 2^{\mathbf{L}}$. ■

THEOREM 15.3 Let $\mathbf{K}$ be the set of feasible solutions of (15.1). If $\mathbf{K} \neq \emptyset$, the n-dimensional volume of $\mathbf{E}_0 \cap \mathbf{K}$ is $\geqq 2^{-(n+1)\mathbf{L}}$.

Proof: Suppose $\mathbf{K} \neq \emptyset$. Since $\mathbf{K}$ is a nonempty open set, it is possible to find a point $x = (x_j)$ in it satisfying $x_j \neq 0$ for all j. By making transformations of the form $x_j = -y_j$ if necessary (these transformations do not affect the volume of any set), we can assume that there exists a solution $\bar{x} = (\bar{x}_j) \in \mathbf{K}$, where $\bar{x}_j > 0$ for all j. Since $\bar{x} \in \mathbf{K}$, we have $A\bar{x} < b$. So, $\bar{x}$ is an interior point of the polyhedron $Ax \leqq b$, $x \geqq 0$. Let $\hat{x}$ be an extreme point of this polyhedron. By Theorem 15.1, $\hat{x}_j < 2^{\mathbf{L}}/\sqrt{n}$ for all j. Thus, it is clear that there exists $0 < \alpha < 1$, such that $\alpha\bar{x} + (1 - \alpha)\hat{x}$ is an interior point of the polytope Γ determined by $Ax \leqq b$, $x \geqq 0$, $x_j \leqq \lfloor 2^{\mathbf{L}}\sqrt{n} \rfloor$ for all j. Since Γ has an interior point, it must have positive n-dimensional volume. Since it is bounded, Γ must have at least one extreme point. If Γ has $\leqq n$ extreme points, since it is the convex hull of its extreme points (because it is bounded, by Theorem 3.6 and Corollary 3.4), it will be on a hyperplane in $\mathbf{R}^n$, contradicting the fact that its n-dimensional volume is positive. So, we can select a set $\mathbf{V} = \{V^0, V', \ldots, V^n\}$ of $n + 1$ extreme points of Γ, all of which do not lie on any hyperplane in $\mathbf{R}^n$. The convex hull of $\mathbf{V}$ is an n-dimensional simplex $\mathbf{S}$. By applying Theorem 15.1 on the system defining Γ, we conclude that each V^i is an integer vector u^i divided by an integer d_i, where $1 \leqq d_i < 2^{\mathbf{L}}/n$. The n-dimensional volume of the simplex $\mathbf{S}$ is (see [15.38]):

$$\frac{1}{n!} \text{determinant} \begin{pmatrix} 1 & \vdots & 1 & \vdots & 1 \\ \cdots & \cdots & \cdots & \cdots & \cdots \\ V^0 & \vdots & V^1 & \vdots & V^n \end{pmatrix}$$

$$= \frac{1}{n!d_0d_1 \ldots d_n} \text{determinant} \begin{pmatrix} d_0 & \vdots & d_n \\ \cdots & \cdots & \cdots \\ u^0 & \vdots & u^n \end{pmatrix}$$

Since all the entries in the determinant on the right-hand-side are integers, the value of that determinant is an integer, and since the n-dimensional volume of the simplex $\mathbf{S}$ is nonzero, this determinant is a nonzero integer, and hence has an absolute value $\geqq 1$. Using this, we have, volume of $\mathbf{S} \geqq 1/(n!d_0 \ldots d_n) \geqq 2^{-(n+1)\mathbf{L}}n^{n+1}/n! \geqq 2^{-(n+1)\mathbf{L}}$. $\mathbf{S} \subset \Gamma$, and from the definition of Γ, we verify that $\Gamma \subset \mathbf{E}_0 \cap \mathbf{K}$. So the volume of $\mathbf{E}_0 \cap \mathbf{K}$ is $\geqq 2^{-(n+1)\mathbf{L}}$. ■

THEOREM 15.4 In Step $r + 1$, assume the D_r is symmetric and positive definite. If $v_r \leqq -1$, $\mathbf{E}_r \cap \{x: A_{i.}x < b_i\} = \emptyset$. If $-1 < v_r \leqq 0$, D_{r+1} is symmetric and positive definite, and $\mathbf{E}_{r+1} \supset \mathbf{E}_r \cap \{x: A_{i.}x \leqq b_i\}$. Also, if Δ_{r+1}, Δ_r are the volumes of $\mathbf{E}_{r+1}$, $\mathbf{E}_r$, then

$$(\Delta_{r+1}/\Delta_r) = (n(1 + v_r)/(n + 1)) \times (n^2(1 - v_r^2)/(n^2 - 1))^{(n-1)/2} < 1 \tag{15.5}$$

We also have $(\Delta_{r+1}/\Delta_r) < e^{-(1/2(n+1))}$, where e is the base of the natural logarithms (e = 2.7 approximately, see [15.18]).

Proof: Let $\mathbf{H} = \{x: A_{i.}x = b_i\}$. Since D_r is positive definite and symmetric, there exists a nonsingular matrix Q_1 of order n satisfying $D_r = Q_1^T Q_1$ (Q_1 can be taken to be the Cholesky factor of D_r). Also there exists a matrix of rotation Q_2 (a square matrix of order n that is a unitary matrix, that is, a matrix satisfying $Q_2^T Q_2 = I$: see references [15.13, 15.21]), such that $(A_{i.}D_r A_{i.}^T)^{-1/2} Q_2 Q_1 A_{i.}^T = (1, 0, \ldots, 0) \in \mathbf{R}^n$. Let $Q = Q_2 Q_1$. Take the nonsingular linear transformation from x to ξ, $x = Q^T\xi + x^r$. Since $(x - x^r)^T D_r^{-1}(x - x^r) = (Q^T\xi)^T(Q_1^T Q_1)^{-1}(Q^T\xi) = \xi^T\xi$, this transforms $\mathbf{E}_r$ into the unit ball $\{\xi: \xi^T\xi \leqq 1\}$ in the ξ-space, which we denote by $\bar{\mathbf{E}}_r$. Since $A_{i.}x - b_i = A_{i.}(Q^T\xi + x^r) - b_i = \xi^T Q A_{i.}^T - \nu_r = \xi_1 - \nu_r$, this transforms $\mathbf{H}$ into the hyperplane $\mathbf{L} = \{\xi: \xi_1 = \nu_r\}$ in the ξ-space. So $\mathbf{E}_r \cap \{x: A_{i.}x < b_i\}$ gets transformed into $\{\xi: \xi^T\xi \leqq 1\} \cap \{\xi: \xi_1 < \nu_r\}$, and this is clearly $\emptyset$ if $\nu_r \leqq -1$. Hence if $\nu_r \leqq -1$, $\mathbf{E}_r \cap \{x: A_{i.}x < b_i\} = \emptyset$ and hence $\mathbf{E}_r \cap \mathbf{K} = \emptyset$.

The symmetry of D_{r+1} follows directly from the updating formula (15.4), since D_r is symmetric.

Let $\bar{D}_{r+1}$ be the diagonal matrix of order n, with diagonal entries $(n(\nu_r+1)/(n+1))^2, n^2(1-\nu_r^2)/(n^2-1), n^2(1-\nu_r^2)/(n^2-1), \ldots, n^2(1-\nu_r^2)/(n^2-1)$. Clearly $\bar{D}_{r+1} = ((1-\nu_r^2)n^2/(n^2-1))(I - (2(1-\nu_r n)/(n+1)(1-\nu_r))(I_{.1}I_{1.}))$. It can be verified that $Q^T\bar{D}_{r+1}Q = D_{r+1}$. Clearly $\bar{D}_{r+1}$ is positive definite, and hence so is D_{r+1}.

Since x^r violates $A_{i.}x < b_i$, $\nu_r \leqq 0$. Suppose $-1 < \nu_r \leqq 0$. Let $\xi^{r+1} = ((n\nu_r - 1)/(n+1), 0, \ldots, 0)^T$. Verify that the inverse transform of the ellipsoid in the ξ-space, $\bar{\mathbf{E}}_{r+1} = \bar{\mathbf{E}}(\xi^{r+1}, \bar{D}_{r+1})$, is the ellipsoid $\mathbf{E}_{r+1} = \mathbf{E}(x^{r+1}, D_{r+1})$ in the x space. It can also be verified that the point $\hat{\xi} = (-1, 0, \ldots, 0)^T$ is a common boundary point of $\bar{\mathbf{E}}_{r+1}$ and $\bar{\mathbf{E}}_r$, and that both these ellipsoids have the same tangent plane at $\hat{\xi}$. Also the intersection $\bar{\mathbf{E}}_r \cap \mathbf{L}$ is the $(n-1)$-dimensional sphere $\{\xi: \xi_1 = \nu_r, \xi_2^2 + \cdots + \xi_n^2 \leqq 1 - \xi_1^2\} = \bar{\mathbf{E}}_{r+1} \cap \mathbf{L}$ (see Figure 15.1).

Since $\nu_r > -1$, $\mathbf{E}_r \cap \mathbf{H}$ is as in Figure 15.2. Move the hyperplane $\mathbf{H}$ parallel to itself away from x^r (i.e., reduce b_i in $A_{i.}x = b_i$) until it becomes a tangent plane to $\mathbf{E}_r$ at the boundary point, $\hat{x}$, say. Point $\hat{x}$ is clearly the inverse transform of $\hat{\xi}$ from the ξ-space. Taking the inverse transform from the ξ-space, we conclude from the above results that $\hat{x}$ is a common boundary point of $\mathbf{E}_r$ and $\mathbf{E}_{r+1}$, that both $\mathbf{E}_r$ and $\mathbf{E}_{r+1}$ have the same tangent plane (which is a translate of $\mathbf{H}$) at $\hat{x}$, and that $\mathbf{E}_r \cap \mathbf{H} = \mathbf{E}_{r+1} \cap \mathbf{H}$ (see Figure 15.2). So, $\mathbf{E}_{r+1} \supset \mathbf{E}_r \cap \{A_{i.}x < b_i\} \supset \mathbf{E}_r \cap \mathbf{K}$. In fact, $\mathbf{E}_{r+1}$ is the smallest volume ellipsoid satisfying the properties that it contain $\mathbf{E}_r \cap \{x: A_{i.}x \leqq b_i\}$, and it has $\hat{x}$ as a boundary point with the translate of $\mathbf{H}$ through $\hat{x}$ as the tangent plane.

The n-dimensional volume of a solid subset of $\mathbf{R}^n$ is obtained by computing a definite multiple intergral over that subset (see [15.18, 15.29, 15.39]). If we make a nonsingular affine transformation of the variables, the multiple integral gets divided by the Jacobian of the transformation. From this, it can be shown that if D is any positive definite matrix, the volume of the elipsoid $\mathbf{E}(x, D)$ is (volume of the unit spherical ball in $\mathbf{R}^n$) ($\sqrt{\text{determinant of } D}$). We also

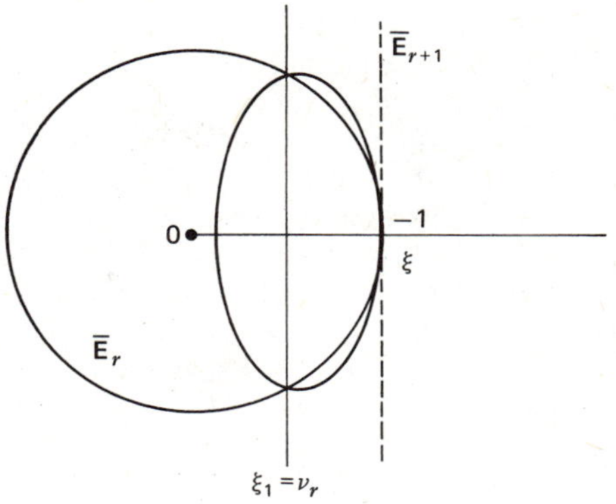

Figure 15.1 $\bar{\mathbf{E}}_{r+1}$ touches $\bar{\mathbf{E}}_r$ at $(-1, 0, \ldots, 0)$ and has the same intersection with the hyperplane $\xi_1 = \nu_r$ as $\bar{\mathbf{E}}_r$.

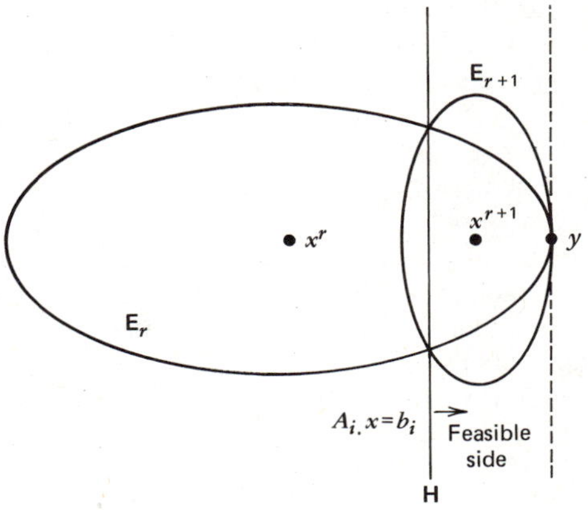

Figure 15.2

have $(\Delta_{r+1}/\Delta_r) =$ (volume of $\bar{\mathbf{E}}_{r+1}$)/(volume of $\bar{\mathbf{E}}_r$). Since $\bar{\mathbf{E}}_r$ is the unit spherical ball in $\mathbf{R}^n$, we conclude from the above that $\Delta_{r+1}/\Delta_r = \sqrt{\text{determinant of } \bar{D}_{r+1}}$, which proves (15.5). It can be verified that in the range $-1 < \nu_r \leqq 0$, the value of the right-hand side of (15.5) increases as ν_r increases to 0. So $(\Delta_{r+1}/\Delta_r) \leqq (n/(n+1))(n^2/(n^2-1))^{(n-1)/2}$. From the properties of e (see [15.18]) we know that $n^2/(n^2-1) = 1 + 1/(n^2-1) < e^{(1/(n^2-1))}$, and $n/(n+1) = 1 - 1/(n+1) < e^{-(1/(n+1))}$. Using these in the above, we get $\Delta_{r+1}/\Delta_r < e^{-(1/2(n+1))}$. ■

THEOREM 15.5 All the matrices D_r obtained in the ellipsoid method are positive definite and symmetric and $\mathbf{E}_r \supset \mathbf{E}_0 \cap \mathbf{K}$ for all r. Also, if the points x^r obtained in the method are infeasible for all $1 \leqq r \leqq 6(n+1)^2\mathrm{L}$, then $\mathbf{K} = \varnothing$.

Proof: $D_0 = 2^{2\mathrm{L}}I$ is positive definite and symmetric. Applying the result in Theorem 15.4 to each step, it follows that D_r has this property for all r. The fact that $\mathbf{E}_r \supset \mathbf{E}_0 \cap \mathbf{K}$ for all r also follows by applying the results in Theorem 15.4 to each step. By Theorem 15.3, if $\mathbf{K} \neq \varnothing$, the volume of $\mathbf{E}_0 \cap \mathbf{K}$ is $\geqq 2^{-(n+1)\mathrm{L}}$. Since $\Delta_{r+1}/\Delta_r < e^{-(1/2(n+1))}$ by Theorem 15.4, $\Delta_0 =$ volume of solid sphere of radius 2^{L} in $\mathbf{R}^n$, we conclude that if $r > 6(n+1)^2\mathrm{L}$, then $\Delta_r < 2^{-(n+1)\mathrm{L}}$. These facts together imply that if $x^r \notin \mathbf{K}$ for all $r = 1$ to $6(n+1)^2\mathrm{L}$, then $\mathbf{K} = \varnothing$. ■

In the updating formula (15.3), the quantity $+\sqrt{A_{i.}D_rA_{i.}^{\mathrm{T}}}$ appears in the denominator. This square root exists and is positive because D_r is positive definite and $A_{i.} \neq 0$. If D_r is not positive definite, we may have $A_{i.}D_rA_{i.}^{\mathrm{T}} \leqq 0$, and in this case (15.3) does not make any sense. Also $(x - x^r)D_r^{-1}(x - x_r) \leqq 1$ represents an ellipsoid in $\mathbf{R}^n$ only if D_r is positive definite. Hence the positive definiteness of D_r for all r is a vital property in proving the validity of this algorithm.

L is the size of the input system (15.1). Each step of the ellipsoid algorithm requires $O(n^2)$ effort. Since it requires at most $6(n+1)^2\mathrm{L}$ steps, the overall computational effort required to process (15.1) by the ellipsoid algorithm is of order $O(n^4\mathrm{L})$. So it is a polynomially bounded algorithm for processing systems of open linear inequalities of the form (15.1).

The updating formulas (15.3) and (15.4) require the computation of a square root, besides other arithmetic operations, and the theorems proved so far are only valid under the assumption of exact arithmetic throughout. In practice one must use finite precision arithmetic. Khachiyan [15.12, 15.19] indicated that computing each quantity to 6 ln L bits of precision would suffice. If each entry in x^{r+1} and D_{r+1} is rounded to this specific number of bits, the ellipsoid $\mathbf{E}_{r+1}$ may not contain $\mathbf{E}_r \cap \{x: A_{i.}x \leqq b_i\}$. However, Khachiyan showed that if D_{r+1} is multiplied by a factor slightly larger than 1 before being rounded (this factor, $(1 + (1/16n^2))$, is known as a *dilation factor*, since multiplying D_{r+1} in (15.4) by it has the effect of enlarging $\mathbf{E}_{r+1}$ slightly), then all the results continue to hold. See [15.11, 15.19, 15.28]. Also, S. Ursic [15.41] considered the problem of controlling round-off errors when taking square roots in the ellipsoid method, and showed that this can be avoided if one considers a variant in which all the computations are performed in exact rational arithmetic using continued fractions.

If the arithmetical operations are not carried out to an adequate degree of precision, after several steps the matrix D_r may accumulate so much round-off error that it may no longer be positive definite. So in writing a computer code for this algorithm, care should be exercised in maintaining adequate degree of precision.

We need the data in the original system of constraints to be all integer for using the ellipsoid algorithm, because the size of the system, L, is needed in the termination criterion for the algorithm. On problems with real but noninteger data, the ellipsoid method has unbounded complexity. For example, consider the problem discussed by J. F. Traub and H. Wozniakowski [14.21] of two inequalities in two unknowns of the form (15.1) with

$$A = \begin{pmatrix} 1 & 0 \\ -1 & 0 \end{pmatrix}, \quad b = \begin{pmatrix} 1 + 1/s \\ 1 \end{pmatrix},$$

where $s \geqq 5/\sqrt{2}$. Let $x^0 = (1 + \sqrt{2}/5,\ 0)^{\mathrm{T}}$, $D_0 = 2I$. It can be verified that the ellipsoid $\mathbf{E}_0 = \mathbf{E}(x^0, D_0)$ satisfies a result corresponding to Theorem 15.3 for this problem. Executing the ellipsoid method on this problem beginning with $\mathbf{E}_0$, using the updating formulas (15.3) and (15.4) with $\nu_r = 0$ in each step, we are led to $x^r = (h_r, 0)^{\mathrm{T}}$,

$$D_r = \begin{pmatrix} f_r & 0 \\ 0 & g_r \end{pmatrix},$$

where $h_0 = 1 + \sqrt{2}/5$, $h_{r+1} = h_r + s_r(\sqrt{2}/3)(2/3)^r$, $s_r = 1$ if $h_r \leqq 1$ or $= -1$ if $h_r \geqq 1 + 1/s$, $f_r = 2(2/3)^{2r}$ and $g_r = 2(4/3)^r$. Hence, as long as h_r is not in the open interval $(1, 1 + 1/s)$, it is equal to 1 +

$(-1)^r(\sqrt{2}/5)(2/3)^r$. Let k be the smallest integer, such that h_k is in the open interval $(1, 1 + 1/s)$. It can be verified that k is approximately $(\log s)/\log(3/2)$. So by choosing s arbitrarily large, we can make the ellipsoid method go through an arbitrarily large number of steps before termination in this problem. Also when s is large and integer, this system can be transformed into an inequality system with integer data by multiplying both sides by s. The size of the resulting integer system, L, is approximately $\log s$. The above analysis shows that the number of steps in the ellipsoid method (and not just the bound) increases as L.

To execute the ellipsoid algorithm, we do not need the entire system of constraints (15.1) explicitly. We need an estimate of the size of the system, L, to determine when to terminate. Then in each step we need to check whether the current center x^r is feasible (i.e., satisfies all the constraints), or find a constraint in the system that x^r violates. If x^r violates the ith constraint, and $\mathbf{K}$ is the feasible region, then the hyperplane corresponding to the ith constraint is a separating hyperplane separating $\mathbf{K}$ from x^r. So the problem of generating such a violated constraint is known as the *separation problem*. If we are given a subroutine for solving this separation problem (for any given point $x \in \mathbf{R}^n$, this subroutine must be able to determine whether $x \in \mathbf{K}$ or generate a hyperplane separating x from $\mathbf{K}$), then we can execute the various steps of the ellipsoid algorithm using this subroutine only, without ever needing the system of constraints (15.1) explicitly.

Most combinatorial optimization problems can be recast as LPs in which the number of defining inequalities is large and hard to find explicity. However, for some of these problems, it is possible to provide an efficient subroutine for the associated separation problem that somehow generates a violated inequality if the current test point is not feasible. In [15.11] M. Grötschel, L. Lovász, and A. Schrijver indicate how to construct efficient separation subroutines for a variety of combinatorial optimization problems, and thereby develop polynomially bounded algorithms for these problems based on the ellipsoid method. Algorithms based on the ellipsoid method have been developed for solving nonlinear convex optimization problems by exploiting the same idea (see references [15.6, 15.8, 15.16] of S. J. Chung and K. G. Murty, J. G. Ecker and M. Kupferschmid, and D. B. Judin and A. S. Nemirovskii).

15.3 HOW TO PROCESS SYSTEMS OF CLOSED LINEAR INEQUALITIES USING THE ELLIPSOID METHOD

Consider the system

$$Ax \leqq b \tag{15.6}$$

where A and b are integer matrices of orders $m \times n$ and $m \times 1$, respectively. L, defined as in (15.2), is the size of this system. We discuss two schemes for processing (15.6) using the ellipsoid method.

Scheme 1 is iterative. It consists of at most minimum $\{n, m\}$ iterations. In each iteration there may be at most m stages. In each stage, the ellipsoid method is used once on an open subsystem of linear inequalities constructed from (15.6). Each iteration either terminates with finding a feasible solution of (15.6), or by identifying a constraint of (15.6) that must hold as an equation at every feasible solution of (15.6). In the second case, a variable is eliminated from the system using that equation, and then the equation itself is eliminated, thereby reducing the system into another in which the number of variables and the number of constraints are both one less. With this reduced system, the scheme moves to the next iteration.

In a stage, we have a solution satisfying a subset of the constraints as strict inequalities (the others are either satisfied as equations or violated). For example, $x^1 = (b_1 - 1)A_{1.}^{\mathrm{T}}/\|A_{1.}\|^2$ satisfies the first constraint in (15.6) as a strict inequality. Suppose we have a solution $\bar{x}$ satisfying the first t constraints in (15.6) as strict inequalities. If $\bar{x}$ satisfies the remaining constraints as equations, we are done, $\bar{x}$ is feasible, and we terminate. Suppose $\bar{x}$ violates the $(t + 1)$th constraint in (15.6). Then apply the ellipsoid method to solve the open system $A_{i.}x < b_i$, $i = 1$ to $t + 1$. If it produces a solution $\hat{x}$ to this system, $\hat{x}$ satisfies at least one more constraint in (15.6) as a strict inequality than $\bar{x}$; with $\hat{x}$ move to the next stage in the same iteration. If the ellipsoid method terminates with the conclusion that the open system of $(t + 1)$ inequalities is infeasible, because the system of the first t constraints in (15.6) treated as strict inequalities has a solution $\bar{x}$, the $(t + 1)$th constraint in (15.6) must hold as an equation, $A_{t+1.}x = b_{t+1}$ at every feasible solution of (15.6). Using this equation, reduce this system to one of lower dimension as discussed above and process that reduced system in the same manner by moving to the next iteration.

For finite termination in the ellipsoid algorithm, the data have to be integral. For eliminating variables using equations in this scheme, we can use the special algorithms of A. Bachem and R. Kannan discussed in [15.4, 15.17]. These algorithms maintain all the coefficients always integral, and guarantee that the number of digits in each coefficient of the reduced system is bounded above by a polynomial in the size of the original system. The computational effort required to carry out this reduction is also polynomially bounded in terms of the size of the original system. Thus this scheme is polynomially bounded.

Scheme 2 from the paper [15.9] of P. Gács and L. Lovász perturbs the closed system (15.6) into the open system $A_{i.}x < b_i + 2^{-\mathbf{L}}$ for $i = 1$ to m, or equivalently

$$2^{\mathbf{L}}(A_{i.}x) < 2^{\mathbf{L}}b_i + 1, \qquad i = 1 \text{ to } m \tag{15.7}$$

expressing all the data in integers. It applies the ellipsoid method on (15.7). It is shown below that (15.7) has a feasible solution iff (15.6) does. Also, a procedure is developed below for generating a feasible solution of (15.6) from a feasible solution of (15.7).

Let $\theta_i(x) = A_{i.}x - b_i$, $i = 1$ to m. So x is feasible to (15.6) iff $\theta_i(x) \leqq 0$ for all i. If $\theta_i(x) > 0$, it provides a measure of the extent of infeasibility of the point x to the ith constraint in (15.6).

THEOREM 15.6 Let $x^0 \in \mathbf{R}^n$ be an arbitrary point. There exists $\bar{x} \in \mathbf{R}^n$ satisfying $\theta_i(\bar{x}) \leqq$ maximum $\{0, \theta_i(x^0)\}$ for each $i = 1$ to m, and that every row vector of A can be expressed as a linear combination of $\{A_{i.}: i$, such that $\theta_i(\bar{x}) \geqq 0\}$.

Proof: The proof is constructive. For convenience in notation, suppose $\{i: \theta_i(x^0) \geqq 0\} = \{1, \ldots, k\}$. Suppose $\{A_{1.}, \ldots, A_{k.}\}$ does not span all the rows A; in particular, suppose $A_{k+1.}$ cannot be expressed as a linear combination of vectors from this set. By a standard result in linear algebra (see Exercise 3.42), the system of equations

$$\begin{aligned} A_{i.}y &= 0, \qquad i = 1 \text{ to } k \\ A_{k+1.}y &= 1 \end{aligned} \tag{15.8}$$

has a solution $y^0 \in \mathbf{R}^n$. Find $\lambda_0 =$ maximum $\{\lambda: \lambda$, such that $\lambda A_{i.}y^0 + \theta_i(x^0) \leqq 0$ for $i = k+1$ to $m\}$. $A_{k+1.}y^0 = 1$ and $\lambda_0 A_{k+1.}y^0 + \theta_{k+1}(x^0) \leqq 0$ imply that $\lambda_0 \leqq -\theta_{k+1}(x^0)$. Also, since $\theta_i(x^0) < 0$ for all $i = k+1$ to m, we know that $\lambda_0 > 0$. So $0 < \lambda_0 \leqq -\theta_{k+1}(x^0)$. Thus λ_0 is a finite positive quantity. Let $x^1 = x^0 + \lambda_0 y^0$. So $\theta_i(x^1) = \theta_i(x^0) + \lambda_0 A_{i.}y^0 = \theta_i(x^0)$ for $i = 1$ to k and $\theta_i(x^1) \leqq 0$ for $i = k+1$ to m by the definition of λ_0. Also $\theta_i(x^1) = 0$ for at least one i between $k+1$ to m (this holds for all i that tie for the maximum in the definition of λ_0). So the set $\{A_{i.}: i$, such that $\theta_i(x^1) \geqq 0\}$ contains at least one more row than $\{A_{i.}: i$, such that $\theta_i(x^0) \geqq 0\}$. If $\{A_i: i$, such that $\theta_i(x^1) \geqq 0\}$ spans all the rows of A, $x^1 = \bar{x}$ does it, otherwise repeat the procedure with x^1. After at most m repetitions, we obtain a point $\bar{x}$ satisfying both conditions of the theorem. ■

THEOREM 15.7 The system (15.6) is feasible iff (15.7) is. Also from a feasible solution of (15.7), a feasible solution of (15.6) can be constructed.

Proof: This proof is also constructive. All feasible solutions for (15.6) are feasible to (15.7). To show the converse, let x^0 be feasible to (15.7). Suppose x^0 is not feasible to (15.6). Define $\theta_i(x^0) = A_{i.}x^0 - b_i$ as in Theorem 15.6. Since x^0 is feasible to (15.7), $\theta_i(x^0) < 2^{-\mathbf{L}}$ for all i. For convenience in notation, assume that $\{i: \theta_i(x^0) \geqq 0\} = \{1, \ldots, k\}$. So $\theta_i(x^0) < 0$ for $i = k+1$ to m and $0 \leqq \theta_i(x^0) < 2^{-\mathbf{L}}$ for $i = 1$ to k. Using the procedure in Theorem 15.6, construct a point $\bar{x}$ satisfying $\theta_i(\bar{x}) \leqq$ maximum $\{0, \theta_i(x^0)\}$, $i = 1$ to m; and $\{A_{i.}: i$, such that $\theta_i(\bar{x}) \geqq 0\}$ spans all the rows of A. Find a maximal linearly independent subset of $\{A_{i.}: i$, such that $\theta_i(\bar{x}) \geqq 0\}$. Suppose this subset is $\{A_{i.}: i \in \mathbf{J} = \{p_1, \ldots, p_r\}\}$. Solve

$$A_{i.}x = b_i, \; i \in \mathbf{J} = \{p_1, \ldots, p_r\} \tag{15.9}$$

We now show that every solution of (15.9) is feasible to (15.6). Suppose $\hat{x}$ is feasible to (15.9). Let $p \in \{1, \ldots, m\}$. By the choice of $\bar{x}$, the system $A_{p.} = \sum_{i \in \mathbf{J}} \lambda_i A_{i.}$ has a solution $(\lambda_i: i \in \mathbf{J})$, and since $\{A_i: i \in \mathbf{J}\}$ is linearly independent, this solution is unique. From Cramer's rule we conclude that each λ_i in this solution is of the form d_i/d, where d_i and d are the values of determinants of appropriate submatrices of A. So $d \neq 0$, and all d_i, d are integers. If $d < 0$, we can multiply all d_i and d by -1, so, without any loss of generality we assume that $d > 0$. By Theorem 15.1, d and $|d_i|$ are all $< 2^{\mathbf{L}}/n$. So we have $d(A_{p.}\hat{x} - b_p) = \sum_{i \in \mathbf{J}} d_i A_{i.}\hat{x} - db_p = \sum_{i \in \mathbf{J}} d_i b_i - db_p$, since $\hat{x}$ solves (15.9). Now $\sum_{i \in \mathbf{J}} d_i b_i - db_p = \sum_{i \in \mathbf{J}} d_i(A_{i.}\bar{x} - \theta_i(\bar{x})) - d(A_{p.}\bar{x} - \theta_p(\bar{x})) = d\theta_p(\bar{x}) - \sum_{i \in \mathbf{J}} d_i\theta_i(\bar{x})$, since $A_{p.} = \sum_{i \in \mathbf{J}} (d_i/d)A_{i.}$. Since $\mathbf{J} \subset \{i: i$, such that $\theta_i(\bar{x}) \geqq 0\}$, we have $\theta_i(\bar{x}) \geqq 0$ for all $i \in \mathbf{J}$. Also $\theta_i(\bar{x}) \leqq 2^{-\mathbf{L}}$ for all i, as discussed above. So $0 \leqq \theta_i(\bar{x}) \leqq 2^{-\mathbf{L}}$ for all $i \in \mathbf{J}$. Using this in the above, since $d > 0$, we have $\sum_{i \in \mathbf{J}} d_i b_i - db_p \leqq d2^{-\mathbf{L}} - \sum_{i \in \mathbf{J}} d_i\theta_i(\bar{x}) \leqq (2^{\mathbf{L}}/n)2^{-\mathbf{L}} - \sum_{i \in \mathbf{J}} d_i\theta_i(\bar{x}) \leqq (1/n) + \sum_{i \in \mathbf{J}} |d_i|\theta_i(\bar{x}) \leqq (1/n) + \sum_{i \in \mathbf{J}} (2^{\mathbf{L}}/n)2^{-\mathbf{L}} < 1$, since all d_i, d are $< 2^{\mathbf{L}}/n$, as discussed above, and $\theta_i(\bar{x}) < 2^{-\mathbf{L}}$. So $\sum_{i \in \mathbf{J}} d_i b_i - db_p < 1$, and it is an integer, so we must have $\sum_{i \in \mathbf{J}} d_i b_i - db_p \leqq 0$. Since $d > 0$, we can divide this by d, leading to $0 \geqq \sum_{i \in \mathbf{J}} (d_i/d)b_i - b_p = \sum_{i \in \mathbf{J}} \lambda_i A_{i.}\hat{x} - b_p = A_{p.}\hat{x} - b_p$. That is $A_{p.}\hat{x} - b_p \leqq 0$ or $A_{p.}\hat{x} \leqq b_p$. Since this holds for any $p \in \{1, \ldots, m\}$, $\hat{x}$ must be feasible to (15.6), and the theorem is proved. ∎

So, to solve (15.6), apply the ellipsoid method on (15.7) and from a solution of it obtain a solution for (15.6) as in Theorem 15.7. L is the size of the system (15.6). The size of the perturbed system (15.7) is $m(n + 1)\mathrm{L} + \mathrm{L}$. So the computational effort required by the ellipsoid algorithm to process the system (15.7) is $\mathrm{O}(mn^5\mathrm{L})$. If a feasible solution of (15.7) is obtained at the termination of the ellipsoid algorithm, to obtain a solution of the original system (15.6) by Scheme 2 requires the solution of at most $(m + 2)$ different systems of linear equations that requires a computational effort of $\mathrm{O}(m^4)$. So the overall computational effort required for solving the system of closed linear inequalities by Scheme 2 is of order $\mathrm{O}(mn^5\mathrm{L}) + \mathrm{O}(m^4)$. It is clear that all the entries in the system are blown up (multiplied by the large number 2^{L}) in the process of constructing the perturbed system. However, the number of digits in each entry in the perturbed system is at most L more than those in the corresponding entry in the original system. Also, since the ellipsoid algorithm has to be applied only once on the perturbed system (15.7), this perturbed system is exact and not an approximation and all the data in it are integer.

15.4 HOW TO SOLVE A LINEAR PROGRAM USING THE ELLIPSOID METHOD

Consider the LP with integer data

$$\begin{aligned} &\text{Minimize} \quad cx \\ &\text{Subject to} \quad Ax \geqq b, \quad x \geqq 0 \end{aligned} \tag{15.10}$$

Introducing the dual variables y from the results in Section 4.6.4, we know that solving (15.10) is equivalent to solving the system

$$Ax \geqq b, \qquad A^{\mathrm{T}}y \leqq c^{\mathrm{T}}, \qquad cx - b^{\mathrm{T}}y \leqq 0, \qquad x, y \geqq 0 \tag{15.11}$$

System (15.11) has integer data. Solve (15.11) by the ellipsoid method. If $(\bar{x}, \bar{y})$ is feasible to (15.11), $\bar{x}$ is an optimum solution of (15.10) and $\bar{y}$ is a dual optimum solution. Suppose (15.11) is infeasible. Then solve the system $Ax \geqq b$, $x \geqq 0$. If this is feasible, then $z(x)$ is unbounded below on the set of feasible solutions of (15.10). This approach provides a polynomially bounded algorithm for solving the LP (15.10). Mathematically, this clearly establishes that LPs belong in the class **P** of problems, which are solvable by polynomially bounded algorithms. This is a major theoretical result.

In Chapter 4 we saw that the weak duality theorem of linear programming provides a very efficient supervisor principle for a pair of primal-dual LPs. But until the ellipsoid method was developed, there was no supervisor principle for an LP by itself (without its dual). The ellipsoid method provides a supervisor principle for an LP by itself without its dual. Given the LP and a feasible solution that is claimed to be optimal for it, we use the complementary slackness theorem, to construct a system of equations and inequalities in the dual variables that must be feasible if the given solution is, in fact, optimal. The feasibility of this dual system can be checked using the ellipsoid method. Since the ellipsoid method is polynomially bounded, this provides a procedure for checking the claimed optimality of a given solution to the LP, whose time complexity is polynomially bounded.

15.5 PERFORMANCE MEASURES FOR ALGORITHMS

In Chapter 14 we discussed two performance measures for the efficiency of an algorithm. One is the *average computational complexity* (this measures the average computational effort required to solve an instance of the problem as a function of its size, average being defined in a statistical sense), and the other is the *worst case computational complexity* (this determines an upper bound on the computational effort to solve any instance of the problem as a function of its size). Until recently, an algorithm was classified as a *good* or *bad algorithm*, depending on whether its worst case computational complexity was polynomially bounded or not. By this classification, the simplex method (with several of the available entering variable choice rules) is a bad algorithm, and the ellipsoid method is a good algorithm, since it has been shown to be polynomially bounded. But in computational tests the ellipsoid method performed very poorly in comparison with the simplex method. For this reason, mathematicians find themselves in an awkward situation; the ellipsoid algorithm, which is considered good by the mathematical criterion of polynomial boundedness in the worst case, is a bad algorithm in practice; and the simplex algorithm, which is considered bad by the same mathematical criterion, is a good algorithm in practice. So practitioners have begun to question the "polynomial boundedness in the worst case" criterion for judging or comparing algorithms. One major side effect of the ellipsoid method is the change in emphasis away from the worst case analysis in judging the computational efficiency of algorithms.

The tremendous excitement that the ellipsoid method generated when it was first announced, waned considerably when computational tests revealed that it is impractical in the form in which it is currently available. The final blow to the ellipsoid method may have been dealt when papers establishing mathematically that the ever popular simplex algorithm is indeed a polynomial growth algorithm on an average [14.2, 14.3, 14.19] began to appear.

15.6 ELLIPSOID-TYPE METHODS USING OTHER CONVEX BODIES

For processing systems of open linear inequalities, it is possible to construct on ellipsoid-type method that uses convex bodies other than ellipsoids. In [15.42] this has been done using the simplex in place of the ellipsoid. We describe this method briefly here. Let the system to be processed be: find $x \in \mathbf{R}^n$ satisfying

$$A'x < b' \tag{15.12}$$

in which all the data are integer. Let L′ be the size of (15.12). From the results in Section 15.2, we know that the set of feasible solutions of (15.12) is nonempty iff the set of feasible solutions of the following system is nonempty.

$$\begin{aligned} A'x &< b' \\ x_j &< 2^{\mathbf{L}'}, \qquad j = 1 \text{ to } n \\ -\sum_{j=1}^{n} x_j &< n2^{\mathbf{L}'} \end{aligned} \tag{15.13}$$

Also, clearly, every feasible solution of (15.13) is feasible to (15.12). So, processing (15.12) can be carried out by processing (15.13). The method works on (15.13). For simplicity, denote (15.13) by

$$Ax < b \tag{15.14}$$

where A is a matrix of order $m \times n$. Clearly $m > n + 1$, and A has rank n.

In the rth step, the method considers the simplex $\mathbf{K}_r = \{x: Q^r(Ax - b) \leqq 0\}$, where Q^r is a nonnegative matrix of order $(n + 1) \times m$ satisfying the property that $\mathbf{K}_r$ is indeed an n-dimensional simplex. So, $\mathbf{K}_r$ has $n + 1$ extreme points. Each extreme point of $\mathbf{K}_r$ is the unique solution to the system consisting of n of the $n + 1$ constraints characterizing $\mathbf{K}_r$ as strict equality constraints. Let x^r be the center of $\mathbf{K}_r$, that is, the average of the $n + 1$ extreme points of $\mathbf{K}_r$. If x^r satisfies (15.14), the method terminates. If x^r violates (15.14), choose a violated constraint, say the ith, that is, $A_{i.}x - b_i < 0$. Let y^r be an extreme point of $\mathbf{K}_r$ that maximizes $A_{i.}x - b_i$ over $\mathbf{K}_r$. By the hypothesis, we have $A_{i.}y^r - b_i > 0$. Let t be such that $Q^r_{t.}(Ax - b) = 0$ is the equation of the facet of $\mathbf{K}_r$ containing all the extreme points of $\mathbf{K}_r$ other than y^r. Obtain Q^{r+1} by just increasing the value of q_{ti} in Q^r by $(Q^r_{t.}(Ay^r - b)/((A_{i.}y^r - b_i)(n^2 - 1))$ and leaving all other entries unchanged. With this Q^{r+1} and the simplex $\mathbf{K}_{r+1} = \{x: Q^{r-1}(Ax - b) \leqq 0\}$, go to Step $r + 1$.

The method can be initiated by taking $Q^0 = (0 \vdots I_{n+1})$ where I_{n+1} is the unit matrix of order $n + 1$, and 0 is the matrix of order $(n + 1) \times (m - n - 1)$ with all zero entries.

Let Δ_r denote the n-dimensional volume of the simplex in the rth step, $\mathbf{K}_r$. It has been shown by B. Yamnitsky and L. A. Levin in [15.42] that $\Delta_r/\Delta_{r+1} \geqq (1 - (1/n^2))^n/(1 - (1/n)) > e^{(1-1/n)/2n^2}$. So the volume of the simplex decreases as the method moves from one step to the next. If L is the size of (15.14), there is a positive constant δ such that either x^r becomes feasible to (15.14) for some $r \leqq \mathrm{L}\delta n^4$; or if the method goes through $\mathrm{L}\delta n^4$ steps, we can conclude that (15.14) is infeasible. There seems to be more room for improvement in this algorithm. It may be possible to construct a variant of this method, in which the volume of the simplex decreases much more rapidly after each step.

Exercises

15.1 For solving (15.1), the ellipsoid methods discussed in Section 15.2 require $0(n^4L)$ arithmetical operations on the data in the worst case. This complexity measure depends on L, the number of digits in the data. Is there an algorithm for solving (15.1) or (15.6) whose worst case computational complexity (in terms of the number of arithmetical operations on the given data) is a polynomial in n and m and does not involve L?

15.2 Discuss an efficient method for computing the dimension of a convex polyhedron specified by a system of linear equations and inequalities, some of which may be sign restrictions on individual variables.

REFERENCES

15.1 M. Akgül, "On Convex Programming via Ellipsoidal Algorithms," Research Report CORR 81-17, Faculty of Mathematics, University of Waterloo, Waterloo, Ontario, Canada, April 1981.

15.2 M. Akgül, "Topics in Relaxation and Ellipsoidal Methods," Ph.D. dissertation, Department of Combinatorics and Optimization, University of Waterloo, Waterloo, Ont., Canada, 1981.

15.3 B. Aspvall and R. E. Stone, "Khachiyan's Linear Programming Algorithm," *Journal of Algorithms 1* (1980), 1–13.

15.4 A. Bachem and R. Kannan, "Applications of Polynomial Smith Normal Form Calculations," in *Numberische Methoden bei graphentheoretischen und Kombinatorischen Problemen*, Band 2, Herausgegeben Von L-Collatz, Gr. Meinardus, W. Wetterling, ISNM Vol. 46, Birkhäuser Verlag Basel, 1979, pp. 9–21.

15.5 R. G. Bland, D. Goldfarb, and M. J. Todd, "The Ellipsoid Method: A Survey," *Operations Research 29*, 6 (November/December 1981), 1039–1091.

15.6 S. J. Chung and K. G. Murty, "Polynomially Bounded Ellipsoid Algorithms for Convex Quadratic Programming," in *Nonlinear Programming 4*, O. L. Mangasarian, R. R. Meyer, and S. M. Robinson (Ed.), Academic, New York, 1981, pp. 439–485.

15.7 P. H. Delsarte, "A Refined Version of Khachian's Algorithm," *Philips Journal of Research 35* (1980), 307–319.

15.8 J. G. Ecker and M. Kupferschmid, "An Ellipsoid Algorithm for Convex Programming," Department of Mathematical Sciences, Rensselaer Polytechnic Institute, Troy, N.Y. April 1981.

15.9 P. Gaćs and L. Lovaśz "Khacian's Algorithm for Linear Programming," *Mathematical Programming Study 14*, North-Holland, Amsterdam, The Netherlands, January 1981, pp. 61–68.

15.10 P. E. Gill and W. Murray, "A Numerical Investigation of Ellipsoid Algorithms for Large-Scale Linear Programming," Technical Report SOL 80-27, Systems Optimization Laboratory, Stanford University, Stanford, Calif. October 1980.

15.11 M. Grötschel, L. Lovász, and A. Schrijver, "The Ellipsoid Method and its Consequences in Combinatorial Optimization," *Combinatorica*. (1981), 169–197.

15.12 L. G. Hačijan "A Polynomial Algorithm in Linear Programming," *Soviet Math-Doklady 20*, 1 (1979), 191–194.

15.13 M. Hall and H. J. Ryser "Normal Completions of Incidence Matrices," *American Journal of Mathematics 16*, (1954), 581–589.

15.14 P. C. Jones and E. S. Marwil, "A Dimensional Reduction Variant of Khachiyan's Algorithm for Linear Programming Problems," EG & G Idaho, Inc., Idaho Falls, Idaho January 1980.

15.15 D. B. Judin and A. S. Nemirovskii, "Evaluation of the Informational Complexity of Mathematical Programming Problems," *Ekonomika i Matematicheskie Metody 12*, 1976, pp. 128–142; translated in Matekon, *Translations of*

Russian and East European Mathematical Economics 13, 2 (Winter 1976), 3–25.

15.16 D. B. Judin and A. S. Nemirovskii, "Informational Complexity and Effective Methods of Solution for Convex Extremal Problems," *Ekonomika i Matematicheskie Metody 12*, 1976, pp. 357–369; translated in Matekon: *Translations of Russian and East European Mathematical Economics 13*, 3 (Spring 1977), 25–45.

15.17 R. Kannan and A. Bachem, "Polynomial Algorithms for Computing the Smith and Hermite Normal Forms of an Integer Matrix," *SIAM Journal on Computing 8*, 4 (November 1979), 499–507.

15.18 W. Kaplan and D. J. Lewis, *Calculus and Linear Algebra*, Vols. 1 and 2, Wiley New York, 1970.

15.19 L. G. Khachiyan, "Polynomial Algorithms in Linear Programming," *Zhurnal Vichislitel'noi Matematiki i Matematicheskoi Fiziki* (in Russian) *20*, 1 (1980), 51–68.

15.20 K. Kubota, private communication, November 1979.

15.21 T. Y. Lam, *The Algebraic Theory of Quadratic Forms*, Benjamin Advanced book program, Reading, Mass., 1973.

15.22 E. L. Lawler, "The Great Mathematical Sputnik of 1979," *The Sciences*, September 1980.

15.23 A. Yu. Levin, "On one Algorithm for Minimization of Convex Functions," *Doklady Academie Nauk SSR 160*, 6 (1965), 286–290. Translated in *Soviet Mathematics Doklady 160*, 6 (1965), 286–290.

15.24 L. Lovász, "A New Linear Programming Algorithm—Better or Worse than the Simplex Method?" *The Mathematical Intelligencer 2*, 3 (1980), 141–146.

15.25 N. Megiddo, "Is Binary Encoding Appropriate for the Problem-Language Relationship," *Theoretical Computer Science 19* (1982), 1–5.

15.26 N. Megiddo, "Towards a Genuinely Polynomial Algorithm for Linear Programming," Technical Report, Statistics Department, Tel Aviv University, Tel Aviv, Israel, 1981.

15.27 N. Megiddo, "Solving Linear Programming in Linear-time When the Dimension Is Fixed," Technical Report, Statistics Department, Tel Aviv University, Tel Aviv, Israel, 1982.

15.28 M. W. Padberg and M. R. Rao, "The Russian Method for Linear Inequalities," and "The Russian Method for Inequalities II: Approximate Arithmetic," Graduate School of Business Administration, New York University, New York, 1980.

15.29 W. Rudin, *Principles of Mathematical Analysis*, 3rd ed., McGraw-Hill, New York, 1976.

15.30 N. Z. Shor, "On the Structure of Algorithms for the Numerical Solution of Optimal Planning and Design Problems," Dissertation, Cybernetics Institute, Academy of Sciences of the Ukrainian SSR, Kiev, Russia, 1964.

15.31 N. Z. Shor, "The Rate of Convergence of the Generalized Gradient Descent Method," *Kibernetika 4*, 3, 1968, 96–99, translated in *Cybernetics 4*, 3 (1968), 79–80.

15.32 N. Z. Shor, "Utilization of the Operation of Space Dilatation in the Minimization of Convex Functions," *Kibernetika 6*, 1, 1970, 6–12, translated in *Cybernetics 6*, 1 (1970), 7–15.

15.33 N. Z. Shor, "Convergence Rate of the Gradient Descent Method with Dilatation of the Space," *Kibernetika 6*, 2, 1970, 80–85, translated in *Cybernetics 6*, 2, 1970, 102–108.

15.34 N. Z. Shor, "Generalized Gradient Methods of Nondifferentiable Function Minimization and their Application to Problems of Mathematical Programming," *Ekonomika i Matematicheskie Metody 12* (1976), 337–356.

15.35 N. Z. Shor, "Cut-off Method with Space Extension in Convex Programming Problems," *Kibernetika 13*, 1, 1977, 94–95, translated in *Cybernetics 13*, 1, (1977), 94–96.

15.36 N. Z. Shor, "New Development Trends in Nondifferentiable Optimization," *Kibernetika 13*, 6, 1977, 87–91, translated in *Cybernetics 13*, 6 (1977), 881–896.

15.37 N. Z. Shor and N. G. Zhurbenko, "A Minimization Method Utilizing the Operation of Space Expansion in the Direction of the Difference of Two Successive Gradients, *Kibernetika 7*, 3, 1971, 51–59, translated in *Cybernetics* 7 (3), 1971.

15.38 D. M. Y. Sommerville, *An Introduction to the Geometry of* N-*Dimensions*, Dover, New York, 1958.

15.39 M. S. Spivak, *Calculus on Manifolds*; *A Modern Approach to Classical Theorems of Advanced Calculus*, Benjamin, New York 1965.

15.40 M. J. Todd, "On Minimum Volume Ellipsoids Containing Part of a Given Ellipsoid," *Mathematics of Operations Research*, 7, 2 (May 1982), 253–261.

15.41 S. Ursic, The Ellipsoid Algorithm for Linear Inequalities in Exact Arithmetic, in "*Proceedings 23rd Annual Symposium on Foundations of Computer Science*," pp. 321–326, IEEE Computer Society, Long Beach, Calif., 1982.

15.42 B. Yamnitsky and L. A. Levin, An old linear programming algorithm runs in polynomial time, in "*Proceedings 23rd Annual Symposium on Foundations of Computer Science*," pp. 327–328, IEEE Computer Society, Long Beach, Calif., 1982.

chapter 16

Iterative Methods For Linear Inequalities and Linear Programs

16.1 INTRODUCTION

The name *iterative method* usually refers to a method that provides a simple formula for computing the $(r+1)$th point as an explicit function of the rth point: $x^{r+1} = f(x^r)$. The method begins with an initial point $x^0 \in \mathbf{R}^n$ (quite often x^0 can be chosen arbitrarily, subject to some simple constraints that may be specified, such as $x^0 \geqq 0$, etc.) and generates the sequence of points $\{x^0, x^1, x^2, \ldots\}$ one after the other using the above formula. The method can be terminated whenever one of the points in the sequence can be recognized as being a solution to the problem under consideration. If finite termination does not occur mathematically, the method has to be continued indefinitely. In some of these methods, it is possible to prove mathematically that the sequence $\{x^r\}$ converges in the limit to a solution of the problem under consideration, or it may be possible to prove that every accumulation point of the sequence $\{x^r\}$ is a solution of the problem. In practice, it is impossible to continue the method indefinitely. In such cases, the sequence is computed to some finite length, and the final solution accepted as an approximate solution of the problem.

So far we have discussed a variety of finite direct pivotal methods such as the simplex method and its variants for solving systems of linear inequalities or LPs. The tremendous practical success of these methods has made them very popular. However, over the years, a number of iterative methods (which are, in general, infinite) have been proposed for solving linear inequalities and linear programs. These iterative methods have never been very popular, and very few textbooks in linear programming even discuss them. This is in surprising contrast with textbooks in numerical linear algebra, most of which usually discuss several iterative methods for the solution of systems of linear equations. If iterative methods make sense for systems of linear equations, they should be of some interest for linear inequalities and linear programs too. We therefore present in this chapter a brief survey of some iterative methods for solving systems of linear inequalities and linear programs. These iterative methods may not be practically attractive in comparison with the simplex method for solving moderate-size LPs with nice special structure, but they do have some advantages worth considering. The iterative methods always work with the original data, do not rely on nondegeneracy or uniqueness of the solution, and do not need matrix inversion. The basic computational steps in iterative methods are extremely simple and very easy to program. The iterative methods are very stable, robust, simple, and seem to provide a viable alternative for tackling large-scale LPs with no special structure. One inconvenience of iterative methods lies in the fact that we have to know a priori that the system of linear inequalities under consideration is feasible (or that the linear program under consideration has an optimum solution). In practical applications, this does not appear to be a serious drawback.

Most of the algorithms in the nonlinear programming area are of the iterative type. Some of the iterative algorithms for linear programs discussed in this chapter are actually specializations of some nonlinear programming algorithm to the linear case.

16.2 RELAXATION METHODS

The relaxation methods for solving systems of linear inequalities are closely related to the relaxation methods for solving systems of linear equations (see references [16.14, 16.28]), and were developed in the early 1950s. These methods are iterative methods that generate a sequence of points. If one of the points in the sequence satisfies all the constraints, it is a feasible solution, and the method terminates. Otherwise, it generates the next point in the sequence, and continues in the same manner. Assuming that the set of feasible solution is nonempty, under some conditions on the system of constraints it can be proved that the method terminates after a finite number of steps with a feasible solution of the system. If the system of constraints does not satisfy these conditions it is possible that the method may continue indefinitely, but under weaker conditions it can be proved that it converges in the limit to a feasible solution.

The relaxation method provides an alternative to the simplex algorithm for solving systems of linear inequalities. By the methods discussed in Section 2.2., every system of linear constraints can be put in standard form, and then using the equations in the standard form, some variables can be eliminated (as discussed in Section 2.5.4), transforming the whole system into an equivalent system of linear inequalities. Also, by the results in Section 4.6.4, we know that every LP can be transformed into a problem of solving a system of linear inequalities. Using these results, every LP can therefore be solved directly by the methods discussed here, without the need for doing any optimization.

Orthogonal Projection of a Point on a Hyperplane

Let $a = (a_1, \ldots, a_n) \neq 0$ and $\mathbf{H} = \{x : ax = d\}$, a hyperplane. If $\bar{x} \notin \mathbf{H}$, the point in $\mathbf{H}$ nearest to $\bar{x}$ in terms of the Euclidean distance is known as the *orthogonal projection* (or just the *projection*) of $\bar{x}$ in $\mathbf{H}$. It is $\hat{x} = \bar{x} + a^{\mathrm{T}}(d - a\bar{x})/||a||^2$. It can be verified that for any $x \in \mathbf{H}$, we have $(\bar{x} - x)^{\mathrm{T}}(\hat{x} - \bar{x}) = 0$. So the straight line joining $\bar{x}$ and $\hat{x}$ is normal to the hyperplane at $\hat{x}$.

The Relaxation Method

Suppose the system to be solved, including any bound restrictions on the variables, is

$$Ax \geqq b \tag{16.1}$$

where A is a matrix of order $m \times n$. Without any loss of generality we assume that each row of A is normalized so that $||A_{i.}|| = 1$ for all $i = 1$ to m. Let $\mathbf{K}$ be the set of feasible solutions of (16.1). Choose the initial point $x^0 \in \mathbf{R}^n$ arbitrarily and go to Step 1.

General Step $r + 1$, for $r \geqq 0$ Let x^r be the present point. If $x^r \in \mathbf{K}$, terminate. If $x^r \notin \mathbf{K}$, the ith constraint in (16.1) is violated by it if $A_{i.}x^r < b_i$, and in this case $b_i - A_i x^r$ is known as the *residual* in this violated inequality at x^r. Select a violated inequality, say the pth. For this selection, three commonly used criteria are given below.

The maximum residual criterion: Here, the chosen pth constraint is the most violated; that is, $b_p - A_{p.}x^r = \text{maximum } \{b_i - A_{i.}x^r : i, \text{ such that } b_i - A_{i.}x^r > 0\}$.

The maximum distance criterion: Here, the selected pth constraint is the one for which the Euclidean distance between x^r and its orthogonal projection on the constraint hyperplane is maximum among all violated constraints. Under the normalization assumption on the rows of A, this criterion is the same as the maximum residual criterion. If the rows of A are not so scaled, it could be different.

The cyclical criterion: Here the constraints are arranged in a fixed order, say the $i_1, \ldots, i_m$th, at the beginning of the method. If the constraint selected in the previous step is the i_tth, in this step select the first violated constraint when the constraints are examined in the order $i_{t+1}, \ldots, i_m, i_1, \ldots, i_{t-1}$.

Let y^r be the orthogonal projection of x^r on the hyperplane of the selected pth constraint. So $y^r = x^r + (A_{p.})^{\mathrm{T}}(b_p - A_{p.}x^r)/||A_{p.}||^2$. Define

$$x^{r+1} = x^r + \lambda_r(y^r - x^r) \tag{16.2}$$

where λ_r is a positive number known as the *relaxation parameter* in this step (see Figure 16.1). With x^{r+1} as the next point in the sequence, go to the next step.

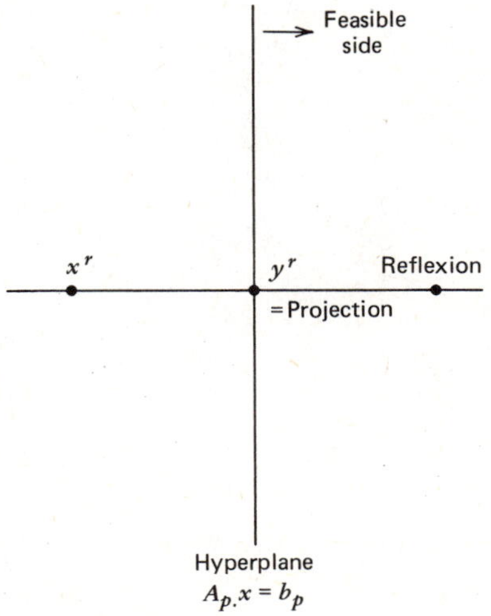

Figure 16.1

The relaxation parameter can be chosen to have different values in different steps. Usually, it is taken to be equal to a positive number λ, where $1 \leqq \lambda \leqq 2$ in all steps. In this case, the next point obtained satisfies the violated constraint, and the procedure is called *the relaxation method* because in each step, it attempts to satisfy only one violated constraint, ignoring or relaxing the others. If $\lambda_r = \lambda$ for all r, then λ is known as the *relaxation parameter*, and the method is normally called by different names, depending on the value of λ.

Value of λ	*Name of the Method*
$\lambda = 1$	Projection method
$\lambda = 2$	Reflexion method
$1 < \lambda < 2$	Overrelaxation method
$\lambda > 2$	Overreflexion method
$0 < \lambda < 1$	Underrelaxation method

It has been proved that if $\mathbf{K} \neq \emptyset$ and $\lambda = 1$, then the method either terminates with a point in $\mathbf{K}$ after a finite number of steps, or the sequence of points $\{x^r\}$ generated converges to a point in $\mathbf{K}$ in the limit. An example where finite termination may not occur is illustrated in Figure 16.2. Here we are trying to solve a system of two inequalities in $\mathbf{R}^2$, $A_{i.}x \geqq b_i$, $i = 1, 2$. On each hyperplane the feasible side is marked by an arrow in Figure 16.2. The initial point x^0 satisfies the second constraint as an equation, but violates the first. The projection method generates a sequence of points x^r, as marked in Figure 16.2, which converges in the limit to the feasible point x^*.

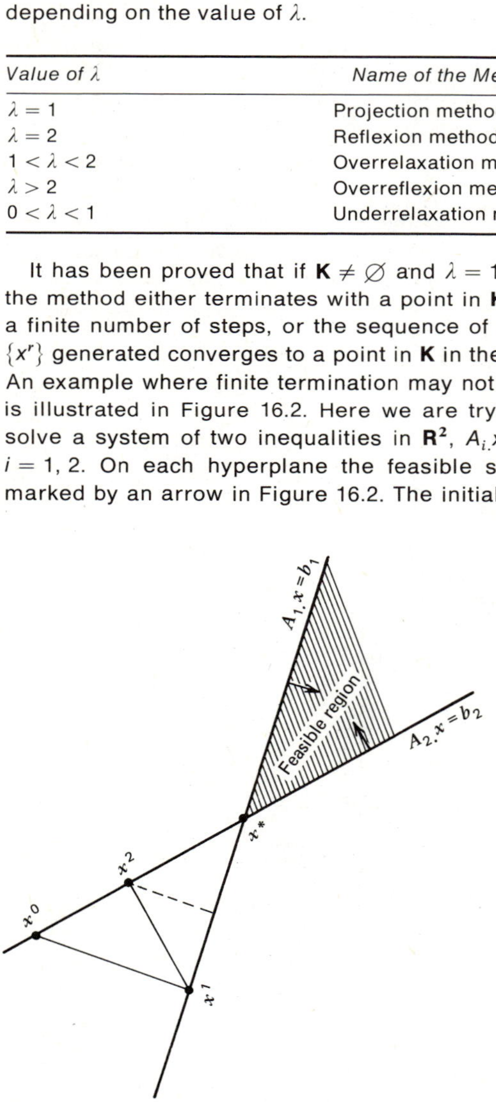

Figure 16.2 The convergence of the projection method.

Let $\alpha = \text{minimum } \{\|x^0 - x\| : x \in \mathbf{K}\}$. In the projection method, if $\{x^r\}$ converges to a point x^* in the limit, it can be proved that there exists a number θ satisfying $0 < \theta < 1$, such that $\|x^r - x^*\| \leqq 2\alpha\theta^r$ for all r, where θ depends only on the matrix A. Hence, the convergence of the sequence $\{x^r\}$ is geometric, which is known as *linear rate of convergence* in nonlinear programming literature.

If $\mathbf{K}$ has dimension n (i.e., $\mathbf{K}$ has full dimension), and if the method implemented with $\lambda_r = \lambda = 2$ for all r, then it can be proved that it terminates in a finite number of steps with a feasible solution (see [16.1, 16.2, 16.6, 16.11, 16.14, 16.16, 16.20, 16.24, 16.25, 16.26, 16.31]).

If it is not known whether $\mathbf{K} = \emptyset$ or not, transform (16.1) into a system of linear equations in nonnegative variables by the methods discussed in Section 2.2. Then construct the Phase I problem corresponding to it by introducing the necessary artificial variables. We know that the Phase I problem has a finite optimum solution. Using the results of Section 4.6.4, transform the Phase I problem into the problem of solving a system of linear inequalities. Any feasible solution of this final system of linear inequalities leads to an optimum solution of the Phase I problem. If all the artificial variables are zero in this solution, this leads to a feasible solution of (16.1). If at least one artificial variable is strictly positive in this solution, we can conclude that $\mathbf{K} = \emptyset$. The final system of linear inequalities has a nonempty set of feasible solutions, and the relaxation method can be used to solve it. However, all these transformations increase the number of constraints and the variables in the system, and thus make it very complicated and possibly impractical.

16.2.1 Finite Versions of the Relaxation Method

When the data in (16.1) are rational or integer, and m, $n \geqq 2$, it is possible to develop a version of the relaxation method that, in a finite number of iterations, finds a feasible solution of (16.1) if one exists, or concludes that it is infeasible. This has been shown by J. Telgen [16.35], and is based on the explicit use of the number theoretic arguments discussed in Chapter 15. The maximum number of iterations in the method can be determined in advance as a function of the size of the system. As a consequence of its finiteness, this version of the relaxation method can also determine the infeasibility of the system.

Assume first that all the data in (16.1) are made integer by multiplying both sides of each constraint by a suitable positive integer, if necessary. Let L be the size of this system, as defined in (15.2) of Section 15.2. Now rescale each constraint so that $A_{i.}A_{i.}^{\mathrm{T}} = 1$ for all i. In the sequel, (16.1) refers to this normalized system.

Given a point $\hat{x} \in \mathbf{R}^n$ and a positive number r, denote the solid ball with $\hat{x}$ as center and r as radius by $\mathbf{B}(\hat{x}, r) = \{x: (x - x)^{\mathrm{T}}(x - \hat{x}) \leqq r^2\}$. This version of the relaxation method can be interpreted as implicitly constructing a series of balls with decreasing radius (notice the similarity with the ellipsoid method of Chapter 15). Let $x^0 = 0$, and define the initial ball to be $\mathbf{B}_0 = \mathbf{B}(x^0, 2^{\mathbf{L}-1}/\sqrt{n})$ with radius $r_0 = 2^{\mathbf{L}-1}/\sqrt{n}$.

Given any $x \in \mathbf{R}^n$, define $\theta(x) = \text{maximum}\,\{-A_{i.}x + b_i: i = 1 \text{ to } m\}$. So, if x is feasible to (16.1), $\theta(x) \leqq 0$. If x is infeasible, $\theta(x)$ provides a (positive) measure of the infeasibility of the point x to the system (16.1). Let $\mathbf{K}$ denote the set of feasible solutions of (16.1). The results of Chapter 15 imply that $\mathbf{B}_0 \cap \mathbf{K} \neq \emptyset$ iff $\mathbf{K} \neq \emptyset$, and that if $\mathbf{K} = \emptyset$, $\theta(x) \geqq 2^{-(\mathbf{L}-1)}$ for all $x \in \mathbf{R}^n$.

Choose the relaxation parameter λ to satisfy $2^{-\mathbf{L}} \leqq \lambda \leqq 2 - 2^{-\mathbf{L}}$. The version of the relaxation method is given by the following iterative scheme. Given x^k, r_k, $\mathbf{B}_k = \mathbf{B}(x^k, r_k)$, apply the following stopping criteria.

1 If x^k is feasible to (16.1), terminate. If x^k is infeasible, but $\theta(x^k) < 2^{-(\mathbf{L}-1)}$, (16.1) has a feasible solution, determine it from x^k using the methods discussed in Theorem 15.7 in Scheme 2 of Section 15.3 and terminate.

2 If (1) does not hold and either $r_k^2 < 2^{-2(\mathbf{L}-1)}$ or $k \geqq \lceil 2^{4\mathbf{L}}/(n\lambda(2 - \lambda))\rceil$, (16.1) is infeasible, terminate.

If neither (1) nor (2) holds, let $1 \leqq \ell \leqq m$ be such that $\theta(x^k) = -A_{\ell.}x^k + b_\ell$. Determine

$$x^{k+1} = x^k + \lambda\theta(x^k)A_{\ell.}^{\mathrm{T}}$$
$$r_{k+1}^2 = r_k^2 - (\theta(x^k))^2\lambda(2 - \lambda)$$
$$\mathbf{B}_{k+1} = \mathbf{B}(x^{k+1}, r_{k+1})$$

with x^{k+1}, r_{k+1}, $\mathbf{B}_{k+1}$, go to the next iteration.

Proofs of Validity

THEOREM 16.1 For all k, if the method did not terminate by iteration k, $\mathbf{B}_{k+1} \supset \mathbf{B}_k \cap \{x: A_{\ell.}x \geqq b_\ell\}$, where ℓ is the index determined as in Step k.

Proof: Let $\bar{x} \in \mathbf{B}_k \cap \{x: A_{\ell.}x \geqq b_\ell\}$. We have to prove that $(x^{k+1} - \bar{x})^{\mathrm{T}}(x^{k+1} - \bar{x}) - r_{k+1}^2 \leqq 0$. Substituting for x^{k+1}, r_{k+1} in terms of x^k, r_k, and using the facts that $A_{\ell.}A_{\ell.}^{\mathrm{T}} = 1$ (by our normalizing assumption.), $(x^k - \bar{x})^{\mathrm{T}}(x^k - \bar{x}) - r_k^2 \leqq 0$ (since $\bar{x} \in \mathbf{B}_k$), $\theta(x^k) \geqq 0$ (since x^k is infeasible to (16.1)), and $b_\ell - A_{\ell.}\bar{x} \leqq 0$ (since $\bar{x} \in \{x: A_{\ell.}x \geqq b_\ell\}$), we have $(x^{k+1} - \bar{x})^{\mathrm{T}}(x^{k+1} - \bar{x}) - r_{k+1}^2 = (x^k - \bar{x})^{\mathrm{T}}(x^k - \bar{x}) + 2\lambda\theta(x^k)A_{\ell.}(x^k - \bar{x}) + \lambda^2(\theta(x^k))^2 - r_k^2 + (\theta(x^k))^2\lambda(2 - \lambda) = [(x^k - \bar{x})^{\mathrm{T}}(x^k - \bar{x}) - r_k^2] + 2\lambda\theta(x^k)(b_\ell - A_{\ell.}\bar{x}) \leqq 2\lambda\theta(x^k)(b_\ell - A_{\ell.}\bar{x}) \leqq 0$. Hence the result. ∎

THEOREM 16.2 If termination did not occur by Step k and $r_k^2 \leqq 2^{-2(\mathbf{L}-1)}$, (16.1) is infeasible.

Proof: If $\mathbf{K} \neq \emptyset$, by the results of Chapter 15 and the choice of $\mathbf{B}_0$, $\mathbf{B}_0 \cap \mathbf{K} \neq \emptyset$. So by applying Theorem 16.1 repeatedly, $\emptyset \neq \mathbf{B}_k \cap \{x: A_{\ell.}x \geqq b_\ell\} \supset \mathbf{B}_0 \cap \mathbf{K}$, if $\mathbf{K} \neq \emptyset$. Since termination did not occur in iteration k we have $\theta(x^k) > 2^{-(\mathbf{L}-1)}$. By the definition of ℓ in Step k, $b_\ell - A_{\ell.}x^k = \theta(x^k) > 2^{-(\mathbf{L}-1)}$. Since $A_{\ell.}A_{\ell.}^{\mathrm{T}} = 1$ by our normalization assumption, it can be verified that the maximum value of $A_{\ell.}x$ over $x \in \mathbf{B}_k$ is $A_{\ell.}x^k + r_k$. So the maximum value of $A_{\ell.}x^k - b_\ell + r_k$ over $x \in \mathbf{B}_k$ is $A_{\ell.}x^k - b_\ell + r_k \leqq r_k - \theta(x^k) < r_k - 2^{-(\mathbf{L}-1)} < 0$ if $r_k^2 \leqq 2^{-2(\mathbf{L}-1)}$. So if $r_k^2 \leqq 2^{-(\mathbf{L}-1)}$, $\mathbf{B}_k \cap \{x: A_{\ell.}x \geqq b_\ell\} = \emptyset$, which implies that $\mathbf{K} = \emptyset$. ∎

THEOREM 16.3 If termination did not occur by iteration $\mathrm{N} = \lceil 2^{4\mathbf{L}}/(n\lambda(2 - \lambda))\rceil$, $\mathbf{K} = \emptyset$.

Proof: If the method goes from iteration k to $k + 1$, we must have $\theta(x^k) > 2^{-(\mathbf{L}-1)}$, and hence $r_{k+1}^2 < r_k^2 - \lambda(2 - \lambda)2^{-2(\mathbf{L}-1)}$. So, each time the method moves from one iteration to the next, the square of the radius of the ball reduces by at least $\lambda(2 - \lambda)2^{-2(\mathbf{L}-1)}$. Also, from the initial choice of $\mathbf{B}_0$, $r_0^2 = 2^{2(\mathbf{L}-1)}/n$. So if the method goes to iteration N without termination, we will have $r_{\mathrm{N}}^2 < 2^{-2(\mathbf{L}-1)}$, which by Theorem 16.2 implies that $\mathbf{K} = \emptyset$. ∎

The convergence of the algorithm is fastest if the reduction in the radius r_k^2 is maximized, that is, if $\lambda = 1$. Also, on infeasible systems, the condition $r_k^2 \leqq 2^{-2(\mathbf{L}-1)}$ may be satisfied long before the Nth iteration. For example, consider the system in $\mathbf{R}^2$: $-x_2 \geqq 0$, $x_2 \geqq 4$. Here, even though N is large, the method can be verified to terminate in 2 iterations.

So, even on infeasible problems, the maximum number of iterations may not be required. Also, the upper bound on the number of iterations in this method is not sufficient to make the method polynomially bounded. Consider the system in $\mathbf{R}^2$: $x_1 - 2^\alpha x_2 \geqq 2^\alpha$, $x_2 \geqq 0$, where α is a positive integer. So $L = 2\alpha + 4$. Taking $\lambda = 1$, it can be verified that generally for k even, $x^{k+2} = (x_1^k + (2^\alpha - x_1^k)/(2^{2\alpha} + 1), \quad 0)^{\mathrm{T}}$. The method stops by satisfying criterion (1). A necessary condition for (1) to hold is $x_1^k > 2^\alpha - 1$. But since the x_1 coordinate increases by at most $2^\alpha/(1 + 2^{2\alpha})$, it can be verified that the method takes at least $2^{2\alpha}$ iterations. Since $L = 2\alpha + 4$, by increasing α among positive integers, we have problem instances on which this method exhibits exponential growth in the number of iterations as a function of problem size. The same worst case behavior can be shown to hold for any λ.

16.2.2 Extensions to Concave Constraints

The relaxation method can be extended directly to find a point $x \in \mathbf{R}^n$ satisfying a system of inequalities, $g_i(x) \geqq 0$, $i = 1$ to m, where each $g_i(x)$ is either an affine function or a nonlinear concave function defined on $\mathbf{R}^n$. See references [16.4, 16.5, 16.7, 16.8, 16.12, 16.13, 16.15, 16.17, 16.18, 16.19, 16.32] for a discussion of various relaxation methods and their convergence properties.

16.3 ITERATIVE METHODS BASED ON A QUADRATIC PERTURBATION

Consider the LP

$$\begin{aligned} \text{Minimize} \quad & cx \\ \text{Subject to} \quad & Ax \geqq b \\ & Fx = d \end{aligned} \tag{16.3}$$

where A and F are given matrices of order $m \times n$ and $k \times n$, respectively. The constraints in (16.3) include all the lower- and upper-bound restrictions on individual variables, if any. Consider the following quadratic perturbation of (16.3):

$$\begin{aligned} \text{Minimize} \quad & (\varepsilon/2)x^{\mathrm{T}}x + cx \\ \text{Subject to} \quad & Ax \geqq b \\ & Fx = d \end{aligned} \tag{16.4}$$

where ε is a positive-valued parameter. It has been shown by O. L. Mangasarian and R. R. Meyer in [16.21] that if (16.3) has an optimum solution, there exists $\bar{\varepsilon} > 0$, such that whenever $0 < \varepsilon < \bar{\varepsilon}$, the unique optimum solution of (16.4) is independent of ε and is also optimum to (16.3). Therefore we can solve (16.3) by solving (16.4) with an appropriately chosen value for ε. Problem (16.4) can be solved by the iterative scheme for convex quadratic programs developed by O. L. Mangasarian in [16.22, 16.23]. This iterative scheme tries to compute the optimum Lagrange multiplier vectors $u = (u_1, \ldots, u_m)^{\mathrm{T}}$, $v = (v_1, \ldots, v_k)^{\mathrm{T}}$ associated with (16.4), and from them the optimum solution for (16.3) is obtained using $x = \varepsilon^{-1}(A^{\mathrm{T}}u + F^{\mathrm{T}}v - c^{\mathrm{T}})$.

If $B = (b_{ij})$ is any square matrix of order N, define the diagonal matrix with diagonal entries $b_{11}, \ldots, b_{\mathrm{NN}}$ to be the *diagonal part of B*. For $j = 1$ to N, the matrix whose jth column is $(0, \ldots, 0, b_{j+1,j}, \ldots, b_{\mathrm{N}j})^{\mathrm{T}}$ is known as the *strict lower triangular part of B*. For $j = 1$ to N, the matrix whose jth column is $(b_{1j}, b_{2j}, \ldots, b_{j-1,j}, 0, \ldots, 0)^{\mathrm{T}}$ is known as the *strict upper triangular part of B*.

Let $y = (y_j) \in \mathbf{R}^{\mathrm{N}}$. For each j, let $y_j^+ =$ maximum $\{0, y_j\}$ and let y^+ be the column vector (y_j^+). It can be verified that y^+ is the nearest point in the nonnegative orthant of $\mathbf{R}^{\mathrm{N}}$ to y.

The Iterative Scheme for (16.4)

Choose a positive value for ε and an initial point (u^0, v^0), where $u^0 \geqq 0$, $u^0 \in \mathbf{R}^m$, $v^0 \in \mathbf{R}^k$. Given (u^r, v^r), determine (u^{r+1}, v^{r+1}) from

$$\begin{pmatrix} u^{r+1} \\ \cdots \\ v^{r+1} \end{pmatrix} = \left(\begin{pmatrix} u^r \\ \cdots \\ v^r \end{pmatrix} - \omega G^{-1} \left(DD^{\mathrm{T}} \begin{pmatrix} u^r \\ \cdots \\ v^r \end{pmatrix} - Dc^{\mathrm{T}} - \varepsilon \begin{pmatrix} b \\ \cdots \\ d \end{pmatrix} + K^r \begin{pmatrix} u^{r+1} - u^r \\ \cdots\cdots \\ v^{r+1} - v^r \end{pmatrix} \right) \right)^* \tag{16.5}$$

where $D = \begin{pmatrix} A \\ \cdots \\ F \end{pmatrix}$; G is the diagonal part of DD^{T}; K^r is either L or U, L, U being the strict lower triangular part and the strict upper triangular part of DD^{T}; ω is a parameter satisfying $0 < \omega < 2$; and $\begin{pmatrix} u \\ \cdots \\ v \end{pmatrix}^* = \begin{pmatrix} u+ \\ \cdots \\ v \end{pmatrix}$.

If $K^r = L$ in (16.5), (u^{r+1}, v^{r+1}) can be computed easily in the order $u_1^{r+1}, \ldots, u_m^{r+1}, v_1^{r+1}, \ldots, v_k^{r+1}$. When K^r is chosen to be equal to U in (16.5), (u^{r+1}, v^{r+1}) can be computed easily in the order $v_k^{r+1}, \ldots, v_1^{r+1}, u_m^{r+1}, \ldots, u_1^{r+1}$. K^r can be chosen to be equal to L for all r, or equal to U for all r, or we can alternate by making $K^r = L$ for odd r and $K^r = U$ for even r. This leads to different versions of the basic iterative scheme (16.5), all of which work well.

In [16.22, 16.23], O. L. Mangasarian has proved that if (16.3) has an optimum solution, there exists $\bar{\varepsilon} > 0$, such that for any $0 < \varepsilon < \bar{\varepsilon}$, each accumulation point $(\bar{u}, \bar{v})$ of the sequence (u^r, v^r) generated by (16.5) leads to an optimum solution of (16.3) through the formula $x = \varepsilon^{-1}(A^T\bar{u} + F^T\bar{v} - c^T)$. He has also proved that if (16.3) has an optimum solution, the system $Ax > b$, $Fx = d$ has a solution $\hat{x}$, and the rows of F form a linearly independent set, then the sequence (u^r, v^r) generated by (16.5) has at least one accumulation point.

Numerical Implementation of this Scheme

In [16.22], Mangasarian reports on the comparative performance of this scheme with the revised simplex method in numerical tests performed on randomly generated dense LPs. It is reported that if the parameters ε, ω are properly chosen, this iterative scheme is competitive with the revised simplex method, and may even be more robust, since it can solve dense unstructed and possibly degenerate problem on which the revised simplex codes perform poorly. Numerical experience indicates that values of ε in the range m to n, and ω in the range 0.5 to 1.5 seem to lead to good results. More numerical experimentation is necessary to determine the optimal choice for ε and ω.

16.4 SPARSITY-PRESERVING ITERATIVE SOR ALGORITHMS FOR LINEAR PROGRAMMING

The iterative successive overrelaxation (SOR) methods discussed in Section 16.3 for LPs require the product of the constraint matrix by its transpose, which can cause the loss of any sparsity that the original problem may have had. So those methods are not suitable for solving large sparse LPs. Here we present sparsity-preserving iterative SOR methods for LPs, which do not require the multiplication of the constraint matrix by its transpose, due to Mangasarian [16.23]. We consider the LP

$$\begin{aligned} \text{Minimize} \quad & cx \\ \text{Subject to} \quad & Ax \geqq b \\ & x \geqq 0 \end{aligned} \tag{16.6}$$

where A is a matrix of order $m \times n$. To avoid trivial cases, we assume that each row and column of A, as well as the vectors b and c, are all nonzero. Let $\pi = (\pi_1, \ldots, \pi_m)$ denote the row vector of dual variables associated with the constraints in (16.6). Let $k = m + n$, and

$$M = \begin{pmatrix} 0 & \vdots & -A^T \\ \cdots & \cdots & \cdots \\ A & \vdots & 0 \end{pmatrix} \qquad q = \begin{pmatrix} c^T \\ \cdots \\ -b \end{pmatrix} \qquad y = \begin{pmatrix} x \\ \cdots \\ \pi^T \end{pmatrix} \tag{16.7}$$

From the results in Sections 4.6.4 and 4.6.5, it is clear that solving (16.6) is equivalent to solving $My + q \geqq 0$, $y \geqq 0$, $q^Ty \leqq 0$. This system may have many solutions, and one way of identifying a specific solution to this system is to solve the quadratic program: minimize $(1/2)y^Ty$, subject to $My + q \geqq 0$, $y \geqq 0$, $-q^Ty \geqq 0$. This gives the optimal y of least norm. Associate the Lagrange multiplier vectors $s \in \mathbf{R}^k$, $t \in \mathbf{R}^k$, $\beta \in \mathbf{R}^1$ to the constraints in this quadratic program. The iterative method discussed below tries to obtain the optimal Lagrange multiplier vectors (s, t, β). From these, the optimal y can be obtained using $y = M^Ts + t - \beta q$.

The iterative method is the following. Choose the initial point $(s^0, t^0, \beta^0) \geqq 0$ arbitrarily, and $0 < \omega < 2$. For $t = 0, 1, 2, \ldots$, given (s^r, t^r, β^r), compute $(s^{r+1}, t^{r+1}, \beta^{r+1})$ as in (16.8).

In [16.23], Mangasarian has proved that if (s, t, β) is any accumulation point of the sequence $\{(s^r, t^r, \beta^r)\}$ generated by (16.8), then the corresponding y obtained as above is the vector of optimum primal and dual solutions associated with (16.6). He has also derived some sufficient conditions under which the

$$\begin{aligned} s_1^{r+1} &= \left(s_1^r - \frac{\omega}{\|M_{1.}\|^2}\left(M_{1.}(M^Ts^r + t^r - \beta^r q) + q_1\right)\right)^+ \\ s_i^{r+1} &= \left(s_i^r - \frac{\omega}{\|M_{i.}\|^2}\left(M_{i.}\left(\sum_{\ell=1}^{i-1}(M_{\ell.})^Ts_\ell^{r+1} + \sum_{\ell=i}^{k}(M_{\ell.})^Ts_\ell^r + t^r - \beta^r q\right) + q_i\right)\right)^+ \qquad i = 2 \text{ to } k \\ t^{r+1} &= (t^r - \omega(M^Ts^{r+1} + t^r - \beta^r q))^+ \\ \beta^{r+1} &= \left(\beta^r + \frac{\omega}{\|q\|^2} q^T(M^Ts^{r+1} + t^{r+1} - \beta^r q)\right)^+ \end{aligned} \tag{16.8}$$

$$u_1^{r+1} = \left(u_1^r - \frac{\omega}{\|A_{1.}\|^2}(A_{1.}(A^{\mathrm{T}}u^r + v^r - c^{\mathrm{T}}) - \varepsilon b_1)\right)^+$$

$$u_i^{r+1} = \left(u_i^r - \frac{\omega}{\|A_{i.}\|^2}\left(A_{i.}\left(\sum_{\ell=1}^{i-1}(A_{\ell.})^{\mathrm{T}}u_\ell^{r+1} + \sum_{\ell=i}^{m}(A_{\ell.})^{\mathrm{T}}u_\ell^r + v^r - c^{\mathrm{T}}\right) - \varepsilon b_i\right)\right)^+ \qquad i = 2 \text{ to } m \qquad (16.10)$$

$$v^{r+1} = (v^r - \omega(A^{\mathrm{T}}u^{r+1} + v^r - c^{\mathrm{T}}))^+$$

sequence $\{(s^r, t^r, \beta^r)\}$ will have an accumulation point.

Another sparsity-preserving version of the iterative SOR method can be developed for solving (16.6) using a quadratic perturbation of it, as discussed in Section 16.3. For this, we consider the quadratic program

$$\begin{aligned} \text{Minimize} \quad & cx + \frac{\varepsilon}{2}x^{\mathrm{T}}x \\ \text{Subject to} \quad & Ax \geqq b \qquad (16.9) \\ & x \geqq 0 \end{aligned}$$

where ε is a small positive number. Let $u \in \mathbf{R}^m$, $v \in \mathbf{R}^n$ be the column vectors of Lagrange multipliers associated with the constraints and sign restrictions in (16.9), respectively. As before, the iterative method tries to obtain the optimum (u, v), and the optimum x is then computed using $x = (1/\varepsilon)\,(A^{\mathrm{T}}u + v - c^{\mathrm{T}})$. Choose $(u^0, v^0) \geqq 0$ arbitrarily, and $0 < \omega < 2$. Given (u^r, v^r), determine (u^{r+1}, v^{r+1}) by (16.10).

It can be shown that if (16.6) has an optimum solution, there exists an $\bar{\varepsilon} > 0$, such that for any $0 < \varepsilon < \bar{\varepsilon}$, each accumulation point of the sequence $\{(u^r, v^r)\}$ generated by (16.10) leads by the above formula to an optimum solution of (16.6). Also, if (16.6) has an optimum solution and $\{x\colon Ax > b, x > 0\} \neq \emptyset$, the sequence $\{(u^r, v^r)\}$ generated by (16.10) can be proved to have at least one accumulation point (see [16.23]).

Unlike the algorithms discussed in Section 16.3, these algorithms do not need AA^{T}, and therefore preserve any sparsity that the matrix A may have.

16.5 AN ITERATIVE METHOD FOR SOLVING A PRIMAL DUAL PAIR OF LINEAR PROGRAMS

This method is due to W. Oettli [16.27]. We consider the LP (16.6). In the notation of Section 16.4, solving (16.6) is equivalent to solving $My + q \geqq 0$ and $q^{\mathrm{T}}y \leqq 0$, $y \geqq 0$. Let $\mathbf{\Gamma} = \{y\colon y \in \mathbf{R}^k, q^{\mathrm{T}}y \leqq 0, y \geqq 0\}$. Given any point $y \in \mathbf{R}^k$, denote by $\mathrm{P}_{\mathbf{\Gamma}}(y)$ the nearest point in $\mathbf{\Gamma}$ to y. It can be shown that $\mathrm{P}_{\mathbf{\Gamma}}(y) = (y - \lambda_0 q)^+$, where λ_0 is the smallest nonnegative value of the parameter λ for which the piecewise linear monotonically decreasing function $q^{\mathrm{T}}(y - \lambda q)^+$ assumes a nonpositive value.

For each $i = 1$ to k, define $\ell_i(y) = \text{minimum}\,\{0, M_{i.}y + q_i\}$, and let $\ell(y) = (\ell_1(y), \ldots, \ell_k(y))^{\mathrm{T}}$. Let $\phi(y) = \|\ell(y)\|$. Let $E(y)$ be the square matrix of order k, where $E_{i.} = M_{i.}$ if $M_{i.}y + q_i \leqq 0$, or 0 otherwise, for $i = 1$ to k. Let $s(y) = (s_1(y), \ldots, s_k(y))^{\mathrm{T}} = (E(y))^{\mathrm{T}}\ell(y)/\|\ell(y)\|$, defined only for y satisfying $\ell(y) \neq 0$. Let $\rho(y) = \phi(y)/\|s(y)\|^2$, defined for $y \in \mathbf{\Gamma}\setminus\{y\colon My + q \geqq 0\}$.

The iterative scheme is the following. Choose $y^0 \in \mathbf{\Gamma}$ arbitrarily as the initial point. Given y^r, if $My^r + q \geqq 0$, terminate, as y^r is the point we are seeking. If $My^r + q \ngeqq 0$, determine $y^{r+1} = \mathrm{P}_{\mathbf{\Gamma}}(y^r - \rho(y^r)s(y^r))$, and continue.

In [16.27], Oettli proved that if (16.6) has an optimum solution and this scheme does not terminate in a finite number of steps, then the sequence generated $\{y^r\}$ converges in the limit to a point $y^* \in \{y\colon y \in \mathbf{\Gamma}, My + q \geqq 0\}$ with a linear rate of convergence. The other tangible aspects of this method are that it is stable, self-correcting, always works with the original data and does not rely on nondegeneracy or uniqueness of the solution. Also, any practical information about the approximate location of an optimum solution can be used here profitably. This property can come in very handy if we have to solve several LPs with only minor changes in the data.

Exercises

16.1 Let A be a matrix of order $m \times n$, with at most two nonzero entries per row, all the entries in which are 0, +1, or −1. Consider the system of linear inequalities $Ax \leqq b$, where b is an integer vector. If this system has a feasible solution, prove that it has a feasible solution that is integral. [*Hint:* Verify the result trivially for $m = 1$. Use the fact that when this method is solved by the reflection method, starting with a feasible integer solution for one constraint, the method converges to an integer solution.]

—[*B. C. Eaves*]

REFERENCES

16.1 S. Agmon, "The Relaxation Method for Linear Inequalities," *Canadian Journal of Mathematics 6* (1954), 382–392.

16.2 D. N. de G. Allen, *Relaxation Methods in Engineering and Science*, McGraw-Hill, New York, 1954.

16.3 D. P. Bertsekas, "Necessary and Sufficient Conditions for a Penalty Method to be Exact," *Mathematical Programming 9* (1975), 87–99.

16.4 L. M. Bregman, "The Method of Successive Projection for Finding a Common Point of Convex Sets," *Soviet Mathematics Doklady* 6, (translation), *6* (1965), 688–692.

16.5 L. M. Bregman, "The Relaxation Method of Finding the Common Point of Convex Sets and Its Application to the Solution of Problems in Convex Programming," *U.S.S.R. Computational Mathematics and Mathematical Physics 3* (1967), 200–217.

16.6 H. D. Block and S. A. Levin, "On the Boundedness of an Iterative Procedure for Solving a System of Linear Inequalities," in *Proceedings of the American Mathematical Society 26* (1970) 229–235.

16.7 P. M. Camerini, L. Fratta, and F. Maffioli, "On Improving Relaxation Methods by Modified Gradient Techniques," in *Non-differentiable Optimization, Mathematical Programming, Study 3*, M. L. Balinski and P. Wolfe (eds.), 1975, 26–34.

16.8 Y. C. Cheng, "Iterative Methods for Solving Linear Complementarity and Linear Programming Problems," Ph.D. dissertation, Computer Science Department, University of Wisconsin, Madison, August 1981.

16,9 V. F. Densyanov and A. M. Rubinov, *Approximate Methods in Optimization Problems*, American Elservier, New York, 1970.

16.10 I. I. Eremin, "An Iterative Method for Chebychev Approximations of Incompatiable Systems of Linear Inequalities," *Soviet Mathematics Doklady 3* (1962), 570–572.

16.11 I. I. Eremin, "A Generalization of the Motzkin-Agmon Relaxation Method," *Uspekhi Matematicheskikh Nauk 20*, 2 (in Russian) (1965) 183–187.

16.12 I. I. Eremin, "The Relaxation Method of Solving Systems of Inequalities with Convex Functions on the Left Sides," *Soviet Mathematics Doklady 6*, (translation), (1965), 219–222.

16.13 I. I. Eremin and V. D. Mazurov, "Iterative Methods for Solving Convex Programming Problems," *Soviet Physics Doklady 11* (1967), 757–759.

16.14 D. K. Faddeev and V. N. Faddeeva, "Computational Methods of Linear Algebra," translated by R. C. Williams, Freeman, San Francisco, Calif., 1963.

16.15 M. A. Germanov and V. S. Spiridonov, "On a Method of Solving Systems of Non-linear Inequalities," *U.S.S.R. Computational Mathematics and Mathematical Physics 2* (1966), 194–196.

16.16 J. L. Goffin, "On the Finite Convergence of the Relaxation Method for Solving Systems of Linear Inequalities" (Ph.D. dissertation), ORC 71–36, Operations Research Center, University of California, Berkeley, December 1971.

16.17 J. L. Goffin, "On the Convergence Rates of Subgradient Optimization Methods," *Mathematical Programming 13* (1977), 329–347.

16.18 J. L. Goffin, "Nondifferentiable Optimization and the Relaxation Method," in *Nonsmooth Optimization* (Proceedings of a IIASA Workshop) C. Lemarechal and R. Mifflin (Eds.), Pergamon, New York, November 1978.

16.19 L. G. Gubin, B. T. Polyok and E. V. Raik, "The Method of Projections for Finding the Common Point of Convex Sets," *U.S.S.R. Computational Mathematics and Mathematical Physics 6* (1967), 1–24.

16.20 R. G. Jeroslow, "Some Relaxation Methods for Linear Inequalities," *Cahiers du C.E.R.O. 21*, 1 (1979), 43–53.

16.21 O. L. Mangasarian and R. R. Meyer, "Nonlinear Perturbation of Linear Programs," *SIAM Journal on Control and Optimization 17* (1979), 743–752.

16.22 O. L. Mangasarian, "Iterative Solutions of Linear Programs," *SIAM Journal on Numerical Analysis 18*, 4 (August 1981), 606–614.

16.23 O. L. Mangasarian, "Sparsity Preserving SOR Algorithms for Separable Quadratic and Linear Programming," Computer Science Technical Report No. 438, University of Wisconsin, Madison, July 1981. (To be published in the *Proceedings of the Third Symposium on Mathematical Programming with Data Perturbations,* A. V. Fiacco (Ed.) May 1981). (Washington, D.C.)

16.24 J. F. Maurras, K. Truemper, and M. Akgül, "Polynomial Algorithms for a Class of Linear Programs," *Mathematical Programming 21* (1981), 121–136.

16.25 Y. I. Merzlyokov, "On a Relaxation Method of Solving Systems of Linear Inequalities," *U.S.S.R. Computational Mathematics and Mathematical Physics* (1964), 504–510.

16.26 T. Motzkin and I. J. Schoenberg, "The Relaxation Method for Linear Inequalities," *Canadian Journal of Mathematics 6* (1954), 393–404.

16.27 W. Oettli, "An Iterative Method, Having Linear Rate of Convergence, for Solving a Pair of Dual Linear Programs," *Mathematical Programming 3*, 3 (December 1972), 302–311.

16.28 J. M. Ortega and W. C. Rheinboldt, *Iterative Solution of Nonlinear Equations in Several Variables*, Academic, New York, 1970.

16.29 B. T. Polyak, "Gradient Methods for Solving Equations and Inequalities," *U.S.S.R. Computational Mathematics and Mathematical Physics 4*, 6 (1964), 17–82.

16.30 B. T. Polyak, "Some Methods for Speeding up the Convergence of Iteration Methods," *U.S.S.R. Computational Mathematics and Mathematical Physics 5* (1964), 1–17.

16.31 B. T. Polyak and N. V. Tretiyakov, "Concerning an Iterative Method for Linear Programming and Its Economic Interpretation," *Economics and Mathematical Methods 8* (in Russian) (1972), 740–751.

16.32 S. Schlechter, "Relaxation Methods for Convex Problems," *SIAM Journal of Numerical Analysis 5*, 3 (1968), 601–612.

16.33 N. Z. Shor, "The Rate of Convergence of the Generalized Gradient Descent Method," *Cybernetics 4*, 3 (1968), 79–80.

16.34 N. Z. Shor, "Generalized Gradient Methods for Non-Smooth Functions and Their Applications to Mathematical Programming Problems," *Ekonomika i Matematicheskie Metody 12*, 2 (in Russian) (1976), 337–356.

16.35 J. Telgen, "On Relaxation Methods for Systems of Linear Inequalities," *European Journal of Operational Research 9* (1982), 184–189.

Vector Minima

chapter 17

17.1 DEFINITIONS AND PROPERTIES OF VECTOR MINIMA

Consider a multiobjective LP. Express the constraints in standard form. Express each objective function in minimization form and eliminate any constant term in it. After these changes, let $c^r x$, $r = 1$ to t be the various objectives, $t \geqq 2$. Let C be the $t \times n$ matrix whose rows are $c^1, \ldots, c^t$. Let $\mathbf{K}$ be the set of feasible solutions. As defined in Section 1.4, $x^0 \in \mathbf{K}$ is said to be a *vector minimum*, (VM) (or an *efficient point*, or a *pareto-optimal point*, or a *nondominated feasible solution*) if there exists no $x \in \mathbf{K}$ satisfying $Cx \leq Cx^0$ (i.e., $c^r x \leqq c^r x^0$ for all $r = 1$ to t and $c^r x < c^r x^0$ for at least one r). The problem is (17.1) given below, where A is a matrix of order $m \times n$. The set of all VM for (17.1) is known as the *efficient frontier* for (17.1)

Vector Minimum for	Cx	
Subject to	$Ax = b$	$x \geqq 0$ (17.1)

THEOREM 17.1 $\bar{x} \in \mathbf{K}$ is a VM for (17.1) iff for each $r = 1$ to t, $\bar{x}$ is an optimum solution of the LP: minimize $c^r x$, subject to $Ax = b$; $c^p x \leqq c^p \bar{x}$ for $p = 1$ to t, $p \neq r$; $x \geqq 0$.

Proof: Follows from the definition of VM. This theorem is taken from the paper [17.15] of C. P. Simon. ■

THEOREM 17.2 Let $\lambda = (\lambda_1, \ldots, \lambda_t) > 0$, $w_\lambda(x) = \lambda C x = \sum_{r=1}^{t} \lambda_r c^r x$. If $\bar{x} \in \mathbf{K}$ minimizes $w_\lambda(x)$ over $x \in \mathbf{K}$, $\bar{x}$ is a VM for (17.1).

Proof: Proof by contradiction. If $\bar{x}$ is not a VM, there must exist an $\hat{x} \in \mathbf{K}$ satisfying $C\hat{x} \leq C\bar{x}$. This implies $w_\lambda(\hat{x}) < w_\lambda(\bar{x})$, a contradiction. So $\bar{x}$ must be a VM for (17.1). ■

THEOREM 17.3 If $\bar{x}$ is a VM for (17.1), there exists a vector $\lambda = (\lambda_1, \ldots, \lambda_t) > 0$, such that $\bar{x}$ minimizes $w_\lambda(x) = \lambda C x$ over $x \in \mathbf{K}$.

Proof: Since $\bar{x}$ is a VM, for each $r = 1$ to t, $\bar{x}$ is optimal for the LP given in Theorem 17.1. By the duality theorem, there exist $\pi^r = (\pi_1^r, \ldots, \pi_m^r)$ and $(\mu_1^r, \ldots, \mu_{r-1}^r, \mu_{r+1}^r, \ldots, \mu_t^r)$, satisfying $\mu_p^r \geqq 0$, $p = 1$ to t, $p \neq r$; $\pi^r A - \sum_{p=1, p \neq r}^{t} \mu_p^r c^p \leqq c^r$; and $\pi^r b - \sum_{p=1, p \neq r}^{t} \mu_p^r c^p \bar{x} = c^r \bar{x}$. Now define $\mu_r^r = 1$, and let $\mu^r = (\mu_1^r, \ldots, \mu_t^r) \in \mathbf{R}^t$. Let $\bar{\pi} = \sum_{r=1}^{t} \pi^r$ and $\bar{\lambda} = \sum_{r=1}^{t} \mu^r$. Clearly $\bar{\lambda} > 0$. From the above facts and the sufficient optimality criterion for LP (Theorem 4.4 of Section 4.5.1) it can be verified that $\bar{x}$, $\bar{\pi}$ are, respectively, primal and dual optima for the problem of minimizing $w_{\bar{\lambda}}(x) = \bar{\lambda} C x$ over $x \in \mathbf{K}$. ■

COROLLARY 17.1 If there exists a VM for (17.1), there exists an extreme point of $\mathbf{K}$ (i.e., a BFS of (17.1)) that is a VM for (17.1).

Proof: Follows from Theorem 17.3 and Theorem 3.3 in Section 3.5.5. ■

THEOREM 17.4 If x^1 and x^* are two extreme points of $\mathbf{K}$ that are both VM for (17.1), there exists an edge path in $\mathbf{K}$ between x^1 and x^* such that all the points along the path are also VM for (17.1).

Proof: By Theorem 17.3, there exist $\lambda^1 > 0$, $\lambda^* > 0$, such that x^1, x^* are optimum solutions for the LP: minimize $w_\lambda(x) = \lambda C x$ over $x \in \mathbf{K}$, when $\lambda = \lambda^1$, λ^*, respectively. Define $\lambda(\alpha) = \lambda^1 + \alpha(\lambda^* - \lambda^1)$ for $0 \leqq \alpha \leqq 1$. Clearly $\lambda(0) = \lambda^1$, $\lambda(1) = \lambda^*$, and $\lambda(\alpha) > 0$ for all $0 \leqq \alpha \leqq 1$. Let $\mathbf{F}^*$ be the face of $\mathbf{K}$ that is the set of optimum solutions of the LP: minimize $\lambda^* C x$ over $x \in \mathbf{K}$. $x^* \in \mathbf{F}^*$, and by Theorem 17.2 every point in $\mathbf{F}^*$ is a VM for (17.1). Consider the parametric cost LP: minimize $\lambda(\alpha) C x$ over $x \in \mathbf{K}$, in which α is the real-valued parameter varying between 0 and 1. Solve this by the parametric cost simplex algorithm beginning with the extreme point x^1 optimal for $\alpha = 0$, until α reaches 1. This traces an edge path in $\mathbf{K}$ from x^1 to an extreme point, say, x^2, in $\mathbf{F}^*$, and since $\lambda(\alpha) > 0$ for all $0 \leqq \alpha \leqq 1$, every point along this path is a VM by Theorem 17.2. Combining this path with an edge path from x^2 to x^* in the face $\mathbf{F}^*$ leads to the edge path from x^1 to x^*, all the points on which are VMs for (17.1). ■

THEOREM 17.5 If x^1 and x^* are two distinct VMs for (17.1), there exists a continuous path in **K** between them, all of whose points are also VM.

Proof: In the notation of the proof of Theorem 17.4, let $\mathbf{F}^1$ be the face of **K**, which is the set of all optimum solutions of the LP: minimize $\lambda^1 Cx$ over $x \in \mathbf{K}$. Every point in the faces $\mathbf{F}^1$ and $\mathbf{F}^*$ is a VM by Theorem 17.2, and by Theorem 17.4 there exists an edge path between an extreme point on $\mathbf{F}^1$ and an extreme on $\mathbf{F}^*$, all the points on which are VM. Since $x^1 \in \mathbf{F}^1$ and $x^* \in \mathbf{F}^*$, this result follows. ■

Mathematically, a subset of $\mathbf{R}^n$ is said to be *simply connected* if between every pair of points in it, there exists a continuous path, all of whose points themselves belong to the subset.

THEOREM 17.6 The set of all VM for (17.1) is a union of faces of **K**, which is simply connected.

Proof: From Theorems 17.2 and 17.3, every VM must lie on a face of **K**, all the points on which are VMs. This theorem follows from this and Theorem 17.5. ■

Example 17.1

This example, taken from reference [17.3] of J. G. Ecker, and I. A. Kouada shows that the set of VM need not be convex. Let $\mathbf{K} = \{(x_1, x_2, x_3): x_1 + x_2 + x_3 \leqq 1,\ x_1, x_2, x_3 \geqq 0\}$, $t = 2$, $c^1 = (4, 2, 1)$, $c_2 = (-4, -3, -2)$. For this problem, it can be verified that the set of VM is as given in Figure 17.1

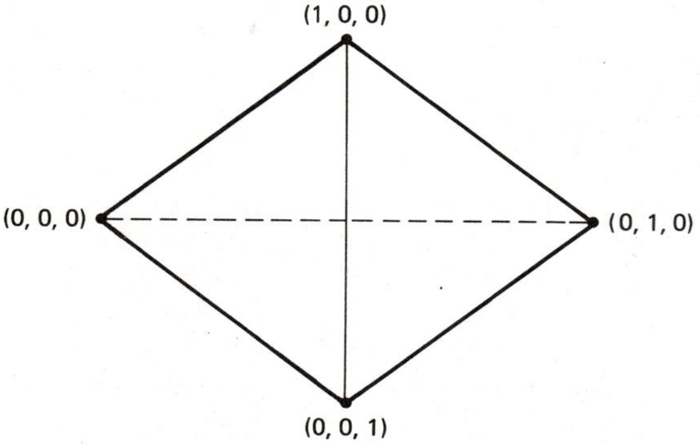

Figure 17.1 Points on the thick edges form the set of VM for the problem in Example 17.1.

THEOREM 17.7 The set of all feasible solutions that are not VMs for (17.1) is a convex set.

Proof: Let $x^1, x^2 \in \mathbf{K}$ be not VMs. By definition, there exist $\bar{x}, \hat{x} \in \mathbf{K}$, satisfying $C\bar{x} \leq Cx^1$ and $C\hat{x} \leq Cx^2$. So, for all $0 \leq \alpha \leq 1$, $C(\alpha\bar{x} + (1-\alpha)\hat{x}) \leq C(\alpha x^1 + (1-\alpha)x^2)$. So, $\alpha x^1 + (1-\alpha)x^2$ is not a VM either. ■

Therefore the set of VMs for (17.1) is the set theoretic difference of two convex sets, which is in general not convex.

Exercises

17.1 Suppose $x^1 \in \mathbf{K}$ is not a VM for (17.1). Prove that $\alpha x^1 + (1-\alpha)x$ is not a VM for all $0 < \alpha \leqq 1$, and $x \in \mathbf{K}$.

—(*P. L. Yu and M. Zeleny* [17.17])

17.2 Let $\mathbf{\Gamma}$ be an arbitrary convex subset of **K**. Prove the following: (1) if a relative interior point of $\mathbf{\Gamma}$, say, x^1, is not a VM, then every point in the relative interior of $\mathbf{\Gamma}$ is not a VM, (2) if a relative interior point of $\mathbf{\Gamma}$, say, x^2, is a VM, then every point in the closure of $\mathbf{\Gamma}$ is a VM.

—(*P. L. Yu and M. Zeleny* [17.17])

17.3 If **K** is bounded, prove that every VM can be expressed as a convex combination of extreme points of **K** that are VMs.

17.4 If the problem of minimizing one of the objective functions, say, $c^r x$, over $x \in \mathbf{K}$ has a unique optimum solution $\bar{x}$, prove that $\bar{x}$ is a VM.

—(*J. Phillips* [17.13])

Necessary and Sufficient Conditions for a VM

For $\bar{x} \in \mathbf{K}$, define $\mathbf{J}(\bar{x}) = \{j: \bar{x}_j = 0\}$.

THEOREM 17.8 $\bar{x} \in \mathbf{K}$ is a VM iff the system $Au = 0$, $u_j \geqq 0$, for all $j \in \mathbf{J}(\bar{x})$, $Cu \leqq 0$, $\sum_{r=1}^{t} c^r u = -1$ has no solution $u \in \mathbf{R}^n$.

Proof: If the system has such a solution u, it can be verified that for α positive and sufficiently small, $\bar{x} + \alpha u \in \mathbf{K}$ and $C(\bar{x} + \alpha u) \leq C\bar{x}$, so $\bar{x}$ is not a VM for (17.1). Conversely, if $\bar{x} \in \mathbf{K}$ is not a VM, there exists $\hat{x} \in \mathbf{K}$, satisfying $C\hat{x} \leq C\bar{x}$, and it can be verified that $u = (\hat{x} - \bar{x})/(\sum_{r=1}^{t} c^r(\bar{x} - \hat{x}))$ is feasible to the above system. ■

THEOREM 17.9 $\bar{x} \in \mathbf{K}$ is a VM for (17.1) iff the system $\pi A + dy + \mu C = 0$, $y \geqq 0$, $\mu \geqq e_t^{\mathrm{T}}$ has a solution $\pi = (\pi_1, \ldots, \pi_m)$, $y = (y_1, \ldots, y_m)^{\mathrm{T}}$, $\mu = (\mu_1, \ldots, \mu_t)$, where e_t^{T} is the row vector of all 1's in $\mathbf{R}^t$ and $d = (d_1, \ldots, d_n)$, with $d_j = 1$ if $j \in \mathbf{J}(\bar{x})$, 0 otherwise.

Proof: This follows by applying Tucker's theorem of the alternatives (Theorem 4.11 of Section 4.6.7) on the result in Theorem 17.8. ■

THEOREM 17.10 Let $\bar{x} \in \mathbf{K}$. Consider the LP

$$\begin{array}{lllr} \text{Maximize} & \sum_{r=1}^{t} s_r & & \\ \text{Subject to} & Ax = b & & \\ & c^r x + s_r = c^r \bar{x} & r = 1 \text{ to } t & (17.2) \\ x \geqq 0 & s_r \geqq 0 & r = 1 \text{ to } t & \end{array}$$

The point $\bar{x}$ is a VM for (17.1) iff the optimum objective value in (17.2) is zero. If the LP (17.2) is unbounded above, (17.1) has no VM. If (17.2) has an optimum solution, in any optimum solution (x^0, s^0) for it, x^0 is a VM for (17.1).

Proof: The point $\bar{x}$ is a VM for (17.1) by definition iff there exists no $x \in \mathbf{K}$ that makes $s = C\bar{x} - Cx \geq 0$. This is also the condition for (17.2) to have the optimum objective value of 0.

If the objective function in (17.2) is unbounded above, it remains unbounded above when we change $C\bar{x}$ in (17.2) to $C\tilde{x}$ for any $\tilde{x} \in \mathbf{K}$ (by Corollary 3.3 in Section 3.7) and by the above $\tilde{x}$ cannot be a VM, too. So, in this case, there are no VMs for (17.1).

Let (x^0, s^0) be optimal to (17.2). Let (π^0, μ^0) be any dual optimum solution corresponding to (17.2). Verify that $(x^0, s = 0)$ and (π^0, μ^0) together satisfy the sufficient optimality criterion for being optimal to the LP (P^0) obtained by changing $C\bar{x}$ in (17.2) to Cx^0. So the optimum objective value in (P^0) is zero, which by the above implies that x^0 is a VM for (17.1). ■

Necessary and Sufficient Conditions for Efficient Edges and Faces

A face of $\mathbf{K}$ is said to be *efficient* iff every point in it is a VM for (17.1). We use the following notation in the sequel. Let $\bar{x}$ be a BFS of (17.1) associated with a basic vector x_B and the corresponding basis B. $\mathbf{N}_B = \{j: x_j \notin x_B\}$, so $|\mathbf{N}_B| = n - m$. C_B, C_D is the corresponding partition by columns of C. $\bar{C}_D$ denotes the corresponding updated matrix (after all the entries in C_B are converted to zero by pivoting out), so it is of order $t \times (n - m)$. For $j \in \mathbf{N}_B$, $\bar{C}_j$ is the column in $\bar{C}_D$ corresponding to x_j. D is the $m \times (n - m)$ matrix of nonbasic columns in A, and $\bar{D} = B^{-1}D$ is the corresponding updated matrix. $\bar{D}_j$ is the column in $\bar{D}$ corresponding to x_j for $j \in \mathbf{N}_B$. If B is a degenerate basis, we let $\bar{D}^0$ denote the matrix consisting of rows of $\bar{D}$ in which the updated right-hand-side constant is 0. $u \in \mathbf{R}^{n-m}$ denotes a column vector of variables u_j for each $j \in \mathbf{N}_B$. If B is nondegenerate, $\mathbf{F}(B, j)$ for $j \in \mathbf{N}_B$ denotes the edge obtained by bringing the nonbasic variable x_j into the basic vector x_B; that is, $\mathbf{F}(B, j) = \{x: x \in \mathbf{K}, \text{ and } x_p = 0 \text{ for all } p \in \mathbf{N}_B \backslash \{j\}\}$. If B is degenerate, the edge $\mathbf{F}(B, j)$ is only defined for $j \in \mathbf{N}_B$ satisfying $\bar{D}_j^0 \leqq 0$ (here $\bar{D}_j^0$ is the column corresponding to x_j in $\bar{D}^0$), as otherwise, the process of bringing x_j into the basic vector x_B leads to a degenerate pivot step and does not lead to an edge. For any $\mathbf{J} \subset \mathbf{N}_B$, in general, where B is degenerate or not, $\mathbf{F}(B, \mathbf{J})$ denotes the face $\{x: x \in \mathbf{K}, x_j = 0 \text{ for all } j \in \mathbf{N}_B \backslash \mathbf{J}\}$.

THEOREM 17.11 Suppose $\bar{x}$ is a VM that is a nondegenerate BFS for (17.1). For $j \in \mathbf{N}_B$, the edge $\mathbf{F}(B, j)$ is efficient iff the optimum objective value is zero in the following LP.

$$\begin{array}{llr} \text{Maximize} & e_t^{\mathrm{T}} s & \\ \text{Subject to} & \bar{C}_D u + s = \bar{C}_j & (17.3) \\ u \geqq 0 & s \geqq 0 & \end{array}$$

Proof: Let u^0 be defined by $u_j^0 = 1$, $u_p^0 = 0$ for all $p \in \mathbf{N}_B \backslash \{j\}$. Then $x^0 = (x_B^0, x_D^0) = (\bar{x}_B - \varepsilon \bar{D}_j, \varepsilon u^0)$ is an interior point in $\mathbf{F}(B, j)$ for ε positive but sufficiently small. If the optimum objective value in (17.3) is 0, there exists no (u, s) feasible to (17.3) satisfying $\bar{C}_D u \leq \bar{C}_D u^0 = \bar{C}_j$. This, in turn, implies that there exists no $x \in \mathbf{K}$ satisfying $Cx \leq Cx^0$ (since for all $x = (x_B, x_D) \in \mathbf{K}$, $Cx = C_B x_B + \bar{C}_D x_D$) and hence x^0 is a VM, and by Exercise 17.2, $\mathbf{F}(B, j)$ is an efficient edge. If the optimum objective value in (17.3) is >0, there exist (u^*, s^*) feasible to (17.3) with $s^* \geq 0$. In this case, define $(x_B^*, x_D^*) = (\bar{x}_B - \varepsilon \bar{D} u^*, \varepsilon u^*)$, and verify that $Cx^* \leq Cx^0$, so x^0 is not a VM, which by Exercise 17.2 implies that $\mathbf{F}(B, j)$ is not efficient. ■

THEOREM 17.12 Suppose $\bar{x}$ is a VM that is a nondegenerate BFS of (17.1). Consider the system (17.4) in row vectors $\gamma = (\gamma_j: j \in \mathbf{N}_B)$ and $\mu = (\mu_1, \ldots, \mu_t)$, and let $\mathbf{\Gamma}_B$ be the set of feasible solutions corresponding to it. Define $j \in \mathbf{N}_B$ to be *nonredundant* if the minimum value of γ_j in $\mathbf{\Gamma}_B$ is zero, redundant otherwise. Let $\mathbf{E}_B = \{j: j \in \mathbf{N}_B, \ j \text{ is nonredundant}\}$.

The edge $\mathbf{F}(B, j)$ is efficient iff $j \in \mathbf{E}_B$.

$$\begin{aligned} \gamma - \mu\bar{C}_D &= e_t^{\mathrm{T}}\bar{C}_D \\ \gamma \geqq 0 \qquad & \mu \geqq 0 \end{aligned} \tag{17.4}$$

Proof: This follows from Theorem 17.11 and by looking at the dual of (17.3). ■

THEOREM 17.13: Suppose $\bar{x}$ is a VM that is a degenerate BFS of (17.1). For $j \in \mathbf{N}_B$ satisfying $\bar{D}_j^0 \leqq 0$, the edge $\mathbf{F}(B, j)$ is efficient iff the optimum objective value in the LP (17.5) is zero.

$$\begin{aligned} \text{Maximize} \quad & e_t^{\mathrm{T}}s \\ \text{Subject to} \quad & \bar{C}_D u + s = \bar{C}_j \\ & \bar{D}^0 u \leqq 0 \\ & u \geqq 0 \qquad s \geqq 0 \end{aligned} \tag{17.5}$$

Proof: Similar to the proof of Theorem 17.11. ■

THEOREM 17.14 Suppose $\bar{x}$ is a VM that is a degenerate BFS of (17.1). Let y be a row vector consisting of a variable associated with each row in $\bar{D}^0$. Let $\boldsymbol{\Gamma}_B$ now be the set of all feasible solutions (μ, y) corresponding to the following system (17.6). Define $j \in \mathbf{N}_B$ to be *nonredundant* iff $\bar{D}_j^0 \leqq 0$ and there exists a $(\mu, y) \in \boldsymbol{\Gamma}_B$ satisfying $(\mu + e_t^{\mathrm{T}})\bar{C}_j = 0$, *redundant* otherwise. Let $\mathbf{E}_B = \{j\colon j \in \mathbf{N}_B,\ \bar{D}_j^0 \leqq 0$ and j is nonredundant$\}$. The edge $\mathbf{F}(B, j)$ is efficient iff $j \in \mathbf{E}_B$.

$$\begin{aligned} \mu\bar{C}_D + y\bar{D}^0 &\geqq -e_t^{\mathrm{T}}\bar{C}_D \\ \mu \geqq 0 \qquad & y \geqq 0 \end{aligned} \tag{17.6}$$

Proof: Similar to the proof of Theorem 17.12. ■

THEOREM 17.15 Let $\bar{x}$ be a VM that is a BFS of (17.1). Define $\mathbf{E}_B$ as in Theorem 17.12 or Theorem 17.14, depending on whether $\bar{x}$ is nondegenerate or degenerate. For any $\mathbf{J} \subset \mathbf{E}_B$, the face $\mathbf{F}(B, \mathbf{J})$ is efficient iff the set $\boldsymbol{\Delta}(B, \mathbf{J}) = \{(\mu, \gamma)\colon \mu = (\mu_1, \ldots, \mu_t) \geqq 0,\ \gamma = (\gamma_j\colon j \in \mathbf{N}_B) \geqq 0,\ \gamma - \mu\bar{C}_D = e_t^{\mathrm{T}}\bar{C}_D,\ \gamma_j = 0$ for each $j \in \mathbf{J}\} \neq \varnothing$.

Proof: If $(\bar{\mu}, \bar{\gamma}) \in \boldsymbol{\Delta}(B, \mathbf{J})$, then $\mathbf{F}(B, \mathbf{J})$ is a subset of the set of optimum solutions for the problem: minimize $(\mu + e_t^{\mathrm{T}})Cx$ over $x \in \mathbf{K}$, and by Theorem 17.2, $\mathbf{F}(B, \mathbf{J})$ is efficient. If $\mathbf{F}(B, \mathbf{J})$ is efficient, let $\hat{x}$ be an interior point in it. So $\hat{x}$ is a VM. By Theorem 17.3 there exists $\lambda = (\lambda_1, \ldots, \lambda_t) > 0$, such that $\hat{x}$ is optimal to the problem: minimize λCx over $x \in \mathbf{K}$. Choose α large and positive so that $\alpha\lambda > e_t^{\mathrm{T}}$, and define μ by $\alpha\lambda - e_t^{\mathrm{T}}$, $\gamma = (\mu + e_t^{\mathrm{T}})\bar{C}_D$, and verify that $(\mu, \gamma) \in \Delta(B, \mathbf{J})$. ■

17.2 AN ALGORITHM TO GENERATE ALL EXTREME POINT VM

Use Theorem 17.10 to generate a VM for (17.1), x^0, if one exists. Use the method discussed in Section 3.5.4 to obtain a BFS $\bar{x}$ starting with x^0. Then $\bar{x}$ is an extreme point of $\mathbf{K}$ that is a VM. If $\bar{x}$ is a nondegenerate BFS, let B be the basis associated with it, and find $\mathbf{E}_B$ as in Theorem 17.12 (this requires solving at most $(n - m)$ LPs over the set (17.4)). Then $\{\mathbf{F}(B, j)\colon j \in \mathbf{E}_B\}$ are all the efficient edges incident at $\bar{x}$.

If $\bar{x}$ is a degenerate BFS, find a basis B associated with it, compute $\mathbf{E}_B$ as in Theorem 17.14, and then $\{\mathbf{F}(B, j)\colon j \in \mathbf{E}_B\}$ are all the efficient edges incident at $\bar{x}$ that can be obtained through the basis B. Repeat this process with every feasible basis for (17.1) associated with $\bar{x}$. (See Section 3.7.1 for a method of generating all such bases.)

Having generated all the efficient edges of $\mathbf{K}$ containing $\bar{x}$, find all the extreme points on these edges, excluding $\bar{x}$. Each of these extreme points is also a VM for (17.1). Repeat the same procedure with each of these extreme points, and continue in the same manner with any new extreme point VM that are generated. By the connectedness property discussed in Theorem 17.4, this method will eventually generate all the extreme points of $\mathbf{K}$ that are VM for (17.1) (see [17.1, 17.3, 17.6, 17.8, 17.11, 17.13, 17.17]).

17.3 AN ALGORITHM FOR GENERATING ALL THE EFFICIENT FACES OF *K*

In many decision problems, the decision maker's utility function may be such that the desired solution is a VM that may not be an extreme point of $\mathbf{K}$. For such situations it may be desirable to generate the set of all VM for the problem. For this, generate a VM $\bar{x}$, which is an extreme point of $\mathbf{K}$, as discussed above. Find a feasible basis B for (17.1) associated with $\bar{x}$. Generate maximal efficient faces of the form $\mathbf{F}(B, \mathbf{J})$ discussed in Theorem 17.15. Repeat with each feasible basis B for (17.1) associated with $\bar{x}$ if it is a degenerate BFS. This generates all the efficient faces containing $\bar{x}$. Repeat this procedure with each adjacent extreme point of $\bar{x}$ that is also a VM, and continue in the same manner with any newly generated extreme point VM that are adjacent to those already investigated. By the simply connectedness property of the union of efficient faces, this method will eventually generate all the efficient faces of **K** (see [17.5]).

Practical Applicability of These Algorithms

When faced with a multiobjective problem in which each objective is required to be minimized, it is very appealing to think of a VM as being a desirable solution for the problem. Clearly, when VMs do exist, any feasible point that is not a VM cannot be attractive, because it is dominated by some other feasible solution. However, usually there are many VM, and practitioners usually find themselves at a loss when they have to choose one feasible solution for implementation from this bewildering choice of efficient faces. A mathematician would consider the vector minimum problem as being completely solved mathematically when the set of all VM is generated. But for this to be practically useful, satisfactory criteria have to be developed to choose one VM for implementation. So far, no such criteria exist. The values of the different objective functions keep varying over different VM. To choose a specific VM for implementation, one should decide on how much of the value of one objective function can be sacrificed for the sake of some gain in the value of another, or, in effect, what kind of trade-off is reasonable among the various objective functions in the model. This kind of information is, in general, hard to provide. Also, when such information is available, we can usually determine positive weights $\lambda_r > 0$ for $r = 1$ to t, such that any optimum solution for the problem of minimizing λCx over $x \in \mathbf{K}$ is a desirable solution for this multiobjective problem (in this case, any such solution is, of course, a VM, but we have not made any use of the algorithms for generating all the VM in resolving this multiobjective problem). When trade-off information among the various objective functions is available, the goal programming type approaches discussed in Section 1.4 may also be more useful for analyzing this multiobjective problem than the algorithms that generate all VM. For all these reasons, the algorithms for computing all the VM have not yet come into wide practical use.

Exercises

17.5 Let $x^0 \in \mathbf{K}$ be a point that is not a VM for (17.1). If VMs do exist, prove that there exists a VM $\bar{x}$ for (17.1) satisfying $C\bar{x} \leqq Cx^0$. Also, for any VM $\bar{x}$ for (17.1), if $C\bar{x} \leqq Cx^0$, prove that $C\bar{x} \leq Cx^0$. Also, if $\mathbf{E}_0$ is the set of efficient points of the problem: vector minimum of Cx, subject to $Ax = b$, $Cx \leqq Cx^0$, $x \geqq 0$, prove that every point in $\mathbf{E}_0$ is also a VM for (17.1).

—(*Evans and Steuer* [17.6])

17.6 Let $\mathbf{K}_0 = \mathbf{K}$, and let $\mathbf{K}_r$ denote the set of all optimum solutions for the problem: minimize $C^r x$ over $x \in \mathbf{K}_{r-1}$, for $r = 1$ to t. If $\mathbf{K}_t \neq \emptyset$, prove that every point in $\mathbf{K}_t$ is a VM for (17.1).

REFERENCES

17.1 S. M. Belenson and K. C. Kapur, "An Algorithm for Solving Multicriterion Linear Problems with Examples," *Operational Research Quarterly 24* (1973), 65–77.

17.2 G. R. Bitran, "Linear Multi-Objective Problems with Interval Coefficients," *Management Science 26-27* (July 1980), 694–706.

17.3 J. G. Ecker and I. A. Kouada, "Finding all Efficient Extreme Points for Multiple Objective Linear Programs," *Mathematical Programming 14* (1978), 249–261.

17.4 J. G. Ecker and N. S. Hegner, "On Computing an Initial Efficient Extreme Point," *Journal of the Operational Research Society 29*, 10 (1978), 1005–1007.

17.5 J. G. Ecker, N. S. Hegner, and I. A. Kouada, "Generating all Maximal Efficient Faces for Multiple Objective Linear Programs," *Journal of Optimization Theory and Applications 30*, 3 (March 1980) 353–381.

17.6 J. P. Evans and R. E. Steuer, "A Revised Simplex Method for Linear Multiple Objective Programs," *Mathematical Programming 5*, 1 (August 1973), 54–72.

17.7 T. Gal and J. Nedoma, "Multiparametric Linear Programming," *Management Science 18* (1972), 406–421.

17.8 T. Gal, "A General Method for Determining the Set of all Efficient Solutions to a Linear Vector Maximum Problem," Institute fur Waitschaftswissenschaften, Report No. 76/12, Aachen, West Germany, 1976.

17.9 A. M. Geoffrion, "Proper Efficiency and the Theory of Vector Maximization," *Journal of*

Mathematical Analysis and Applications 22 (1968), 618–630.

17.10 C. L. Hwang and A. S. Masud, *Multiple Objective Decision Making Methods and Applications: A State of the Art Survey*, Springer, New York, 1979.

17.11 H. Isermann, "The Enumeration of the Set of all Efficient Solutions to a Linear Vector Multiple Objective Program," *Operational Research Quarterly 28* (1977), 711–725.

17.12 J. S. H. Kornbluth, "A Survey of Goal Programming," *Omega 1*, 2 (1973), 193–205.

17.13 J. Philip, "Algorithms for the Vector Maximization Problem," *Mathematical Programming 2*, 2 (April 1972), 207–229.

17.14 B. Roy, "Problems and Methods with Multiple Objective Functions," *Mathematical Programming 1* (1971), 239–266.

17.15 C. P. Simon, "Scalar and Vector Maximization: Calculus Techniques with Economics Applications," in S. Reiter (Ed.), *Studies in Mathematical Economics*, Mathematical Association of America, to be published.

17.16 M. Starr and M. Zeleny (Eds.) *Multiple Criteria Decision Making*, TIMS Studies in Management Science, vol. 6, North Holland, Amsterdam, The Netherlands, 1977.

17.17 P. L. Yu and M. Zeleny, "The Set of all Nondominated Solutions in Linear Cases and a Multiple Criteria Simplex Method," *Journal of Mathematical Analysis and Applications 49* (1975), 430–468.

17.18 M. Zeleny, *Linear Multiobjective Programming*, Springer, New York, 1974.

17.19 M. Zeleny, *Multiple Criteria Decision Making*, McGraw-Hill, New York, 1981.

Index